Fundamentals of Technical Thermodynamics

Martin Dehli · Ernst Doering ·
Herbert Schedwill

Fundamentals of Technical Thermodynamics

Textbook for Engineering Students

Martin Dehli
Hochschule Esslingen
Esslingen, Germany

Ernst Doering
Hochschule Esslingen
Esslingen, Germany

Herbert Schedwill
Hochschule Esslingen
Esslingen, Germany

ISBN 978-3-658-38909-3   ISBN 978-3-658-38910-9 (eBook)
https://doi.org/10.1007/978-3-658-38910-9

© The Editor(s) (if applicable) and The Author(s), under exclusive license to Springer Fachmedien Wiesbaden GmbH, part of Springer Nature 2023
Extended translation of the 9. German original edition published by Springer Fachmedien Wiesbaden, Wiesbaden, Germany, 2020
This work is subject to copyright. All rights are solely and exclusively licensed by the Publisher, whether the whole or part of the material is concerned, specifically the rights of translation, reprinting, reuse of illustrations, recitation, broadcasting, reproduction on microfilms or in any other physical way, and transmission or information storage and retrieval, electronic adaptation, computer software, or by similar or dissimilar methodology now known or hereafter developed.
The use of general descriptive names, registered names, trademarks, service marks, etc. in this publication does not imply, even in the absence of a specific statement, that such names are exempt from the relevant protective laws and regulations and therefore free for general use.
The publisher, the authors, and the editors are safe to assume that the advice and information in this book are believed to be true and accurate at the date of publication. Neither the publisher nor the authors or the editors give a warranty, expressed or implied, with respect to the material contained herein or for any errors or omissions that may have been made. The publisher remains neutral with regard to jurisdictional claims in published maps and institutional affiliations.

Responsible Editor: Eric Blaschke
This Springer imprint is published by the registered company Springer Fachmedien Wiesbaden GmbH, part of Springer Nature.
The registered company address is: Abraham-Lincoln-Str. 46, 65189 Wiesbaden, Germany

Foreword

'Technical Thermodynamics' is one of the fundamental areas of knowledge in mechanical engineering, energy engineering, automotive engineering, building services engineering, environmental engineering, chemical process engineering and other fields of engineering. This textbook is intended to help students gain access to areas of knowledge in technical thermodynamics that are often experienced as difficult.

The content focuses on the basic thermodynamic concepts, the first law of thermodynamics, the second law of thermodynamics, ideal gases, real gases and vapors, thermal machines, cyclic processes, exergy, heat transfer, moist air, combustion and chemical thermodynamics. The basic thermodynamic facts are illustrated by means of graphical representations - in particular also by state diagrams.

In addition, further areas of knowledge are imparted to the readers: For example, section 7 presents a new approach to the generalisation of thermodynamic cyclic processes with the help of additional evaluation parameters. Also, reference is made to a form of heat (called temperature change heat) as – besides reversible heat (called entropy change heat), volume change work and pressure change work – a fourth process variable which has hardly been used so far in engineering.

The late founder of the book, Prof. Dipl.-Phys. Ernst Doering, developed guidelines for a practical presentation of thermodynamics and, among other things, pursued the goal of treating friction and substitute processes as fundamental phenomena which are typical for all technical processes. Prof. Dr.-Ing. Herbert Schedwill maintained and deepened this approach: E.g. in section 3, the treatment of the second law of thermodynamics deals with the model of the reversible substitute process. In addition, the treatment of irreversible changes of state in comparison to reversible changes of state affects section 6 (Ideal and Real Machines), section 7 (Ideal and Real Cyclic Processes) and section 8 (Exergy). Herbert Schedwill also rendered great services in the extension and update of the book. This new edition contains additional sections on gas dynamics and chemical thermodynamics.

The book originated, among other things, from lectures for thermodynamics, heat and mass transfer, gas technology, air-conditioning technology, refrigeration technology, energy technology as well as combustion technology and heat economy at the Esslingen University of Applied Sciences (HE). The LaTeX typesetting and most of the figures were produced by the authors themselves.

Parallel to this book, the use of the book "Aufgabensammlung Technische Thermodynamik" (2nd edition, ISBN 978-3-658-22943-6, published by Springer Vieweg) is recommended. On more than 380 pages, it contains numerous problems from energy and thermotechnical practice as well as from exercises and examinations, each with detailed solutions. The collection of exercises can be used to independently check one's own learning success in thermodynamics, as its structure is based on that of the book "Fundamentals of Technical Thermodynamics". The tasks are designed in such a way that they can also show suitable solutions, especially to engineers who deal with questions of thermodynamics in their professional work.

Esslingen, Germany, autumn 2022

Prof. Dr.-Ing. Martin Dehli

Table of Contents

1 Basic Thermodynamic Terms 1
  1.1 Applications of Thermodynamics 1
  1.2 System 4
  1.3 State, State Variables, Changes of State 5
  1.4 Process, Process Variables 8

2 The First Law of Thermodynamics 9
  2.1 The Principle of Conservation of Energy 9
  2.2 Potential Energy 10
  2.3 Kinetic Energy 13
  2.4 Work 14
    2.4.1 Volume Change Work 14
    2.4.2 Coupling Work 16
    2.4.3 Shift Work 16
    2.4.4 Pressure Change Work 17
    2.4.5 Friction Work 19
  2.5 Thermal Energy 20
    2.5.1 Internal Energy 21
    2.5.2 Heat 22
    2.5.3 Enthalpy 23
  2.6 Energy Balances 24
    2.6.1 Energy Balance for the Closed System 24
    2.6.2 Energy Balance for the Open System 26
  2.7 Heat Capacity 30
    2.7.1 Specific Heat Capacity 31
    2.7.2 The Specific Heat Capacity of Gases 34
  2.8 Fluid Mechanics 35
    2.8.1 General Aspects 35
    2.8.2 Flow Shapes 36
    2.8.3 Friction and Roughness 36
    2.8.4 Individual Resistances 38
    2.8.5 Equivalent Pipe Length 39
    2.8.6 Pressure Loss or Pressure Gain Due to Density Difference
          between Flowing Fluid and Ambient Fluid 39
    2.8.7 Total Pressure Difference During Fluid Transfer 39

3 The Second Law of Thermodynamics 43
  3.1 The Statement of the Second Law 43
    3.1.1 Reversible and Irreversible Processes 44
    3.1.2 Quasi-Static Changes of State 45
  3.2 Irreversible Processes 46
    3.2.1 Friction 46
    3.2.2 Temperature Equalisation 47
    3.2.3 Pressure Equalisation 47
    3.2.4 Throttling 48

3.3 Entropy 49
   3.3.1 Reversible Substitute Processes of Adiabatic Processes 50
   3.3.2 The Calculation of the Entropy Change 52
   3.3.3 Entropy as a State Variable, Total Differential 52
3.4 The Entropy Change of Irreversible Processes 55
   3.4.1 Friction 55
   3.4.2 Temperature Equalisation 56
   3.4.3 Pressure Equalisation 58
   3.4.4 Throttling 59
3.5 Non-Adiabatic Process and Reversible Substitute Process 59
   3.5.1 Isentropic Change of State; Interpretations of Entropy 61
   3.5.2 Entropy Diagrams 62
   3.5.3 Circular Integral, Thermodynamic Temperature 64
   3.5.4 Dissipative Energy 66

4 Ideal Gases 69
4.1 Thermal Equation of State 69
   4.1.1 Law of *Boyle* and *Mariotte* 69
   4.1.2 Law of *Gay-Lussac* 69
   4.1.3 Physical Norm State 70
   4.1.4 Gas Thermometer 71
   4.1.5 Specific Gas Constant 72
   4.1.6 Universal Gas Constant 74
4.2 Caloric State Variables of Ideal Gases 74
   4.2.1 Internal Energy 75
   4.2.2 Enthalpy 75
   4.2.3 Entropy 76
4.3 Changes of State 77
   4.3.1 Isochoric Change of Statee 77
   4.3.2 Isobaric Change of State 79
   4.3.3 Isothermal Change of State 82
   4.3.4 Isentropic Change of State 84
   4.3.5 Polytropic Change of State 89
   4.3.6 Changes of State With Variable Mass 96
4.4 Specific Thermal Energy and Specific Work in the T, s Diagram 97
4.5 Mixtures of Ideal Gases 99
   4.5.1 The Mixing Process in the Closed System 102
   4.5.2 The Mixing Process without Total Volume Change 105
   4.5.3 The Mixing Process without Temperature Change, Pressure Change and Total Volume Change 106
   4.5.4 The Mixing Process in the Open System 108
4.6 Dynamics of Ideal Gases: Compressible Stationary Gas Flow 111
   4.6.1 Introduction 111
   4.6.2 Velocity of Sound and Propagation of Sound 112
   4.6.3 Energy Equation and *Bernoulli* Equation of Compressible One-Dimensional Ideal Gas Flow 115
   4.6.4 Stagnation State Variables and Critical State 118
   4.6.5 The Velocity Diagram of the Specific Energy Equation 121

		4.6.6 Flow Function 122

- 4.6.6 Flow Function 122
- 4.6.7 Isentropic Gas Flow in Nozzles and Orifices 124
- 4.6.8 Accelerated Compressible Flow 126
- 4.6.9 Compression Shock 142

5 Real Gases and Vapors 151
5.1 Properties of Vapors 151
5.1.1 Phase Transitions 151
5.1.2 Two-Phase Regions 152
5.1.3 Boiling and Condensing 153
5.1.4 Evaporation and Thawing 155
5.1.5 Liquid 157
5.1.6 Two-Phase Liquid-Vapor State 160
5.1.7 Superheated Steam 163
5.2 State Diagrams 164
5.2.1 The p, v, T Surface 164
5.2.2 The T, s Diagram 167
5.2.3 The h, s Diagram 170
5.3 Thermal Equations of State 172
5.3.1 The *van der Waals* Equation 172
5.3.2 The Boundary Curve and the *Maxwell* Relation 174
5.3.3 The reduced *van der Waals* Equation 176
5.3.4 Different Approaches 177
5.3.5 Virial Coefficients 181
5.4 Calculation of State Variables; Property Tables 183
5.4.1 The Caloric State Variables 184
5.4.2 The Specific Heat Capacities c_p and c_v 189
5.4.3 The Isentropic Exponent and the Isothermal Exponent 191
5.4.4 The *Clausius-Clapeyron* Equation 193
5.4.5 Free Energy and Free Enthalpy 197
5.4.5.1 General 197
5.4.5.2 A g, s Diagram for Water and Steam 203
5.4.6 The *Joule-Thomson* Effect 207

6 Thermal Machines 213
6.1 Classification and Types of Machines 213
6.1.1 Classification According to the Direction of Energy Conversion 213
6.1.2 Classification According to the Constructionn of the Machines 214
6.1.3 Classification According to the Type of Process Taking Place 214
6.2 Ideal Machines 214
6.2.1 Compression and Expansion in Ideal Machines 215
6.2.2 Multi-Stage Compression and Expansion 216
6.2.3 The Energy Balance for Flow Machines 218
6.2.4 The Energy Balance for Displacement Machines 220
6.3 Energy Balances for Real Machines 223
6.3.1 Internal or Indexed Work 224
6.3.2 Total Work 226
6.3.3 Total Enthalpy 226

- 6.4 Real Machines 226
  - 6.4.1 The Uncooled Compressor 226
  - 6.4.2 The Cooled Compressor 229
  - 6.4.3 Piston Compressor 231
  - 6.4.4 Turbo Compressor 231
  - 6.4.5 Gas and Steam Turbines 231
- 6.5 Efficiencies 235
  - 6.5.1 Comparative Processes 236
  - 6.5.2 The Internal Efficiency 236
  - 6.5.3 The Mechanical Efficiency 238
  - 6.5.4 The Total Efficiency 238
  - 6.5.5 The Isentropic Efficiency 238
  - 6.5.6 The Isothermal Efficiency 239
  - 6.5.7 The Polytropic Efficiency 239

7 Cyclic Processes 245
- 7.1 Cyclic Process Work, Heat Input and Heat Output 245
- 7.2 Right-Hand and Left-Hand Cyclic Processes 251
- 7.3 The Theory of Right-Hand Cyclic Processes 252
  - 7.3.1 Conversion of Thermal Energy to Mechanical Energy 253
  - 7.3.2 Thermal Efficiency 254
  - 7.3.3 Right-Hand *Carnot* Process 255
  - 7.3.4 Effect of Irreversible Processes 256
  - 7.3.5 *Carnot* Factor 258
- 7.4 Technically Used Right-Hand Cyclic Processes 260
  - 7.4.1 *Seiliger* Process, *Otto* Process, *Diesel* Process,
    Generalised *Diesel* Process 261
  - 7.4.2 *Joule* Process 265
  - 7.4.3 *Ericsson* Process 268
  - 7.4.4 *Stirling* Process 270
  - 7.4.5 Single-Polytropic *Carnot* Process 272
  - 7.4.6 Gas Expansion Process 273
  - 7.4.7 *Clausius-Rankine* Process 274
- 7.5 Comparative Evaluation of Right-Hand Cyclic Processes 277
  - 7.5.1 Process Variables and Cyclic Processes 278
  - 7.5.2 Mechanical Effort Ratios and Thermal Effort Ratios 279
  - 7.5.3 Evaluation Criteria for Important Thermodynamic Cyclic Processes . 283
    - 7.5.3.1 General Thermodynamic Relations 283
    - 7.5.3.2 Examples 285
    - 7.5.3.3 Graphical Representation of the Thermodynamic Relations .. 299
    - 7.5.3.4 Cyclic Process Calculations for Real Fluids 311
- 7.6 Left-Hand Cyclic Processes 317
  - 7.6.1 Performance Number 318
  - 7.6.2 Left-Hand *Carnot* Process 319
  - 7.6.3 Left-Hand *Joule* Process 320
  - 7.6.4 Gas Expansion Process as a Left-Hand Cycle Process 321
  - 7.6.5 Cold Vapor Compression Process 325

8 Exergy ... 333
8.1 Energy and Exergy ... 333
8.1.1 Exergy of Heat ... 335
8.1.2 Exergy of Bound Energy ... 337
8.1.3 Exergy of Temperature Change Heat ... 338
8.1.4 Exergy of Volume Change Work ... 340
8.1.5 Exergy of Shift Work ... 341
8.1.6 Exergy of Pressure Change Work ... 342
8.1.7 Exergy of Internal Energy ... 343
8.1.8 Exergy of Enthalpy ... 347
8.1.9 Exergy of Free Energy ... 351
8.1.10 Exergy of Free Enthalpy ... 351
8.1.11 Difference between EU and EF ... 354
8.1.12 Difference between EH and EG ... 355
8.1.13 Free Energy and Free Enthalpy as Thermodynamic Potentials ... 355
8.2 Exergy and Anergy ... 358
8.2.1 Anergy in a p,V Diagram and in a T,S Diagram ... 360
8.2.2 Anergy-Free Energies ... 363
8.3 Exergy Loss ... 363
8.3.1 Irreversibility and Exergy Loss ... 363
8.3.2 Exergy Loss and Anergy Gain ... 367
8.3.3 Exergetic Efficiencies ... 370

9 Heat Transfer ... 375
9.1 Heat Radiation ... 375
9.1.1 *Stefan-Boltzmann* Law ... 375
9.1.2 *Kirchhoff*'s Law ... 375
9.1.3 *Planck*'s Radiation Law ... 376
9.1.4 *Wien*'s Displacement Law ... 377
9.1.5 *Lambert*'s Cosine Law ... 378
9.1.6 Irradiance Number ... 378
9.2 Radiation Exchange ... 383
9.2.1 Cavity Method ... 384
9.2.2 Envelopment of One Surface by Another ... 385
9.2.3 Two Parallel Surfaces of Equal Size ... 386
9.2.4 Matrix Representation ... 386
9.3 Stationary One-Dimensional Heat Conduction ... 389
9.3.1 Plane Wall ... 389
9.3.2 Pipe Wall ... 390
9.4 Instationary One-Dimensional Heat Conduction ... 391
9.4.1 Plane Single-Layer Wall ... 392
9.4.2 Semi-Infinite Body ... 394
9.4.3 Contact Temperature ... 395
9.5 Heat Transfer by Convection ... 396
9.5.1 Heat Transfer Coefficient ... 397
9.5.2 Similarity Theory ... 398
9.5.3 *Reynolds* Analogy ... 401
9.5.4 *Prandtl* Analogy ... 402

    9.5.5 Power Number Approaches for Laminar and Turbulent Flow 405
    9.5.6 Approaches for Phase Transitions 411
 9.6 Over-all Heat Transfer 414
    9.6.1 Over-all Heat Transfer Coefficient 415
    9.6.2 Fin Efficiency and Area Efficiency 416
    9.6.3 Mean Temperature Difference 417
    9.6.4 Operating Characteristic (Effectiveness) 417
 9.7 Finned Heat Transfer Surfaces 418
    9.7.1 Straight Fin With Rectangular Cross-Section 419
    9.7.2 Circular Fin With Rectangular Cross-Section 419
 9.8 Partition Wall Heat Exchangers 421
    9.8.1 Unidirectional Flow Heat Exchanger 421
    9.8.2 Counterflow Heat Exchanger 422
    9.8.3 Crossflow Heat Exchanger 424
    9.8.4 Heat Transfer with Phase Transition in a Heat Exchanger 428
 9.9 Evaluation and Design 429
    9.9.1 Correction Factor for a Crossflow Heat Exchanger 430
    9.9.2 Representation of the Operating Characteristic 432
    9.9.3 Longitudinal Heat Conduction in a Plane Partition Wall 434
    9.9.4 Design Diagram 437

10 Humid Air 443
 10.1 State Variables of Humid Air 443
    10.1.1 Relative Humidity 443
    10.1.2 Humidity Ratio and Saturation 443
    10.1.3 Specific Enthalpy 445
    10.1.4 Specific Volume and Density 446
 10.2 Changes of State of Humid Air 446
    10.2.1 Temperature Change 447
    10.2.2 Humidification and Dehumidification 447
    10.2.3 Mixing of Two Humid Air Quantities 448
 10.3 The h,x Diagram of *Mollier* 449
    10.3.1 Temperature Change 451
    10.3.2 Humidification and Dehumidification 452
    10.3.3 Mixing of Two Humid Air Quantities 452
 10.4 Evaporation Model 452
    10.4.1 Evaporation Coefficient 452
    10.4.2 Energy Balances 453
    10.4.3 *Lewis* Relationship 454
 10.5 Cooling Limit 455
 10.6 Evaporation and Dew Precipitation 457
 10.7 Water Vapour Diffusion Through Walls 458

11 Combustion 465
 11.1 Fuels 466
    11.1.1 Gaseous Fuels 466
    11.1.2 Solid and Liquid Fuels 470
    11.1.3 Composition of the Combustion Gas, Combustion Triangles,

	Combustion Control	473
11.2	Technical Aspects of Combustion	481
	11.2.1 Initiation and Progression of Combustion	481
	11.2.2 Complete and Incomplete Combustion	481
	11.2.3 Dew Point of Combustion Gases	484
	11.2.4 Chimney Draught	484
11.3	Upper Calorific Value and Lower Calorific Value	485
11.4	Theoretical Combustion Temperature	487

12 Chemical Thermodynamics ... 497

- 12.1 Systems Involving Chemical Reactions ... 497
- 12.2 Reaction Turnover and Reaction Rate ... 499
- 12.3 Molar Enthalpies of Reaction and Standard Molar Enthalpies of Formation; Theorem of *Hess* ... 502
  - 12.3.1 Molar Enthalpies of Reaction ... 502
  - 12.3.2 Standard Molar Enthalpies of Formation; Theorem of *Hess* ... 506
- 12.4 Absolute Molar Entropies; Third Law of Thermodynamics ... 512
- 12.5 The Importance of the Second Law for Chemical Reactions ... 516
- 12.6 Chemical Exergies ... 523
- 12.7 Fuel Exergies ... 526
- 12.8 Chemical Potentials ... 531
- 12.9 The Law of Mass Action ... 533
- 12.10 Pressure and Temperature Dependence of the Constants of the Law of Mass Action; Law of *Le Chatelier* and *Braun* ... 538
- 12.11 Model of Isothermal-Isobaric Reversible Chemical Reactions ... 543
  - 12.11.1 Model of Reversible Oxiation of Hydrogen ... 543
  - 12.11.2 Model of Arbitrary Homogeneous Reversible Chemical Reactions of Ideal Gases ... 547
  - 12.11.3 Reversible Storage of Heat and Work in the Form of Chemical Energy ... 548
- 12.12 Fuel Cells ... 550

Appendix ... 555

References ... 592

Index ... 600

Important Formula Characters

A	buoyancy, area, cross-section, first virial coefficient, anergy	$\Delta^R G_m$	molar reaction *Gibbs* function
a	specific anergy, thermal diffusivity, absorption ratio, constant, parameter, velocity of sound, mass fraction ash	Ga	*Galilei* number
		Gr	*Grashof* number
		g	gravitational acceleration, specific free enthalpy
a_{qs}	thermal effort ratio	H	enthalpy, height, incident area related radiation
a_{qT}	thermal effort ratio		
a_{wp}	mechanical effort ratio	$\dot{H}$	enthalpy flow
a_{wv}	mechanical effort ratio	H_{Amax}	maximum enthalpy of combustion gas (maximum enthalpy of exhaust gas)
a_g	total effort ratio		
B	second virial coefficient, emitted area related radiation, width	$H_{m,0i}^{f\square}$	standard molar enthalpy of formation
		H_i	specific lower calorific value
b	heat penetration coefficient, constant, parameter	$H_{i,0}$	norm volume related lower calorific value
C	heat capacity, norm volume related heat capacity, third virial coefficient, radiation coefficient, *Euler* constant, water vapor concentration	$H_{m,0i}$	molar enthalpy
		H_s	specific upper calorific value
		$H_{s,0}$	norm volume related upper calorific value
$\dot{C}$	time-related heat capacity of a fluid flow (heat capacity flow)	H_t	total enthalpy
		$\Delta^A H_m$	molar activation enthalpy
C_m	molar heat capacity	$\Delta^R H_m$	molar enthalpy of reaction
c	velocity, specific heat capacity, radiation constant, speed of light, mass fraction carbon	ΔH_T	reduced work
		ΔH_V	additional work
		h	specific enthalpy, height, *Planck* constant, mass fraction hydrogen, norm volume related enthalpy
D	fourth virial coefficient, diffusion coefficient		
d	differential	I	intensity
d	diameter, transmittance ratio	i	number of stages, running variable
E	energy, energy content of a substance flow, exergy, emission, elastic modulus	j	running variable
		K	constant, function
$\dot{E}$	energy flow	k	over-all heat transfer coefficient, *Boltzmann* constant, roughness, proportionality factor
E_k	kinetic energy		
$E_{m,0i}$	molar energy		
E_p	potential energy	L	state of air, length
ΔE_T	energy recovery	l	length, norm volume related air requirement
ΔE_V	heat loss		
$Ex_{m,0i}$	molar exergy	l^*	specific air requirement
$Ex_{m,0i}^{\square}$	standard molar exergy	l'	equivalent pipe length
$Ex_{m,B}$	molar fuel exergy	l_a	air content
e	specific exergy, basis of natural logarithms, unit matrix	ln	natural logarithm
		lg	logarithm to base 10
F	force, liquid, free energy, function	M	molar mass, torque, *Mach* number
		m	mass
F_R	friction force	$\dot{m}$	mass flow
F_T	tangential force	N_A	*Avogadro* number
f	specific free energy, functional	Nu	*Nußelt* number
G	weight, free enthalpy	n	molar quantity, polytrope exponent, rotational speed, mass fraction nitrogen
$G_{m,0i}$	molar free enthalpy, molar *Gibbs* function		
		$\dot{n}$	molar quantity flow
$G_{m,0i}^{\square}$	standard molar *Gibbs* function	n_S	isentropic exponent

n_T	isothermal exponent	V_0	physical norm volume
o	norm volume related oxygen requirement, mass fraction oxygen	$\Delta^R V_m$	molar reaction volume
		v	specific volume
o^*	specific oxygen demand	v_0	norm volume related combustion volume, (norm volume related exhaust gas volume)
P	power		
Pe	Peclet number		
Pr	Prandtl number	v_0^*	specific combustion gas volume (specific exhaust gas volume)
p	pressure		
Δp	pressure difference	W	work, displacement work, time related heat capacity of a fluid flow, (heat capacity flow, water value)
Q	heat		
$\dot{Q}$	heat flow (heat output)		
Q_{rev}	reversible heat (reversible substitute heat)	Wa	van der Waals number
		W_e	coupling work (shaft work, technical work)
q	specific heat		
q_{rev}	specific reversible heat (specific reversible substitute heat)	W_i	internal work (indexed work)
		W_{Kreis}	cyclic process work
q_s	specific entropy change heat (specific reversible heat, specific reversible substitute heat)	$(W_m)_{rev}$	reversible molar reaction work
		W_p	pressure change work
		W_R	friction work
q_T	specific temperature change heat	W_{RA}	external friction work
R	specific gas constant, radius	W_{RI}	internal friction work
R_a	heat transfer resistance outside	W_t	total work
R_D	over-all heat transfer resistance	W_V	volume change work
R_i	heat transfer resistance inside	w	specific work, velocity, mass fraction water
R_L	thermal resistance		
Ra	Raleigh number	w_{Kreis}	specific cyclic process work
Re	Reynolds number	w_p	specific pressure change work
R_m	universal gas constant	w_R	specific friction work
r	volume fraction, specific enthalpy of vaporisation (specific heat of vaporisation), reflection ratio, radius, matrix	w_{RA}	specific external friction work
		w_{RI}	specific internal friction work
		w_V	specific volume change work
S	entropy, vapor conduction resistance	x	quality, humidity ratio (moisture content, water content), coordinate
$\dot{S}$	entropy flow		
$S_{m,0i}$	molar entropy	y	coordinate
$S_{m,0i}^{\square}$	standard molar entropy	z	coordinate, real gas factor, fraction, reaction turnover
$\Delta^R S_m$	molar reaction entropy		
St	Stanton number	$\dot{z}$	reaction rate, turnover rate
s	specific entropy, path, thickness, mass fraction sulphur, length		
		α	heat transfer coefficient, angle
T	absolute temperature (thermodynamic temperature, Kelvin temperature), depth	β	thermal expansion coefficient, angle
		γ	specific weight, angle
		δ	differential
t	Celsius temperature	δ	reduced density (normalised density), isenthalpic throttle coefficient (Joule-Thomson coefficient), wall thickness
t_{Amax}	theoretical combustion temperature (adiabatic combustion temperature)		
U	internal energy, circumference	ϵ	performance number, compression ratio, emission ratio
$\dot{U}$	internal energy flow		
u	specific internal energy, dimensionless ratio	ζ	exergetic efficiency, resistance number (individual resistance coefficient)
V	volume	η	efficiency, dynamic viscosity
$\dot{V}$	volume flow	ϑ	temperature, reduced temperature

	(normalised temperature)
κ	isentropic exponent, c_p/c_v
λ	thermal conductivity, wave length, pipe friction coefficient, air ratio
μ	mass fraction, diffusion resistance factor
ν	kinematic viscosity, nitrogen characteristic
Π	product
π	reduced pressure (normalised pressure), circumferential number
ρ	density
Σ	sum
σ	evaporation coefficient, specific heat of melting (specific melting enthalpy), oxygen requirement characteristic
τ	duration of time, shear stress
Φ	operating characteristic, matrix of irradiation numbers
φ	injection ratio, irradiance number, relative humidity, angle
ψ	pressure increase ratio, flow function, degree of saturation
Ψ	ratio
Ω	solid angle
ω	angular velocity, acentric factor, auxiliary size

Indices (subscript)

A	combustion gas (exhaust gas)
a	outer, exit, outlet
ab	released
Af	moist combustion gas
amb	ambient
At	dry combustion gas
B	fuel, fuel gas
C	Carnot
D	throttling, steam, water vapor
E	energy, ice, water ice
e	effective, shaft, entry, inlet
el	electric
F	liquid
G	total, base
g	total
gl	equivalent
H	stroke, heating
h	isenthalpic, hydraulic
i	inside, stage number, component, running variable
id	ideal
is	inside, isentropic
it	inside, isothermal
j	running variable
K	piston, compression, refrigeration system, equilibrium constant, contact
k	critical, critical point
$Kreis$	cyclic, cycle
L	air, length
lim	limit
M	mixture of gas
m	mean, molar, mechanical
max	maximum
min	minimum
$min\,f$	minimum moist
$min\,t$	minimum dry
N	wet vapor
n	polytropic
p	pressure, isobaric, polytropic, partial pressure
Qu	source
R	friction, rib, fin
RA	external friction
RI	internal friction
r	reduced
rev	reversible
S	saturation, boiling state, boiling point, black body, face area, slag, ash
Se	sink
s	isentropic
T	turbine, triple point
Tr	triple point
t	isothermal, total
th	thermal
U	ambient
$\ddot{u}$	over
V	volume, compressor, loss, vaporisation
v	isochoric
W	water, heat, heat pump, heat transfer
Wd	wall
Z	intermediate state
zu	supplied
0	physical norm state, reference state, ambient state, initial quantity
1	initial, start, entry, inlet
2	end, exit, outlet

Indices (superscript)

af	exhaust gas moist (combustion gas moist)
at	exhaust gas dry (combustion gas dry)
$'$	boiling liquid
$''$	saturated vapor
$*$	critical state of dynamik flow

Authors Vita

Prof. Dr.-Ing. Martin Dehli, *1948, studied mechanical engineering and energy technology and obtained his doctorate on canonical equations of state in thermodynamics at the Technical University of Stuttgart. After working for three years in an engineering company, he worked for fourteen years in a large energy supply company - among other things as head of department for decentralised energy technologies and for fundamental issues. Since 1991 at the Esslingen University of Applied Sciences (Hochschule Esslingen). Teaching areas: Thermodynamics, Energy Technology, Gas Technology, Combustion etc. Co-author of this textbook since 2003.

Prof. Dipl.-Phys. Ernst Doering, *1925, †1982, studied at the Technical University of Karlsruhe. After six years in industry in the fields of fluid mechanics and drying technology, he worked from 1958 at the Staatliche Ingenieurschule, now Esslingen University of Applied Sciences. 1968 Author of the first edition of this textbook.

Prof. Dr.-Ing. Herbert Schedwill *1927, †2015, after apprenticeship as a mechanic, studied at the Technical University of Stuttgart. Then three years of engineering work in design, development and testing in the field of heat transfer. He obtained his doctorate on longitudinal heat conduction in the plane partition wall of recuperative heat exchangers. From 1970 to 1989 at the Staatliche Ingenieurschule, now Esslingen University of Applied Sciences (Hochschule Esslingen). Until 2015 co-author of this textbook, whereby he gave essential impulses for its further development.

1 Basic Thermodynamic Terms

Thermodynamics is concerned with the conversion and transfer of energy, insofar as a form of energy is involved, which is commonly referred to as "heat". In addition, properties of matter are also covered. The term "heat" has undergone a change of meaning in the last decades. What in the development of the classical natural sciences was meant by the term "heat" is today largely described by the term "internal energy". Thus, heat is no longer the central energy form of thermodynamics. The designation of the field of natural science has followed this change in meaning, and "heat theory" became thermodynamics.

The relationships of the different forms of energy to each other and the laws that occur during their transformations are described by three main theorems, which form the theoretical framework of thermodynamics. The confirmation of their validity is one of the most important discoveries in the history of natural science. The first law represents the application of the law of conservation of energy, which has unlimited validity in classical physics, to the various problems of thermodynamics. The second law allows statements about the direction of the energy transformation, which always runs in such a way that the high-grade forms of energy (e.g. mechanical energy, electrical energy) decrease in favor of the inferior ones (e.g. internal energy of the environment). The high-grade energy is also referred to as exergy and can be described, by the second law, as the principle of the reduction of exergy. The third main theorem describes the behavior of entropy - a state variable, which, among other things, can be interpreted as the degree of disorder of a substance or a system - as zero at the zero point of the thermodynamic temperature.

1.1 Applications of Thermodynamics

The task of technical thermodynamics is the application of scientific knowledge to problems of technology. The greatest difficulty is generally the transfer of theoretical methods to the requirements of practice. In theoretical science, the problem to be considered is idealized, environmental influences are reduced or eliminated, and the "friction" of the process is often neglected. Then one arrives at manageable interrelationships and aspects. A technical problem is complicated, several processes overlap, and some of the variables required for the solution are unknown. An exact solution is sometimes possible, but it would require an unjustifiably high expenditure of time and money. In many cases, an approximate solution will suffice. The technical problem must be simplified in such a way that it allows a solution. Here it must be decided which influences are significant, and which are insignificant. This decision is often a difficult task in the computational solution of technical problems. In the meantime, comprehensive tools have been developed with which thermodynamic problems - e.g. in plant and building services engineering - can be treated by simulation calculations with the help of approximate solutions.

An approximate solution obtained by calculation will not always suffice. Therefore experimental methods are used to obtain the desired results from a small-scale model or a large-scale model; here, too, the knowledge of the scientific correlations is required.

There are only a few areas of technology that are not affected by thermodynamics. For some areas it forms the scientific basis. Without claiming to be complete and to express an order of priority by the sequence, the following areas are mentioned:

Energy technology: Generation of electrical energy from the energy sources coal, oil, gas, biomass or nuclear fuels (Figures 1.1 and 1.2)

Heating and process heat technology: Generation of heat from gas, oil, coal, biomass, electricity or solar energy (Figures 1.3 to 1.5)

Energy conversion in machines: Combustion of fuels in machines to drive vehicles, to generate electrical energy, or e.g. for the compression of gases or liquids (Fig. 1.6)

Refrigeration: Cooling of substances or rooms below ambient temperature (Fig. 1.7)

Heat engineering: e.g. heating or cooling of substances in heat exchangers (Fig. 1.8)

Ventilation and air-conditioning technology: e.g. air handling, humidification or dehumidification, air management in factory halls and rooms for living (Fig. 1.9)

Combustion processes; chemical process engineering: combustion of gaseous, liquid and solid energy sources, chemical process technology from an energetic and exergetic point of view (Fig. 1.10)

Figure 1.1 Hard coal fired power plant; conventional power plant units; gas and steam turbine combined heat and power plant; internal combustion engine power plant (from left to right)

Figure 1.2 Combined heat and power plant; ORC plant; fuel cell power plant (from left to right)

Figure 1.3 Natural gas condensing boiler; electricity-generating heating system with fuel cell and condensing boiler; circulation pumps; vacuum solar collector (from left to right)

Figure 1.4 Boiler house with various heat generators; steam boiler system; waste heat boiler with auxiliary firing; high-temperature furnaces (from left to right)

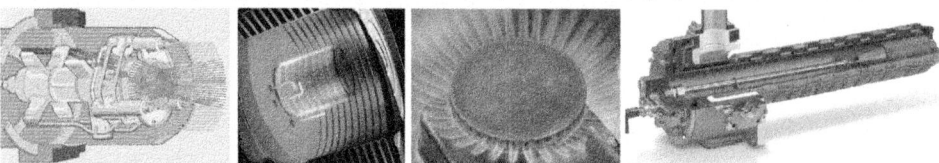

Figure 1.5 Multi-stage industrial burner; burner for different natural gas compositions; cooker burner; burner with recuperative heat recovery (from left to right)

Figure 1.6 *Diesel* engine; gas turbine; *Stirling* engine; liquid pump (from left to right)

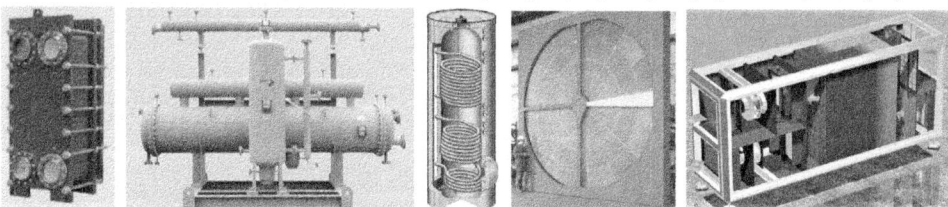

Figure 1.7 Plate heat exchanger; double-tube heat exchanger; stratified storage tank with coils; rotary heat exchanger; regenerative heat exchanger (from left to right)

Figure 1.8 Chiller; liquid chiller in outdoor installation; absorption chiller; industrial heat pump (from left to right)

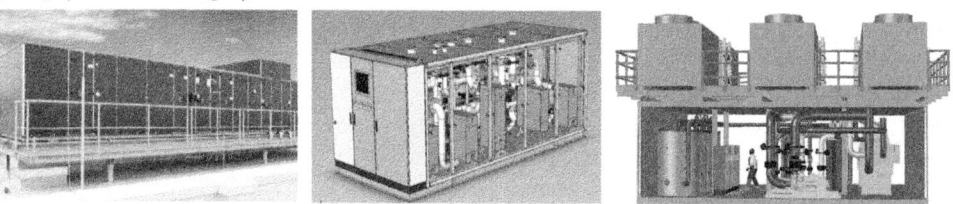

Figure 1.9 Ventilation and air-conditioning technology: central air-conditioning unit for roofs; heat pump for heating and cooling of large buildings; combined heat and power unit, absorption heat pump and recooling plant in a data centre (from left to right)

Figure 1.10 Fluid transport in a chemical plant; chemical process technology; sea-based natural gas production and processing plant (from left to right)

Research into the thermodynamic processes that take place in machines and appliances requires the introduction of concepts and terms such as system and environment, state and change of state, state variables, process and process variables.

1.2 System

The part of a machine or plant that is to be the subject of thermodynamic investigation is called a system. It is delimited by the system boundary from the environment, which lies outside from the system.

The systems to be investigated can have very different sizes and contents. For example, a system can include an entire thermal power plant, if the investigation comprises the conversion of energy released by the combustion of a fuel via energy of flowing steam via mechanical energy by a turbine into electrical energy by a generator (Figure 1.11). If, on the other hand, the energy conversion in a steam turbine is considered, it makes sense to make only the steam turbine the content of the system. Finally, the energy transformation in the blade channel of the steam turbine becomes clear if a volume element of the steam flowing in it forms the system content.

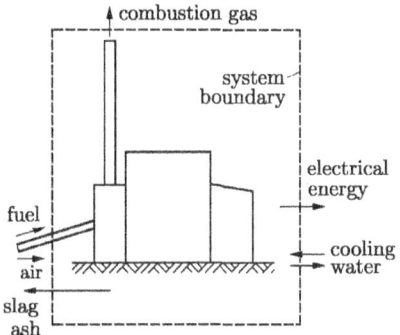 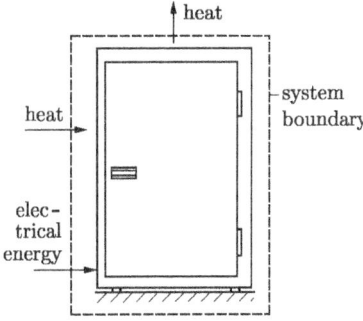

Figure 1.11 Thermal power plant as an example of an open system

Figure 1.12 Refrigerator as an example of a closed system

If the amount of substance in a system remains constant, it is called a closed system. The volume of the amount of substance can be variable. There is no mass transport across the system boundary. Energy transport across the system boundary is possible. A refrigerator, for example, is a closed system when the door is closed (Figure 1.12) although heat constantly flows through the insulating layer around the cooling chamber and thus through the system boundary into the system and although the condenser constantly transfers heat to the surrounding air. Electrical energy is supplied to the

system to drive the compressor, but no mass flow passes through the system boundary. Circulating refrigerant, cooled air and food remain within the system at all times.

An open system is defined by a specific space. Across the system boundary mass and energy can pass the system boundary. For example, the thermal power plant described before is an open system: Here, not only e.g. electrical energy, but also substances (e.g. fuel, water, air) cross the system boundary.

The compressor of a refrigerator is also an open system: During operation, refrigerant flows into the system and leaves it with increased pressure. The driving electric motor draws electrical energy from the electric grid.

Two special cases are worth noting: If the system boundary is not penetrated by either material or energy flows, it is referred to as an isolated system. A system that is heat-tight with respect to the environment is called an adiabatic system.

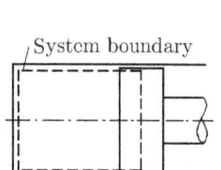
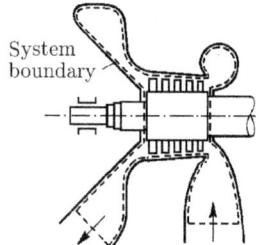
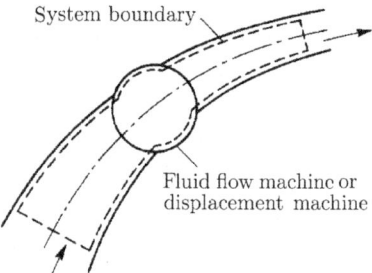

Figure 1.13 Piston machine (simple closed system)

Figure 1.14 Gas turbine (simple open system)

Figure 1.15 General view of a simple open system

To illustrate the thermodynamic laws, in the following we will use simple systems that are easier to understand. These simple systems contain parts (e.g. shafts, impellers) and a uniform homogeneous fluid (gas or liquid).

The simple closed system is approximately the homogeneous content of a container with a constant volume or of a cylinder which is closed by a movable piston. The simple open system can be a flowing fluid in a pipe, a flowing fluid in a piston machine or in a heat exchanger.

Non-simple systems are, for example, water and water vapor (two phases) in a steam boiler or the – locally differently composed – combustion gases in the combustion chamber of a boiler or in the cylinder of a combustion engine (combustion processes).

1.3 State, State Variables, Changes of State

All measurable properties of a system represent the state of a system. The description of a state is made by state variables, which always take on the same value when the state of the system is the same again.

In mechanics, e. g. the external state of a body is examined. To characterise the external state of a body information about the spatial coordinates, velocity or acceleration is used.

On the other hand, in thermodynamics one is interested in the internal state, which, in the case of simple systems is described e.g. by the mass m, the pressure p, the volume V and the temperature T. These properies are called internal state variables.

T is the absolute or thermodynamic temperature on the *Kelvin* scale[1] (sections 3.5.3 and 4.1.2). The relationship with the temperature t on the *Celsius* scale[2] is given by the relation

$$T = T_0 + t \,, \tag{1.1}$$

where

$$T_0 = 273.15 \text{ K} \tag{1.2}$$

on the Kelvin scale is the zero point of the Celsius scale. The zero point of the Kelvin scale $T = 0$ K is the absolute zero. With $dT = dt$ the temperature change

$$T_2 - T_1 = t_2 - t_1 \tag{1.3}$$

is the same on both scales (Figure 1.16).

The state variables pressure p, volume V and temperature T are called thermal state variables. These are important for the designer of a machine or system. The volume or flow rate determines the size, the pressure determines the wall thickness and the temperature determines the choice of material for the construction to be designed.

In thermodynamics, also caloric state variables are known, which are discussed in sections 2.5 and 3.3. The thermal state variables are linked at constant substance quantity by the following relationship, which is referred to as the thermal equation of state (Sect. 4.1.5, 4.1.6 and 5.3):

$$F(p, V, T) = 0 \tag{1.4}$$

If all thermal state variables describing the state of a closed system remain constant in time, the system is in equilibrium. The state of a closed system can be changed by external influences, e.g. by heating or cooling the substance or by changing the volume of the system. As a rule, the state variables will change in the process.

Since state variables only depend on the state of the system, it follows as an important consequence, the way in which these final values are reached from an initial state to a final state is irrelevant. State variables are path-independent, i.e. the final value of a state variable does not depend on the way in which one arrives at this final value. If the final state reached by the system is the same as the initial state, the state variables have the same values as at the beginning of the change of state.

The quality of being path-independent is not self-evident for all variables. If a quantity of gas is heated from a temperature T_1 to a temperature T_2 at constant pressure, while a second, identical quantity of gas is heated from the same initial state via different intermediate states with changes in pressure, finally to the same final state with the temperature T_2 and the initial pressure, then the ways used have no influence on all final state variables.

If, however, one asks about the heat required for these changes of state or the work to be done, these two changes of state show clear differences. Heat and work are not state variables, but process variables, since their description also includes the indication of the ways used.

---

[1] Sir *William Thomson*, 1824 to 1907, since 1892 Lord *Kelvin*, Professor of Physics in Glasgow
[2] *Anders Celsius*, 1701 to 1744, Swedish astronomer, director of the observatory in Uppsala

The amount of substance of the fluid in a closed system can be described by the mass m or the molar quantity n. Between the two state variables m and n there is the relation

$$m = Mn, \qquad (1.5)$$

where M is called the molar mass. The mass m can be represented as the product of the density ρ and the volume V: $\quad m = \rho V$

In an open system, the substance flow takes the place of the quantity of substance; it is described by the mass flow $\dot{m}$ or the molar flow $\dot{n}$:

$$\dot{m} = \frac{m}{\tau} \qquad\qquad \dot{n} = \frac{n}{\tau} \qquad (1.6)$$

τ is the time in which the mass m or the molar quantity n flows through a flow cross-section. If the mass or molar quantity – flowing in a considered cross-section – is constant in time, then the flow process is called stationary. The mass flow rate $\dot{m}$ can be calculated as the product of the density ρ and the volume flow rate $\dot{V}$: $\quad \dot{m} = \rho \dot{V}$

In the case of continuous operation of machines, stationary processes generally occur. This case can be described by Eqs. (1.6). During start-up or shutdown, the material flow increases or decreases, and the process is unsteady. Eqs. (1.6) are changed to

$$\dot{m} = \frac{dm}{d\tau} \qquad\qquad \dot{n} = \frac{dn}{d\tau}. \qquad (1.7)$$

In an open system, stationary processes are to be assumed in the following.

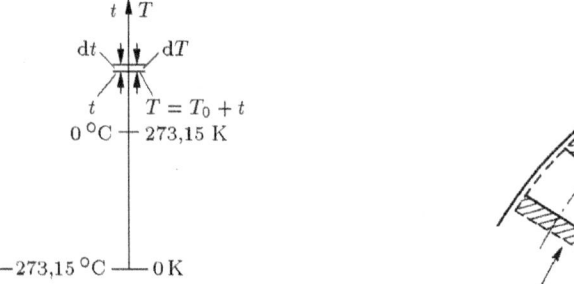

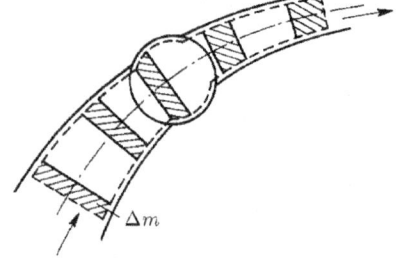

Figure 1.16 *Celsius* and *Kelvin* temperature scale

Figure 1.17 Moving closed system: substance quantity Δm flowing through an open system

In a closed system, a change of state is based on a temporal sequence. The numbers on the state variables indicate a temporal sequence. Thus the system, for example, first has the temperature T_1 and then, at the same place, the temperature T_2. In the open system, on the other hand, the numbers indicate the local sequence of the states of the mass flow, whereby the counting method is determined by the flow direction of the fluid. Thus, a mass flow at the inlet of the system has e.g. the temperature T_1 and at the same time has the temperature T_2 at the outlet of the system.

Both approaches can be combined if one follows a delimited quantity of substance with mass Δm as it passes through an open system. This delimited quantity of substance can be regarded as a moving closed system (Fig. 1.17). In the quantity of substance Δm under consideration, the state variables change in time succession between two states occurring simultaneously at different locations.

Changes of state in which certain state variables are kept constant, have been given special names:

$V =$ const	$dV = 0$	Isochoric change of state
$p =$ const	$dp = 0$	Isobaric change of state
$T =$ const	$dT = 0$	Isothermal change of state

1.4 Process, Process Variables

The action on a system is called a process, whereby in the general case a change of state of the quantity of substance is caused. It is possible that the same change of state is caused by different processes.

This becomes clear in the following example: The statement that the temperature of a certain amount of water is risen from 10 °C to 20 °C in an open container (and thus isobaric) is sufficient to describe the change of state: $t_2 - t_1 = 20\,°C - 10\,°C$. The process to which this change of state belongs is not yet defined. It can take place in such a way that the heating is effected by a supply of heat from the outside. The same change of state can also occur when an agitator rotates in the water, whereby the temperature change takes place through friction work.

Even if a change of state is generally associated with the running of a process, the two terms must nevertheless be strictly distinguished. The indication of the process describes more precisely, what happens. The change of state is only a part of the process description. From the above example we can conclude that the energy quantities heat and friction work cannot be described by the change of state alone. Rather, knowledge of the process is necessary to make a statement about the heat or friction work required for the change of state. Heat and friction work are therefore referred to as process variables and are not state variables. In contrast to the state variables, the process variables are path-dependent. They act over the process duration and thus during the change of state on the system.

Tasks For Section 1

1. In gas welding, ethine (acetylene) C_2H_2 and oxygen O_2 are reacting with each other to produce high temperatures. These gases are contained under pressure in two gas cylinders and have the masses 3.418 kg and 10.500 kg respectively. What are the molar quantities n of both gases if their molar masses M are 26.04 kg/kmol and 32.00 kg/kmol?

2. 0.04 kmol/min of liquid bioethanol (ethyl alcohol) C_2H_5OH is flowing into a tank; its molar mass M is 46.08 kg/kmol. What is the mass flow rate? How many hours does it take for the tank to contain 221 kg of bioethanol?

3. For transport by sea in tankers, natural gas is liquefied (Liquefied Natural Gas LNG) and stored as liquid methane CH_4 at ambient pressure at a cryogenic temperature of $t = -161.5\,°C$. What is the *Kelvin* temperature of the LNG?

2 The First Law of Thermodynamics

2.1 The Principle of Conservation of Energy

In mechanics, the principle of the conservation of energy is applied to two types of energy which occur in the frictionless motion of bodies in the earth's gravitational field: potential and kinetic energy. It states that in a closed mechanical system, the sum of both energies remains constant. In thermodynamics, the principle of the conservation of energy is applied to the following types of energy:

> mechanical energy (potential energy, kinetic energy)
> work
> thermal energy (internal energy, heat, enthalpy)
> electrical energy
> chemical energy
> nuclear energy

This comprehensive energy principle is called the first law of thermodynamics. In the formulation of the first law, only mechanical and thermal energy or work appear. The other types of energy are replaced by mechanical or thermal energy or work.

If one wishes to trace the historical development that led to the formulation of this theorem one must bear in mind that the concept of heat in the recent thermodynamics has been given a more restrictive definition than in earlier times of science. In the following description, the more comprehensive term of thermal energy is used today in place of the formerly common term heat.

For a long time, the view prevailed that thermal energy was not a form of energy but a material quantity. It was at the end of the 18th and in the 19th century that the concept of the energetic nature of this quantity became established.

In the years from 1842 to 1850 *Joule*[1] demonstrated experimentally that the conversion of mechanical energy into thermal energy always produced the same ratio of values. Independently of this, *Mayer*[2] had calculated this in 1842 through theoretical considerations and pointed out that the reverse energy conversion is also possible. In 1847 *Helmholtz*[3] developed the extended principle of the conservation of energy, which has since been called the first law of thermodynamics:

---

[1] *James Prescott Joule* (pronounced dschuhl), 1818 to 1889, owner of a brewery, was engaged in experimental investigations of electromagnetic processes and in the relationship between heat and work.

[2] *Robert Mayer*, 1814 to 1878, doctor in Heilbronn (South West Germany), since 1841. His research on the equivalence of work and heat did not receive the necessary recognition from contemporary physicists.

[3] *Hermann von Helmholtz*, 1821 to 1894, studied medicine and physiology, in 1849 became professor of physiology in Königsberg, came to Berlin as professor of physics via Bonn and Heidelberg in 1871. In 1888 he was elected president of the newly founded Physikalisch-Technische Reichsanstalt in Charlottenburg. In his treatise " Über die Erhaltung der Kraft" ("On the Conservation of Force"), which formulated the energy principle, Mayer's achievements are not mentioned. The term "force" was previously used in the sense of "energy".

> In an isolated system, the total amount of energy can neither be increased nor decreased. Only the different types of energy can be converted into each other.

According to the first law, a "perpetuum mobile" is impossible. One does not understand a "perpetuuml mobile" as a device that remains in constant movement without external drive, as the name actually implies, but a machine that creates energy in form of work out of nothing. A perpetual movement without drive is possible in the absence of friction and does not contradict the energy principle. An example is the movement of the planets around the sun. Since a "perpetuum mobile" contradicts the first law, it can also be formulated as follows:

> A perpetum mobile of the first kind is impossible.

The first law of thermodynamics is a principle (axiom) and cannot be proven. Its validity, however, is assured by many experiments.

In the following, the different types of energy occurring in thermodynamics are to be considered to find a formulation of the energy transformations for the closed and the open system according to the first law In doing so, it is necessary to describe the quantities "heat" and "work" more precisely. We make this agreement with respect to the system and determine according to Fig. 2.1:

> Work and heat are positive when they are supplied to the system and negative when they are released from the system.

Figure 2.1 Sign agreement and representation of heat and work in figures

2.2 Potential Energy

The forces in the earth's gravitational field are the cause of the potential energy. In the earth's gravitational field, two forces act on every body: the weight G and the buoyancy A. The static buoyancy or Archimedean buoyancy A, which occurs when a body is immersed in a gaseous or liquid medium, may be less than, equal to or even greater than the weight G of the body. The buoyancy is calculated from the weight of the gaseous or liquid medium displaced by the body, surrounding the body. The weight G of a body or a quantity of substance with the mass m, the volume V, the density ϱ and the specific weight γ is

$$G = mg = V\varrho g = V\gamma , \qquad (2.1)$$

where g is the acceleration due to gravity.

$$g = 9.80665 \text{ m/s}^2 = 9.80665 \text{ N/kg} \qquad (2.2)$$

In technical calculations, the approximate value 9.81 m/s² often is used.

The buoyancy A of a body or a quantity of matter, which is characterised by a volume V, in an environment with the density ϱ_U and the specific weight γ_U is

$$A = V \varrho_U g = V \gamma_U . \qquad (2.3)$$

If, in the superposition of the weight and the buoyancy, the weight predominates, then the excess weight or the resulting weight is positive. If the buoyancy predominates, then the excess weight is negative:

$$F = G - A = V(\varrho - \varrho_U) g = V(\gamma - \gamma_U) \qquad (2.4)$$

In the case of solid bodies or liquids that are in a gas, one can generally neglect the buoyancy A. In the case of a gas surrounded by another gas, this neglection is not possible, since weight and buoyancy are of the same order of magnitude.

If one imagines a partial air volume of the same temperature delimited in an air space (e.g. in a hot air balloon before take-off), then it follows $G = A$ and $F = 0$ for this partial air volume. On a partial air volume which is warmed up compared to its environment (e.g. in a hot air balloon after the take-off) with the specific weight γ and a volume V, according to Eq. (2.4) acts a negative excess weight. In the heated partial air volume and in the environment the same pressure prevails. The partial air volume, which fills the volume V in the heated state, occupied the volume V_U before heating. The weight G of this has not changed:

$$G = V \gamma = V_U \gamma_U \qquad (2.5)$$

In the air space, the following relationship applies between the specific weight and the absolute temperature:

$$\frac{\gamma}{\gamma_U} = \frac{T_U}{T} \qquad (2.6)$$

With the thermal expansion coefficient of an ideal gas (section 4.1.2)

$$\beta = \frac{1}{T} \qquad \beta_U = \frac{1}{T_U} \qquad (2.7)$$

follows from Eq. (2.6)

$$\frac{\gamma}{\gamma_U} = \frac{\beta}{\beta_U} . \qquad (2.8)$$

With the excess temperature of the heated air compared to the environment

$$\Delta t = T - T_U \qquad (2.9)$$

one obtains from Eqs. (2.4) to (2.9) the excess weight

$$F = -V \gamma_U \beta \Delta t = -V \gamma \beta_U \Delta t = -V_U \gamma_U \beta_U \Delta t . \qquad (2.10)$$

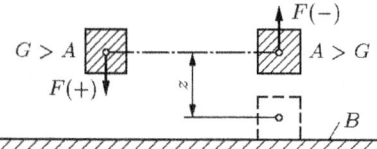

Figure 2.2 Potential energy
B reference plane

The potential energy of a body or a quantity of matter is

$$E_p = F z = V(\varrho - \varrho_U) g z , \qquad (2.11)$$

where z is the height measured from a reference plane B (Fig. 2.2). It corresponds to the work required to lift the body without friction from the reference plane B to the height z. For thermodynamics, the change in potential energy of a flow of substance between the inlet and the outlet cross-section of an open system is important, if the height position changes.

12 2 The First Law of Thermodynamics

If the variability of the densities ϱ and ϱ_U as a function of z is negligible, the following applies
$$E_{p2} - E_{p1} = V(\varrho - \varrho_U) g (z_2 - z_1) . \tag{2.12}$$

If one can disregard the buoyancy, then Eq. (2.12) is simplified to
$$E_{p2} - E_{p1} = V \varrho g (z_2 - z_1) = m g (z_2 - z_1) . \tag{2.13}$$

The path along which the change in altitude occurs has no influence. The potential energy is a state variable.

Example 2.1 The difference in height between the two water reservoirs of a storage power station is 220 m (Fig. 2.3). What power can be gained in a turbine plant if a flow of water of 20 m³/s flows from the upper to the lower basin and friction can be neglected?

$$\begin{aligned}\dot{E}_{p2} - \dot{E}_{p1} &= \dot{m} g (z_2 - z_1) = \dot{V} \varrho g (z_2 - z_1) \\ &= 20 \,\frac{\mathrm{m}^3}{\mathrm{s}} \cdot 10^3 \,\frac{\mathrm{kg}}{\mathrm{m}^3} \cdot 9.81 \,\frac{\mathrm{m}}{\mathrm{s}^2} \cdot (-220 \text{ m}) \cdot \frac{\mathrm{N\,s}^2}{\mathrm{kg\,m}} \,\frac{\mathrm{MJ}}{10^6 \,\mathrm{N\,m}} = -43.2 \text{ MW}\end{aligned}$$

If the difference in potential energy could be fully utilised, a power output of 43.2 MW could be obtained.

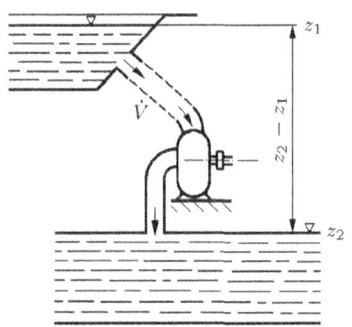

← **Figure 2.3** Illustration to example 2.1

↓ **Figure 2.4** Illustration to example 2.2

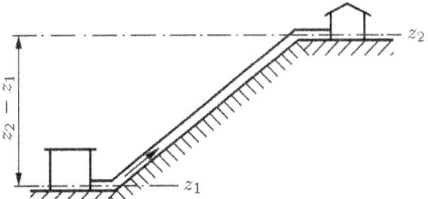

Example 2.2 A gas supply area is located 80 m above the gas plant, in which natural gas with a density of $\varrho = 0.742$ kg/m³ is fed into the pipeline network (Fig. 2.4). The ambient air has a density of $\varrho_U = 1.16$ kg/m³. What is the difference in potential energy between the gas plant and the supply area for 1 m³ of natural gas?

Gl. (2.12):
$$E_{p2} - E_{p1} = 1 \text{ m}^3 \cdot (0.742 - 1.16) \,\frac{\mathrm{kg}}{\mathrm{m}^3} \cdot 9.81 \,\frac{\mathrm{m}}{\mathrm{s}^2} \cdot 80 \text{ m} \cdot \frac{\mathrm{N\,s}^2}{\mathrm{kg\,m}} \,\frac{\mathrm{J}}{\mathrm{N\,m}} = -328 \text{ J}$$

The difference in potential energy can be used to overcome pipe friction.

Table 2.1 Relationships between pressure units

	Pa	bar	Torr*	atm*	at*
1 Pa =	1	10^{-5}	$7.50062 \cdot 10^{-3}$	$9.86923 \cdot 10^{-6}$	$1.01972 \cdot 10^{-5}$
1 bar =	10^5	1	750.062	0.986923	1.01972
1 Torr* =	133.322	$1.33322 \cdot 10^{-3}$	1	$1.31579 \cdot 10^{-3}$	$1.35951 \cdot 10^{-3}$
1 atm* =	101325	1.01325	760	1	1.03323
1 at* =	98066.5	0.980665	735.559	0.967841	1

* Units have no longer been permitted since 1.1.1978.   1 Pa = 1 N/m²
1 MPa = 10 bar   1 Torr = 1 mm Hg   1 at = 1 kp/cm²   1 mm WS = 1 kp/m²

Table 2.2 Relationships between energy units

		J	kJ	kW h	kcal*	kp m*
1 J	=	1	10^{-3}	$2.7778 \cdot 10^{-7}$	$2.3885 \cdot 10^{-4}$	1.10197
1 kJ	=	10^3	1	$2.7778 \cdot 10^{-4}$	0.23885	1019.7
1 kW h	=	$3.6 \cdot 10^6$	$3.6 \cdot 10^3$	1	859.85	367098
1 kcal*	=	4186.8	4.1868	$1.1630 \cdot 10^{-3}$	1	426.935
1 kp m*	=	9.80665	$9.80665 \cdot 10^{-3}$	$2.7241 \cdot 10^{-6}$	$2.3423 \cdot 10^{-3}$	1

* Units have no longer been permitted since 1.1.1978.
$1 \text{ J} = 1 \text{ N m} = 1 \text{ kg m}^2/\text{s}^2 = 1 \text{ W s} \quad 1 \text{ N} = 1 \text{ kg m}/\text{s}^2$

Table 2.3 Relationships between poweer units

		W	kW	kcal/h*	kp m/s*
1 W	=	1	10^{-3}	0.85985	0.10197
1 kW	=	10^3	1	859.85	101.97
1 kcal/h*	=	1.1630	$1.1630 \cdot 10^{-3}$	1	0.11859
1 kp m/s*	=	9.80665	$9.80665 \cdot 10^{-3}$	8.4322	1

* Units have no longer been permitted since 1.1.1978.

2.3 Kinetic Energy

If a body with a mass of m is to be accelerated from rest to a speed of c a certain amount of work must be done on the body. This work is stored in the moving body and is called kinetic energy. According to the laws of mechanics, the kinetic energy is

$$E_k = m \frac{c^2}{2} . \qquad (2.14)$$

If a fluid enters an open system with velocity c_1 and the exit velocity is c_2, then the change in the kinetic energy of the fluid is

$$E_{k2} - E_{k1} = \frac{m}{2} (c_2^2 - c_1^2) . \qquad (2.15)$$

The velocity of the particles in a real flow is locally different in a certain cross-section; therefore, the average flow velocity c is used to calculate the kinetic energy. It is obtained by dividing the volume flow $\dot{V}$ by the cross-sectional area A:

$$c = \frac{\dot{V}}{A} \qquad (2.16)$$

The kinetic energy is a state variable.

Example 2.3 In a waterfall, water flowing at a velocity of 1.2 m/s, falls freely from a height of 14 m. What increase in kinetic energy does 1 m³ of water achieve? What speed does the water reach?

$$E_{k2} - E_{k1} = \frac{m}{2}(c_2^2 - c_1^2) = m\,g\,h = 1000 \text{ kg} \cdot 9.81 \frac{m}{s^2} \cdot 14 \text{ m} \cdot \frac{\text{N s}^2}{\text{kg m}} \frac{\text{kJ}}{1000 \text{ N m}} = 137 \text{ kJ}$$

$$c_2 = \sqrt{2\,g\,h + c_1^2} = \sqrt{2 \cdot 9.81 \text{ m/s}^2 \cdot 14 \text{ m} + (1.2 \text{ m/s})^2} = 16.6 \text{ m/s}$$

2.4 Work

The concept of work is borrowed from mechanics. Mechanical work is done by forces whose mechanical center shifts in the direction of the force. Here work equals force multiplied with distance. When work is done on a body, its energy increases. The increase in energy has its cause in the work done; if friction can be neglected, then the increase in energy can be completely converted back into work. The increase in energy can therefore also be explained as work capacity.

Energy and work have the same dimension. This means that they are of the same type of magnitude and are measured in the same units. Energy is a generic term that also includes work as well. On the other hand, one must also distinguish between energy and work.

With the description of the energy of a body, one describes a state, while work is a process that takes place over a period of time. One can speak of the energy content of a body but there is no such thing as the work content. Work is a form of energy transfer.

In thermodynamics, work occurs in several different forms, which are given different names to make it easier to distinguish between them.

2.4.1 Volume Change Work

To explain the volume change work, let us consider a simple closed system with a variable volume (Fig. 2.5). The cylinder, which is closed by a frictionless piston contains a gas which exerts a pressure p on the inner boundary of the system and thus on the piston. The pressure p causes a inner force $F_p = p\,A$. A normal force F from outside acts as a counterforce on the piston. We will assume that the forces F and F_p acting on the piston are in equilibrium:

$$\vec{F} + \vec{F_p} = 0 \qquad \vec{F} = -\vec{F_p} \qquad (2.17)$$

Since the lines of action of the forces are equal, the following equation also applies:

$$F = -p\,A \qquad (2.18)$$

To compress the gas without friction, we move the piston by the displacement element ds. The force F does the work

$$dW_V = F\,ds\ . \qquad (2.19)$$

This work changes the volume of the gas, so it is called volume change work. The path element ds can be expressed by the piston area A and the volume change dV:

$$dW_V = -p\,A\,ds = -p\,A\,\frac{dV}{A} = -p\,dV \qquad (2.20)$$

Eq. (2.20) satisfies the sign agreement established in section 2.1. During the compression of the gas volume work is supplied to the system, it must therefore become positive; during an expansion (dV positive) work is released from the system, the volume change work becomes negative. The volume change work that is done on the way from the initial state 1 to the final state 2, is obtained by integrating equation (2.20). It is usual to write only the numbers 1 and 2 instead of the integration limits V_1 and V_2:

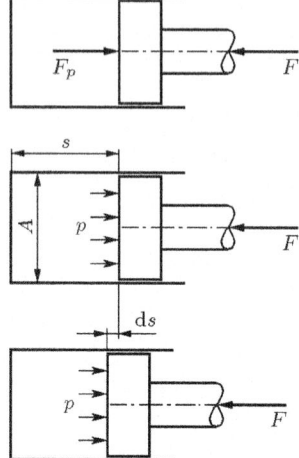

$$W_{V12} = -\int_1^2 p\,dV \qquad (2.21)$$

The integral is given by the course of the state curve between points 1 and 2 in the p,V diagram and is therefore dependent on the path. The volume change work is not a state variable, but a process variable. This is also expressed by the notation:

$$\int_1^2 dW_V = W_{V12} \qquad (2.22)$$

The notation $W_{V2} - W_{V1}$ would be wrong.

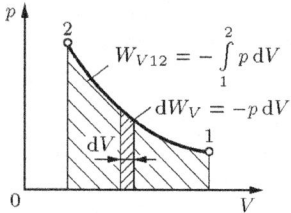

Figure 2.5 Representation of the volume change work in the p,V diagram

Example 2.4 1 kg of the refrigerant R 134 a with a temperature of 20 °C and a volume of $V_1 = 0.2344$ m³ is to be isothermally compressed to a volume $V_2 = 0.04296$ m³. According to example 5.9, the pressure during the isothermal change of state depends on the volume as follows:

$$p = \frac{\varphi_0}{V} + \frac{\varphi_1}{V^2} + \frac{\varphi_2}{V^3}$$

$\varphi_0 = 0.23888$ m³ bar $\varphi_1 = -0.0010558$ m⁶ bar $\varphi_2 = 0.00000087285$ m⁹ bar

What is the volume change work to be done?

$$W_{V12} = -\int_1^2 p\,dV = -\varphi_0 \int_1^2 \frac{dV}{V} - \varphi_1 \int_1^2 \frac{dV}{V^2} - \varphi_2 \int_1^2 \frac{dV}{V^3}$$

$$= -\varphi_0 \ln\frac{V_2}{V_1} + \varphi_1\left(\frac{1}{V_2} - \frac{1}{V_1}\right) + \frac{\varphi_2}{2}\left(\frac{1}{V_2^2} - \frac{1}{V_1^2}\right)$$

$$= 0.38548\,\text{m}^3\,\text{bar} \cdot \frac{10^5\,\text{N}}{\text{bar}\,\text{m}^2} \cdot \frac{\text{kJ}}{10^3\,\text{N}\,\text{m}} = 0.38548\,\text{m}^3\,\text{bar} \cdot \frac{100\,\text{kJ}}{\text{m}^3\,\text{bar}} = 38.548\,\text{kJ}$$

The sign is positive: The volume change work is supplied to the system.

2.4.2 Coupling Work

If the thermodynamic system includes a machine, which is connected via a machine shaft to a second machine outside of the system, then the machine shaft is cut by the system boundary as a component of energy transfer (Fig. 2.6). Via this shaft, work can either be supplied to the system (e.g. in the case of a compressor) or work can be delivered from the system to a second system (e.g. in the case of a turbine). This type of work is called coupling work (or shaft work, effective energy, technical work) W_e. If the torque occurring on the shaft is designated as M, the angle of rotation as φ, the duration of the energy transfer as τ and the angular velocity as $\omega = d\varphi/d\tau$, then the following equation applies:

$$dW_e = M(\tau)\, d\varphi = M(\tau)\, \omega(\tau)\, d\tau \tag{2.23}$$

In the stationary case is

$$W_e = M\omega\tau \quad . \tag{2.24}$$

The coupling work is a process variable.

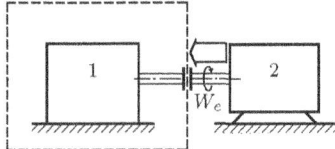

Figure 2.6 Example of the occurrence of coupling work: Drive of a machine 1 by an electric motor 2

Example 2.5 A compressor requires a coupling power of 20 kW to drive it. What is the torque occurring at the shaft at a speed of $n = 1200$ 1/min? What coupling work occurs during one revolution? The speed n can be understood as the reciprocal value of time τ, if one considers that during one revolution the angle $\varphi = 2\pi$ is passed through. Thus:

$$P_e = \dot{W}_e = \frac{W_e}{\tau} = \frac{M\varphi}{\tau} = M\omega = M\,2\pi n \qquad M = \frac{P_e}{2\pi n} = \frac{20 \text{ kW}}{2\pi\, 20 \text{ 1/s}} \cdot \frac{1000 \text{ N m}}{1 \text{ kW s}} = 159 \text{ N m}$$

$$W_e = \frac{P_e}{n} = \frac{20 \text{ kW}}{20 \text{ 1/s}} \cdot \frac{1 \text{ kJ}}{1 \text{ kW s}} = 1 \text{ kJ}$$

The sign is positive: The coupling work is supplied to the system.

2.4.3 Shift Work

If a quantity of substance is transported across a system boundary, the shift work occurs. We consider a simple open system (Fig. 2.7). A delimited mass m with the volume V_1 flows across the system boundary in the inlet cross-section A_1. If the pressure of the fluid at this point is p_1, then on entry into the system, work W_1 must be done on the fluid against this pressure.

$$W_1 = F_1\, s_1 = p_1\, A_1\, \frac{V_1}{A_1} = p_1\, V_1 \tag{2.25}$$

In the stationary process, at the same time, an equal amount of substance of mass m with volume V_2 leaves the outlet cross-section A_2 and has to perform work W_2 against the pressure p_2 prevailing there, i.e. work W_2 is done by the fluid against the pressure p_2, i.e. work is released by the fluid.

$$W_2 = F_2\, s_2 = p_2\, A_2\, \frac{V_2}{A_2} = p_2\, V_2 \tag{2.26}$$

often only the difference of the shift work is of interest:

$$W_2 - W_1 = p_2 V_2 - p_1 V_1 \tag{2.27}$$

Since the shift work depends only on the entry or exit state and not on the change of state that the mass m undergoes when passing through the open system. the shift work is a state variable.

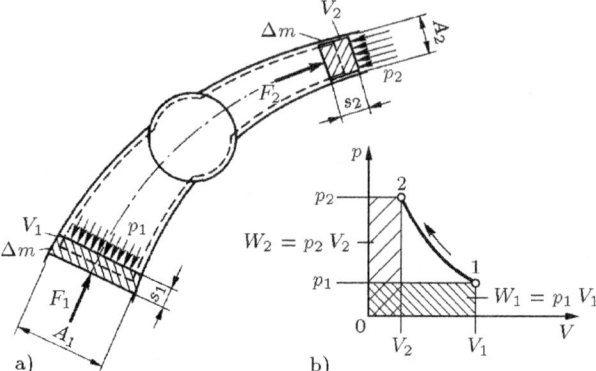

Figure 2.7 Representation of the shift work a) in an open system (view from above) b) in the p, V diagram

Example 2.6 In the compression mentioned in example 2.4, the pressure p_1 is 1 bar and the pressure p_2 is 5 bar. What shift work occurs on entry into the open system, and when flowing out of the open system?

$$W_1 = p_1 V_1 = 1 \text{ bar} \cdot 0.2344 \text{ m}^3 = 0,2344 \text{ bar m}^3 \cdot \frac{100 \text{ kJ}}{\text{m}^3 \text{ bar}} = 23.440 \text{ kJ} \qquad 1 \text{ m}^3 \text{bar} = 100 \text{ kJ}$$

$$W_2 = p_2 V_2 = 5 \text{ bar} \cdot 0.04296 \text{ m}^3 = 0,21480 \text{ bar m}^3 \cdot \frac{100 \text{ kJ}}{\text{m}^3 \text{ bar}} = 21.480 \text{ kJ}$$

2.4.4 Pressure Change Work

To understand this work, which occurs in open systems, we consider a compressor in the form of a turbo machine (Fig. 2.7). This compresses a gas flow from a low-pressure vessel at pressure p_1 into a high-pressure vessel at pressure p_2. The gas flow is compressed in the process. The space inside the compressor housing through which the gas flows forms an open system. The gas flowing through the system is supplied with work by the impeller of the compressor. This work is called pressure change work. It results from the volume change of the gas during compression $W_{V\,12}$ and the shift works that occur at the system boundary W_1 and W_2. The impeller of the compressor must apply the volume change work $W_{V\,12}$ and the shift work W_2 when the fluid is leaving the open system, while the shift work W_1 on entry into the open system is not performed by the impeller but by the low-pressure vessel:

$$W_{p12} = W_{V\,12} + W_2 - W_1 \ . \tag{2.28}$$

With Eqs. (2.21), (2.25) and (2.26) one obtains

$$W_{p12} = p_2 V_2 - p_1 V_1 - \int_1^2 p \, dV \ . \tag{2.29}$$

A p, V diagram according to Fig. 2.8 shows that the pressure change work can also be represented as an integral:

$$W_{p12} = \int_1^2 V\,dp \qquad (2.30)$$

$$dW_p = V\,dp = d(pV) - p\,dV \qquad (2.31)$$

Since the integrals in Eqs. (2.29) and (2.30) depend on the course of the state curve e.g. in a p, V diagram and thus depend on the path, the pressure change work is a process variable.

Example 2.7 In examples 2.4 and 2.6, what is the pressure change work during compression?

$$W_{p12} = W_{V12} + W_2 - W_1 = 38.548\text{ kJ} + 21.480\text{ kJ} - 23.440\text{ kJ} = 36.588\text{ kJ}\ .$$

It can also be calculated by the expression according to Eq. (2.30). For 1 kg of the refrigerant R 134 a with a temperature of 20 °C the following equation applies:

$$V = \frac{\psi_0}{p} + \psi_1 + \psi_2 p$$

$\psi_0 = 0.23882$ m³ bar $\qquad \psi_1 = -0.0043404$ m³ $\qquad \psi_2 = -0.000093042$ m³/ bar

$$W_{p12} = \psi_0 \ln\frac{p_2}{p_1} + \psi_1(p_2 - p_1) + \frac{\psi_2}{2}(p_2^2 - p_1^2)$$

With $p_1 = 1$ bar and $p_2 = 5$ bar, with equation (2.30) the above result is obtained. In the case of compression the pressure change work is positive.

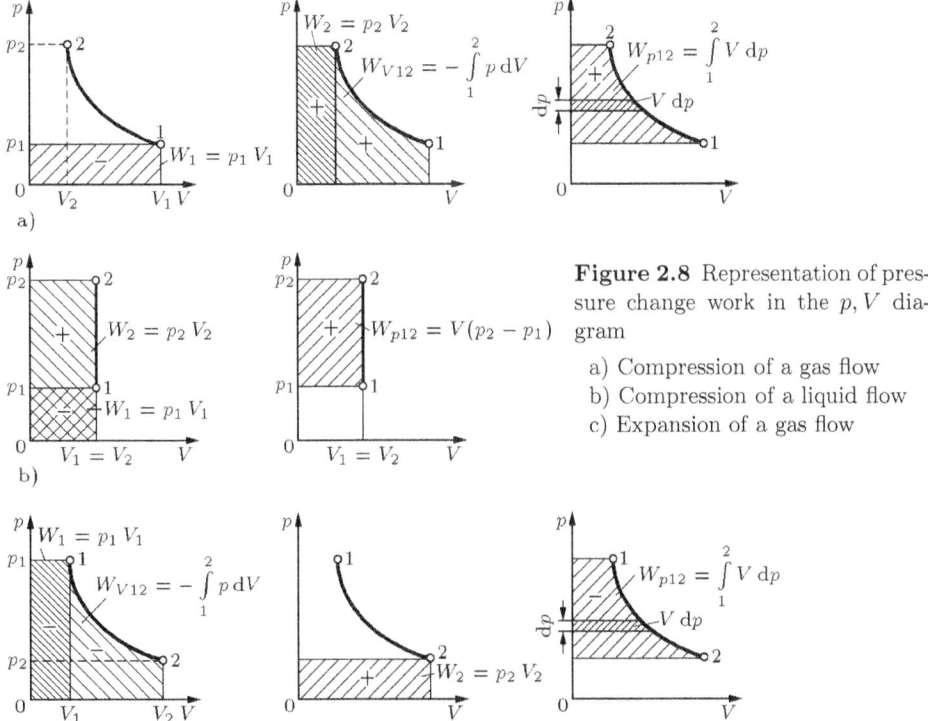

Figure 2.8 Representation of pressure change work in the p, V diagram

a) Compression of a gas flow
b) Compression of a liquid flow
c) Expansion of a gas flow

2.4.5 Friction Work

When a body slides on a solid surface, a friction force F_R occurs, which tries to stop the sliding movement. The tangential force F_T must overcome the friction force F_R, acting in the same line of action and directed in the opposite direction (Fig. 2.9). The magnitude of F_T is equal to the frictional force in the case of a uniform movement. By analogous application of Eqs. (2.17) and (2.19), the friction work results from the product of the friction force and the path.

This idea of friction work originates from mechanics. It is also applicable to a thermodynamic system, e.g. in the case of friction in the bearing of a shaft. For thermodynamic systems, however, friction is also important in fluid flows and has its cause in the viscous forces. These forces have the effect, for example, that in the cross-section of a flowing through an open system, locally different velocities are formed or, in other words, a velocity profile (Fig. 2.10) is created. Friction also occurs when vortices form in the flow due to changes in the cross-section or deflections.

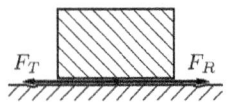
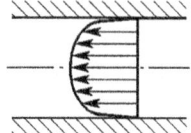

Figure 2.9 Sliding of a solid body: F_R Friction force F_T Tangential force

Figure 2.10 Velocity profile in a fluid

Friction work never exceeds the system boundary. It can occur inside or outside of a system. Friction work inside a system is called internal friction work. Friction work in the environment of the system is called external friction work. Internal friction work occurs, for example, when a fluid flows through an open system or when an agitator, which is driven from the outside, rotates in a liquid or a gas and increases the temperature and thus the internal energy of the liquid or gas. If a machine shaft is mounted outside of a system, external frictional work arises in the bearing.

The position of the system boundary determines whether a friction work is to be regarded as internal or external friction work. In Fig. 2.11 a and Fig. 2.11 b, the internal friction work is represented by the friction when a gas is flowing along the blades and vanes of a turbo machine (flow losses). The external friction work occurs at the bearing of the shaft. If the system boundary is set differently (Fig. 2.11 c) and e.g. forced cooling of the bearing is used (Fig. 2.11 d), the bearing friction is internal friction work, or it increases the internal energy of the coolant flow. In the case of a piston machine, the internal friction work occurs by the turbulence of the fluid as well as by a part of the sliding friction of the piston rings, which is transferred to the fluid (Fig. 2.11 e). The other part of the sliding friction is transferred through the cylinder wall to the environment or by oil as external friction work. External friction work also appears in the process of engine lubrication.

Friction work initially increases the internal energy of the material elements involved. For internal friction work, this means that temperature differences occure in a system and energy balancing flows are set in motion in a system. If the temperature fields in

a non-adiabatic system exceed the system boundary, part of the internal friction work can be lost out of the system as heat.

If the internal friction work is denoted by W_{RI} and the external friction work by W_{RA}, the total friction work done is W_R. Since the friction work does not exceed the system boundary, a sign, indicating an input to the system or an output from the system to the environment, is meaningless. When friction work is used in an energy balance and in the formulation of the first law of thermodynamics, one always uses the absolute amounts:

$$|W_R| = |W_{RI}| + |W_{RA}| \tag{2.32}$$

The friction work is a process variable.

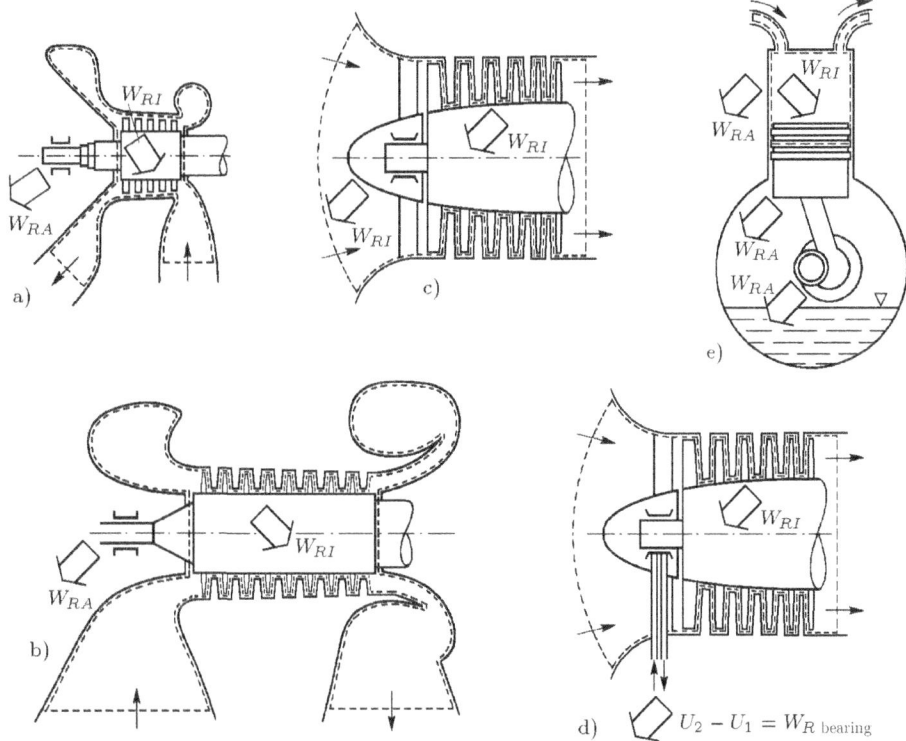

Figure 2.11 Internal and external friction work in machines
a), c) and d) gas turbine   b) turbo compressor   e) piston engine

2.5 Thermal Energy

In order to describe the concept of thermal energy, one has to use the idea of the structure of matter developed within the framework of classical physics. In this concept, the smallest particles that still have the properties of the quantity of matter under consideration are called molecules or - e.g. in the case of noble gases - atoms.

In solids as well as in liquids or gases, these constituting forms of matter are not at rest, but are in motion. In the crystalline solid, these movements consist of oscillations

around a position of rest; in a gas, the molecules move freely in space at high speeds. Besides oscillation and translation, in molecules that are made up of several atoms, there can also be rotational movements of the atoms in the molecule.

The total of the potential energies and kinetic energies of all molecules of a quantity of substance is called thermal energy. A characteristic feature of this thermal energy is that it is distributed in an unordered manner among the molecules of the quantity of substance. In addition, this distribution is steadily changed by exchange processes and can only be indicated by statistical laws. To measure the average thermal energy of a quantity of substance in a simple system, the state variable temperature is used.

2.5.1 Internal Energy

The thermal energy stored in the substance quantities of a system is referred to as the internal energy U of the system. An increase in the internal energy has the following effect: The system (without a phase transition) is reflected in an increase in the temperature of the substance. Modern thermodynamics does not use the illustrative concept of thermal energy given in section 2.5, which presupposes a model of the matter (phenomenological approach); it makes use of a definition, which also retains its unrestricted validity outside the field of classical physics.

A system that is thermally impermeable to its environment is called an adiabatic system (section 1.2). If we perform work on an adiabatic system, which we want to denote by W_{12ad}, we supply energy to the system. According to the energy principle, this energy must be stored in the system; we increase the internal energy of the system and can define

$$W_{12ad} = U_2 - U_1 \,. \tag{2.33}$$

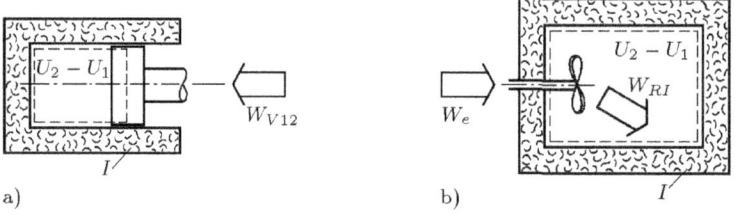

Figure 2.12 Increasing the internal energy of an adiabatic system by adding
a) volume change work W_{V12} b) coupling work W_e (I heat insulation)

If the adiabatic system gives off work, its internal energy decreases; at the same time its temperature also decreases.

The internal energy proves to be a state variable, because the final state U_2 can be reached by various processes from the initial state U_1. In a closed system, for example, W_{12ad} could be a volume change work W_{V12}, which increases the internal energy through compression (Fig. 2.12 a):

$$W_{V12} = U_2 - U_1 \tag{2.34}$$

However, W_{12ad} could just as well be a coupling work W_e, which is converted by an agitator into internal friction work $|W_{RI}|$ (Fig. 2.12 b):

$$W_e = |W_{RI}| = U_2 - U_1 \qquad (2.35)$$

In technical problems, it is often not the absolute value of the internal energy that is of interest, but only its difference. Statements about the absolute value of the internal energy are therefore not necessary in most cases.

2.5.2 Heat

The concept of heat is defined in thermodynamics in a very special and restricted way compared to the general use of language. To explain we consider a simple closed system that has a different temperature compared with the temperature of its environment. Let the temperature of the system be t_S, the temperature of the environment t_U. If the system is non-adiabatic, i.e. lacks any thermal insulation, then a temperature gradient will develop across the system boundary. This temperature gradient causes a transport of thermal energy across the system boundary, which is called heat Q_{12} (Fig. 2.13). Heat is therefore the energy that crosses the system boundary under the effect of a temperature gradient and therefore represents a form of energy transfer. Heat always flows by itself in the direction of the temperature gradient. It is not a state variable, heat is a process variable (section 3.3.3).

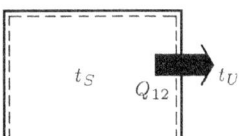

Figure 2.13 Heat transferred from a system with the temperature t_S to the environment with the temperature t_U. For $t_S > t_U$, according to the sign agreement, the heat Q_{12} is negative

The heat related to time represents the heat flux (heat flow) $\dot{Q}_{12}$:

$$\dot{Q}_{12} = \frac{Q_{12}}{\tau} \qquad (2.36)$$

τ is the time period in which the heat Q_{12} in a stationary process exceeds the system boundary. A transient process can be described as follows:

$$\dot{Q}_{12} = \frac{dQ}{d\tau} \qquad (2.37)$$

An equation for the heat flow $\dot{Q}_{12}$ can be derived from the fact that heat always flows by itself in the direction of a temperature gradient (Fig. 2.14). If one considers two systems separated by a wall, whereby the system with the higher temperature t transfers the heat flow $d\dot{Q}$ through the wall element dA to the system with the lower temperature t', then

$$d\dot{Q} = k\,dA\,(t - t') \,. \qquad (2.38)$$

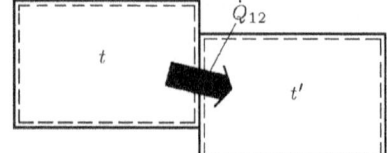

Figure 2.14 Heat flow from a system with temperature t to a system with temperature t'

The proportionality factor k is the heat transfer coefficient. When integrating over the total heat transfer area A, it is important whether the temperatures t and t' and thus the temperature difference $(t - t')$ are constant or variable. In the case of stationary heat transfer, temperatures can be assumed to be locally unchanged, as is the case, for example, with a house wall in the context of the calculation of the heat demand of buildings, the following result is obtained:

$$\dot{Q}_{12} = k\,A\,(t - t')\,. \tag{2.39}$$

If one has to consider that in the stationary case the temperature difference $t - t'$ is locally different, e.g. in the case of heat exchangers without a phase transition, then the mean temperature difference $(t - t')_m$ is used:

$$\dot{Q}_{12} = k\,A\,(t - t')_m \tag{2.40}$$

Since during heat transfer one system gives off heat and to the other system this heat is supplied, a distinction between positive and negative heat flow does not make sense. In this case, the heat flow is always calculated as positive.

2.5.3 Enthalpy

Enthalpy H is the sum of the internal energy U and the shift work pV; it is defined as follows:

$$H = U + pV \tag{2.41}$$

Since both the internal energy U and the shift work pV are state variables, the enthalpy H is also a state variable. It is for isobaric changes of state in closed and open systems and in the treatment of the work of adiabatic machines of particular importance. As with the internal energy, here is usually only the difference in enthalpy between the initial and final state of interest:

$$H_2 - H_1 = U_2 - U_1 + p_2 V_2 - p_1 V_1 \tag{2.42}$$

The differential of the enthalpy is

$$dH = dU + d(pV)\,. \tag{2.43}$$

Since pV is a function of two variables - namely p and V - $d(pV)$ is a total differential (Section 3.3.3):

$$d(pV) = \left(\frac{\partial(pV)}{\partial p}\right)_V dp + \left(\frac{\partial(pV)}{\partial V}\right)_p dV = V\,dp + p\,dV \tag{2.44}$$

Eq. (2.44) agrees with Eq. (2.31). From Eqs. (2.43) and (2.44) follows

$$dH = dU + V\,dp + p\,dV . \tag{2.45}$$

2.6 Energy Balances

The various energy transformations and energy transfers that occur in machines and apparatuses can be formulated by equations. The validity of the principle of the conservation of energy can always be assumed.

2.6.1 Energy Balance for the Closed System

The first law of thermodynamics for a closed system can be expressed as follows:

> Heat and work, which as forms of energy transfer are supplied to a closed system at rest, cause an increase of the internal energy of the system.

$$Q_{12} + W_{12} = U_2 - U_1 \tag{2.46}$$

If the work W_{12} supplied to the system consists of volume change work W_{V12} and internal friction work W_{RI} (Fig. 2.15), then Eq. (2.46) becomes

$$Q_{12} + W_{V12} + |W_{RI}| = U_2 - U_1 \tag{2.47}$$

and in differential form

$$dQ + dW_V + |dW_{RI}| = dU . \tag{2.48}$$

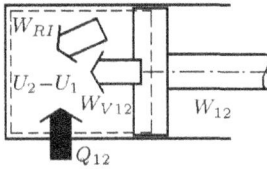

Figure 2.15 Representation of the first law of thermodynamics for closed systems

If no friction occurs, this becomes

$$Q_{12} + W_{V12} = U_2 - U_1 \tag{2.49}$$

or

$$Q_{12} - \int_1^2 p\,dV = U_2 - U_1 \tag{2.50}$$

and in differential notation

$$dQ - p\,dV = dU . \tag{2.51}$$

With frictionless isochoric change of state and with $dV = 0$, then Eq. (2.48) becomes

$$dQ = dU . \tag{2.52}$$

For frictionless adiabatic change of state, and with $dQ = 0$, Eq. (2.51) becomes

$$-p\,dV = dU . \tag{2.53}$$

We apply the statements of Eq. (2.46) to the simple systems introduced in section 1.2.

To a gas enclosed in a container of constant volume, work can be supplied in the form of coupling work W_e. If, for example, in a bearing external friction work W_{RA} occurs, only the difference $W_e - |W_{RA}|$ reaches the system boundary (Fig. 2.16). If this is converted into internal friction work W_{RI} via a paddle wheel, the following applies:

$$W_{12} = W_e - |W_{RA}| = |W_{RI}| \tag{2.54}$$

Eq. (2.46) becomes in this case

$$Q_{12} + W_e - |W_{RA}| = U_2 - U_1 \tag{2.55}$$

or

$$Q_{12} + |W_{RI}| = U_2 - U_1 . \tag{2.56}$$

Added heat and internal friction work increase the internal energy.

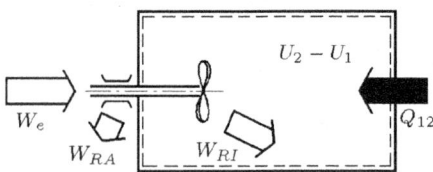
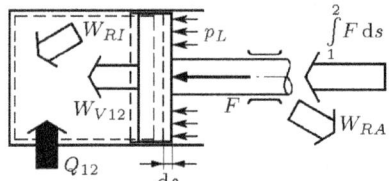

Figure 2.16 Energy balance for a container with constant volume

Figure 2.17 Energy balance for the compression of a gas

To compress a gas that is in a cylinder, work must be supplied to the piston from outside. The piston force F acts on the rod of the piston. The transmitted work $\int_1^2 F\,ds$ can again be reduced by the external frictional work occurring in a bearing. In addition, the air pressure p_L does the work $p_L(V_1 - V_2)$ during the piston movement (Fig. 2.17). Thus the work that is supplied to the system is

$$W_{12} = \int_1^2 F\,ds + p_L(V_1 - V_2) - |W_{RA}| . \tag{2.57}$$

In the system this work is used to compress the gas by the volume change work W_{V12}, but part of it can be transformed into internal friction work W_{RI}:

$$W_{12} = W_{V12} + |W_{RI}| = -\int_1^2 p\,dV + |W_{RI}| \tag{2.58}$$

Eq. (2.46) reads in this case

$$Q_{12} + \int_1^2 F\,ds + p_L(V_1 - V_2) - |W_{RA}| = U_2 - U_1 \tag{2.59}$$

or

$$Q_{12} - \int_1^2 p\,dV + |W_{RI}| = U_2 - U_1 . \tag{2.60}$$

Example 2.8 A cylindrical disc gas tank has a circular base area of $d = 16$ m in diameter and a height of $h = 20$ m. It is sealed gas-tight by a sliding lid. The lid has a mass of $m = 45\,320$ kg, the air pressure is $p_L = 1$ bar $= 100\,000$ N/m² $=$ const. Solar radiation causes the gas volume to increase while the air pressure and the gas pressure remain constant, and the lid rises frictionlessly and very slowly by $s = \Delta h = $ const $= 0.48$ m. How great is the work W_{L12} done by the gas to move the outside air, and the entire volume change work W_{V12} done by the gas, which causes not only the shift of the outside air but also the lift of the lid? What change in internal energy $U_2 - U_1$ occurs if the solar radiation is $Q_{12} = 40\,690$ kJ?

The base area of the tank is

$$A = \frac{\pi}{4} d^2 = 201.06 \text{ m}^2 \ .$$

When the volume of the gas increases, outside air is moved away. The work done by the gas against the air pressure is

$$W_{L12} = -\int_1^2 p_L \, dV = -p_L \int_1^2 dV = -p_L(V_2 - V_1) = -p_L\, s\, A$$

$$= -1000 \text{ mbar} \cdot 0.48 \text{ m} \cdot 201.06 \text{ m}^2 \cdot \frac{1 \text{ bar}}{1000 \text{ mbar}} \cdot \frac{100 \text{ kJ}}{\text{m}^3 \text{ bar}} = -9651 \text{ kJ}.$$

The volume change work, since the gas pressure remains constant, is

$$W_{V12} = W_{V12} = -\int_1^2 p\, dV = -p(V_2 - V_1) = -(p_L + \frac{G}{A})s\,A = -(p_L + \frac{m\,g}{A})s\,A$$

$$= -(100\,000\,\frac{\text{N}}{\text{m}^2} + \frac{45\,320\text{ kg} \cdot 9.81\text{ m}}{201.06\text{ m}^2\text{ s}^2} \cdot \frac{\text{N s}^2}{\text{kg m}}) \cdot 0.48\text{ m} \cdot 201.06\text{ m}^2 \frac{\text{kJ}}{1000\text{ N m}} = -9864 \text{ kJ} \ .$$

The difference between W_{V12} and W_{L12} is the work done by the gas on the lid, which is lifted by the distance $s = \Delta h$.

The change in internal energy is

$$U_2 - U_1 = Q_{12} + W_{V12} = 40\,690 \text{ kJ} - 9864 \text{ kJ} = 30\,826 \text{ kJ}.$$

As the internal energy of the gas increases, its temperature also increases.

2.6.2 Energy Balance for the Open System

The first law of thermodynamics for an open system states:

> **Heat and work, which are forms of energy transfer to an open system, cause an increase in the energy content of the material flow.**

$$Q_{12} + W_{12} = E_2 - E_1 \tag{2.61}$$

For the simple open system introduced in section 1.2, the energy content of the substance flow is the sum of internal, kinetic and potential energy. Neglecting the buoyancy, the energy content is

$$E = U + \frac{m}{2}c^2 + m\,g\,z \ . \tag{2.62}$$

The increase in the energy content of the material flow is thus

$$E_2 - E_1 = U_2 - U_1 + \frac{m}{2}(c_2^2 - c_1^2) + m\,g\,(z_2 - z_1) \ . \tag{2.63}$$

The work supplied to the system is made up of the shift work W_1 and the coupling work W_e, reduced by the external friction work W_{RA}. The shift work W_2 is delivered by the system (Fig. 2.18).

$$W_{12} = W_1 - W_2 + W_e - |W_{RA}| = p_1 V_1 - p_2 V_2 + W_e - |W_{RA}| \tag{2.64}$$

Eq. (2.61) takes the form

$$Q_{12} + W_e - |W_{RA}| + W_1 - W_2 = E_2 - E_1 \tag{2.65}$$

or

$$Q_{12} + W_e - |W_{RA}| + p_1 V_1 - p_2 V_2 = U_2 - U_1 + \frac{m}{2}(c_2^2 - c_1^2) + m g (z_2 - z_1) . \tag{2.66}$$

With Eq. (2.42) for the enthalpy we get

$$Q_{12} + W_e - |W_{RA}| = H_2 - H_1 + \frac{m}{2}(c_2^2 - c_1^2) + m g (z_2 - z_1) . \tag{2.67}$$

The open system can be understood as a moving closed system according to section 1.3. Eq. (2.47) applies: $Q_{12} + W_{V12} + |W_{RI}| = U_2 - U_1$ If this equation is inserted into Eq. (2.66), one obtains

$$W_e - |W_{RA}| + p_1 V_1 - p_2 V_2 = -\int_1^2 p \, dV + |W_{RI}| + \frac{m}{2}(c_2^2 - c_1^2) + m g (z_2 - z_1) . \tag{2.68}$$

Using Eqs. (2.29) and (2.32) it becomes

$$W_e = W_{p12} + \frac{m}{2}(c_2^2 - c_1^2) + m g (z_2 - z_1) + |W_R| . \tag{2.69}$$

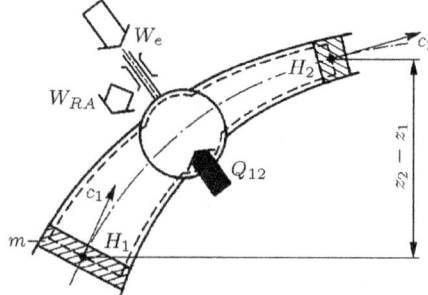

Figure 2.18 Representation of the first law of thermodynamics for open systems (view from the side)

From Eqs. (2.67) and (2.69) the final result is

$$Q_{12} + W_{p12} + |W_{RI}| = H_2 - H_1 \tag{2.70}$$

or with Eq. (2.30)

$$Q_{12} + \int_1^2 V \, dp + |W_{RI}| = H_2 - H_1 \tag{2.71}$$

and in differential notation

$$dQ + V \, dp + |dW_{RI}| = dH . \tag{2.72}$$

If the system is frictionless, then it becomes

$$Q_{12} + \int_1^2 V \, dp = H_2 - H_1 \tag{2.73}$$

and in differential notation
$$dQ + V\,dp = dH \,. \tag{2.74}$$
With frictionless isobaric change of state ($|dW_{RI}| = 0$ and $dp = 0$), it follows
$$dQ = dH \,. \tag{2.75}$$
For frictionless adiabatic change of state with $dQ = 0$ one obtains
$$V\,dp = dH \,. \tag{2.76}$$

In many cases it is advantageous to draw up a power balance instead of an energy balance. The following quantities then occur:

Temporal change of the potential energy $\quad \dot{E}_{p2} - \dot{E}_{p1} = \dot{m}\,g\,(z_2 - z_1)$

temporal change of the kinetic energy $\quad \dot{E}_{k2} - \dot{E}_{k1} = \dfrac{\dot{m}}{2}(c_2^2 - c_1^2)$

Coupling power $\quad \dot{W}_e = P_e = M\omega$

Frictional power $\quad \dot{W}_R = P_R \quad \dot{W}_{RI} = P_{RI} \quad \dot{W}_{RA} = P_{RA}$

Heat flow $\quad \dot{Q}_{12}$

Difference of enthalpy flow $\quad \dot{H}_2 - \dot{H}_1$

Eq. (2.67) results in
$$\dot{Q}_{12} + P_e - |P_{RA}| = \dot{H}_2 - \dot{H}_1 + \frac{\dot{m}}{2}(c_2^2 - c_1^2) + \dot{m}\,g\,(z_2 - z_1) \,. \tag{2.77}$$

The formulation of the first law of thermodynamics for an adiabatic flow process without supply or output of coupling power results in
$$\dot{H}_2 - \dot{H}_1 + \frac{\dot{m}}{2}(c_2^2 - c_1^2) + \dot{m}\,g\,(z_2 - z_1) = 0 \,. \tag{2.78}$$

The flow through a pipeline forms an open system in which coupling work, heat and external friction work are equal to zero. Let us assume that the differences in kinetic and potential energy between two considered cross-sections are negligible, then it follows

$$W_e = 0 \qquad W_{RA} = 0 \qquad c_2 = c_1 \qquad z_2 = z_1 \qquad Q_{12} = 0 \,.$$

According to Eq. (2.69), then, with Eq. (2.32), the result is
$$W_{p12} = -|W_{RI}| \tag{2.79}$$
or with Eq. (2.30)
$$|W_{RI}| = -\int_1^2 V\,dp \,. \tag{2.80}$$

If there is internal friction work (section 2.4.5) in the pipeline, the pressure in the direction of flow must decrease, otherwise the right hand side of equation (2.80) cannot be positive. In the case of flow through heat exchangers of all types (e.g. heaters, coolers, evaporators, condensers, boilers) in general the same conditions apply as for pipes, except for the heat transfer. Eq. (2.70) gives with Eq. (2.79)
$$Q_{12} = H_2 - H_1 = U_2 - U_1 + p_2 V_2 - p_1 V_1 \,. \tag{2.81}$$

2.6 Energy Balances

The heat transferred can be calculated from the change in enthalpy of the substance flow. Since one strives to keep the internal friction work small, the pressure drop Δp measured against the absolute pressure p is generally so small that the flow can be considered isobaric. For incompressible fluids ($V = $ const) it follows that

$$Q_{12} = U_2 - U_1 . \tag{2.82}$$

Example 2.9 At a water turbine, in the inlet cross-section with a diameter of $d_1 = 400$ mm at a gage pressure of $p_e = 23.2$ bar a velocity of $c_1 = 4.6$ m/s is measured. In the outlet cross-section 4.2 m below, the water flows with a velocity of $c_2 = 8.3$ m/s without gage pressure but only with atmospheric pressure of the outside air. This atmospheric pressure is $p_L = 0.98$ bar. The temperature increase of the water between the inlet cross-section and the outlet cross-section is determined as $\Delta t = 0.24$ K; the specific heat capacity of the water is $c = 4.19$ kJ/(kg K); its density is $\rho = 1000$ kg/m^3 = const. The flow through the turbine is adiabatic. How large is the mass flow $\dot{m}$? What is the coupling power P_e of the turbine if there is no external friction power $|P_{RA}|$? What is the internal friction power $|P_{RI}|$? What coupling power $(P_e)_{id}$ would the turbine deliver if there were no internal friction $|P_{RI}|$?

In the given pressure range, water can be assumed to be incompressible: $\dot{V}_1 = \dot{V}_2 = \dot{V} = \dot{m}/\rho$

The mass flow rate $\dot{m}$ is given by

$$\dot{m} = \dot{V}\rho = A_1 c_1 \rho = \frac{\pi}{4} d_1^2 c_1 \rho = \frac{3.14159}{4} \cdot 0.4^2 \text{ m}^2 \cdot 4.6 \frac{\text{m}}{\text{s}} \cdot 1000 \frac{\text{kg}}{\text{m}^3} = 578 \frac{\text{kg}}{\text{s}}$$

In Eq. (2.77), $\dot{Q}_{12} = 0$ and $|P_{RA}| = 0$ can be set:

$$P_e = \dot{H}_2 - \dot{H}_1 + \frac{\dot{m}}{2}(c_2^2 - c_1^2) + \dot{m} g (z_2 - z_1)$$

$$= \dot{U}_2 - \dot{U}_1 + p_2 \dot{V}_2 - p_1 \dot{V}_1 + \frac{\dot{m}}{2}(c_2^2 - c_1^2) + \dot{m} g (z_2 - z_1)$$

Using Eq. (2.84) one obtains:

$$P_e = \dot{m}[c\Delta t + \frac{p_2 - p_1}{\rho} + \frac{1}{2}(c_2^2 - c_1^2) + g(z_2 - z_1)]$$

$$P_e = 578 \text{ kg/s} \cdot [4.19 \text{ kJ/(kg K)} \cdot 0.24 \text{ K} + \frac{(0.98 - 24.18) \cdot 10^5 \text{ N/m}^2}{1000 \text{ kg/m}^3} \frac{\text{kJ}}{1000 \text{ Nm}} +$$

$$+ \frac{1}{2}(8.3^2 - 4.6^2)\frac{\text{m}^2}{\text{s}^2}\frac{\text{N s}^2}{\text{kg m}}\frac{\text{kJ}}{1000 \text{ Nm}} + 9.81 \frac{\text{m}}{\text{s}^2} \cdot (-4.2 \text{ m})\frac{\text{N s}^2}{\text{kg m}}\frac{\text{kJ}}{1000 \text{ Nm}}]$$

$$= 578 \text{ kg/s} \cdot [1.0056 - 2.32 + 0.023865 - 0.041202] \text{ kJ/kg} = -770 \text{ kW}.$$

The internal friction power is obtained by considering the open system as a moving closed system. By Eq. (2.47), since $\dot{Q}_{12} = 0$ and due to incompressibility, also $P_{V12} = 0$, becomes

$$|P_{RI}| = \dot{U}_2 - \dot{U}_1 = \dot{m} c \Delta t = 581 \text{ kW} .$$

If no friction power occurs, the water temperature remains constant: $\dot{U}_2 = \dot{U}_1$. The coupling power in the ideal case is

$$(P_e)_{id} = \dot{m}[\frac{p_2 - p_1}{\rho} + \frac{1}{2}(c_2^2 - c_1^2) + g(z_2 - z_1)]$$

$$(P_e)_{id} = 578 \text{ kg/s} \cdot [\frac{(0.98 - 24.18) \cdot 10^5 \text{ N/m}^2}{1000 \text{ kg/m}^3} \frac{\text{kJ}}{1000 \text{ Nm}} +$$

$$+ \frac{1}{2}(8.3^2 - 4.6^2)\frac{\text{m}^2}{\text{s}^2}\frac{\text{N s}^2}{\text{kg m}}\frac{\text{kJ}}{1000 \text{ Nm}} + 9.81 \frac{\text{m}}{\text{s}^2} \cdot (-4.2 \text{ m})\frac{\text{N s}^2}{\text{kg m}}\frac{\text{kJ}}{1000 \text{ Nm}}]$$

$$= 578 \text{ kg/s} \cdot [-2.32 + 0.023865 - 0.041202] \text{ kJ/kg} = -1351 \text{ kW}.$$

Example 2.10 In a heat exchanger as part of the water cooling system of a truck, a mass flow of $\dot{m} = 3.64$ kg/s of a water/glycol mixture with an inlet temperature of $t_1 = 75.0\,°C$ is cooled down to $t_2 = 32.0\,°C$. The specific heat capacity is $c = 3.687$ kJ/(kg K). Before entering the heat exchanger, a velocity of $c_1 = 2.1$ m/s and a pressure of $p_1 = 1.23$ bar of the coolant are measured. After leaving the heat exchanger, the velocity of the coolant is $c_2 = 1.8$ m/s, and the pressure drop in the heat exchanger is $\Delta p = p_2 - p_1 = $ -5200 N/m². The liquid has the density $\rho = 1000$ kg/m³.

What heat flow $\dot{Q}_{12}$ is transferred to the coolant if the outlet measuring point is $\Delta z = z_2 - z_1 = 0.6$ m above the inlet measuring point?

With $P_e = 0$ and $P_{RA} = 0$, Eq. (2.77) yields

$$\begin{aligned}
\dot{Q}_{12} &= \dot{U}_2 - \dot{U}_1 + p_2\dot{V}_2 - p_1\dot{V}_1 + \frac{\dot{m}}{2}(c_2^2 - c_1^2) + \dot{m}\,g\,(z_2 - z_1) \\
&= \dot{m}\left[c\,\Delta t + \frac{p_2 - p_1}{\varrho} + \frac{1}{2}(c_2^2 - c_1^2) + g(z_2 - z_1)\right] \\
&= 3.64\text{ kg/s} \cdot \left[3.687\text{ kJ/(kg K)} \cdot (-43.0\text{ K}) - \frac{5200\text{ N m}^3}{\text{m}^2\,1000\text{ kg}}\frac{\text{kJ}}{1000\text{ Nm}}\right. \\
&\quad \left.+ \frac{1}{2}(1.8^2 - 2.1^2)\frac{\text{m}^2}{\text{s}^2}\frac{\text{N s}^2}{\text{kg m}}\frac{\text{kJ}}{1000\text{ Nm}} + 9.81\frac{\text{m}}{\text{s}^2} \cdot 0.6\text{ m}\frac{\text{N s}^2}{\text{kg m}}\frac{\text{kJ}}{1000\text{ Nm}}\right] \\
&= 3.64\text{ kg/s} \cdot (-158.541 - 0.0052 - 0.000585 + 0.005886)\text{ kJ/kg} = -577\text{ kW} .
\end{aligned}$$

Neglecting the pressure loss, the change in kinetic energy and the change in potential energy, practically the same result is obtained.

2.7 Heat Capacity

If heat is added to a solid substance without any work occurring and without changing its state of aggregation, its temperature increases. The increase in temperature depends on the heat capacity of the substance: with the same heat input, a small temperature increase is associated with a large heat capacity, and a large increase in temperature is associated with a small heat capacity. In the case of isobaric melting or evaporation of a pure substance, the temperature remains the same despite the heat input. This means an infinitely large heat capacity during melting or evaporation. The same effect as a heat supply has the internal friction work.

Friction work is done e.g. when a solid body rubs against a rough surface. If one considers the solid body as a system and the surface of the solid body as a system boundary, the frictional work performed between the stationary surface and the moving body is partly external and partly internal friction work. That part of the friction work which causes an increase in the temperature of the stationary surface is external friction work. Internal friction work is the part that causes the temperature of the moving body to rise. The internal friction work thus has the same effect as an addition of heat. The temperature increase also depends on the heat capacity of the body under consideration.

Between the sum of heat input dQ and internal friction work dW_{RI} as well as the temperature increase dt there is a linear relationship, whereby the heat capacity C is the proportionality factor:

$$dQ + |dW_{RI}| = C\,dt = m\,c\,dt \tag{2.83}$$

2.7 Heat Capacity

c is the specific heat capacity (Section 2.7.1). For solids and liquids the change in volume associated with the increase in temperature can be practically neglected. Thus the volume change work is $W_{V12} = 0$, and the first law for closed systems takes the form of Eq. (2.48)

$$dQ + |dW_{RI}| = dU = C\,dt = m\,c\,dt\ . \tag{2.84}$$

If one relates all quantities to time in a fluid, then it follows

$$d\dot{Q} + |d\dot{W}_{RI}| = \dot{C}\,dt = W\,dt\ . \tag{2.85}$$

The time-related heat capacity of the fluid flow $\dot{C}$ can be understood as the heat capacity flow. $\dot{C}$ is also called the water value and is denoted by the letter W. This is related to the original definition of the kilocalorie, according to which the kilocalorie is the heat required to heat one kilogram of water by one degree centigrade. The water value is then the substitute water flow of a fluid flow which, with the same heat flow, results in the same temperature change.

2.7.1 Specific Heat Capacity

The heat capacity C related to the mass m is the specific heat capacity c

$$c = \frac{C}{m} \tag{2.86}$$

or with Eq. (2.84)

$$c = \frac{dQ + |dW_{RI}|}{m\,dt}\ . \tag{2.87}$$

The specific heat capacity is generally a function of temperature. It increases with increasing temperature. The specific heat capacity according to Eq. (2.87) is called the true specific heat capacity.

The integration of Eq. (2.83) with respect to Eq. (2.86) gives

$$Q_{12} + |W_{RI}| = m\int_1^2 c\,dt\ . \tag{2.88}$$

For practical use, it is more convenient to write Eq. (2.88) with the mean specific heat capacity c_m:

$$Q_{12} + |W_{RI}| = m\,c_m(t_2 - t_1) \tag{2.89}$$

According to the right part of Fig. 2.19, Eqs. (2.88) and (2.89) lead to

$$c_m = \frac{1}{t_2 - t_1}\int_1^2 c\,dt\ . \tag{2.90}$$

Averaging according to Eq. (2.90) for each application of Eq. (2.89) is practically not feasible. An anticipation of all integrations and a presentation of the results

$$c_m = f(t_1, t_2) \tag{2.91}$$

in a table becomes extensive because Eq. (2.91) is a function of two variables - namely t_1 and t_2. With a simple mathematical operation, the problem of averaging can be solved more easily. With the relation

$$\int_1^2 c\,dt = \int_{t_1}^{t_2} c\,dt = \int_{t_1}^{0} c\,dt + \int_0^{t_2} c\,dt = \int_0^{t_2} c\,dt - \int_0^{t_1} c\,dt \qquad (2.92)$$

Eq. (2.90) can be transformed as follows:

$$c_m = \frac{1}{t_2 - t_1}\left(\int_0^{t_2} c\,dt - \int_0^{t_1} c\,dt\right) \qquad (2.93)$$

In order that each integration appears as an averaging according to Eq. (2.90), Eq. (2.93) is extended, where the round brackets should clearly emphasise the mean values:

$$c_m = \frac{1}{t_2 - t_1}\left[\left(\frac{1}{t_2 - 0}\int_0^{t_2} c\,dt\right)t_2 - \left(\frac{1}{t_1 - 0}\int_0^{t_1} c\,dt\right)t_1\right] \qquad (2.94)$$

The mean specific heat capacity in a certain temperature range is obtained according to Eq. (2.94) by a simple arithmetic operation and with the help of two mean values, where the lower limits coincide. With the simplified notation for Eq. (2.90)

$$c_m = \frac{1}{t_2 - t_1}\int_1^2 c\,dt = c_m|_{t_1}^{t_2} \qquad (2.95)$$

Eq. (2.94) takes the following form:

$$c_m|_{t_1}^{t_2} = \frac{1}{t_2 - t_1}(c_m|_{0\,°C}^{t_2} \cdot t_2 - c_m|_{0\,°C}^{t_1} \cdot t_1) \qquad (2.96)$$

To calculate the mean specific heat capacity using the important equation (2.96) (left part of Fig. 2.19) ()one needs a representation

$$c_m|_{0\,°C}^{t} = f(t) \ . \qquad (2.97)$$

Eq. (2.97) is a function of only one single variable, namely t. A table according to Eq. (2.97) has a much smaller dimension than a table according to Eq. (2.91).

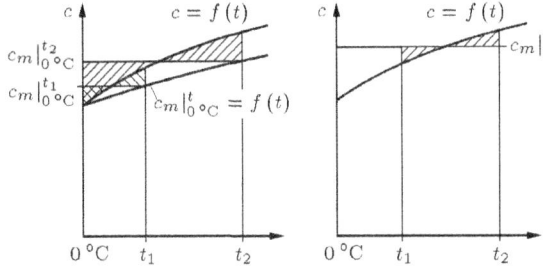

Figure 2.19 True specific heat capacities $c = f(t)$ at t_1 and t_2, mean specific heat capacity $c_m|_{t_1}^{t_2}$ (right); mean specific heat capacities $c_m|_{0\,°C}^{t_1} < c_m|_{0\,°C}^{t_2} < c_m|_{t_1}^{t_2}$ (left)

In many cases, in the designation of the mean specific heat capacity $c_m|_{t_1}^{t_2}$ the index m and the temperatures t_1 and t_2 are omitted. It then means

$$c = c_m|_{t_1}^{t_2}. \qquad (2.98)$$

2.7 Heat Capacity

If the specific heat capacity c has the meaning of an average value, then from Eq. (2.88) follows:

$$Q_{12} + |W_{RI}| = m\,c \int_1^2 dt = m\,c\,(t_2 - t_1) \qquad (2.99)$$

In the treatment of changes of state, the mean specific heat capacity is usually used in this sense.

Table 2.4 True specific heat capacity c and mean specific heat capacity $c_m|_{0\,°C}^t$ in kJ/(kg K) of iron according to [98]

t °C	0	20	100	200	300	400	500	600	700	800	1000
c	0.440	0.452	0.486	0.532	0.582	0.628	0.678	0.754			
$c_m|_{0\,°C}^t$	0.440	0.444	0.465	0.486	0.511	0.532	0.557	0.582	0.628	0.670	0.703

Table 2.5 True specific heat capacity c in kJ/(kg K) of water according to [128]

↓ p bar	t in °C → 0	10	20	30	40	50	60	70	80
1	4.2194	4.1955	4.1848	4.1800	4.1786	4.1796	4.1828	4.1881	4.1955
50	4.1955	4.1773	4.1698	4.1670	4.1667	4.1684	4.1720	4.1775	4.1849
100	4.1723	4.1595	4.1551	4.1541	4.1549	4.1573	4.1613	4.1670	4.1744
150	4.1501	4.1425	4.1410	4.1417	4.1435	4.1466	4.1509	4.1568	4.1642
200	4.1290	4.1262	4.1274	4.1297	4.1325	4.1361	4.1409	4.1468	4.1542
250	4.1090	4.1107	4.1144	4.1181	4.1218	4.1260	4.1311	4.1372	4.1445
300	4.0899	4.0958	4.1018	4.1069	4.1115	4.1162	4.1215	4.1278	4.1351

Table 2.6 Mean specific heat capacity $c_m|_{0\,°C}^t$ in kJ/(kg K) of water according to [128]

↓ p bar	t in °C → 0	10	20	30	40	50	60	70	80
1	4.2194	4.2058	4.1976	4.1924	4.1891	4.1870	4.1860	4.1859	4.1866
50	4.1955	4.1850	4.1790	4.1754	4.1732	4.1720	4.1717	4.1721	4.1732
100	4.1723	4.1648	4.1609	4.1587	4.1576	4.1573	4.1576	4.1585	4.1600
150	4.1501	4.1455	4.1435	4.1427	4.1427	4.1431	4.1440	4.1454	4.1473
200	4.1290	4.1270	4.1269	4.1274	4.1283	4.1295	4.1310	4.1328	4.1350
250	4.1090	4.1095	4.1110	4.1127	4.1145	4.1164	4.1184	4.1206	4.1231
300	4.0899	4.0927	4.0958	4.0986	4.1013	4.1038	4.1063	4.1089	4.1117

Example 2.11 What is the mean specific heat capacity of iron in the temperature range from 150 °C to 350 °C?

By linear interpolation of the mean specific heat capacities of Table 2.4 one obtains

$$c_m|_{0\,°C}^{150\,°C} = 0.4755 \text{ kJ/(kg K)} \qquad c_m|_{0\,°C}^{350\,°C} = 0.5215 \text{ kJ/(kg K)}$$

According to Eq. (2.96), the mean specific heat capacity sought is

$$c_m|_{150\,°C}^{350\,°C} = \frac{0.5215 \text{ kJ/(kg K)} \cdot 350\,°C - 0.4755 \text{ kJ/(kg K)} \cdot 150\,°C}{350\,°C - 150\,°C} = 0.556 \text{ kJ/(kg K)}\,.$$

Example 2.12 A hardening bath contains 0.5 m³ of oil with the density $\rho_{oil} = 900$ kg/m³ and a temperature of $t_{oil} = 20\,°C$. Steel parts are to be hardened in this bath, which are brought into the bath with a temperature of $t_{st} = 800\,°C$. What is the largest batch mass m_{st} being permissible if the mixing temperature t_M is to remain 10 K below the flame temperature of the oil of $t_f = 200\,°C$?

$$c_{m\,oil}\big|_{20\,°C}^{190\,°C} = 2.294 \text{ kJ/(kg K)} \quad \text{is given.}$$

According to Table 2.4 with Eq. (2.96) one gets

$$c_{m\,st}\big|_{190\,°C}^{800\,°C} = 0.728 \text{ kJ/(kg K)}$$

Both quantities of substance together form a closed system after the steel parts have been inserted. This is a system whose internal energy remains constant. The internal energies of the individual substances change. For the oil one obtains

$$U_{2\,oil} - U_{1\,oil} = m_{oil}\,c_{m\,oil}\,(t_M - t_{oil}) = V_{oil}\,\rho_{oil}\,c_{m\,oil}\,(t_M - t_{oil})$$

and for the steel parts

$$U_{2\,st} - U_{1\,st} = m_{st}\,c_{m\,st}\,(t_M - t_{st})\,.$$

It follows

$$U_{2\,oil} - U_{2\,st} - U_{1\,oil} - U_{1\,st} = 0\,.$$

The transformation according to m_{St} results in

$$m_{st} = \frac{0.5 \text{ m}^3 \cdot 900 \text{ kg/m}^3 \cdot 2.294 \text{ kJ/(kg K)} \cdot (190 - 20)\,°C}{0.728 \text{ kJ/(kg K)} \cdot (800 - 190)\,°C} = 395 \text{ kg}\,.$$

2.7.2 The Specific Heat Capacity of Gases

The specific heat capacity of gases depends

1. on the type of gas,
2. on the temperature,
3. on the pressure,
4. on the type of change of state.

In the special case of ideal gases, there is no dependence on pressure. The dependence on the change of state is illustrated by a comparison of isochoric change of state and isobaric change of state. At constant volume, the specific heat capacity is c_v (specific isochoric heat capacity). According to Eq. (2.48) and Eq. (2.87) it applies for an isochoric change of state with $dV = 0$

$$dQ + |dW_{RI}| = dU = m\,du = m\,c_v\,dt \qquad c_v = \left(\frac{\partial u}{\partial t}\right)_v, \qquad (2.100)$$

where

$$u = \frac{U}{m} \qquad (2.101)$$

means the specific internal energy.

At constant pressure, the specific heat capacity is denoted by c_p (specific isobaric heat capacity). From Eqs. (2.72) and (2.87) it follows for an isobaric change of state with $dp = 0$

$$\mathrm{d}Q + |\mathrm{d}W_{RI}| = \mathrm{d}H = m\,\mathrm{d}h = m\,c_p\,\mathrm{d}t \qquad c_p = \left(\frac{\partial h}{\partial t}\right)_p , \qquad (2.102)$$

where

$$h = \frac{H}{m} \qquad (2.103)$$

is the specific enthalpy.

However, in addition, other types of changes of state are possible in which the temperature of the gas changes. The specific heat capacity can become negative if the temperature decreases when heat is added and more energy in form of work is removed. The specific heat capacity can become negative, too, if the temperature rises when heat is released, which is the case when more energy in form of work is supplied at the same time.

2.8 Fluid Mechanics

2.8.1 General Aspects

In order to be able to formulate simple laws for the flow of fluids (of liquids and gases), the following assumptions are first useful:

1. The fluid flow is stationary (unchanging flow pattern).

2. The fluid is incompressible (examples: incompressible liquids; gases whose density does not change significantly during the flow process – e.g. in a pipe).

3. The flow is frictionless in the fluid itself and on the inner wall of the flow-carrying structure (such as a pipe).

4. The flow is adiabatic, i.e. heat is neither added nor removed; the temperature does not change.

For steady-state flow, the continuity equation is given by

$$\dot{m} = A\,c\,\varrho = \text{const} . \qquad (2.104)$$

It takes the following form for incompressible fluids at constant temperature according to Eq. (2.16):

$$\dot{V} = A\,c = \text{const} \qquad (2.105)$$

Using the *Bernoulli*[4] equation for a steady and frictionless flow of incompressible fluids, the total energy is the sum of potential energy, pressure energy and kinetic energy, where the total energy remains constant:

$$m\,g\,z + m\,\frac{p}{\varrho} + m\,\frac{c^2}{2} = \text{const} \qquad (2.106)$$

After dividing by m and multiplying by ϱ, the sum elements of Eq. (2.106) obtain the measure of pressure:

---

[4] *Daniel Bernoulli*, 1700 to 1782, mathematician and natural scientist from Basel, and Johann Bernoulli, 1667 to 1748, mathematician and physician, father of Daniel Bernoulli, investigated laws of fluid mechanics; they also achieved groundbreaking discoveries in mathematics.

$$\varrho g z + p + \frac{\varrho}{2} c^2 = \text{const} \qquad (2.107)$$

The following terms are commonly used for the individual elements of this equation:

The first element $\varrho g z$ represents the pressure due to the gravity of the fluid; it is therefore called "gravity pressure".

The second element p is the internal pressure of the fluid; it is given as "static pressure", in the literature often named p_{st}. With a pressure measurement device, this static pressure is recorded as gage pressure to the ambient pressure (being a resting pressure or a flow pressure p_e).

The third element $\varrho c^2/2$ indicates the pressure generated by the kinetic energy; this "kinetic pressure" is often denoted by p_{kin} in the literature.

The sum of static, kinetic and, if to be considered, gravity pressure is called "total pressure".

If Eq. (2.107) is divided by the expression ϱg, the *Bernoulli* equation is obtained in the form of a "height equation" with height z, which is common, for example, in the flow of liquids - such as in water supply:

$$z + \frac{p}{\varrho g} + \frac{c^2}{2 g} = \text{const} \qquad (2.108)$$

2.8.2 Flow Shapes

The forms of flow observed are laminar flow and turbulent flow. In laminar flow, flow filaments are formed that appear next to each other and parallel to the direction of flow. A prerequisite for laminar flow is a comparatively low flow velocity or a high viscosity of the fluid. When the 'critical velocity' is exceeded, laminar flow changes to turbulent flow.

The two flow forms can be distinguished from each other with the help of the similarity laws established by *Reynolds*[5]. The dimensionless parameter Re named after him is derived from the flow velocity c, in the case of pipes from the internal pipe diameter d and from the kinematic viscosity ν:

$$Re = \frac{c \, d}{\nu} \qquad (2.109)$$

2.8.3 Friction and Roughness

The assumption of an imaginary frictionless flow formulated in section 2.8.1 does not apply in practice. The viscosity of the fluid and the nature of the pipe walls result in energy conversions due to friction, which lead to pressure losses.

Due to regularities recognised by *Hagen* and *Poiseuille*[6], the equation of *Darcy* and *Weisbach*[7] applies to the pressure loss caused by friction in the straight pipe for incompressible fluids:

$$\Delta p_R = \lambda \frac{l}{d} \frac{1}{2} \varrho c^2 \qquad (2.110)$$

---

[5] *Osborne Reynolds*, 1842 to 1912, Irish-English physicist, worked on the fields of hydraulics and soil mechanics as well as heat transfer.

[6] *Gotthilf Heinrich Ludwig Hagen*, 1797 to 1884, German hydraulic engineer
Jean Leonard Marie Poiseuille, 1797 to 1869, French natural scientist

[7] *Philibert Gaspard Henry Darcy*, 1803 to 1858, French engineer, investigated flows through porous structures.
Julius Ludwig Weisbach, 1806 to 1871, German mathematician and engineer

λ is called the pipe friction coefficient.

In laminar flow, λ is independent of the roughness of the pipe wall k and is only a function of the *Reynolds* number Re (Table 2.7, Eq. (1)).

For turbulent flow, a distinction must be made between
1. hydraulically smooth pipes, where $k = 0$ is assumed and λ depends only on Re (Table 2.7, Eqs. (2) to (4)), and
2. hydraulically rough pipes, where λ depends only on the pipe diameter d and the pipe wall roughness k (Table 2.7, Eq. (5)), and
3. pipes that lie in the transition region between smooth and rough, where λ depends on Re, d and k (Table 2.7, Eq. (6)).

The three areas mentioned above are visible in Fig. 2.20.

Table 2.7 Equations for the determination of the pipe friction coefficient λ (cf. Figure 2.20)[8]

$\lambda = Re/64$	(1)	*Hagen, Poiseuille*	$Re < 2320$	lam.
$\lambda = 0.3164 \cdot Re^{-0.25}$	(2)	*Blasius*	$Re = 2320 \ldots 10^5$	turb. g
$\lambda = 0.0032 + 0.221 \cdot Re^{-0.237}$	(3)	*Nikuradse*	$Re = 10^5 \ldots 5 \cdot 10^6$	turb. g
$1/\sqrt{\lambda} = 2\lg(Re\sqrt{\lambda}) - 0.8$	(4)	*Prandtl, Karman*	$Re > 10^6$	turb. g
$1/\sqrt{\lambda} = -2\lg\left[\dfrac{k}{3.71\,d}\right] = 2\lg\left(\dfrac{d}{k}\right) + 1.14$	(5)	*Nikuradse*	$Re > \dfrac{200\,d}{\sqrt{\lambda}\,k}$	turb. r
$1/\sqrt{\lambda} = -2\lg\left[\dfrac{2.51}{Re\sqrt{\lambda}} + \dfrac{k}{3.71\,d}\right]$	(6)	*Prandtl, Colebrook*	$Re < \dfrac{200\,d}{\sqrt{\lambda}\,k}$	turb. Ü
$\lambda = \left[-2\lg\left[2.7\dfrac{(\lg Re)^{1.2}}{Re} + \dfrac{k}{3.71\,d}\right]\right]^{-2}$	(7)	*Zanke*	$Re > 2320$	turb.

lam. : laminar turb. : turbulent g : hydraulic smooth r : hydraulic rough Ü: transition range

The transition law for pipes with technical roughness (Eq. 6), which is frequently used in practical calculations, links the laws found by *Prandtl* (Eq. 4) and *Nikuradse* (Eq. 5) for the limiting cases of hydraulically smooth and hydraulically rough on the basis of a proposal by *Colebrook* and *White* by superimposing both influences in the argument of the logarithm. According to *Zanke*, the implicit equation (6), which can only be solved iteratively, can be replaced with very good accuracy by the explicit equation (7), which applies in the entire turbulent range.

Besides the calculation of the pipe friction coefficient λ according to Eqs. (1) to (7) in Table 2.7, λ can be taken from the pipe friction diagram (also called *Moody* diagram or λ, Re diagram for pipes) (Fig. 2.20). The partially required parameter d/k is formed with the pipe roughness k according to Table 2.8.

---

[8] *Heinrich Blasius*, 1883 to 1970, German physicist
Johann Nikuradse, 1894 to 1979, German engineer and physicist, born in Georgia
Ludwig Prandtl, 1875 to 1953, German engineer, Professor in Hannover and Göttingen, made significant contributions to fluid mechanics, boundary layer theory and supersonic flow.
Theodore von Karman, 1881 to 1963, North American physicist and aviation engineer of Hungarian origin, worked in the fields of aerodynamics, aviation and space research.
Cyril Frank Colebrook, 1910 to 1997, British physicist, made essential contributions to fluid mechanics.
Cedric Masey White, 1898 to 1993, British physicist, worked on fluid mechanics and hydraulics.
Lewis Ferry Moody, 1880 to 1953, North American mechanical engineer, hydraulics professor

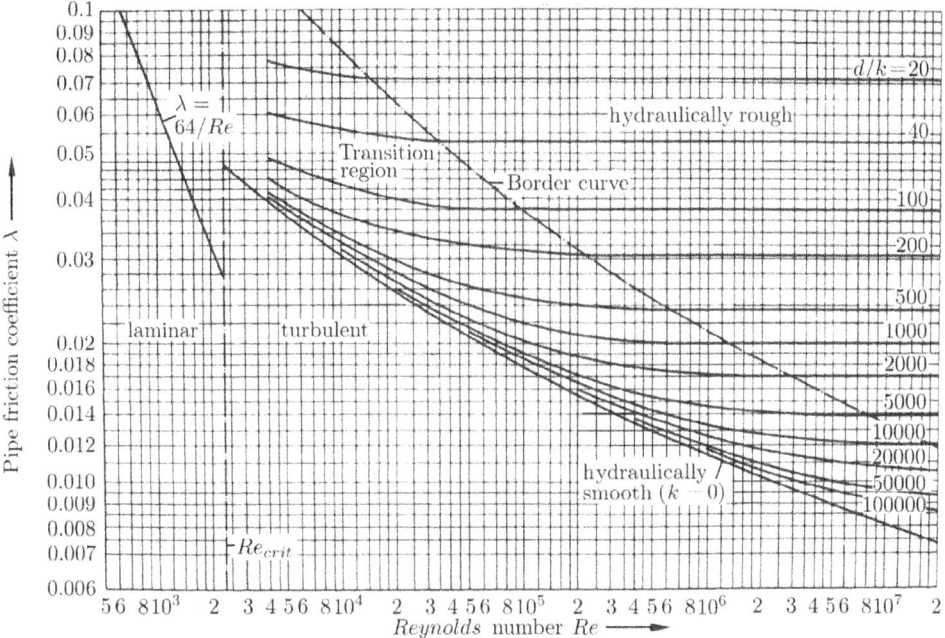

Figure 2.20 Pipe friction diagram (also called *Moody* diagram or λ, Re diagram for pipes)

Table 2.8 Roughness values k of piping

Steel, polished, chrome-plated	0.0014 ... 0.0015
Stainless steel	0.0015
Copper pipe, new	0.0015
Plastic pipe	0.0015
PE material	0.003 ... 0.028
PE pipes, PB pipes	0.007
Steel, plastic coated	0.010 ... 0.032
Steel, not welded but drawn	0.01 ... 0.06
Steel, welded	0.02 ... 0.10
Copper, aged	up to 0.03
Commercial steel pipe	0.045
Steel pipe, welded, new	0.04 ... 0.10
Gas pipe, new	0.020 ... 0.065
Gas pipe, old	0.10 ... 0.14
Steel, galvanised	0.06 ... 0.30

Table 2.9 Individual resistance coefficient values ζ

Reduction	0.5
Storey arch	0.5
Angle 90°	1.5
Angle 45°	0.7
Arch 90°	0.4
Arch 45°	0.3
Cross piece 90° branch	1.5
Arch-T passage	0
Bend-T branch	1.3
Ball shut-off valve	0.2
Gate valve	0.5
Butterfly valve	0.2
Safety valve	3.0
Shut-off device (TAE)	2.0
Gas flow monitor (GS)	5 ... 10

2.8.4 Individual Resistances

Pipelines for fluid transport have fittings as well as shaped and connecting component parts in which additional pressure losses occur as a result of flow deflections, turbulences and separations. These losses are usually determined by tests and specified with the aid of individual resistance coefficients ζ (reference values see Table 2.9), which can be added up to $\Sigma \zeta$, for example. Thus, the additional pressure loss due to fittings and shaped and connecting component parts Δp_F is determined as follows:

$$\Delta p_F = \Sigma \zeta \frac{1}{2} \varrho c^2 \qquad (2.111)$$

2.8.5 Equivalent Pipe Length

In straight pipelines as well as in pipe fittings, the pressure loss is proportional to $\varrho c^2/2$. Therefore, the pressure loss in fittings as well as in other components can be given according to Eq. (2.110) by the equal pressure loss of an equivalent straight pipe section l' $\Delta p_F = \lambda l' \varrho c^2/(2d)$ as follows:

$$l' = \frac{\zeta}{\lambda} d \tag{2.112}$$

2.8.6 Pressure Loss or Pressure Gain Due to Density Difference between Flowing Fluid and Ambient Fluid

In non-horizontal pipelines – for example in inhouse gas installation systems – buoyancy has to be taken into account due to the difference in density between the air and the gas (cf. example 2.2) The pressure difference depends on the difference in densities and heights (cf. Eq. (2.4)):

$$\Delta p_A = (\varrho - \rho_U)(z_2 - z_1) g \tag{2.113}$$

In those cases where the gas has the lower density compared to the surrounding gas (usually air), there is a pressure gain (negative pressure drop) in rising pipelines and a pressure drop in falling pipelines. For fluids to be transported with greater densities compared to the surrounding fluid, the reverse process can be observed. At higher pressures of the fluid in the pipe, the buoyancy is negligible as long as the height differences are insignificant.

2.8.7 Total Pressure Difference During Fluid Transfer

When the fluid is conveyed, the pressure is reduced by the pressure loss due to pipe friction in the straight pipe Δp_R as well as the pressure loss due to the valves and fittings Δp_F; in addition, the pressure influence due to buoyancy (geodetic height differences, density differences) Δp_A changes the pressure. The total pressure difference is thus

$$\Delta p_{ges} = \Delta p_R + \Delta p_F + \Delta p_A = \lambda \frac{l}{d}\frac{1}{2}\varrho c^2 + \Sigma \zeta \frac{1}{2} \varrho c^2 + (\varrho - \varrho_U)(z_2 - z_1) g . \tag{2.114}$$

If instead of the total pressure difference Δp_{ges} the total height difference Δz_{ges} is to be calculated, this equation takes the form

$$\Delta z_{ges} = \Delta z_R + \Delta z_F + \Delta z_A = \lambda \frac{l}{d}\frac{1}{2g} c^2 + \Sigma \zeta \frac{1}{2g} c^2 + \frac{\varrho - \varrho_U}{\varrho}(z_2 - z_1) . \tag{2.115}$$

Example 2.13 Natural gas H is distributed through a local gas network operating at medium pressure. In the network, a pipe section of 2.8 km length is flowed through by the volume flow $\dot{V}_B = 1700$ m³/h (volume flow at operation state). The pipe is made of polyethylene and has the internal diameter $d_i = 248.2$ mm; the pipe roughness has the value $k = 0.007$ mm. The flow pressure is $p_e = 0.80$ bar (gage pressure); the ambient pressure is $p_{amb} = p_L = 1.0$ bar; the dry natural gas H has the temperature $t = 12$ °C; the kinematic viscosity is $\nu = 7.835 \cdot 10^{-6}$ m²/s. The natural gas can be taken as an ideal gas with the practically invariable density during the flow through the pipe $\varrho_B = 1.33243$ kg/m³.

a) The cross-sectional area A of the pipe and the flow velocity c are to be determined.

b) What is the *Reynolds* number Re? What is the value of the quotient d_i/k? Is the flow laminar or turbulent? The pipe friction coefficient λ is to be determined with the λ, Re diagram and checked with a suitable equation. How high is the pressure loss Δp_R due to pipe friction under the simplified assumption of constant density?

c) There are four $90°$ angles in the pipe, each with the individual resistance coefficient $\zeta = 0.4$, five $45°$ angles, each with $\zeta = 0.3$ and six gate valves, each with $\zeta = 0.5$. What are the sum of the individual resistance coefficients $\Sigma \zeta$ and the resulting additional pressure loss Δp_F?

d) The pipe is laid in the direction of flow as a rising pipe with a height increase over the entire length of $z_2 - z_1 = 30$ m. The additional pressure loss Δp_A is to be calculated. (Note: For air, the density $\varrho_L = 1.239$ kg/m^3 can be assumed at about 12 °C).

e) The total pressure drop $p_1 - p_2 = \Delta p_{ges}$ is to be determined.

a) $A = \dfrac{\pi}{4} d_i^2 = \dfrac{3.14159}{4} 0.2482^2 \text{ m}^2 = 0.04838 \text{ m}^2$

$c = \dfrac{\dot{V}_B}{A} = \dfrac{1700 \text{ m}^3}{\text{h } 0.04838 \text{ m}^2} \dfrac{1 \text{ h}}{3600 \text{ s}} = 9.760 \dfrac{\text{m}}{\text{s}}$

b) $Re = \dfrac{cd}{\nu} = \dfrac{9.760 \text{ m} \cdot 0.2482 \text{ m s}}{\text{s } 7.8353 \cdot 10^{-6} \text{ m}^2} = 3.092 \cdot 10^5 = 309200 \quad \rightarrow \quad$ turbulent flow

$d/k = (248.2 \text{ mm})/(0.007 \text{ mm}) = 35457 \quad \rightarrow \quad \lambda = 0.015$

$\rightarrow$ Check with the equation of *Zanke* for the turbulent flow range $Re > 2320$:

$\lambda = \left[-2 \lg\left[2.7 \dfrac{(\lg Re)^{1.2}}{Re} + \dfrac{k}{3.71d}\right]\right]^{-2} = \left[-2 \lg\left[2.7 \dfrac{(\lg 309200)^{1.2}}{309200} + \dfrac{0.007}{3.71 \cdot 248.2}\right]\right]^{-2} = 0.0147$

$\Delta p_R = \lambda \dfrac{l}{d} \dfrac{1}{2} \varrho c^2 = 0.0147 \dfrac{2800 \text{ m}}{0.2482 \text{ m}} \cdot \dfrac{1}{2} \cdot 1.33243 \dfrac{\text{kg}}{\text{m}^3} 9.760^2 \dfrac{\text{m}^2}{\text{s}^2} \dfrac{\text{N s}^2}{\text{kg m}} \dfrac{\text{mbar m}^2}{100 \text{ N}} = 105.24 \text{ mbar}$

c) $\Sigma \zeta = 4 \cdot 0.4 + 5 \cdot 0.3 + 6 \cdot 0.5 = 6.1$

$\Delta p_F = \Sigma \zeta \dfrac{1}{2} \varrho c^2 = 6.1 \cdot \dfrac{1}{2} \cdot 1.33243 \dfrac{\text{kg}}{\text{m}^3} 9.760^2 \dfrac{\text{m}^2}{\text{s}^2} \dfrac{\text{N s}^2}{\text{kg m}} \dfrac{\text{mbar m}^2}{100 \text{ N}} = 3.871 \text{ mbar}$

d) $\Delta p_A = (\varrho - \varrho_U)(z_2 - z_1) g =$

$= (1.33243 - 1.239) \dfrac{\text{kg}}{\text{m}^3} (30 \text{ m}) \cdot 9.81 \dfrac{\text{m}}{\text{s}^2} \dfrac{\text{N s}^2}{\text{kg m}} \dfrac{\text{mbar m}^2}{100 \text{ N}} = 0.275 \text{ mbar}$

e) $\Delta p_{ges} = \Delta p_R + \Delta p_F + \Delta p_A = (105.24 + 3.871 + 0.275) \text{ mbar} = 109.388 \text{ mbar} \approx 110 \text{ mbar}$

The calculation result shows that the pressure loss Δp_{ges}, related to the initial pressure $p_1 = p_{amb} + p_e = 1.80$ bar, is about 6 % and therefore the condition of an almost incompressible flow only roughly applies, so that Eq. (2.110) only applies approximately. The exact calculation of the pressure loss of ideal gases when taking their compressibility into account is dealt with in section 4.6, where Eq. (4.282) results.

Example 2.14 For the electricity supply of an island, $\dot{V}_B = 17$ m^3/h fuel oil of quality specification M with density $\varrho = 920$ kg/m^3 at temperature $t_B = 10$ °C and air density $\varrho_L = \varrho_0 = 1.247$ kg/m^3 is conveyed from the storage tank in the port to a power station; there, several large power generating *Diesel* units are operated with fuel oil M. The pipeline of $d = 125$ mm internal diameter has a length of $l = 1600$ m and overcomes a height difference of $z_2 - z_1 = z_{stat} = 25.0$ m; the roughness of the pipeline is $k = 0.1$ mm. A rounded inlet from the oil tank, eighteen $45°$ angles and eight open gate valves in the pipeline must be taken into account; the sum of the loss height, caused by the individual resistance coefficients, is $\Sigma \Delta z_F = 0.027$ m. The kinematic viscosity of the fuel oil M is $\nu = 68 \cdot 10^{-6}$ m^2/s.

a) The mean flow velocity c that occurs is to be calculated.

b) What is the Reynolds number Re? Is the flow laminar or turbulent? The pipe friction coefficient λ and thus the loss height in the pipe Δz_R are to be determined.

c) The total height loss Δz_{ges} including the static height z_{stat} are to be calculated.

a) $c = \dfrac{\dot{V}_B}{A} = \dot{V}_B \dfrac{4}{\pi d^2} = \dfrac{17\,\text{m}^3 \cdot 4}{\text{h}\, 3.14159 \cdot 0.125^2\,\text{m}^2} \dfrac{1\,\text{h}}{3600\,\text{s}} \approx 0.3848 \dfrac{\text{m}}{\text{s}}$

b) $Re = \dfrac{c\,d}{\nu} = \dfrac{0.3848\,\text{m} \cdot 0.125\,\text{m}\,\text{s}}{\text{s}\,68 \cdot 10^{-6}\,\text{m}^2} = 7.07 \cdot 10^2 \rightarrow$ laminar flow $\rightarrow \lambda = 0.09$

→ Check with the equation of *Hagen-Poiseuille* for the laminar flow range $Re < 2320$:

$\lambda = 64/Re = 64/707 = 0.0905$

c) . $\quad \Delta z_R = \lambda \dfrac{l}{d} \dfrac{1}{2g} c^2 = 0.0905 \dfrac{1600\,\text{m}}{0.125\,\text{m}} \dfrac{1}{2 \cdot 9.81} \dfrac{\text{s}^2}{\text{m}} 0.3848^2 \dfrac{\text{m}^2}{\text{s}^2} = 8.7424\,\text{m}$

d) As $\varrho = 920\,\text{kg/m}^3$ and $\varrho_L = \varrho_0 = 1.247\,\text{kg/m}^3$ are given and thus it becomes $\varrho \gg \varrho_U$, a good approximation is

$\Delta z_A = \dfrac{\varrho - \varrho_U}{\varrho}(z_2 - z_1) \approx (z_2 - z_1) = 25.0\,\text{m}$

e) $\Delta z_{ges} = \Delta z_R + \Sigma \Delta z_F + \Delta z_A = (8.7424 + 0.027 + 25.0)\,\text{m} = 33.7694\,\text{m} \approx 33.8\,\text{m}$

Tasks for Section 2

1. The coupling power of an engine is to be measured with a water brake. In the water brake, the coupling power is supplied by friction to a cooling water flow of 6.59 kg/s, and the temperature increases from 10 °C to 50 °C. What is the coupling power of the engine?

2. 80 litres of water are to be heated from 12 °C to 57 °C in a water heater which has an electric power of 3 kW. The density at a gage pressure of 5 bar is 994 kg/m^3. In what time is the water heated up?

3. In a waterfall, water with a temperature of 10 °C falls 200 metres into the depth. In the process, its total difference in potential energy leads to an increase in internal energy. The process is adiabatic with respect to the environment. What temperature increase does the water experience? What temperature increase would occur if mercury were to fall instead? (The specific heat capacity of mercury is $c = 0.1393$ kJ/(kg K) = const.)

4. A high-pressure pump is used in a valley to supply water to a mountain village. The circular inlet nozzle has a diameter $D_1 = 0.1$ m; the water flowing through it has an absolute pressure $p_1 = 1.0$ bar, and the flow velocity is $c_1 = 2.0$ m/s. In the outlet nozzle, which is 1.2 m higher, the water flows with a velocity of $c_2 = 4.0$ m/s with the absolute pressure $p_2 = 19.0$ bar into the pressure pipe. At the adiabatically operating pump, the water temperature increases as a result of internal friction power $|P_{RI}|$ from $t_1 = 10.00$ °C to $t_2 = 10.08$ °C. The external frictional power $|P_{RA}|$ can be neglected. The density of the water is to be taken as $\varrho = 1000$ kg/m^3 = const. The specific heat capacity of water is $c = 4.19$ kJ/(kg K)= const. The volume flow $\dot{V}_1 = \dot{V}$ = const and the mass flow $\dot{m}_1 = \dot{m}$ = const of the water at the inlet cross-section are to be determined. What coupling power P_e is required for the drive? What coupling power $(P_e)_{id}$ would be required for the drive if no internal friction power $|P_{RI}|$ would occur?

5. In a chemical plant, oxygen O_2 with a mass flow rate of 10 kg/s at an ambient pressure of 1 bar is heated isobarically from $t_1 = 50$ °C to $t_2 = 600$ °C and without friction before it enters a combustion chamber as an oxidant. The mean specific isobaric heat capacity $c_m|_{t_1}^{t_2}$ of oxygen in the temperature interval is to be considered on the basis of the $c_m|_{0\,°C}^t$ values of the corresponding table in the appendix. How large is the heat flow that is transferred to the

oxygen stream? What heat flow must be supplied if the oxygen stream is not heated from $t_1 = 50\,°C$ to $t_2 = 600\,°C$, but even further to $700\,°C$ isobarically and frictionless?

6. A constant heat flow passes through a heat exchanger that is used to collect geothermal heat by a heat pump. For this, a constant mass flow of 1.1 kg/s of a liquid water/glycol mixture flows through the heat exchanger. The density is 1102 kg/m^3 = const; the average specific heat capacity is 3.85 kJ/(kg K).

At the inlet to the heat exchanger the temperature $t_1 = -3\,°C$, the velocity $c_1 = 2.0$ m/s and the pressure $p_1 = 4.5$ bar are measured. At the outlet, the temperature $t_1 = 1\,°C$, the velocity $c_2 = 2.0$ m/s and the pressure $p_2 = 1.2$ bar are measured. The inlet measuring point is 1.2 m above the outlet measuring point.

a) What heat flow is transferred from the soil to the water/glycol mixture? (For the calculation the general formulation of the first law of thermodynamics for open systems with inclusion of the kinetic and potential energies is to be used).

b) How many heat exchangers of the same design are necessary to gain a heat flow of 32 kW out of the soil?

3 The Second Law of Thermodynamics

According to the first law of thermodynamics, any energy conversion is conceivable. For example, any form of energy could be converted into any other. However, if we take a closer look at the processes that occur in nature and technology, we find that in many processes a complete conversion into the desired energy can never be achieved. For example, a water turbine or an electric motor works with a certain degree of efficiency, which indicates what proportion of the energy supplied is converted into the desired form of energy.

A turbine has the task of converting the potential energy and the kinetic energy of a quantity of water into coupling work, which can be taken off the shaft of the turbine. Part of the energy is used to cover the friction "losses" that occur at the bearings, but also in the fluid itself as it flows through the machine. An electric motor is designed to convert electrical energy into coupling work. However, part of the energy supplied is transformed into heat by the various mechanical and electrical "losses" and is released into the environment by the cooling air (or another coolant).

In both cases, the energy that is not converted into the desired useful energy is referred to as "loss". What justification do we have to call this energy a "loss"? According to the first law of thermodynamics, we know that energy can never disappear. However, the engineer understands "loss" not as a disappearance of energy, but a transformation into an undesirable form of energy. Such a conversion occurs in the processes described. The conversion of this "lost" energy back into the original energy encounters particular difficulties. Both the friction "losses" in the turbine and the electrical and mechanical "losses" of the electric motor, lead to an increase of the internal energy of the environment (air, cooling water). There is no process with which the originally available amount of mechanical or electrical energy can be completely recovered by removing the internal energy of the environment, provided that no permanent changes occur in the system or in the environment compared to the original state.

Due to the generation of the energy "loss", the considered processes of energy conversion in the turbine and in the electric motor have become irreversible or non-reversible. The reversal of the described processes fails because the extraction of internal energy from the environment alone is not possible. Additional facilities would be required for this, in which permanent changes occur compared to the original state.

3.1 The Statement of the Second Law

The investigation of further processes leads to the conclusion that all real transformations or transmissions of energy in machines and apparatuses, but also in nature are subject to the same restriction and are therefore not reversible - they are irreversible. These experimental findings are summarised in a law, which is called the second law of thermodynamics:

> **All natural and technical processes are irreversible.**

3.1.1 Reversible and Irreversible Processes

We can now formulate what we want to understand by a reversible or irreversible process: If a system has gone through a process and reached a final state that is different from the initial state, and can we restore the initial state without leaving any changes in the system or in the environment, and the original state is fully restored, then the process is reversible.

Since, according to the second law, all natural and technical processes are irreversible, there are in fact no reversible processes. Nevertheless, a reversible process is not only interesting from a physical point of view, but also as a borderline case of technical realisation, it is also of technical importance.

Example 3.1 A water turbine 1 according to Fig. 3.1 is supplied with water from an upper reservoir. After flowing through the turbine, the water flows into a lower reservoir or river. Friction work occurs in the connecting pipes 2, in the turbine and in the bearings. According to Eq. (2.69), the coupling work gained is thus less than it would be without friction work under otherwise identical conditions. The friction work increases the internal energy of the water and, to a lesser extent, the internal energy of the ambient air.

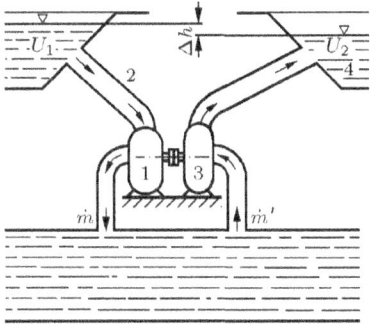

Figure 3.1 Irreversible flow through a water turbine: $\dot{m}$ mass flow of water in turbine and pump, $U_2 - U_1$ increase of the internal energy of the water and the ambient air

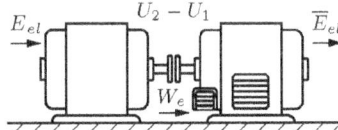

Figure 3.2 Irreversible operation of an electric motor

A centrifugal pump 3, which is driven by the generated coupling work, is capable to pump water into the basin 4, which is of the same size and the same height as the reservoir. But the pump cannot manage in the basin 4 the water level of the reservoir to be reached, even if the pump itself were to operate without any "losses" and the pipes leading to basin 4 were free of friction.

An additional machine does not exist, which, without permanent changes to the system or the environment, can manage the level difference by extracting the internal energy of the water and the ambient air. The process taking place in the water turbine is thus recognised as irreversible. The smaller the friction "losses" can be made, the closer one comes to a reversible process.

Example 3.2 An electric motor according to Fig. 3.2 is supplied with the electric energy E_{el}; the machine converts part of it into coupling work W_e. The energy difference not used increases the internal energy of the ambient air $(U_2 - U_1)$ via heating the motor winding. The generator, which is connected to the engine via a clutch, is not able to restore the electric energy E_{el} completely. The electrical energy $\overline{E}_{el}$ generated is less than the electric energy E_{el}. Since the internal energy of the environment cannot be converted back into electrical energy without permanent changes in the environment, this process is also irreversible.

3.1.2 Quasi-Static Changes of State

The initial point of a change of state is considered to be a state of equilibrium. In order for a change of state to be initiated, a disturbance of the equilibrium must occur. A change of state therefore leads via non-equilibrium states. All thermal equations of state and state diagrams only apply to equilibrium states, i.e. only for the initial and for the final state. If the perturbances of the equilibrium are small and if the change of state approximately consists of a sequence of equilibrium states, then the intermediate states also can be assumed as being in an equilibrium state. Such a change of state is called quasi-static. Reversible processes are only conceivable if the changes of state associated with the process are quasi-static.

Figure 3.3 Compression and expansion of a quantity of gas
a) quasi-static change of state
b) real change of state; pressure curve in the cylinder corresponding with the piston position

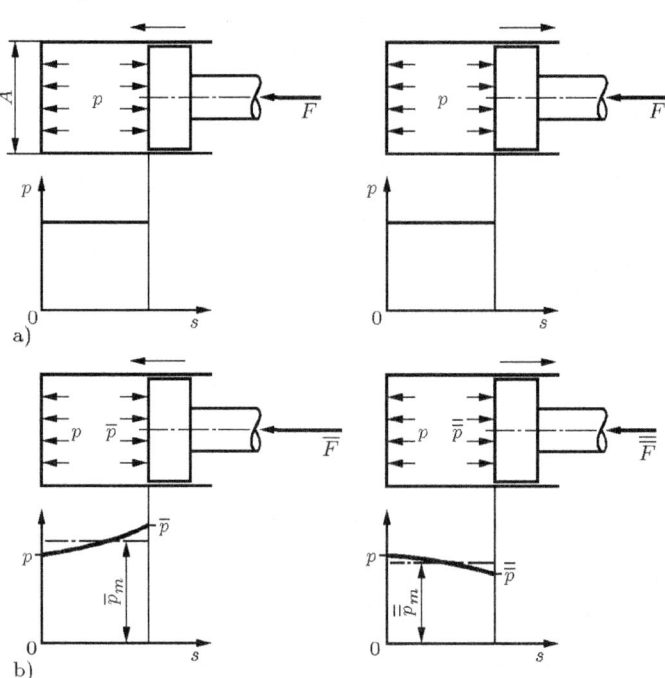

Example 3.3 If a gas is expanded or compressed in a cylinder without friction and infinitely slowly, the forces acting on the piston are in equilibrium in each piston position (Fig. 3.3 above). The change of state is quasi-static:
$$F = p\,A$$
During the actual compression process (Fig. 3.3 below), a gas pressure $\bar{p}$ results at the piston head which is greater than the pressure p at the bottom of the cylinder and greater than the mean pressure $\bar{p}_m$ in the gas chamber:
$$\bar{p} > \bar{p}_m > p$$
The rapid movement of the piston first acts on the gas molecules at the head of the piston. As a result of their inertia, this gas quantity is compressed and thus obtains an increased gas pressure $\bar{p}$.

Conversely, if the gas expands, the gas pressure $\bar{p}$ at the piston head becomes smaller than the pressure p at the bottom of the cylinder and less than the mean pressure $\bar{\bar{p}}_m$ in the gas chamber:

$$\bar{\bar{p}} < \bar{\bar{p}}_m < p$$

The real changes of state are no longer quasi-static, the process of compression and relaxation becomes irreversible.

3.2 Irreversible Processes

The irreversibility of thermodynamic processes is based on processes that always run only in one direction and whose reversal is impossible. They can be divided into two groups: friction processes and equalisation processes.

To describe them in more detail it is helpful to use simple adiabatic systems. The restriction to adiabatic systems really does not imply any restriction. It is always possible to choose a system boundary so large that no heat transfer with the environment occurs.

3.2.1 Friction

We use the example described in section 2.6.1. A gas enclosed in a container of constant volume is supplied with coupling work (Fig. 3.4).

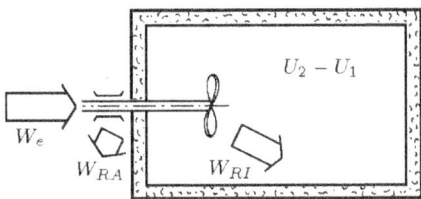 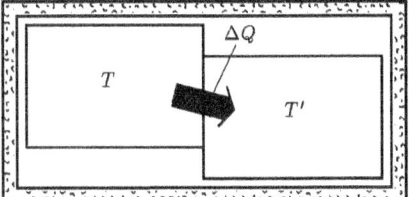

Figure 3.4 Supply of internal friction work; closed adiabatic system

Figure 3.5 Change of temperature to reach thermal equilibrium between two subsystems of different temperatures; closed adiabatic system

Inside the system, on the shaft there is an agitator - a paddle wheel - which performs internal friction work on the gas in the system. The internal friction work causes an increase of the internal energy. If $Q_{12} = 0$ applies to the heat, it follows from Eqs. (2.54) to (2.56)

$$W_e - |W_{RA}| = |W_{RI}| = U_2 - U_1 . \tag{3.1}$$

The notation of Eq. (3.1) indicates the energy flow from left to right. A reversal of this sequence of energy transformations is not possible according to experience. *M. Planck*[1] used this experience to formulate the second law of thermodynamics:

> **It is impossible to construct a periodically working machine, that does nothing more than lift a load and continuously extract heat from a heat container.**

---

[1] *Max Planck*, 1858 to 1947, professor of physics in Kiel in 1885, in Berlin in 1889. He founded quantum theory in 1900 and was awarded the Nobel Prize for Physics in 1918. In today's linguistic expression the term "heat" in this formulation of the second law would be replaced by "internal energy".

3.2.2 Temperature Equalisation

If there is a heat-conducting connection between two bodies of different temperatures, heat flows from the body with the higher temperature to the body with the lower temperature (Fig. 3.5). One can formulate this everyday experience in general terms as follows: Two neighbouring systems have a common heat-permeable boundary. Both systems are the content of an adiabatic overall system. The systems have different temperatures.

Under the effect of the temperature gradient, the heat ΔQ flows from the system of higher temperature T across the system boundary into the system of lower temperature T'. The process runs by itself and is finished when a temperature balance between the two systems has taken place and thermal equilibrium is reached.

The reverse process is impossible according to experience: No heat could flow by itself to change an initial state of temperature balance to a state of different temperatures T and T'.

Rudolf Clausius[2] used this observation to formulate the second law of thermodynamics:

> **Heat can never flow by itself from a body of lower temperature to a body of higher temperature.**

3.2.3 Pressure Equalisation

An adiabatic system consisting of two vessels of equal size 1 and 2 with a connecting pipe 3 that is shut-off, contains a gas under high pressure in one vessel, while in the other vessel there is no gas or, respectively, only a very small quantity of the same gas at the same temperature. The pressure in the second vessel is therefore zero respectively much lower than in the first vessel. If the connecting pipe is opened, gas flows from the vessel with high pressure into the vessel with low pressure until the pressure is balanced in both vessels (Fig. 3.6).

If one measures the temperature after the equalisation process, the temperature in ideal gases is the same as the temperature in both vessels before the pressure equalisation. If the laws of ideal gases cannot describe the behaviour of the gas, then the temperature after the pressure equalisation is different compared with the temperature before the connecting pipe was opened. This gas is called a real gas.

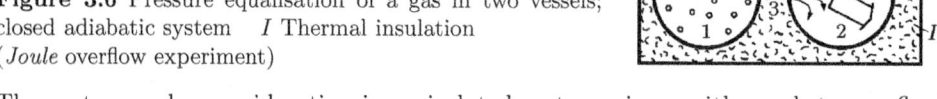

Figure 3.6 Pressure equalisation of a gas in two vessels; closed adiabatic system *I* Thermal insulation
(*Joule* overflow experiment)

The system under consideration is an isolated system, since neither substances flow nor energies exceed the system boundary. Accordinmg to

$$Q_{12} = W_{12} = 0 , \qquad (3.2)$$

---

[2] *Rudolf Clausius*, 1822 to 1888, German Professor of physics in Zurich in 1855, in Bonn in 1869. His fundamental "Abhandlungen über die mechanische Wärmetheorie" (Treatises on the Mechanical Theory of Heat), in which entropy as a quantity of state was introduced and elaborated, was published between 1864 and 1867.

it follows from the first law of thermodynamics with Eq. (2.46) $U_2 - U_1 = 0$ or
$$U_2 = U_1 \ . \tag{3.3}$$

As a form of thermal energy, internal energy is a function of temperature. The question arises as to whether the internal energy is only a function of the temperature, or whether there is, in addition, a dependence on another state variable. For the gas after the change of state in both vessels, the process of pressure equalisation is associated with an increase in volume.

If in a gas in an isolated adiabatic system with constant internal energy, no change in temperature can be measured, then the gas is ideal, and the internal energy is only a function of temperature and independent of volume:

$$T_2 = T_1 \qquad U = f(T) \qquad U \neq f(T,V) \tag{3.4}$$

If, in a gas that undergoes a volume increase with constant internal energy in an isolated system, a temperature change is measured, then the gas is real and the internal energy is not only a function of the temperature but also depends on the volume:

$$T_2 \neq T_1 \qquad U = f(T,V) \tag{3.5}$$

Since no work exceeds the system boundary, it follows from Eq. (2.58)

$$W_{12} = W_{V12} + |W_{RI}| = 0$$

or

$$-W_{V12} = |W_{RI}| \ . \tag{3.6}$$

According to the sign agreement, the volume change work in an expansion process is negative. The volume change work done during pressure equalisation is converted into friction work. With Eq. (2.21) it follows from Eq. (3.6)

$$\frac{|dW_{RI}|}{dV} = p \ . \tag{3.7}$$

In Eq. (3.7) the pressure p can only be positive. Thus the left side of Eq. (3.7) must also be positive. This is only the case if the volume change dV is also positive. If the process were reversed, dV would be negative, which is therefore impossible. Eq. (3.7) is the mathematical formulation of the statement that the pressure equalisation is irreversible.

3.2.4 Throttling

Throttling is an adiabatic flow process associated with a pressure reduction through a constriction (for example a valve in a pipeline) (Fig. 3.7). In the energy balance for the open system according to Eq. (2.67), the quantities on the left side are zero. If one neglects the change in potential energy, then it becomes

$$H_2 - H_1 + \frac{m}{2}(c_2^2 - c_1^2) = 0 \ . \tag{3.8}$$

In many cases, one can also neglect the change in kinetic energy. Then equation (3.8) becomes $H_2 - H_1 = 0$ or

$$H_2 = H_1 \ . \tag{3.9}$$

With Eq. (3.9) it follows from the first law according to Eq. (2.70)

$$W_{12} = W_{p12} + |W_{RI}| = 0 \quad \text{or} \quad -W_{p12} = |W_{RI}|. \qquad (3.10)$$

The pressure change work W_{p12} can be represented as an integral according to Eq. (2.30). Eq. (3.10) can thus also be written as follows:

$$-\int_1^2 V\,dp = |W_{RI}| \qquad (3.11)$$

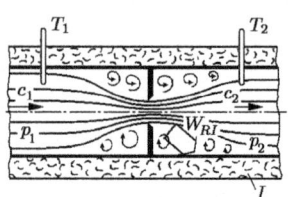

Figure 3.7 Throttling of a fluid; open adiabatic system   *I* Thermal insulation (experiment by *Joule und Thomson*)

According to section 2.4.4, the pressure change work in an expansion process is negative. Since a throttling means a pressure decrease of the fluid, dp is negative. This is also represented by Eqs. (3.10) and (3.11). The pressure change work performed by the fluid is converted into internal friction work. From Eq. (3.11) one obtains

$$\frac{|dW_{RI}|}{dp} = -V. \qquad (3.12)$$

In Eq. (3.12) dp can only be negative. With a reversal of the throttling, dp would be positive. Eq. (3.12) states that the throttling process is irreversible.

According to Eq. (2.41), the enthalpy is the sum of the internal energy and the shift work, and thus according to Eqs. (3.4) and (3.5) it is a function of temperature. Whether a dependence on a further state variable is present can be clarified by temperature measurements during the throttling process. One finds that the temperatures at a greater distance before and after the throttling point are the same for ideal gases and different for real gases. If, in the case of a gas with constant enthalpy, a pressure drop does not cause a change in temperature, then the gas is ideal, and the enthalpy is only a function of the temperature and independent of the pressure:

$$T_2 = T_1 \qquad H = f(T) \qquad H \neq f(T,p) \qquad (3.13)$$

If a reduction in pressure causes a change in temperature in a gas with constant enthalpy, then the gas is real and the enthalpy is not only a function of the temperature, but also a function of the pressure:

$$T_2 \neq T_1 \qquad H = f(T,p) \qquad (3.14)$$

This phenomenon in real gases is called the *Joule-Thomson* effect (cf. also section 5.4.6).

3.3 Entropy

All processes of energy conversion or energy transfer occurring in nature and technology contain one or more of the friction or equalisation processes listed in section 3.2. These have the effect that with every energy conversion or energy transfer one has to reckon

with a devaluation of energy, which manifests itself in the fact that the energies present at the end cannot be completely converted back into the originally existing types of energy.

In order to be able to compare the irreversibility of the different processes, one needs a measure for the devaluation of the energy through irreversibility. It must allow to compare such completely different processes as heat transfer and throttling with each other. But if it is available, one can also test more complicated processes in terms of their irreversibility and identify the greatest "loss" points. There, one will then try to improve the process in terms of energy e.g. by trying to reduce friction or equalisation processes.

Before we can pursue the question of a state variable that allows the irreversibility of a process to be calculated, the concept of a reversible substitute process must be introduced. First, we will restrict the consideration to adiabatic systems and refer to the processes occurring in adiabatic systems as adiabatic processes. In section 3.5, the consideration is extended to the non-adiabatic processes occurring in non-adiabatic systems.

3.3.1 Reversible Substitute Processes of Adiabatic Processes

All real processes are irreversible because they involve friction respectively equalisation processes. For a closed adiabatic system, the first law states that

$$-\int_1^2 p\,dV + |W_{RI}| = U_2 - U_1 \ . \tag{3.15}$$

The first law for an open adiabatic system gives

$$\int_1^2 V\,dp + |W_{RI}| = H_2 - H_1 \ . \tag{3.16}$$

The irreversible real processes are to be replaced by reversible substitute processes, whose initial and final state is the same as in the real processes. For the reversible substitute process to be formed, we can use any model of thought which does not need to have any claim to technical realisation. It is only necessary that the substance(s) involved - starting from the same initial state - reversibly reach the same final state that they reach in the irreversible process.

A reversible process must be frictionless. If, in the energy balance of a real process, the internal friction work $|W_{RI}|$ would be omitted, then the process is changed, and the final state of that changed process will not be the same as the final state of the real process. If one removes the internal friction work $|W_{RI}|$, then something must be introduced as a substitute that acts like the inner friction work, but is not friction work itself. This substitute is a reversibly supplied heat $(Q_{12})_{rev}$, which corresponds to the amount of internal friction work in Eqs. (3.15) and (3.16) respectively. In the case of a reversibly supplied heat, the temperature gradient required for this is zero, i.e. the heat can also flow in the opposite direction (cf. sections 3.2.2 and 3.4.2). In place of Eqs. (3.15) and (3.16) for the real adiabatic processes, the following equations occur

for the reversible substitute processes:

$$(Q_{12})_{rev} - \int_1^2 p\,dV = U_2 - U_1 \tag{3.17}$$

$$(Q_{12})_{rev} + \int_1^2 V\,dp = H_2 - H_1 . \tag{3.18}$$

The reversible substitutes for the real adiabatic processes are no longer adiabatic. Since $(Q_{12})_{rev}$ is the substitute heat for $|W_{RI}|$, it must be a heat supply. In differential notation, from Eqs. (3.17) and (3.18) follows

$$dQ_{rev} - p\,dV = dU \tag{3.19}$$

$$dQ_{rev} + V\,dp = dH . \tag{3.20}$$

If $(Q_{12})_{rev}$ is a heat input, the differential dQ_{rev} has a positive sign.

A distinction can be done between adiabatic processes with friction, adiabatic processes without friction and impossible adiabatic processes. This can be made with the help of the equivalent heat $(Q_{12})_{rev}$ of the corresponding reversible substitute processes. The differential notation according to the equations (3.19) and (3.20) is particularly suitable for this.

For the substitute process of an irreversible adiabatic process is

$$dQ_{rev} > 0 . \tag{3.21}$$

For the substitute process of a reversible adiabatic process on can write

$$dQ_{rev} = 0 . \tag{3.22}$$

For the substitute process of an impossible adiabatic process is

$$dQ_{rev} < 0 . \tag{3.23}$$

The heat $(Q_{12})_{rev}$ has the disadvantage that it is not a state variable. For its calculation it is not sufficient to know the initial point and the end point of the change of state, one also needs the path. However, one obtains a state variable, if dQ_{rev} is divided by the absolute temperature T (proof: section 3.3.3). This quantity is called entropy S according to the proposal of *Rudolf Clausius*. Its differential is

$$dS = \frac{dQ_{rev}}{T} . \tag{3.24}$$

Since the absolute temperature is always positive, the signs of the differentials dS and dQ_{rev} coincide. Therefore, the same rules can be formulated for dS as for dQ_{rev}.

For the substitute process of an irreversible adiabatic process is

$$dS > 0 . \tag{3.25}$$

For the substitute process of a reversible adiabatic process is

$$dS = 0 . \tag{3.26}$$

For the substitute process of an impossible adiabatic process is

$$dS < 0 . \tag{3.27}$$

3.3.2 The Calculation of the Entropy Change

The integration of Eq.(3.24) yields

$$S_2 - S_1 = \int_1^2 \frac{dQ_{rev}}{T} . \tag{3.28}$$

If several substances or bodies are involved in an adiabatic process, a distinction must be made between the entropy change of a single substance involved in the process and the total entropy change of all substances involved in the process.

The entropy change of a single substance can be positive or negative, depending on whether the heat has to be added or has to be removed in the reversible substitute process (cf. section 3.5). The total entropy change of all the substances involved provides a statement about the reversibility of the adiabatic process. The total entropy change can be positive, in which case the process is irreversible, or zero, in which case it is reversible.

3.3.3 Entropy as a State Variable, Total Differential

In simple systems, a state variable is a state function of two variables:

$$z = z(x, y) \tag{3.29}$$

The differential of a function of two variables is a total differential:

$$dz = \left(\frac{\partial z}{\partial x}\right)_y dx + \left(\frac{\partial z}{\partial y}\right)_x dy \tag{3.30}$$

The mixed second derivatives of the function according to Eq. (3.29) are equal; this follows from the mathematical quality of a state function:

$$\frac{\partial^2 z}{\partial x \, \partial y} = \frac{\partial^2 z}{\partial y \, \partial x} \tag{3.31}$$

It is therefore irrelevant whether a state function $z = z(x, y)$ is first derived according to x at constant y and then to y at constant x, or whether a state function $z = z(x, y)$ is first derived according to y at constant x and then according to x at constant y.

The individual expressions of Eq. (3.30) can be interpreted geometrically:

The function $z = z(x, y)$ represents a surface in space. At a point $P_1(x_1|y_1)$ a tangential plane is placed to the surface. On the tangential plane in the neighbourhood of the point $P_1(x_1|y_1)$ lies the point $P_2(x_2|y_2)$. Between the coordinates of the points P_1 and P_2 there is the relation

$$x_2 = x_1 + dx \tag{3.32}$$
$$y_2 = y_1 + dy \tag{3.33}$$
$$z_2 = z_1 + dz. \tag{3.34}$$

In Fig. 3.8, the surface $P_1CP_2DP_1$ is part of the tangential plane. With the lines

$$\overline{P_1A} = dx \qquad (3.35)$$
$$\overline{P_1B} = dy \qquad (3.36)$$

and the angles

$$\sphericalangle AP_1C = \arctan\left(\frac{\partial z}{\partial x}\right)_y \qquad (3.37)$$

$$\sphericalangle BP_1D = \arctan\left(\frac{\partial z}{\partial y}\right)_x \qquad (3.38)$$

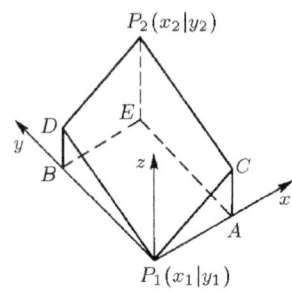

Figure 3.8 For explanation of the total differential

one obtains the distances

$$\overline{AC} = \left(\frac{\partial z}{\partial x}\right)_y dx \qquad (3.39)$$

$$\overline{BD} = \left(\frac{\partial z}{\partial y}\right)_x dy . \qquad (3.40)$$

The addition of these distances gives

$$\overline{AC} + \overline{BD} = \overline{EP_2} = dz . \qquad (3.41)$$

From Eqs. (3.39) to (3.41) follows Eq. (3.30). If one has an expression

$$dz = P(x,y)\, dx + Q(x,y)\, dy , \qquad (3.42)$$

then this expression represents a total differential with the meaning

$$P(x,y) = \left(\frac{\partial z}{\partial x}\right)_y \qquad Q(x,y) = \left(\frac{\partial z}{\partial y}\right)_x , \qquad (3.43)$$

if according to Eq. (3.31) the relation

$$\frac{\partial^2 z}{\partial x\, \partial y} = \frac{\partial^2 z}{\partial y\, \partial x} = \left(\frac{\partial P(x,y)}{\partial y}\right)_x = \left(\frac{\partial Q(x,y)}{\partial x}\right)_y \qquad (3.44)$$

holds. If Eq. (3.44) does not hold, then Eqs. (3.43) do not hold, and Eq. (3.42) is then not a total differential.

Example 3.4 It has to be checked whether the two expressions

$$dz = 2x y^3\, dx + 3 x^2 y^2\, dy \qquad \text{und} \qquad dz = 2 x^3 y\, dx + 3 x y\, dy$$

represent total differentials.

If one assumes that the first expression is a total differential, then the following applies

$$P(x,y) = \left(\frac{\partial z}{\partial x}\right)_y = 2xy^3 \qquad \left(\frac{\partial}{\partial y} 2xy^3\right)_x = 6xy^2$$

$$Q(x,y) = \left(\frac{\partial z}{\partial y}\right)_x = 3x^2y^2 \qquad \left(\frac{\partial}{\partial x} 3x^2y^2\right)_y = 6xy^2 \ .$$

Eq. (3.31) is satisfied since both mixed second derivatives are equal in magnitude. The first expression is indeed a total differential. If the second expression were also a total differential, then the following would apply

$$P(x,y) = \left(\frac{\partial z}{\partial x}\right)_y = 2x^3 y \qquad \left(\frac{\partial}{\partial y} 2x^3 y\right)_x = 2x^3$$

$$Q(x,y) = \left(\frac{\partial z}{\partial y}\right)_x = 3xy \qquad \left(\frac{\partial}{\partial x} 3xy\right)_y = 3y \ .$$

Since here the two mixed second derivatives are not equal in magnitude, Eq. (3.31) is not fulfilled. The second expression is therefore not a total differential.

The proof that the heat $(Q_{12})_{rev}$ is a process variable, while the entropy S is a state variable, still has to be done. For the sake of simplicity relations are used, which are only valid for ideal gases.

An expression according to Eq. (3.42) is provided by the first law according to Eq. (3.19):

$$dQ_{rev} = dU + p\,dV \qquad (3.45)$$

A comparison with Eqs. (3.42) to (3.44) gives

$$P(x,y) \rightarrow 1$$
$$dx \rightarrow dU$$
$$Q(x,y) \rightarrow p$$
$$dy \rightarrow dV$$

$$\left(\frac{\partial P(x,y)}{\partial y}\right)_x \rightarrow \left(\frac{\partial 1}{\partial V}\right)_U = 0 \qquad (3.46)$$

$$\left(\frac{\partial Q(x,y)}{\partial x}\right)_y \rightarrow \left(\frac{\partial p}{\partial U}\right)_V \neq 0 \ . \qquad (3.47)$$

To understand the expression (3.47) at constant volume, note that the internal energy is a function of temperature. One can therefore ask what $\partial p/\partial T$ gives at constant volume. Since at variable temperature and constant volume the pressure is not constant, $\partial p/\partial T \neq 0$ follows. It can be seen that Eq. (3.44) is not fulfilled and that Eq. (3.43) does not apply. dQ_{rev} according to Eq. (3.45) is not a total differential, and $(Q_{12})_{rev}$ is not a state variable. $(Q_{12})_{rev}$ is a process variable.

According to Eqs. (3.24) and (3.45) the differential of entropy is

$$dS = \frac{dQ_{rev}}{T} = \frac{dU + p\,dV}{T} \ . \qquad (3.48)$$

It is to be shown that dS is a total differential. For this purpose, in anticipation of the treatment of ideal gases (Sections 4.2 and 4.1) the following equations are used:

$$dU = m\,c_v\,dT \qquad (3.49)$$

$$pV = mRT \tag{3.50}$$

c_v is the specific heat capacity at constant volume according to Eq. (2.100). For ideal gases it is only a function of temperature. R is the gas constant. Eqs. (3.49) and (3.50) are inserted into Eq. (3.48):

$$dS = \frac{m\,c_v}{T}\,dT + \frac{m\,R}{V}\,dV \tag{3.51}$$

A comparison with Eqs. (3.42) to (3.44) gives

$$P(x,y) \rightarrow \frac{m\,c_v}{T}$$

$$dx \rightarrow dT$$

$$Q(x,y) \rightarrow \frac{m\,R}{V}$$

$$dy \rightarrow dV$$

$$\left(\frac{\partial P(x,y)}{\partial y}\right)_x \rightarrow \left(\frac{\partial}{\partial V}\frac{m\,c_v}{T}\right)_T = 0 \tag{3.52}$$

$$\left(\frac{\partial Q(x,y)}{\partial x}\right)_y \rightarrow \left(\frac{\partial}{\partial T}\frac{m\,R}{V}\right)_V = 0\,. \tag{3.53}$$

Note that the partial derivative according to Eq. (3.52) is formed at constant temperature and the partial derivative according to Eq. (3.53) is formed at constant volume. From Eqs. (3.52) and (3.53) it follows that Eq. (3.44) is fulfilled. Thus Eqs. (3.43) are valid, dS is a total differential, and the entropy S is a state variable.

3.4 The Entropy Change of Irreversible Processes

All calculations carried out in connection with a technical process refer to the substitute process. The technical process is irreversible, the substitute process is reversible. In many cases, some of the components of the system in which the real process takes place, have to be changed for the substitute process. In the surroundings of the substitute system, heat sources and additional equipment must be assumed.

The substitute process is generally no longer adiabatic, the heat transfer from the heat sources of the surroundings to the substitute process must be reversible. The formation of the substitute processes and the calculation of the entropy change is based on the irreversible processes discussed in section 3.2. The reversible substitute process not only serves to calculate the irreversible real process, but also can show in which direction the real process can be improved.

3.4.1 Friction

For the processes described in section 3.2.1, the following applies according to Eq. (3.1):

$$|W_{RI}| = U_2 - U_1 \tag{3.54}$$

In the reversible substitution process, the internal friction work $|W_{RI}|$ is replaced by the heat $(Q_{12})_{rev}$ (Fig. 3.9). The first law for the replacement system is

$$(Q_{12})_{rev} = U_2 - U_1\,. \tag{3.55}$$

The reversible substitute heat $(Q_{12})_{rev}$ comes from a heat source in the surroundings of the substitute system. Then one must assume constant temperatures T of both systems to make sure that the heat transfer ist reversible. The entropy change is

$$S_2 - S_1 = \int_1^2 \frac{dQ_{rev}}{T} \ . \tag{3.56}$$

As dQ_{rev} is positive, $S_2 - S_1$ becomes also positive.

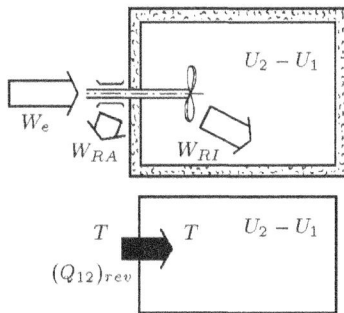

Figure 3.9 Supply of friction work and reversible substitute process

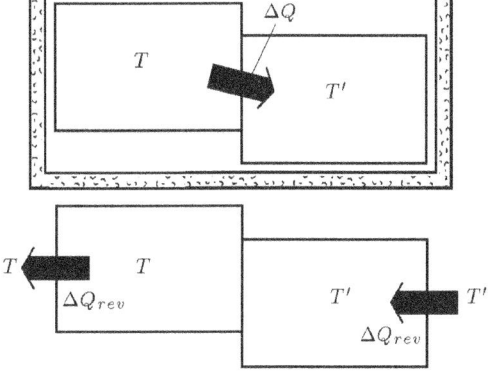

Figure 3.10 Temperature equalisation and reversible substitute process

Example 3.5 Which change in entropy flow $\dot{S}_2 - \dot{S}_1$ occurs when water flows through the water turbine according to example 2.9? The water temperature at the entrance of the turbine is 12 °C. The heat flow $(\dot{Q}_{12})_{rev}$ is equal to the internal friction power $|P_{RI}|$.

$$(\dot{Q}_{12})_{rev} = |P_{RI}| = \dot{U}_2 - \dot{U}_1 = \dot{m}\, c\, \Delta t = 581 \text{ kW}$$

This heat flow must be supplied to the turbine in the substitute process at a nearly constant temperature of 12 °C. The temporal change in entropy is therefore

$$\dot{S}_2 - \dot{S}_1 = \frac{(\dot{Q}_{12})_{rev}}{T} = \frac{581 \text{ kW}}{285 \text{ K}} = 2.04 \text{ kW/K} \ .$$

3.4.2 Temperature Equalisation

A small amount of heat ΔQ is transferred from a system with a higher temperature to a system with a lower temperature. This process of temperature equalisation can be described as an adiabatic process if the two systems are considered as an adiabatic overall system (Fig. 3.10).

The process of temperature equalisation is divided into two substitute processes. In the first substitute process heat is reversibly released from the system with the higher temperature. The second substitute process involves the reversible supply of heat to the system with the lower temperature. If one considers a small amount of heat ΔQ_{rev},

one can assume that the different temperatures of the two systems T and T' remain constant each for itself. The entropy change in the first substitute process is

$$\Delta S_T = \frac{\Delta Q_{rev}}{T} . \qquad (3.57)$$

This expression is negative because ΔQ_{rev} is negative. For the sake of clarity one writes

$$\Delta S_T = -\frac{|\Delta Q_{rev}|}{T} . \qquad (3.58)$$

The entropy change of the second substitute process is

$$\Delta S_{T'} = \frac{\Delta Q_{rev}}{T'} . \qquad (3.59)$$

This expression is positive because ΔQ_{rev} is positive. To make the sign perceptible, one writes

$$\Delta S_{T'} = \frac{|\Delta Q_{rev}|}{T'} . \qquad (3.60)$$

The total entropy change is

$$\Delta S = \Delta S_T + \Delta S_{T'} = |\Delta Q_{rev}|\left(\frac{1}{T'} - \frac{1}{T}\right) = |\Delta Q_{rev}|\frac{T - T'}{TT'} . \qquad (3.61)$$

Since there is $T > T'$, one obtains $\Delta S > 0$. If $T = T'$, there is $\Delta S = 0$.

The formation of the substitute process can be imagined in different ways. One possibility is that heat is reversibly transferred from the system with the higher temperature to a heat sink outside of the substitute system, while the system with the lower temperature receives the same heat reversibly from a heat source outside the substitute system.

Another possibility arises if, instead of a heat sink and a heat source, a frictionless piston engine is assumed as an auxiliary unit instead of a heat sink and a heat source. During the first part of an expansion of the working gas in the cylinder it reversibly receives heat from the system with the higher temperature. During the second part of the expansion the cylinder is adiabatic, whereby the temperature of the working gas in the cylinder drops from the temperature T to the temperature T'.

During the subsequent compression the working gas reversibly transfers the previously received heat to the system with the lower temperature. The work gained during the expansion is stored and is available for compression. The frictionless machine is located as an additional device outside the substitute system. The entropy change in this machine is not taken into account in the calculation of the total entropy change.

Example 3.6 In the furnace of a steam boiler, by combustion a heat flow of $\Delta \dot{Q}_{rev} = 300$ kW at $t = 1400\,°\text{C}$. is produced. It is supplied to the steam to be provided at a temperature of $t' = 200\,°\text{C}$. What increase in entropy over time $\dot{S}_2 - \dot{S}_1$ occurs as a result of this process?

$$\dot{S}_2 - \dot{S}_1 = |\Delta \dot{Q}_{rev}|\frac{T - T'}{TT'} = 300 \text{ kW} \frac{1673 \text{ K} - 473 \text{ K}}{1673 \text{ K} \cdot 473 \text{ K}} = 0.455 \text{ kW/K}$$

3.4.3 Pressure Equalisation

For the pressure equalisation between two gas vessels in a closed adiabatic system (Fig. 3.11) Eq. (3.6) is

$$-W_{V12} = |W_{RI}| \ . \tag{3.62}$$

In the reversible substitute process, the internal friction work $|W_{RI}|$ is replaced by the reversibly supplied heat $(Q_{12})_{rev}$. With Eq. (2.21) one obtains

$$(Q_{12})_{rev} = \int_1^2 p \, dV \tag{3.63}$$

or in differential notation

$$dQ_{rev} = p \, dV \ . \tag{3.64}$$

In the surroundings of the reversible substitute system, a heat source must be assumed as an additional device. The heat dQ_{rev} which is transferred over the boundary of the substitute system causes the entropy change

$$dS = \frac{dQ_{rev}}{T} = \frac{p}{T} dV \ . \tag{3.65}$$

The integration gives

$$S_2 - S_1 = \int_1^2 \frac{p}{T} dV \ . \tag{3.66}$$

On the gas, the pressure equalisation has the effect of an increase in volume. In the substitute process, the two gas vessels are replaced by a cylinder in which a piston encloses the gas under high pressure at the beginning. If the piston moves outwards, volume change work W_{V12} is performed and given off to the surroundings. It is equal to the heat $(Q_{12})_{rev}$ supplied from outside. If dV is positive, the entropy change becomes positive according to Eq. (3.66).

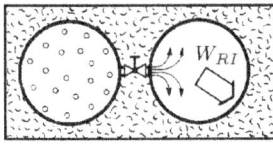

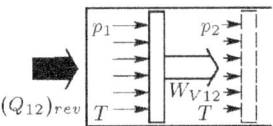

Figure 3.11 Pressure equalisation and reversible substitute process (for an ideal gas: isothermal expansion)

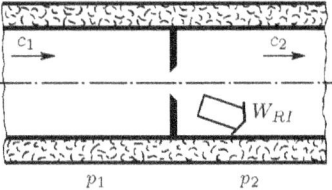

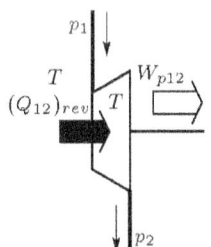

Figure 3.12 Throttling and reversible substitute process (for an ideal gas: isothermal expansion)

3.4.4 Throttling

According to Eq. (3.10), the following applies to the process of throttling, which is the adiabatic expansion of a gas in an open adiabatic system (Fig. 3.12):

$$-W_{p12} = |W_{RI}| \tag{3.67}$$

In the reversible substitute process, the internal friction work $|W_{RI}|$ is replaced by the reversibly supplied heat $(Q_{12})_{rev}$. If one uses the integral representation of the pressure change work according to Eq. (2.30), the first law gives for the substitute process

$$(Q_{12})_{rev} = -\int_1^2 V\,dp \tag{3.68}$$

or in differential notation

$$dQ_{rev} = -V\,dp\,. \tag{3.69}$$

In the substitute process, the function of the throttling point (e.g. a valve) is taken over by a frictionless turbine. As an additional device of the substitute system, a heat source is required from which heat is reversibly supplied to the turbine via the system boundary of the turbine. The turbine must therefore not be adiabatic. The pressure change work W_{p12} performed in the turbine and transferred to the surroundings is equal to the heat W_{p12} supplied by the heat source $(Q_{12})_{rev}$. The differential entropy change according to Eqs. (3.24) and (3.69) is

$$dS = \frac{dQ_{rev}}{T} = -\frac{V}{T}\,dp\,. \tag{3.70}$$

The integration gives

$$S_2 - S_1 = -\int_1^2 \frac{V}{T}\,dp\,. \tag{3.71}$$

Since dp is negative at throttling, the entropy change becomes positive according to Eq. (3.71).

3.5 Non-Adiabatic Process and Reversible Substitute Process

Substitute processes have so far only been formed for adiabatic processes (section 3.3.1). The reversible substitute processes can also be formed for non-adiabatic processes. According to Eq. (2.47), the following applies to a technical process in a closed system:

$$Q_{12} + W_{V12} + |W_{RI}| = U_2 - U_1 \tag{3.72}$$

For the reversible substitute process in a closed system, we obtain

$$(Q_{12})_{rev} + W_{V12} = U_2 - U_1\,. \tag{3.73}$$

According to Eq. (2.70), the following holds for a technical process in an open system:

$$Q_{12} + W_{p12} + |W_{RI}| = H_2 - H_1 \tag{3.74}$$

For the reversible substitute process in an open system, one obtains

$$(Q_{12})_{rev} + W_{p12} = H_2 - H_1 \ . \tag{3.75}$$

The differential forms of Eqs. (3.73) and (3.75) for the reversible substitute process are

$$dQ_{rev} - p\,dV = dU \qquad\qquad dQ_{rev} + V\,dp = dH \ . \tag{3.76}$$

Using Eq. (3.24) for entropy, Eqs. (3.76) can be written as follows:

$$T\,dS - p\,dV = dU \qquad\qquad T\,dS + V\,dp = dH \ . \tag{3.77}$$

The fundamental equations (3.77) represent a combination of the first and second law of thermodynamics. They summarise essential statements of these two generalisations of experimental findings for simple systems in a condensed form.

In Section 1.3, the thermal state variables p, V and T were introduced. The state variables entropy S, internal energy U and enthalpy H are called caloric state variables.

From Eqs. (3.72) and (3.73) and from Eqs. (3.74) and (3.75) follows

$$Q_{12} + |W_{RI}| = (Q_{12})_{rev} \ . \tag{3.78}$$

Eq. (3.78) can be used to transform Eq. (3.24):

$$dS = \frac{dQ_{rev}}{T} = \frac{dQ}{T} + \frac{|dW_{RI}|}{T} \tag{3.79}$$

This equation illustrates the two possible causes of a change in entropy: a heat input or heat output, and friction. While in adiabatic processes is

$$dS \geq 0 \ , \tag{3.80}$$

in non-adiabatic processes when heat is released, the entropy change can also be negative.

The left-hand sides of Eqs. (3.78) and (2.88) respective (2.89) are identical. The right-hand sides of the equations mentioned give with Eq. (1.3)

$$(Q_{12})_{rev} = m\int_1^2 c\,dt = m\int_1^2 c\,dT \tag{3.81}$$

$$(Q_{12})_{rev} = m\,c_m(t_2 - t_1) = m\,c_m(T_2 - T_1) \ . \tag{3.82}$$

According to section 2.7.1, c is the true specific heat capacity. It is a function of temperature. c_m is the mean specific heat capacity. According to Eq. (2.98), one often writes only c for the mean specific heat capacity c_m and omits the index m. Eq. (3.82) then results in

$$(Q_{12})_{rev} = m\,c\,(t_2 - t_1) = m\,c\,(T_2 - T_1) \ . \tag{3.83}$$

According to Eq. (3.78), the heat $(Q_{12})_{rev}$ occurring in the reversible substitute process comprises the heat transferred in the original process Q_{12} and the internal friction work $|W_{RI}|$. The following special cases arise:

a) adiabatic change of state with
$$Q_{12} = 0 \qquad (Q_{12})_{rev} = |W_{RI}| . \qquad (3.84)$$
b) frictionless change of state with
$$|W_{RI}| = 0 \qquad (Q_{12})_{rev} = Q_{12} . \qquad (3.85)$$
c) isentropic change of state with
$$(Q_{12})_{rev} = 0 . \qquad (3.86)$$

3.5.1 Isentropic Change of State; Interpretations of Entropy

Substances involved in a process undergo changes of state, which in the reversible substitute process must be reversible. The changes of state must be quasi-static according to Section 3.1.2.

If the entropy remains constant, the change of state is called an isentropic change of state. According to Eq. (3.24) this is represented by

$$dS = \frac{dQ_{rev}}{T} = 0 \qquad (3.87)$$

equal to
$$dQ_{rev} = 0 . \qquad (3.88)$$

For a state variable X, $dX = 0$ means that $X = $ const. For a process variable X, $dX = 0$ means that $X = 0$. Since heat is a process variable, the statements of Eqs. (3.86) and (3.88) represent an identical meaning.

According to Eq. (3.78), a change of state is isentropic if it is adiabatic and frictionless:

$$Q_{12} = 0 \qquad |W_{RI}| = 0 \qquad (Q_{12})_{rev} = 0 \qquad (3.89)$$

However, an isentropic change of state is also possible at $Q_{12} \neq 0$ and $|W_{RI}| \neq 0$ if

$$Q_{12} = -|W_{RI}| \qquad (Q_{12})_{rev} = 0 \qquad (3.90)$$

is valid. Therefore, according to Eqs. (3.89) a reversible adiabatic change of state is called an isentropic change of state. The inversion of this statement is not generally valid. In the case of Eq. (3.90) it does not apply.

Interpretations of the entropy term have led to different formulations describing different aspects of entropy. *Rudolf Plank* stated in 1926: "The entropy term has a significance that extends far beyond the scope of specialised technical studies, and the second law belongs to the most important foundations of the knowledge of nature." By *Rudolf Clausius* entropy was first called "Äquivalenzwert einer Verwandlung" (equivalent value of a transformation), before in 1865 he imprinted the term "entropy". Entropy was understood by *Ludwig Boltzmann* as a measure of disorder, by *Max Planck* as a measure of probability. In 1948, Claude Shannon understood entropy as a measure of the observer's ignorance.

In the previous sections, it was shown that entropy is a state variable which increases in an irreversible adiabatic process and remains constant in a reversible adiabatic process; a decrease of entropy in an adiabatic process is impossible.

Adiabatic processes with friction, temperature equalisation, pressure equalisation or throttling are irreversible. In an adiabatic process with friction, disorder occurs. In

the adiabatic processes of pressure equalisation and throttling, the disorder grows in connection with the expansion of the gas and the friction that occurs at the same time. The disorder also increases in the adiabatic process of temperature equalisation; this is shown in a reduced ability of the heat that occurs in this process to be able to gain valuable work with a cyclic process.

If a substance that exists as a solid in a well-ordered state is heated and then liquefied, its entropy - i.e. its disorder - increases. When heated further, the liquid finally changes to a gaseous state and becomes even more disordered. When the gas is heated further, its disorder increases even more. The name 'gas' refers to its state as 'chaos' as a result of the very rapid, disordered movement of its particles.

3.5.2 Entropy Diagrams

For the representation of quasi-static changes of state, the T,S diagram is particularly suitable with the temperature T as ordinate and the entropy S as abscissa. In this diagram, the heat $(Q_{12})_{rev} = Q_{12} + |W_{RI}|$ is represented as the area between the state curve and the abscissa. This can be seen from Eq. (3.24) if written as follows:

$$dQ_{rev} = T\,dS \tag{3.91}$$

The integration gives

$$(Q_{12})_{rev} = \int_1^2 T\,dS \ . \tag{3.92}$$

The sign of the entropy change determines whether the heat $(Q_{12})_{rev}$ is to be added ($dS > 0$) or removed ($dS < 0$). Because an area in the T,S diagram has the meaning of a reversible substitute heat, the T,S diagram is a heat diagram in which the internal friction work can also be represented (Fig. 3.13).

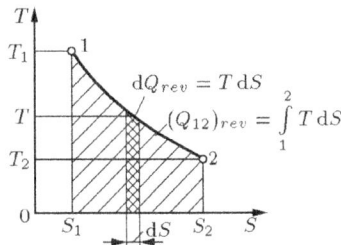

Figure 3.13 T,S diagram and representation of the reversible substitute heat $(Q_{12})_{rev}$

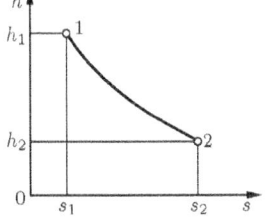

Figure 3.14 h,s diagram

The H,S diagram – with specific values

$$h = \frac{H}{m} \tag{3.93}$$

$$s = \frac{S}{m} \tag{3.94}$$

as h, s diagram of *Mollier*[3] (Fig. 3.14) – is primarily used to represent changes of state in steady-state flow processes. It has gained special importance for the calculation of steam power plants. Since in technical calculations one often only has to calculate the change in entropy, it is generally not of importance which zero point is chosen for the entropy scale on the abscissa. The change in entropy is independent of the choice of the zero point.

In a T, S diagram or a H, S diagram, an isentropic change of state is represented by a vertical line. The description of a change of state as adiabatic or frictionless does not yet provide a definite curve in one of the diagrams mentioned. In the case of a frictionless change of state, any curve progression is possible. If a change of state is adiabatic and internal friction work occurs, the change of state can only take place in the direction of increasing entropy ($dS > 0$). Since the term "adiabatic" does not definitely characterise a change of state, it is often avoided in connection with a change of state and is used only to describe a system. Under no circumstances should the terms adiabatic and isentropic be confused.

Example 3.7 The change of state of the water when flowing through the turbine according to example 2.9 or example 3.5 is to be represented in a T, s diagram.

We calculate the change in specific entropy:

$$s_2 - s_1 = \frac{\dot{S}_2 - \dot{S}_1}{\dot{m}} = \frac{2.04 \text{ kW s}}{\text{K } 578 \text{ kg}} \cdot \frac{1000 \text{ W}}{\text{kW}} \cdot \frac{\text{J}}{\text{W s}} = 3.53 \frac{\text{J}}{\text{kg K}}$$

Arbitrarily, we set $s_1 = 0$ and obtain the representation according to Fig. 3.15. The specific reversible substitute heat corresponding to the specific friction work appears as a rectangular area

$$(q_{12})_{rev} = |w_{RI}| = u_2 - u_1 = \frac{\dot{U}_2 - \dot{U}_1}{\dot{m}} = \frac{581 \text{ kW s}}{578 \text{ kg}} = 1005 \frac{\text{J}}{\text{kg}}.$$

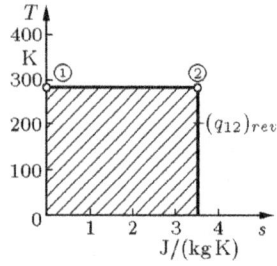

Figure 3.15 To example 3.7

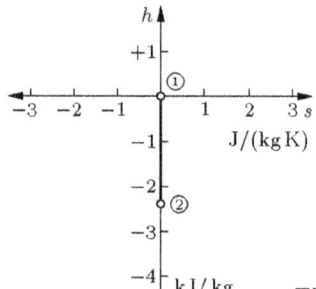

Figure 3.16 To example 3.8

Example 3.8 The change of state of the water flowing through the turbine in example 2.9, is to be shown in the h, s diagram for the case that no friction work occurs.

When there is no friction, the coupling power is $P_e = (P_e)_{id}$. From the equations given in Example 2.9

$$(P_e)_{id} = \dot{H}_2 - \dot{H}_1 + \frac{\dot{m}}{2}(c_2^2 - c_1^2) + \dot{m} g (z_2 - z_1)$$

---

[3] *Richard Mollier*, 1863 to 1935, professor at the TH Dresden, designed in 1904 the h, s diagram and later the h, x diagram for humid air.

and follows

$$(P_e)_{id} = \dot{m}\left[\frac{p_2 - p_1}{\varrho} + \frac{1}{2}(c_2^2 - c_1^2) + g(z_2 - z_1)\right]$$

$$\dot{H}_2 - \dot{H}_1 = \dot{m}\frac{p_2 - p_1}{\varrho}$$

and after division by the mass flow

$$h_2 - h_1 = \frac{p_2 - p_1}{\varrho} = \frac{(0.98 - 24.18) \cdot 10^5 \text{ N m}^3}{\text{m}^2 \ 1000 \text{ kg}} \cdot \frac{\text{kJ}}{10^3 \text{ N m}} = -2.32 \frac{\text{kJ}}{\text{kg}}.$$

Since there is no friction, $s_2 = s_1$ follows. The change of state is represented as a vertical line. We arbitrarily set $s_1 = 0$ and $h_1 = 0$ and obtain the representation in Fig. 3.16.

3.5.3 Circular Integral, Thermodynamic Temperature

For the process variable heat in a reversible substitute process, the following applies:

$$\int_1^2 dQ_{rev} = (Q_{12})_{rev} \tag{3.95}$$

For the state variable entropy is

$$\int_1^2 dS = S_2 - S_1 . \tag{3.96}$$

The integrals in Eqs. (3.95) and (3.96) can be interpreted as line integrals, which is especially useful when a sequence of different changes of state is to be considered.

A line integral has the general form

$$\int_1^2 (P\,dx + Q\,dy) . \tag{3.97}$$

It is a certain integral to be calculated along a curve or a section of a curve from initial point 1 to end point 2. Here P and Q are functions of x and y:

$$P = P(x, y) \qquad\qquad Q = Q(x, y) \tag{3.98}$$

The question is to be clarified under which condition the integral (3.97) depends only on the initial point 1 and on the end point 2 and is independent of the path from 1 to 2. This question can be answered in such a way that the integral (3.97) is independent of the path if for the solution the equation

$$\int_1^2 (P\,dx + Q\,dy) = z(x_2, y_2) - z(x_1, y_1) \tag{3.99}$$

holds. The right-hand side of equation (3.99) can be thought of as arisen as follows:

$$\int_1^2 dz = [z(x,y)]_1^2 = z(x_2, y_2) - z(x_1, y_1) \tag{3.100}$$

dz is the expression according to Eq. (3.42) and the total differential according to Eq. (3.30). The line integral according to Eq. (3.97) is independent of the path if Eq. (3.44) or Eq. (3.31) holds.

3.5 Non-Adiabatic Process and Reversible Substitute Process

If the integration takes place over a closed path, i.e. a path that leads back to the starting point, then the line integral (3.97) becomes the circular integral $\oint (P\,dx + Q\,dy)$. If the integral is independent of the path, then the circular integral becomes

$$\oint dz = \oint (P\,dx + Q\,dy) = 0 . \tag{3.101}$$

This can be seen from Eq. (3.100) when the upper and lower boundaries are equal. If one forms a circular integral with Eq. (3.95), then with Eq. (3.92) follows

$$\oint dQ_{rev} = \oint T\,dS = (Q_{zu})_{rev} - |(Q_{ab})_{rev}| . \tag{3.102}$$

In closed-loop integration, there are regions with $dS > 0$ and $dS < 0$. All positive and negative partial results are combined to $(Q_{zu})_{rev}$ and to $|(Q_{ab})_{rev}|$, respectively, where the overall result is the difference of both expressions $(Q_{zu})_{rev} - |(Q_{ab})_{rev}|$.

A circular integral with Eq. (3.96) yields according to Eq. (3.101)

$$\oint dS = 0 . \tag{3.103}$$

Correspondingly, Eq. (2.44) gives $d(pV) = V\,dp + p\,dV$ and $d(TS) = T\,dS + S\,dT$:

$$\oint d(pV) = \oint V\,dp + \oint p\,dV = 0 \qquad \oint V\,dp = -\oint p\,dV \tag{3.104}$$

$$\oint d(TS) = \oint T\,dS + \oint S\,dT = 0 \qquad -\oint T\,dS = \oint S\,dT \tag{3.105}$$

For the expressions according to Eqs. (3.104) and (3.105) the term cyclic process work W_{Kreis} will be introduced later. Eq. (3.105) can be understood as area calculation of the circled area in the T, S diagram and Eq. (3.104) as area calculation of the circled area in the p, V diagram.

Eqs. (3.102) and (3.103) shall be applied to a special path in the T, S diagram according to Fig. 3.17 (*Carnot* process, see section 7.3.3). Heat is transferred on the path from 1 to 2 and from 3 to 4. These are isothermal changes of state with temperatures T_1 and T_3:

$$(Q_{zu})_{rev} = (Q_{12})_{rev} = \int_1^2 T\,dS = T_1(S_2 - S_1) \tag{3.106}$$

$$|(Q_{ab})_{rev}| = -(Q_{34})_{rev} = -\int_3^4 T\,dS = -T_3(S_4 - S_3) = T_3(S_2 - S_1) \tag{3.107}$$

The total heat transferred is

$$(Q_{zu})_{rev} - |(Q_{ab})_{rev}| = (T_1 - T_3)(S_2 - S_1) . \tag{3.108}$$

On the same paths where heat is transferred, the entropy also changes:

$$S_2 - S_1 = \frac{(Q_{12})_{rev}}{T_1} = \frac{(Q_{zu})_{rev}}{T_1} \tag{3.109}$$

$$S_4 - S_3 = \frac{(Q_{34})_{rev}}{T_3} = -\frac{|(Q_{ab})_{rev}|}{T_3} \tag{3.110}$$

According to Eq. (3.103), the entropy change on the total path is

$$\oint dS = S_2 - S_1 + S_4 - S_3 = \frac{(Q_{zu})_{rev}}{T_1} - \frac{|(Q_{ab})_{rev}|}{T_3} = 0 . \tag{3.111}$$

Figure 3.17 Illustration of the path for the calculation of the circular integral according to Eq. (3.111)

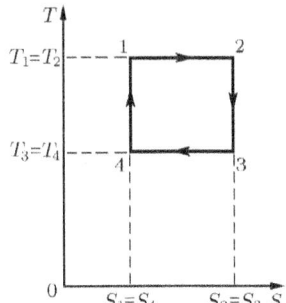

Eq. (3.111) can be used to define a temperature scale called absolute temperature, thermodynamic temperature, universal temperature or *Kelvin* temperature.

If one assumes that the *Kelvin* scale for the absolute temperature is still unknown and only the *Celsius* scale is available, then the following approach can be made for T:

$$T = t + T_0 \tag{3.112}$$

t is the temperature on the known *Celsius* scale, T_0 is still unknown. The values for $(Q_{zu})_{rev}$ and $|(Q_{ab})_{rev}|$ can be calculated for an ideal gas at temperatures t_1 and t_3, using relations for the isothermal change of state of ideal gases (section 4.3.3, Eqs. (4.80) and (4.86)):

$$(Q_{zu})_{rev} = (Q_{12})_{rev} = \int_1^2 T\,dS = \int_1^2 p\,dV = \int_1^2 \frac{m\,R\,T_1}{V}dV = p_1 V_1 \ln \frac{V_2}{V_1} = p_1 V_1 \ln \frac{p_1}{p_2} \tag{3.113}$$

$$|(Q_{ab})_{rev}| = -(Q_{34})_{rev} = -\int_3^4 T\,dS = -\int_3^4 p\,dV = -\int_3^4 \frac{m\,R\,T_3}{V}dV = -p_3 V_3 \ln \frac{V_4}{V_3} = -p_3 V_3 \ln \frac{p_3}{p_4} \tag{3.114}$$

Using the approach given by Eq. (3.112), Eq. (3.111) becomes

$$\frac{(Q_{zu})_{rev}}{t_1 + T_0} - \frac{|(Q_{ab})_{rev}|}{t_3 + T_0} = 0 . \tag{3.115}$$

The transformation of Eq. (3.115) according to the unknown T_0 yields

$$T_0 = \frac{|(Q_{ab})_{rev}|\,t_1 - (Q_{zu})_{rev}\,t_3}{(Q_{zu})_{rev} - |(Q_{ab})_{rev}|} . \tag{3.116}$$

There are no substance values in Eq. (3.116); thus, it is independent of the choice of a substance. Since T_0 can therefore be calculated from measurements of thermal state variables and of thermal process variables without using substance values, the temperature T, which is defined with equation (3.112) and calculated with the help of

$$T_0 = 273{,}15 \text{ K} , \tag{3.117}$$

is also called absolute temperature, thermodynamic temperature, universal temperature or *Kelvin* temperature. It coincides with the temperature of the ideal gas thermometer [54] (also called only gas thermometer), which is discussed in section 4.1.4.

3.5.4 Dissipative Energy

A force, acting on a body whose position in a coordinate system is described by its positional coordinates, changes the position of the body in the coordinate system, and work is done on the body. If F is the force and dx is a differential change of the spatial coordinate in the direction of the force effect, then the mechanical work W_m on the way from the initial point 1 to the end point 2 is

$$W_m = \int_1^2 F\,dx \ . \tag{3.118}$$

Because the mechanical work is connected with a change of the spatial coordinate x, one can call the spatial coordinate also a work coordinate of the mechanical work. If one considers the volume change work

$$W_{V12} = -\int_1^2 p\,dV \ , \tag{3.119}$$

then the volume V is to be regarded as the work coordinate of the volume change work. For the pressure change work

$$W_{p12} = \int_1^2 V\,dp \tag{3.120}$$

the pressure p can be interpreted as the work coordinate of the pressure change work.

There is a certain relationship between work and heat as forms of energy transfer, so that analogue to the term work coordinate the term heat coordinate can be used. The term heat coordinate is obvious. From the equation

$$(Q_{12})_{rev} = \int_1^2 T\,dS \tag{3.121}$$

we see that the entropy S can be understood as a heat coordinate. According to Eq. (3.78)

$$Q_{12} + |W_{RI}| = (Q_{12})_{rev} \tag{3.122}$$

not only the heat Q_{12} is transferred via the heat coordinate. Also the internal friction work $|W_{RI}|$ uses the heat coordinate instead of a work coordinate. Internal friction work $|W_{RI}| = |W_{diss}|$ is therefore an energy that has "gone astray", that has got onto a "wrong path". It is a dissipated energy that can no longer be converted back into usable work.

This is expressed in the term dissipative energy. When one says energy dissipates, one means that energy undergoes a transformation that is not desired. The energy is not lost, it just cannot be used in the way one actually wanted. If the term internal friction work is still used, it is in the sense that the meaning of the term dissipative energy is included.

Tasks for Section 3

1. For the determination of the mechanical heat equivalent, *Joule* used the schematic arrangement shown in Fig. 3.18. An agitator in a vessel is driven by the sinking of the mass m. The vessel contains a liquid that is heated due to the internal friction work performed. The vessel is thermally insulated from the surroundings. What temperature increase is measured when 0.5 kg of mercury of 14 °C with the specific heat capacity $c = 0.1393$ kJ/(kg K) is heated without heat loss by a mass of 5 kg sinking by 1.5 m? What entropy change occurs during this process?

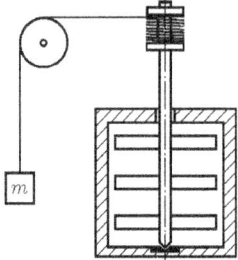

Figure 3.18 Arrangement of *Joule* for the determination of the mechanical heat equivalent

2. According to task 2, section 2, the water is heated during a longer time perid of 1.8 hours due to the fact that the insulation of the water heater is partly damaged and a heat transfer to the surroundings must be accepted. The ambient temperature can be assumed to be constant at 18 °C, and for the heat source, a constant temperature of 200 °C can be assumed. What total entropy change occurs?

3. In a pressure reducing valve, a water flow of 60 ℓ/min with a temperature of 12 °C is throttled from a gage pressure of 6 bar to a gage pressure of 2.5 bar. What entropy change occurs in one hour during this process?

4. It has to be checked whether the two expressions

$$dz = 6\,x\,y^5\,dx + 15\,x^2\,y^4\,dy \quad \text{and} \quad dz = 4\,x^3\,y^6\,dx + 6\,x^4\,y^5\,dy$$

represent total differentials or not.

5. A heat pump is used to draw 8 kW of heat from the ground, which has an average temperature of 10 °C. This heat output is transferred to a water/glycol mixture flowing in a borehole heat exchanger at an average temperature of -1 °C. What is the increase in entropy over time associated with this process?

4 Ideal Gases

In a gas, due to the thermal energy of the molar substance, the molecules move in a disordered manner and with great velocity in the available space. If we compare the mean distance between molecules with the "diameter" of the molecules (for the sake of simplicity, let us introduce this term), it can be found that under normal conditions the ratio between the intrinsic volume of the molecules and the total volume of the gas is very small.

If one neglects the intrinsic volume of the molecules and thus imagines the molecules as mass points and, in addition, neglects the interaction forces between the molecules, important statements can be made about the behaviour, and thus about the properties, of the gas. Such an idealised gas is called an ideal or perfect gas. If these simplifications are not possible, one speaks of a real gas.

4.1 Thermal Equation of State

According to Eq. (1.4), between the thermal variables of state pressure p, volume V and temperature T there exists the thermal equation of state

$$F(p, V, T) = 0 \; . \tag{4.1}$$

This equation is particularly easy to formulate for ideal gases. In the following, the thermal equation of state for ideal gases will be discussed.

4.1.1 Law of *Boyle* and *Mariotte*

According to *Boyle*[1] and *Mariotte*[2], for ideal gases, the product of pressure and volume is constant at unchanged temperature:

$$p_1 V_1 = p_2 V_2 \qquad pV = \text{const} \tag{4.2}$$

The numerical value of the constant on the right-hand side of Eq. (4.2) depends on the temperature, the mass and the type of the gas. For a given mass of a given gas, the higher the temperature, the greater is the constant. Eq. (4.2) represents a hyperbola in a p, V diagram.

4.1.2 Law of *Gay-Lussac*

According to *Gay-Lussac*[3], the volume of an ideal gas at constant pressure changes linearly with temperature. According to Fig. 4.1 holds

$$\frac{V}{T_0 + t} = \frac{\overline{V_0}}{T_0} \qquad V = \overline{V_0} \left(1 + \frac{t}{T_0}\right) . \tag{4.3}$$

---

[1] *Robert Boyle*, 1627 to 1691, Irish scientist
[2] *Edme Mariotte*, 1620 to 1684, French scientist
[3] *Joseph Louis Gay-Lussac*, 1778 to 1850. The measurement of the coefficient of expansion of gases was first discovered not by *Gay-Lussac*, but by *Guillaume Amontons* (1663 to 1705).

© The Author(s), under exclusive license to
Springer Fachmedien Wiesbaden GmbH, part of Springer Nature 2023
M. Dehli et al., *Fundamentals of Technical Thermodynamics*,
https://doi.org/10.1007/978-3-658-38910-9_4

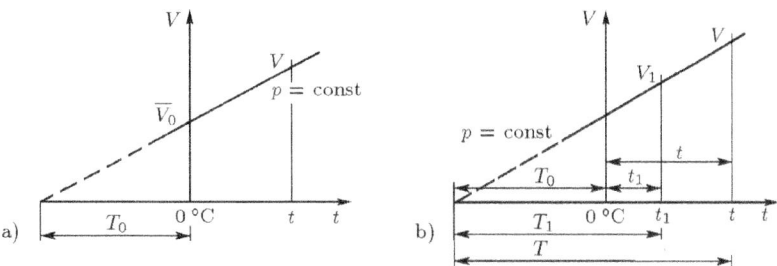

Figure 4.1 a) Change of volume of an ideal gas with temperature at constant pressure
b) Diagram to derive the coefficient of volume expansion

With the thermal expansion coefficient (volume expansion coefficient) β (at the physical norm state β_0, cf. Eq. (4.12))

$$\beta = \frac{1}{V}\left(\frac{\partial V}{\partial T}\right)_p = \frac{1}{V}\left(\frac{V}{T}\right)_p = \frac{1}{T} \qquad \beta_0 = \frac{1}{V_0}\left(\frac{V_0}{T_0}\right)_p = \frac{1}{T_0} \qquad (4.4)$$

one obtains

$$V = \overline{V_0}\,(1 + \beta_0\, t)\;. \qquad (4.5)$$

$\overline{V_0}$ is the volume at $0\,°C$ and pressure p. The increase in volume ΔV during isobaric heating from $0\,°C$ to a temperature t is $\Delta V = \overline{V_0}\,\beta_0\, t$. The coefficient of volume expansion β_0 has the value valid for all ideal gases

$$\beta_0 = \frac{1}{273{,}15\ \text{K}}\;. \qquad (4.6)$$

Note that $\overline{V_0}$ and β_0 refer to a temperature of $0\,°C$.

If at pressure p and temperature t_1 the initial volume is V_1, then for the final volume V at isobaric heating to the temperature t according to Figure 4.1 b the relation is

$$\frac{V}{T_0 + t} = \frac{V_1}{T_0 + t_1}$$

$$V = V_1\!\left(1 + \frac{t - t_1}{T_0 + t_1}\right). \qquad (4.7)$$

With the thermal expansion coefficient

$$\beta_1 = \frac{1}{T_0 + t_1} = \frac{1}{T_1} \qquad (4.8)$$

Eq. (4.7) becomes

$$V = V_1[1 + \beta_1(t - t_1)]\;. \qquad (4.9)$$

The increase in volume is $\Delta V = V_1\,\beta_1(t - t_1)$.

According to Figure 4.1, from measurement results that can be represented as a straight line in the V, t diagram, the constant T_0 and thus the absolute or thermodynamic temperature T can be found by extrapolation - i.e. by extending the line to the t axis (cf. Sections 1.3 and 3.5.3).

4.1.3 Physical Norm State

Eq. (4.5) is multiplied by the pressure p which is constant with the temperature change:

$$pV = p\,\overline{V_0}\,(1 + \beta_0\, t) \qquad (4.10)$$

According to the *Boyle-Mariotte* law Eq. (4.2), at $0\,°C$ follows

$$p\overline{V_0} = \bar{p}_0 V = p_0 V_0 \; . \tag{4.11}$$

$\bar{p}_0$ is the pressure that occurs at the volume V at $0\,°C$. p_0 is an agreed normalised pressure, called physical norm pressure or norm pressure. Correspondingly, the temperature T_0 is called physical norm temperature or norm temperature, which is the zero point of the *Celsius* scale according to Eq. (1.2).

$$\boxed{\begin{array}{l} p_0 = 1\;\text{atm} = 1.01325\;\text{bar} = 760\;\text{torr} = 10\,332\;\text{kp/m}^2 \\ T_0 = 273.15\;\text{K} \quad \text{resp.} \quad t_0 = 0\,°C \end{array}} \tag{4.12}$$

p_0 and $T-$ = represent the physical norm state. V_0 is the volume associated with the physical norm pressure p_0 at $0\,°C$. It is called the norm volume or physical norm volume.

The norm volume V_0 is a volume quantity, but, as the state (p_0, T_0) is determined as the physical norm state, it is possible to determine with V_0 the quantity of a gas. The norm volume V_0 thus indicates the volume of the gas quantity in the norm state and is at the same time a measure of the gas quantity itself. The norm volume V_0 is therefore equivalent to the mass m and to the molar quantity n.

Example 4.1 What mass flow corresponds to a norm volume flow of 3000 m^3/h of natural gas whose density in the physical norm state is $\varrho_0 = 0.716$ kg/m^3?

$$\dot{m} = \dot{V}_0\,\varrho_0 = 3000\;\text{m}^3/\text{h} \cdot 0.716\;\text{kg/m}^3 = 2148\;\text{kg/h}$$

4.1.4 Gas Thermometer

From Eqs. (4.10) and (4.11) follows

$$p = \bar{p}_0(1 + \beta_0\, t) \; . \tag{4.13}$$

Using Eqs. (1.1) and (4.4), Eq. (4.13) becomes

$$p = \bar{p}_0 \frac{T}{T_0} \; . \tag{4.14}$$

Eqs. (4.13) and (4.14) are valid for constant volume. They show that by measuring pressure at constant volume with the help of a gas thermometer (Fig. 4.2) temperatures can be determined. It can be shown that the measurement results obtained with the gas thermometer are independent of the nature of the ideal gas (see sections 4.1.5 and 4.1.6). The temperature scale of a gas thermometer working with an ideal gas agrees with the thermodynamic temperature scale (section 3.5.3) [54].

According to Eq. (4.14), the zero point of the *Celsius* scale is the fixed point of the *Kelvin* scale for the absolute temperature with the numerical values for $\bar{p}_0$ and T_0. Since 1954, instead of the zero point of the *Celsius* scale, the triple point of water has been used as the fixed point. At a temperature of $0.01\,°C = 273.16$ K and a pressure of 6.112 mbar = 62.33 kp/m^2 = 4.58 Torr, ice, water and water vapor coexist in a state of equilibrium, suggesting the name triple point. With $T_{Tr} = 273.16$ K as the temperature of the triple point on the *Kelvin* scale, Eq. (4.14) becomes

$$p = p^* \frac{T}{T_{Tr}} \ . \tag{4.15}$$

p^* is the pressure when the gas occupies a volume V at the temperature $t_{Tr} = 0.01\,°\mathrm{C}$. The volume V is given by Eqs. (4.14) and (4.15). From Eq. (4.14) it follows:

$$p^* = \bar{p}_0 \frac{T_{Tr}}{T_0} \tag{4.16}$$

Eq. (4.15) in the form

$$T = \frac{p}{p^*} T_{Tr} \tag{4.17}$$

traces the temperature measurement back to a pressure measurement.

The gas thermometer is unsuitable for technical temperature measurements. Thus, the direct realisation of the thermodynamic temperature scale is also difficult. It is therefore replaced by a practical temperature scale with a series of well-reproducible fixed points. With the help of these fixed points, fixed standard thermometers are calibrated. The "International Practical Temperature Scale of 1968" (IPTS-68) and the "International Practical Temperature Scale of 1990" (ITS-90) [60], [64] were chosen so that a temperature value determined in these scales approximates very accurately the thermodynamic temperature corresponding to it. The accuracy of the approximation depends on the state of the art of measurement.

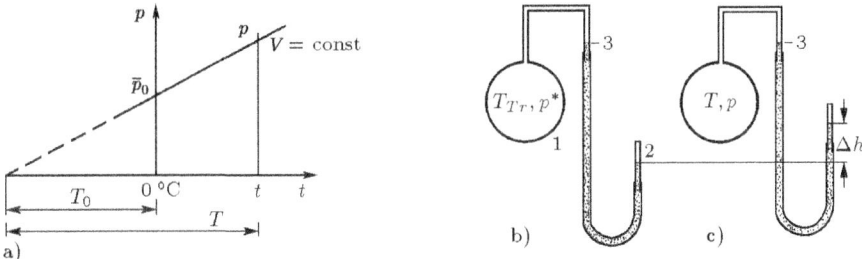

Figure 4.2 a) Change of pressure of an ideal gas with temperature at constant volume. Gas thermometer: b) at the triple point of water   c) at higher temperature
1 spherical vessel, 2 rubber hose with glass tube, 3 marking of the upper limit of the liquid column in the left leg of the U-shaped tube. In figure c) the right leg of the U-shaped tube is modified in such a way that the mark 3 is reached again. Δh with $p - p^* = \varrho g \Delta h$ corresponds to a measured temperature difference according to Eq. (4.17).

4.1.5 Specific Gas Constant

According to Eqs. (4.10) and (4.11) is

$$pV = p_0 V_0 (1 + \beta_0 t) \ . \tag{4.18}$$

Using Eqs. (4.4) and (1.1) one obtains

$$pV = \frac{p_0 V_0}{T_0} T \ . \tag{4.19}$$

In Sections 1.3 and 2.2, reference has already been made to the density ϱ:

$$\varrho = \frac{m}{V} \tag{4.20}$$

The reciprocal of the density ϱ is the specific volume v:

$$v = \frac{1}{\varrho} = \frac{V}{m} \qquad (4.21)$$

From Eqs. (4.20) and (4.21) it follows:

$$\varrho_0 = \frac{m}{V_0} \qquad v_0 = \frac{V_0}{m} \qquad (4.22)$$

ϱ_0 and v_0 are density and specific volume at $0\,°C$ and norm pressure 1013.25 mbar. Eq. (4.19) gives with Eq. (4.22)

$$pV = m\frac{p_0 v_0}{T_0} T \ . \qquad (4.23)$$

The expression $p_0 v_0 / T_0$ is a characteristic quantity of a particular gas. The expression

$$R = \frac{p_0 v_0}{T_0} \qquad (4.24)$$

is called the specific gas constant of an ideal gas. Eq. (4.24) is used to transform Eq. (4.23) to

$$\boxed{pV = mRT} \ . \qquad (4.25)$$

Eq. (4.25) is the gas equation or thermal equation of state of ideal gases. With Eq. (4.21), the division by the mass results in

$$pv = RT \ . \qquad (4.26)$$

By transformation we get

$$\varrho = \frac{p}{RT} \ . \qquad (4.27)$$

Eq. (4.27) is used to calculate the density of an ideal gas.

If two different ideal gases have the same temperature T and the same pressure p, Eq. (4.27) gives the relation

$$\varrho_1 : \varrho_2 = R_2 : R_1 \ . \qquad (4.28)$$

The densities behave inversely to the specific gas constants.

At constant mass, it follows from Eq. (4.25)

$$d(pV) = mR\,dT = pV\frac{dT}{T} \qquad (4.29)$$

or

$$\frac{d(pV)}{pV} = \frac{dT}{T} \ . \qquad (4.30)$$

With Eq. (2.44) this becomes

$$\frac{dp}{p} + \frac{dV}{V} = \frac{dT}{T} \ . \qquad (4.31)$$

In place of $\frac{dV}{V}$, in Eq. (4.31) $\frac{dv}{v}$ or $-\frac{d\varrho}{\varrho}$ can also be inserted.

4.1.6 Universal Gas Constant

A mole is a quantity of substance consisting of a certain number of molecules or atoms. The number of molecules or atoms is chosen so that 1 mole of the carbon $^{12}_{6}C$ has the mass of exactly 12 g. The carbon isotope $^{12}_{6}C$ is said to have the molar mass $M = m/n = 12$ g/mol $= 12$ kg/kmol. According to the law of Avogadro[4], the molar quantity (molar substance) n of each ideal gas at the same temperature and pressure occupies the same volume. The volume per molar quantity n is called the molar volume V_m. Analogous to Eq. (4.21), one can define

$$V_m = \frac{V}{n} \quad \text{and see it in parallel to} \quad v = \frac{V}{m} = \frac{V}{Mn} = \frac{V_m}{M}. \tag{4.32}$$

At $0\,°C$ and 1013.25 mbar, the molar volume for all ideal gases is

$$(V_m)_0 = 22.41410 \text{ m}^3/\text{kmol [25]}. \tag{4.33}$$

The number of molecules or atoms in a kilomole is given by the Avogadro number N_A:

$$N_A = 6.0221367 \cdot 10^{26} \text{ molecules/kmol or atoms/kmol [20]} \tag{4.34}$$

To derive a general gas equation, Eq. (4.26) is to be multiplied by the molar mass M:

$$pvM = MRT \tag{4.35}$$

At the same pressure and temperature, the molar volume $V_m = vM$ is the same for all ideal gases. Since in this case the left-hand side of Eq. (4.35) is the same for all ideal gases, the right-hand side must also be the same for all ideal gases. Thus MR must also give a value that is the same for all ideal gases:

$$R_m = MR = 8{,}314510 \text{ kJ/(kmol K) [20]} \tag{4.36}$$

R_m is the molar or universal gas constant. Using Eqs. (4.32) and (4.36), Eq. (4.35) becomes

$$pV_m = R_m T \quad \text{resp.} \quad pV = nR_m T. \tag{4.37}$$

Eq. (4.37) is the thermal equation of state that holds for all ideal gases.

4.2 Caloric State Variables of Ideal Gases

According to section 3.5, internal energy U, enthalpy H and entropy S are summarised under the term caloric state variables.

---

[4]Count *Amedeo Avogadro di Quaregna e di Cerreto*, 1776 to 1856, first a jurist, dedicated himself to the natural sciences and in 1820 was appointed to the chair of physics in his hometown of Turin.

4.2.1 Internal Energy

In section 3.2.3 we found that the internal energy of the ideal gases is only a function of temperature (Fig. 4.3):

$$U = f(T) \tag{4.38}$$

The function is obtained from Eq. (2.100):

$$dU = m\, c_v\, dT \tag{4.39}$$

With c_v as the mean value according to Eq. (2.98), Eq. (4.39) leads to

$$U_2 - U_1 = m\, c_v (T_2 - T_1). \tag{4.40}$$

This is true for all changes of state of ideal gases.

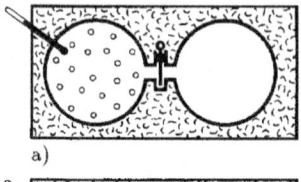
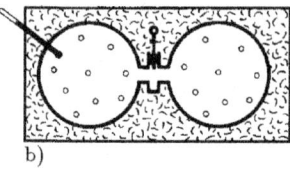

Figure 4.3 Pressure equalisation of an ideal gas between two vessels according to Fig. 3.6
a) before pressure equalisation
b) after pressure equalisation

4.2.2 Enthalpy

According to section 3.2.4, the enthalpy of ideal gases depends only on temperature (Fig. 4.4):

$$H = f(T) \tag{4.41}$$

Eqs. (4.38) and (4.41) are called the caloric equations of state of ideal gases. The function given by Eq. (4.41) is obtained from Eq. (2.102).

$$dH = m\, c_p\, dT \tag{4.42}$$

With c_p as the mean value according to Eq. (2.98), Eq. (4.42) results in

$$H_2 - H_1 = m\, c_p (T_2 - T_1). \tag{4.43}$$

Eq. (4.43) is also valid for all changes of state of ideal gases. With the abbreviation κ, which is called the isentropic exponent (cf. Section 4.3.4)

$$\kappa = \frac{c_p}{c_v}, \tag{4.44}$$

follows from Eqs. (4.39) and (4.42) respectively (4.40) and (4.43) a relation between enthalpy and internal energy:

$$dH = \kappa\, dU \qquad H_2 - H_1 = \kappa\, (U_2 - U_1) \tag{4.45}$$

If one inserts Eq. (4.42) into Eq. (2.43), one gets

$$m\, c_p\, dT = dU + d(pV). \tag{4.46}$$

Using Eqs. (4.39) and (4.29), Eq. (4.46) leads to

$$m\, c_p\, dT = m\, c_v\, dT + m\, R\, dT. \tag{4.47}$$

This gives a relationship between the specific heat capacities and the specific gas constant:
$$c_p - c_v = R \tag{4.48}$$
From this it follows with Eq. (4.44)
$$c_p = \frac{\kappa}{\kappa - 1} R \tag{4.49}$$
$$c_v = \frac{1}{\kappa - 1} R \ . \tag{4.50}$$

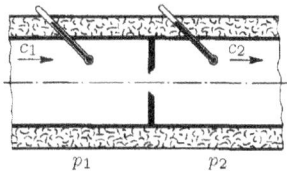

Figure 4.4 Throttling of an ideal gas according to Fig. 3.7

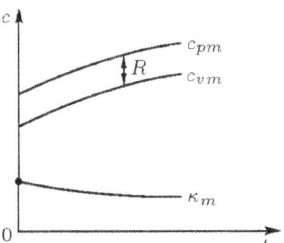

Figure 4.5 Temperature dependence of the specific heat capacities of ideal gases $c_{pm}|_{0\,°C}^{t}$ and $c_{vm}|_{0\,°C}^{t}$, respectively of κ_m

For the same temperature increase, isobaric heating requires a greater heat input than isochoric heating. This is due to the fact that in the isobaric change of state a volume change work has to be done, which is not the case in the isochoric change of state. Thus $c_p > c_v$ holds, and κ is always greater than 1.

The caloric equations of state of the ideal gases Eqs. (4.38) and (4.41) only show a dependence on the temperature. Therefore, the specific heat capacities of the ideal gases and κ according to Eq. (4.44) are also only functions of the temperature and independent of the pressure. The difference of the specific heat capacities c_p and c_v according to Eq. (4.48), on the other hand, is independent of temperature (Fig. 4.5).

4.2.3 Entropy

Equations can be developed for the entropy change of ideal gases according to Eq. (3.24) and to Eq. (3.28), respectively. According to Eqs. (3.76) and (3.79) is

$$dS = \frac{dU + p\,dV}{T} \qquad dS = \frac{dH - V\,dp}{T} \ . \tag{4.51}$$

For the internal energy and enthalpy, Eqs. (4.39) and (4.42) can be used. With

$$\frac{p}{T} = \frac{mR}{V} \tag{4.52}$$

and

$$\frac{V}{T} = \frac{mR}{p} \tag{4.53}$$

according to Eq. (4.25) one gets

$$dS = m\,c_v\frac{dT}{T} + m\,R\frac{dV}{V} \tag{4.54}$$

and

$$dS = m\,c_p\frac{dT}{T} - m\,R\frac{dp}{p}\,. \tag{4.55}$$

Eq. (4.54) results with Eqs. (4.31) and (4.48) in

$$dS = m\,c_p\frac{dV}{V} + m\,c_v\frac{dp}{p}\,. \tag{4.56}$$

The integration yields, if one simplifies the specific heat capacities as temperature-independent or as average specific heat capacities,

$$S_2 - S_1 = m\,c_v\ln\frac{T_2}{T_1} + m\,R\ln\frac{V_2}{V_1} \tag{4.57}$$

$$S_2 - S_1 = m\,c_p\ln\frac{T_2}{T_1} - m\,R\ln\frac{p_2}{p_1} \tag{4.58}$$

$$S_2 - S_1 = m\,c_p\ln\frac{V_2}{V_1} + m\,c_v\ln\frac{p_2}{p_1}\,. \tag{4.59}$$

Eqs. (4.57) to (4.59) are applicable to all changes of state of ideal gases.

4.3 Changes of State

In the following, the thermodynamic relations of the special changes of state of ideal gases will be examined in more detail:

Isochoric change of state	$dV = 0$	$V = \text{const}$	(4.60)
Isobaric change of state	$dp = 0$	$p = \text{const}$	(4.61)
Isothermal change of state	$dT = 0$	$T = \text{const}$	(4.62)
Isentropic change of state	$dS = 0$	$S = \text{const}$	(4.63)
	$dQ_{rev} = 0$	$(Q_{12})_{rev} = 0$	(4.64)

In the treatment of changes of state, the specific heat capacity is understood as a mean value according to Eq. (2.98).

4.3.1 Isochoric Change of State

If one applies the thermal equation of state (4.25) for the initial state and the final state of a a process at constant volume (Fig. 4.6), one can write

$$p_1 V = m\,R\,T_1 \qquad\qquad p_2 V = m\,R\,T_2\,. \tag{4.65}$$

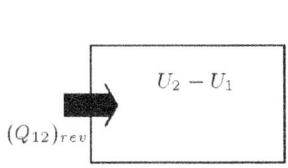

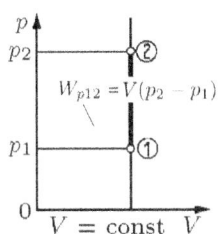

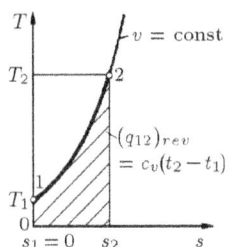

Figure 4.6 Isochoric heating of a gas volume

Figure 4.7 Isochoric heating in the p, V diagram

Figure 4.8 Isochoric heating in the T, s diagram. Explanation of the curve in example 4.5

From Eqs. (4.65) follows

$$\frac{p_1}{T_1} = \frac{p_2}{T_2} \qquad \frac{p}{T} = \text{const} . \qquad (4.66)$$

In the p, V diagram the isochores are represented as vertical straight lines (Fig. 4.7). The volume change work according to Eq. (2.21) and the pressure change work according to Eq. (2.30) are

$$W_{V12} = -\int_1^2 p\,dV = 0 . \qquad (4.67)$$

$$W_{p12} = \int_1^2 V\,dp = V\,(p_2 - p_1) . \qquad (4.68)$$

The heat transferred in a reversible substitute process with isochoric change of state is given by Eqs. (2.88) and (3.80) with c_v as the specific heat capacity at constant volume

$$dQ_{rev} = m\,c_v\,dT \qquad (Q_{12})_{rev} = m\,c_v(T_2 - T_1) . \qquad (4.69)$$

The first law for a closed system according to Eq. (3.76) and Eq. (4.60) is given by

$$dQ_{rev} = dU \qquad (Q_{12})_{rev} = U_2 - U_1 . \qquad (4.70)$$

The entropy change is given by Eqs. (4.57) and (4.59) with Eq. (4.60) (see Fig. 4.8):

$$S_2 - S_1 = m\,c_v \ln \frac{T_2}{T_1} = m\,c_v \ln \frac{p_2}{p_1} \qquad (4.71)$$

Example 4.2 A gas cylinder of 80 dm³ capacity contains CO_2 under a pressure of 4.6 bar at a temperature of 18 °C.

a) What is the gage pressure in the cylinder when the ambient air pressure is 990 mbar?

b) What is the gage pressure when the temperature of the gas rises to 212 °C as a result of a fire in the building that leads to an ambient temperature in the surroundings of the gas cylinder up to 850 °C?

c) What is the total entropy change due to the increase in temperature of the gas?
d) After the fire has been extinguished and the ambient temperature has cooled down to 18 °C, the gas in the cylinder also cools down to 18 °C, releasing heat to the surroundings. What is the total entropy change associated with this process?

a) $p_{1\ddot{u}} = p_1 - p_L = 4.6$ bar $- 0.99$ bar $= 3.61$ bar

b) $p_2 = \dfrac{T_2}{T_1} p_1 = \dfrac{485 \text{ K}}{291 \text{ K}} \cdot 4.6$ bar $= 7.67$ bar $p_{2\ddot{u}} = p_2 - p_L = (7.67 - 0.99)$ bar $= 6.68$ bar

c) $m = \dfrac{pV}{RT} = \dfrac{4.6 \text{ bar} \cdot 0.08 \text{ m}^3 \text{ kg K}}{0.18892 \text{ kJ} \cdot 291 \text{ K}} \cdot \dfrac{100 \text{ kJ}}{\text{m}^3 \text{ bar}} = 0.669$ kg

According to Eq. (2.96), the mean specific heat capacity c_{pm} and according to Eq. (4.48) c_{vm} is calculated:

$c_{pm}|_{0\,°C}^{18\,°C} = 0.8262$ kJ/(kg K) $\qquad\qquad c_{pm}|_{0\,°C}^{212\,°C} = 0.9171$ kJ/(kg K)

$c_{pm}|_{18\,°C}^{212\,°C} = 0.9255$ kJ/(kg K) $\qquad\qquad c_{vm}|_{18\,°C}^{212\,°C} = 0.7366$ kJ/(kg K)

$S_{G2} - S_{G1} = m\, c_{vm} \ln \dfrac{T_2}{T_1} = 0.669 \text{ kg} \cdot 0.7366 \text{ kJ/(kg K)} \cdot \ln \dfrac{485 \text{ K}}{291 \text{ K}} = 0.252$ kJ/K

$(Q_{12})_{rev} = m\, c_{vm}(t_2 - t_1) = 0.669 \text{ kg} \cdot 0.7366 \text{ kJ/(kg K)} \cdot 194 \text{ K} = 95.60$ kJ

$S_{U2} - S_{U1} = -\dfrac{(Q_{12})_{rev}}{T_U} = -\dfrac{95.60 \text{ kJ}}{1123 \text{ K}} = -0.085$ kJ/K

The total entropy change is $(S_{G2} - S_{G1}) + (S_{U2} - S_{U1}) = 0.167$ kJ/K

d) $S_{G2} - S_{G1} = -0.252$ kJ/K $\qquad\qquad S_{U2} - S_{U1} = \dfrac{95.60 \text{ kJ}}{291 \text{ K}} = 0.329$ kJ/K

$(S_{G2} - S_{G1}) + (S_{U2} - S_{U1}) = 0.077$ kJ/K

4.3.2 Isobaric Change of State

For the initial and final states of an isobaric change of state (Fig. 4.9), one can write:

$$pV_1 = m R T_1 \qquad\qquad pV_2 = m R T_2 \qquad (4.72)$$

This leads to
$$\dfrac{V_1}{T_1} = \dfrac{V_2}{T_2} \qquad\qquad \dfrac{V}{T} = \text{const}. \qquad (4.73)$$

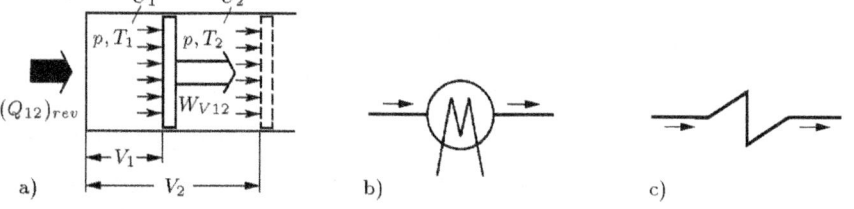

Figure 4.9 Examples of isobaric changes of state
a) isobaric heating of a gas in a piston engine
b) isobaric cooling of a gas flow in a cooler
c) isobaric heating of a gas flow in a heater

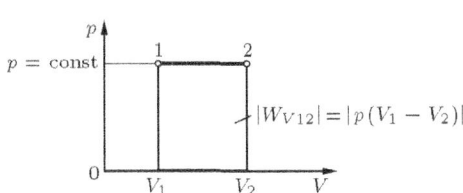
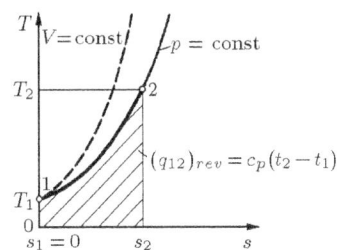

Figure 4.10 Isobaric heating in a p, V diagram

Figure 4.11 Isobaric heating in a T, s diagram

In a p, V diagram the isobars are represented as horizontal straight lines (Fig. 4.10). The volume change work is

$$W_{V12} = -\int_1^2 p\,dV = -p(V_2 - V_1) = p(V_1 - V_2) \,. \tag{4.74}$$

It is equal to the difference of two shift operations. Since the pressure is constant, the pressure change work is given by Eq. (2.30)

$$W_{p12} = \int_1^2 V\,dp = 0 \,. \tag{4.75}$$

The heat transferred in a reversible substitute process with isobaric change of state is given by Eqs. (2.88) and (3.80). If c_p is the specific heat capacity at constant pressure, one finds

$$dQ_{rev} = m\,c_p\,dT \tag{4.76}$$

$$(Q_{12})_{rev} = m\,c_p(T_2 - T_1) \,. \tag{4.77}$$

Applying the first law for an open system according to Eq. (3.76) to an isobaric change of state yields with Eq. (4.61)

$$dQ_{rev} = dH \qquad (Q_{12})_{rev} = H_2 - H_1 \,. \tag{4.78}$$

The entropy change is given by Eqs. (4.58) and (4.59) with Eq. (4.61) (Fig. 4.11)

$$S_2 - S_1 = m\,c_p \ln\frac{T_2}{T_1} = m\,c_p \ln\frac{V_2}{V_1} \,. \tag{4.79}$$

The changes of state in heat exchangers are treated as isobaric in thermodynamics. A fluid that is cooled or heated in a heat exchanger will only flow if there is a pressure gradient. However, this is generally so small compared to the absolute pressure of the fluid that approximately an isobaric change of state can be assumed.

Example 4.3 On the type plate of a fan heater, the electrical power is stated as 1.5 kW. Of this, 0.05 kW is the connected power of the motor for the fan; with 1.45 kW, the largest part is converted into heat flow in the heating wires. In the fan heater, the electrical power supplied to the motor is changed in friction power as well as the electrical power in the heating wires.

4.3 Changes of State

a) What mass of air is heated per hour if the air enters the fan heater with a temperature of 16 °C and leaves it with a temperature of 31 °C?

b) What increase in air volume occurs during heating?

c) What is the entropy change per second of the heated air?

a) $(\dot{Q}_{12})_{rev} = \dot{Q}_{12} + |P_{RI}| = \dot{H}_2 - \dot{H}_1 = \dot{m}\, c_{pm}(t_2 - t_1)$ $c_{pm}|_{16\,°C}^{31\,°C} = 1.0042$ kJ/(kg K)

$$\dot{m} = \frac{(\dot{Q}_{12})_{rev}}{c_{pm}(t_2 - t_1)} = \frac{1{,}5 \text{ kW kg K}}{1{,}0042 \text{ kJ} \cdot 15 \text{ K}} \cdot \frac{\text{kJ}}{\text{kW s}} = 0.0996\,\frac{\text{kg}}{\text{s}} = 358.5\,\frac{\text{kg}}{\text{h}}$$

b) $\dfrac{V_2}{V_1} = \dfrac{T_2}{T_1} = \dfrac{304 \text{ K}}{289 \text{ K}} = 1.052$ $\Delta V = 0.052 \approx 5\,\%$

c) $\dot{S}_2 - \dot{S}_1 = \dot{m}\, c_{pm} \ln \dfrac{T_2}{T_1} = 0.00506$ kW/K

Example 4.4 It is to be shown that the change of state in Example 2.8 can be treated with relations for ideal gases.
The change of state is an isobaric one. According to Eqs. (4.45) and (4.78), it must hold:

$(Q_{12})_{rev} = H_2 - H_1 = 40\,690$ kJ $U_2 - U_1 = 30\,826$ kJ $\kappa = \dfrac{H_2 - H_1}{U_2 - U_1} = \dfrac{40\,690 \text{ kJ}}{30\,826 \text{ kJ}} = 1.32$

For the temperature change during expansion, $T_2 = \dfrac{V_2}{V_1} T_1 = 1{,}024\, T_1$ holds.

From this follows $T_2 - T_1 = 0.024\, T_1$.   The amount of gas is given by $m = \dfrac{p_1 V_1}{R T_1}$.

Thus the heat input according to Eq. (4.77) is $(Q_{12})_{rev} = m\, c_p (T_2 - T_1) = \dfrac{c_p}{R} 0.024\, p_1 V_1$.

With Eq. (4.49), it results in

$(Q_{12})_{rev} = \dfrac{\kappa}{\kappa - 1} 0.024\, p_1 V_1 = \dfrac{1.32}{0.32} 0.024 \cdot 1.02211 \text{ bar} \cdot 4021.2 \text{ m}^3 = 40\,690$ kJ.

Example 4.5 Construction of constant-pressure lines and constant-volume lines in a T, s diagram.
If instead of the pair of values T_2, s_2 one introduces the variable state point T, s into Eqs. (4.71) and (4.79) and sets $s_1 = 0$, then this leads to the functions

with isobaric change of state : $T = T_1 e^{\frac{s}{c_p}}$ (1)

with isochoric change of state : $T = T_1 e^{\frac{s}{c_v}}$ (2)

The first function does not include the magnitude of the constant pressure p for the isobaric change of state. It follows that the function values in the T, s diagram are congruent for all isobars. The function curves for different pressures are obtained by a shift parallel to the abscissa axis. The shift amount is obtained by setting $T = $ const in Eq. (4.58):

$$s_2 - s_1 = -R \ln \frac{p_2}{p_1}$$

Since for $p_2 > p_1$ the difference $s_2 - s_1$ is negative, the isobars of higher pressure lie to the left of the isobars of lower pressure. The shift amount is independent of the selected temperature and is therefore the same at all points of the isobars (condition for the congruence of the function curves).

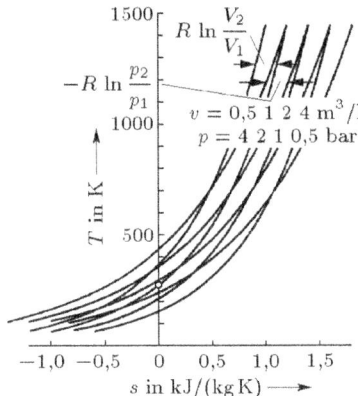

For the isochoric change of state the same construction procedure results. The shift amount is here according to Eq. (4.57)

$$s_2 - s_1 = R \ln \frac{V_2}{V_1} = R \ln \frac{v_2}{v_1}.$$

The isochores with larger volume are to the right of the isochores with smaller volume. Fig. 4.12 shows the T, s diagram for air with isobars and isochores drawn in. For simplicity, the specific heat capacities are assumed to be constant

Figure 4.12 Construction of the isobars and isochores in the T, s diagram for air

4.3.3 Isothermal Change of State

For the volume change work of an isothermal change of state (Fig. 4.13), it follows from Eqs. (2.21) and (4.25)

$$W_{V12} = -\int_1^2 p\,dV = -mRT\int_1^2 \frac{dV}{V} = mRT\ln\frac{V_1}{V_2}. \tag{4.80}$$

With Eq. (4.2)

$$p_1 V_1 = p_2 V_2 \qquad pV = \text{const} \tag{4.81}$$

one obtains

$$W_{V12} = mRT\ln\frac{p_2}{p_1}. \tag{4.82}$$

Eqs. (4.80) and (4.82) give with Eq. (4.25) (Fig. 4.14)

$$W_{V12} = p_1 V_1 \ln\frac{V_1}{V_2} = p_1 V_1 \ln\frac{p_2}{p_1}. \tag{4.83}$$

Eqs. (4.39), (4.42) and (4.62) lead to

$$dU = dH = 0, \tag{4.84}$$

and Eqs. (3.76) result in

$$dQ_{rev} = p\,dV = -V\,dp. \tag{4.85}$$

This means according to Eqs. (2.21) and (2.30)

$$(Q_{12})_{rev} = -W_{V12} = -W_{p12}. \tag{4.86}$$

Thus, all equations for the volume change work W_{V12} also apply to the pressure change work W_{p12}; if one changes the sign in these equations, they also apply to the reversible heat $(Q_{12})_{rev}$.

With Eqs. (4.80) and (4.82) one gets

$$(Q_{12})_{rev} = mRT\ln\frac{V_2}{V_1} = mRT\ln\frac{p_1}{p_2} \tag{4.87}$$

and with Eqs. (4.25) and (4.87)

$$(Q_{12})_{rev} = p_1 V_1 \ln\frac{V_2}{V_1} = p_1 V_1 \ln\frac{p_1}{p_2}. \tag{4.88}$$

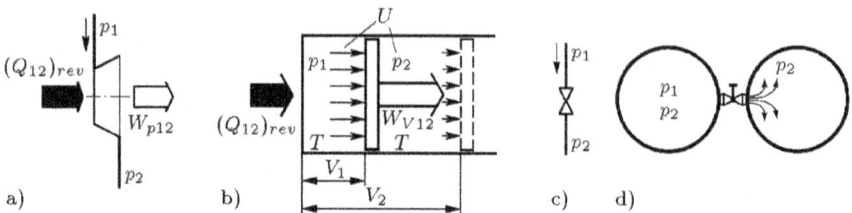

Figure 4.13 Examples of isothermal changes of state of ideal gases
a) isothermal expansion of a gas flow in a turbine
b) isothermal expansion of a gas in a piston engine
c) throttling in an adiabatic system
d) pressure equalisation in an adiabatic system

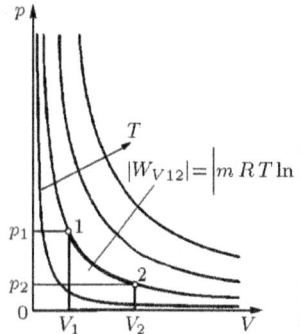

Figure 4.14 Isothermal expansion in a p, V diagram

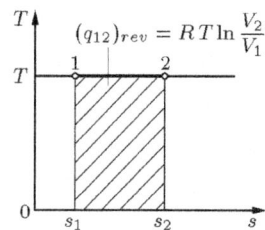

Figure 4.15 Isothermal expansion in a T, s diagram

From this follows for the entropy change (Fig. 4.15)

$$S_2 - S_1 = \frac{(Q_{12})_{rev}}{T} = m R \ln \frac{V_2}{V_1} = m R \ln \frac{p_1}{p_2} \qquad (4.89)$$

and with the thermal equation of state (4.25)

$$S_2 - S_1 = \frac{p_1 V_1}{T} \ln \frac{V_2}{V_1} = \frac{p_1 V_1}{T} \ln \frac{p_1}{p_2}. \qquad (4.90)$$

In all equations containing $p_1 V_1$, according to Eq. (4.81) $p_1 V_1$ can be replaced by $p_2 V_2$.

Example 4.6 The friction resistance caused by a valve in an adiabatic natural gas pipeline is described by the dimensionless factor ζ. If p_1, ϱ_1 and c_1 are pressure, density and velocity upstream of the valve, then the pressure drop according to Eq. (2.111) is given by

$$p_1 - p_2 = \zeta \frac{1}{2} \varrho_1 c_1^2 .$$

With a pipe diameter $d = 25.4$ mm, a temperature $t = 12\,°C$ and a pressure $p_1 = 1.032$ bar, the gas flow has a velocity $c_1 = 10$ m/s. In the physical norm state ($T_0 = 273.15$ K, $p_0 = 1.01325$ bar) the gas has the density $\varrho_0 = 0.717$ kg/m^3. For the fully opened valve of a certain design, $\zeta = 4.2$ has been determined. What friction power $|P_{RI}|$ is effective in the gas flow? What entropy change per second occurs as a result of this process? The change of state can be assumed to be isothermal, the gas is an ideal gas.

84   4 Ideal Gases

According to Eqs. (3.78) and (4.88), since no heat flow $\dot{Q}_{12}$ is added or removed, it follows:

$$|\dot{P}_{RI}| = (\dot{Q}_{12})_{rev} = p_1 \dot{V}_1 \ln \frac{p_1}{p_2} \qquad \dot{V}_1 = \frac{\pi d^2}{4} c_1 = 5.067 \cdot 10^{-3} \text{ m}^3/\text{s}$$

$$\varrho_1 = \varrho_0 \frac{p_1}{p_0} \frac{T_0}{T_1} = 0.717 \text{ kg/m}^3 \cdot \frac{1.032 \text{ bar}}{1.01325 \text{ bar}} \cdot \frac{273.15 \text{ K}}{285.15 \text{ K}} = 0.6995 \text{ kg/m}^3$$

$$p_1 - p_2 = 4.2 \cdot \frac{1}{2} \cdot 0.6995 \frac{\text{kg}}{\text{m}^3} \cdot 10^2 \frac{\text{m}^2}{\text{s}^2} \cdot \frac{1 \text{ N s}^2}{\text{kg m}} \cdot \frac{1 \text{ bar m}^2}{10^5 \text{ N}} = 0.00147 \text{ bar} = 147 \text{ Pa}$$

$$p_2 = 1.03200 \text{ bar} - 0.00147 \text{ bar} = 1.03053 \text{ bar}$$

$$|\dot{P}_{RI}| = 1.032 \text{ bar} \cdot 5.067 \cdot 10^{-3} \frac{\text{m}^3}{\text{s}} \cdot \frac{10^5 \text{ J}}{\text{m}^3 \text{ bar}} \cdot \frac{1 \text{ W s}}{\text{J}} \cdot \ln \frac{1.03200 \text{ bar}}{1.03053 \text{ bar}} = 0.7454 \text{ W}$$

According to Eq. (4.90) is

$$\dot{S}_2 - \dot{S}_1 = \frac{p_1 \dot{V}_1}{T} \ln \frac{p_1}{p_2} = \frac{|\dot{P}_{RI}|}{T} = \frac{0.7454 \text{ W}}{285.15 \text{ K}} = 2.614 \cdot 10^{-3} \text{ W/K}$$

Example 4.7 A balloon for weather observation is filled with hydrogen gas on the ground, but only partially inflated. During the ascent, it can expand frictionlessly to a spherical volume of 6.8 m^3. The pressure and temperature of the gas are constantly equal to the outside air pressure and outside air temperature. The outside air pressure drops from 1 bar to 0.3 bar during the ascent, the outside air temperature 8 °C remains constant during the ascent. How great is the volume change work, what heat is transferred? What is the total entropy change?

$$W_{V12} = m R T \ln \frac{p_2}{p_1} = p_2 V_2 \ln \frac{p_2}{p_1} = 0.3 \text{ bar} \cdot 6.8 \text{ m}^3 \cdot \frac{100 \text{ kJ}}{\text{m}^3 \text{ bar}} \cdot \ln \frac{0.3 \text{ bar}}{1 \text{ bar}} = -246 \text{ kJ}$$

$$(Q_{12})_{rev} = -W_{V12} = 246 \text{ kJ}$$

The entropy change of the gas is

$$(S_2 - S_1)_G = \frac{(Q_{12})_{rev}}{T} = \frac{246 \text{ kJ}}{281 \text{ K}} = 0.875 \text{ kJ/K}.$$

The heat $(Q_{12})_{rev}$ is given off to the environment at the same temperature. Therefore

$$(S_2 - S_1)_U = -\frac{(Q_{12})_{rev}}{T} = -0.875 \text{ kJ/K} \quad \text{holds.}$$

The total entropy change is thus

$$S_2 - S_1 = (S_G + S_U)_2 - (S_G + S_U)_1 = 0 .$$

The process is reversible.

4.3.4 Isentropic Change of State

In the case of an isentropic change of state (Fig. 4.15), according to Eqs. (4.63) and (4.64) are

$$dS = 0 \quad \text{resp.} \quad S_2 - S_1 = 0 \quad \text{(Fig. 4.18)} \quad \text{and}$$

$$dQ_{rev} = 0 \quad \text{resp.} \quad (Q_{12})_{rev} = 0 .$$

Thus, it follows from Eqs. (3.76) for reversible substitute processes and Eqs. (4.39) and (4.42) for the internal energy and the enthalpy

$$dU = -p\,dV = m\,c_v\,dT \tag{4.91}$$

$$dH = V\,dp = m\,c_p\,dT . \tag{4.92}$$

4.3 Changes of State

The transformation of Eqs. (4.91) and (4.92) to $m\,dT$ and equating them yields

$$\frac{c_p}{c_v}\frac{dV}{V} = -\frac{dp}{p}. \tag{4.93}$$

Eq. (4.93) is integrated. With the isentropic exponent κ according to Eq. (4.44) and κ = const is

$$\kappa \ln \frac{V_2}{V_1} = \ln\left(\frac{V_2}{V_1}\right)^\kappa = -\ln\frac{p_2}{p_1} = \ln\frac{p_1}{p_2}$$

$$\left(\frac{V_2}{V_1}\right)^\kappa = \frac{p_1}{p_2} \qquad\qquad p_1 V_1^\kappa = p_2 V_2^\kappa \tag{4.94}$$

$$p V^\kappa = \text{const}. \tag{4.95}$$

Eq. (4.95) illustrates, why κ is called the isentropic exponent.
The thermal equation of state for the ideal gas (4.25) is applied to the initial state 1 and the final state 2. By eliminating $m R$ one obtains

$$\frac{p_1 V_1}{T_1} = \frac{p_2 V_2}{T_2} \tag{4.96}$$

$$\frac{p_1}{p_2} = \frac{V_2 T_1}{V_1 T_2} \qquad\qquad \frac{V_2}{V_1} = \frac{p_1 T_2}{p_2 T_1}. \tag{4.97}$$

Eqs. (4.96) and (4.97) apply in general, i.e. not only to a special change of state. From Eqs. (4.94) and (4.97) follows

$$\left(\frac{V_2}{V_1}\right)^{\kappa-1} = \frac{T_1}{T_2} \qquad\qquad T_1 V_1^{\kappa-1} = T_2 V_2^{\kappa-1} \tag{4.98}$$

$$T V^{\kappa-1} = \text{const} \tag{4.99}$$

$$\left(\frac{p_1}{p_2}\right)^{\kappa-1} = \left(\frac{T_1}{T_2}\right)^\kappa \qquad\qquad \frac{T_2}{T_1} = \left(\frac{p_2}{p_1}\right)^{\frac{\kappa-1}{\kappa}} \tag{4,100}$$

$$\frac{T^\kappa}{p^{\kappa-1}} = \text{const}. \tag{4.101}$$

If one wants to integrate Eq. (2.21) for the volume change work, a relation between pressure and volume is needed. It is obtained from Eq. (4.94) by replacing the end point with index 2 by any point without index to get a variable between the initial point 1 and the end point 2:

$$p = p_1 V_1^\kappa \frac{1}{V^\kappa} = p_1 V_1^\kappa V^{-\kappa} \tag{4.102}$$

Eq. (2.21) gives with Eq. (4.102) (Fig. 4.17):

$$W_{V12} = -\int_1^2 p\,dV = -p_1 V_1^\kappa \int_1^2 V^{-\kappa} dV = -\frac{p_1 V_1 V_1^{\kappa-1}}{1-\kappa}\left[V_2^{1-\kappa} - V_1^{1-\kappa}\right] = \frac{p_1 V_1}{\kappa - 1}\left[\left(\frac{V_1}{V_2}\right)^{\kappa-1} - 1\right] \tag{4.103}$$

Using the equation for the ideal gas (4.25), one can replace $p_1 V_1$ by $m R T_1$:

$$W_{V12} = \frac{m R T_1}{\kappa - 1}\left[\left(\frac{V_1}{V_2}\right)^{\kappa-1} - 1\right] \tag{4.104}$$

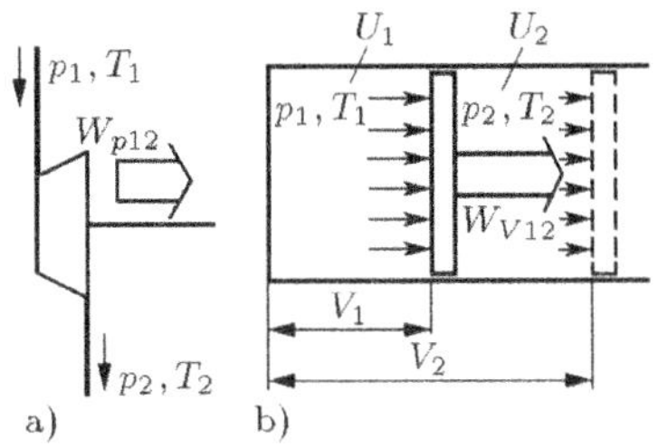

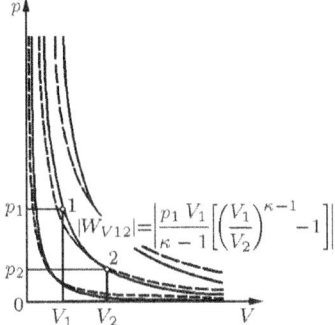
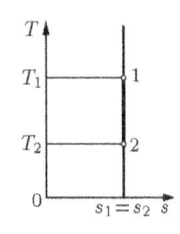

Figure 4.16 Examples of isentropic changes of state
a) isentropic expansion of a gas flow in a turbine
b) isentropic expansion of a volume of gas in a piston engine

Figure 4.17 Isentropic expansion in a p, V diagram

Figure 4.18 Isentropic expansion in a T, s diagram

Eq. (4.94) expresses the volume ratio by the pressure ratio. From Eqs. (4.103) and (4.104) becomes

$$W_{V12} = \frac{p_1 V_1}{\kappa - 1}\left[\left(\frac{p_2}{p_1}\right)^{\frac{\kappa-1}{\kappa}} - 1\right] = \frac{mRT_1}{\kappa - 1}\left[\left(\frac{p_2}{p_1}\right)^{\frac{\kappa-1}{\kappa}} - 1\right]. \qquad (4.105)$$

From Eq. (4.94) also follows

$$\left(\frac{V_1}{V_2}\right)^{\kappa-1} = \frac{p_2 V_2}{p_1 V_1}.$$

If this equation is inserted into Eq. (4.103), the result is

$$W_{V12} = \frac{1}{\kappa - 1}(p_2 V_2 - p_1 V_1). \qquad (4.106)$$

Further equations for the volume change work are obtained by inserting Eq. (4.98) into Eqs. (4.103) and (4.104):

$$W_{V12} = \frac{p_1 V_1}{\kappa - 1}\left(\frac{T_2}{T_1} - 1\right) = \frac{mR}{\kappa - 1}(T_2 - T_1) = m c_v (T_2 - T_1) \qquad (4.107)$$

From Eqs. (4.91) and (2.20) as well as Eqs. (4.92) and (2.31) it follows:

$$dU = dW_V \qquad\qquad U_2 - U_1 = W_{V12} \qquad (4.108)$$

$$dH = dW_p \qquad\qquad H_2 - H_1 = W_{p12} \qquad (4.109)$$

Furthermore, with Eq. (4.45) one obtains

$$dW_p = \kappa\, dW_V \qquad\qquad W_{p12} = \kappa\, W_{V12}. \qquad (4.110)$$

From this follow, among others, with Eqs. (4.104), (4.105), (4.106) and (4.107):

$$W_{p12} = \frac{\kappa}{\kappa - 1} p_1 V_1\left[\left(\frac{V_1}{V_2}\right)^{\kappa-1} - 1\right] = \frac{\kappa}{\kappa - 1} p_1 V_1\left[\left(\frac{p_2}{p_1}\right)^{\frac{\kappa-1}{\kappa}} - 1\right] =$$

$$= \frac{\kappa}{\kappa - 1}(p_2 V_2 - p_1 V_1) = \frac{\kappa}{\kappa - 1} mR(T_2 - T_1) = m c_p (T_2 - T_1) \qquad (4.111)$$

As already shown in section 3.5.1, an adiabatic and frictionless change of state is also an isentropic change of state. The abbreviation κ according to Eq. (4.44), which is used in the formulation of the relationship between the thermal state variables pressure, volume and

4.3 Changes of State

temperature in the case of an isentropic change of state in Eqs. (4.94), (4.95), (4.98) to (4.108) and (4.111) is called isentropic exponent, as mentioned before. The isentropic exponent κ is a function of temperature.

Example 4.8 A horizontally arranged shock absorber consists of a cylinder of $d = 80$ mm diameter and a piston closing off a volume of air of $s = 200$ mm length. The air in the cylinder is at ambient state $t_1 = 18\,°C$ and $p_1 = 949.3$ mbar. What impact work can the shock absorber absorb if the piston moves $\Delta s = 150$ mm in, and what pressure, what gage pressure and what temperature does the air in the cylinder then reach? The change of state can be assumed to be adiabatic and frictionless (Fig. 4.19).

For κ we first insert the value for $0\,°C$: $\kappa = 1.40$

$$V_1 = \frac{\pi}{4} d^2 s = 1.0053 \cdot 10^{-3}\,\mathrm{m}^3 \qquad V_2 = \frac{\pi}{4} d^2 (s - \Delta s) = 0.2513 \cdot 10^{-3}\,\mathrm{m}^3 \qquad \frac{V_1}{V_2} = 4$$

$$T_2 = T_1 \left(\frac{V_1}{V_2}\right)^{\kappa - 1} = 291.15\,\mathrm{K} \cdot 4^{0.40} = 506.92\,\mathrm{K}$$

With the calculated temperature $t_2 = 233.77\,°C \approx 234\,°C$, κ is corrected:

$$\kappa_m = \frac{c_{pm}}{c_{pm} - R}$$

$$c_{pm}\big|_{18\,°C}^{234\,°C} = 1.0141\,\mathrm{kJ/(kg\,K)}$$

$$\kappa_m = \frac{1.0141\,\mathrm{kJ/(kg\,K)}}{1.0141\,\mathrm{kJ/(kg\,K)} - 0.28706\,\mathrm{kJ/(kg\,K)}} = 1.3948$$

$$T_2 = 291.15\,\mathrm{K} \cdot 4^{0.3948} = 503.28\,\mathrm{K} \qquad t_2 = 230.13\,°C$$

$$p_2 = p_1 \left(\frac{V_1}{V_2}\right)^{\kappa_m} = 0.9493\,\mathrm{bar} \cdot 4^{1.3948} = 6.564\,\mathrm{bar}$$

$$p_{2\ddot{u}} = p_2 - p_L = 6.564\,\mathrm{bar} - 0.949\,\mathrm{bar} = 5.615\,\mathrm{bar}$$

The compression of the gas is given by the impact work $\int_1^2 F\,ds$ and the air pressure work $p_L(V_1 - V_2)$:

$$\int_1^2 F\,ds + p_L(V_1 - V_2) = W_{V12}$$

According to Eq. (4.103) one finds

$$W_{V12} = \frac{p_1 V_1}{\kappa - 1}\left[\left(\frac{V_1}{V_2}\right)^{\kappa - 1} - 1\right] = \frac{0.9493\,\mathrm{bar} \cdot 1.0053 \cdot 10^{-3}\,\mathrm{m}^3}{1.3948 - 1} \cdot \frac{10^5\,\mathrm{J}}{\mathrm{m}^3\,\mathrm{bar}} \cdot (4^{0.3948} - 1) = 176.12\,\mathrm{J}$$

$$p_L(V_1 - V_2) = 0.9493\,\mathrm{bar} \cdot (1.0053 - 0.2513) \cdot 10^{-3}\,\mathrm{m}^3 \cdot \frac{10^5\,\mathrm{J}}{\mathrm{m}^3\,\mathrm{bar}} = 71.58\,\mathrm{J}$$

$$\int_1^2 F\,ds = 176.12\,\mathrm{J} - 71.58\,\mathrm{J} = 104.54\,\mathrm{J}\ .$$

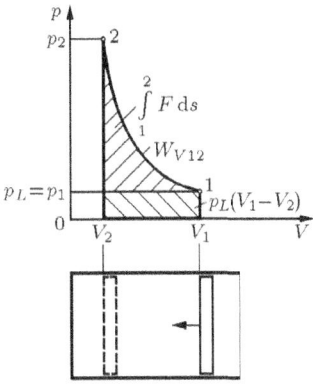

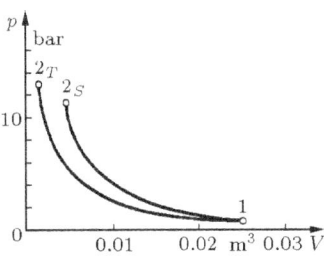

Figure 4.19 To example 4.8

Figure 4.20 To example 4.9

Example 4.9 A vehicle with a mass of $m = 32\,000$ kg hits an air brake integrated in the vehicle with a speed of $c = 0.5$ m/s and comes to a standstill. The brake cylinder has a diameter $d = 250$ mm and a length $s = 500$ mm. Before hitting the air brake, the air filling is in the ambient state and has the air temperature $t_1 = 20\,°C$, the air pressure $p_1 = 1$ bar and the isentropic exponent $\kappa = 1.4 = $ const. What are the braking distance and the maximum cylinder pressure for isothermal respectively isentropic compression (Fig. 4.20)?

If the friction work is neglected, according to Eqs. (2.57) and (2.58) one can write

$$\int_1^2 F\,ds + p_L(V_1 - V_2) = W_{V12}$$

$$p_L(V_1 - V_2) = p_L V_1 \left(1 - \frac{V_2}{V_1}\right) = p_L V_1 (1 - \varepsilon)\ .$$

$\varepsilon = \dfrac{V_2}{V_1}$ is the compression ratio. $\quad V_1 = \dfrac{\pi}{4} d^2 s = 0.02454$ m^3

Isothermal compression: $\quad W_{V12} = p_1 V_1 \ln \dfrac{V_1}{V_2} = p_L V_1 \ln \dfrac{1}{\varepsilon_T}$

Isentropic compression: $\quad W_{V12} = \dfrac{p_1 V_1}{\kappa - 1}\left[\left(\dfrac{V_1}{V_2}\right)^{\kappa-1} - 1\right] = \dfrac{p_L V_1}{\kappa - 1}\left[\left(\dfrac{1}{\varepsilon_S}\right)^{\kappa-1} - 1\right]$

The impact work is equal to the kinetic energy:

$$\int_1^2 F\,ds = \frac{1}{2} m c^2 = \frac{1}{2} \cdot 32\,000\ \text{kg} \cdot 0.5^2\,\frac{\text{m}^2}{\text{s}^2} \cdot \frac{\text{N s}^2}{\text{kg m}} \cdot \frac{1\,\text{kJ}}{1000\,\text{N m}} = 4.0\ \text{kJ}$$

With $\quad p_L V_1 = 1$ bar $\cdot\ 0.02454$ m$^3 \cdot \dfrac{100\,\text{kJ}}{\text{m}^3\,\text{bar}} = 2.454$ kJ $\quad$ and $\quad \dfrac{1}{p_L V_1}\int_1^2 F\,ds = 1.6300$

one obtains the following determining equations for the compression ratio

- for isothermal compression: $\quad 1.63 = \ln \dfrac{1}{\varepsilon_T} + \varepsilon_T - 1 \quad \ln\dfrac{1}{\varepsilon_T} + \varepsilon_T - 2.63 = 0$

$\varepsilon_T = 0.0779 \quad V_2 = V_1 \varepsilon_T = 0.001913$ m$^3 \quad p_2 = p_1 \dfrac{V_1}{V_2} = \dfrac{p_1}{\varepsilon_T} = 12.83$ bar

braking distance: $\quad (1 - \varepsilon_T)\cdot 500$ mm $= 461$ mm

- for isentropic compression: $\quad 1.63 = \dfrac{1}{0.4}\left[\left(\dfrac{1}{\varepsilon_S}\right)^{0.4} - 1\right] + \varepsilon_S - 1 \quad 2.5\left(\dfrac{1}{\varepsilon_S}\right)^{0.4} + \varepsilon_S - 5.13 = 0$

$\varepsilon_S = 0.1814 \quad V_2 = V_1 \varepsilon_S = 0.00445$ m$^3 \quad p_2 = p_1 \left(\dfrac{V_1}{V_2}\right)^{\kappa} = \dfrac{p_1}{\varepsilon_S^{\kappa}} = 10.91$ bar

braking distance: $\quad (1 - \varepsilon_S)\cdot 500$ mm $= 409$ mm

4.3.5 Polytropic Change of State

The special changes of state discussed so far can be characterised as follows:

Isochoric change of state	$V_1 = V_2$	(4.112)
Isobaric change of state	$p_1 = p_2$	(4.113)
Isothermal change of state	$p_1 V_1 = p_2 V_2$	(4.114)
Isentropic change of state	$p_1 V_1^\kappa = p_2 V_2^\kappa$	(4.115)

An equation is to be looked for which satisfies Eqs. (4.112) to (4.115) as special cases, and with which one can also find all the changes of state between the special cases (Fig. 4.21). The following equation fulfils the above requirements:

$$p_1 V_1^n = p_2 V_2^n \qquad p V^n = \mathrm{const} \qquad (4.116)$$

The general change of state defined by Eq. (4.116) is the polytropic change of state. The exponent n is called the polytropic exponent. It takes the following numerical values for the special changes of state:

Isentropic change of state	$n_S = \kappa$	(4.117)
Isothermal change of state	$n_T = 1$	(4.118)
Isobaric change of state	$n_p = 0$	(4.119)
Isochoric change of state	$n_V = \infty$	(4.120)

To understand Eq. (4.120), one can also write Eq. (4.116) as follows:

$$\frac{V_1}{V_2} = \sqrt[n]{\frac{p_2}{p_1}} \qquad (4.121)$$

The transition of Eq. (4.121) to Eq. (4.112) is fullfilled if

$$\sqrt[n]{\frac{p_2}{p_1}} = 1 \qquad (4.122)$$

becomes. For $p_2/p_1 \neq 1$, this is the case for $n \to \infty$:

$$\lim_{n \to \infty} \sqrt[n]{\frac{p_2}{p_1}} = 1 \qquad (4.123)$$

As can be seen from Eqs. (4.115) to (4.117), in a whole series of equations of the section 4.3.4 about the isentropic change of state, on can substitute the constant isentropic exponent κ by the polytropic exponent n, which is constant for the respective change of state. In this way, one obtains the corresponding equations for the polytropic change of state.

This is the case for the equations describing the relationships between the thermal state variables, as well as for the equations for the volume change work and the pressure change work.

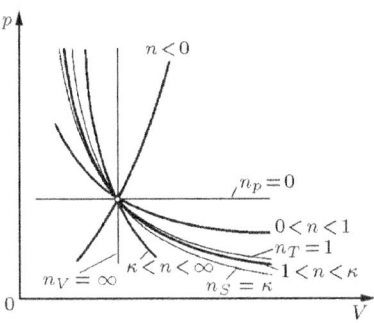 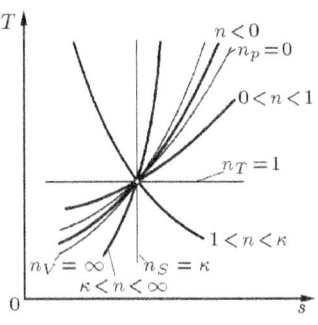

Figure 4.21 Polytropes in a p, V diagram and in a T, s diagram; thinner lines for special changes of state

The equations for the volume change work are obtained from Eqs. (4.103) to (4.105), (4.107) and (4.108). They are compiled in Table 4.1.

Table 4.1 Equations for the volume change work in a polytropic change of state

$$W_{V12} = \frac{p_1 V_1}{n-1}\left[\left(\frac{V_1}{V_2}\right)^{n-1} - 1\right] = \frac{m R T_1}{n-1}\left[\left(\frac{V_1}{V_2}\right)^{n-1} - 1\right] \quad (1)$$

$$W_{V12} = \frac{p_1 V_1}{n-1}\left[\left(\frac{p_2}{p_1}\right)^{\frac{n-1}{n}} - 1\right] = \frac{m R T_1}{n-1}\left[\left(\frac{p_2}{p_1}\right)^{\frac{n-1}{n}} - 1\right] \quad (2)$$

$$W_{V12} = \frac{1}{n-1}(p_2 V_2 - p_1 V_1) \quad (3)$$

$$W_{V12} = \frac{p_1 V_1}{n-1}\left(\frac{T_2}{T_1} - 1\right) = \frac{m R}{n-1}(T_2 - T_1) \quad (4)$$

From the equations for the volume change work according to Table 4.1 and Eq. (4.128), the equations for the pressure change work follow according to Table 4.2.

Table 4.2 Equations for the pressure change work in a polytropic change of state

$$W_{p12} = \frac{n}{n-1}p_1 V_1\left[\left(\frac{V_1}{V_2}\right)^{n-1} - 1\right] = \frac{n}{n-1}m R T_1\left[\left(\frac{V_1}{V_2}\right)^{n-1} - 1\right] \quad (1)$$

$$W_{p12} = \frac{n}{n-1}p_1 V_1\left[\left(\frac{p_2}{p_1}\right)^{\frac{n-1}{n}} - 1\right] = \frac{n}{n-1}m R T_1\left[\left(\frac{p_2}{p_1}\right)^{\frac{n-1}{n}} - 1\right] \quad (2)$$

$$W_{p12} = \frac{n}{n-1}(p_2 V_2 - p_1 V_1) \quad (3)$$

$$W_{p12} = \frac{n}{n-1}p_1 V_1\left(\frac{T_2}{T_1} - 1\right) = \frac{n}{n-1}m R (T_2 - T_1) \quad (4)$$

From Eqs. (4.98) to (4.101) it follows directly

$$\left(\frac{V_2}{V_1}\right)^{n-1} = \frac{T_1}{T_2} \qquad T_1 V_1^{n-1} = T_2 V_2^{n-1} \quad (4.124)$$

$$TV^{n-1} = \text{const} \tag{4.125}$$

$$\left(\frac{p_1}{p_2}\right)^{n-1} = \left(\frac{T_1}{T_2}\right)^n \qquad \frac{T_2}{T_1} = \left(\frac{p_2}{p_1}\right)^{\frac{n-1}{n}} \tag{4.126}$$

$$\frac{T^n}{p^{n-1}} = \text{const}. \tag{4.127}$$

Eq. (4.111) can be generalised in the same simple way:

$$\mathrm{d}W_p = n\,\mathrm{d}W_V \qquad W_{p12} = n\,W_{V12} \tag{4.128}$$

In section 2.7.2, the dependence of the specific heat capacity of the gases on the type of change of state was pointed out. The type of change of state is described for ideal gases by the polytrope exponent n. c_n is a generalised specific heat capacity, also called specific polytropic heat capacity. In c_n, the index n is an indication of the polytropic exponent, on which the specific heat capacity of an ideal gas depends. According to Eq. (3.81), the following applies to the heat in a reversible substitute process for a polytropic change of state:

$$\mathrm{d}Q_{rev} = m\,c_n\,\mathrm{d}T \tag{4.129}$$

$$(Q_{12})_{rev} = m\,c_n\,(T_2 - T_1) \tag{4.130}$$

In Eq. (3.73) for the reversible substitute process of a closed system

$$(Q_{12})_{rev} + W_{V12} = U_2 - U_1,$$

Eq. (4.130) is used for the reversible substitute heat, Eq. (4.40) for the difference in internal energy and Eq. (4) of Table 4.1 for the volume change work:

$$m c_n (T_2 - T_1) + \frac{m R}{n-1}(T_2 - T_1) = m c_v (T_2 - T_1)$$

Dividing by $m(T_2 - T_1)$ gives (Fig. 4.22):

$$c_n = c_v - \frac{R}{n-1} \tag{4.131}$$

In Eq. (3.75) for the reversible substitute process of an open system

$$(Q_{12})_{rev} + W_{p12} = H_2 - H_1$$

Eq. (4.130) is used for the reversible heat, Eq. (4.43) for the difference in enthalpy and Eq. (4) of Table 4.2 for the pressure change work:

$$m\,c_n\,(T_2 - T_1) + \frac{n m R}{n-1}(T_2 - T_1) = m\,c_p\,(T_2 - T_1)$$

Dividing by $m(T_2 - T_1)$ gives:

$$c_n = c_p - \frac{n}{n-1} R. \tag{4.132}$$

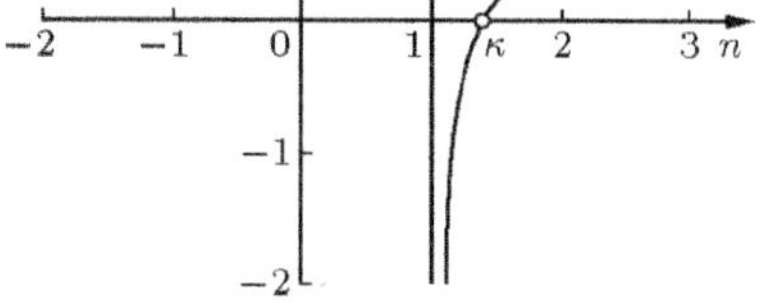

Figure 4.22 Dependence of the specific polytropic heat capacity c_n on the polytropic exponent n for air

Eqs. (4.131) and (4.132), combined with Eqs. (4.49) and (4.50), result in

$$\frac{c_n}{R} = \frac{1}{\kappa - 1} - \frac{1}{n-1} \tag{4.133}$$

$$\frac{c_n}{R} = \frac{\kappa}{\kappa - 1} - \frac{n}{n - 1} \tag{4.134}$$

$$\frac{c_n}{R} = \frac{n - \kappa}{(\kappa - 1)(n - 1)} \tag{4.135}$$

$$c_n = c_v \frac{n - \kappa}{n - 1} \tag{4.136}$$

$$c_n = \frac{c_p}{\kappa} \frac{n - \kappa}{n - 1}. \tag{4.137}$$

In the range $1 < n < \kappa$, c_n is negative.

In this range, if a system is supplied with heat and at the same time releases a larger amount of work, this results in a decrease of temperature. If a system releases heat and at the same time is supplied with a larger amount of work, this results in an increase of temperature. These two changes of state are correlated with a negative specific polytropic heat capacity c_n.

A relation between the heat and the volume change work is given by Eq. (4) according to Table 4.1, Eq (4.130) and Eq. (4.135):

$$\frac{(Q_{12})_{rev}}{W_{V12}} = \frac{c_n}{R}(n - 1) = \frac{n - \kappa}{\kappa - 1} \tag{4.138}$$

From Eqs. (1) to (4) of Table 4.1 and Eq. (4.138), the Eqs. of Table 4.3 follow.

Table 4.3 Equations for the reversible substitute heat $(Q_{12})_{rev}$ in a polytropic change of state

$(Q_{12})_{rev} = \dfrac{n - \kappa}{(\kappa - 1)(n - 1)} p_1 V_1 \left[\left(\dfrac{V_1}{V_2}\right)^{n-1} - 1\right] = \dfrac{n - \kappa}{(\kappa - 1)(n - 1)} m R T_1 \left[\left(\dfrac{V_1}{V_2}\right)^{n-1} - 1\right]$	(1)
$(Q_{12})_{rev} = \dfrac{n - \kappa}{(\kappa - 1)(n - 1)} p_1 V_1 \left[\left(\dfrac{p_2}{p_1}\right)^{\frac{n-1}{n}} - 1\right] = \dfrac{n - \kappa}{(\kappa - 1)(n - 1)} m R T_1 \left[\left(\dfrac{p_2}{p_1}\right)^{\frac{n-1}{n}} - 1\right]$	(2)
$(Q_{12})_{rev} = \dfrac{n - \kappa}{(\kappa - 1)(n - 1)} (p_2 V_2 - p_1 V_1)$	(3)
$(Q_{12})_{rev} = \dfrac{n - \kappa}{(\kappa - 1)(n - 1)} p_1 V_1 \left(\dfrac{T_2}{T_1} - 1\right) = \dfrac{n - \kappa}{(\kappa - 1)(n - 1)} m R (T_2 - T_1)$	(4)

A relation between the internal energy and the volume change work is obtained from Eq. (4) of Table 4.1, Eq. (4.40) and Eq. (4.50):

$$\frac{U_2 - U_1}{W_{V12}} = \frac{c_v}{R}(n - 1) = \frac{n - 1}{\kappa - 1} \tag{4.139}$$

Eqs. (1) to (4) of Table 4.1 and Eq. (4.139) result in the equations of Table 4.4.

The generally valid equation (4.45)

$$H_2 - H_1 = \kappa (U_2 - U_1)$$

and the Eqs. (1) to (4) of Table 4.4 give the equations of Table 4.5.

Table 4.4 Equations for the change of internal energy in a polytropic change of state

$$U_2 - U_1 = \frac{p_1 V_1}{\kappa - 1}\left[\left(\frac{V_1}{V_2}\right)^{n-1} - 1\right] = \frac{m R T_1}{\kappa - 1}\left[\left(\frac{V_1}{V_2}\right)^{n-1} - 1\right] \quad (1)$$

$$U_2 - U_1 = \frac{p_1 V_1}{\kappa - 1}\left[\left(\frac{p_2}{p_1}\right)^{\frac{n-1}{n}} - 1\right] = \frac{m R T_1}{\kappa - 1}\left[\left(\frac{p_2}{p_1}\right)^{\frac{n-1}{n}} - 1\right] \quad (2)$$

$$U_2 - U_1 = \frac{1}{\kappa - 1}(p_2 V_2 - p_1 V_1) \quad (3)$$

$$U_2 - U_1 = \frac{p_1 V_1}{\kappa - 1}\left(\frac{T_2}{T_1} - 1\right) = \frac{m R}{\kappa - 1}(T_2 - T_1) \quad (4)$$

Table 4.5 Equations for the change of enthalpy in polytropic change of state

$$H_2 - H_1 = \frac{\kappa}{\kappa - 1}p_1 V_1\left[\left(\frac{V_1}{V_2}\right)^{n-1} - 1\right] = \frac{\kappa}{\kappa - 1}m R T_1\left[\left(\frac{V_1}{V_2}\right)^{n-1} - 1\right] \quad (1)$$

$$H_2 - H_1 = \frac{\kappa}{\kappa - 1}p_1 V_1\left[\left(\frac{p_2}{p_1}\right)^{\frac{n-1}{n}} - 1\right] = \frac{\kappa}{\kappa - 1}m R T_1\left[\left(\frac{p_2}{p_1}\right)^{\frac{n-1}{n}} - 1\right] \quad (2)$$

$$H_2 - H_1 = \frac{\kappa}{\kappa - 1}(p_2 V_2 - p_1 V_1) \quad (3)$$

$$H_2 - H_1 = \frac{\kappa}{\kappa - 1}p_1 V_1\left(\frac{T_2}{T_1} - 1\right) = \frac{\kappa}{\kappa - 1}m R(T_2 - T_1) \quad (4)$$

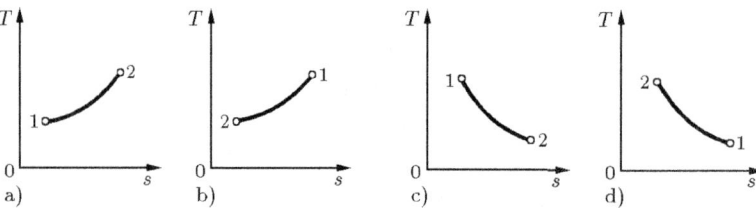

Figure 4.23 Interpretation of polytropes in a T, s diagram
a) c_n positive, ds positive, heat input: temperature increases
b) c_n positive, ds negative, heat output: temperature decreases
c) c_n negative, ds positive, heat input, greater work output, too: temperature decreases
d) c_n negative, ds negative, heat output, greater work input, too: temperature increases

If Eq. (4.129) is inserted into Eq. (3.24) for the differential of entropy, this gives

$$dS = \frac{dQ_{rev}}{T} = m\, c_n\, \frac{dT}{T}. \quad (4.140)$$

With $c_n = $ const, the ingetration results in

$$S_2 - S_1 = m\, c_n \ln \frac{T_2}{T_1} \quad (4.141)$$

(Fig. 4.23).

In addition to Eq. (4.141), Eqs. (4.57) to (4.59) can be used to calculate the entropy change in a polytropic change of state.

If two of the thermal state variables pressure p, volume V and temperature T are known for the initial point 1 and for the end point 2 of a polytropic change of state, the polytropic exponent n can be calculated. From Eqs. (4.116), (4.124) and (4.126) one obtains

$$n = \frac{\ln \dfrac{p_2}{p_1}}{\ln \dfrac{V_1}{V_2}} \qquad (4.142)$$

$$n = \frac{\ln \dfrac{V_2}{V_1} + \ln \dfrac{T_1}{T_2}}{\ln \dfrac{V_2}{V_1}} \qquad (4.143)$$

$$n = \frac{\ln \dfrac{p_2}{p_1}}{\ln \dfrac{p_2}{p_1} - \ln \dfrac{T_2}{T_1}} \qquad (4.144)$$

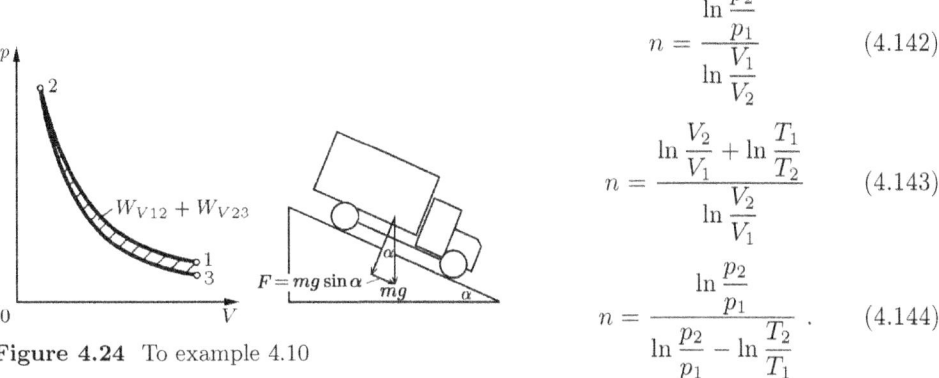

Figure 4.24 To example 4.10

The natural logarithm in Eqs. (4.142) to (4.144) can be replaced by another logarithm.

Example 4.10 A motor vehicle with a *Diesel* engine drives down a hill with the fuel supply switched off, but the gear is engaged (Fig. 4.24). The piston of the *Diesel* engine is driven by the wheels and compresses the air in the cylinder, which has at the initial point 1 the volume $V_1 = 2.817 \ \ell$ and is under pressure $p_1 = 0.95$ bar and at the temperature $t_1 = 20\,°C$. At the point 2, the compressed air has the volume $V_2 = 0.217 \ \ell$. The compression is characterised by the polytropic exponent $n_K = 1.35$. Subsequently, the air expands again from the point 2 to the initial volume $V_3 = V_1 = 2.817 \ \ell$ with the polytropic exponent $n_E = 1.45$. The processes during the air exchange can be neglected. Which pressures and temperatures are reached after compression, which after expansion? What braking power does the engine achieve when 3000 compression and 3000 expansion processes per minute take place? What speed does the vehicle reach if wheel friction and air resistance can be neglected, the mass of the vehicle is 3 t and the gradient of the road is 10 %?

$$p_2 = p_1 \left(\frac{V_1}{V_2}\right)^{n_K} = 30.25 \text{ bar} \qquad p_3 = p_2 \left(\frac{V_2}{V_1}\right)^{n_E} = 0.74 \text{ bar}$$

$$T_2 = T_1 \left(\frac{V_1}{V_2}\right)^{n_K-1} = 719 \text{ K} \qquad T_3 = T_2 \left(\frac{V_2}{V_1}\right)^{n_E-1} = 227 \text{ K}$$

$$W_{V12} = \frac{p_1 V_1}{n_K - 1}\left[\left(\frac{V_1}{V_2}\right)^{n_K-1} - 1\right] = 1111 \text{ J}$$

$$W_{V23} = \frac{p_2 V_2}{n_E - 1}\left[\left(\frac{V_2}{V_1}\right)^{n_E-1} - 1\right] = -998 \text{ J}$$

Differential work: $\quad \Delta W = W_{V12} + W_{V23} = 113$ J

Braking power at uniform speed: $\Delta W/\tau = \Delta W n = 113 \text{ J} \cdot 50 \text{ s}^{-1} = 5.65$ kW

Braking power and braking speed: $\quad \Delta W/\tau = F\,s/\tau = F\,c = m\,g\,\sin\alpha\,c$

$c = (\Delta W/\tau)/(m\,g\,\sin\alpha) \quad \sin\alpha = 0.10 \quad c = 1.92$ m/s $= 6.91$ km/h

4.3 Changes of State

Example 4.11 The decrease of air pressure and air temperature in the atmosphere with increasing height is defined in the standard DIN 5450. For the decrease of the pressure p with increasing height h the following approach applies (cf. Fig. 4.25):

$$(p + dp)\, dA - p\, dA + dG = 0$$

$$dG = \varrho g\, dA\, dh$$

$$-dp = \varrho g\, dh$$

The relation between temperature and pressure is given by a polytropic change of state:

$$\frac{T}{T_0} = \left(\frac{p}{p_0}\right)^{\frac{n-1}{n}}$$

T_0 and p_0 are temperature and pressure at the ground. What is the polytropic exponent n to be used? The vertical temperature gradient γ in the range from -5000 m to $11\,000$ m is given by measurements as follows:

$$\gamma = \frac{dT}{dh} = -6.5 \cdot 10^{-3}\, \text{K/m}$$

Which functions describe the dependence of pressure p, temperature T and density ϱ on the height h?

With Eq. (4.27) calculations lead to

$$\varrho = \frac{p}{RT} = \frac{p}{RT_0}\left(\frac{p_0}{p}\right)^{\frac{n-1}{n}} \qquad dp = -\frac{g}{RT_0}\, p_0^{\frac{n-1}{n}}\, p^{\frac{1}{n}}\, dh$$

$$p^{-\frac{1}{n}}\, dp = -\frac{g}{RT_0}\, p_0^{\frac{n-1}{n}}\, dh \qquad \int_{p_0}^{p} p^{-\frac{1}{n}}\, dp = -\frac{g}{RT_0}\, p_0^{\frac{n-1}{n}} \int_0^h dh$$

$$p = p_0\left(1 - \frac{n-1}{n}\frac{gh}{RT_0}\right)^{\frac{n}{n-1}} \qquad T = T_0\left(1 - \frac{n-1}{n}\frac{gh}{RT_0}\right) \qquad \varrho = \varrho_0\left(1 - \frac{n-1}{n}\frac{gh}{RT_0}\right)^{\frac{1}{n-1}}$$

With

$$\varrho_0 = \frac{p_0}{RT_0} \qquad T = T_0 + \gamma h \qquad \gamma = -\frac{n-1}{n}\frac{g}{R} \qquad n \text{ results in} \qquad n = \frac{1}{1 + \frac{R\gamma}{g}} = 1.235\ .$$

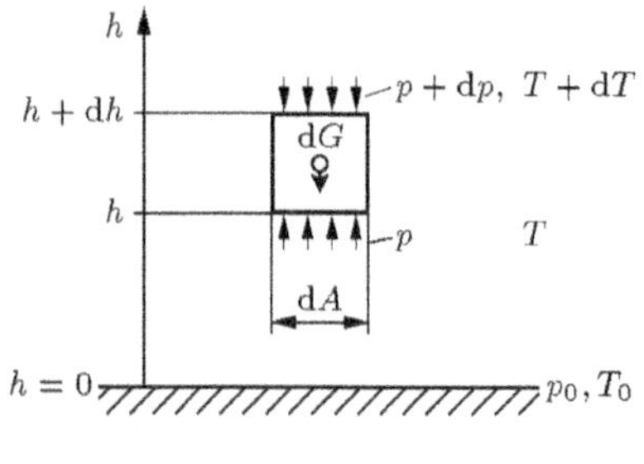

Figure 4.25 To example 4.11

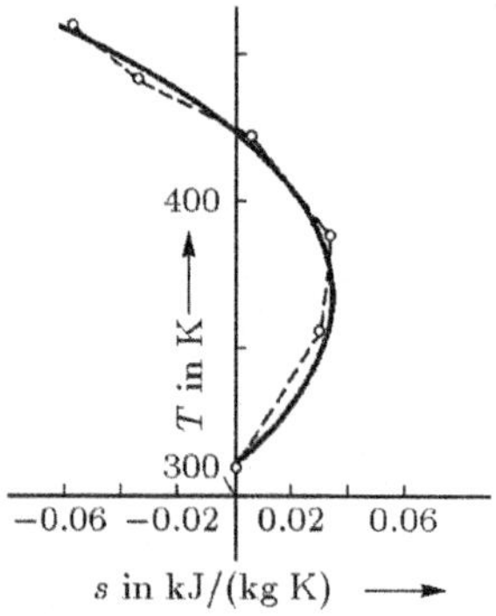

Figure 4.26 To example 4.12. The dashed curves represent the calculated polytropic changes of state.

Example 4.12 The indicator diagram of a single-stage, air-cooled piston compressor [71] gives the following changes of state for the compression of the intake air volume:

p_i in bar	1.01	1.5	2	3	4	5
V_i in ℓ	6.12	4.75	3.88	2.82	2.21	1.84

96   4 Ideal Gases

What is the mass m of the air? What temperatures T_i are reached when the initial temperature is $t_1 = 36\,°\text{C}$? What values are obtained for the mean specific isobaric heat capacities $c_{pm}|_{t_i}^{t_{i+1}}$, the mean isentropic exponents κ_m, the polytropic exponents n and the specific polytropic heat capacities c_n between two state points? What reversible substitute heats $Q_{rev} = (Q_{i,i+1})_{rev}$ are transferred? The changes of state are to be represented in a T,s diagram.

The mass of air m and the temperatures T_i are obtained from Eq. (4.25), the temperature-dependent specific heat capacities $c_{pm}|$ from Eq. (2.96) and the mean isentropic exponents κ_m from Eq. (4.44). The polytropic exponents n are given by Eq. (4.142), the specific heat capacities c_n by Eq. (4.132). The reversible substitute heats $Q_{rev} = (Q_{i,i+1})_{rev}$ and the changes in the specific entropy Δs are given by Eqs. (4.130) and (4.141).

$$m = \frac{p_1 V_1}{R T_1} = 6.9651 \cdot 10^{-3}\ \text{kg} \qquad T_i = \frac{p_i V_i}{m R} \qquad c_{pm}|_{t_i}^{t_{i+1}} = \frac{1}{t_{i+1} - t_i}\left(c_{pm}|_{0\,°C}^{t_{i+1}} t_{i+1} - c_{pm}|_{0\,°C}^{t_i} t_i\right)$$

$$\kappa_m = \frac{c_{pm}}{c_{vm}} \approx \frac{c_{pm}}{c_{pm} - R} \qquad n = \frac{\ln\frac{p_{i+1}}{p_i}}{\ln\frac{V_i}{V_{i+1}}} \qquad c_n = c_{pm} - \frac{n}{n-1} R$$

$$(Q_{i,i+1})_{rev} = Q_{i,i+1} + |W_{RI}| = m\,c_n\,(T_{i+1} - T_i) \qquad \Delta s = s_{i+1} - s_i = c_n \ln\frac{T_{i+1}}{T_i}$$

p_i bar	V_i ℓ	T_i K	c_{pm} kJ/(kg K)	κ_m	n	c_n kJ/(kg K)	Q_{rev} J	Δs kJ/(kg K)
1.01	6.12	309.15	1.0063	1.3991	1.5607	0.2073	68.159	0,0295
1.5	4.75	356.35	1.0097	1.3973	1.4220	0.0423	9.366	0.0036
2	3.88	388.11	1.0135	1.3952	1.2707	−0.3342	−81.490	−0.0289
3	2.82	423.12	1.0172	1.3932	1.1803	−0.8624	−114.155	−0.0379
4	2.21	442.13	1.0200	1.3916	1.2179	−0.5847	−73.327	−0.0233
5	1.84	460.13						

From the table and from Fig. 4.26 it can be seen that for the reversible substitute heat $Q_{rev} = (Q_{i,i+1})_{rev} = Q_{i,i+1} + |W_{RI}|$ the internal friction work $|W_{RI}|$ initially outweighs the heat given off $Q_{i,i+1}$. From the change of state from 2 to 3, from 3 to 4 and from 4 to 5 is the absolute amount of heat given off greater than the internal friction work, and the reversible substitute heat becomes negative.

4.3.6 Changes of State with Variable Mass

If the state of a gas, which is in a space (e.g in a vessel or in a room) of constant volume, is changed in such a way that a part of the gas can escape from this space, there is a change of state with variable mass.

Example 4.13 The air in a room which has the volume 100 m³ is changed from the initial temperature $3\,°\text{C}$ to the final temperature $23\,°\text{C}$. The air pressure is 1 bar. What heat is supplied to the air in the room?

Since air can escape through existing leaks when heated, the mass of the air to be heated diminishes during the heating process. The heating is isobaric.

The expression $m = \dfrac{pV}{RT}$ is inserted into Eq. (4.76): $\qquad dQ_{rev} = m\,c_p\,dT = \dfrac{pV}{R}\,c_p\,\dfrac{dT}{T}$

The temperature-dependent specific heat capacity c_p is replaced by the mean specific heat capacity c_{pm}. This is followed by the integration:

$$dQ_{rev} = \frac{pV}{R}\,c_{pm}\,\frac{dT}{T} \qquad\qquad c_{pm}|_{3\,°C}^{23\,°C} = 1.0036\ \text{kJ/(kg K)}$$

$$(Q_{12})_{rev} = \frac{pV}{R} c_{pm} \ln \frac{T_2}{T_1} = \frac{1 \text{ bar} \cdot 100 \text{ m}^3 \text{ kg K}}{0.28706 \text{ kJ}} \cdot \frac{100 \text{ kJ}}{\text{m}^3 \text{ bar}} \cdot 1.0036 \frac{\text{kJ}}{\text{kg K}} \cdot \ln \frac{296 \text{ K}}{276 \text{ K}} = 2446 \text{ kJ}$$

Example 4.14 The envelope of a balloon has an opening at the bottom. Through this opening part of the filling gas can escape, when expanding during the ascent of the balloon. The balloon is fully inflated at the launch site, so its volume remains constant during ascent. The pressure and temperature of the gas are constantly equal to the ambient air pressure and ambient air temperature. For the change of the air density ϱ_L with the height H the following numerical equation applies according to example 4.11:

$$\varrho_L = 1.225 \left(\frac{288.15 - 6.5\,H}{288.15} \right)^{4.255} \text{ in kg/m}^3 \text{ with } H \text{ in km}$$

How does the acceleration force of the balloon change with altitude if the balloon has a radius of 10 m, hydrogen is used as the filling gas and the weight of the envelope and the nacelle is negligible?

According to Eq. (2.4), the acceleration force is equal to the negative excess weight.

$$F = G - A = V g (\varrho_G - \varrho_L) = V g \varrho_L \left(\frac{\varrho_G}{\varrho_L} - 1 \right)$$

According to Eq. (4.28) $\quad \varrho_G = \dfrac{p}{R_G T} \quad$ and $\quad \varrho_L = \dfrac{p}{R_L T} \quad$ hold. Thus results

$$F = V g \varrho_L \left(\frac{R_L}{R_G} - 1 \right) = 4188.8 \text{ m}^3 \cdot 9.81 \text{ m/s}^2 \left(\frac{287.06 \text{ J/(kg K)}}{4124.5 \text{ J/(kg K)}} - 1 \right) \varrho_L = -38\,232 \text{ m}^4/\text{s}^2 \cdot \varrho_L$$

H in km	0	0.2	0.4	0.6	0.8	1.0
F in N	$-46\,834$	$-45\,942$	$-45\,062$	$-44\,196$	$-43\,342$	$-42\,501$

4.4 Specific Thermal Energy and Specific Work in the T,s Diagram

For frictionless changes of state of ideal gases the specific quantities

$$(q_{12})_{rev}, \quad h_2 - h_1, \quad u_2 - u_1, \quad w_{V12} \quad \text{and} \quad w_{p12}$$

in the T,s diagram are shown as areas. According to Eqs. (3.73) and (3.75) is

$$(q_{12})_{rev} + w_{V12} = u_2 - u_1 \tag{4.145}$$
$$(q_{12})_{rev} + w_{p12} = h_2 - h_1 . \tag{4.146}$$

For ideal gases, specific internal energy and specific enthalpy depend only on temperature, and their changes are thus constant for any change of state between the same temperature limits. According to Eqs. (4.69) and (4.70) it follows for isochoric changes of state

$$u_2 - u_1 = c_v(T_2 - T_1) = (q_{12})_{rev} = \int_1^2 T\,ds. \tag{4.147}$$

According to Eqs. (4.77) and (4.78), in the case of isobaric changes of state follows

$$h_2 - h_1 = c_p(T_2 - T_1) = (q_{12})_{rev} = \int_1^2 T\,ds . \tag{4.148}$$

The change of the specific internal energy in the T,s diagram thus corresponds to the area under the curve for an isochoric change of state, the change of specific enthalpy corresponds to the area under the curve for an isobaric change of state. Since all isochores or isobars can be obtained by shifting in the direction of the s axis (see example 4.5), any isochore or isobar between the same temperature limits T_2 and T_1 can be used for the representation. In Fig. 4.27 and Fig. 4.28, the isochores or isobars passing through state point 2 are mostly chosen.

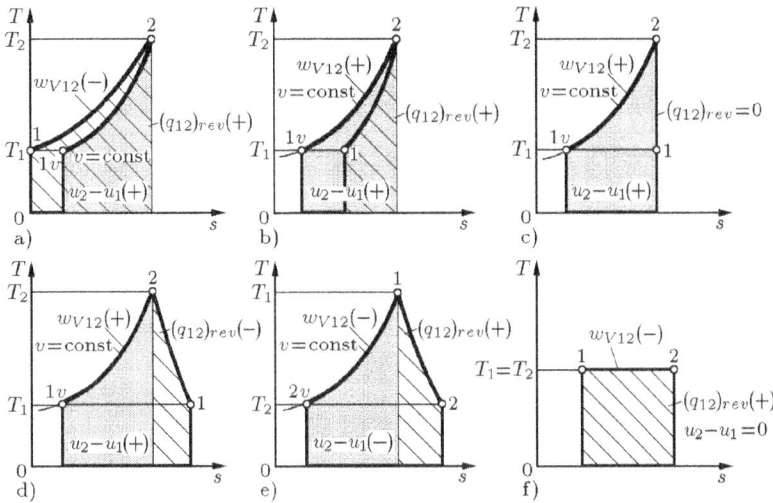

Figure 4.27 Specific quantities in the T, s diagram (changes of state between state points 1 and 2)
a) isobaric expansion b) polytropic compression $\kappa < n < \infty$ c) isentropic compression
d) polytropic compression $1 < n < \kappa$ e) polytropic expansion $1 < n < \kappa$ f) isothermal expansion

$(q_{12})_{rev}$ $u_2 - u_1$ w_{V12}

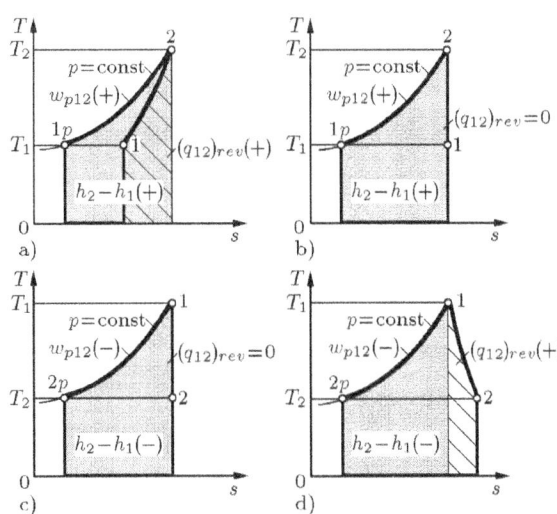

Figure 4.28 Specific quantities in the T, s diagram (changes of state between state points 1 and 2)
a) polytropic compression $\kappa < n < \infty$ b) isentropic compression c) isentropic expansion
d) polytropic expansion $1 < n < \kappa$

$(q_{12})_{rev}$ $h_2 - h_1$ w_{p12}

4.5 Mixtures of Ideal Gases

In technical processes with gases, often not pure gases but gas mixtures are handled, e.g. air, natural gas H, natural gas L, coke oven gas, water gas, luminous gas, blast furnace gas, landfill gas, sewage gas, biogas or combustion gas.

The behaviour of several ideal gases in the same space that do not react chemically with each other is described by the law of *Dalton*:[5]

> If a gas consists of a mixture of several ideal gases, then each individual gas spreads out evenly throughout the room and exerts a pressure that is as great as if the other gases in the mixture were not present.

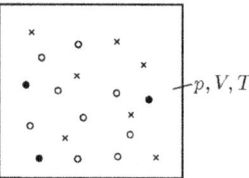

Figure 4.29 Schematic figure of a gas mixture

The pressure of a single gas in a given mixture (Fig. 4.29) is expressed as the partial pressure p_{pi} of that gas. The sum of the partial pressures of all n individual gases is equal to the total pressure p of the mixture:

$$p = \sum_{i=1}^{n} p_{pi} \qquad (4.149)$$

From Eq. (4.149) it follows

$$\sum_{i=1}^{n} \frac{p_{pi}}{p} = 1 \; . \qquad (4.150)$$

For example, the total pressure p of the air in the free atmosphere – the barometric pressure – is given by the sum of the partial pressures p_{pi} of the gaseous individual components, where the partial pressure fractions of the individual gases p_{pi}/p are as follows: Nitrogen 0.78097, Oxygen 0.20944, Argon 0.00916, Carbon Dioxide 0.00040 and other gases 0.00003. As will be shown in section 4.5.3, for air these partial pressure fractions are also equal to the volume fractions (space fractions) r_i.

The gas equation (4.25) can be applied to each individual gas of the gas mixture, where the temperatures of all individual gases and the gas mixture are equal:

$$p_{pi} V = m_i R_i T \qquad (4.151)$$

By addition over all gases in the mixture, Eq. (4.149) gives

$$\sum_{i=1}^{n} p_{pi} V = \sum_{i=1}^{n} m_i R_i T \; . \qquad (4.152)$$

The right-hand side of Eq. (4.152) is multiplied by m and the sum is divided by m at the same time:

$$pV = m \sum_{i=1}^{n} \frac{m_i}{m} R_i T \qquad (4.153)$$

---

[5] *John Dalton*, 1766 to 1844, English physicist and chemist

The sum on the right-hand side of Eq. (4.153) is called the specific gas constant of the ideal gas mixture R:

$$R = \sum_{i=1}^{n} \frac{m_i}{m} R_i \tag{4.154}$$

Thus one obtains the thermal equation of state for the ideal gas mixture

$$pV = mRT \ . \tag{4.155}$$

It has the same form as the gas equation of single ideal gases (4.25). Likewise, the following applies to the density of the ideal gas mixture

$$\varrho = \frac{m}{V} = \frac{p}{RT} \ . \tag{4.156}$$

The quotient

$$\mu_i = \frac{m_i}{m} \tag{4.157}$$

is called the mass fraction of the single gas i in the ideal gas mixture. For the mass m as a conservation variable holds:

$$m = \sum_{i=1}^{n} m_i \tag{4.158}$$

If the left and right side of Eq. (4.158) is divided by m, then

$$\sum_{i=1}^{n} \mu_i = 1 \tag{4.159}$$

is given. Thus Eq. (4.454) can be written as follows:

$$R = \sum_{i=1}^{n} \mu_i R_i \tag{4.160}$$

With $R = R_m/M$ and $R_i = R_m/M_i$ (Eq. (4.36)) it follows for the molar mass of the mixture of ideal gases

$$\frac{1}{M} = \sum_{i=1}^{n} \frac{\mu_i}{M_i} \ . \tag{4.161}$$

From $p_{pi} V = m_i R_i T$ (Eq. (4.151)) and $V = mRT/p$ (Eq. (4.155)) follows $p_{pi} mRT/p = m_i R_i T$ and therefore for the partial pressures of the gas components

$$p_{pi} = \mu_i \frac{R_i}{R} p = \mu_i \frac{M}{M_i} p \tag{4.162}$$

For the change of state of a gas mixture between the initial point 1 and the final point 2, the following relations apply according to the law of energy conservation to the internal energy, enthalpy, entropy and the reversible substitute heat:

$$U_2 - U_1 = \sum_{i=1}^{n} (U_2 - U_1)_i \tag{4.163}$$

$$H_2 - H_1 = \sum_{i=1}^{n} (H_2 - H_1)_i \tag{4.164}$$

$$S_2 - S_1 = \sum_{i=1}^{n} (S_2 - S_1)_i \tag{4.165}$$

4.5 Mixtures of Ideal Gases

$$(Q_{12})_{rev} = \sum_{i=1}^{n}(Q_{12i})_{rev} \tag{4.166}$$

For the specific caloric variables

$$u = \frac{U}{m}, \quad h = \frac{H}{m}, \quad s = \frac{S}{m}, \quad q = \frac{Q}{m}$$

also applies

$$u_2 - u_1 = \sum_{i=1}^{n}\mu_i(u_2 - u_1)_i = c_v(T_2 - T_1) = \sum_{i=1}^{n}\mu_i c_{vi}(T_2 - T_1) \tag{4.167}$$

$$h_2 - h_1 = \sum_{i=1}^{n}\mu_i(h_2 - h_1)_i = c_p(T_2 - T_1) = \sum_{i=1}^{n}\mu_i c_{pi}(T_2 - T_1) \tag{4.168}$$

$$s_2 - s_1 = \sum_{i=1}^{n}\mu_i(s_2 - s_1)_i = c_n \ln\frac{T_2}{T_1} = \sum_{i=1}^{n}\mu_i c_{ni} \ln\frac{T_2}{T_1} \tag{4.169}$$

$$(q_{12})_{rev} = \sum_{i=1}^{n}\mu_i(q_{12i})_{rev} = c_n(T_2 - T_1) = \sum_{i=1}^{n}\mu_i c_{ni}(T_2 - T_1). \tag{4.170}$$

From Eqs. (4.167) to (4.170), the specific heat capacity of a mixture of ideal gases is given by

$$c = \sum_{i=1}^{n}\mu_i c_i. \tag{4.171}$$

Here, the specific heat capacities c and c_i, respectively, can be c_v, c_p, c_n and c_{vi}, c_{pi}, c_{ni}, respectively.

If Eq. (4.50) is inserted into the equation for c_v, an equation for the calculation of the isentropic exponent of a mixture of ideal gases is obtained:

$$\frac{1}{\kappa - 1} = \sum_{i=1}^{n}\mu_i \frac{R_i}{R}\frac{1}{\kappa_i - 1} \tag{4.172}$$

Example 4.15 A coke oven generator gas has the composition in mass fractions

0.9 % H_2 30.6 % CO   0.3 % CH_4 8.6 % CO_2 59.6 % N_2.

What are the gas constant, the molar mass and the partial pressures of the gas mixture in the physical norm state?

In the physical norm state the pressure is $p = 1013$ mbar.

Gas	μ_i	R_i J/kg K	M_i kg/kmol	$\mu_i R_i$ J/(kg K)	μ_i/M_i kmol/kg	$p_{pi} = \mu_i \dfrac{R_i}{R} p$ mbar
H_2	0.009	4124.4	2.02	37.12	0.00446	117
CO	0.306	296.8	28.01	90.82	0.01092	285
CH_4	0.003	518.3	16.04	1.55	0.00019	5
CO_2	0.086	188.9	44.01	16.25	0.00195	51
N_2	0.596	296.8	28.01	176.89	0.02128	555
				322.63	0.03880	1013

$R = 322.63 \text{ J/(kg K)}$ $\quad\quad M = 1/(0.03880 \text{ kmol/kg}) = 25.77 \text{ kg/kmol}$

For control one calculates: $\quad M = \dfrac{R_m}{R} = \dfrac{8314.51 \text{ J/(kmol K)}}{322.63 \text{ J/(kg K)}} = 25.77 \text{ kg/kmol}$

4.5.1 The Mixing Process in the Closed System

One can always think of a gas mixture as being created by an adiabatic mixing process (Fig. 4.30). The mixing process is irreversible, an equalisation process takes place for each individual gas (section 3.2).

The sum of the internal energies of individual ideal gases before the mixing is at the respective temperatures of the individual gases T_i

$$\sum_{i=1}^{n} U_i = \sum_{i=1}^{n}(m_i\, u_i)_{T_i} = \sum_{i=1}^{n} m_i\, c_{vi}\, T_i \ . \tag{4.173}$$

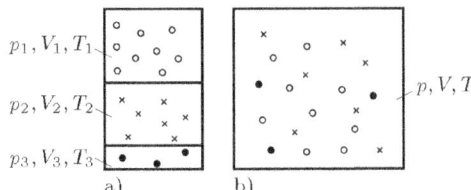

Figure 4.30 Mixture of three gases in a closed system ($\Sigma V_i \neq V$)
a) before mixing
b) after mixing

After the mixture, all individual gases have the same temperature T. Using Eq. (4.167), it follows

$$U = \sum_{i=1}^{n}(m_i\, u_i)_T = \sum_{i=1}^{n} m_i\, c_{vi}\, T \ . \tag{4.174}$$

If the quantity of individual gases forms an isolated closed system during the mixing process, neither heat nor work is exchanged with the environment. The internal energy of this system then remains constant: The sum of the internal energies of the individual gases before mixing is equal to the internal energy of the mixture after mixing:

$\sum_{i=1}^{n} U_i = U \quad$ By equating the right-hand sides of Eqs. (4.173) and (4.174) it follows for the mixture temperature T

$$T = \dfrac{\sum_{i=1}^{n} m_i\, c_{vi}\, T_i}{\sum_{i=1}^{n} m_i\, c_{vi}} \ . \tag{4.175}$$

If p_i, V_i and T_i are pressure, volume and temperature of every individual gas before mixing, then the thermal equation of state of ideal gases applies to the product $p_i V_i$ and the density ϱ_i of every individual gas i in the state before the mixture

$$p_i V_i = m_i R_i T_i \tag{4.176}$$

$$\varrho_i = \dfrac{m_i}{V_i} = \dfrac{p_i}{R_i T_i} \ . \tag{4.177}$$

According to Eqs. (4.154) and (4.176), before the mixture is

$$mR = \sum_{i=1}^{n} m_i R_i = \sum_{i=1}^{n} \frac{p_i V_i}{T_i} . \quad (4.178)$$

With Eq. (4.155) the resulting pressure p after the mixture is

$$p = \frac{T}{V} \sum_{i=1}^{n} \frac{p_i V_i}{T_i} . \quad (4.179)$$

Example 4.16 Two vessels, the first of which has exhaust gas of $p_A = 0.93$ bar and $t_A = 378\,°C$ in a volume of $V_A = 0.03\text{ m}^3$, the second containing air of $p_L = 0.21$ bar and $t_L = 20\,°C$ in a volume of $V_L = 0.056\text{ m}^3$, are connected with a third, completely empty vessel of $V_{empty} = 0.046\text{ m}^3$. What mixing temperature T and what mixing pressure p arise if there is no heat exchange with the environment, and if the exhaust gas has a density in the physical norm state of $\varrho_{0A} = 1.203\text{ kg/m}^3$ and a specific heat capacity $c_{vA} = 0.634\text{ kJ/(kg K)}$ in the temperature range in question?

With $\quad R_A = \dfrac{p_{0A}}{\varrho_{0A}\, T_{0A}}$, $\quad$ it follows: $\quad m_A = \dfrac{p_A V_A}{R_A T_A} = \dfrac{p_A\, T_{0A}}{p_{0A}\, T_A} V_A\, \varrho_{0A} = 0.01389\text{ kg}$

$m_L = \dfrac{p_L V_L}{R_L T_L} = 0.01398\text{ kg}$

With $c_{vL} = 0.724\text{ kJ/(kg K)}$ for the air in the temperature range T_L to T
and $\quad V = V_A + V_L + V_{empty} = 0.132\text{ m}^3 \quad$ are

$T = \dfrac{m_A\, c_{vA}\, T_A + m_L\, c_{vL}\, T_L}{m_A\, c_{vA} + m_L\, c_{vL}} = 459\text{ K} \quad$ and $\quad p = \dfrac{T}{V}\left(\dfrac{p_A V_A}{T_A} + \dfrac{p_L V_L}{T_L}\right) = 0.289$ bar.

In order to obtain further equations for the density ϱ, the specific volume v, the individual gas constant R and the molar mass M of a gas mixture, the volume fraction (space fraction) r_i is introduced:

$$r_i = \frac{V_i}{V} \quad (4.180)$$

Using the relation for mass m as the product of volume V and density ϱ

$$m = V\varrho \qquad\qquad m_i = V_i\, \varrho_i \quad (4.181)$$

one obtains with $\quad m = \sum_{i=1}^{n} m_i \quad$ (Eq. (4.158))

$$\varrho = \sum_{i=1}^{n} r_i\, \varrho_i . \quad (4.182)$$

With mass m as quotient of volume V and specific volume v

$$m = \frac{V}{v} \qquad\qquad m_i = \frac{V_i}{v_i} \quad (4.183)$$

one gets

$$\frac{1}{v} = \sum_{i=1}^{n} \frac{r_i}{v_i} . \quad (4.184)$$

Inserting $\quad m_i = \dfrac{p_{pi} V}{R_i T} \quad$ and $\quad m = \dfrac{pV}{RT} \quad$ (transformed Eqs. (4.151) and (4.155))

into $\quad m = \sum_{i=1}^{n} m_i \quad$ (Eq. (4.158)), $\quad$ the result is

$$\frac{1}{R} = \sum_{i=1}^{n} \frac{p_{pi}}{p} \frac{1}{R_i} \quad (4.185)$$

and with Eq. (4.36)
$$M = \sum_{i=1}^{n} \frac{p_{pi}}{p} M_i \ . \tag{4.186}$$

Expressing the mass m as the product of molar mass M and molar quantity n according to Eq. (1.5) gives
$$m = M n \qquad\qquad m_i = M_i n_i \ . \tag{4.187}$$

From this follows with Eq. (4.158) and the respective mole fraction n_i/n
$$M = \sum_{i=1}^{n} \frac{n_i}{n} M_i \tag{4.188}$$

$$\frac{1}{R} = \sum_{i=1}^{n} \frac{n_i}{n} \frac{1}{R_i} \ . \tag{4.189}$$

A comparison of Eqs. (4.186) and (4.188) respective (4.185) and (4.189) shows, that is
$$\frac{n_i}{n} = \frac{p_{pi}}{p} \ . \tag{4.190}$$

The ratio of the molar quantities of the individual gases in a gas mixture is equal to the ratio of the partial pressures. For example, if the ideal gas mixture consists of three individual gases, the following applies:
$$n_1 : n_2 : n_3 = p_{p1} : p_{p2} : p_{p3} \tag{4.191}$$

With respect to section 4.1.6, the ratio of the molar quantities of the individual gases in a gas mixture is also equal to the ratio of the number of molecules of these gases.

According to Eq. (4.165), the entropy change of the mixing process ΔS is the sum of the entropy changes of the individual gases ΔS_i. With Eq. (4.57) one obtains
$$\Delta S = \sum_{i=1}^{n} \Delta S_i = \sum_{i=1}^{n} (m_i \, c_{vi} \ln \frac{T}{T_i} + m_i \, R_i \ln \frac{V}{V_i}) \ . \tag{4.192a}$$

If dividing by the mass m and using Eq. (4.180), one gets
$$\Delta s = \sum_{i=1}^{n} \mu_i (c_{vi} \ln \frac{T}{T_i} - R_i \ln r_i) \ . \tag{4.192b}$$

Since the mixing process is adiabatic, Δs has the meaning of an overall change in specific entropy. According to section 3.3.2, always $\Delta s > 0$ will be.

By dividing Eqs. (4.181), a relation can be derived which enables to convert volume fractions into mass fractions and vice versa:
$$\mu_i = r_i \frac{\varrho_i}{\varrho} = r_i \frac{v}{v_i} \tag{4.193}$$

Eqs. (4.156) and (4.177) finally give
$$\mu_i = r_i \frac{p_i}{p} \frac{T}{T_i} \frac{R}{R_i} = r_i \frac{p_i}{p} \frac{T}{T_i} \frac{M_i}{M} \tag{4.194}$$

$$\mu_i = \frac{p_{pi}}{p} \frac{R}{R_i} = \frac{p_{pi}}{p} \frac{M_i}{M} \tag{4.195}$$

Eq. (4.195) is the transformed Eq. (4.162). With Eq. (4.190) becomes

$$\mu_i = \frac{n_i}{n}\frac{M_i}{M} = \frac{n_i}{n}\frac{R}{R_i}. \qquad (4.196)$$

In the relation according to Eq. (4.190) another expression can be included:

$$\frac{n_i}{n} = \frac{p_{pi}}{p} = r_i \frac{p_i}{p}\frac{T}{T_i} \qquad (4.197)$$

Analogous to Eqs. (4.150) and (4.159) it can be determined:

$$\sum_{i=1}^{n} \frac{n_i}{n} = 1 \qquad (4.198)$$

$$\sum_{i=1}^{n} r_i \frac{p_i}{p}\frac{T}{T_i} = 1 . \qquad (4.199)$$

Example 4.17 In example 4.16, what are the volume fractions, the molar quantities and the partial pressures of the individual gases in the mixture, and what entropy change occurs during the mixing process?

$$r_L = \frac{V_L}{V} = 0.42 \qquad r_A = \frac{V_A}{V} = 0.23$$

With $M_L = 28.96$ kg/kmol and $M_A = \dfrac{R_m}{R_A} = \dfrac{8314.51 \text{ J/(kmol K)}}{308.35 \text{ J/(kg K)}} = 26.96$ kg/kmol is

$$n_L = \frac{m_L}{M_L} = 0.483 \text{ mol} \qquad \text{and} \qquad n_A = \frac{m_A}{M_A} = 0.515 \text{ mol}.$$

According to Eq. (4.190), the partial pressures are $p_{pL} = 0.140$ bar and $p_{pA} = 0.149$ bar. The entropy change is

$$\begin{aligned}
m\,\Delta s = \Delta S &= m_A(c_{vA}\ln\frac{T}{T_A} - R_A\ln r_A) + m_L(c_{vL}\ln\frac{T}{T_L} - R_L\ln r_L) \\
&= 0.01389 \text{ kg} \cdot (0.634 \text{ kJ/(kg K)}) \cdot \ln\frac{459 \text{ K}}{651 \text{ K}} - 308.35 \text{ J/(kg K)} \cdot \ln 0.23) \\
&\quad + 0.01398 \text{ kg} \cdot (0.724 \text{ kJ/(kg K)}) \cdot \ln\frac{459 \text{ K}}{293 \text{ K}} - 287.06 \text{ J/(kg K)} \cdot \ln 0.42) \\
&= 3.2172 \text{ J/K} + 8.0247 \text{ J/K} = 11.2419 \text{ J/K} .
\end{aligned}$$

4.5.2 The Mixing Process Without Total Volume Change

For the mixing process with total volume unchanged (Fig. 4.31) one finds

$$V = \sum_{i=1}^{n} V_i . \qquad (4.200)$$

With $r_i = V_i/V$ (Eq. (4.180)) one obtains

$$\sum_{i=1}^{n} r_i = 1 . \qquad (4.201)$$

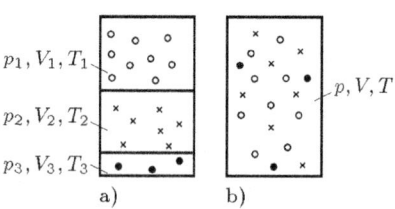

Figure 4.31 Mixing of three gases with total volume unchanged ($\Sigma V_i = V$)
a) before mixing b) after mixing

Note that Eq. (4.201) only holds in this section and did not yet hold in example 4.17.

Example 4.18 The two vessels in Example 4.16, containing exhaust gas and air, are connected only with each other (without the involvement of a third vessel). What are the resulting mixing temperature and mixing pressure? What are the volume fractions of the individual gases in the mixture?

The mixing temperature is the same as in example 4.16, the mixing pressure is $p = 0.443$ bar, the volume fractions are $r_L = 0.65$ and $r_A = 0.35$; the sum of the volume fractions is $\Sigma r_i = 1$.

4.5.3 The Mixing Process Without Temperature Change, Pressure Change and Total Volume Change

If the individual gases all have the same temperature $T_i = T$ before mixing, the special case is of particular interest, that also all pressures $p_i = p$ are the same (Fig. 4.32).
With (Eq.4.176) for the state of the individual gases before the mixture

$$p_i V_i = m_i R_i T$$

and (Eq.4.151) for the state of the individual gases after the mixture

$$p_{pi} V = m_i R_i T$$

becomes

$$p_i V_i = p V_i = p_{pi} V . \tag{4.202}$$

With Eqs. (4.180) and (4.190), Eq. (4.202) becomes

$$\frac{p_{pi}}{p} = \frac{n_i}{n} = r_i . \tag{4.203}$$

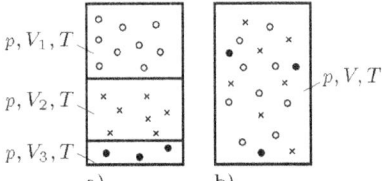

Figure 4.32 Mixing of three gases without temperature and pressure change ($\Sigma V_i = V$)
a) before mixing   b) after mixing

For Eqs. (4.188), (4.189) and (4.194) follows:

$$M = \sum_{i=1}^{n} r_i M_i \tag{4.204}$$

$$\frac{1}{R} = \sum_{i=1}^{n} \frac{r_i}{R_i} \tag{4.205}$$

$$\mu_i = r_i \frac{R}{R_i} = r_i \frac{M_i}{M} \tag{4.206}$$

The analysis of a gas mixture is carried out under the conditions $T_i = T$, $\Sigma V_i = V$ (Eq. (4.200)) and $p_i = p$, since the gas mixture can be imagined to have been mixed without a change in temperature, with an unchanged total volume and with the same pressures of the individual components before the mixture and the same pressure of the gas mixture after the mixing process. Thus also $\Sigma r_i = 1$ (Eq. (4.201)) is fulfilled.

Example 4.19 By an analysis carried out at constant temperature, the composition of water gas in volume fractions is measured as

4.5 Mixtures of Ideal Gases

49 % H_2 42 % CO 0.5 % CH_4 5 % CO_2 3.5 % N_2

What are the molar mass, the composition in mass fractions and the specific gas constant of water gas? What is the density of water gas at a temperature of 12 °C and a pressure of 0.963 bar? What is the density of water gas in its physical norm state? What is the molar ratio of the individual gases? What are the partial pressures?

Gas	r_i	M_i kg/kmol	$r_i M_i$ kg/kmol	$\mu_i = r_i M_i / M$
H_2	0.49	2.02	0.99	0.062
CO	0.42	28.01	11.76	0.735
CH_4	0.005	16.04	0.08	0.005
CO_2	0.05	44.01	2.20	0.137
N_2	0.035	28.01	0.98	0.061

$$M = 16.01 \text{ kg/kmol}$$

The specific gas constant is

$$R = \frac{R_m}{M} = \frac{8.31451 \text{ kJ/(kmol K)}}{16.01 \text{ kg/kmol}} = 519 \text{ J/(kg K)}$$

The density in the given state is

$$\varrho = \frac{p}{RT} = \frac{0.963 \text{ bar}}{519 \text{ J/(kg K)} \cdot 285 \text{ K}} = 0.651 \text{ kg/m}^3 .$$

The density in the physical norm state is

$$\varrho_0 = \frac{p_0}{RT_0} = 0.715 \text{ kg/m}^3 .$$

The molar quantity ratio is given by Eq. (4.203) ($CH_4 \triangleq 1$)

$$n_{H_2} : n_{CO} : n_{CH_4} : n_{CO_2} : n_{N_2} = 98 : 84 : 1 : 10 : 7 .$$

The partial pressures are given by $p_{pi} = r_i p$:

$p_{pH_2} = 0.472$ bar $p_{pCO} = 0.404$ bar $p_{pCH_4} = 0.005$ bar
$p_{pCO_2} = 0.048$ bar $p_{pN_2} = 0.034$ bar.

Example 4.20 What is the total entropy change if the individual gases of example 4.19 are mixed without a temperature change and 1 kg of water gas is produced?

For each individual gas, Eq. (4.192) holds:

$$S_{2i} - S_{1i} = -m_i R_i \ln r_i = -m_i R_i \ln \frac{V_i}{V} = m_i R_i \ln \frac{p}{p_{pi}}$$

Gas	m_i kg	R_i J/(kg K)	$\ln \frac{p}{p_{pi}}$	$S_{2i} - S_{1i}$ J/K
H_2	0.062	4124.5	0.7131	182.3
CO	0.735	296.8	0.8686	189.5
CH_4	0.005	518.3	5.2606	13.6
CO_2	0.137	188.9	2.9989	77.6
N_2	0.061	296.8	3.3437	60.5
				523.5

The total entropy change is $S_2 - S_1 = 523.5$ J/K.

The analysis can also refer to the physical norm state. For this purpose, from Eq. (4.180) follows

$$r_i = \frac{V_i}{V} = \frac{V_{0i}}{V_0} , \qquad (4.207)$$

where V_{0i} is the physical norm volume of the single individual gas i before the mixture and V_0 is the physical norm volume of the mixture. Thus, from Eq. (4.193) one gets

$$\mu_i = r_i \frac{\varrho_{0i}}{\varrho_0} = r_i \frac{v_0}{v_{0i}} . \qquad (4.208)$$

ϱ_{0i} and v_{0i} are density and specific volume of the single individual gas i in the physical norm state before mixing, ϱ_0 and v_0 are density and specific volume of the mixture in the physical norm state. It is particularly interesting to apply Eq. (4.208) to Eq. (4.171):

$$c = \sum_{i=1}^{n} \mu_i c_i = \sum_{i=1}^{n} r_i \frac{\varrho_{0i}}{\varrho_0} c_i \tag{4.209}$$

$$C = \frac{c}{v_0} = c\varrho_0 = \sum_{i=1}^{n} r_i \varrho_{0i} c_i \tag{4.210}$$

C is the volume-related heat capacity in the the physical norm state. With this quantity, the heat $(Q_{12})_{rev}$, which occurs in a reversible substitute process with e.g. isobaric change of state can be calculated as follows:

$$H_2 - H_1 = m\, c_p\, (t_2 - t_1) = (Q_{12})_{rev} = V_0\, C_p (t_2 - t_1) \tag{4.211}$$

Example 4.21 A volume flow of $\dot{V}_0 = 1300$ m^3/h of exhaust gas in the physical norm state with the composition in volume fractions r_i

6.6 % H_2O 11.3 % CO_2 2.4 % O_2 5.9 % CO 73.8 % N_2

is cooled in a heat exchanger from 430 °C to 220 °C. What heat flow does the exhaust gas give off if the internal friction power can be neglected?

Using Eqs. (4.211) and (2.96) one obtains

$$(\dot{Q}_{12})_{rev} = \dot{m}\, c_p (t_2-t_1) = \dot{V}_0\, C_p(t_2-t_1) = \dot{V}_0 \sum_{i=1}^{n} r_i\, \varrho_{0i} \left(c_{pi}|_{0°C}^{t_2}\cdot t_2 - c_{pi}|_{0°C}^{t_1}\cdot t_1\right)(t_2-t_1) = \dot{V}_0 \sum_{i=1}^{n} X_i$$

Gas	r_i	ϱ_{0i}	$c_{pi}|_{0°C}^{220°C}$	$c_{pi}|_{0°C}^{430°C}$	X_i
		kg/m^3	kJ/(kg K)	kJ/(kg K)	kJ/m^3
H_2O	0.066	0.804	1.897	1.955	-22.463
CO_2	0.113	1.965	0.920	0.994	-49.964
O_2	0.024	1.428	0.938	0.969	-7.208
CO	0.059	1.250	1.048	1.067	-16.833
N_2	0.738	1.251	1.044	1.058	-207.969
					-304.437

$$(\dot{Q}_{12})_{rev} = \dot{V}_0 \sum_{i=1}^{n} X_i = 1300\, \frac{m^3}{h} \cdot \frac{1\,h}{3600\,s} \cdot (-304.437)\, \frac{kJ}{m^3} \cdot \frac{kW\,s}{kJ} = -109.936\, kW$$

4.5.4 The Mixing Process in the Open System

Gas flows that are mixed together form an open system (Fig. 4.33). Let the volume flows of the individual gases $\dot{V}_i$ have the temperatures T_i and the pressures p_i. Then, in the stationary operating condition, the following applies to the enthalpy flows of the individual gases before mixing:

$$\sum_{i=1}^{n} \dot{H}_i = \sum_{i=1}^{n} (\dot{m}_i\, h_i)_{T_i} = \sum_{i=1}^{n} \dot{m}_i\, c_{pi}\, T_i \tag{4.212}$$

After the mixture, using Eq. (4.164) the total enthalpy flow becomes

$$\dot{H} = \sum_{i=1}^{n} (\dot{m}_i\, h_i)_T = \sum_{i=1}^{n} \dot{m}_i\, c_{pi}\, T\,, \tag{4.213}$$

where T is the temperature of the mixture. If neither heat nor work exceeds the system boundary of the mixture device ($\dot{Q}_{12} = 0$, $P_e = 0$, $P_{RA} = 0$) and the changes of the kinetic and potential energy are negligible, according to Eq. (2.77) $\sum_{i=1}^{n} \dot{H}_i = \dot{H}$ is valid. Then, using Eqs. (4.212) and (4.213), it can be written

$$\sum_{i=1}^{n} \dot{m}_i c_{pi} T_i = \sum_{i=1}^{n} \dot{m}_i c_{pi} T . \qquad (4.214)$$

For the mixing temperature, Eq. (4.214) gives the relation

$$T = \frac{\sum_{i=1}^{n} \dot{m}_i c_{pi} T_i}{\sum_{i=1}^{n} \dot{m}_i c_{pi}} . \qquad (4.215)$$

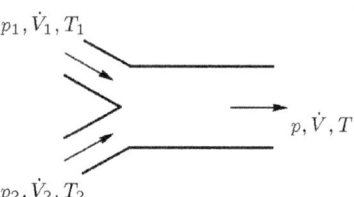

Figure 4.33 Mixing of two individual gas flows in the open system

For each individual gas, before mixing, the thermal equation of state is

$$p_i \dot{V}_i = \dot{m}_i R_i T_i . \qquad (4.216)$$

After the mixing process, the following applies analogously to Eq. (4.155)

$$p \dot{V} = \dot{m} R T . \qquad (4.217)$$

For the mass flows, Eq. (4.154) applies

$$\dot{m} R = \sum_{i=1}^{n} \dot{m}_i R_i . \qquad (4.218)$$

Using Eqs. (4.216) and (4.217), the volume flow of the mixture is

$$\dot{V} = \frac{T}{p} \dot{m} R = \frac{T}{p} \sum_{i=1}^{n} \dot{m}_i R_i = \frac{T}{p} \sum_{i=1}^{n} \frac{p_i \dot{V}_i}{T_i} . \qquad (4.219)$$

According to the laws of fluid mechanics, a pressure drop takes place in the pipes of the individual gases, which results from the friction resistances. The pressure p at the point of union is always lower than the pressures p_i.

Example 4.22 In a pipe with a circular cross-section of 450 mm diameter, at a temperature of 180 °C the velocity of an exhaust gas flow ($R_A = 293.88$ J/(kg K)) is measured to be 2.573 m/s at a gage pressure of 6300 Pa. The ambient air pressure is 1 bar. The exhaust gas flow is mixed with an air flow ($R_L = 287.06$ J/(kg K)), for which in a pipe with 300 mm diameter at a temperature of 20 °C and a gage pressure of 460 Pa, a velocity of 5.287 m/s is measured. At the point of union, the pressure is equal to the air pressure of 1 bar. What is the mixing temperature and the volume flow of the mixture?

First the mass flows are calculated:

For the exhaust gas flow one gets

$$\dot{m}_A = \frac{p_A \dot{V}_A}{R_A T_A} = \frac{p_A \pi D_A^2 c_A}{4 R_A T_A} = \frac{106\,300 \text{ N} \cdot \pi \text{ kg K} \cdot 0.450^2 \text{ m}^2 \cdot 2.573 \text{ m}}{\text{m}^2 \, 4 \cdot 293.88 \text{ J} \cdot 453.15 \text{ K s}} \cdot \frac{1 \text{ J}}{\text{N m}} = 0.327 \text{ kg/s}.$$

For the air flow we get

$$\dot{m}_L = \frac{p_L \dot{V}_L}{R_L T_L} = \frac{p_L \pi D_L^2 c_L}{4 R_L T_L} = \frac{100\,460 \text{ N} \cdot \pi \text{ kg K} \cdot 0.300^2 \text{ m}^2 \cdot 5.287 \text{ m}}{\text{m}^2 \, 4 \cdot 287.06 \text{ J} \cdot 293.15 \text{ K s}} \cdot \frac{1 \text{ J}}{\text{N m}} = 0.446 \text{ kg/s}.$$

To determine the specific heat capacities, the mixing temperature is first estimated to be 90 °C. From this it follows with Eq. (2.96) for the mean specific isobaric heat capacities:

$$c_{pA}|_{90\,°C}^{180\,°C} = 0.976 \text{ kJ/(kg K)} \qquad\qquad c_{pL}|_{20\,°C}^{90\,°C} = 1.004 \text{ kJ/(kg K)}$$

$$T = \frac{\sum_{n=1}^{n} \dot{m}_i c_{pi} T_i}{\sum_{n=1}^{n} \dot{m}_i c_{pi}}$$

$$= \frac{0.327 \text{ kg/s} \cdot 0.976 \text{ kJ/(kg K)} \cdot 453 \text{ K} + 0.446 \text{ kg/s} \cdot 1.004 \text{ kJ/(kg K)} \cdot 293 \text{ K}}{0.327 \text{ kg/s} \cdot 0.976 \text{ kJ/(kg K)} + 0.446 \text{ kg/s} \cdot 1.004 \text{ kJ/(kg K)}}$$

$$= 360 \text{ K}$$

$$\dot{V} = \frac{T}{p} \sum_{i=1}^{n} \dot{m}_i R_i$$

$$= \frac{360 \text{ K}}{1.0 \text{ bar}} \cdot \frac{\text{bar m}^3}{10^5 \text{ J}} \cdot (0.327 \frac{\text{kg}}{\text{s}} \cdot 293.88 \frac{\text{J}}{\text{kg K}} + 0.446 \frac{\text{kg}}{\text{s}} \cdot 287.06 \frac{\text{J}}{\text{kg K}}) = 0.807 \frac{\text{m}^3}{\text{s}}$$

Note that $\dot{V}$ does not result from the addition of $\dot{V}_A$ and $\dot{V}_L$, since
$\dot{V}_A + \dot{V}_L = 0.409 \text{ m}^3/\text{s} + 0.374 \text{ m}^3/\text{s} = 0.783 \text{ m}^3/\text{s}$.

4.6 Dynamics of Ideal Gases: Compressible Stationary Gas Flow

4.6.1 Introduction

In the following, a steady-state flow of an ideal gas in only one direction with variable velocity c is considered (one-dimensional gas flow), taking into account its compressibility. Thus, the density ϱ is variable along the length x, and the velocity is not only a function of the length x but also of the density: $c = c(x, \varrho)$. The density ϱ changes in a fluidic continuum depending on the pressure p, on the velocity c and on the temperature T.

The laws of gas dynamics enable the calculation of subsonic and supersonic flows in the blade grids of gas and steam turbines as well as axial and radial compressors, in fans, in special supersonic nozzles as well as in gas pipes and gas burners. Furthermore, outflow processes in pressure vessels – e.g. also in case of leakages – can be described. In addition, flow processes on aircraft wings and engines, on rockets and ballistic projectiles as well as on ground vehicles can be treated [47].

At very low pressures of $p \leq 0.1$ Pa (almost vacuum), the gas is no longer a continuum, but is to be understood as a free molecular flow according to the laws of gas kinetics.

Since in the continuity equation according to Eq. (2. 104) $\dot{m}_1 = \dot{m}_2 = \varrho_1 A_1 c_1 = \varrho_2 A_2 c_2$ the density ϱ is linked to the velocity c and to the flow cross-section A, the change of a state variable or the change of the flow cross-section also results in the change of the respective state variable $\varrho(c, A), c(\varrho, A), A(\varrho, c)$:

$$\mathrm{d}\dot{m} = \mathrm{d}(\varrho \dot{V}) = \mathrm{d}(\varrho A c) \tag{4.220}$$

The change of state of the ideal gas occurring in this process can be captured with the equations of thermodynamics in order to establish the conservation laws of compressible gas flow in this way. With Eq. (4.21) $v = 1/\varrho$ and Eq. (3.77) the specific internal energy is $\mathrm{d}u = T\,\mathrm{d}s - p\,\mathrm{d}v = T\,\mathrm{d}s - p\,\mathrm{d}(1/\varrho)$ and the specific enthalpy is $\mathrm{d}h = T\,\mathrm{d}s + v\,\mathrm{d}p = T\,\mathrm{d}s + (1/\varrho)\,\mathrm{d}p$. Further, the thermal equation of state of the ideal gas $\varrho = p/(RT)$ (Eq. (4.27)) is relevant.

The *Bernoulli* equation (2.106) is valid for incompressible flows; this is extended to adiabatic flows of compressible gas in section 4.6.3. It should be noted that although reversible adiabatic flow is isentropic flow, isentropic flow can be both reversible and irreversible, i.e. the distinguishable cases of frictionless adiabatic flow as well as frictional flow with the property $q_{12} = -|w_{RI}|$ can occur; here the relationships explained in section 3.5.1 should be noted.

In the following considerations, the rest state 0 (also called resting state 0 or stagnation state 0) at $c = c_0 = 0$ is represented by p_0, T_0 and ϱ_0.

In addition, the concept of a fluidically "critical" state is introduced, whereby this is not to be confused with the critical state in the sense of the critical point as a material property. This fluidically "critical" state is characterised by p^*, T^*, ϱ^*, M^* at $c^* = a^*$, where a is the velocity of sound according to section 4.6.2.

In addition, the state variables after a compression shock (also called only shock or compaction shock) are denoted by $\hat{p}$, $\hat{T}$, $\hat{\varrho}$, $\hat{M}$, where M is the *Mach*[6] number according to Section 4.6.2.

---
[6] *Ernst Mach*, 1838 to 1916, Austrian physicist and philosopher

112 4 Ideal Gases

4.6.2 Velocity of Sound and Propagation of Sound

Velocity of Sound

Sound is characterised by small pressure fluctuations in the range of mPa (Fig. 4.34); it propagates in an elastic continuum in the form of longitudinal waves with the velocity of sound a.

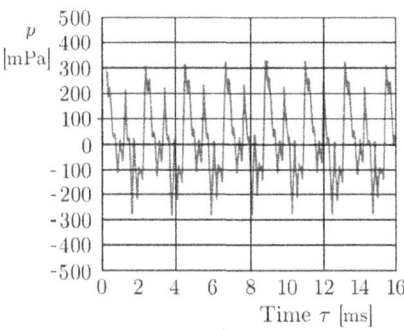

Figure 4.34 Sound: Pressure oscillation of a violin tone

Because of the connection of the pressure p with the density ϱ and the velocity c, the pressure fluctuations lead to fluctuations of the density and the velocity. If the damping of the small pressure fluctuations by friction is neglected and it is taken into account that with respect to the environment no heat emission or supply takes place, the process can be treated as frictionless and adiabatic – i.e. isentropic.

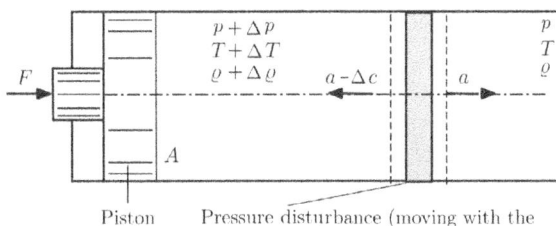

Figure 4.35 Propagation of a pressure disturbance in the fluid of a cylinder

The gas flow in the co-moving control space shown in Fig. 4.35 can be represented as follows, taking into account the continuity equation along a streamline at constant flow cross-section $A = $ const:

$$\dot{m} = \varrho \dot{V} = \varrho c A = \text{const} \qquad (4.221)$$

The flow density ϱc is for $A = $ const

$$\frac{\dot{m}}{A} = \varrho c = \text{const} . \qquad (4.222)$$

Eq. (4.221) can also be formulated in logarithmic form:

$$\ln \dot{m} = \ln(\varrho c A) = \ln \text{const} = \ln \varrho + \ln c + \ln A \qquad (4.223)$$

With the derivatives $d(\ln \varrho)/d\varrho = 1/\varrho$, $d(\ln c)/dc = 1/c$ and $d(\ln A)/dA = 1/A$, the differential form of Eq. (4.223)

$$d(\ln \varrho + \ln c + \ln A) = d(\ln \varrho) + d(\ln c) + d(\ln A) = 0$$

leads to the continuity equation in its differential form:

$$\frac{d\varrho}{\varrho} + \frac{dc}{c} + \frac{dA}{A} = 0 \qquad (4.224)$$

4.6 Dynamics of Ideal Gases: Compressible Stationary Gas Flow

After differentiation of Eq. (4.222) one gets:
$$\varrho\, dc + c\, d\varrho = 0 \quad \text{resp.} \quad c\, d\varrho = -\varrho\, dc \quad (4.225)$$
The momentum set along the horizontal streamline in Figure 4.35 is with the force F and the acceleration b
$$F + pA = \text{const} = mb + pA = \text{const} = \dot{m}c + pA = \text{const}. \quad (4.226)$$
After dividing by A corresponding to $A = \dot{m}/\varrho c$, this gives Eq. (4.227):
$$\varrho c^2 + p = \text{const} \quad (4.227)$$
By differentiation of this equation it follows:
$$d(\varrho c^2) + dp = 2\varrho c\, dc + c^2\, d\varrho + dp = 2\varrho c\, dc + cc\, d\varrho + dp = 0 \quad (4.228)$$
Substituting Eq. (4.225) $c\, d\varrho = -\varrho\, dc$ gives the equations
$$\varrho c\, dc + dp = 0 \quad (\textit{Euler's equation of motion})^7 \quad (4.229)$$
and
$$-c^2\, d\varrho + dp = 0 \quad \text{or after division by } d\varrho \quad c^2 = a^2 = \frac{dp}{d\varrho}. \quad (4.230)$$

For the assumed isentropic change of state, for the propagation velocity c of the pressure perturbation, denoted as the velocity of sound a, Eq. (4.230) also can be written as follows:
$$a^2 = \left(\frac{\partial p}{\partial \varrho}\right)_s \quad (4.231)$$
The isentropic equation according to Eq. (4.95) $pv^\kappa = \text{const} = p/\varrho^\kappa = \text{const}$ assumes $\kappa = \text{const}$. It can be written after a logarithmisation on both sides as
$$\ln(p/\varrho^\kappa) = \ln p - \ln \varrho^\kappa = \ln p - \kappa \ln \varrho = \ln \text{const}.$$
The differential form $d(\ln p) - d(\kappa \ln \varrho) = d(\ln p) - \kappa\, d(\ln \varrho)$ with $d(\ln p) = dp/p$ as well as $d(\ln \varrho) = d\varrho/\varrho$ leads to
$$\frac{dp}{p} - \kappa \frac{d\varrho}{\varrho} = 0 \quad \text{resp.} \quad \frac{dp}{d\varrho} = \left(\frac{\partial p}{\partial \varrho}\right)_s = \kappa \frac{p}{\varrho}. \quad (4.232)$$
With the thermal equation of state of the ideal gases $p/\varrho = RT$, at constant isentropic exponent κ or constant specific heat capacities c_p and c_v, respectively, it follows from Eqs. (4.231) and (4.232) an equation for the calculation of the velocity of sound a for ideal gases with isentropic propagation [47]:
$$a = \sqrt{\left(\frac{\partial p}{\partial \varrho}\right)_s} = \sqrt{\kappa \frac{p}{\varrho}} = \sqrt{\kappa RT} \quad (4.233)$$
The velocity of sound a can also be given for – reversibly and elastically behaving – solids for which the law of $Hooke^8$ $d\sigma_d = dp = E\, d\varrho/\varrho$ holds and the value of the elastic modulus E is known [47]
$$a = \sqrt{\left(\frac{\partial p}{\partial \varrho}\right)_s} = \sqrt{\frac{E}{\varrho}}. \quad (4.234)$$

Propagation of Sound
Sound sources generally emit a periodic sequence of pressure fluctuations that propagate spherically in space in the form of waves of sound. The particles excited by a wave of sound oscillate around their invariable location (Fig. 4.36).

---
[7] *Leonhard Euler*, 1707 to 1783, was a Swiss mathematician, physicist, astronomer, logician and engineer, whose work led to outstanding scientific discoveries.
[8] *Robert Hooke*, 1635 to 1703, English scientist

The sound source (interference source) can be stationary or moving - for example, if it is in a moving motor vehicle. Consequently, waves of sound propagate differently, with respect to the amount of movement of the disturbance source.

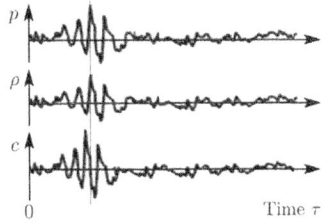

Figure 4.36 Pressure, density and velocity perturbation in a pipe flow

Figure 4.37 shows the cases $c = 0$, $c < a$, $c = a$ and $c > a$. The quotient of velocity c and velocity of sound propagation a is called *Mach* number

$$M = c/a \ . \tag{4.235}$$

Depending on the velocity of the pressure disturbance c or the incident flow velocity of the pressure disturbance, the *Mach* number has values from $M = 0$ (fluid at rest) to $M = \infty$ (at $c = c_{max}$). This results in regions of

$M = 0$: fluid at rest: aerostatics and thermodynamics
$M < 1$: subsonic flow
$M = 1$: sonic flow, transonic flow
$M > 1$: supersonic flow
$M > 5$: hypersonic flow

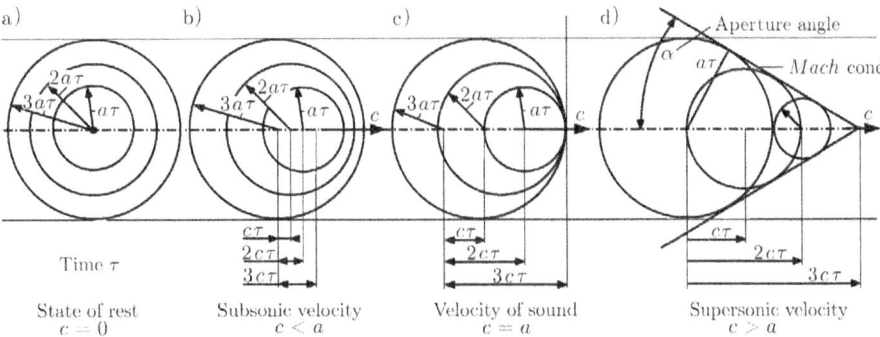

Figure 4.37 Propagation of pressure disturbances at different velocities of the disturbance source. The disturbance source travels the path $s = c\tau$ and the wave of sound travels the path $s = a\tau$ in time τ [47].
a) Propagation in the gas at rest: state of rest denoted by 0
b) The disturbance source moves with subsonic velocity $c < a$.
c) The disturbance source moves with velocity of sound $c = a$.
d) The disturbance source moves with supersonic velocity $c > a$.

The sound propagation lines in Fig. 4.37 are spherical surfaces which, for a supersonic flow with $M > 1.0$ in Fig. 4.37 d, are surrounded by the *Mach* cone. This cone is the more stretched, the higher the velocity of motion of the disturbance source c is, i.e. the larger the *Mach* number is. Half the aperture angle α of the *Mach* cone is

$$\sin \alpha = \frac{a\tau}{c\tau} = \frac{a}{c} = \frac{1}{M} \ , \quad \text{also it is} \quad M = \frac{c}{a} = \frac{1}{\sin \alpha} \ . \tag{4.236}$$

Example 4.23 A military aircraft flies horizontally over an observer or listener at an altitude of $z = 11$ km at an air temperature of $-56\,°C$ with the *Mach* number $M = 1.5$. The speed of the aircraft and *Mach*'s aperture angle are to be calculated. What time has elapsed before the observer perceives the sonic boom and what is the distance of the aircraft measured horizontally from the observer? The specific gas constant of air is $R = 0.2872$ kJ/(kg K). The isentropic exponent is $\kappa = 1.4$.

Sound velocity according to Eq. (4.233):

$$a = \sqrt{\kappa R T} = \sqrt{1.4 \cdot 0.2872 \cdot 217.15 \, \frac{\text{kJ K}}{\text{kg K}} \, \frac{1000\,\text{N m}}{\text{kJ}} \, \frac{\text{kg m}}{\text{N s}^2}} = 295.49 \, \frac{\text{m}}{\text{s}}$$

Mach's aperture angle according to Eq. (4.236):

$$\sin \alpha = \frac{1}{M} = 1/1.5 = 0.66667 \quad \text{respective} \quad \alpha = 41.81°$$

Speed of the aircraft according to Eq. (4.235):

$$c = M a = 1.5 \cdot 295.49 \, \frac{\text{m}}{\text{s}} = 443.23 \, \frac{\text{m}}{\text{s}} = 1595.63 \, \frac{\text{km}}{\text{h}}$$

Horizontal distance of the aircraft from the observer:

$$s = \frac{z}{\tan \alpha} = \frac{11\,\text{km}}{0.8944} = 12.299\,\text{km}$$

Time:

$$\tau = \frac{s}{c} = \frac{12299\,\text{m s}}{443.23\,\text{m}} = 27.75\,\text{s}$$

4.6.3 Energy Equation and *Bernoulli* Equation of Compressible One-Dimensional Ideal Gas Flow

The *Euler* equation of motion (4.229) was derived in Section 4.6.2 for horizontal steady flow. Because there are no restrictive statements about the density ϱ, it is also valid for the compressible gas flow. If the dependence on the height z is also taken into account, it takes the following form:

$$c\,dc + \frac{dp}{\varrho} + g\,dz = 0 \qquad (4.237\,\text{a})$$

For the further considerations, a reasonable assumption has to be made about the type of change of state of the ideal gas; here, the reversible-adiabatic – i.e. isentropic – change of state is chosen as a special case of the general polytropic change of state, in which heat is neither added to nor removed from the ideal gas.

Because in gas-dynamics, gas density is small, the height coordinate z has only a very small influence in gas-dynamic considerations, if meteorological processes with large height differences can be neglected, Eq. (4.237 a) simplifies to

$$c\,dc + \frac{dp}{\varrho} = 0 \qquad (4.237\,\text{b})$$

If Eq. (4.237 b) is integrated, the *Bernoulli* equation of gas dynamics is obtained. This has – starting from an initial point 0 and leading to a variable state – the following form:

$$\int_{c_0}^{c} c\,dc + \int_{p_0}^{p} \frac{dp}{\varrho} = 0 \qquad (4.238)$$

Assuming 0 as the rest state for an isentropic change of state, the pressure ratio can be converted to

$$\frac{p}{p_0} = \left(\frac{\varrho}{\varrho_0}\right)^\kappa \quad \text{resp.} \quad \frac{1}{\varrho} = \frac{1}{\varrho_0}\left(\frac{p_0}{p}\right)^{\frac{1}{\kappa}}. \tag{4.239}$$

This relation is inserted into Eq. (4.238) and integration is performed on this:

$$\int_{c_0}^{c} c\, dc + \frac{1}{\varrho_0} \int_{p_0}^{p} \left(\frac{p_0}{p}\right)^{-\frac{1}{\kappa}} dp = \frac{c^2}{2} - \frac{c_0^2}{2} + \frac{\kappa}{\kappa-1}\frac{p_0}{\varrho_0}\left(\frac{p_0}{p}\right)^{\frac{\kappa-1}{\kappa}} - \frac{\kappa}{\kappa-1}\frac{p_0}{\varrho_0} = 0 \tag{4.240}$$

This equation is applied to the outflow of an ideal gas from a container, where in the container the quiescent state of the gas is denoted by p_0, ϱ_0, T_0 (cf. Figure 4.43 in Example 4.24 below). There, the velocity is $c_0 = 0$; the state at the outflow opening is denoted by p_2, ϱ_2, T_2. Thus it follows Eq. (4.241):

$$\frac{c_2^2}{2} = -\frac{\kappa}{\kappa-1}\frac{p_0}{\varrho_0}\left(\frac{p_2}{p_0}\right)^{\frac{\kappa-1}{\kappa}} + \frac{\kappa}{\kappa-1}\frac{p_0}{\varrho_0} = \frac{\kappa}{\kappa-1}\frac{p_0}{\varrho_0}\left[1 - \left(\frac{p_2}{p_0}\right)^{\frac{\kappa-1}{\kappa}}\right]. \tag{4.241}$$

The outflow velocity from the container thus becomes

$$c_2 = \left[\frac{2\kappa}{\kappa-1}\frac{p_0}{\varrho_0}\left[1 - \left(\frac{p_2}{p_0}\right)^{\frac{\kappa-1}{\kappa}}\right]\right]^{\frac{1}{2}}. \tag{4.242}$$

Eq. (4.242) illustrates that the maximum outflow velocity is reached at $p_2 = 0$, i.e. when the ideal gas flows out of the pressure tank into an absolute vacuum. Here the maximum velocity is

$$c_{max} = \left[\frac{2\kappa}{\kappa-1}\frac{p_0}{\varrho_0}\right]^{\frac{1}{2}} = \left[\frac{2\kappa}{\kappa-1}RT_0\right]^{\frac{1}{2}} \quad c_{max} = \left[\frac{2}{\kappa-1}a_0^2\right]^{\frac{1}{2}} = \left[2c_pT_0\right]^{\frac{1}{2}} \tag{4.243}$$

making use of the thermal equation of state (4.27) $p_0/\varrho_0 = RT_0$ and equation (4.49) $c_p = \kappa R/(\kappa-1)$ in the transformation [47]. Eq. (4.243) expresses that the total specific energy contained in the tank $c_p T_0 = (\kappa p_0/[(\kappa-1)\varrho_0])$ compared to the vacuum is converted into the specific kinetic energy $c_{max}^2/2$ of the velocity c_{max}.

The specific energy in the tank is given by the specific enthalpy of rest $h_0 = c_p T_0$ of the gas in the tank. It follows from this that the specific energy equation for compressible gas flows according to Eq. (4.241) with the specific enthalpy $dh = Tds + dp/\varrho = du + d(p/\varrho)$ can also be given in the following form:

$$u_1 + \frac{p_1}{\varrho_1} + \frac{c_1^2}{2} = u_2 + \frac{p_2}{\varrho_2} + \frac{c_2^2}{2} \tag{4.244}$$

With the specific enthalpy $h = u + pv = u + p/\varrho$, it becomes

$$h_1 + \frac{c_1^2}{2} = h_2 + \frac{c_2^2}{2}, \tag{4.245}$$

where h is the specific enthalpy of the flow filament for which holds: $h = c_p T = (c_p/R)(p/\varrho) = (\kappa/(\kappa-1))(p/\varrho)$.

The sum of the specific enthalpy h and the specific kinetic energy $c^2/2$ is also called specific total enthalpy $h_t = h + c^2/2$ (cf. Eq. (6.46)), which during a flow process on the flow filament remains constant.

Eq. (4.245) also illustrates that the specific energy equation of gas flow applies to both reversible-adiabatic flows and frictional flows.

The expression $\kappa p / \varrho$ represents the square of the velocity of sound a^2 according to Eq. (4.233). This gives the relation $h = (\kappa /(\kappa - 1))(p/\varrho) = a^2/(\kappa - 1)$ for h, which is inserted into Eq. (4.245):

$$\frac{a_1{}^2}{\kappa - 1} + \frac{c_1{}^2}{2} = \frac{a_2{}^2}{\kappa - 1} + \frac{c_2{}^2}{2} \qquad (4.246)$$

Eq. (4.246) represents the specific energy equation for a jet flow or nozzle flow (cf. Fig. 4.38). If the *Mach* number $M = c/a$ is used, this specific energy equation becomes

$$\frac{a_1{}^2}{\kappa - 1}\left[1 + \frac{\kappa - 1}{2}M_1{}^2\right] = \frac{a_2{}^2}{\kappa - 1}\left[1 + \frac{\kappa - 1}{2}M_2{}^2\right]. \qquad (4.247)$$

If this is used to calculate outflows from tanks, $c_0 = 0$ is set. Further, the resting velocity of sound $a_0 = \sqrt{\kappa R T_0}$ and the square of the resting *Mach* number $M_0{}^2 = c_0{}^2/a_0{}^2 = 0$ are inserted into Eq. (4.247). From this one gets

$$\frac{a_0{}^2}{\kappa - 1} = \frac{\kappa}{\kappa - 1}\frac{p_0}{\varrho_0} = c_p T_0 = \frac{a_2{}^2}{\kappa - 1}\left[1 + \frac{\kappa - 1}{2}M_2{}^2\right]. \qquad (4.248)$$

This can be used to give the exit velocity of sound a_2 and the exit *Mach* number M_2 at the exit port of a tank.

To determine the effect of a gas flow on the temperature of bodies surrounded by gas, the specific enthalpy $h = c_p T$ is used in Eq. (4.245), from which it follows after a transformation:

$$T_1 + \frac{c_1{}^2}{2 c_p} = T_2 + \frac{c_2{}^2}{2 c_p} \qquad (4.249)$$

Eq. (4.249) shows that a gas flow cools down when it is accelerated; for example, in the case of a gas containing water vapor, unwanted icing may occur. Conversely, the temperature rises when a gas flow is decelerated. If, for example, it is isentropically decelerated to $c_2 = 0$ at the stagnation point of a body, the expression $c_1^2/(2c_p)$ reflects the temperature increase occurring there [47].

Eqs. (4.244) to (4.249) are equivalent specific energy equations of gas dynamics. The *Bernoulli* equation can also be represented in the following form:

$$\frac{a^2}{\kappa - 1} + \frac{c^2}{2} = \text{const} \qquad (4.250)$$

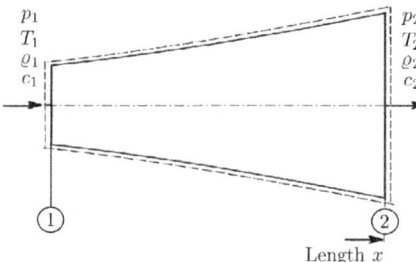

Figure 4.38 System boundaries of a jet flow or a nozzle flow

4 Ideal Gases

If the velocity c reaches the local velocity of sound a, it is called the critical velocity; this also defines the critical $Mach$ number $M^* = 1.0$. This is obtained at the critical velocity of $c^* = a^*$, which is equal to the critical velocity of sound a^*.

If the general representation of Eq. (4.246)

$$\frac{a^2}{\kappa - 1} + \frac{c^2}{2} = \text{const}$$

is to be applied to the rest state – say, in a tank with $c_0 = 0$ – as well as to the flow state in an exit orifice with velocity c, this gives

$$\frac{a_0^2}{\kappa - 1} = \frac{a^2}{\kappa - 1} + \frac{c^2}{2}. \tag{4.251}$$

If the ideal gas flows into a container with the absolute vacuum of $p = 0$ and the absolute temperature $T = 0$, then the outflow velocity of sound is $a = \sqrt{\kappa RT} = 0$. Therefore, according to Eq. (4.251), it follows $a_0^2/(\kappa - 1) = c^2/2$, and the velocity reaches its maximum value $c = c_{max}$. This has been described by Eq. (4.243), which is confirmed: $c_{max} = \sqrt{2 a_0^2/(\kappa - 1)}$.

If air has the state variables $t = 20\,°C$ as well as $\kappa = 1.40$ and thus $a_0 = 343.32$ m/s and flows into an absolute vacuum, the theoretically achievable maximum outflow velocity results in $c_{max} = 767.69$ m/s $= 2763.69$ km/h. Since an absolute vacuum with $p = 0$ and $T = 0$ cannot be realised, this velocity is practically unattainable. In hypersonic wind tunnels, a maximum speed of $c_{max} = 670$ m/s $= 2412$ km/h has been achieved so far [47].

For the critical state $c^* = a^*$ and $a = a^*$, Eq. (4.252) gives the specific energy equation:

$$\frac{a_0^2}{\kappa - 1} = \frac{a^{*2}}{(\kappa - 1)} + \frac{a^{*2}}{2} = \frac{2 a^{*2}}{2(\kappa - 1)} + \frac{(\kappa - 1) a^{*2}}{2(\kappa - 1)} = \frac{2 a^{*2} - a^{*2} + \kappa a^{*2}}{2(\kappa - 1)} = \frac{1}{2} \frac{(\kappa + 1)}{(\kappa - 1)} a^{*2} \tag{4.252}$$

It follows:

$$a^{*2} = 2 a_0^2/(\kappa + 1) \tag{4.253 a}$$

If one forms the quotient of a state variable of the gas flow at any point on the streamline (e.g. a, p, ϱ, T) to the corresponding rest quantity of the streamline (e.g. a_0, p_0, ϱ_0, T_0), then this quotient depends only on the substance – represented by κ – and on the $Mach$ number at the point under consideration. This will be discussed in the following.

4.6.4 Stagnation State Variables and Critical State

When considering the outflow from a tank (cf. Fig. 4.39), Eq. (4.248), applied to the rest state 0 (also called the stagnation state 0) and to the general outflow state, results in

$$\frac{a_0^2}{\kappa - 1} = \frac{a^2}{\kappa - 1}\left[1 + \frac{\kappa - 1}{2}\left(\frac{c}{a}\right)^2\right]. \tag{4.254}$$

According to Eq. (4.233), with $a^2 = \kappa RT$ and $a_0^2/a^2 = T_0/T$ for the quotient of the velocities of sound, Eq. (4.254) becomes

$$\left(\frac{a_0}{a}\right)^2 = \frac{T_0}{T} = 1 + \frac{\kappa - 1}{2} M^2. \tag{4.255}$$

4.6 Dynamics of Ideal Gases: Compressible Stationary Gas Flow

With the isentropic equations $p_0/p = (T_0/T)^{\frac{\kappa}{\kappa-1}}$ and $\varrho_0/\varrho = (T_0/T)^{\frac{1}{\kappa-1}}$, the ratios of the rest variables (stagnation variables) for the pressure p_0 and the density ϱ_0 related to the local variables p and ϱ can be determined:

$$\frac{p_0}{p} = \left(\frac{T_0}{T}\right)^{\frac{\kappa}{\kappa-1}} = \left[1 + \frac{\kappa-1}{2} M^2\right]^{\frac{\kappa}{\kappa-1}} \quad (4.256)$$

$$\frac{\varrho_0}{\varrho} = \left(\frac{T_0}{T}\right)^{\frac{1}{\kappa-1}} = \left[1 + \frac{\kappa-1}{2} M^2\right]^{\frac{1}{\kappa-1}} \quad (4.257)$$

From Eq. (4.257) it can be seen that the density reduction $\Delta\varrho = \varrho_0 - \varrho$ of a flowing ideal gas becomes stronger the greater is the velocity c or the *Mach* number of the flow.

For practical purposes, it is of interest which error results when the density change is not taken into account for gas flow velocities of about $c = 10$ m/s to 70 m/s. The error is given in Table 4.6, which lists the errors of the related density and pressure changes in air depending on the *Mach* number in the range of $M = 0.1$ to 1.0 or depending on the velocity c in the range of 34.32 m/s to 343.26 m/s [47].

Table 4.6 Error size of density and pressure in calculations for air at $p = 1.0$ bar, $T = 293.16$ K, $R = 0.2876$ kJ/(kg K) as an incompressible gas [47]

Mach number M	0,1	0,2	0,4	0,5	0,8	1,0
Velocity c m/s	34.32	68.64	137.42	205.92	274.56	343.26
Density change $\Delta\varrho/\varrho_0$	0.005	0.020	0.076	0.160	0.260	0.366
Pressure change $\Delta p/p_0$	0.007	0.028	0.104	0.216	0.344	0.472

From Table 4.6 it can be concluded that, for example, fans with low total pressure ratios up to $p_{tD}/p_{tS} = 1.18$ and low velocities up to $c = 40$ m/s can be designed without consideration of the density changes — i.e. with neglect of the compressibility. On the other hand, compressibility must be taken into account when designing high-pressure fans with total pressure ratios of $p_{tD}/p_{tS} > 1.20$ [47].

The stagnation variables of a fluid, like the fluidic critical variables (cf. e.g. Eq. (4.253 a)), are characteristic for the value of the constant on a flow filament. Therefore, the ratio of these two constants again results in a constant value. This has the same value for all streamlines, i.e. it is independent of the flow and depends solely on the substance size κ of the fluid.

The following Eqs. (4.258) to (4.260), which are derived with the help of Eq. (4.253 b) → Eq. (4.258) for the stagnation state (resting state) and for the critical state, give the ratio values in the fluidic critical state and in the resting state:

Sound velocity ratio: $\quad \left(\dfrac{a^*}{a_0}\right)^2 = \dfrac{2}{\kappa+1} \quad$ (4.253 b)

Temperature ratio: $\quad \left(\dfrac{T^*}{T_0}\right) = \left(\dfrac{a^*}{a_0}\right)^2 = \dfrac{2}{\kappa+1} \quad$ (4.258)

Pressure ratio: $\quad \left(\dfrac{p^*}{p_0}\right) = \dfrac{2}{\kappa+1}^{\frac{\kappa}{\kappa-1}} \quad$ (4,259)

Density ratio: $\quad \left(\dfrac{\varrho^*}{\varrho_0}\right) = \dfrac{2}{\kappa+1}^{\frac{1}{\kappa-1}} \quad$ (4.260)

4 Ideal Gases

Table 4.7 Critical state variables of some gases and vapors [47]

Gas type	κ	$\kappa/(\kappa-1)$	p^*/p_0	ϱ^*/ϱ_0	T^*/T_0
Helium	1.66	2.515	0.488	0.649	0.752
N_2; O_2	1.40	3.50	0.528	0.634	0.833
Air	1.40	3.50	0.528	0.634	0.833
Superheated steam	1.33	4.030	0.540	0.629	0.858
Saturated steam	1.135	8.407	0.577	0.616	0.937

Table 4.7 contains critical ratio values for important ideal gases and vapors [47]. For example, for air and other diatomic ideal gases with $\kappa = 1.4$, the following values result: $p^*/p_0 = 0.528$; $\varrho^*/\varrho_0 = 0.634$ and $T^*/T_0 = 0.833$.

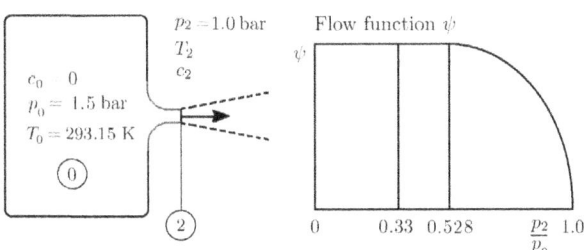

Figure 4.39 Outflow from a pressure tank

Example 4.24 From a pressure tank filled with air with $p_0 = 1{,}5$ bar, $R = 0.2872$ kJ/(kg K), $T_0 = 273.15$ K, $c_p = 1.004$ kJ/(kg K) and $\kappa = 1.40$ the outflow velocity c of air into the free atmosphere with $p_{amb} = p_2 = 1.0$ bar is to be calculated (see Fig. 4.43).

Air density in the tank:

$$\varrho_0 = \frac{p_0}{RT_0} = \frac{1.5 \text{ bar kg K}}{0.2872 \text{ kJ} \cdot 273.15 \text{ K}} \frac{10^5 \text{ N}}{\text{bar m}^2} \frac{\text{kJ}}{10^3 \text{ N m}} = 1.9121 \frac{\text{kg}}{\text{m}^3}$$

Pressure ratio from Table 4.7:

$$\frac{p_2}{p_0} = \frac{1.0 \, bar}{1.5 \, bar} = 0.6667 > \frac{p^*}{p_0} = 0.528 \quad \rightarrow \quad \text{subcritical outflow}$$

Outflow velocity (Eq. (4.242)):

$$c_2 = \left[\frac{2\kappa}{\kappa-1}\frac{p_0}{\varrho_0}\left[1-\left(\frac{p_2}{p_0}\right)^{\frac{\kappa-1}{\kappa}}\right]\right]^{\frac{1}{2}} = \left[\frac{2.8}{0.4}\frac{1.5 \text{ bar m}^3}{1.9121 \text{ kg}}\frac{10^5 \text{ N}}{\text{bar m}^2}\frac{\text{kg m}}{\text{N s}^2}\left[1-\left(\frac{1.0}{1.5}\right)^{\frac{0.4}{1.4}}\right]\right]^{\frac{1}{2}} = 245.089 \frac{\text{m}}{\text{s}}$$

Outlet temperature (Eq. (4.249)):

$$T_2 = T_0 - \frac{c_2{}^2}{2c_p} = 273.15 \text{ K} - \frac{245.089^2 \text{ m}^2}{2 \cdot 1.004 \text{ s}^2}\frac{\text{kg K}}{\text{kJ}}\frac{\text{kJ}}{1000 \text{ N m}}\frac{\text{N s}^2}{\text{kg m}}$$

$$= 273.15 \text{ K} - 29.91 \text{ K} = 243.24 \text{ K}$$

Outlet Mach number, Eq. (4.235):

$$M_2 = \frac{c_2}{a_2} = \frac{c_2}{\sqrt{\kappa R T_2}} = \frac{245.089 \frac{\text{m}}{\text{s}}}{\sqrt{1.4 \cdot 0.2872 \frac{\text{kJ}}{\text{kg K}}\frac{10^3 \text{ N m}}{\text{kJ}}\frac{\text{kg m}}{\text{N s}^2} \cdot 243.24 \text{ K}}} = 0.7837$$

The isentropic exponent κ of real gases can be plotted, for example, as a function of pressure and temperature. Fig. 4.40 shows the isentropic exponent of air; it becomes clear that at temperatures above about 273,15 K and pressures of up to about 5 bar, κ

becomes $\kappa \approx 1{,}4$. Therefore, with good approximation, ideal gas behaviour and $\kappa = 1{,}4$ can be assumed.

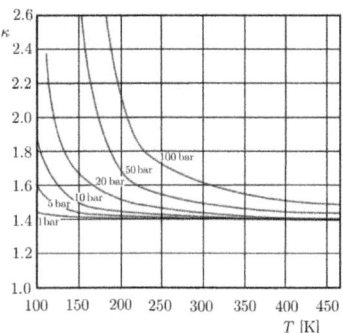

Figure 4.40 Isentropic exponent κ of air as a function of temperature and pressure [2]

4.6.5 The Velocity Diagram of the Specific Energy Equation

The specific energy equation according to (Eq. (4.251)) is valid for the rest state and for any flow state. If it is transformed and related to the velocity of sound a_0 as well as to the maximum velocity c_{max}, an elliptic equation with the coordinate sections $a_0 = \sqrt{\kappa R T_0} = \sqrt{\kappa p_0/\varrho_0}$ and $a_0 \sqrt{2/(\kappa - 1)}$ according to Eq. (4.261 a) follows:

$$\left(\frac{a}{a_0}\right)^2 + \left[\frac{c}{a_0 \sqrt{\frac{2}{\kappa - 1}}}\right]^2 = 1 \qquad (4.261\,\text{a})$$

The expression $a_0 \sqrt{2/(\kappa - 1)}$ is, according to Eq. (4.243), the maximum outflow velocity c_{max} from a pressure tank into an absolute vacuum. Thus, the elliptic equation can also be formulated as a specific energy equation in the following way:

$$\left(\frac{a}{a_0}\right)^2 + \left(\frac{c}{c_{max}}\right)^2 = 1 \qquad (4.261\,\text{b})$$

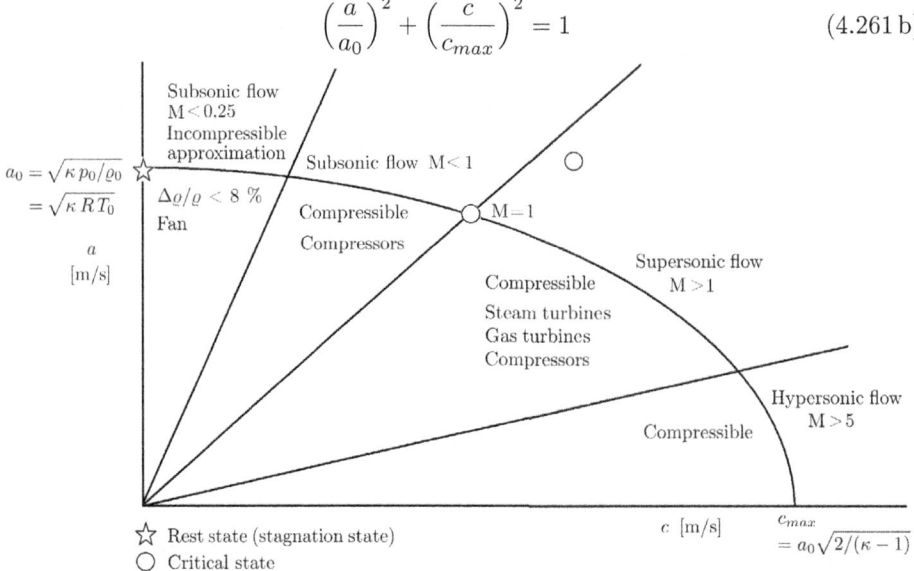

Figure 4.41 Velocity ellipse with all areas of compressible flow

The course of this ellipse in the first mathematical quadrant is given in Fig. 4.41, where the regions of compressible gas flow are shown.

Both the subsonic range as well as the supersonic range are covered with the following flow regions:

Subsonic flow:
- incompressible approximation with $M < 0.25$
- subsonic flow with $M < 1$
- critical flow with $M = M^* = 1$

Supersonic flow:
- supersonic flow with $M > 1$
- hypersonic flow with $M > 5$

Example 4.25 Air with specific isobaric heat capacity $c_p = 1.004 \text{ kJ}/(\text{kg K})$ and temperature $T = 283.15 \text{ K}$ flows around a body with velocity $c_1 = 125 \text{ m/s}$. At the stagnation point of the body being flowed around, where is $c_2 = 0 \text{ m/s}$, the temperature increase $\Delta T = T_2 - T_1$ and the temperature T_2 are to be determined.

Transforming the *Bernoulli* equation (Eq. (4.249)) gives:

$$\Delta T = T_2 - T_1 = \frac{c_2^2}{2 c_p} = \frac{125^2 \frac{\text{m}^2}{\text{s}^2}}{2 \cdot 1.004 \frac{\text{kJ}}{\text{kg K}} \frac{10^3 \text{ N m}}{\text{kJ}} \frac{\text{kg m}}{\text{N s}^2}} = 7.78 \text{ K}$$

$T_2 = 283.15 \text{ K} + 7.78 \text{ K} = 290.93 \text{ K}$ (17.78 °C)

4.6.6 Flow Function

The theoretical mass flow $\dot{m}$ exiting a pressure tank can be determined using the continuity equation $\dot{m} = \varrho c A$ by inserting the outflow velocity c_2 according to Eq. (4. 242) as well as the density ϱ_2 according to the transformed isentropic equation (Eq. (4.239)) for state 2 $\varrho_2 = \varrho_0 \left(\frac{p_2}{p_0}\right)^{\frac{1}{\kappa}}$:

$$\dot{m} = \varrho_2 c_2 A_2 = \left[\varrho_0^2 \left(\frac{p_2}{p_0}\right)^{\frac{2}{\kappa}}\right]^{\frac{1}{2}} c_2 A_2 = \left[\varrho_0^2 \left(\frac{p_2}{p_0}\right)^{\frac{2}{\kappa}}\right]^{\frac{1}{2}} \left[\frac{2\kappa}{\kappa - 1} \frac{p_0}{\varrho_0} \left[1 - \left(\frac{p_2}{p_0}\right)^{\frac{\kappa-1}{\kappa}}\right]\right]^{\frac{1}{2}} A_2 \quad (4.262)$$

Thus becomes, with the generalised pressure $p = p_2$,

$$\dot{m} = A_2 \left[\frac{2\kappa}{\kappa - 1} p_0 \varrho_0 \left[\left(\frac{p}{p_0}\right)^{\frac{2}{\kappa}} - \left(\frac{p}{p_0}\right)^{\frac{\kappa+1}{\kappa}}\right]\right]^{\frac{1}{2}}. \quad (4.263)$$

If this equation is represented by the form

$$\dot{m} = \psi A_2 \sqrt{2 p_0 \varrho_0}, \quad (4.264)$$

then the newly introduced quantity ψ in Eq. (4.264) can be written as

$$\psi = \left[\frac{\kappa}{\kappa - 1} \left[\left(\frac{p}{p_0}\right)^{\frac{2}{\kappa}} - \left(\frac{p}{p_0}\right)^{\frac{\kappa+1}{\kappa}}\right]\right]^{\frac{1}{2}}. \quad (4.265)$$

ψ is called the flow function or outflow function. If this is known, Eq. (4.264) can be used to determine the outflowing theoretical mass flow $\dot{m}$.

With regard to the consideration of irreversibilities during the outflow process, the jet contraction (see Fig. 4.46) with $\alpha = f(Re, (d_2/d_1)) = 0.60$ to 1.20 or the pressure loss coefficient ζ can be taken into account. Then the outflowing mass flow results in

$$\dot m = \alpha\,\psi\,A_2\sqrt{2\,p_0\,\varrho_0}\;. \qquad (4.266)$$

As Eq. (4.265) shows, the flow function ψ depends only on the isentropic exponent κ (i.e. on the type of gas) and on the pressure ratio p/p_0: $\psi = f\,(\kappa,\,p/p_0)$.

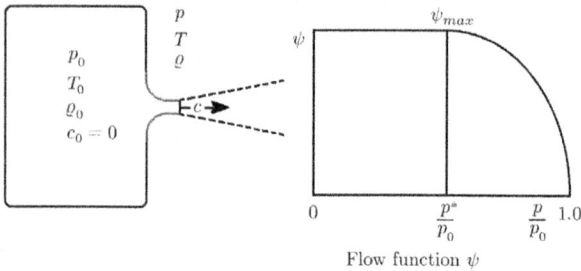

Figure 4.42 Tank with jet expansion (left); outflow function for supercritical outflow without supersonic nozzle (right)

If the pressure ratio p/p_0 decreases, the flow function ψ increases and reaches the critical pressure ratio p^*/p_0 as a maximum.

Afterwards, ψ decreases again because it does not have any dependence on the cross-sectional area A of the gas flow. The influence of A is found in the calculation of the mass flow $\dot m$ according to equation (4.264).

In reality, however, the achieved maximum value of the flow function ψ and also the achieved maximum mass flow $\dot m$ remains unchanged in the entire pressure range p/p_0 below the critical pressure ratios $(p/p_0) < (p^*/p_0)$ because the cross-sectional area A of the gas flow exiting the tank opening widens, as Figure 4.42 shows.

In Fig. 4.43, for two ideal gases and for two states of steam,, the flow function ψ, which remains constant below the critical pressure ratio p^*/p_0, is shown by a solid horizontal line, while the purely mathematical course of ψ is shown by a dashed line.

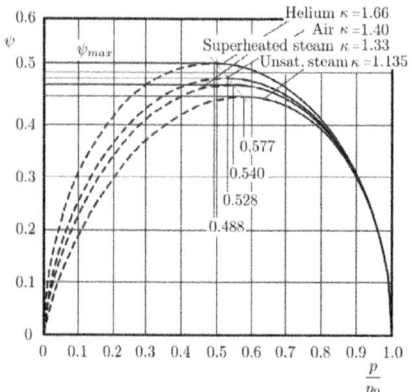

Figure 4.43 Flow function ψ for two ideal gases and two states of steam

As it will be shown in a subsequent section, the flow rate c, the flow function ψ and the mass flow rate $\dot m$ passing through can be further increased below the critical pressure ratio $(p/p_0) < (p^*/p_0)$ if a special nozzle – the supersonic nozzle (referred to as a Laval[9] nozzle) – is used for the outflow.

---

[9] *Carl Gustaf de Laval*, 1845 to 1913, Swedish engineer, one of the inventors of the steam turbine

The maximum value of the flow function ψ depends on the critical pressure ratio $p^*/p_0 = [2/(\kappa+1)]^{\frac{\kappa}{\kappa-1}}$ according to Eq. (4.259). This can be used to calculate the maximum value of ψ

$$\psi_{max} = \left(\frac{2}{\kappa+1}\right)^{\frac{1}{\kappa-1}} \sqrt{\frac{\kappa}{\kappa+1}}. \qquad (4.267)$$

The maximum theoretical mass flow $\dot{m}_{max}$ is thus given by

$$\dot{m}_{max} = \alpha\, A_2 \sqrt{2\, p_0\, \varrho_0}\left[\left(\frac{2}{\kappa+1}\right)^{\frac{1}{\kappa-1}} \sqrt{\frac{\kappa}{\kappa+1}}\right]. \qquad (4.268)$$

If the pressure ratio is less than the critical pressure ratio $(p/p_0 < p^*/p_0)$, then the jet expands and widens after exiting the orifice (Figure 4.46).

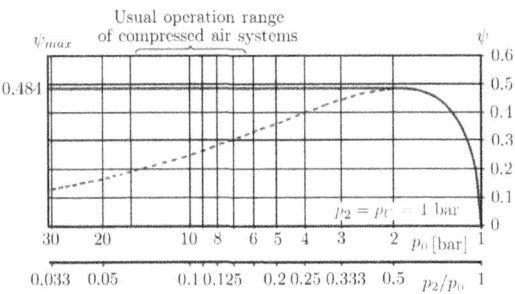

Figure 4.44 Outflow function ψ for air from a tank in logarithmic representation at constant external pressure $p_2 = p_U = 1.0$ bar

Fig. 4.44 shows the outflow function ψ for air from a pressure tank in logarithmic form. The usual operating range of compressed air systems with $p_0 \approx 6 \ldots 16$ bar is marked in the diagram. The external pressure is the ambient pressure of the air ($p_2 = p_U = 1$ bar).

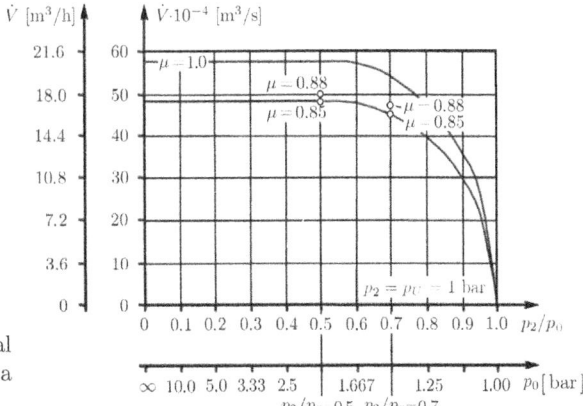

Figure 4.45 Theoretical and real outflow volume flow from a tank as a function of the pressure ratio p_2/p_0

Fig. 4.45 shows the theoretical and the real outflow volume flow of air $\dot{V}$ from a pressure tank for ratios of the real mass flow to the theoretical mass flow of $\mu = 1$, of $\mu = 0.88$ as well as of $\mu = 0.85$ as a function of the pressure ratio p_2/p_0. The external pressure is the ambient pressure of the air ($p_2 = p_U = 1$ bar).

4.6.7 Isentropic Gas Flow in Nozzles and Orifices

The outflow function ψ is also used to be able to handle the flow through orifices and nozzles (Figure 4.46). It can be used to measure, among other things, volume flows and mass flows in nozzles and orifices, which can be standardised for this purpose.

If the mass flow is to be determined precisely, the nozzle coefficient α or the jet contraction must also be taken into account. Fig. 4.46 illustrates the jet contraction for some orifices.

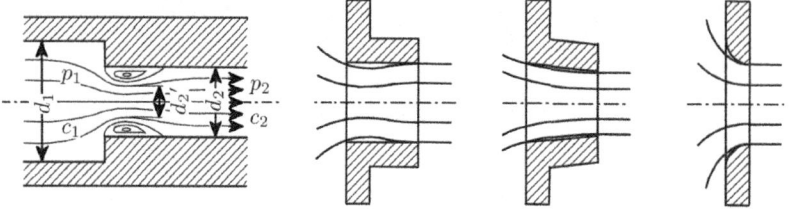

Figure 4.46 Jet contraction in different outlets and flow-throughs

The equation for the mass flow $\dot{m}$ is

$$\dot{m} = \alpha \psi A_2 \sqrt{2 p_0 \varrho_0} \ . \tag{4.269}$$

Example 4.26 Superheated water vapor, to be taken as an ideal gas with $R = 0.46152$ kJ/(kg K) and $\kappa = 1.33 = \text{const}$, is in a tank with constant absolute pressure $p_0 = 9.0$ bar and temperature $t_0 = 450\ °C$. It flows isentropically through an opening with the diameter d = 50 mm into a technical system with the absolute pressure $p_2 = 5.5$ bar, whereby the nozzle coefficient is to be set at $\alpha = 0.8$.

The mass flow and the maximum value of the flow function are to be determined. Furthermore, the steam temperature, the density and the velocity of the outlet flow, the local and the critical *Mach* number as well as the cooling of the steam during the outlet expansion shall be calculated.

For a subcritical flow, according to Table 4.7, must be:

$$\frac{p_2}{p_0} > \frac{p^*}{p_0} \qquad \frac{p_2}{p_0} = 0.61111 > 0.540 \quad \rightarrow \quad \text{subcritical flow}$$

Flow function, theoretical mass flow, real mass flow and maximum value of flow function:
At first, the density ϱ_0 has to be determined:

$$\varrho_0 = \frac{p_0}{R T_0} = \frac{9.0\ \text{bar kg K}}{0.46152\ \text{kJ} \cdot 723.15\ \text{K}} \cdot \frac{\text{kJ}}{10^3\ \text{N m}} \cdot \frac{10^5\ \text{N}}{\text{bar m}^2} = 2.69664\ \frac{\text{kg}}{\text{m}^3}$$

$$\psi = \left[\frac{\kappa}{\kappa-1}\left[\left(\frac{p_2}{p_0}\right)^{\frac{2}{\kappa}} - \left(\frac{p_2}{p_0}\right)^{\frac{\kappa+1}{\kappa}}\right]\right]^{\frac{1}{2}} = \left[\frac{1.33}{0.33}\left[\left(\frac{5.5}{9.0}\right)^{\frac{2}{1.33}} - \left(\frac{5.5}{9.0}\right)^{\frac{2.33}{1.33}}\right]\right]^{\frac{1}{2}} = 0.47011$$

$$\dot{m} = \psi A_2 \sqrt{2 \varrho_0 p_0}$$

$$= 0.47011 \cdot \frac{3.14159}{4} \cdot 0.050^2\ \text{m}^2 \sqrt{2 \cdot 2.69664\ \frac{\text{kg}}{\text{m}^3} \cdot 9.0\ \text{bar} \cdot \frac{10^5\ \text{N}}{\text{bar m}^2} \cdot \frac{\text{kg m}}{\text{N s}^2}} = 2.0336\ \frac{\text{kg}}{\text{s}}$$

$$\dot{m}_{re} = \alpha \psi A_2 \sqrt{2 \varrho_0 p_0} = \alpha \dot{m} = 0.8 \cdot 2.0336\ \frac{\text{kg}}{\text{s}} = 1.6269\ \frac{\text{kg}}{\text{s}}$$

$$\psi_{max} = \left(\frac{2}{\kappa+1}\right)^{\frac{1}{\kappa-1}} \sqrt{\frac{\kappa}{\kappa+1}} = \left(\frac{2}{2,33}\right)^{\frac{1}{0.33}} \cdot \sqrt{\frac{1.33}{2.33}} = 0.47561$$

Outlet temperature according to the isentropic equation (Eq. (4.256)):

$$T_2 = T_0 \left(\frac{p_2}{p_0}\right)^{\frac{\kappa-1}{\kappa}} = 723.15\ \text{K} \cdot 0.61111^{0.2481} = 639.97\ \text{K}$$

Vapor density at the outlet:

$$\varrho_2 = \frac{p_2}{RT_2} = \frac{5.5\,\text{bar}\,\text{kg}\,\text{K}}{0.46152\,\text{kJ} \cdot 639.97\,\text{K}} \frac{\text{kJ}}{10^3\,\text{N}\,\text{m}} \frac{10^5\,\text{N}}{\text{bar}\,\text{m}^2} = 1.862\,\frac{\text{kg}}{\text{m}^3}$$

Exit velocity of vapor from orifice, Eq. (4.249):

$$c_2 = \left[2\,c_p\,(T_0 - T_2)\right]^{\frac{1}{2}} = \left[\frac{2\,\kappa}{\kappa - 1} R\,(T_0 - T_2)\right]^{\frac{1}{2}}$$

$$c_2 = \left[\frac{2 \cdot 1.33}{0.33} 0.46152\,\frac{\text{kJ}}{\text{kg}\,\text{K}} \frac{10^3\,\text{N}\,\text{m}}{\text{kJ}} \frac{\text{kg}\,\text{m}}{\text{N}\,\text{s}^2} (723.15\,\text{K} - 639.97\,\text{K})\right]^{\frac{1}{2}} = 556.27\,\frac{\text{m}}{\text{s}}$$

Local *Mach* number according to Eq. (4.235):

$$M_2 = \frac{c_2}{a_2} = \frac{c_2}{\sqrt{\kappa\,R\,T_2}} = \frac{556.27\,\frac{\text{m}}{\text{s}}}{\sqrt{1.33 \cdot 0.46152\,\frac{\text{kJ}}{\text{kg}\,\text{K}} \frac{10^3\,\text{N}\,\text{m}}{\text{kJ}} \frac{\text{kg}\,\text{m}}{\text{N}\,\text{s}^2} \cdot 639.97\,\text{K}}} = 0.8875$$

Critical velocity of sound, Eq. (4.257):

$$a^* = \sqrt{\frac{2}{\kappa + 1}}\,a_0\quad\, ^2 = \sqrt{\frac{2\,\kappa}{\kappa + 1} R\,T_0}$$

$$a^* = \sqrt{\frac{2 \cdot 1.33}{1.33 + 1} \cdot 0.46152\,\frac{\text{kJ}}{\text{kg}\,\text{K}} \frac{10^3\,\text{N}\,\text{m}}{\text{kJ}} \frac{\text{kg}\,\text{m}}{\text{N}\,\text{s}^2} \cdot 639.97\,\text{K}} = 580.68\,\frac{\text{m}}{\text{s}}$$

Comparison with the flow-critical *Mach* number $M^* = 1$ (cf. Eq. (4.235)):

$$M = \frac{c_2}{a^*} = \frac{556.27\,\text{m/s}}{580.68\,\text{m/s}} = 0.958$$

Temperature drop of water vapor in isentropic expansion flow:

$$\Delta T = T_0 - T_2 = 723.125\,\text{K} - 639.97\,\text{K} = 83.18\,\text{K}$$

4.6.8 Accelerated Compressible Flow

Frictional Compressible Pipe Flow

In a frictional steady pipe flow, the friction work that occurs is supplied to the gas as dissipation energy. This energy supply – with a simultaneous decrease of the pressure energy and thus with a decrease of the pressure – contributes to a decrease of the density.

Since at the same time the mass flow remains unchanged, the volume flow increases with a constant pipe cross-section; this leads to an acceleration of the flow.

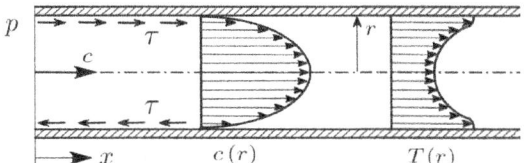

Figure 4.47 Velocity and temperature profile for a frictional pipe flow

The *Euler* equation of motion for the frictional flow of a *Newton*[10] fluid, neglecting the gravitational component and taking into account the frictional component, is given in accordance with Eq. (4.229) as shown in Fig. 4.47:

---

[10] *Isaac Newton*, 1643 to 1727, English scientist, made important contributions to physics, astronomy and mathematics.

4.6 Dynamics of Ideal Gases: Compressible Stationary Gas Flow

$$c\,dc + \frac{dp}{\varrho} + \frac{d\tau}{\varrho} = 0 \tag{4.270}$$

Here, $d\tau$ is the differential of the shear stress (and not the differential of time). The thermal equation of state of ideal gases $p/\varrho = RT$ can be written in the differential notation as $d(p/\varrho) = R\,dT = (p\,dT)/(\varrho T) = (dp\,\varrho - d\varrho\,p)/\varrho^2 = dp/\varrho - (d\varrho\,p)/\varrho^2$. If this equation is multiplied on both sides by ϱ/p, it follows

$$\frac{dp}{p} - \frac{d\varrho}{\varrho} = \frac{dT}{T} \tag{4.271}$$

(cf. Eq. (4.31)). Further, the continuity equation $\dot{m} = \varrho c A$ in its differential form $d\varrho/\varrho + dc/c + dA/A = 0$ (Eq. (4.224)) for the pipeline at constant pipe cross-section ($A = $ const, $dA = 0$) leads to

$$\frac{d\varrho}{\varrho} + \frac{dc}{c} = 0\,. \tag{4.272}$$

If this equation is inserted into Eq. (4.271) and the pressure p is replaced there using the thermal equation of state of the ideal gas $p = \varrho RT$, then one gets

$$\frac{dp}{\varrho RT} + \frac{dc}{c} = \frac{dT}{T} \quad \text{as well as} \quad \frac{dp}{\varrho} = RT\left(\frac{dT}{T} - \frac{dc}{c}\right). \tag{4.273}$$

Thus, Eq. (4.273) gives a relationship for the pressure change dp due to a temperature change dT and a velocity change dc. If this is used in Eq. (4.270), and inserting the equation for the specific friction energy $\tau/\varrho = \lambda(c^2/2)(x/d_h)$ with $d = d_h$ as the inner diameter of the pipe, this leads to the equation for the compressible frictional flow of an ideal gas

$$c\,dc + R\,dT - RT\frac{dc}{c} + \lambda\frac{c^2}{2}\frac{dx}{d_h} = 0\,. \tag{4.274}$$

If one knows the temperature course $T(x)$ along the longitudinal pipe axis x, the velocity course along the longitudinal pipe axis $c(x)$ can be determined with this equation. If first an isothermal pipe flow ($T = $ const) is assumed, then Eq. (4.274) can be calculated in a simple way.

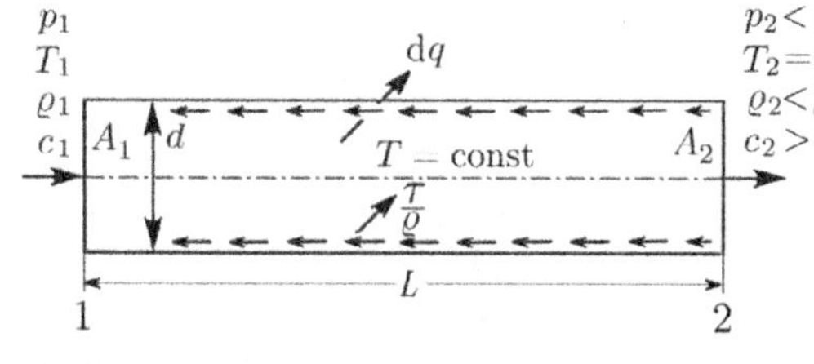

Figure 4.48 Frictional compressible isothermal pipe flow

Frictional Compressible Isothermal Pipe Flow

In long, ground-bedded gas pipelines such as, in particular, long-distance gas transmission pipelines and local gas supply pipelines, the pipelines and the gas have practically the constant temperature of the ground, and a heat release dq is possible (Fig. 4.48).

With $T = $ const, Euler's equation of motion for compressible frictional pipe flow of an ideal gas (Eq. (4.274)) becomes

$$\left(1 - \frac{RT}{c^2}\right)c\,dc + \lambda\frac{c^2}{2}\frac{dx}{d_h} = 0\,. \tag{4.275}$$

The pipe friction coefficient λ (cf. section 2.8) does not depend on the Mach number for an incompressible flow, and also does not depend on the Mach number for a compressible subsonic flow ($M < 1.0$).

λ depends only on the *Reynolds* number Re as well as on the ratio of the pipe diameter to the pipe roughness d/k (cf. section 2.8): $\lambda = f(Re, d/k)$

λ can be determined e.g. from the λ, Re diagram (Figure 2.20) and does not depend on the flow path x. Thus Eq. (4.275) – after division by $c^2/2$ – can be calculated in the limits 1 and 2 according to Fig. 4.48:

$$\frac{2}{c}\,\mathrm{d}c - \frac{2\,RT}{c^3}\,\mathrm{d}c + \lambda\,\frac{1}{d_h}\,\mathrm{d}x = 0 \qquad (4.276)$$

The integration becomes

$$2\int_1^2 c^{-1}\,\mathrm{d}c - 2\,RT\int_1^2 c^{-3}\,\mathrm{d}c + \lambda\,\frac{1}{d_h}\int_1^2 \mathrm{d}x = 2\ln\frac{c_2}{c_1} + RT\left[\frac{1}{c_2{}^2} - \frac{1}{c_1{}^2}\right] + \lambda\,\frac{L}{d_h} \,. \qquad (4.277)$$

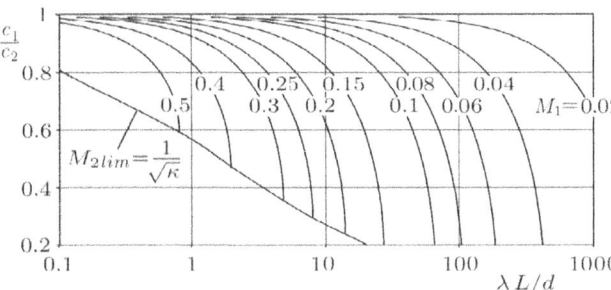

Figure 4.49 Velocity profile for compressible isothermal pipe flow for air and other diatomic gases with $\kappa = 1.4$ [47]

If the square of the velocity of sound, divided by κ, expressed as $a_1{}^2/\kappa = RT = RT_1$, and the *Mach* number at the beginning of the pipe $M_1^2 = c_1{}^2/a_1{}^2$ are considered, the following equation with *Mach* number M_1 can be derived from Eq. (4.277):

$$2\ln\frac{c_1}{c_2} + \frac{1}{\kappa\,M_1{}^2}\left[1 - \frac{c_1{}^2}{c_2{}^2}\right] = \lambda\,\frac{L}{d_h} \qquad (4.278)$$

The first expression of Eq. (4.278) $2\ln(c_1/c_2)$ represents the acceleration of the gas as a consequence of its compressible influence. In the iterative solution of the equation, it can initially be assumed to be approximately zero.

With the help of the continuity equation $\dot{m} = \varrho_1\,c_1\,A_1 = \varrho_2\,c_2\,A_2$ for $A = \text{const}$ and with the thermal equation of state of the ideal gas $p_1 = \varrho_1\,RT_1$ and $p_2 = \varrho_2\,RT_1$ for $T = \text{const}$, the ratios ϱ_2/ϱ_1 and p_2/p_1 as well as the state variables at the end of the pipe can be calculated. For a constant pipe cross-section $A = \text{const}$ and $T = \text{const}$ they become

$$\frac{\varrho_2}{\varrho_1} = \frac{c_1}{c_2} \qquad \text{and} \qquad \frac{p_2}{p_1} = \frac{\varrho_2}{\varrho_1} = \frac{c_1}{c_2} \,. \qquad (4.279)$$

An iterative solution of the complete Eq. (4.278) leads to results which can be clearly represented in a diagram depending on the pipe geometry $\lambda L/d_h = \lambda L/d$ and the *Mach* number M.

Fig. 4.49 shows the velocity ratio c_1/c_2 for the compressible isothermal pipe flow of air and other diatomic gases with $\kappa = 1.4$ [47].

In Fig. 4.49, for different *Mach* numbers of M_1 between 0.02 and 0.50, the ratio of the velocity at the beginning and at the end of the pipe c_1/c_2 is given in the range from 0.2 to 1 over $\lambda L/d$ with values from 0.1 to 1000.

Furthermore, Fig. 4.49 contains the course of the respective limit $Mach$ number. This is derived with the help of a limit value consideration of Eq. (4.278) for

$$\left(\frac{c_1}{c_2}\right)_{lim} = \left(\frac{M_1}{M_2}\right)_{lim} = \sqrt{\kappa}\, M_1 , \qquad (4.280)$$

where this becomes [47]

$$M_{2Gr\,lim} = \frac{1}{\sqrt{\kappa}} . \qquad (4.281)$$

Fig. 4.49 illustrates the acceleration of the gas flow due to the simultaneous decrease in pressure and density caused by friction. If very long pipelines or very large values of $\lambda L/d > 10$ to 10^3 are to be considered, such a gas flow can lead to a compression shock with a corresponding increase in pressure and a corresponding decrease in velocity. Hereafter, the isothermal frictional flow process can find a new beginning.

In order to solve Eq. (4.278), it is convenient to assume the acceleration term $2 \ln c_1/c_2$ to be zero. Thus, the solution of the simplified Eq. (4.278) makes it possible to calculate the pressure loss Δp for the frictional compressible flow of an ideal gas:

$$\Delta p = p_1 - p_2 = p_1\left(1 - \sqrt{1 - \lambda \frac{L}{d} \frac{\varrho_1}{p_1} c_1^{\,2}}\right) \qquad (4.282)$$

Eq. (4.282) shows that the pressure drop Δp for pipe flows with variable density is always higher than the pressure drop for incompressible frictional flow with the same pipe parameters and initial conditions according to (Eq. (4.283)) and Eq. (2.110), respectively.

For compressible gas flows with $Mach$ numbers of $M_1 > 0.2$, the pressure loss must always be determined with consideration of the gas compressibility with the help of Eq. (4.282) or Eq. (4.278). Here it applies that for the velocity ratio c_1/c_2 and for the pipe length L limit values can occur which must not be ignored if undesirable compression shocks are to be avoided.

Another limit value for the calculation of pressure losses of compressible gases in pipelines is the ratio of the pressure loss Δp to the absolute inlet pressure p_1, which results in $\Delta p/p_1 \geq 0.08$.

A mathematical series transformation of Eq. (4.282) leads to the pressure drop for incompressible pipe flow (cf. section 2.8):

$$\Delta p = \lambda \frac{L}{d} \frac{1}{2} \varrho\, c_1^{\,2} \qquad (4.283)$$

Example 4.27 A high-pressure hydrogen pipeline is planned to supply heat to a pulp mill. A physical norm volume flow of $\dot{V}_0 = 15000$ m^3/h is to be transported in a pipeline section of $l = 12.0$ km length with an inner diameter $d_i = 327.2$ mm (compressible gas with variable density). The pipeline consists of polyethylene plastic pipes with a pipe roughness of $k = 0.007$ mm.

The absolute pressure at the inlet cross-section of the pipe is $p_1 = 3.50$ bar; the hydrogen H$_2$ has the temperature $t_1 = 12\,°$C and the dynamic viscosity $\eta = 8.8 \cdot 10^{-6}$ kg/(m s). The physical norm density of hydrogen is $\varrho_0 = 0.08989$ kg/m^3. In the pressure range under consideration, hydrogen can be regarded as an ideal gas.

a) What is the volume flow $\dot{V}_1$ at operating condition at the inlet cross-section of the pipe? The velocity c_1 of the hydrogen is to be determined. What is the value of the density of the hydrogen ϱ_1?

130 4 Ideal Gases

The kinematic viscosity ν, the parameter d/k and the Reynolds number Re are to be calculated. With the help of the λ, Re diagram the pipe friction coefficient λ is to be determined and checked with a suitable equation. Using these values, the pressure loss Δp and the pressure p_2 at the end of the pipe are to be calculated.

b) What final pressure p_2 would result if this were calculated according to section 2.8 without taking the compressibility of the gas into account?

c) For a comparison, the supply of the pulp mill through the same pipeline with natural gas H is to be assumed, whereby the same heat output – related to the calorific value – is to be provided. A physical norm volume flow of natural gas H of $\dot{V}_0 = 4350 \ \text{m}^3/\text{h}$ is sufficient for this. Natural gas H with the pressure $p_1 = 3.50$ bar at the inlet cross-section can be treated as an ideal gas; it has the dynamic viscosity $\eta = 10.8 \cdot 10^{-6}$ kg/(m s) and the physical norm density $\varrho_0 = 0.7885 \ \text{kg/m}^3$. The pressure drop Δp and the pressure p_2 at the end of the pipe shall be determined.

a) $\dot{V}_1 = \dot{V}_0 \dfrac{p_0}{p_1} \dfrac{T_1}{T_0} = 15000 \dfrac{\text{Nm}^3}{\text{h}} \dfrac{1.01325 \ \text{bar}}{3.50 \ \text{bar}} \dfrac{285.15 \ \text{K}}{273.15 \ \text{K}} = 4533.274 \dfrac{\text{m}^3}{\text{h}}$

$c_1 = \dfrac{\dot{V}_1}{A} = \dot{V}_1 \dfrac{4}{\pi d^2} = \dfrac{4533.274 \ \text{m}^3 \cdot 4}{\text{h} \ 3.14159 \cdot 0.3272^2 \ \text{m}^2} \dfrac{1 \ \text{h}}{3600 \ \text{s}} = 14.976 \dfrac{\text{m}}{\text{s}}$

$\varrho_1 = \varrho_0 \dfrac{p_1}{p_0} \dfrac{T_0}{T_1} = 0.08989 \dfrac{\text{kg}}{\text{Nm}^3} \dfrac{3.50 \ \text{bar}}{1.01325 \ \text{bar}} \dfrac{273.15 \ \text{K}}{285.15 \ \text{K}} = 0.2974 \dfrac{\text{kg}}{\text{m}^3}$

b) $\nu = \dfrac{\eta}{\varrho_1} = \dfrac{8.8 \cdot 10^{-6} \ \text{kg m}^3}{0.2974 \ \text{m s kg}} = 2.9586 \cdot 10^{-5} \dfrac{\text{m}^2}{\text{s}}$ $\dfrac{d}{k} = \dfrac{327.2 \ \text{mm}}{0.007 \ \text{mm}} = 46743$

$Re = \dfrac{c \, d}{\nu} = \dfrac{14.976 \ \text{m} \cdot 0.3272 \ \text{m s}}{\text{s} \ 2.9586 \cdot 10^{-5} \ \text{m}^2} = 1.6562 \cdot 10^5$ $\rightarrow$ turbulent flow

$\lambda = 0.0163$ $\rightarrow$ nearly hydraulically smooth flow condition

Verification with the equation of *Nikuradze* in the range $Re = 10^5 \dots 5 \cdot 10^6$:

$\lambda = 0.0032 + 0.221 \cdot Re^{-0.237} = 0.0032 + 0.221 \cdot (1.6562 \cdot 10^5)^{-0.237} = 0.0160$

$\Delta p_R = p_1 - p_2 = p_1 \left(1 - \sqrt{1 - \lambda \dfrac{l}{d} \dfrac{\varrho_1}{p_1} c_1^2} \right)$

$= 3.50 \ \text{bar} \left(1 - \sqrt{1 - 0.0163 \dfrac{12000 \ \text{m}}{0.3272 \ \text{m}} \dfrac{0.2974 \ \text{kg}}{\text{m}^3 \ 3.50 \ \text{bar}} 14.976^2 \dfrac{\text{m}^2}{\text{s}^2} \dfrac{\text{bar m}^2}{10^5 \ \text{N}} \dfrac{\text{N s}^2}{\text{kg m}}} \right)$

$= 3.50 \ \text{bar} \left(1 - \sqrt{1 - 0.11393} \right) = 0.2054 \ \text{bar} \approx 0.205 \ \text{bar}$

$p_2 \approx 3.295 \ \text{bar}$

c) The calculation as incompressible gas flow according to section 2.8 leads to the pressure loss

$\Delta p = \lambda \dfrac{l}{d} \dfrac{1}{2} \varrho c_1^2 = 0.0163 \dfrac{12000 \ \text{m}}{0.272 \ \text{m}} \dfrac{1}{2} 0.2974 \dfrac{\text{kg}}{\text{m}^3} 14.976^2 \dfrac{\text{m}^2}{\text{s}^2} \dfrac{\text{bar m}^2}{10^5 \ \text{N}} \dfrac{\text{N s}^2}{\text{kg m}} = 0.1994 \ \text{bar}$

$\approx 0.199 \ \text{bar}$

$p_2 \approx 3.301 \ \text{bar}$

This calculation results in a somewhat lower pressure loss. However, at considerably higher input pressures p_1, the deviations would be considerably greater when assuming an incompressible gas flow, so that the more accurate approaches for the compressible gas flow must then be used.

4.6 Dynamics of Ideal Gases: Compressible Stationary Gas Flow

d) $\dot{V}_1 = \dot{V}_0 \dfrac{p_0}{p_1} \dfrac{T_1}{T_0} = 4350 \dfrac{\text{Nm}^3}{\text{h}} \dfrac{1.01325 \text{ bar}}{3.50 \text{ bar}} \dfrac{285.15 \text{ K}}{273.15 \text{ K}} = 1314.650 \dfrac{\text{m}^3}{\text{h}}$

$c_1 = \dfrac{\dot{V}_1}{A} = \dot{V}_1 \dfrac{4}{\pi d^2} = \dfrac{1314.650 \text{ m}^3 \cdot 4}{\text{h} \; 3.14159 \cdot 0.3272^2 \text{ m}^2} \dfrac{1 \text{ h}}{3600 \text{ s}} = 4.343 \dfrac{\text{m}}{\text{s}}$

$\varrho_1 = \varrho_0 \dfrac{p_1}{p_0} \dfrac{T_0}{T_1} = 0.7885 \dfrac{\text{kg}}{\text{Nm}^3} \dfrac{3.50 \text{ bar}}{1.01325 \text{ bar}} \dfrac{273.15 \text{ K}}{285.15 \text{ K}} = 2.609 \dfrac{\text{kg}}{\text{m}^3}$

$\nu = \dfrac{\eta}{\varrho_1} = \dfrac{10.8 \cdot 10^{-6} \text{ kg m}^3}{2.609 \text{ m s kg}} = 4.1395 \cdot 10^{-6} \dfrac{\text{m}^2}{\text{s}} \quad \dfrac{d}{k} = \dfrac{327.2 \text{ mm}}{0.007 \text{ mm}} = 46743$

$Re = \dfrac{c d}{\nu} = \dfrac{4.343 \text{ m} \cdot 0.3272 \text{ m s}}{\text{s} \; 4.1395 \cdot 10^{-6} \text{ m}^2} = 3.433 \cdot 10^5 \quad \rightarrow \quad \text{turbulent flow}$

$\lambda = 0.0140 \quad \rightarrow \quad$ almost hydraulically smooth flow condition

Verification with the equation of *Nikuradze* in the range $Re = 10^5 \ldots 5 \cdot 10^6$:

$\lambda = 0.0032 + 0.221 \cdot Re^{-0.237} = 0.0032 + 0.221 \cdot (3.433 \cdot 10^5)^{-0.237} = 0.0140$

$\Delta p_R = p_1 - p_2 = p_1 \left(1 - \sqrt{1 - \lambda \dfrac{l}{d} \dfrac{\varrho_1}{p_1} c_1^2}\right)$

$= 3.50 \text{ bar} \left(1 - \sqrt{1 - 0.014 \dfrac{12000 \text{ m}}{0.3272 \text{ m}} \dfrac{2.609 \text{ kg}}{\text{m}^3 \; 3.50 \text{ bar}} 4.343^2 \dfrac{\text{m}^2}{\text{s}^2} \dfrac{\text{bar m}^2}{10^5 \text{ N}} \dfrac{\text{N s}^2}{\text{kg m}}}\right)$

$= 3.50 \text{ bar} \left(1 - \sqrt{1 - 0.07219}\right) = 0.12870 \text{ bar} \approx 0.129 \text{ bar}$

$p_2 \approx 3.371 \text{ bar}$

The calculation as incompressible flow according to section 2.8 gives the pressure loss

$\Delta p = \lambda \dfrac{l}{d} \dfrac{1}{2} \varrho c_1^2 = 0.014 \dfrac{12000 \text{ m}}{0.3272 \text{ m}} \dfrac{1}{2} 2.609 \dfrac{\text{kg}}{\text{m}^3} 4.343^2 \dfrac{\text{m}^2}{\text{s}^2} \dfrac{\text{bar m}^2}{10^5 \text{ N}} \dfrac{\text{N s}^2}{\text{kg m}} = 0.12633 \text{ bar}$

$\approx 0.126 \text{ bar}$

$p_2 \approx 3.374 \text{ bar}$

Frictional Compressible Adiabatic Pipe Flow

For the transport of gases in thermally insulated long gas pipelines as well as in short, non-thermally insulated supply lines, the heat transfer through the pipe wall can be assumed to be zero or approximately neglected (Fig. 4.50).

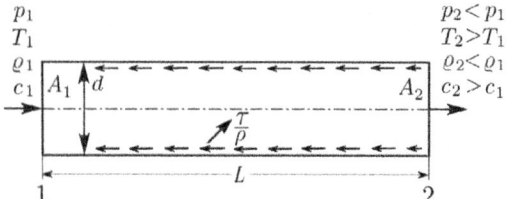

Figure 4.50 Frictionally adiabatic pipe flow

Using the approach for the total temperature of a flow based on Eq. (4.249) $T_t = T + c^2/(2 c_p)$ and the specific isobaric heat capacity $c_p = \kappa R/(\kappa - 1)$, the gas temperature can be given as follows:

$$T = T_t - \dfrac{c^2}{2 c_p} = T_t - \dfrac{\kappa - 1}{2 \kappa} \dfrac{c^2}{R} \qquad (4.284)$$

From Eq. (4.284), after a differentiation according to dT/dc, follows

$$dT = -\frac{\kappa - 1}{\kappa R} c\, dc . \qquad (4.285)$$

If this temperature change dT as well as T according to Eq. (4.284) are inserted in the equation of the compressible frictional pipe flow Eq. (4.274) and then divided by $c^2/2$, the result is

$$c\,dc - \frac{\kappa-1}{\kappa} c\,dc - R\left[T_t - \frac{\kappa-1}{2\kappa}\frac{c^2}{R}\right]\frac{dc}{c} + \lambda \frac{c^2}{2}\frac{dx}{d_h} = 0 ,$$

$$(1-\frac{\kappa-1}{\kappa}+\frac{\kappa-1}{2\kappa})c\,dc - \frac{RT_t}{c}dc + \lambda \frac{c^2}{2}\frac{dx}{d_h} = 0 \quad \text{and} \quad \frac{\kappa+1}{\kappa}\frac{dc}{c} - 2RT_{t1}\frac{dc}{c^3} + \lambda\frac{dx}{d_h} = 0 .$$

$$(4.286)$$

The pipe friction coefficient λ shows no dependence on the length x not only for isothermal but also for adiabatic pipe flow and depends only on the *Reynolds* number Re and the quotient d/k: $\lambda = f(Re, d/k)$. Integrating Eq. (4.286) between states 1 and 2 according to Figure 4.50 leads to Eq. (4.287):

$$\frac{\kappa+1}{\kappa}\int_1^2 \frac{dc}{c} - 2RT_{t1}\int_1^2 \frac{dc}{c^3} + \int_1^2 \frac{\lambda}{d_h}dx = \frac{\kappa+1}{\kappa}\ln\left(\frac{c_2}{c_1}\right) + RT_{t1}\left[\frac{1}{c_2^2}-\frac{1}{c_1^2}\right] + \lambda\frac{L}{d_h} = 0$$

$$\frac{\kappa+1}{\kappa}\ln\left(\frac{c_2}{c_1}\right) - \frac{RT_{t1}}{c_1^2}\left[1-\left(\frac{c_1}{c_2}\right)^2\right] + \lambda\frac{L}{d_h} = 0 \qquad (4.287)$$

A transformation of Eq. (4.284) leads to

$$\frac{T_t}{T}-1 = \frac{\kappa-1}{2}\frac{c^2}{\kappa RT} \quad \text{and} \quad \left(\frac{T_t}{T}-1\right)\frac{\kappa RT}{c^2} = \frac{\kappa RT_t}{c^2} - \frac{\kappa RT}{c^2} = \frac{\kappa-1}{2}. \qquad (4.288)$$

Thus the expression RT_{t1}/c_1^2 in Eq. (4.287) can be represented by $\kappa RT_1 = a_1^2$ as follows:

$$\frac{RT_{t1}}{c_1^2} = \frac{\kappa-1}{2\kappa} + \frac{a_1^2}{\kappa c_1^2} \quad \text{and} \quad \frac{RT_{t1}}{c_1^2} = \frac{1}{\kappa M_1^2} + \frac{\kappa-1}{2\kappa} \qquad (4.289)$$

Inserting Eq. (4.289) into Eq. (4.287) yields an equation for the velocity ratio c_1/c_2 as a function of the gas type expressed by κ, the *Mach* number M_1, the pipe friction coefficient λ, the pipe length L and the hydraulic pipe diameter d_h:

$$\left[\frac{1}{\kappa M_1^2}+\frac{\kappa-1}{2\kappa}\right]\left[1-\left(\frac{c_1}{c_2}\right)^2\right]+\frac{\kappa+1}{\kappa}\ln\left(\frac{c_1}{c_2}\right)-\lambda\frac{L}{d_h} = 0 \qquad (4.290)$$

This transcendental equation can be solved iteratively if the pipe geometry $d_h = d$ and L is given, λ is known, the initial conditions p_1, T_1 and c_1 are given, and the isentropic exponent κ is known [47].

The results obtained with Eq. (4.290) for the adiabatic pipe flow of an ideal gas can be reproduced depending on the two quantities λ and L/d of the pipe geometry as well as on the inlet *Mach* number M_1. Fig. 4.51 shows the velocity ratios for the compressible adiabatic pipe flow, following the structure of Fig. 4.49.

The ratio of velocities c_1/c_2 is shown for numerical values of $c_1/c_2 = 0.2$ to 1 over $\lambda L/d$, where $\lambda L/d$ spans numerical values from 0.1 to 1000. For *Mach* numbers of $M_1 = 0.02$ to 0.5, increasing values of $\lambda L/d$ result with decreasing numerical values of c_1/c_2 according to Fig. 4.51.

In addition, the course of the limiting $Mach$ number with $M_{2Gr} = 1$ is shown in the diagram.

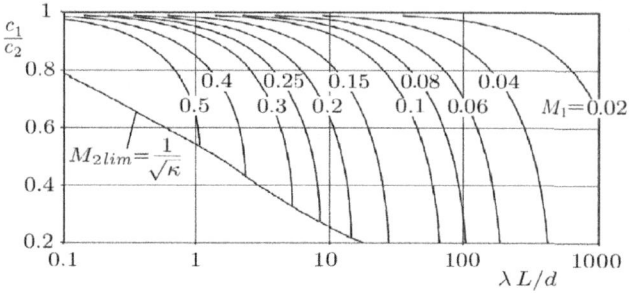

Figure 4.51 Velocity profile for compressible adiabatic pipe flow for air and other diatomic gases with $\kappa = 1.4$ [47]

For lower velocities, the acceleration term $((\kappa+1)/\kappa) \ln(c_1/c_2)$ in Eq. (4.290) can be set to approximate zero. Thus, after a transformation to the velocity ratio c_1/c_2, Eq. (4.290) becomes

$$\frac{c_1}{c_2} = \left[1 - \frac{2\lambda \dfrac{L}{d_h} \kappa M_1^2}{2 + (\kappa - 1)(M_1)^2}\right]^{\frac{1}{2}}. \tag{4.291}$$

For the pipe flow, the temperature ratio T_2/T_1 is obtained via links of Eqs. (4.284) and (4.288) to

$$\frac{T_2}{T_1} = 1 + \frac{\kappa-1}{2} \frac{c_1^2}{\kappa R T_1}\left[1 - \left(\frac{c_2}{c_1}\right)^2\right] = 1 + \frac{\kappa-1}{2} M_1^2 \left[1 - \left(\frac{c_2}{c_1}\right)^2\right]. \tag{4.292}$$

The density ratio is obtained by integrating Eq. (4.272) $(d\varrho/\varrho) + (dc/c) = 0$ to $\ln(\varrho_2/\varrho_1) = -\ln(c_2/c_1) = \ln(c_1/c_2)$ and

$$\frac{\varrho_2}{\varrho_1} = \frac{c_1}{c_2}, \tag{4.293a}$$

respectively. The pressure ratio is obtained by dividing $p_2 = \varrho_2 R T_2$ by $p_1 = \varrho_1 R T_1$:

$$\frac{p_2}{p_1} = \frac{\varrho_2 T_2}{\varrho_1 T_1} = \frac{c_1 T_2}{c_2 T_1} \tag{4.293b}$$

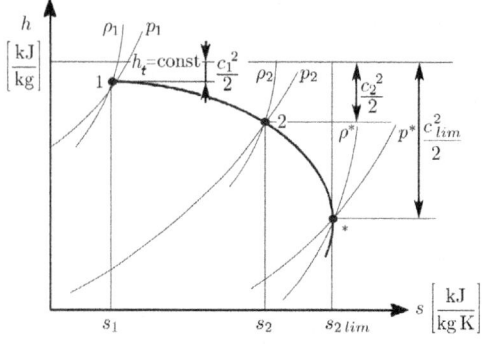

Figure 4.52 Change of state of adiabatic pipe flow in a h,s diagram (*Fanno-*curve) for gases with $\kappa = 1.4$

The course of the states of the friction-affected adiabatic gas flow in the pipeline can be clearly shown in a h, s diagram. This results in the so-called *Fanno* curve in Fig. 4.52. At the critical pressure p^* the velocity c reaches the possible limit value c_{lim}.

134   4 Ideal Gases

At this limit point of Fig. 4.52, the velocity ratio (cf. Eq. (4.291)) is

$$\left(\frac{c_1}{c_2}\right)_{lim} = \left[\frac{(\kappa+1)\,M_1{}^2}{2+(\kappa-1)\,(M_1)^2}\right]^{\frac{1}{2}}. \tag{4.294}$$

Example 4.28 A thermally insulated pipeline containing superheated steam has the internal diameter $d = 107.1$ mm and the length 1250 m. At the inlet cross-section 1, the velocity of the steam is $c_1 = 23$ m/s, the temperature $t_1 = 275$ °C and the pressure $p_1 = 10.0$ bar. The steel pipe has the surface roughness $k = 0.1$ mm. For $q = 0$, the velocity ratio c_1/c_2, the temperature ratio T_2/T_1, the density ratio ϱ_2/ϱ_1, the density ϱ_2, the pressure ratio p_2/p_1 and the pressure loss $\Delta p = p_1 - p_2$ are to be calculated, related to state 2 at the end of the pipe.

The superheated steam, which is to be treated as an ideal gas, has the kinematic viscosity $\nu = 4.6 \cdot 10^{-6}$ m²/s, the isentropic exponent $\kappa = 1.33$ and the specific gas constant $R = 0.46152$ kJ/(kg K).

Reynolds number of the steam flow:

$$Re = \frac{cd}{\nu} = \frac{23\,\mathrm{m} \cdot 0.1071\,\mathrm{m\,s}}{\mathrm{s}\,4.6 \cdot 10^{-6}\,\mathrm{m}^2} = 535500$$

Relative surface roughness:

$$\frac{d}{k} = \frac{107.1\,\mathrm{mm}}{0.1\,\mathrm{mm}} = 1071$$

Pipe friction coefficient λ from the λ, Re diagram:

$$\lambda = f(Re, d/k) = 0.020$$

Sound velocity of superheated steam, Eq. (4.233):

$$a = \sqrt{\kappa\,R\,T} = \sqrt{1.33 \cdot 0.46152\,\frac{\mathrm{kJ}}{\mathrm{kg\,K}}\,\frac{1000\,\mathrm{N\,m}}{\mathrm{kJ}}\,\frac{\mathrm{kg\,m}}{\mathrm{N\,s}^2}\,548.15\,\mathrm{K}} = 580.06\,\frac{\mathrm{m}}{\mathrm{s}}$$

Mach number, Eq. (4.235):

$$\frac{c}{a} = \frac{23\,\mathrm{m\,s}}{\mathrm{s}\,580.06\,\mathrm{m}} = 0.03965$$

Velocity ratio using Eq. (4.291) and final velocity c_2 when the acceleration term is neglected:

$$\frac{c_1}{c_2} = \left[1 - \frac{2\lambda\dfrac{L}{d_h}\kappa\,M_1{}^2}{2+(\kappa-1)\,(M_1)^2}\right]^{\frac{1}{2}} = \left[1 - \frac{2 \cdot 0.020\,\dfrac{1250\,\mathrm{m}}{0.1071\,\mathrm{m}}\,1.33 \cdot 0.03965^2}{2+(1.33-1)\,0.03965^2}\right]^{\frac{1}{2}} = 0.71558$$

$$c_2 = \frac{c_2}{c_1}\,c_1 = \frac{1}{0.71558} \cdot 23\,\frac{\mathrm{m}}{\mathrm{s}} = 32.142\,\frac{\mathrm{m}}{\mathrm{s}}$$

Temperature ratio and final temperature T_2, Eq. (4.292):

$$\frac{T_2}{T_1} = 1 + \frac{\kappa-1}{2}\,M_1{}^2\left[1-\left(\frac{c_2}{c_1}\right)^2\right] = 1 + \frac{1.33-1}{2}\,0.03965^2\left[1-\left(\frac{1}{0.71558}\right)^2\right] = 0.99975$$

$$T_2 = \frac{T_2}{T_1}\,T_1 = 0.99975 \cdot 548.15\,\mathrm{K} = 548.015\,\mathrm{K}$$

Density ratio and density ϱ_2, Eq. (4.293):

$$\frac{\varrho_2}{\varrho_1} = \frac{c_1}{c_2} = 0.71558$$

$$\varrho_2 = \frac{\varrho_2}{\varrho_1}\,\varrho_1 = \frac{\varrho_2}{\varrho_1}\,\frac{p_1}{R\,T_1} = 0.71558 \cdot \frac{10.0\,\mathrm{bar\,kg\,K}}{0.46152\,\mathrm{kJ} \cdot 548.15\,\mathrm{K}}\,\frac{\mathrm{kJ}}{1000\,\mathrm{N\,m}}\,\frac{10^5\,\mathrm{N}}{\mathrm{bar\,m}^2} = 2.829\,\frac{\mathrm{kg}}{\mathrm{m}^3}$$

Pressure ratio, final pressure p_2 and pressure drop, Eq. (4.293 b):

$$\frac{p_2}{p_1} = \frac{\varrho_2}{\varrho_1}\frac{T_2}{T_1} = 0.71558 \cdot 0.99975 = 0.71540$$

$$p_2 = \frac{p_2}{p_1} p_1 = 0.71540 \cdot 10.0\,\text{bar} = 7.1540\,\text{bar}$$

$$\Delta p = p_1 - p_2 = 10.0\,\text{bar} - 7.1540\,\text{bar} = 2.8460\,\text{bar} \approx 2.85\,\text{bar}$$

Area-Velocity Relationship

In a compressible frictionless accelerated gas flow in nozzles, the state variables p, T, ϱ, c as well as the flow cross-section A change, depending on the velocity c or the *Mach* number along the path x.

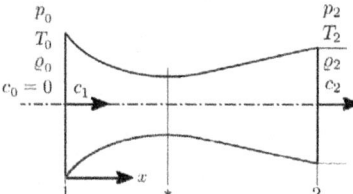

Figure 4.53 One-dimensional flow in a *Laval* nozzle

Gas flows of this type are formed in supersonic *Laval* nozzles (Fig. 4.53 and Fig. 4.54); they also occur in blade grids of gas and steam turbines, which are used, among other aspects, to generate high velocities with large specific energy flows. To capture these processes, a one-dimensional isentropic flow along a flow filament is investigated in the following according to Fig. 4.53.

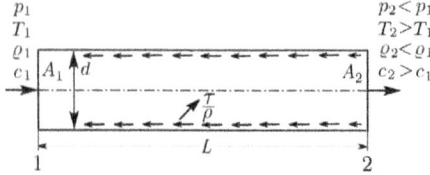

Figure 4.58 Guide apparatus with *Laval* nozzles of a steam turbine [47]

The continuity equation for the compressible gas flow $\dot{m} = \varrho c A$ is in its differential form according to Eq. (4.224): $d\varrho/\varrho + dc/c + dA/A = 0$. Eq. (4.224) shows that a desired change in velocity also requires the change in density of the flow and the flow cross-section.

The density change in turn leads to a pressure and temperature change, as illustrated by the thermal equation of state of the ideal gas in its differential form according to Eq. (4.271):

$$\frac{dp}{p} - \frac{d\varrho}{\varrho} - \frac{dT}{T} = 0 \qquad (4.271)$$

Further, the simplified equation of motion (4.237) is required in the form $(p/\varrho)(dp/p) + c\,dc = 0$. Then, the isentropic relation according to Eq. (4.100) in the form $p/p_0 = (T/T_0)^{(\kappa/(\kappa-1))}$ is used, from which with a logarithmisation of both sides $\ln(p/p_0) = \kappa/(\kappa-1) \ln(T/T_0)$ follows and the differential notation $dp/p = \kappa/(\kappa-1)\,dT/T$

results, from which with the transformed Eq. (4.271) $dp/p = \kappa/(\kappa-1)(dp/p - d\varrho/\varrho)$ becomes.

This relation is inserted into the above form of the equation of motion, resulting in the differential form of the energy equation of gas dynamics [47]:

$$\frac{\kappa}{\kappa-1}\frac{p}{\varrho}\left(\frac{dp}{p} - \frac{d\varrho}{\varrho}\right) + c\,dc = 0 \qquad (4.295)$$

If the pressure change dp/p is separated from this and $dp/p = \kappa\,d\varrho/\varrho$ is also separated from the isentropic equation, equations for the density-velocity relation and the area-velocity relation are obtained. From Eq. (4.295) and with the square of the velocity of sound $a^2 = \kappa p/\varrho$ and the $Mach$ number $M = c/a$ follows

$$\frac{dp}{p} = -\frac{\kappa-1}{\kappa}\frac{\varrho}{p}c\,dc + \frac{d\varrho}{\varrho} = \kappa\frac{d\varrho}{\varrho} \quad \text{as well as} \quad -\frac{\kappa-1}{a^2}c\,dc = (\kappa-1)\frac{d\varrho}{\varrho}, \qquad (4.296)$$

and from this the density-velocity relationship

$$\frac{d\varrho}{\varrho} = -\frac{c^2}{a^2}\frac{dc}{c} = -M^2\frac{dc}{c}. \qquad (4.297)$$

Inserting Eq. (4.297) into the continuity equation $d\varrho/\varrho + dc/c + dA/A = 0$ (Eq. (4.224)) leads to the area-velocity relationship [47]

$$\frac{dA}{A} = (M^2 - 1)\frac{dc}{c}. \qquad (4.298)$$

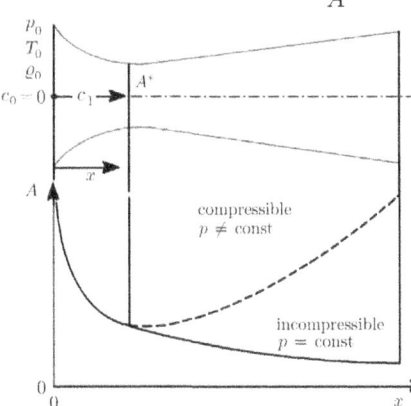

Figure 4.55 Cross-sectional profile in a nozzle with compressible and incompressible flow

The numerical value of the $Mach$ number affects the density change $d\varrho/\varrho$ and the cross-sectional change dA/A of the nozzle, as can be seen from Eqs. (4.297) and (4.298) and from Table 4.8. Fig. 4.55 shows the cross-section of a nozzle as a function of the nozzle length x for compressible and incompressible flow.

Table 4.8 Effect of the magnitude of the $Mach$ number on the density change and the cross-sectional change of flows

Rest state (stagnation state)	$M = 0$	$\|d\varrho/\varrho\| = 0$	$dA/A = -dc/c = 0$
Subsonic velocity	$M < 1$	$\|d\varrho/\varrho\| < \|dc/c\|$	$dA/A < dc/c$
Critical state	$M = M^* = 1$	$\|d\varrho/\varrho\| = \|dc/c\|$	$dA/A = 0$
Supersonic velocity	$M > 1$	$\|d\varrho/\varrho\| > \|dc/c\|$	$dA/A > dc/c$

For a constant *Mach* number M, the area-velocity relationship can be easily integrated according to Eq. (4.298)); the cross-section ratio A_2/A_1 of a supersonic nozzle results from this to

$$\ln \frac{A_2}{A_1} = (M^2 - 1) \ln \frac{c_2}{c_1} = \ln \left[\left(\frac{c_2}{c_1}\right)\right]^{(M^2-1)} \qquad (4.299)$$

and further

$$\frac{A_2}{A_1} = \left(\frac{c_2}{c_1}\right)^{(M^2-1)} \qquad (4.300)$$

and

$$\frac{A}{c^{(M^2-1)}} = \text{const} \ . \qquad (4.301)$$

Eq. (4.300) can also be written for the subsonic range ($M < 1.0$) in the form

$$\frac{A_2}{A_1} = \frac{c_1^{(1-M_1^2)}}{c_2^{(1-M_2^2)}} \ . \qquad (4.302)$$

As Figs. 4.53 to 4.56 show, in the subsonic region the cross-sectional constriction and in the supersonic region the cross-sectional expansion of the nozzle causes the acceleration of the gas flow.

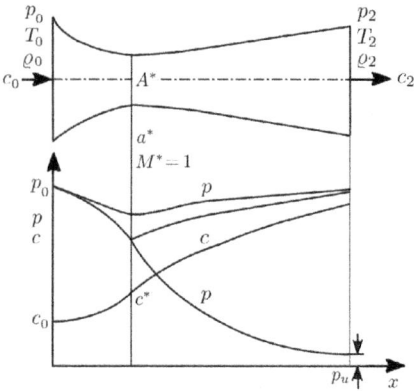

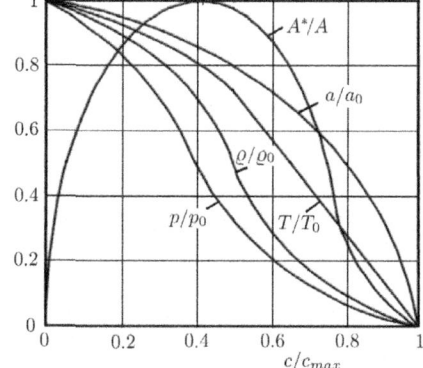

Figure 4.56 Pressure and velocity profile in a *Laval* nozzle at subcritical, critical and supercritical expansion

Figure 4.57 Related state variables as a function of the velocity ratio c/c_{max} for an isentropic flow of air with $\kappa = 1.4$ from a tank through a *Laval* nozzle into absolute vacuum

If Eqs. (4.297) and (4.298) are resolved to dc/c and equated to it, this leads to

$$\frac{d\varrho}{\varrho} = -\frac{M^2}{M^2-1} \frac{dA}{A} \ . \qquad (4.303)$$

If Eq. (4.303) is inserted into the isentropic relation, follows

$$\frac{dp}{p} = \kappa \frac{d\varrho}{\varrho} = -\frac{\kappa M^2}{M^2-1} \frac{dA}{A} \ . \qquad (4.304)$$

From this, after integration and a subsequent de-logarithmisation, the pressure ratio of the *Laval* nozzle is

$$\frac{p_2}{p_1} = \left(\frac{A_1}{A_2}\right)^{\frac{\kappa M^2}{M^2-1}} , \qquad (4.305\,\text{a})$$

which in the case of a circular cross-section with $A = \pi d^2/4$ becomes

$$\frac{p_2}{p_1} = \left(\frac{d_1}{d_2}\right)^{\frac{2\kappa M^2}{M^2-1}}. \qquad (4.305\,b)$$

Fig. 4.56 shows that in the left narrowing region of the *Laval* nozzle, the pressure of the gas flow decreases and the velocity increases, continuing beyond the critical velocity c^* in the narrowest cross-section A^*.

However, if the critical velocity c^* is not reached in the narrowest cross-section A^*, the right widening part of the nozzle acts as a diffuser; in this case, the pressure p increases back to the initial value p_0 in the case of frictionless flow (Fig. 4.56).

Eq. (4.243) shows the achievable maximum velocity when flowing out into the complete vacuum. Using the elliptic equation (4.261 b) $(a/a_0)^2 + (c/c_{max})^2 = 1$, of the equations for the velocities of sound $a^2 = \kappa R T$ and $a_0^2 = \kappa R T_0$, the isentropic relations between T/T_0 and ϱ/ϱ_0, respective between T/T_0 and p/p_0, as well as the relations for the ratio values in the resting state (Eqs. (4.253 b) to (4.260)), the changes of the ratio values of the state variables a/a_0, T/T_0, ϱ/ϱ_0 and p/p_0 depending on the velocity ratio c/c_{max} can be represented in equation form:

$$\left(\frac{c}{c_{max}}\right)^2 = 1 - \left(\frac{a^2}{a_0^2}\right) = 1 - \frac{\kappa R T}{\kappa R T_0} = 1 - \frac{T}{T_0} = 1 - \left(\frac{\varrho}{\varrho_0}\right)^{\kappa-1} = 1 - \left(\frac{p}{p_0}\right)^{\frac{\kappa-1}{\kappa}} \qquad (4.306)$$

The corresponding curves are shown in Fig. 4.57. There it becomes visible, for example, that the local velocity of sound a, respective the ratio a/a_0, decreases with increasing velocity c from the value $a/a_0 = 1.0$ in the resting state and at the maximum velocity $c = c_{max}$, which is reached in absolute vacuum at $p = 0$, $\varrho = 0$ and $T = 0$, to the value $a = \sqrt{\kappa R T} = 0$ with $a/a_0 = 0$.

This progression of a/a_0 makes it clear that different velocities of sound are formed in flows which are to be distinguished from each other:

Thus, the resting velocity of sound is given by

$$a_0 = \sqrt{\kappa \frac{p_0}{\varrho_0}}. \qquad (4.307)$$

The local velocity of sound a at a point on the streamline, according to $R T_0 = p_0/\varrho_0$ and $T/T_0 = (p/p_0)^{((\kappa-1)/\kappa)}$, is given by

$$a = \sqrt{\kappa \frac{p}{\varrho}} = \sqrt{\kappa R T} = \sqrt{\kappa \frac{p_0}{\varrho_0}\left(\frac{p}{p_0}\right)^{\frac{\kappa-1}{\kappa}}}, \qquad (4.308)$$

and the critical velocity of sound a^* is obtained according to Eq. (4.253 b) for a^* and Eq. (4.243) as well as Eq. (4.233) for a_0 to

$$a^* = \sqrt{\frac{2}{\kappa+1}}\, a_0 = \sqrt{\frac{\kappa-1}{\kappa+1}}\, c_{max} = \sqrt{\frac{2\kappa}{\kappa+1}\frac{p_0}{\varrho_0}}. \qquad (4.309)$$

The Rest *Mach* number at $c = 0$ is always $M_0 = 0$. With the different velocities of sound according to Eqs. (4.308) and (4.309), two different *Mach* numbers can be introduced. The local *Mach* number is

$$M = c/a = \frac{c}{\sqrt{\kappa R T}},$$

4.6 Dynamics of Ideal Gases: Compressible Stationary Gas Flow

and the critical $Mach$ number according to Eq. (4.309) for a^* and Eq. (4.306) for $c/c_{max} = c^*/c_{max}$ is

$$M^* = \frac{c^*}{a^*} = \left[\frac{\kappa+1}{\kappa-1} \left[1 - \left(\frac{p}{p_0}\right)^{\frac{\kappa-1}{\kappa}} \right] \right]^{\frac{1}{2}}. \tag{4,310}$$

Using this equation, applying the isentropic equation $p/p_0 = (\varrho/\varrho_0)^\kappa = (T/T_0)^{(\kappa/(\kappa-1))}$ and using Eq. (4.256), this leads to the representation of the pressure ratio p/p_0, to which the exponent $(\kappa-1)/\kappa$ is given:

$$\left(\frac{p}{p_0}\right)^{\frac{\kappa-1}{\kappa}} = \left(\frac{\varrho}{\varrho_0}\right)^{\kappa-1} = \frac{1}{1 + \frac{\kappa-1}{2} M^2}. \tag{4.311}$$

If this equation is inserted into Eq. (4.242), and if the continuity equation (Eq. (4.221)) for any point on the streamline and for the critical condition $\dot{m} = \varrho c A = \varrho^* c^* A^*$ is considered, the cross-section ratio in the critical cross-section $A^*/A = (\varrho/\varrho^*)(c/c^*) = (\varrho/\varrho_0)(\varrho_0/\varrho^*)(c/c^*)$ as well as the density ratio ϱ^*/ϱ in the critical cross-section of a supersonic nozzle can be reproduced mathematically.

The cross-section ratio A^*/A thus becomes

$$\frac{A^*}{A} = \frac{\varrho}{\varrho^*} M^* = M^* \left[\frac{1 + \frac{\kappa-1}{\kappa+1} M^{*2}}{1 + \frac{\kappa-1}{\kappa+1}} \right]^{\frac{1}{\kappa-1}} \tag{4.312}$$

and the density ratio ϱ/ϱ^* from this is

$$\frac{\varrho}{\varrho^*} = \frac{1}{M^*} \frac{A^*}{A} = \left[\frac{1 + \frac{\kappa-1}{\kappa+1} M^{*2}}{1 + \frac{\kappa-1}{\kappa+1}} \right]^{\frac{1}{\kappa-1}}. \tag{4.313}$$

For the critical $Mach$ number $M^* = 1$, Eq. (4.312) gives the cross-section ratio $A^*/A = 1.0$. This states that at the critical point, the function A^*/A for the cross-section of a supersonic nozzle – expressed mathematically – passes through an inflection point between the converging and the following diverging part of the supersonic nozzle depending on the spatial coordinate x (maximum point at $M^* = 1$ in Fig. 4.58). In the supersonic region of the $Laval$ nozzle with $M > 1.0$, it becomes $A^*/A < 1$, thus expressing an extension of the cross-section.

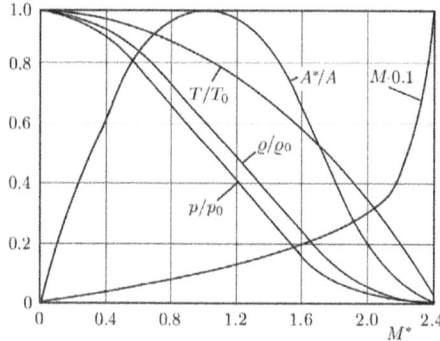

Figure 4.58 Related state variables as a function of the critical $Mach$ number M^* for an isentropic flow of air with $\kappa = 1.4$ from a tank through a $Laval$ nozzle into absolute vacuum

The relation between the critical $Mach$ number M^* and the local $Mach$ number M is given via the velocities of sound a_0 and a^* of Eq. (4.253 b) and the definition of the $Mach$ numbers $M^* = c^*/a^*$ and $M = c/a$ (Eq. (4.235)) to

$$M^* = \frac{c}{a^*} = \frac{c}{\left[\dfrac{2}{\kappa+1}\right]^{\frac{1}{2}} a_0} = \left[\frac{(\kappa+1) M^2}{2 + (\kappa-1) M^2}\right]^{\frac{1}{2}}. \qquad (4.314)$$

The ratio values of the state variables p/p_0, T/T_0 and ϱ/ϱ_0 can be represented not only as a function of c/c_{max} according to Fig. 4.57, but can also be reproduced according to Fig. 4.58 as a function of the critical $Mach$ number M^*, the maximum value of which results in $M^* = 2.45$ [47].

In [93], among other aspects, the equations for the local $Mach$ numbers M, for the critical $Mach$ numbers M^* as well as for the gas-dynamic ratios a/a_0, T/T_0, p/p_0 and ϱ/ϱ_0 can be found for the isentropic flow of ideal gases with constant specific isobaric heat capacity $c_p = \text{const.}$

Example 4.29 In the steam turbine of an industrial cogeneration plant, superheated steam is accelerated in a $Laval$ nozzle. The steam state at the inlet cross section is $p_1 = 15.0$ bar, $T_1 = 773.15$ K and $\varrho_1 = 4.2038$ kg/m³. The superheated steam should be treatable as an ideal gas with $R = 0.46152$ kJ/(kg K) and $\kappa = 1.33$ as a first approximation. The steam is to be expanded in the $Laval$ nozzle to $p_2 = 2.0$ bar and $T_2 = 511.00$ K, with the steam density reaching the value $\varrho_2 = 0.8480$ kg/m³. The nozzle has as its narrowest cross-section the value $A^* = 0{,}0026$ m² and a cross-section ratio of $A_2/A^* = 25$.

a) Is the $Laval$ nozzle operated in the supercritical region?
b) What are the values of the state variables at the nozzle outlet cross-section c_2, a_2 and M_2 during isentropic expansion?
c) Determine the volume flow $\dot{V}_2$ at the nozzle outlet cross-section, the mass flow $\dot{m}$ and the outlet cross-section A_2.
d) How do the required cross-section ratio A_1/A_2' of the $Laval$ nozzle and the outlet cross-section A_2' change if the pressure at the outlet is lowered to $p_2 = 1.3$ bar with otherwise the same parameters?

a) Expansion pressure ratio:

$p_2/p_1 = (2.0 \text{ bar})/(15.0 \text{ bar}) = 0.13333 < 0.577 \quad \rightarrow \quad$ supercritical area

b) Nozzle exit velocity, Eq. (4.242):

$$c_2 = \left[\frac{2\kappa}{\kappa-1} \frac{p_1}{\varrho_1}\left[1 - \left(\frac{p_2}{p_1}\right)^{\frac{\kappa-1}{\kappa}}\right]\right]^{\frac{1}{2}}$$

$$c_2 = \left[\frac{2 \cdot 1.33}{1.33 - 1} \frac{15.0 \text{ bar m}^3}{4.2038 \text{ kg}} \frac{10^5 \text{ N}}{\text{bar m}^2} \frac{\text{kg m}}{\text{N s}^2}\left[1 - \left(\frac{2.0}{15.0}\right)^{\frac{1.33-1}{1.33}}\right]\right]^{\frac{1}{2}} = 1063.76 \frac{\text{m}}{\text{s}}$$

Sound velocity at the outlet cross-section, Eq. (4.233):

$$a_2 = \sqrt{\kappa R T_2} = \sqrt{1.33 \cdot 0.46152 \frac{\text{kJ}}{\text{kg K}} \frac{10^3 \text{ N m}}{\text{kJ}} \frac{\text{kg m}}{\text{N s}^2} 511.00 \text{ K}} = 560.06 \frac{\text{m}}{\text{s}}$$

$Mach$ number at the outlet cross-section, Eq. (4.235):

$$M = \frac{c_2}{a_2} = \frac{1063.76 \text{ m/s}}{560.06 \text{ m/s}} = 1.899$$

c) Volume flow at the outlet cross-section, Eq. (4.222):

$$\dot V_2 = A_2\, c_2 = \frac{A_2}{A^*}\, A^*\, c_2 = 25 \cdot 0.0026\,\text{m}^2 \cdot 1063.76\,\frac{\text{m}}{\text{s}} = 69.1444\,\frac{\text{m}^3}{\text{s}}$$

Mass flow, Eq. (4.221):

$$\dot m = \varrho_2\, \dot V_2 = \varrho_2\, \frac{A_2}{A^*}\, A^*\, c_2 = 0.8480\,\frac{\text{kg}}{\text{m}^3} \cdot 25 \cdot 0.0026\,\text{m}^2 \cdot 1063.76\,\frac{\text{m}}{\text{s}} = 58.634\,\frac{\text{kg}}{\text{s}}$$

$$A_2 = \frac{A_2}{A^*}\, A^* = 25 \cdot 0.0026\,\text{m}^2 = 0.065\,\text{m}^2$$

d)
$$c_2 = \left[\frac{2\kappa}{\kappa - 1}\, \frac{p_1}{\varrho_1}\left[1 - \left(\frac{p_2}{p_1}\right)^{\frac{\kappa-1}{\kappa}}\right]\right]^{\frac{1}{2}}$$

$$c_2 = \left[\frac{2 \cdot 1.33}{1.33 - 1} \cdot \frac{15.0\,\text{bar m}^3}{4.2038\,\text{kg}}\, \frac{10^5\,\text{N}}{\text{bar m}^2}\, \frac{\text{kg m}}{\text{N s}^2}\left[1 - \left(\frac{1.3}{15.0}\right)^{\frac{1.33-1}{1.33}}\right]\right]^{\frac{1}{2}} = 1143.87\,\frac{\text{m}}{\text{s}}$$

Sound velocity $a_2' = a_2$ for $T_2' = T_2$:

Mach number at the outlet cross-section: $\quad M_2 = \dfrac{c_2}{a_2} = \dfrac{1143.87\,\text{m/s}}{560.06\,\text{m/s}} = 2.0424$

Cross-section ratio: $\quad \dfrac{A_1}{A_2} = \left(\dfrac{p_2}{p_1}\right)^{\frac{M_2^2 - 1}{\kappa M_2^2}} = \left(\dfrac{1.3\,\text{bar}}{15.0\,\text{bar}}\right)^{\frac{2.0424^2 - 1}{1.33 \cdot 2.0424^2}} = 0.2471$

Steam density: $\quad \varrho_2' = \dfrac{p_2}{R T_2} = \dfrac{1.3\,\text{bar kg K}}{0.46152\,\text{kJ} \cdot 511.00\,\text{K}}\, \dfrac{100\,\text{kJ}}{\text{bar m}^3} = 0.55123\,\dfrac{\text{kg}}{\text{m}^3}$

New outlet cross-section A_2', Gl. (4.222):

$$A_2' = \frac{\dot m}{\varrho_2'\, c_2'} = \frac{58.634\,\text{kg m}^3\,\text{s}}{\text{s}\, 0.55123\,\text{kg} \cdot 1143.87\,\text{m}} = 0.0930\,\text{m}^2$$

Operational Behaviour of Supersonic Nozzles

Laval nozzles must be operated at the design point, i.e. at the specified pressure ratio p_1/p_2. If this is not ensured, in the case of a too high back pressure p_2', a compression shock (also called a compaction shock or only a shock) will form with jet detachment in the nozzle; the intended final velocity will not be achieved. If the back pressure p_2' is too low, a jet expansion occurs at the outlet, whereby the pressure drops abruptly to the outlet pressure p_2' (Fig. 4.59).

If the pressure in a *Laval* nozzle in the supercritical range does not form in such a way that the pressure curve $p(x)$ according to Eq. (4.305 a) is reached, the gas flow in the nozzle cannot approach the outlet state p_2 isentropically. Instead, it changes abruptly to the pressure p_2', the density ϱ_2' and the temperature T_2'.

If the pressure p_2' at the nozzle outlet cross-section falls below the design pressure p_2, i.e. if $p_2' < p_2$ applies, the flow at the nozzle outlet expands. If, on the other hand, the pressure at the nozzle outlet is higher than the design pressure $p_2' > p_2$, a compression shock is formed and the flow datachs from the nozzle wall (Fig. 4.59). Such a surge causes irreversible flow processes associated with an increase in specific entropy $s_2 - s_1$:

$$s_2 - s_1 = c_p \ln \frac{T_2}{T_1} - R \ln \frac{p_2}{p_1} \qquad (4{,}315)$$

The dimensionless pressure curve p/p_0, the curve of the *Mach* number M and the curve of the increase of the specific entropy $s_2 - s_1$ in a supersonic nozzle at increased pressure behind the nozzle $p_2' > p_2$ are shown in Fig. 4.60.

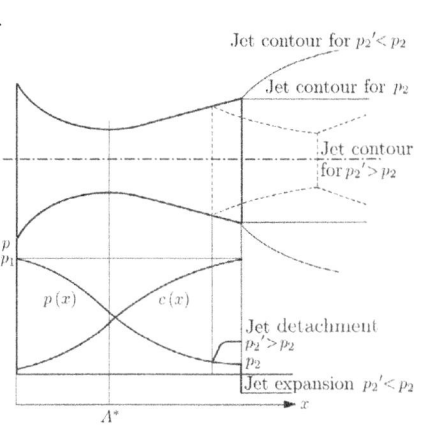
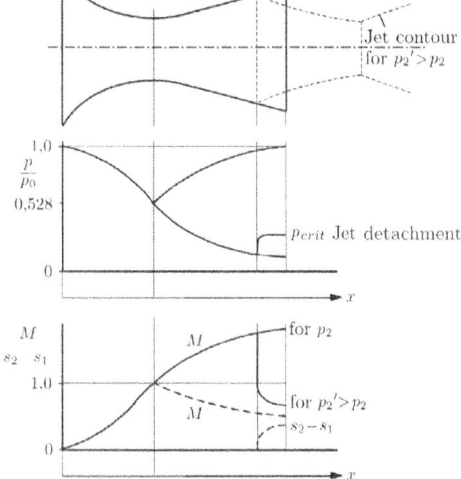

Figure 4.59 (above) Pressure change in *Laval* nozzles at variable backpressure p_2

Figure 4.60 (right) Variation of *Mach* number and specific entropy in *Laval* nozzles at too high backpressure with detachment and compression shock

Supersonic nozzles can be used to greatly reduce the pressure of the gas flow. If almost vacuum is achieved in such a strongly diluted gas, a molecular flow at gas densities of $\varrho \leq 1.21 \cdot 10^{-6}$ kg/m^3 is formed at pressures of $p \leq 10^{-6}$ bar (0.1 Pa). This follows the laws of the kinetic theory of gases, whereby the free path length l of the gas molecules lies in areas of $l/d > 0.5$, i.e. is larger than the radius of the pipe. In the transition region of viscous flow to molecular flow, the *Reynolds* number is only $Re = 0.12$ [47].

4.6.9 Compression Shock
Normal Compression Shock

In supersonic flows, the compression shock is a characteristic process. For its thermodynamic description, a one-dimensional, compressible and isentropic gas flow in a pipe with constant cross-section A in the control space between 1 and 2 is considered according to Fig. 4.61.

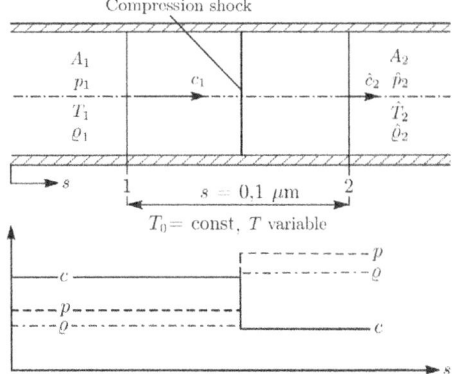

Figure 4.61 Change of state of a supersonic flow during a right-angle compression shock [47]

It is assumed that the distance between the inlet and outlet surfaces of the control space is very small and only has the free path length l of a few molecules. The free

4.6 Dynamics of Ideal Gases: Compressible Stationary Gas Flow

path length of the molecules has, for example, at air of $p_0 = 1$ bar and $t_0 = 20\,°C$, a value of about $l = 0.1\,\mu m$.

The expansion of the shock front, in which the temperature increases abruptly from T_1 to $\hat{T}_2$, depends on the pressure ratio during the compression shock; it leads to values of about $s = 0.07\,\mu m$ to $0.50\,\mu m$.

If the flow is frictionless and incompressible, the solution can be obtained from the continuity equation for $A = $ const, $c_2 = c_1$ and $\varrho_2 = \varrho_1$, which can also be applied to compressible flows at low velocity. In the case of compressible fluids, another solution is obtained with $c_2 \neq c_1$ and $p_2 \neq p_1$ [47].

The state variables after the compression shock are denoted in the following by $\hat{c}_2$, $\hat{p}_2$, $\hat{\varrho}_2$, $\hat{T}_2$, $\hat{a}_2$, $\hat{M}_2$, etc.

The following three balance equations are used to describe the rectangular compression shock (also called normal compression shock):

Continuity equation (Eq. (4.222)) for constant flow cross-section $A = A_1 = A_2$

$$\varrho_1 c_1 = \hat{\varrho}_2 \hat{c}_2 = \frac{\dot{m}}{A} \tag{4.316}$$

Momentum equation according to Eq. (4.227)

$$\varrho_1 c_1^2 + p_1 = \hat{\varrho}_2 \hat{c}_2^2 + \hat{p}_2 \tag{4.317}$$

Specific energy equation (*Bernoulli* equation)

$$\frac{\kappa}{\kappa - 1} \frac{p_1}{\varrho_1} + \frac{c_1^2}{2} = \frac{\kappa}{\kappa - 1} \frac{\hat{p}_2}{\hat{\varrho}_2} + \frac{\hat{c}_2^2}{2} \tag{4.318}$$

This form of the *Bernoulli* equation follows from Eq. (4.245) $h_1 + c_1^2/2 = \hat{h}_2 + \hat{c}_2^2/2 = c_p T_1 + c_1^2/2 = c_p \hat{T}_2 + \hat{c}_2^2/2$ if the thermal equation of state of the ideal gas Eq. (4.26) $T = p/(R\varrho)$ and Eq. (4.49) $c_p = \kappa R/(\kappa - 1)$ are considered and merged to $h = c_p T = c_p p/(R\varrho) = (\kappa/(\kappa-1)) \cdot (p/\varrho)$.

If the flow state before the shock with p_1, c_1, T_1 and ϱ_1 is known, the three balance equations can be used to derive the impact relations for $\hat{p}_2/p_1$, $\hat{c}_2/c_1$ and $\hat{\varrho}_2/\varrho_1$ as solutions.

From the continuity equation (Eq. (4.316)), the momentum equation (Eq. (4.317)) and the specific energy equation (Eq. (4.318)), the thermal equation of state of the ideal gas $p/\varrho = RT$ and the *Mach* number $M = c/a = c/\sqrt{\kappa RT}$ yield the pressure ratio for the shock process

$$\frac{\hat{p}_2}{p_1} = 1 + \frac{2\kappa}{\kappa + 1}(M_1^2 - 1) \ . \tag{4.319}$$

If $M_1 > 1.0$ is given, then $\hat{p}_2/p_1 > 1.0$ follows; the pressure therefore increases (compression shock). The continuity equation (Eq. (4.316)) and the specific energy equation (Eq. 4.318)) lead to the second and third shock relations:

$$\frac{\hat{c}_2}{c_1} = \frac{\varrho_1}{\hat{\varrho}_2} = 1 - \frac{2}{\kappa + 1}\frac{M_1^2 - 1}{M_1^2} \tag{4.320}$$

As a result of the compression shock with $M_1 > 1.0$, the velocity decreases: $\hat{c}_2 < c_1$ resp. $\hat{M}_2 < 1$. For example, for an air flow of $t = 20\,°C$, $\kappa = 1.4$ and $M_1 = 1.6$, the velocity ratio decreases to $\hat{c}_2/c_1 = 0.492$, i.e. to about half the value; thus the density increases to about two times.

4 Ideal Gases

With $a^2 = \kappa RT$, with the thermal equation of state for ideal gases $p/\varrho = RT$ and with the relations according to Eqs. (4.319) and (4.320), the temperature ratio for the normal compression shock leads to

$$\frac{\hat{T}_2}{T_1} = \frac{\hat{a}_2^2}{a_1^2} = \frac{\hat{p}_2}{p_1}\frac{\varrho_1}{\hat{\varrho}_2} = \left[1 + \frac{2\kappa}{\kappa+1}(M_1^2 - 1)\right] \cdot \left[1 - \frac{2}{\kappa+1}\frac{M_1^2 - 1}{M_1^2}\right] . \quad (4.321)$$

This fourth shock relation expresses that in the compression shock the temperature $\hat{T}_2$ increases in the ratio of the pressures $\hat{p}_2/p_1$ or with the decreased velocity ratio $\hat{c}_2/c_1$. The temperature ratio is $\hat{T}_2/T_1 > 1$ for $M_1 > 1$ and the ratio of the velocities of sound $\hat{a}_2/a_1 > 1$ also increases. The ratio of the Mach numbers after and before the compression shock can be given as follows:

$$\frac{\hat{M}_2}{M_1} = \frac{\hat{c}_2}{c_1^2}\frac{a_1}{\hat{a}_2} < 1 , \quad (4.322)$$

because $\hat{c}_2/c_1 < 1$ and $a_1/\hat{a}_2 < 1$.

Thus, for the normal compression shock, the supersonic Mach number M_1 before the shock transforms to the subsonic range. With Eqs. (4.320) and 4.321), the equation for the Mach number $\hat{M}_2$, which is in the subsonic range $\hat{M}_2 < 1$, results in

$$\frac{\hat{M}_2}{M_1} = \left[\frac{\frac{2}{M_1^2} + (\kappa - 1)}{2\kappa M_1^2 - (\kappa - 1)}\right]^{\frac{1}{2}} . \quad (4.323)$$

The ratio of the total pressure (total pressure: see section 2.8.1) after the compression shock $\hat{p}_{t2}$ to the pressure $\hat{p}_2$ is given by

$$\frac{\hat{p}_{t2}}{\hat{p}_2} = \left[1 + \frac{\kappa - 1}{2}\hat{M}_2^2\right]^{\frac{\kappa}{\kappa-1}} . \quad (4,324)$$

The specific entropy increase in the right-angle compression shock is given by Eqs. (4.58) and (4.49):

$$\hat{s}_2 - s_1 = R\left[\frac{\kappa}{\kappa - 1}\ln\frac{\hat{T}_2}{T_1} - \ln\frac{\hat{p}_2}{p_1}\right] \geq 0 \quad (4.325)$$

If in Eq. (4.325) the temperature ratio $\hat{T}_2/T_1$ and the pressure ratio $\hat{p}_2/p_1$ are calculated according to Eqs. (4.321) and (4.319) by the Mach number M_1, the specific entropy change can also be expressed in the following way:

$$\hat{s}_2 - s_1 = R\ln\left[\frac{\left[1 + \frac{2\kappa}{\kappa+1}(M_1^2 - 1)\right]^{\frac{1}{\kappa-1}}}{\left[\frac{(\kappa+1)M_1^2}{2 + (\kappa-1)M_1^2}\right]^{\frac{\kappa}{\kappa-1}}}\right] \quad (4.326)$$

Since, according to the second law of thermodynamics, the specific entropy change $\hat{s}_2 - s_1$ for adiabatic real – i.e. irreversible – flow processes only can increase, and because for the normal compression shock the temperature ratio $\hat{T}_2/T_1$ is > 1 and the pressure ratio is also $\hat{p}_2/p_1 > 1$, in Eq. (4.325) the following must be fulfilled:

$$\frac{\kappa}{\kappa - 1}\ln\frac{\hat{T}_2}{T_1} > \ln\frac{\hat{p}_2}{p_1} \quad (4,327)$$

Table 4.9 contains the ratio values of the relevant state variables in the normal compression shock. The pressure $\hat{p}_2$ and the density $\hat{\varrho}_2$ increase abruptly, and the velocity $\hat{c}_2$ of the gas decreases; besides, the temperature $\hat{T}_2$ and thus also the velocity of sound $\hat{a}_2 = \sqrt{\kappa R \hat{T}_2}$ increase.

It is also relevant that the rest pressure (also called static pressure) $\hat{p}_{02}$ is lower after the compression shock, while the rest temperature has not changed: $\hat{T}_{02} = T_{01}$.

Table 4.9 State variables after a normal compression shock [47]

Pressure ratio	$\hat{p}_2/p_1 > 1 \rightarrow \hat{p}_2 > p_1$
Velocity ratio	$\hat{c}_2/c_1 < 1 \rightarrow \hat{c}_2 < c_1$
Density ratio	$\hat{\varrho}_2/\varrho_1 > 1 \rightarrow \hat{\varrho}_2 > \varrho_1$
Temperature ratio	$\hat{T}_2/T_1 > 1 \rightarrow \hat{T}_2 > T_1$
Velocity of sound ratio	$\hat{a}_2/a_1 > 1 \rightarrow \hat{a}_2 > a_1$
Mach number	$M_1 > 1, \hat{M}_2 < 1$
Critical Mach number	$\hat{M}_2^* = 1/M_1^*$
Specific entropy change	$\hat{s}_2 - s_1 > 0 \rightarrow \hat{s}_2 > s_1$
Rest pressure ratio	$\hat{p}_{02}/p_{01} = \hat{\varrho}_{02}/\varrho_{01} < 1$
	$\rightarrow \hat{p}_{02} < p_{01}$
Rest density ratio	$\hat{\varrho}_2/\varrho_1 < 1 \rightarrow \hat{\varrho}_2 < \varrho_1$
Rest temperature ratio	$\hat{T}_{02}/T_{01} = \hat{a}_{02}/a_{01} = 1$
	$\rightarrow \hat{T}_{02} = T_{01}$
Rest velocity of sound	$\hat{a}_{02} = a_{01}$

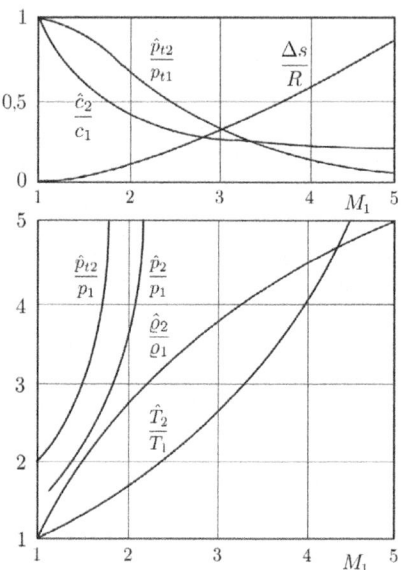

Figure 4.62 (right) Changes of state during a normal compression shock as a function of the incoming flow Mach number M_1

Fig. 4.62 gives the ratio values of the shock relations as functions of the incoming flow Mach number before the compression shock in the range of M_1 between 1 and 5.

The pressure ratio $\hat{p}_2/p_1$, the pressure ratio $\hat{p}_{t2}/p_1$, the temperature ratio $\hat{T}_2/T_1$ and the density ratio $\hat{\varrho}_2/\varrho_1$ increase with an increasing incoming flow Mach number M_1 in the normal compression shock (lower part of Fig. 4.62), while the velocity ratio $\hat{c}_2/c_1$ and the total pressure ratio $\hat{p}_{t2}/p_{t1}$ decrease (upper part of Fig. 4.62).

The specific energy equation (Eq. (4.318)) and its constant sum value (expressed by the left and right sides of the equation) are valid throughout the compression shock. From this condition, the rest quantities $\hat{p}_0, \hat{\varrho}_0, \hat{T}_0$ and $\hat{a}_0$ can also be derived after the compression shock. If this constancy is observed, the following quantities remain unchanged: Rest enthalpy $h_0 = c_p T_0$, rest temperature T_0 and rest velocity of sound $a_0 = \sqrt{\kappa R T_0}$.

The ratio of the rest pressures is equal to the ratio of the rest densities: $\hat{p}_{02}/p_{01} = \hat{\varrho}_{02}/\varrho_{01} < 1$. The ratio of rest pressures in the right-angle compression shock is thus

$$\frac{p_{01}}{\hat{p}_{02}} = \frac{\varrho_{01}}{\hat{\varrho}_{02}} = \left[1 + \frac{2\kappa}{\kappa+1}(M_1^2 - 1)\right]^{\frac{1}{\kappa-1}} \cdot \left[1 - \frac{2}{\kappa+1}\frac{(M_1^2 - 1)}{M_1^2}\right]^{\frac{\kappa}{\kappa-1}}. \quad (4.328)$$

As mentioned, the change of state of the gas flow by a compression shock is a non-reversible process. The increase of the specific entropy is small at a low incoming flow Mach number, but always larger than in the case of the non-rectangular (non-normal)

compression shock. The non-normal compression shock is described in more detail in [47], for example.

Example 4.30 In a supersonic flow of air with $M_1 = 1.5$, $T_1 = 293.15$ K, $p_1 = 1.90$ bar, $\kappa = 1.4$ and $R = 0.2872$ kJ/(kg K), a normal compression shock occurs. What is the pressure ratio $\hat{p}_2/p_1$ and the velocity ratio $\hat{c}_2/c_1$ as well as the pressure $\hat{p}_2$ and the velocity $\hat{c}_2$ after the compression shock?

According to Eq. (4.319), the pressure ratio is

$$\frac{\hat{p}_2}{p_1} = 1 + \frac{2\kappa}{\kappa+1}(M_1{}^2 - 1) = 1 + \frac{2 \cdot 1.4}{2.4}(1.5^2 - 1) = 2.458 \ .$$

$\hat{p}_2 = 2.458 \cdot 1.90$ bar $= 4.671$ bar

The velocity ratio is according to Eq. (4.320)

$$\frac{\hat{c}_2}{c_1} = \frac{\varrho_1}{\hat{\varrho}_2} = 1 - \frac{2}{\kappa+1}\frac{(M_1{}^2 - 1)}{M_1{}^2} = 1 - \frac{2}{1.4+1}\frac{(1.5^2 - 1)}{1.5^2} = 0.537 \ .$$

$$\hat{c}_2 = \frac{\hat{c}_2}{c_1}c_1 = \frac{\hat{c}_2}{c_1}M_1 a_1 = \frac{\hat{c}_2}{c_1}M_1 \sqrt{\kappa R T_1}$$

$$\hat{c}_2 = 0.537 \cdot 1.5 \cdot \sqrt{1.4 \cdot 0.2872 \frac{\text{kJ}}{\text{kg K}} \frac{1000 \, \text{Nm}}{\text{kJ}} \frac{\text{kg m}}{\text{N s}^2} \cdot 293.15 \, \text{K}} = 276.546 \ \frac{\text{m}}{\text{s}}$$

Example 4.31 A supersonic wind tunnel is operated with a $Mach$ number $M_1 = 1.7$ in the measuring section. The static pressure in the air jet has the value $p_1 = 1.04$ bar, the temperature is $t_1 = 25\ °$C ($T_1 = 298.15$ K). The isentropic exponent is $\kappa = 1.4$.

A normal compression shock is formed in the experimental section. The static pressure behind the compression shock $\hat{p}_2$, the $Mach$ number $\hat{M}_2$, the total pressure ratio $\hat{p}_{t2}/p_1$, the total pressure $\hat{p}_{t2}$ and the total temperature $\hat{T}_{t2}$ are to be calculated.

Pressure ratio according to Eq. (4.319):

$$\frac{\hat{p}_2}{p_1} = 1 + \frac{2\kappa}{\kappa+1}(M_1{}^2 - 1) = 1 + \frac{2 \cdot 1.4}{2.4}(1.7^2 - 1) = 3.205$$

$\hat{p}_2 = 3.205 \cdot 1.04$ bar $= 3.333$ bar

$Mach$ number $\hat{M}_2$ behind the compression shock, Eq. (4.323):

$$\frac{\hat{M}_2}{M_1} = \left[\frac{\frac{2}{M_1{}^2}+(\kappa-1)}{2\kappa M_1{}^2 - (\kappa-1)}\right]^{\frac{1}{2}} \to \hat{M}_2 = \left[\frac{2+(\kappa-1)M_1{}^2}{2\kappa M_1{}^2 - (\kappa-1)}\right]^{\frac{1}{2}} = \left[\frac{2+(1.4-1)1.7^2}{2 \cdot 1.4 \cdot 1.7^2 - (1.4-1)}\right]^{\frac{1}{2}} = 0.6405$$

Total pressure ratio at isentropic flow behind the compression shock, Eql. (4.324):

$$\frac{\hat{p}_{t2}}{\hat{p}_2} = \left[1 + \frac{\kappa-1}{2}\hat{M}_2{}^2\right]^{\frac{\kappa}{\kappa-1}} = \left[1 + \frac{1.4-1}{2}0.6405^2\right]^{\frac{1.4}{1.4-1}} = 1.318$$

$\hat{p}_{t2} = \hat{p}_2 \dfrac{\hat{p}_{t2}}{\hat{p}_2} = 3.3333$ bar $\cdot 1.318 = 4.393$ bar

Temperature ratio, Eq. (4.255):

$$\frac{\hat{T}_{t2}}{T_1} = 1 + \frac{\kappa-1}{2}M_1{}^2 = 1 + \frac{1.4-1}{2}1.7^2 = 1.578$$

$\hat{T}_{t2} = T_1 \dfrac{\hat{T}_{t2}}{T_1} = 298.15$ K $\cdot 1.578 = 470.48$ K $\quad (197.33\ °$C$)$

Tasks for Section 4

1. On an oxygen cylinder, it is remarked the empty weight 70 kg[11] and the volume 40 ℓ. A weight check of the cylinder described by the welder as empty gives the weight 72.5 kg at a temperature of 20 °C. What is the pressure of the oxygen in the cylinder?

2. A natural gas consumption of 47.6 m³ has been read on a gas meter. During consumption, the natural gas (molar mass 16.03 kg/kmol) was on average under a gage pressure of 20 mbar. The ambient air had the pressure 961 mbar and the temperature 9 °C. The consumer paid 0.30 Euro for 1 m³ of gas.

a) How much is 1 kg of natural gas?

b) By how many Euro does the price of 1 kg of natural gas increase if the given volume is used under a gage pressure of 16 mbar at the same ambient air pressure of 961 mbar but at an ambient air temperature of 34 °C?

3. A balloon filled with helium is to reach a summit altitude of 6000 m where the barometric pressure is 500 mbar and the temperature is 0 °C. The balloon is not fully inflated at the launch site and just reaches spherical shape at 6000 m altitude. During the ascent, the pressure and temperature in the balloon are constantly equal to the ambient air pressure and ambient air temperature. The mass of the balloon envelope and the load is 1350 kg.

a) What diameter does the balloon reach at 6000 m altitude?

b) What mass of gas is required to fill it?

c) What volume does the balloon occupy at the launch site at 1013 mbar and 30 °C?

d) What is the buoyancy in the hovering state on the ground and at 6000 m altitude?

4. With an electric heating device, a gas flow of 3000 m³/h of carbon dioxide used for a chemical process is heated from 20 °C and 933 mbar to 150 °C.

a) What connected load is required if 8 % of the electrical power is dissipated to the environment as heat loss (temperature 20 °C)?

b) What would be the gas outlet temperature if no heat loss occurred?

5. The altitude corrector of a carburettor (Fig. 4.63) is used to adjust the mixture formation to atmospheric conditions. It consists of 14 barodoses (1 to 14) which are rigidly connected to each other. The cylindrical cans with $D = 20$ mm in diameter and $s_1 = 5$ mm in height are filled with a sealed volume of air and can change their height smoothly. The adjustment change of the cans is taken by a stamp 15 whose diameter is $d = 2$ mm. This stamp adjusts the fuel supply to the carburettor with increasing local height. The air filling of the cans is in a state of 15 °C and 1013 mbar. With a change in altitude to 3000 m, the ambient air pressure drops to 701 mbar and the ambient air temperature to -4.5 °C.

a) By which polytropes can the temperature and pressure decrease of the ambient air be represented?

b) What adjustment can be taken from the stamp and used for fuel control if the air in the cans is exposed to the same change of state as the ambient air?

c) What heat transfer must occur between the ambient air and the air in the cans?

---

[11] According to the German Implementing Ordinance to the Law on Units of Measurement of June 26th, 1970, § 7 (4), units of weight as a designation used in commercial transactions for the indication of quantities of goods for mass are the units of mass.

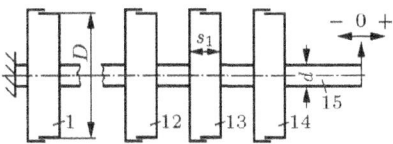
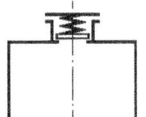
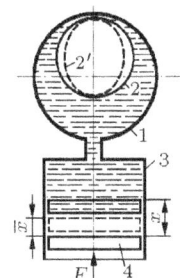

Figure 4.63 To task 5   **Figure 4.64** To task 6

Figure 4.65 To task 7

6. A cylinder (Fig. 4.64) with a capacity of 600 ℓ equipped with a safety valve, contains carbon dioxide of 4 bar gage pressure and 20 °C. The valve is set at 12 bar gage pressure. The ambient air pressure is 1027 mbar. Heat is applied until the safety valve responds.

a) What is the temperature reached and the heat supplied?

b) While the valve blows off, heating continues until the cylinder content reaches the temperature 600 °C. How much gas has escaped?

c) What heat was supplied when the valve blew off?

d) After reaching 600 °C, the heat supply stops, the valve closes the cylinder. Slowly the carbon dioxide content of the cylinder cools down to the ambient temperature. What is the gage pressure in the cylinder now?

e) What heat is given off to the surroundings?

7. A hydraulic storage (accumulator) consists of a spherical container 1 which contains nitrogen in a plastic bladder 2. The remaining space is filled with hydraulic oil (Fig. 4.65). In contrast to hydraulic oil, the nitrogen filling is highly compressible (bubble 2') and can be used as an energy storage.

The type of a hydraulic accumulator e.g. is used as a hydropneumatic spring. A hydraulic accumulator of 4 ℓ capacity is connected to a hydraulic cylinder 3 of 106 mm diameter. The plastic bladder of the reservoir contains 2.2 ℓ of nitrogen under 5.5 bar gage pressure at an ambient air pressure of 0.95 bar. The force $F = 13\,500$ N suddenly acts on piston 4, causing an increase in pressure of the oil and compressing the bladder.

a) What change of state describes the process?

b) How high does the pressure rise?

c) What stroke x does the piston of the cylinder make?

d) What temperature does the nitrogen reach if the temperature at the beginning was 20 °C?

e) After absorbing the volume change work caused by the impact, the volume of nitrogen expands until a state of equilibrium is reached. What stroke $\bar{x}$ to the equilibrium state does the piston perform?

8. An analysis of dry atmospheric air gives 78.04 % N_2, 20.99 % O_2, 0.93 % Ar, 0.03 % CO_2 and 0.01 % H_2.[12]

a) What are the density in the physical norm state and the specific gas constant?

b) Often, for simplification, all gases except oxygen are added to nitrogen, and the composition is given as 21 % O_2 and 79 % N_2. What density in the physical norm state must be used for the nitrogen in this case if the density of the air is to reach the value calculated above?

---

[12] Without special reference, the composition of a gas mixture is given in volume fractions r_i. The volume fraction for CO_2 in task 8 corresponds to the value in 1925; in contrast, the value in 2021 is 0.0417 %.

9. The exhaust gas of an older stationary *Otto* engine at idle contains 74 % N_2, 9 % H_2O, 7 % CO_2, 5 % CO, 4 % H_2 and 1 % O_2. It is passed through an absorber to reduce pollutants, which lowers the CO content to 1 %.

a) What amount of CO is absorbed per hour when 7 kg/h of exhaust gas is passed through the absorber?

b) What is the composition of the exhaust gas after pollution reduction?

10. In the exhaust gas pipe of an old steam boiler whose diameter is 350 mm, the flow velocity is measured to be 3.5 m/s. At the measuring point the temperature of the exhaust gas is 220 °C and the pressure is 960 mbar. The analysis yields the following composition of the dry exhaust gas (steam not included): 12.1 % CO_2, 2.6 % O_2, 6.3 % CO, 79.0 % N_2. A separately performed determination of the water content results in a water vapor content of 53 g H_2O per m^3 of moist exhaust gas in the physical norm state. What quantity of exhaust gas flows through the pipe?

11. Endogas can be used as a protective gas in a process of treating steel. It is produced, for example, from coke oven gas by complete catalytic decomposition of CO_2 and CH_4 at temperatures between 750 and 950 °C. In one plant, 750 m^3/h of coke oven gas in the physical norm state with the composition 5.1 % CO_2, 17.4 % CO, 50.8 % H_2, 17.2 % CH_4, 9.5 % N_2 are processed with air. The following reactions take place completely:

$$CO_2 + CH_4 = 2\ CO + 2\ H_2 \qquad\qquad 2\ CH_4 + O_2 = 2\ CO + 4\ H_2$$

a) What amount of air (21 % O_2, 79 % N_2) must be supplied to meet the oxygen demand?

b) What amount of endogas is produced?

c) What is the composition of the resulting endogas?

12. A vessel of 1 m^3 capacity filled with air is evacuated to an absolute pressure of 0.8 bar. The ambient air pressure is 1070 mbar, the temperature 20 °C. After this, the vessel is filled with water gas (composition 49 % H_2, 43 % CO, 5 % CO_2, 3 % N_2) until the pressure of 1070 mbar is reached again.

a) What mass of gas is then in the vessel?

b) What is the composition of the gas?

13. In a forging furnace, 13 m^3/h of exhaust gas in the physical norm state is produced. The gas has the temperature 650 °C. The exhaust gas gives off some of its energy in an air preheater and cools down to 430 °C in the process (Fig. 4.70). Leakage increases the volume of the exhaust gas to 15.5 m^3/h in the physical norm state, because air from the environment with 20 °C is sucked in. The exhaust gas consists of 8 % H_2O, 14 % CO_2, 8 % CO and 70 % N_2. In the air preheater, 9.5 m^3/h of combustion air in the physical norm condition is preheated with an inlet temperature of 20 °C for the forging furnace. To what temperature can the air be heated? The internal friction power is negligible.

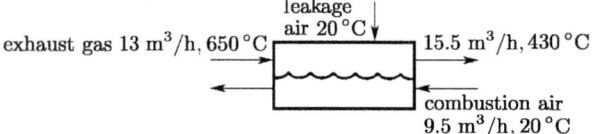

Figure 4.70 To task 13

14. A car tyre with a volume of 20 litres is filled with air, the temperature of which is 18 °C and the pressure of which is 2.9 bar. While driving, the temperature increases to 30 °C and the volume increases to 1.03 times. What is the mass and the molar quantity of air in the tyre? To what value does the pressure increase?

150   4 Ideal Gases

15. A 250 mm high glass cylinder filled with air and having an inner diameter of 100 mm is closed at the top with a weightless and frictionless movable piston. The air has a temperature of 20 °C and a pressure of 0.95 bar (state 1). A mass with a weight G is attached to the piston. This moves the piston downwards and after some time it comes to a standstill at a height of 150 mm, whereby an isothermal change of state of the air to state 2 is to be assumed. What pressure results? How great is the weight force G? Now the glass cylinder is extended and turned around, whereby the air reaches state 3 as a result of an additional isothermal change of state. What is the pressure and what is the minimum length of the cylinder to make sure that the piston does not fall out? What is the mass of the enclosed air and how many molecules does it contain?

16. A *Zeppelin* airship has a volume of 250 000 m^3 filled with the noble gas helium. At an altitude of about 500 m, at an air pressure of 0.95 bar, it suddenly moves from an area of low air temperature with 7 °C into a warm air region of the same pressure with an air temperature of 20 °C. How does the buoyancy – and thus the load force – of the airship change? (It should be noted that the temperature of the helium in the airship cannot immediately equalise with the temperature of the ambient air and therefore remains constant.)

17. Natural gas (i.e. predominantly methane CH$_4$) is stored in a cylindrical disc gas tank. The thermodynamic properties of methane are $c_p = 2.156$ kJ/(kg K) = const; $c_v = 1.638$ kJ/(kg K) = const; $M = 16.043$ kg/kmol. The cylinder has a base area of 25 m diameter and a movable concrete lid with the mass $m_D = 80\,000$ kg. During the day, in state 1, the gas volume has a height of 70 m at a temperature of 15 °C. During the night, the height decreases to 69.0 m (state 2) due to isobaric cooling of the natural gas. The change of state is without friction. The atmospheric pressure of the environment remains unchanged at $p_{1L} = p_{2L} = 1.0$ bar = const.

a) What is the difference $V_2 - V_1$? The shift work that the ambient air adds to the gas is to be calculated.

b) What is the weight force G_D of the lid and the pressure p_D exerted by the lid on the gas? What is the total pressure p_{ges} on the gas? What is the volume change work done on the gas?

c) Determine the specific gas constant R and the isentropic exponent κ.

d) What is the mass $m_1 = m_2$ of the gas? How large is t_2?

e) The differences of the internal energy $U_2 - U_1$, of the enthalpy $H_2 - H_1$ and of the entropy $S_2 - S_1$ are to be calculated. What is the heat $Q_{12} = (Q_{12})_{rev}$ given off by the gas to the surroundings?

18. Experiments are carried out on the *Otto* engine of a combined heat and power plant to separate CO$_2$ from the exhaust gas with the aid of a cooling, compression and subsequent pressure water scrubbing process. The hot exhaust gas contains the following gaseous components in volume fractions r_i: 72 % N$_2$, 14 % H$_2$O, 10 % CO$_2$, 1 % CO, 3 % O$_2$. The exhaust gas can be considered as an ideal gas.

a) Determine the specific gas constant R and the density ϱ_0 of the exhaust gas in the physical norm state. What are the mass fractions μ_i of all components in the exhaust gas?

b) During cooling, compression and pressurised water scrubbing, not only the CO$_2$ but also the water vapor H$_2$O is separated from the exhaust gas. What quantity of CO$_2$ and what quantity of H$_2$O are separated every hour when 8 kg/h of exhaust gas leave the *Otto* engine? What is the composition in volume fractions of the CO$_2$–free and H$_2$O–free exhaust gas?

5 Real Gases and Vapors

Any pure substance can occur in three different phases: solid, liquid and gas. The phases differ in their physical properties such as density, specific heat capacity, refraction index. In the gaseous phase, the substance is called a real gas or a vapor. There is no difference between a real gas and a vapor. However, it has become common usage to refer to real gases near their liquefaction as vapors.

With decreasing pressure p and increasing specific volume v, the real gas approaches the state of the ideal gas.

5.1 Properties of Vapors

A vapor can occur in complete isolation and fill a space on its own, but it can also exist together with the associated liquid or solid. A vapor can also be a component of a mixture of gases. Thus, water vapor in the atmosphere is a component of the gas mixture moist air.

In order to describe the properties of a vapor, it is helpful to include the liquid and the solid phases also in the consideration.

5.1.1 Phase Transitions

In a state diagram the different phases are assigned to certain regions, as shown schematically in Fig. 5.1. The individual regions are separated from each other by boundary lines. Between the regions of the liquid and of the vapor phase lies the vapor pressure

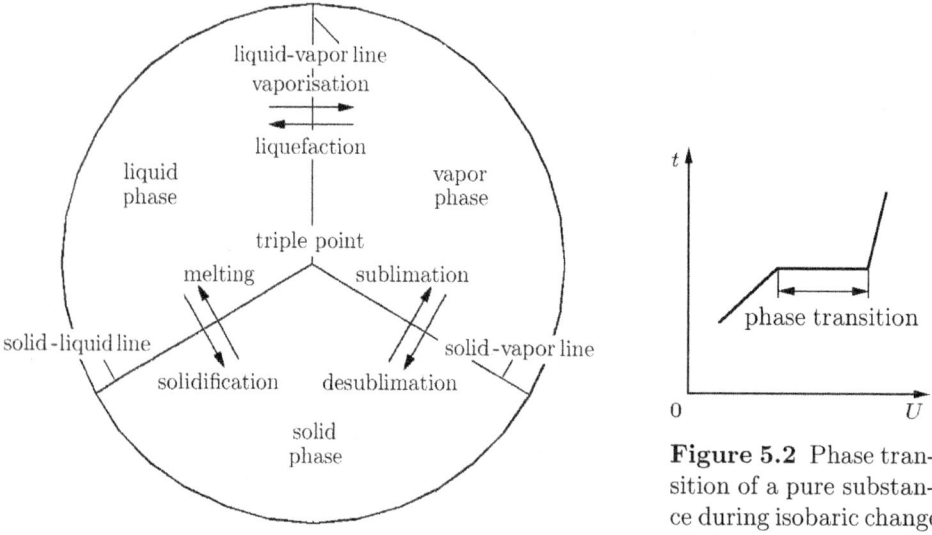

Figure 5.1 Schematic representation of the phase regions

Figure 5.2 Phase transition of a pure substance during isobaric change of state

curve, also called liquid-vapor line. The transition from the liquid to the gaseous phase of a substance is called vaporisation. The reverse process is called liquefaction or con-

© The Author(s), under exclusive license to
Springer Fachmedien Wiesbaden GmbH, part of Springer Nature 2023
M. Dehli et al., *Fundamentals of Technical Thermodynamics*,
https://doi.org/10.1007/978-3-658-38910-9_5

densation. Between the regions of the solid and the gas lies the sublimation pressure curve (also called solid-vapor line) as a boundary. The transition from the solid to the gas phase is called sublimation, the transition in the opposite direction desublimation. The regions of the solid and the liquid phases are separated by the melting pressure curve (also called solid-liquid line). Melting and solidification (freezing) denote the transitions from the solid to the liquid phase and vice versa.

The three boundary curves meet at the triple point. It denotes the state where all three phases can coexist at the same time.

Different internal energies are associated with the individual phases of a pure substance. The solid phase has the smallest internal energy U. It increases more and more during the transitions to the liquid phase and finally to the gas phase. Initially, with isobaric heat input, the internal energy of a phase can be increased by increasing the temperature. This is only possible up to a certain limit. After that, the phase transition begins and ends at a constant temperature. Only after this transition has been completed can the temperature continue to rise. During the phase transition, the change of state is both isobaric and isothermal (Fig. 5.2).

5.1.2 Two-Phase Regions

A phase transition does not occur simultaneously in all parts of a substance. Thus, between the beginning and the end of the phase transition, the old and the new phase are in variable proportions in relation to each other in a two-phase region. Therefore, in addition to an approach according to Fig. 5.1, one can also consider a phase transition according to Fig. 5.3, in which the regionss of the pure phases are separated from each other by two-phase regions.

The two-phase regions represent transition areas, and the dividing lines between pure phases and two-phase regions mark the beginning and the end of the phase transition, respectively.

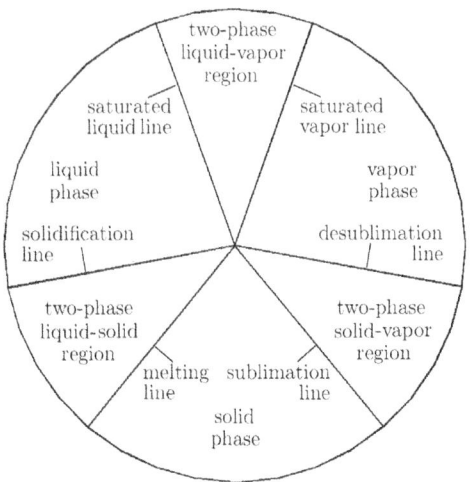

Figure 5.3 Schematic representation of the two-phase regions and separation lines:

  two-phase solid-liquid region with
    melting line and
    solidification line (freezing line)

  two-phase liquid-vapor region with
    boiling line (vaporisation line or
    saturated liquid line) and
    dew line (liquefaction line or saturated vapor line)

  two-phase solid-vapor region with
    sublimation line and
    desublimation line

In pure substances, the region of two-phase liquid-vapor lies between the liquid and the vapor phase. The boundary with respect to the liquid phase is the boiling line

(vaporisation line or saturated liquid line), and the boundary with respect to the vapor phase is the dew line (liquefaction line or saturated vapor line). The regions of the liquid and the solid phase are separated by the melting region with the solidification line (freezing line) and the melting line as the boundaries. At very low pressures there is also a direct transition from the solid to the gaseous phase; in this case one passes through the sublimation region by first crossing the sublimation line and then the desublimation line.

A comparison of the names for the boundary lines of the two-phase regions with the names of the phase transitions shows that the names of the boundary lines according to Fig. 5.3 originate from the phase transitions which lead from the regions of the pure phases into the two-phase regions. Boiling line and dew line are an exception. They should actually be called vaporisation line or saturated liquid line and liquefaction line or saturated vapor line.

5.1.3 Boiling and Condensing

Of the above-mentioned two-phase areas of pure substances, the region of two-phase liqid-vapor is the most important in technical machines and installations. The highest temperature that a liquid can reach during isobaric heating is the boiling temperature (saturated liquid temperature). It is a function of the pressure: The boiling temperature increases with increasing pressure. The boiling liquid (saturated liquid) has the maximum internal energy that a liquid in stable equilibrium can have at a given pressure. Boiling is an evaporation process in turbulence. When heat is applied to the liquid from below, the liquid surface below is not sufficient for the phase transition. The liquid creates further phase interfaces inside in the form of vapor bubbles. These rise to the upper liquid surface and swirl through the liquid.

The vapor that is created during the phase transition is called saturated vapor. It has the minimum internal energy that a vapor in stable equilibrium can have at a given pressure. Saturated liquid and saturated vapor have the same temperature and are under the same pressure. This means that saturated liquid and saturated vapor are in thermodynamic equilibrium. The pressure of saturated vapor is the saturation pressure. It is a function of temperature. The saturation pressure as a function of temperature and the saturated temperature as a function of saturation pressure represent the inversions of the same function: It is the vapor pressure curve mentioned in section 5.1.1, which is also called the solid-liquid line. Instead of saturation pressure one can also say boiling pressure and instead of saturation temperature one can also say boiling temperature. In analogy to saturated vapor, one therefore also speaks of saturated liquid instead of boiling liquid.

In order for a liquid to turn into vapor, the vaporisation heat must be added to it. The vaporisation heat is greater than the difference between the internal energies of saturated vapor and saturated liquid because vaporisation is also associated with a considerable increase in volume. If one considers isobaric vaporisation, vaporisation heat is actually an enthalpy of vaporisation.

The term saturated vapor can be explained as follows. When a liquid is brought into a completely evacuated volume, the liquid boils immediately. The boiling process does not stop until the vapor pressure in the space above the liquid surface has reached the saturation pressure. Then the room is "satiated", it cannot hold more vapor. The space is filled with saturated vapor.

Liquid and vapor in a saturated state are collectively called wet vapor or two-phase liquid-vapor. Saturated liquid and saturated vapor are in a state of equilibrium in the same volume. For thermodynamic purposes, it is irrelevant in which form the liquid is present in the volume: whether it is uniformly distributed and mixed with the saturated vapor (condensation droplets), or it exists as a coherent liquid (bottom body).

The temperature of the wet vapor cannot be increased by isobaric heat supply. This causes an increase of the amount of saturated vapor and a decrease of the amount of the saturated liquid. If the wet vapor has been completely transformed into saturated vapor, a further isobaric heat supply will cause a temperature increase. Superheated vapor is obtained.

If superheated vapor is isobarically cooled, saturated vapor is reached at the saturation temperature. The dew line (liquefaction line or saturated vapor line) is reached and condensation of the saturated vapor begins. At a constant temperature, the amount of saturated vapor decreases and the amount of saturated liquid increases until condensation ends when the boiling line (vaporisation line or saturated liquid line) is reached. Further heat extraction leads to cooling of the liquid.

If, according to Fig. 5.4, a liquid volume is closed off by a piston which exerts a constant pressure on the liquid, the vaporisation can only take place at the temperature at which the vapor pressure is equal to the pressure applied from the outside. This is the boiling temperature or saturation temperature t_S. As long as the temperature is lower than t_S, there is only liquid in the cylinder. Its specific volume, specific enthalpy and specific entropy shall be denoted by v_F, h_F, s_F.

If the liquid is heated to the boiling temperature (saturation temperature) t_S, vaporisation begins at all points in the liquid, the liquid boils. The specific state variables of the boiling liquid are denoted by v', h', s'.

If heat is added further, a vapor space is formed. The liquid and vapor continue to be at boiling temperature and the volume increases extensively. For example, the increase in volume during the vaporisation of water at a pressure of 1 bar is more than 1600 times. The sum of liquid and vapor is called wet vapor or two-phase liquid-vapor and its specific state variables are denoted by v_N, h_N, s_N.

When the last drop of liquid is vaporised, the entire volume is filled with saturated vapor. Its specific state variables are v'', h'', s''. At the slightest withdrawal of heat, the saturated vapor begins to condense. The liquid can be deposited in the form of very fine droplets in the vapor space itself or on cold boundary walls.

If the saturated vapor is heated further, the superheated vapor, unsaturated vapor or hot vapor is formed. Its temperature is always higher than the saturation temperature t_S. The specific state variables are $v_{\ddot{u}}, h_{\ddot{u}}, s_{\ddot{u}}$. The superheated vapor is at the same time a real gas.

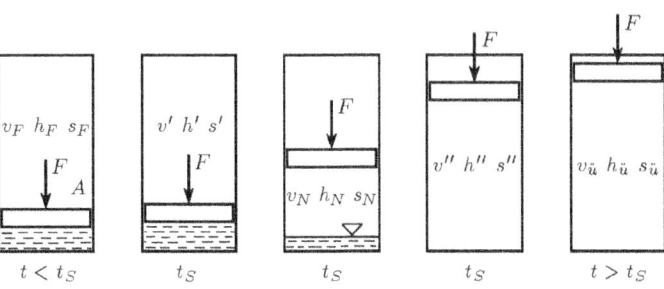

Figure 5.4 Boiling the liquid under constant pressure $p = F/A$

Example 5.1 In 1997, the previous 'Properties of Water and Steam in SI-Units' [113] were replaced by the 'Industrial Standard IAPWS IF97' [128]. The vapor pressure curve (liquid-vapor line, Fig. 5.5) of water p_S/p^* and T_S/T^* according to [128] are given.

$$\frac{p_S}{p^*} = \left[\frac{2C}{-B + (B^2 - 4AC)^{0,5}}\right]^4 \qquad \frac{T_S}{T^*} = \frac{n_{10} + D - \left[(n_{10} + D)^2 - 4(n_9 + n_{10}D)\right]^{0,5}}{2}$$

$p^* = 1\,\text{MPa}$ $1\,\text{MPa} = 10\,\text{bar}$ $T^* = 1\,\text{K}$ $\vartheta = \dfrac{T_S}{T^*} + \dfrac{n_9}{(T_S/T^*) - n_{10}}$ $\beta = (p_S/p^*)^{0,25}$

$A = \vartheta^2 + n_1\vartheta + n_2$ $B = n_3\vartheta^2 + n_4\vartheta + n_5$ $C = n_6\vartheta^2 + n_7\vartheta + n_8$

$D = \dfrac{2G}{-F - (F^2 - 4EG)^{0,5}}$ $E = \beta^2 + n_3\beta + n_6$ $F = n_1\beta^2 + n_4\beta + n_7$ $G = n_2\beta^2 + n_5\beta + n_8$

The constants n_1 to n_{10} have the following values:

$n_1 = 0.11670521452767 \cdot 10^4$
$n_2 = -0.72421316703206 \cdot 10^6$
$n_3 = -0.17073846940092 \cdot 10^2$
$n_4 = 0.12020824702470 \cdot 10^5$
$n_5 = -0.32325550322333 \cdot 10^7$
$n_6 = 0.14915108613530 \cdot 10^2$
$n_7 = -0.48232657361591 \cdot 10^4$
$n_8 = 0.40511340542057 \cdot 10^6$
$n_9 = -0.23855557567849$
$n_{10} = 0.65017534844798 \cdot 10^3$

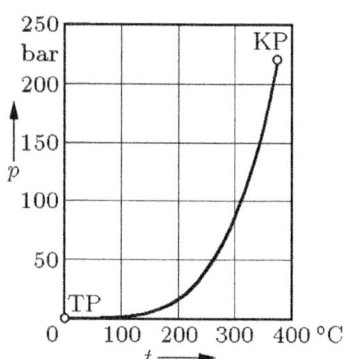

Figure 5.5 Liquid-vapor line of water
TP triple point
$T_{TP} = 273.16\,\text{K}$
$p_{TP} = 611.657\,\text{Pa}$
KP critical point
$T_{KP} = 647.096\,\text{K}$
$p_{KP} = 220.64\,\text{bar}$

5.1.4 Evaporation and Thawing

The previous considerations referred to a system of a pure substance. If a liquid substance that is not represented in the gas phase is brought into a gas space, then the liquid will try to saturate the gas space above the liquid surface with its vapor. Vaporisation begins, which is called evaporation, whereby a gas mixture of the originally present gas and the vapor of the evaporated liquid is formed in the gas space. Such a system is a multi-substance system. Of the possible multi-substance systems, only the case is considered here where the liquid can be regarded as a pure substance while the gas phase consists of a mixture of several substances and also contains the gas phase of the evaporated liquid. For example, water vapor is a component of moist air that is above a liquid water surface.

The evaporation process of the multi-substance system differs from that of the single-substance system. This will be illustrated by the following points:

1) While a phase transition from the liquid phase to the gas phase is only possible at the boiling temperature (satuaration temperature) in a single-substance system, evaporation in a multi-substance system can take place at all temperatures of the liquid phase. Water boils at normal atmospheric pressure at 100 °C. Evaporation temperature of water in air can be any temperature between ice point and boiling point.

2) In an open system in which an isobaric change of state takes place (section 2.6.2), the temperature of a boiling liquid cannot be changed. An evaporating liquid in a multi-substance system can be heated as well as cooled: Increasing the temperature of the liquid intensifies the evaporation, decreasing the temperature decreases the evaporation.

3) Boiling is a vaporisation in turbulence which, when heat is supplied from below, takes place not only on the surface but also inside the liquid. This is expressed in the terms bubble vaporisation and film vaporisation when describing the process from the point of view of heat transfer. In contrast, evaporation in a multi-substance system is a slow evaporation that occurs only at the liquid surface.

4) The pressure of the saturated vapor is equal to the total pressure in boiling. During evaporation in a multi-substance system, the vapor pressure is less than the total pressure.

5) There is thermodynamic equilibrium between the liquid and the saturated vapor during boiling, and only thermal equilibrium during evaporation in a multi-substance system. If instead of the saturated vapor one considers the entire gaseous phase near the surface of the liquid, thermodynamic equilibrium also exists in the case of evaporation. In thermodynamic equilibrium, temperature and pressure are the same. In thermal equilibrium, only the temperatures are equal.

6) Boiling is a transition process, caused by the increase of the specific internal energy above the maximum specific internal energy of the liquid. Evaporation in a multi-substance system is an equalisation process in which the liquid has a partial pressure gradient between the surface of the liquid and the gas space in order to achieve saturation everywhere in the gas space. However, this saturation state can only be achieved if the quantity of liquid evaporating in relation to the gas space is sufficiently large.

Vaporisation is a generic term that generally describes the phase transition from the liquid to the gas phase. Boiling is a phase transition in the single-substance system, evaporation is the corresponding phase transition in the multi-substance system. Liquefaction as the reversal of vaporisation is again a generic term. Condensing is the reversal of boiling in a single-substance system. Thawing refers to a multi-substance system and applies to the process of crossing the saturated vapor line during the cooling of a gas mixture saturated with vapor.

Table 5.1 Phase transitions between liquid and gas

	Vaporisation	Liquefaction
Single-substance system	Boiling	Condensing
Multi-substance system	Evaporation	Thawing

Naming the boundary curves which limit the two phase liquid-vapor region as saturated liquid line and saturated vapor line makes sense. In contrary, naming them as the boiling line and the dew line is not entirely consistent, since the term boiling belongs to the single-substance system and the term dew to the multi-substance system. It also would make sense, naming the boundary curves as the vaporisation line and the liquefaction line. The two phase liquid-vapor region is also called vapor dome.

If one places an open vessel partially filled with water on a gas flame, an evaporation process begins on the surface of the liquid water, which intensifies as the temperature rises. It cannot be seen because water vapor in the air is invisible. In a boundary layer directly at the surface of the liquid, the saturation state is reached first. However, it is not stable because convection and diffusion processes constantly move vapor away from the liquid surface into areas of lower water vapor concentration in the air. On the other hand, air masses with low moisture content enter the boundary layer region (section 10.5).

As the boiling temperature is approached, the vapor pressure increases more and more and the water vapor increasingly displaces the air from the immediate vicinity of the water surface. The originally existing multi-substance system changes at the water surface in the direction of the single-substance system. The beginning of boiling can be recognised by the restless water surface and the formation of bubbles. The now visible vapor clouds are caused by saturated vapor entering a colder environment, by condensation and the formation of tiny liquid droplets. This is perceived by the observer as mist. If the gas flame is extinguished and thus the heat supply to the water is stopped, the boiling process comes to an end. The vaporisation takes on the form of evaporation again.

A loose lid on the vessel partially filled with water supports the formation of a saturated boundary layer at the water surface. The air is displaced more quickly from the vicinity of the water surface, the attainment of the saturation state and the transition from the multi-substance system to the single-substance system is accelerated. If the gas flame is extinguished, the saturation state is maintained in the gas space closed by the lid. The vaporisation stops. When the vessel cools down, the saturation temperature and saturation pressure drop, so that more and more moisture precipitates out and air enters the vessel from the outside. The saturation temperature is equal to the dew point temperature when humid air is cooled.

5.1.5 Liquid

Since technical processes often lead into the liquid region, it is necessary to make statements about the state variables in the liquid region. The specific heat capacity of a liquid c_F usually represents the specific heat capacity at constant pressure c_p. The specific heat capacity at constant volume c_v is usually unknown for liquids. The following comparison of the values for water at a pressure of 1 bar shows the differences [128].

Table 5.2 Specific heat capacities c_p and c_v of water

	0 °C	20 °C	40 °C	60 °C	80 °C	
c_p	4.2194	4.1848	4.1786	4.1828	4.1955	kJ/(kg K)
c_v	4.2170	4.1574	4.0725	3.9736	3.8703	kJ/(kg K)
c_p/c_v	1.0006	1.0066	1.0260	1.0526	1.0840	

For the heat transferred and the changes in enthalpy and entropy per unit mass in the case of isobaric change of state one obtains

$$\mathrm{d}q = \mathrm{d}h = c_F \, \mathrm{d}T = c_F \, \mathrm{d}t \tag{5.1}$$

$$\mathrm{d}s = c_F \frac{\mathrm{d}T}{T} \, . \tag{5.2}$$

With c_F as a constant mean value, integration yields

$$h_F - h_0 = c_F(t - t_0) \tag{5.3}$$

$$s_F - s_0 = c_F \ln \frac{T}{T_0} . \tag{5.4}$$

At this point, the question arises as to the zero point of the scales for enthalpy and entropy. For the caloric state variables, one determines a common zero point in the entire state area of solids, liquids and gases. According to the system of equations by *Wagner* and *Kruse* [128], which has become the international industrial standard for the thermodynamic properties of water and water vapor (IAPWS-IF97), for liquid water at the triple point at $0.01\,°C$ and 0.00611657 bar, the specific internal energy and the specific entropy are equal to zero:

$$u'_{TP} = 0 \qquad\qquad s'_{TP} = 0 \tag{5.5}$$

The index TP means triple point, the dash refers to the saturation state of the liquid. The specific enthalpy of the boiling liquid at the triple point is

$$h'_{TP} = 0.000611783 \ \frac{\text{kJ}}{\text{kg}} . \tag{5.6}$$

According to Eq. (2.41), the specific volume of the boiling liquid at the triple point is

$$v'_{TP} = \frac{h'_{TP} - u'_{TP}}{p_{TP}} = \frac{0.000611783 \text{ kJ/kg}}{0.00611657 \text{ bar}} = 0.1000206 \ \frac{\text{kJ}}{\text{kg bar}}. \tag{5.7}$$

With the relation

$$1 \text{ bar m}^3 = 100 \text{ kJ} \tag{5.8}$$

results in the following for the specific volume of the boiling liquid at the triple point

$$v'_{TP} = 0.001000206 \ \frac{\text{m}^3}{\text{kg}} . \tag{5.9}$$

In engineering calculations, the distinction between the ice point at $0\,°C$ and the triple point at $0.01\,°C$ is often irrelevant, so that the zero point of the enthalpy and entropy scales can be approximated at $0\,°C$.

The specific heat capacity c_F is a function of temperature and pressure:

$$c_F = c_F(t, p) \tag{5.10}$$

The state variables of the liquid

$$v_F = v_F(t, p) \tag{5.11}$$

$$h_F = h_F(t, p) \tag{5.12}$$

$$s_F = s_F(t, p) \tag{5.13}$$

can be taken as functions of temperature and pressure from the state tables of the substance. For water, these values can be taken from the tables of the 'Industrial

Standard IAPWS-IF97' [128]. In many cases, such tables exist only for the state of the boiling liquid:

$$v' = v'(t) \quad \text{or} \quad v' = v'(p) \tag{5.14}$$
$$h' = h'(t) \quad \text{or} \quad h' = h'(p) \tag{5.15}$$
$$s' = s'(t) \quad \text{or} \quad s' = s'(p) \tag{5.16}$$

A fluid can be approximately regarded as incompressible in the case of an increase in pressure. Under this condition it follows

$$v_F = v'(t) . \tag{5.17}$$

Considering the specific heat capacity c_F only as a function of temperature it follows

$$c_F = c_F(t) , \tag{5.18}$$

so enthalpy and entropy are also only temperature-dependent. One can directly use the values for the boiling liquid in the whole liquid region:

$$h_F = h'(t) \tag{5.19}$$
$$s_F = s'(t) \tag{5.20}$$

If one takes into account Eqs. (5.3) and (5.4), one must note that the mean values of c_F in both equations are to be formed in different ways. When calculating the enthalpy change the following equation shall be applied:

$$c_{Fm} = \frac{1}{t_2 - t_1} \int_{t_1}^{t_2} c_F \, dt \tag{5.21}$$

When calculating the entropy change the following equation shall be applied:

$$\bar{c}_{Fm} = \frac{1}{\ln \frac{T_2}{T_1}} \int_{T_1}^{T_2} c_F \frac{dT}{T} \tag{5.22}$$

Example 5.2 Water is heated from 0 °C to 300 °C at constant pressure 120 bar. What are c_{Fm} and $\bar{c}_{Fm}$?

According to the 'Industrial Standard IAPWS-IF97' [128], specific enthalpy and specific entropy in the initial state are

$$h_1 = 12.074 \text{ kJ/kg} \qquad s_1 = 0.00039314 \text{ kJ/(kg K)}$$

and in the final state

$$h_2 = 1340.9 \text{ kJ/kg} \qquad s_2 = 3.2397 \text{ kJ/(kg K)} .$$

From this, the average specific heat capacities can be determined:

$$c_{Fm} = \frac{h_2 - h_1}{300 \text{ K}} = 4.4295 \text{ kJ/(kg K)}$$

$$\bar{c}_{Fm} = \frac{s_2 - s_1}{\ln \frac{573.15}{273.15}} = 4.3708 \text{ kJ/(kg K)}$$

The following table contains the true specific heat capacities of water at 120 bar and the indicated temperatures [128].

t °C	$c_F = c_p$ kJ/(kg K)	t °C	$c_F = c_p$ kJ/(kg K)	t °C	$c_F = c_p$ kJ/(kg K)	t °C	$c_F = c_p$ kJ/(kg K)
0	4.1633	80	4.1703	160	4.2992	240	4.6779
10	4.1526	90	4.1793	170	4.3271	250	4.7652
20	4.1494	100	4.1902	180	4.3590	260	4.8688
30	4.1491	110	4.2028	190	4.3953	270	4.9936
40	4.1503	120	4.2174	200	4.4369	280	5.1467
50	4.1530	130	4.2340	210	4.4847	290	5.3398
60	4.1571	140	4.2530	220	4.5397	300	5.5924
70	4.1629	150	4.2746	230	4.6035		

The average properties c_{Fm} and $\bar{c}_{Fm}$ according to Eqs. (5.21) and (5.22) need an integration which can be done approximately by using the *Simpson* formula

$$\int_a^b y\, dx = \frac{b-a}{3n}(y_0 + 4y_1 + 2y_2 + 4y_3 + 2y_4 + \cdots + 2y_{n-2} + 4y_{n-1} + y_n),$$

where n is the even number of subdivisions (Fig. 5.6). With $n = 30$ and the given c_F values one obtains

$$c_{Fm} = \frac{1}{300\,\text{K}} \int_{0\,°C}^{300\,°C} c_F\, dt = 4.4294 \text{ kJ/(kg K)}$$

$$\bar{c}_{Fm} = \frac{1}{\ln\frac{573,15}{273,15}} \int_{273,15\,\text{K}}^{573,15\,\text{K}} c_F \frac{dT}{T} = 4.3708 \text{ kJ/(kg K)}$$

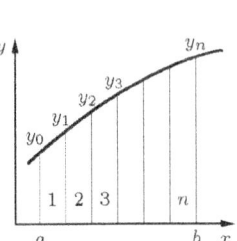

Figure 5.6 Subdivision of an integration area between the boundaries a and b into n sections

The congruence with the values determined from the 'Industrial Standard IAPWS-IF97' [128] is very good. The example shows the necessity of the different averaging methods when applying Eqs. (5.3) and (5.4).

5.1.6 Two-Phase Liquid-Vapor State

The two-phase liquid-vapor state (also called wet vapor state) is a state of equilibrium that extends from the start of boiling to the vaporisation of the last drop of liquid. In the two-phase liquid-vapor state region, in which the wet vapor can have very different compositions, the pressure and temperature remain unchanged in an open system. It is sufficient to specify the pressure or the temperature, because both quantities depend on each other via the vapor pressure curve (liquid-vapor line). However, pressure and temperature do not unambiguously describe the two-phase liquid-vapor state, because the proportions of saturated vapor and saturated liquid can still be arbitrary. Therefore, for an unambiguous description of the two-phase liquid-vapor mixture, one needs an additional state variable named quality x. If m_N is the mass of the two-phase liquid-vapor mixture, m_F the mass of the saturated liquid and m_D the mass of the saturated

vapor, then is the quality x defined as

$$x = \frac{m_D}{m_N} = \frac{m_D}{m_F + m_D} \,. \tag{5.23}$$

For the specific volume of the two-phase liquid-vapor mixture one gets

$$v_N = \frac{V_N}{m_N} = \frac{V_F + V_D}{m_F + m_D} \,, \tag{5.24}$$

where V_N is the volume of the two-phase liquid-vapor mixture, V_F is the volume of the saturated liquid, and V_D is the volume of the saturated vapor. With the specific volume of the saturated liquid v' and the specific volume of the saturated steam v''

$$v' = \frac{V_F}{m_F} \qquad\qquad v'' = \frac{V_D}{m_D} \,, \tag{5.25}$$

one obtains the specific volume of the two-phase liquid-vapor mixture

$$v_N = (1-x)v' + x\,v'' = v' + x(v'' - v') \,. \tag{5.26}$$

From Eq. (5.26) it follows

$$x = \frac{v_N - v'}{v'' - v'} \,. \tag{5.27}$$

For the specific internal energy u_N, the specific enthalpy h_N and the specific entropy s_N of the two-phase liquid-vapor mixture, corresponding relations can be derived:

$$u_N = (1-x)u' + x\,u'' = u' + x(u'' - u') \tag{5.28}$$

$$h_N = (1-x)h' + x\,h'' = h' + x(h'' - h') \tag{5.29}$$

$$s_N = (1-x)s' + x\,s'' = s' + x(s'' - s') \tag{5.30}$$

$$x = \frac{u_N - u'}{u'' - u'} \tag{5.31}$$

$$x = \frac{h_N - h'}{h'' - h'} \tag{5.32}$$

$$x = \frac{s_N - s'}{s'' - s'} \tag{5.33}$$

The difference between the specific enthalpy of the saturated vapor h'' and the specific enthalpy of the saturated liquid h' is the specific enthalpy of vaporisation r (specific heat of vaporisation at constant pressure):

$$r = h'' - h' \tag{5.34}$$

Eq. (5.29) gives with Eq. (5.34)

$$h_N = h' + x\,r \,. \tag{5.35}$$

According to Eq. (5.8) the specific enthalpy of the saturated liquid h' and the specific enthalpy of the saturated vapor h'' are

$$h' = u' + p\,v' \qquad\qquad h'' = u'' + p\,v'' \,. \tag{5.36}$$

5 Real Gases and Vapors

For the specific heat of vaporisation according to Eq. (5.34) it follows that

$$r = u'' - u' + p(v'' - v') . \qquad (5.37)$$

$u'' - u'$ is the specific internal heat of vaporisation. It is used to increase the thermal energy of the molecules.

$p(v'' - v')$ is the specific external heat of vaporisation. It is used to apply the specific volume change work, which at constant pressure is equal to the difference of the specific shift work.

The difference of the specific entropies of the saturated vapor s'' and the saturated liquid s' can be calculated according th Eq. (3.56) by the added specific reversible substitute heat

$$q_{rev} = \frac{Q_{rev}}{m} , \qquad (5.38)$$

which is needed for the vaporisation of one mass unit of liquid, and the boiling temperature T_S, because isobaric vaporisation is also isothermal:

$$s'' - s' = \frac{q_{rev}}{T_S} \qquad (5.39)$$

Since in isobaric vaporisation there is $q_{rev} = h'' - h' = r$, one gets

$$s'' - s' = \frac{r}{T_S} . \qquad (5.40)$$

Eq. (5.30) gives with Eq. (5.40)

$$s_N = s' + x \frac{r}{T_S} . \qquad (5.41)$$

For water, the values for the specific quantities of the saturated liquid v', h', s' and the saturated vapor v'', h'', s'' can be taken from the 'Industrial Standard IAPWS-IF97' [128].

Example 5.3 What is the specific internal and the specific external heat of vaporisation of water at a pressure of 1.2 bar?

From the 'Industrial Standard IAPWS-IF97' for the saturation state at 1.2 bar we get: $t = 104.78\,°C$

$$v' = 0.0010473 \text{ m}^3/\text{kg} \qquad v'' = 1.428445 \text{ m}^3/\text{kg} \qquad r = 2243.8 \text{ kJ/kg}$$

The specific external heat of vaporisation, with 1 bar m^3 = 100 kJ, is

$$p(v'' - v') = 171.3 \text{ kJ/kg} .$$

The specific internal heat of vaporisation is

$$u'' - u' = r - p(v'' - v') = 2072.5 \text{ kJ/kg} .$$

5.1.7 Superheated Vapor

Vapor with a temperature higher than the saturation temperature is superheated vapor. Thus, the water vapor in unsaturated humid air is also superheated vapor. The temperature given by the thermal equation of state

$$F(p, v, T) = 0 \tag{5.42}$$

must be determined experimentally. The simple Eq. (4.26) of ideal gases

$$pv = RT \tag{5.43}$$

is no longer sufficient. With the help of a correction factor z, Eq. (5.43) can be adapted to the laws applicable here. One calls z the real gas factor (also named compressibility factor) and represents it as a function of two thermal state variables:

$$\frac{pv}{RT} = z(p, T) = z(v, T) \tag{5.44}$$

The relation between the specific heat capacities c_p and c_v and the special gas constant R according to Eq. (4.48) no longer holds, c_p and c_v can no longer be calculated by the simple Eqs. (4.49) and (4.50). The differential of the specific internal energy

$$du = c_v \, dT \tag{5.45}$$

is no longer independent of the type of change of state and, according to $c_v = \left(\frac{\partial u}{\partial T}\right)_v$, is only applicable to an isochoric change of state. Also the differential of the specific enthalpy

$$dh = c_p \, dT \tag{5.46}$$

is no longer generally valid, but according to $c_p = \left(\frac{\partial h}{\partial T}\right)_p$ only refers to an isobaric change of state. If one continues to use the abbreviation

$$\kappa = \frac{c_p}{c_v} \tag{5.47}$$

according to Eq. (4.44), however, this is no longer the isentropic exponent n_S. The following relation now applies to it

$$n_S = n_T \kappa \,, \tag{5.48}$$

whose definition can be derived according to Eqs. (5.187) and (5.190). n_T is the isotherm exponent, where in general is $n_T \neq 1$ (section 5.4.3).

Working with real gases and vapors requires knowledge of complicated thermodynamic relationships. Vapor tables and state diagrams are important aids for the engineer.

Example 5.4 For the real gas factor z of air between $0\,°C$ and $200\,°C$ in the pressure range from 0 bar to 100 bar, according to *Holborn* and *Otto* [58] the values of Table 5.3 can be used:

164 5 Real Gases and Vapors

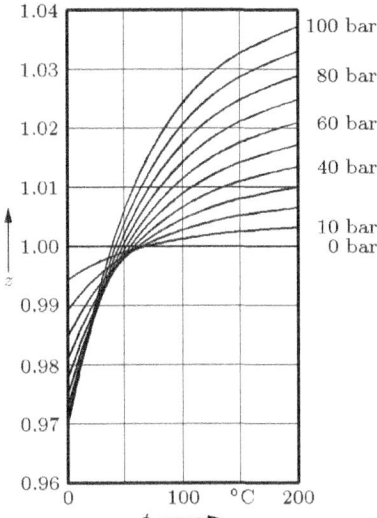

Figure 5.7 Real gas factor for air
$$z(T,p) = \frac{pv}{RT}$$

Table 5.3 Values of the real gas factor $z(T,p)$ for air

	$t=0\,°C$	$50\,°C$	$100\,°C$	$150\,°C$	$200\,°C$
$p=0$	1.0000	1.0000	1.0000	1.0000	1.0000
10	0.9943	0.9990	1.0013	1.0025	1.0032
20	0.9893	0.9984	1.0028	1.0052	1.0065
30	0.9848	0.9981	1.0046	1.0080	1.0099
40	0.9809	0.9982	1.0066	1.0110	1.0135
50	0.9776	0.9986	1.0089	1.0142	1.0171
60	0.9749	0.9994	1.0115	1.0176	1.0209
70	0.9728	1.0005	1.0143	1.0211	1.0248
80	0.9712	1.0021	1.0174	1.0248	1.0288
90	0.9703	1.0039	1.0207	1.0286	1.0329
100 bar	0.9699	1.0061	1.0242	1.0327	1.0372

The real gas factor can be calculated by the equation

$$z(T,p) = \sum_{i=0}^{4}\sum_{j=0}^{2} a_{ij} T^{-i} p^{j}$$

If one inserts the pressure p in bar and applies the absolute temperature T, then the factors a_{ij} have the following numerical values:

$a_{00} = 1$ $a_{01} = -1.141097858 \cdot 10^{-3}$ $a_{02} = 5.167838557 \cdot 10^{-5}$
$a_{10} = 0$ $a_{11} = 2.319871842$ $a_{12} = -7.853777953 \cdot 10^{-2}$
$a_{20} = 0$ $a_{21} = -1.236139494 \cdot 10^{3}$ $a_{22} = 4.404939194 \cdot 10^{1}$
$a_{30} = 0$ $a_{31} = 2.794319527 \cdot 10^{5}$ $a_{32} = -1.072732180 \cdot 10^{4}$
$a_{40} = 0$ $a_{41} = -2.833385469 \cdot 10^{7}$ $a_{42} = 9.728593264 \cdot 10^{5}$

The graphical representation of the real gas factor of air is shown in Fig. 5.7. z increases with higher temperatures and higher pressures.

5.2 State Diagrams

The change of thermodynamic properties with the change of initial values can be followed particularly easily in state diagrams. If they allow a sufficient reading accuracy as working diagrams, technical problems can be solved graphically and the numerical values of interest can be extracted directly.

5.2.1 The p,v,T Surface

For pure substances, the thermal equation of state (5.42) represents in the entire state region of all three phases and of the two-phase regions a surface according to Fig. 5.8, if a three-dimensional application in a rectangular p, v, T-coordinate system is used.

In the two-phase regions, the isobars and the isotherms coincide. Therefore, the two-phase regions are cylindrical surfaces that can be unwound in a plane. The boundary curves between the two-phase regions and the state regions of the pure phases each mark a kink in the p, v, T surface. Only at the critical point is there no kink.

By projecting the three-dimensional p, v, T surface in p-direction, v-direction or T-direction one obtains a two-dimensional T, v diagram, p, T diagram or p, v diagram.

The p,v diagram of Fig. 5.9 and the T,v diagram of Fig. 5.10 correspond to a representation according to Fig. 5.3. The regions of the pure phases are separated by two-phase regions. The two-phase liquid-vapor region as a transition region between liquid phase and gas phase does not separate these two regions completely. At the critical point as the intersection of the critical isobar, the critical isochore and the critical isotherm, the saturated liquid line and the saturated vapor line merge continuously. Above the critical point, no change can be observed in the transition from the liquid phase to the gas phase. This means that in this range the two phases cannot be distinguished. Below the critical point, the transition from liquid to superheated vapur passes through the two-phase liquid-vapor region. Thereby a visible interface — the liquid surface — is formed between the liquid and the vapor.

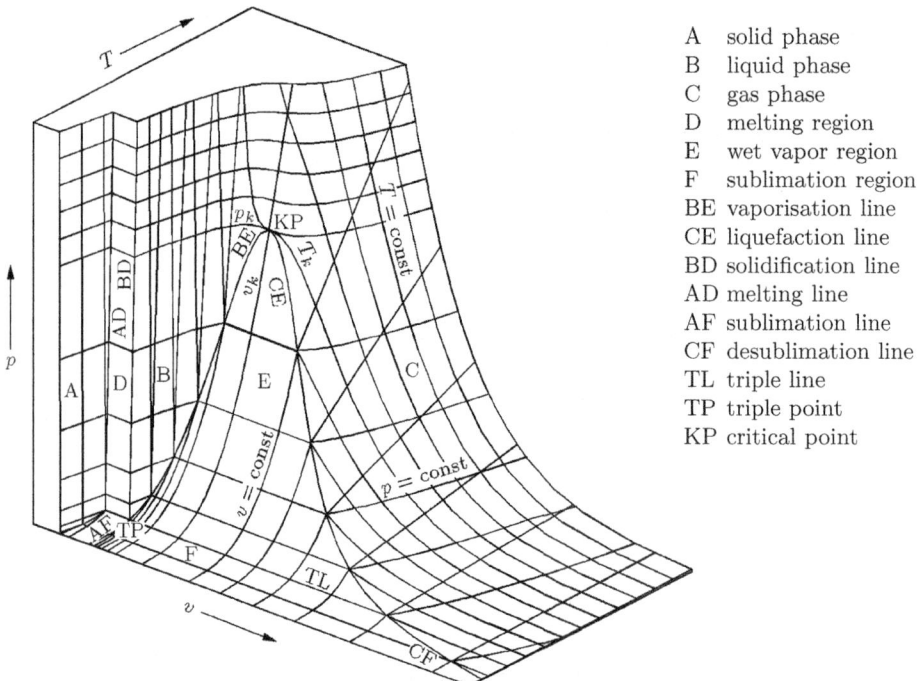

A solid phase
B liquid phase
C gas phase
D melting region
E wet vapor region
F sublimation region
BE vaporisation line
CE liquefaction line
BD solidification line
AD melting line
AF sublimation line
CF desublimation line
TL triple line
TP triple point
KP critical point

Figure 5.8 Principle representation of the p,v,T surface of a pure substance (e.g. CO_2). The specific volume v is plotted logarithmically

The p,T diagram in Fig. 5.11 corresponds to a representation according to Fig. 5.1. It is called a phase diagram. The regions of the pure phases are separated by the boundary curves. Vaporisation line (saturated liquid line) and liquefaction line (saturated vapor line) overlap and appear as a liquid-vapor line (vapor pressure curve). Solidification line and melting line result in the solid-liquid line (melting pressure curve). Sublimation line and desublimation line form the solid-vapor line (sublimation pressure curve). Since the two-phase regions in the p,v,T representation are cylindrical surfaces with the axis in the v-direction, in the p,T diagram they coincide with the respective boundary curve. The liquid-vapor line as a boundary curve between liquid and gas phase only partially separates these two regions. It starts at the triple point and ends at the critical point.

Fig. 5.5 represents a part of Fig. 5.11. Above the critical point, both phase areas merge directly into each other without a boundary.

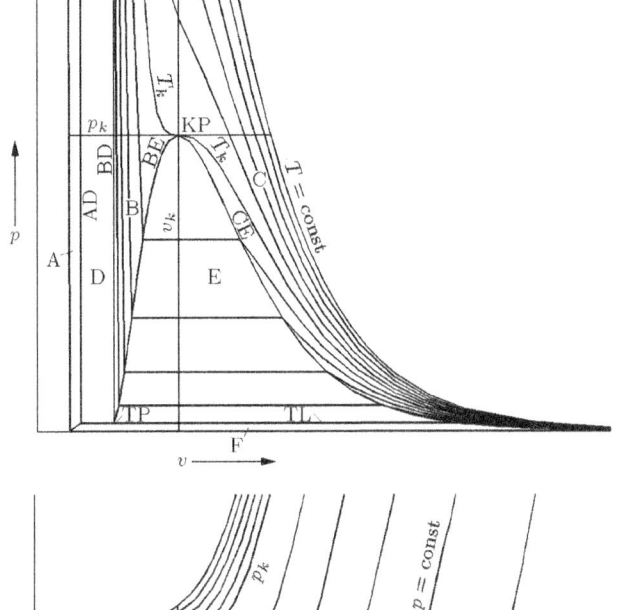

Figure 5.9 p,v diagram with logarithmic plot of the specific volume. The critical isotherm has a terrace point at the critical point. Designation as in Fig. 5.8

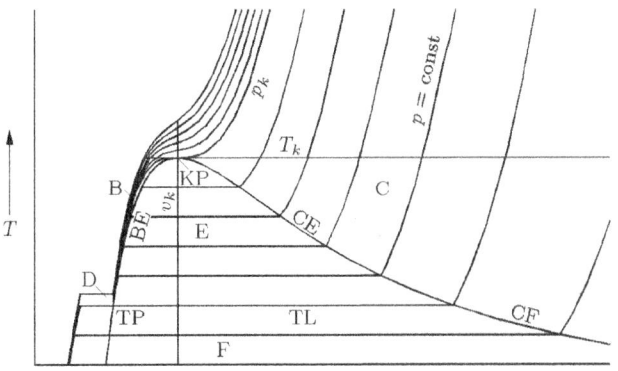

Figure 5.10 T,v diagram with logarithmic plot of the specific volume. The critical isobar has a terrace point at the critical point. Designation as in Fig. 5.8

Figure 5.11 p,T diagram showing the three boundary curves between the three phases: AB solid-liqid line, BC liqid-vapor line, AC solid-vapor line Other designations as in Fig. 5.8

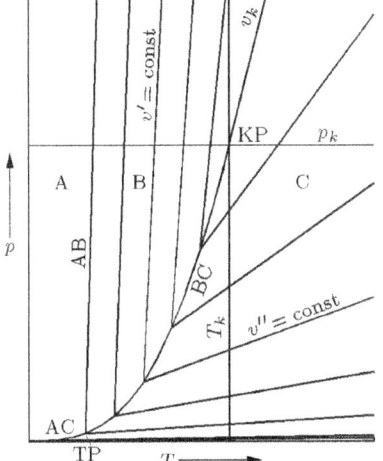

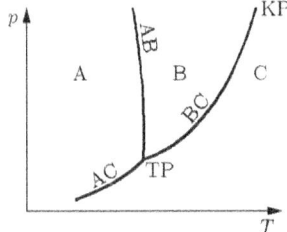

Figure 5.12 p,T diagram for water. Designations as in Fig. 5.8 and Fig. 5.11

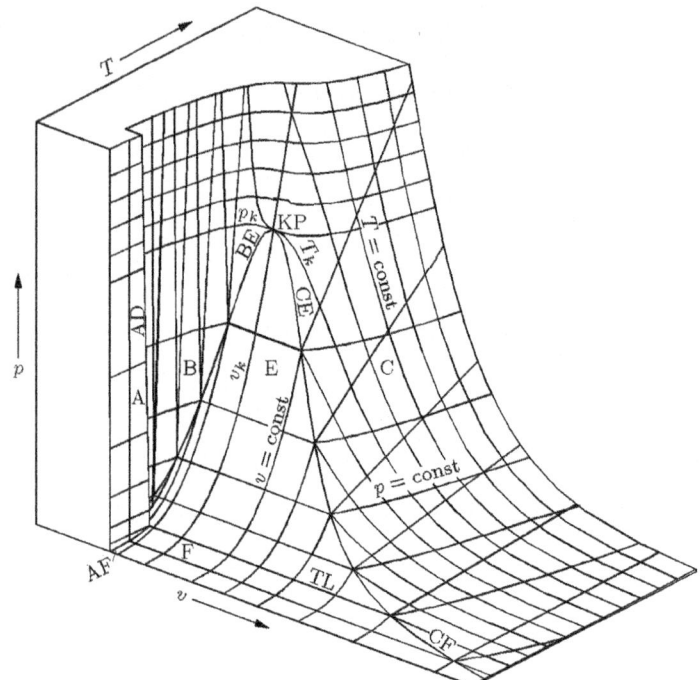

Figure 5.13 Principle representation of the p, v, T surface of water. The specific volume v is plotted logarithmically. Designations as in Fig. 5.8

At the triple point where the solid-liquid line, the liquid-vapor line and the solid-vapor line meet, the transition from the liquid-vapor line to the solid-vapor line is associated with a slight bend. This is not visible in Fig. 5.11, but can be seen in Fig. 5.12, a schematic sketch for water, which also shows that the solid-liquid line has a negative slope.

In the normal case (e.g. CO_2), the slope of the solid-liquid line is positive, as can be seen in Fig. 5.11. Water is an exception, which can be seen, for example, in the fact that the specific volume of the ice is greater than the specific volume of the liquid. Such anomalies are also found in antimony, gallium and bismuth. Because an iceberg floats on water, Fig. 5.8 does not apply to water. The principle representation of the p, v, T area for water is Fig. 5.13. The various modifications [101] in which ice occurs are not taken into account.

5.2.2 The T, s Diagram

In a T, s Diagram according to Fig. 5.14, the saturated liquid line and the saturated vapor line together give a bell-shaped boundary curve, which separates the two-phase liquid-vapor region from the liquid region and the superheated vapor region. The critical point is the highest point of the boundary curve and a terrace point of the critical isobar. For a pressure below the critical pressure, the isobars (p = const) extend through the liquid region, the two-phase liquid-vapor region and the superheated vapor region.

The course of the isobars in the liquid region is almost indistinguishable from the course of the saturated liquid line in the lower part. In the two-phase liquid-vapor region, the isotherms (T = const) and the isobars coincide. In the superheated vapor region the isobars are steeply rising left-hand curves.

168   5 Real Gases and Vapors

In the lower part of the liquid region, the isochores ($v = $ const) are hardly distinguishable from the saturated liquid line. In the two-phase liquid-vapor region, the isochores are curved slightly to the right, diverging fan-like from the lower part of the saturated liquid line. They pass through the saturated vapor line with a bend and in the superheated vapor region are left-hand curves with a greater slope than the isobars.

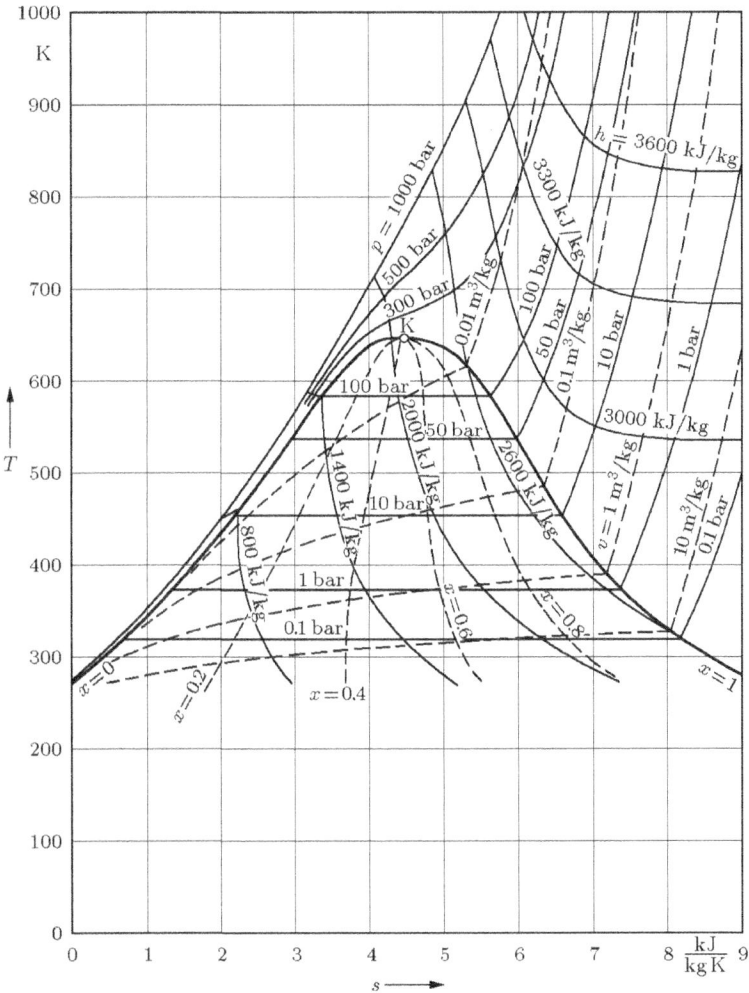

Figure 5.14 T, s diagram for water and steam

In the two-phase liquid-vapor region, the curves of equal quality (equal vapor content) x are found. On the saturated liquid line the quality is $x = 0$, on the saturated vapor line the quality is $x = 1$. In between, the curves $x = $ const are obtained by linear interpolation according to Eq. (5.33).

The curves of constant specific enthalpy ($h = $ const) – the isenthalps – show a hyperbola-like course. In the region of superheated vapor they approach more and more the isotherms with increasing entropy. This is an indication of the approximation of the behaviour of superheated vapor to the behaviour of an ideal gas.

Since according to Eq. (3.92) an area in the T,s diagram represents a specific substitute heat $(q_{12})_{rev}$ and in the case of an isobaric change of state at pressure p, the specific substitute heat $(q_{12})_{rev}$ according to Eq. (4.148) is equal to the specific enthalpy difference $h_2 - h_1$, the specific enthalpy appears in the T,s diagram as an area below of an isobar (Fig. 5.15). The specific enthalpy of the superheated vapor $h_{\ddot{u}}$ can be divided into three parts using Eq. (5.34):

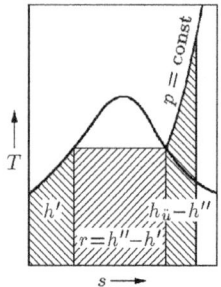

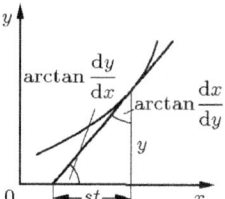

Figure 5.16 To explain the sub-tangent st

Figure 5.15 Plot of specific enthalpy in a T,s diagram

$$h_{\ddot{u}} = h' + r + (h_{\ddot{u}} - h'') \tag{5.49}$$

The specific enthalpy of the boiling liquid h' is formed as the area between the isobar p (which practically coincides with the boiling line) and the abscissa. The specific heat of vaporisation r is found as a rectangle between the saturated liquid line and the saturated vapor line. This is followed by the specific heat in the superheated vapor region $h_{\ddot{u}} - h''$ under the steeply rising isobar.

The subtangents of the isobars and isochores in the T,s diagram represent the specific heat capacities c_p and c_v. This can be seen from the following two equations, which will be discussed later (section 5.4.1):

$$c_p = T \left(\frac{\partial s}{\partial T} \right)_p \tag{5.50}$$

$$c_v = T \left(\frac{\partial s}{\partial T} \right)_v \tag{5.51}$$

It can be seen that in the two-phase liqid-vapor region and at the critical point $c_p \to \infty$ becomes. Eqs. (5.50) and (5.51) can be used to obtain the true specific heat capacity approximately. For this, the aid of properties of specific entropy and temperature is needed which can be taken from a T,s diagram or a vapor table:

If y is a function of x with derivative

$$\frac{\mathrm{d}y}{\mathrm{d}x} = y', \tag{5.52}$$

then the length of the sub-tangent st according to Fig. 5.16 is

$$st = y \frac{\mathrm{d}x}{\mathrm{d}y} = \frac{y}{y'}. \tag{5.53}$$

Replacing x and y by the specific entropy s and the thermodynamic temperature T and the derivative y' by the partial derivative of the temperature T to the entropy s, then with Eq.

(5.53) Eqs. (5.50) and (5.51) are obtained. This consideration is not only valid for isobaric and isochoric changes of state, but in general for all changes of state. If the sub-tangent in the T, s diagram lies to the right of the perpendicular to the abscissa, the specific heat capacity c_n is negative.

Example 5.5 The true specific isobaric heat capacity of water at 120 bar and 300 °C is to be determined approximately.

The slope of the isobar $p = 120$ bar at $t = 300$ °C is replaced by the slope of the connecting straight line between the points at $t_1 = 299$ °C and $t_2 = 301$ °C on the isobar $p = 120$ bar in the T, s diagram:

$$c_p = (t + T_0)\frac{s_2 - s_1}{t_2 - t_1}$$

Using the equations given in the 'Industrial Standard IAPWS-IF97' [128], $t_1 = 299$ °C gives the specific entropy $s_1 = 3.229978$ kJ/(kg K) and at $t_2 = 301$ °C the specific entropy $s_2 = 3.249493$ kJ/(kg K). The evaluation gives $c_p = 5.5924$ kJ/(kg K). This value agrees with the value given in example 5.2.

5.2.3 The h, s Diagram

Enthalpy is of particular importance in engineering calculations. The specific enthalpy appears in the T, s diagram as an area under an isobar. It can be determined by planimetry, but this is a cumbersome procedure. Easier than planimetry is the tapping of a distance. This possibility is offered by the h, s diagram, in which the specific enthalpy appears as the ordinate (Fig. 5.17).

In the h, s diagram for water — according to the agreement on the scale zero point (section 5.1.5) — the beginning of the saturated liquid line can be placed in the origin of the coordinates. This runs steadily upwards as a left-hand curve and merges into the saturated vapor line at the critical point without a kink. The critical point is a turning point of the boundary curve consisting of the saturated liquid line and the saturated vapor line. Since the highest specific enthalpy of saturated vapor occurs at about 236 °C, the saturated vapor line has its highest point here.

The isotherms and isobars in the two-phase liquid-vapor region coincide and are straight lines. According to Eqs. (2.75), (3.91) and (4.78), for an isobaric change of state one can write

$$dH = T\,dS \qquad\qquad dh = T\,ds\ . \qquad (5.54)$$

Applying Eq. (5.54) to the two-phase liquid-vapor region, the temperature T is equal to the boiling temperature T_S:

$$\left(\frac{\partial h}{\partial s}\right)_p = T_S \qquad (5.55)$$

Since the boiling temperature T_S remains constant in the case of isobaric change of state in the two-phase liquid-vapor region, Eq. (5.55) is the mathematical formulation of the statement that the isobars and isotherms in the two-phase liquid-vapor region are straight lines.

The isobars and isotherms approach the boiling line almost tangentially in their lower part in the two-phase liquid-vapor region. However, they are not tangents. At the transition from the two-phase liquid-vapor region to the superheated vapor region, the isobars and isotherms separate. The isobars pass through the saturated vapor line without a kink and run as left-hand curves in the superheated vapor region. The

isotherms bend to the right at the saturated vapor line and almost coincide with the isenthalps at some distance from the saturated vapor line. This corresponds to the approximation of the behaviour of real gases to the behaviour of ideal gases in the superheated vapor region.

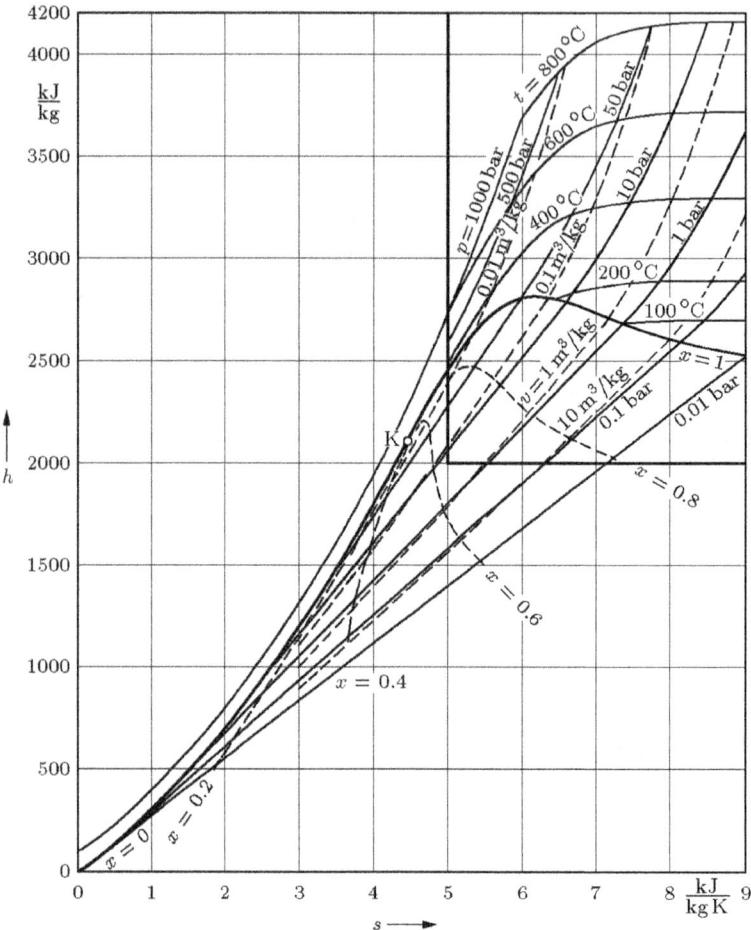

Figure 5.17 h, s diagram for water and steam. The practically important part at the top right is highlighted.

The isochores are steeper than the isobars in the two-phase liquid-vapor region and in the superheated vapor region. At the saturated vapor line they have a kink. The points of equal quality (equal vapor content) x in the two-phase liquid-vapor region are connected by curves.

So that the h, s diagram for water does not become unmanageably large for practical use, one can limit oneself to a section. The section of the superheated vapor region and a part of the two-phase liquid-vapor region has been chosen (see Fig. 5.17). The saturated liquid line, the critical point and the region of the liquid are missing. In these regions it is easy to work with a table containing thermodynamic properties of the fluid (property table, vapor table).

5.3 Thermal Equations of State

The p, v, T surface in a rectangular p, v, T coordinate system is an illustration of the thermal equation of state. The shape of the p, v, T surface makes it clear that it cannot be a simple equation if also the liquid region and the two-phase liquid-vapor region are to be included. Attempts to solve this problem are based partly on physically justified approaches and partly on empirical approaches. Some equations are not generally applicable, but refer only to partial regions.

5.3.1 The *van der Waals* Equation

An equation of state of real gases which, in a first approximation, takes into account those influencing variables by which real gases and ideal gases differ, was given by J. D. van der Waals[1]. By introducing two correction terms, the thermal equation of state of ideal gases is transformed. van der Waals started from the following considerations:

A quantity of gas is compressed by pressure. The greater the pressure is chosen, the smaller the volume of the gas becomes. Eventually, one reaches a limit that is determined by the intrinsic volume of the molecules. A further increase in pressure can no longer reduce the volume, because otherwise the molecules would have to penetrate each other. This limit volume in relation to the gas quantity is denoted by b. The space available for the unit of gas quantity is smaller by this limiting volume than the measured specific gas volume v, it is equal to $(v - b)$.

The second correction takes into account the attractive forces acting between the molecules. Molecules located in the outer surface of the gas are pulled inwards by the attractive forces of the molecules located further inwards, while these forces are balanced on average for a molecule inside the gas. The pressure force of the gas on its boundary surfaces arises from the collisions of the outer molecules on this surface. The impact force of the outer molecules on the boundary surface is reduced by the effect of the attractive forces. The reduction of the impact force of a molecule will be proportional to the number of neighbouring molecules and thus to the gas density ϱ. The effect of all the molecules hitting the boundary surface will also be proportional to the number of molecules and thus to the density ϱ of the gas. One will therefore have to apply a correcting element $a\,\varrho^2 = a/v^2$ to the measured gas pressure p. In comparison to the gas pressure prevailing inside the gas which is decisive for the equation of state, the measured gas pressure on the boundary surfaces p is smaller by $a\,\varrho^2 = a/v^2$. Thus, one must calculate with the term $p + a/v^2$.

According to these considerations, the equation of state of a real gas is

$$(p + \frac{a}{v^2})(v - b) = RT \tag{5.56}$$

with a and b as constants.

There are several ways to determine the constants a, b and R. The first variant is to calculate the three constants from the critical data. The critical data are the pressure p_k, the specific volume v_k and the temperature T_k at the critical point. The critical

---
[1] J. D. van der Waals, 1837 to 1923, Dutch physicist

5.3 Thermal Equations of State

point in the p, v diagram according to Fig. 5.9 is a terrace point (turning point with horizontal tangent) of the critical isotherm. If one develops Eq. (5.56) according to p

$$p = \frac{RT}{v-b} - \frac{a}{v^2}, \tag{5.57}$$

so the conditions for the considered terrace point are given by the first partial derivative and the second partial derivative

$$\left(\frac{\partial p}{\partial v}\right)_T = -\frac{RT_k}{(v_k - b)^2} + \frac{2a}{v_k^3} = 0 \tag{5.58}$$

$$\left(\frac{\partial^2 p}{\partial v^2}\right)_T = \frac{2RT_k}{(v_k - b)^3} - \frac{6a}{v_k^4} = 0. \tag{5.59}$$

From Eqs. (5.58) and (5.59) follows

$$\frac{2a}{v_k^3} = \frac{RT_k}{(v_k - b)^2} \tag{5.60}$$

$$\frac{6a}{v_k^4} = \frac{2RT_k}{(v_k - b)^3}. \tag{5.61}$$

Eq. (5.60) is divided by Eq. (5.61):

$$v_k = 3b \tag{5.62}$$

Inserting Eq. (5.62) into Eq. (5.60) follows:

$$T_k = \frac{8a}{27bR} \tag{5.63}$$

Applying Eq. (5.57) to the critical point, Eqs. (5.62) and (5.63) lead to

$$p_k = \frac{a}{27b^2}. \tag{5.64}$$

From Eqs. (5.62) to (5.64) it follows

$$b = \frac{v_k}{3} \tag{5.65}$$

$$a = 3v_k^2 p_k \tag{5.66}$$

$$R = \frac{8}{3}\frac{p_k v_k}{T_k}. \tag{5.67}$$

(For comparison: for ideal gases, $R = p_k v_k / T_k$ would apply).
In the second variant, one uses the special gas constant of the ideal gas for R and determines v_k, a and b from R, p_k and T_k: If Eq. (5.67) is transformed to the critical specific volume

$$v_k = \frac{3}{8}\frac{RT_k}{p_k} \tag{5.68}$$

and this expression is inserted into Eqs. (5.65) and (5.66), then one obtains

$$b = \frac{1}{8}\frac{RT_k}{p_k} \tag{5.69}$$

$$a = \frac{27}{64}\frac{R^2 T_k^2}{p_k}. \tag{5.70}$$

Example 5.6 The constants a and b of the *van der Waals* equation are to be calculated for water with the properties for the critical pressure, the critical temperature and the specific gas constant of the ideal gas.

Specific gas constant, critical pressure and critical temperature are according to the 'Industrial Standard IAPWS-IF97' [128]

$R = 461.526$ J/(kg K) $= 0.00461526$ bar m^3/(kg K) $\qquad p_k = 220.64$ bar $\qquad T_k = 647.096$ K .

According to Eq. (5.68), the modified critical volume is $v_k = 0.0050647$ m^3/kg .

$\qquad a = 0.017022$ bar m^6/kg^2 $\qquad\qquad b = 0.0016882$ m^3/kg .

5.3.2 The Boundary Curve and the *Maxwell* Relation

The *van der Waals* equation is not applicable to the solid region, the two-phase solid-liquid region and the two-phase solid-vapor region. However, since it is valid for the liquid region and the superheated vapor region, a statement about the two-phase liquid-vapor region and thus about the saturated liquid line and the saturated vapor line are of interest (Fig. 5.18).

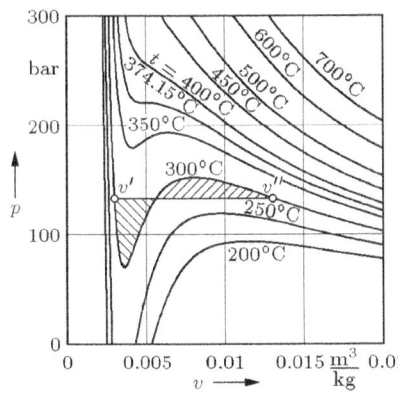

Figure 5.18 p, v diagram with isotherms according to the *van der Waals* equation. Numerical values according to examples 5.6 and 5.7

In the two-phase liquid-vapor region, an isotherm according to the *van der Waals* equation is wave-shaped with a low point and a high point, but a real isotherm coincides with an isobar. According to a consideration of *Maxwell*[2], the real isotherm must have a form in such a way that the two areas which are separated in a p, v diagram by the theoretical isotherm and by the real isotherm are of equal size [82]. Otherwise, a

---
[2] *James Clerk Maxwell*, 1831 to 1879, English physicist who worked extensively on electromagnetic fields as well as on thermodynamic questions

reversible cyclic process with heat input during vaporisation and output of the same heat during the subsequent liquefaction at the same temperature would yield a work output. The *Maxwell* relationship can be expressed by the following equation:

$$\int_{v'}^{v''} p\,dv - p\,(v'' - v') = 0 \tag{5.71}$$

With Eq. (5.57) one obtains

$$RT \ln \frac{v'' - b}{v' - b} + a \left(\frac{1}{v''} - \frac{1}{v'} \right) - p\,(v'' - v') = 0 \,. \tag{5.72}$$

The specific volume of the saturated liquid v' and the specific volume of the saturated vapor v'' are solutions of the transformed Eq. (5.56):

$$v^3 - \left(b + \frac{RT}{p} \right) v^2 + \frac{a}{p} v - \frac{a\,b}{p} = 0 \tag{5.73}$$

Eq. (5.73) is a third degree equation with three real solutions for v in the two-phase liquid-vapor region. v' is the smallest and v'' the largest of the three v values. In Eqs. (5.72) and (5.73), the pressure p must be chosen so that for given values of R, T, a and b Eq. (5.72) is fulfilled. The sequence of the pairs of values of temperature and pressure, which can be derived from Eqs. (5.72) and (5.73) iteratively, represents the liquid vapor line (vapor pressure curve).

If one wants to adapt the constants a, b and R of the *van der Waals* equation to given values of the saturated liquid line and of the saturated vapor line, one starts from Eqs. (5.72) and (5.73). Eq. (5.73) applies to both v' and v''. First, one calculates the specific gas constant R with an estimated b:

$$R = \frac{p[v''^2 + v'^2 - b\,(v'' + v')]}{T \left(v'' + v' - \dfrac{v''\,v'}{v'' - v'} \ln \dfrac{v'' - b}{v' - b} \right)} \tag{5.74}$$

Then the constant a is obtained from

$$a = \left(\frac{RT}{v'' - b} - p \right) v''^2 \tag{5.75}$$

and an improved value of the constant b:

$$b = v' - \frac{RT}{p + \dfrac{a}{v'^2}} \tag{5.76}$$

The calculation is repeated until the old and new b values agree.

The course of the theoretical isotherm in the two-phase liquid-vapor region between the saturated liquid line and the minimum and, in addition, between the saturated vapor line and the maximum, can be interpreted physically (Fig. 5.18): In the region between the saturated liquid line and the minimum, the liquid is superheated. It has a higher temperature than the

boiling temperature which is associated with its pressure. In the region between the saturated vapor line and the maximum, the vapor is subcooled. It has a lower temperature than the saturation temperature which belongs to its pressure. The actual temperatures are taken from the isotherms in both cases. The temperature, which actually belongs to the pressure of the fluid, results in the p, v diagram from the intersection of the horizontal line through the state point with the boundary curve. The region between the minimum and the maximum represents an unstable state.

Example 5.7 For water of 300 °C in the saturation state, the pressure, the specific volume of the saturated liquid and the specific volume of the saturated vapor have to be calculated, using the *van der Waals* equation and the constants according to example 5.6.

The mathematical formulation of the problem leads to a zero problem according to Eq. (5.72) with the additional constraint according to Eq. (5.73). The result is

$$p = 133.44 \text{ bar} \qquad v' = 0.0029719 \text{ m}^3/\text{kg} \qquad v'' = 0.012940 \text{ m}^3/\text{kg}.$$

Example 5.8 Water of 300 °C in the saturation state, according to the 'Industrial Standard IAPWS-IF97' [128], is described by

$$p = 85.877 \text{ bar} \qquad v' = 0.001404 \text{ m}^3/\text{kg} \qquad v'' = 0.02166 \text{ m}^3/\text{kg}.$$

The constants a, b and R of the *van der Waals* equation are to be adjusted to these values. Eqs. (5.74) to (5.76) give

$$a = 0.010559 \text{ bar m}^6/\text{kg}^2 \qquad b = 0.00099241 \text{ m}^3/\text{kg} \qquad R = 0.39088 \text{ kJ}/(\text{kg K}).$$

The modified critical data are

$$p_k = 397.09 \text{ bar} \qquad v_k = 0.0029772 \text{ m}^3/\text{kg} \qquad T_k = 806.53 \text{ K}.$$

5.3.3 The Reduced *van der Waals* Equation

If one sets Eqs. (5.65), (5.66) and (5.67) into Eq. (5.56), one obtains

$$\left[\frac{p}{p_k} + 3 \left(\frac{v_k}{v} \right)^2 \right] \left(3 \frac{v}{v_k} - 1 \right) = 8 \frac{T}{T_k}. \tag{5.77}$$

With the reduced state variables

$$\frac{p}{p_k} = p_r \qquad \frac{v}{v_k} = v_r \qquad \frac{T}{T_k} = T_r, \tag{5.78}$$

from Eq. (5.77) is obtained

$$\left(p_r + \frac{3}{v_r^2} \right) (3 v_r - 1) = 8 T_r. \tag{5.79}$$

Eq. (5.79) is the reduced or normalised *van der Waals* equation. It does not contain any substance values and is therefore generally valid for all substances. In a p_r, v_r diagram, the saturated liquid line and the saturated vapor line are obtained according to the same considerations as in the previous section. The equations paralleled to Eqs. (5.72) and (5.73) are as follows

$$\frac{8}{3} T_r \ln \frac{v_r'' - \frac{1}{3}}{v_r' - \frac{1}{3}} + 3 \left(\frac{1}{v_r''} - \frac{1}{v_r'} \right) - p_r (v_r'' - v_r') = 0 \tag{5.80}$$

$$v_r^3 - \frac{1}{3}\left(1 + 8\frac{T_r}{p_r}\right) v_r^2 + \frac{3}{p_r} v_r - \frac{1}{p_r} = 0 \ . \tag{5.81}$$

For a fixed value of the reduced temperature T_r, the reduced pressure p_r is varied until Eq. (5.80) is satisfied with the values v_r' and v_r'' which are obtained from Eq. (5.81). Also the liquid-vapor line, as an condition of the reduced saturation pressure and the reduced saturation temperature, is valid for all substances.

Eq. (5.79) as a generalized equation, which contains only dimensionless quantities and no individual substance constants, is an expression of the law of coincident states. In this context, states are called coincident if they have equal values p_r, v_r and T_r respectively. There is only one equation for states of this type. The coincidence of measured values with Eq. (5.79) is not satisfactory. This is because the *van der Waals* equation is too simple and the definition of coincident states by equal values of p_r, v_r and T_r is not sufficient. For this reason, further parameters were introduced and more complicated equations were set up, which, however, still today do not satisfy the law of the coincident states validly and accurately for all substances.

The efforts for an extended correspondence principle have led to partial successes [102]. For example, *Rombusch* [103] has formulated an equation of state that is valid in certain regions and is particularly useful when only few measurement data are available for a substance. With its help, e.g. vapor tables and state diagrams were calculated for a number of refrigerants [105], [106], [117], [37].

The 'acentric factor' introduced by *Pitzer* [97] is used to formulate thermal equations of state, which are called generalised equations of state [8]. Two generalised equations of state are, for example, the equation of *Redlich-Kwong-Soave* and the equation of *Peng-Robinson*. The equation of *Redlich-Kwong-Soave* is presented in the next section.

Figures 5.8 to 5.11 and 5.13, in the liquid region and in the superheated vapor region, are designed with values calculated with the reduced *van der Waals* equation.

5.3.4 Different Approaches

The *van der Waals* equation is qualitatively correct but quantitatively unsatisfactory. It describes the liquid region and the superheated vapor region correctly in principle. In the two-phase liquid-vapor region, it does not take into account that isobar and isotherm coincide, but the isotherms in the two-phase liquid-vapor region near the boundary curves are physically meaningful: They describe the special cases of superheating of the liquid and subcooling of the saturated vapor. The saturated liquid line and the saturated vapor line can be determined from the *van der Waals* equation. It allows the transformation into a reduced or normalised form, which corresponds to a generalisation, and the derivation of the principle of coincident states. Even though it is a third-degree equation in v, it is a simple equation that, apart from the specific gas constant, it contains only two other constants. These constants are physically interpretable, and their disappearance causes the transition into the thermal equation of state of ideal gases.

The inaccuracies in the application in a larger range and the necessity of the permanent adjustment of the constants indicates that the number of constants is too small. For this reason, attempts have been made to meet the requirements of practice with more

complicated approaches. Some examples of such equations follow. First of all, some very simple equations shall be pointed out which, like the *van der Waals* equation contain only two other constants apart from the specific gas constant. Such an equation was proposed by *Redlich* and *Kwong* [99]:

$$p = \frac{RT}{v - b} - \frac{a}{T^{0,5} v (v + b)} \tag{5.82}$$

According to research, it is superior to the *van der Waals* equation [125]. Applications of the *Redlich-Kwong* equation in the evaluation of thermodynamic measurements and the graphical representation of a p, v, T surface can be found in [63], [126].

A further development of this type of equation comes from *Soave* [121]. He inserted a temperature function and the acentric factor of *Pitzer* [97] into the *Redlich-Kwong*-equation. The *Redlich-Kwong-Soave* equation has the form

$$p = \frac{RT}{v - b} - \frac{a\,\alpha(T)}{v(v + b)} \cdot \tag{5.83}$$

In it is

$$\alpha(T) = [1 + (0,480 + 1{,}574\,\omega - 0{,}176\,\omega^2)(1 - \sqrt{T_r})]^2 = [1 + \beta(1 - \sqrt{T_r})]^2 \tag{5.84}$$

T_r is the reduced temperature according to Eq. (5.78). ω is the acentric factor:

$$\omega = -\lg\left(\frac{p_s}{p_k}\right)_{T_r = 0{,}7} - 1 \tag{5.85}$$

lg is the logarithm of ten and p_s/p_k is the reduced saturation pressure at the reduced temperature $T_r = 0{,}7$. The constants a and b are calculated from the critical data as in the *van der Waals* equation.

$$a = \frac{1}{9(2^{1/3} - 1)} \frac{R^2 T_k^2}{p_k} \tag{5.86}$$

$$b = \frac{1}{3}(2^{1/3} - 1) \frac{R T_k}{p_k} \tag{5.87}$$

R is the specific gas constant of the ideal gas. At the critical point is

$$z_k = \frac{p_k v_k}{R T_k} = \frac{1}{3} \cdot \tag{5.88}$$

As a precursor of modern development, one can consider the equation of *Beattie* and *Bridgman* [12]:

$$p = \frac{RT(1 - \varepsilon)}{v^2}(v + B) - \frac{A}{v^2} \tag{5.89}$$

$$A = A_0\left(1 - \frac{a}{v}\right) \qquad B = B_0\left(1 - \frac{b}{v}\right) \qquad \varepsilon = \frac{c}{v T^3} \tag{5.90}$$

Numerical values for the constants can be found in [73].

5.3 Thermal Equations of State

An improvement of the equation of *Beattie* and *Bridgman* was given by *Benedict*, *Webb* and *Rubin* [15], [16]. They proposed:

$$p = \frac{RT}{v} + \frac{B_0 RT - A_0 - \frac{C_0}{T^2}}{v^2} + \frac{bRT - a}{v^3} + \frac{a\alpha}{v^6} + \frac{c\left(1 + \frac{\gamma}{v^2}\right)}{v^3 T^2} e^{-\frac{\gamma}{v^2}} \quad (5.91)$$

For some substances, numerical values of the constants are given in [73]. In the papers [35], [76] the constants for the refrigerants R 12, R 22, R 23 and R 115, in [1] the values for argon, nitrogen and oxygen are published.

Some more constants were added and the thermal state variables were normalised (reduced). This led to the extended *Benedict-Webb-Rubin* equation:

$$\pi = Wa\,\vartheta\,\delta + \left(A_2 + B_2\vartheta + \frac{C_2}{\vartheta^2} + \frac{D_2}{\vartheta^4}\right)\delta^2 + \left(A_3 + B_3\vartheta + \frac{C_3}{\vartheta^2}\right)\delta^3 +$$

$$+ \left(\frac{A_4}{\vartheta^2} + \frac{A_5}{\vartheta^4}\right)\delta^3\left(1 + \beta\delta^2\right)e^{-\beta\delta^2} + A_6\delta^6 \quad (5.92)$$

π, ϑ and δ are the normalised pressure, normalised temperature and normalised density:

$$\pi = \frac{p}{p_k} \qquad \vartheta = \frac{T}{T_k} \qquad \delta = \frac{\varrho}{\varrho_k} = \frac{v_k}{v} \quad (5.93)$$

Wa is the *van der Waals* number:

$$Wa = \frac{RT_k \varrho_k}{p_k} = \frac{RT_k}{p_k v_k} \quad (5.94)$$

Papers [35], [36], [70], [86], [87], [88] give the numerical values of the constants for refrigerants R 22, R 23 and R 717.

With Eqs. (5.89), (5.91) and (5.92) the saturation state has not been captured. It is described by special equations for the liquid-vapor line and the density of the saturated liquid. If the *Benedict-Webb-Rubin* equation should allow statements about the saturation state with the help of the *Maxwell* relationship (section 5.3.2), the number of constants has to be increased again. This equation is called the *Bender* equation:

$$p = \varrho T\left[R + B\varrho + C\varrho^2 + D\varrho^3 + E\varrho^4 + F\varrho^5 + (G + H\varrho^2)\varrho^2 e^{-a_{20}\varrho^2}\right] \quad (5.95)$$

The constants B, C, D, E, F, G and H are independent of the density ϱ but depend on the temperature T:

$$B = a_1 - \frac{a_2}{T} - \frac{a_3}{T^2} - \frac{a_4}{T^3} - \frac{a_5}{T^4} \quad (5.96)$$

$$C = a_6 + \frac{a_7}{T} + \frac{a_8}{T^2} \quad (5.97)$$

$$D = a_9 + \frac{a_{10}}{T} \quad (5.98)$$

$$E = a_{11} + \frac{a_{12}}{T} \quad (5.99)$$

$$F = \frac{a_{13}}{T} \quad (5.100)$$

$$G = \frac{a_{14}}{T^3} + \frac{a_{15}}{T^4} + \frac{a_{16}}{T^5} \tag{5.101}$$

$$H = \frac{a_{17}}{T^3} + \frac{a_{18}}{T^4} + \frac{a_{19}}{T^5} \ . \tag{5.102}$$

Values of the constants for methane, ethylene, propylene, ethane, propane, n-butane, n-pentane and carbon dioxide are given in the papers [13], [14], [118] and [124]. A *Benedict-Webb-Rubin* equation, which contains 33 constants in addition to the specific gas constant, is used in [83] for the formulation of a thermal equation of state for methane.

Another thermal equation of state based on a proposal by *R. Plank* is given by *Martin* and *Hou* [79], [80]:

$$p = \frac{RT}{v-b} + \frac{A_2 + B_2 T + C_2 \, e^{-\frac{KT}{T_k}}}{(v-b)^2} + \frac{A_3 + B_3 T + C_3 \, e^{-\frac{KT}{T_k}}}{(v-b)^3} +$$

$$+ \frac{A_4}{(v-b)^4} + \frac{A_5 + B_5 T + C_5 \, e^{-\frac{KT}{T_k}}}{(v-b)^5} \tag{5.103}$$

The constants for the refrigerants R 14, R 23, R 114, R 500, R 503, R 504 and for a mixture of the refrigerants R 31 and R 114 can be taken from [59], [84], [85], [89], [90], [119], [120]. An extended *Martin-Hou* equation has been used in [81] to describe the thermodynamic properties of the refrigerant R 502.

Example 5.9 The refrigerant R 134 a (Tetrafluoroethane $F\,H_2\,C$–$C\,F_3$) has according to *Baehr* and *Kabelac* [10] the specific gas constant $R = 0.081489$ kJ/(kg K) and the critical data $p_k = 40.58$ bar and $T_k = 374.25$ K. According to *Heckenberger* [55], the acentric factor is 0.326. With the *Redlich-Kwong-Soave* equation (5.83) the functions

$$p = \frac{\varphi_0}{v} + \frac{\varphi_1}{v^2} + \frac{\varphi_2}{v^3} \qquad\qquad v = \frac{\psi_0}{p} + \psi_1 + \psi_2 \, p$$

for a temperature of 20 °C between the pressures 1 bar and 5 bar in the region of the super-heated vapor can be determined.

The two equations can be written as follows

$$p\,v = \varphi_0 + \frac{\varphi_1}{v} + \frac{\varphi_2}{v^2} \qquad\qquad p\,v = \psi_0 + \psi_1\,p + \psi_2\,p^2$$

or in general
$$y = a_0 + a_1\,x + a_2\,x^2 \ .$$

The coefficients a_0, a_1 and a_2 can be determined via a balancing polynomial according to the method of sum of least squares of errors. The following values can be used:

p bar	v m³/kg	p bar	v m³/kg	p bar	v m³/kg	p bar	v m³/kg
1.0	0.234396	2.1	0.109184	3.2	0.0699931	4.3	0.0507997
1.1	0.212672	2.2	0.104006	3.3	0.0677225	4.4	0.0495278
1.2	0.194567	2.3	0.0992773	3.4	0.0655849	4.5	0.0483120
1.3	0.179247	2.4	0.0949418	3.5	0.0635689	4.6	0.0471486
1.4	0.166114	2.5	0.0909525	3.6	0.0616644	4.7	0.0460342
1.5	0.154731	2.6	0.0872694	3.7	0.0598623	4.8	0.0449657
1.6	0.144771	2.7	0.0838585	3.8	0.0581545	4.9	0.0439404
1.7	0.135981	2.8	0.0806906	3.9	0.0565337	5.0	0.0429556
1.8	0.128166	2.9	0.0777405	4.0	0.0549935		
1.9	0.121174	3.0	0.0749865	4.1	0.0535279		
2.0	0.114880	3.1	0.0724095	4.2	0.0521316		

The following values are obtained:

$$\varphi_0 = 0.23888 \; \frac{\text{m}^3 \, \text{bar}}{\text{kg}} \qquad \varphi_1 = -0.0010558 \; \frac{\text{m}^6 \, \text{bar}}{\text{kg}^2} \qquad \varphi_2 = 0.000087285 \; \frac{\text{m}^9 \, \text{bar}}{\text{kg}^3}$$

$$\psi_0 = 0.23882 \; \frac{\text{m}^3 \, \text{bar}}{\text{kg}} \qquad \psi_1 = -0.0043404 \; \frac{\text{m}^3}{\text{kg}} \qquad \psi_2 = -0.000093042 \; \frac{\text{m}^3}{\text{kg} \, \text{bar}}$$

These equations are used in examples 2.4 and 2.7.

5.3.5 Virial Coefficients

The real gas factor (compressibility factor) z according to Eq. (5.44) can be represented by the following equation:

$$z = z(v,T) = A + \frac{B}{v} + \frac{C}{v^2} + \frac{D}{v^3} + \ldots \qquad (5.104)$$

The coefficients $A, B, C, D, \ldots$ — the so-called virial coefficients — are functions of the temperature T. The first virial coefficient is set to

$$A = 1 \, . \qquad (5.105)$$

For the second virial coefficient one makes the following approach:

$$B = b_0 + \frac{b_1}{T} + \frac{b_2}{T^2} + \frac{b_3}{T^3} + \ldots \, . \qquad (5.106)$$

Corresponding equations apply to the other virial coefficients. In many cases, one can limit oneself to the first three virial coefficients and then obtain according to Eq. (5.44) the thermal equation of state in the form

$$p = \frac{RT}{v} \left(1 + \frac{B}{v} + \frac{C}{v^2} \right) \, . \qquad (5.107)$$

For $B = C = 0$, Eq. (5.107) is transformed to the thermal equation of state of ideal gases. After the first virial coefficient is determined by Eq. (5.105), the further virial

coefficients can be derived from p, v, T-measured values. From Eqs. (5.104) and (5.105) follows:

$$(z - 1) v = y = B + \frac{C}{v} + \frac{D}{v^2} + \ldots \tag{5.108}$$

y is clearly determined by the measured values and the specific gas constant. The right-hand side of Eq. (5.108) is an interpolation or balancing polynomial for the y values at the same temperature. If one has determined the virial coefficients for different temperatures, one calculates the constants of Eq. (5.106) for the second virial coefficient and for the following virial coefficients.

The term virial coefficient is justified by statistical thermodynamics. According to this, the properties of a gas are determined by the interaction forces between the molecules. [vires (lat.) – forces].

The second virial coefficient describes the mutual influence of two molecules. The third virial coefficient expresses the interaction between three molecules. If another molecule enters the sphere of influence, the series of virial coefficients continues accordingly. If all interaction forces between the molecules cease to exist, the second and all further virial coefficients become zero: The thermal equation of state of ideal gases applies.

The *van der Waals* equation (5.57) can also be converted into the virial form. First it can be wtritten as

$$\frac{pv}{RT} = \frac{v}{v-b} - \frac{1}{RT}\frac{a}{v}. \tag{5.109}$$

With the series expansion

$$\frac{v}{v-b} = \frac{1}{1 - \frac{b}{v}} = 1 + \frac{b}{v} + \frac{b^2}{v^2} + \frac{b^3}{v^3} + \ldots \tag{5.110}$$

one gets

$$\frac{pv}{RT} = 1 + \left(b - \frac{a}{RT}\right)\frac{1}{v} + \frac{b^2}{v^2} + \frac{b^3}{v^3} + \ldots \tag{5.111}$$

For the second, third and fourth virial coefficients, one has

$$B = b - \frac{a}{RT} \qquad C = b^2 \qquad D = b^3. \tag{5.112}$$

The second virial coefficient increases with rising temperature. Attractive and repulsive forces act between the molecules. The attractive forces predominate at low temperatures and cause a decrease in pressure compared to the ideal gas. The second virial coefficient is negative. At high temperatures the repulsive forces predominate, the pressure of the gas on the surroundings is greater than in the case of the ideal gas; the second virial coefficient becomes positive. Virial coefficients for oxygen can be found in [127].

In many cases, a different series expansion of the real gas factor z is more favourable:

$$z = z(p, T) = A' + B' p + C' p^2 + D' p^3 + \ldots \tag{5.113}$$

In this case, too, 0ne speaks of virial coefficients, although the physical interpretation given earlier in the connection with the two- or multi-body problem is now no longer possible. Analogously follows

$$A' = 1 \tag{5.114}$$

$$B' = b'_0 + \frac{b'_1}{T} + \frac{b'_2}{T^2} + \frac{b'_3}{T^3} + \ldots \tag{5.115}$$

If one restricts himself to the first three virial coefficients, the following thermal equation of state is obtained:
$$v = \frac{RT}{p}(1 + B'p + C'p^2) \tag{5.116}$$

Using the equation
$$(z-1)\frac{1}{p} = y' = B' + C'p + D'p^2 + \ldots , \tag{5.117}$$

one can determine the second and the following virial coefficients from p, v, T-measurements. In the form of Eq. (5.116), two equations are given by *Wukalowitsch* and co-workers [134], [135] for water vapour at 800 °C to 1500 °C and pressures up to 1000 bar.

Example 5.10 The equation for the real gas factor (compressibility factor) of air according to Example 5.4 yields a thermal equation of state according to Eq. (5.116). With the factors a_{ij} given in Example 5.4, the following applies to the second and third virial coefficients

$$B' = a_{01} + \frac{a_{11}}{T} + \frac{a_{21}}{T^2} + \frac{a_{31}}{T^3} + \frac{a_{41}}{T^4}$$

$$C' = a_{02} + \frac{a_{12}}{T} + \frac{a_{22}}{T^2} + \frac{a_{32}}{T^3} + \frac{a_{42}}{T^4} .$$

A comparison with the values given by *Baehr* and *Schwier* [2] for the thermodynamic properties of air shows good agreement.

5.4 Calculation of State Variables; Property Tables

A vapor table (steam table) generally contains the thermal state variables
      pressure p
      temperature T
      specific volume v,
perhaps the
      density ϱ
and the caloric state variables
      specific enthalpy h
      specific entropy s,
possibly the
      specific heat of vaporisation r
      specific internal energy u.

It contains these properties for the saturation state, i.e. for the saturated liquid line and the saturated vapor line, for the liquid and for the superheated vapor. If one considers the thermal state variables as measured values and thus as given, caloric state variables can be calculated. The relationship between the thermal and the caloric state variables is described in the following sections.

5.4.1 The Caloric State Variables

The caloric state variables internal energy U, enthalpy H and entropy S can be represented as functions of two variables. The independent variables are the thermal state variables pressure p, volume V and temperature T. If one takes the specific values, i.e. the values related to the mass, then the following functions are involved:

$$u = u(T, v) \tag{5.118}$$

$$h = h(T, p) \tag{5.119}$$

$$s = s(T, v) \tag{5.120}$$

$$s = s(T, p) \tag{5.121}$$

Two modes of representation are discussed for the specific entropy because both are of practical importance. Since in Eqs. (5.118) to (5.121), specific state variables apply according to Eqs. (3.29) and (3.30), the total differentials are

$$du = \left(\frac{\partial u}{\partial T}\right)_v dT + \left(\frac{\partial u}{\partial v}\right)_T dv \tag{5.122}$$

$$dh = \left(\frac{\partial h}{\partial T}\right)_p dT + \left(\frac{\partial h}{\partial p}\right)_T dp \tag{5.123}$$

$$ds = \left(\frac{\partial s}{\partial T}\right)_v dT + \left(\frac{\partial s}{\partial v}\right)_T dv \tag{5.124}$$

$$ds = \left(\frac{\partial s}{\partial T}\right)_p dT + \left(\frac{\partial s}{\partial p}\right)_T dp \ . \tag{5.125}$$

The partial derivatives in Eqs. (5.122) to (5.125) can be calculated with the help of the formulations of the first and second law of thermodynamics: According to Eqs. (3.77) are

$$T\,ds - p\,dv = du \tag{5.126}$$

$$T\,ds + v\,dp = dh \ . \tag{5.127}$$

In the case of an isochoric change of state ($dv = 0$), Eqs. (5.122), (5.124) and (5.126) lead to

$$du = \left(\frac{\partial u}{\partial T}\right)_v dT \tag{5.128}$$

$$ds = \left(\frac{\partial s}{\partial T}\right)_v dT \tag{5.129}$$

$$T\,ds = du \ . \tag{5.130}$$

According to Eq. (4.69), for an isochoric change of state, the change in specific internal energy is equal to the differential specific substitute heat:

$$du = dq_{rev} = c_v\,dT \tag{5.131}$$

From Eqs. (5.128) to (5.131) follows

$$c_v = \left(\frac{\partial u}{\partial T}\right)_v \tag{5.132}$$

$$c_v = T\left(\frac{\partial s}{\partial T}\right)_v. \tag{5.133}$$

Eqs. (5.123), (5.125) and (5.127) take the following form for an isobaric change of state $(dp = 0)$:

$$dh = \left(\frac{\partial h}{\partial T}\right)_p dT \tag{5.134}$$

$$ds = \left(\frac{\partial s}{\partial T}\right)_p dT \tag{5.135}$$

$$T\,ds = dh\,. \tag{5.136}$$

According to Eq. (4.78), the change in specific enthalpy for an isobaric change of state is equal to the differential specific substitute heat:

$$dh = dq_{rev} = c_p\,dT \tag{5.137}$$

Eqs. (5.134) to (5.137) give

$$c_p = \left(\frac{\partial h}{\partial T}\right)_p \tag{5.138}$$

$$c_p = T\left(\frac{\partial s}{\partial T}\right)_p. \tag{5.139}$$

In Eqs. (5.126) and (5.127), which are transformed to ds, Eqs. (5.122) and (5.123) are inserted:

$$ds = \frac{1}{T}\left(\frac{\partial u}{\partial T}\right)_v dT + \frac{1}{T}\left[\left(\frac{\partial u}{\partial v}\right)_T + p\right] dv \tag{5.140}$$

$$ds = \frac{1}{T}\left(\frac{\partial h}{\partial T}\right)_p dT + \frac{1}{T}\left[\left(\frac{\partial h}{\partial p}\right)_T - v\right] dp \tag{5.141}$$

Comparing Eqs. (5.124) and (5.140), and likewise (5.125) and (5.141), one obtains the relations

$$\left(\frac{\partial s}{\partial v}\right)_T = \frac{1}{T}\left[\left(\frac{\partial u}{\partial v}\right)_T + p\right] \tag{5.142}$$

$$\left(\frac{\partial s}{\partial p}\right)_T = \frac{1}{T}\left[\left(\frac{\partial h}{\partial p}\right)_T - v\right]. \tag{5.143}$$

Using the partial differentiation of c_v with respect to v at constant temperature, it follows from Eqs. (5.132) and (5.133)

$$\frac{\partial^2 s}{\partial T\,\partial v} = \frac{1}{T}\frac{\partial^2 u}{\partial T\,\partial v}. \tag{5.144}$$

5 Real Gases and Vapors

If Eq. (5.142) is differentiated according to T at constant v, it follows

$$\frac{\partial^2 s}{\partial v\, \partial T} = -\frac{1}{T^2}\left[\left(\frac{\partial u}{\partial v}\right)_T + p\right] + \frac{1}{T}\left[\frac{\partial^2 u}{\partial v\, \partial T} + \left(\frac{\partial p}{\partial T}\right)_v\right]. \qquad (5.145)$$

From the equality of the second partial derivatives

$$\frac{\partial^2 s}{\partial T\, \partial v} = \frac{\partial^2 s}{\partial v\, \partial T} \qquad (5.146)$$

follows

$$\left(\frac{\partial u}{\partial v}\right)_T = T\left(\frac{\partial p}{\partial T}\right)_v - p. \qquad (5.147)$$

Accordingly, one obtains from Eqs. (5.138), (5.139) and (5.143)

$$\left(\frac{\partial h}{\partial p}\right)_T = v - T\left(\frac{\partial v}{\partial T}\right)_p. \qquad (5.148)$$

Eqs. (5.147) and (5.148) are inserted into Eqs. (5.142) and (5.143):

$$\left(\frac{\partial s}{\partial v}\right)_T = \left(\frac{\partial p}{\partial T}\right)_v \qquad (5.149)$$

$$\left(\frac{\partial s}{\partial p}\right)_T = -\left(\frac{\partial v}{\partial T}\right)_p \qquad (5.150)$$

Eqs. (5.122) to (5.125) give with Eqs. (5.132), (5.133), (5.138), (5.139) and (5.147) to (5.150)

$$du = c_v\, dT + \left[T\left(\frac{\partial p}{\partial T}\right)_v - p\right] dv \qquad (5.151)$$

$$dh = c_p\, dT + \left[v - T\left(\frac{\partial v}{\partial T}\right)_p\right] dp \qquad (5.152)$$

$$ds = c_v\frac{dT}{T} + \left(\frac{\partial p}{\partial T}\right)_v dv \qquad (5.153)$$

$$ds = c_p\frac{dT}{T} - \left(\frac{\partial v}{\partial T}\right)_p dp. \qquad (5.154)$$

Eqs. (5.151) to (5.154) are integrated in two steps. The integration starts at a state characterised by the thermal state variables p_1, v_1 and T_1. At this point and during the first integration step, where only the temperature is variable ($dv = 0$ or $dp = 0$), the gas phase is said to be in the state of the ideal gas. Let the initial value p_1 be very small and the initial value v_1 be very large. Thus, the thermal equation of state of the ideal gases (4.26) is valid:

$$p_1 v_1 = R T_1 \qquad (5.155)$$

5.4 Calculation of State Variables; Property Tables

The specific heat capacities $c_p(T,p)$ and $c_v(T,v)$ in the state of the ideal gas are only dependent on temperature (section 2.7.2). This shall be expressed in the notation

$$c_p^0 = c_p(T) \tag{5.156}$$

$$c_v^0 = c_v(T). \tag{5.157}$$

In the state of the ideal gas, Eq. (4.48) applies:

$$c_p^0 - c_v^0 = R \tag{5.158}$$

During the second integration step (v or p, respectively, is hereby the variable) T remains constant ($dT = 0$). For $u(T,v)$ and $u_1(T_1,v_1)$ only u and u_1 are written for simplification. The specific enthalpy h and the specific entropy s are written accordingly. The integration of Eqs. (5.151) to (5.154) yields the following equations:

$$u = u_1 + \int_{T_1}^{T} c_v^0 \, dT + \int_{v_1}^{v} \left[T \left(\frac{\partial p}{\partial T} \right)_v - p \right] dv \tag{5.159}$$

$$h = h_1 + \int_{T_1}^{T} c_p^0 \, dT + \int_{p_1}^{p} \left[v - T \left(\frac{\partial v}{\partial T} \right)_p \right] dp \tag{5.160}$$

$$s = s_1 + \int_{T_1}^{T} c_v^0 \frac{dT}{T} + \int_{v_1}^{v} \left(\frac{\partial p}{\partial T} \right)_v dv \tag{5.161}$$

$$s = s_1 + \int_{T_1}^{T} c_p^0 \frac{dT}{T} - \int_{p_1}^{p} \left(\frac{\partial v}{\partial T} \right)_p dp \tag{5.162}$$

Example 5.11 According to [10], the following applies to the refrigerant R 134 a for the state of the ideal gas:

$$c_p^0 = c_v^0 + R = c_0 \frac{T_k}{T} + c_1 + c_2 \frac{T}{T_k} + c_3 \left(\frac{T}{T_k} \right)^2$$

$c_0 = 0.042276$ kJ/(kg K) $c_1 = -0.00526$ kJ/(kg K)

$c_2 = 1.23374$ kJ/(kg K) $c_3 = -0.282552$ kJ/(kg K)

Using the *Redlich-Kwong-Soave* equation (5.83) and the data given in Example 5.9, the equations for the specific enthalpy and the specific entropy are to be determined.

The application of Eq. (5.160) is difficult in this case because the thermal equation of state is of the form $p = p(T,v)$. It is recommended to first calculate the change of the specific internal energy with Eq. (5.159). With Eq. (2.41) is then

$$h = u + pv.$$

For the calculation of the first integral in Eq. (5.159) one sets

$$T = T_k \frac{T}{T_k} \qquad\qquad dT = T_k \, d\left(\frac{T}{T_k}\right)$$

and, using the equation for the specific heat capacity c_v^0, obtains

$$\int_{T_1}^{T} c_v^0 \, dT = T_k \left[c_0 \ln \frac{T}{T_k} + (c_1 - R)\frac{T}{T_k} + \frac{c_2}{2}\left(\frac{T}{T_k}\right)^2 + \frac{c_3}{3}\left(\frac{T}{T_k}\right)^3 \right] - u_1^* .$$

From Eq. (5.83) it follows for the second integral

$$\left(\frac{\partial p}{\partial T}\right)_v = \frac{R}{v-b} + \frac{a}{v(v+b)}\beta\sqrt{\frac{\alpha(T)}{T_k T}}$$

$$T\left(\frac{\partial p}{\partial T}\right)_v - p = \frac{a}{v(v+b)}(1+\beta)\sqrt{\alpha(T)}$$

$$\int_{v_1}^{v} \left[T\left(\frac{\partial p}{\partial T}\right)_v - p\right] dv = -\frac{a}{b}(1+\beta)\sqrt{\alpha(T)} \ln\left(1 + \frac{b}{v}\right) - u_2^* .$$

The constants u_1^* and u_2^* result with the lower limits of the integrals. For the change of the specific internal energy according to Eq. (5.159) one obtains

$$u = T_k \left[c_0 \ln \frac{T}{T_k} + (c_1 - R)\frac{T}{T_k} + \frac{c_2}{2}\left(\frac{T}{T_k}\right)^2 + \frac{c_3}{3}\left(\frac{T}{T_k}\right)^3 \right] - \frac{a}{b}(1+\beta)\sqrt{\alpha(T)} \ln\left(1 + \frac{b}{v}\right) + u^*$$

u^* is a constant obtained by combining the constants u_1, u_1^* and u_2^*.

The final formula for the specific enthalpy is

$$h = T_k \left[c_0 \ln \frac{T}{T_k} + (c_1 - R)\frac{T}{T_k} + \frac{c_2}{2}\left(\frac{T}{T_k}\right)^2 + \frac{c_3}{3}\left(\frac{T}{T_k}\right)^3 \right] - \frac{a}{b}(1+\beta)\sqrt{\alpha(T)} \ln\left(1 + \frac{b}{v}\right)$$

$$+ \frac{RTv}{v-b} - \frac{a\,\alpha(T)}{v+b} + h^* .$$

An equation for the specific entropy is obtained from Eq. (5.161). The first integral gives

$$\int_{T_1}^{T} c_v^0 \frac{dT}{T} = -c_0 \frac{T_k}{T} + (c_1 - R)\ln\frac{T}{T_k} + c_2\frac{T}{T_k} + \frac{c_3}{2}\left(\frac{T}{T_k}\right)^2 - s_1^* .$$

The second integral gives

$$\int_{v_1}^{v} \left(\frac{\partial p}{\partial T}\right)_v dv = R \ln \frac{v-b}{v_1-b} - \frac{a}{b}\beta\sqrt{\frac{\alpha(T)}{T_k T}} \ln\left(1 + \frac{b}{v}\right) - s_2^* .$$

The first expression on the right-hand side can be transformed as follows:

$$R \ln \frac{v-b}{v_1-b} = R \ln(v-b) - R \ln(v_1-b) + R \ln(1-b) - R \ln(1-b) = R \ln \frac{v-b}{1-b} + R \ln \frac{1-b}{v_1-b}$$

The expression $R \ln[(1-b)/(v_1-b)]$ is included into the integration constant. As a final formula for the specific entropy one obtains

$$s = -c_0 \frac{T_k}{T} + (c_1 - R)\ln\frac{T}{T_k} + c_2\frac{T}{T_k} + \frac{c_3}{2}\left(\frac{T}{T_k}\right)^2 + R \ln \frac{v-b}{1-b} - \frac{a}{b}\beta\sqrt{\frac{\alpha(T)}{T_k T}} \ln\left(1 + \frac{b}{v}\right) + s^*$$

The values for h^* and s^* result from the choice of the zero points or reference points for the scales of the specific enthalpy and specific entropy. Since a point on the saturated liquid line is usually chosen as the reference point and statements about the boundary curve and the saturated vapor region are not yet available, the calculation of the constants h^* and s^* is carried out in example 5.14. With the data in example 5.9 for R, p_k, v_k, T_k and ω, the numerical values for a, b, α and ω result according to Equations (5.86), (5.87) and (5.84).

5.4.2 The Specific Heat Capacities c_p and c_v

The specific heat capacities at constant pressure and at constant volume c_p and c_v can also be expressed by the thermal state variables. To derive a suitable relation for the calculation of c_p, Eq. (5.139) can be used. The partial derivation according to the pressure at constant temperature yields

$$\left(\frac{\partial c_p}{\partial p}\right)_T = T\frac{\partial^2 s}{\partial T\, \partial p}\;. \tag{5.163}$$

The partial derivative of Eq. (5.150) with respect to the temperature at constant pressure

$$\frac{\partial^2 s}{\partial p\, \partial T} = -\left(\frac{\partial^2 v}{\partial T^2}\right)_p \tag{5.164}$$

results because of

$$\frac{\partial^2 s}{\partial T\, \partial p} = \frac{\partial^2 s}{\partial p\, \partial T} \tag{5.165}$$

with Eq. (5.163) in

$$\left(\frac{\partial c_p}{\partial p}\right)_T = -T\left(\frac{\partial^2 v}{\partial T^2}\right)_p\;. \tag{5.166}$$

Eq. (5.166) is integrated between the pressure p_1 at which the gas is in the state of the ideal gas and the pressure p:

$$c_p = c_p^0 - T\int_{p_1}^{p}\left(\frac{\partial^2 v}{\partial T^2}\right)_p dp \tag{5.167}$$

Starting from Eq. (5.133), using Eqs. (5.146) and (5.149), one obtains

$$\left(\frac{\partial c_v}{\partial v}\right)_T = T\left(\frac{\partial^2 p}{\partial T^2}\right)_v \tag{5.168}$$

$$c_v = c_v^0 + T\int_{v_1}^{v}\left(\frac{\partial^2 p}{\partial T^2}\right)_v dv \tag{5.169}\;.$$

Another interesting relation is for the difference $c_p - c_v$. From the equation $h = u + pv$ it follows with Eq. (2.44)

$$dh - du = d(pv) = v\,dp + p\,dv\;. \tag{5.170}$$

This equation, with the difference from Eqs. (5.151) and (5.152), results in

$$dh - du = (c_p - c_v)\,dT + d(pv) - T\left(\frac{\partial v}{\partial T}\right)_p dp - T\left(\frac{\partial p}{\partial T}\right)_v dv$$

$$c_p - c_v = T\left(\frac{\partial v}{\partial T}\right)_p \frac{dp}{dT} + T\left(\frac{\partial p}{\partial T}\right)_v \frac{dv}{dT}\;. \tag{5.171}$$

5 Real Gases and Vapors

The total differential (also called exact differential) of pressure $dp = (\partial p/\partial T)_v \, dT + (\partial p/\partial v)_T \, dv$ is divided by the differential dT

$$\frac{dp}{dT} = \left(\frac{\partial p}{\partial T}\right)_v + \left(\frac{\partial p}{\partial v}\right)_T \frac{dv}{dT} \tag{5.172}$$

and inserted in Eq. (5.171):

$$c_p - c_v = T \left(\frac{\partial v}{\partial T}\right)_p \left(\frac{\partial p}{\partial T}\right)_v + T \left[\left(\frac{\partial v}{\partial T}\right)_p \left(\frac{\partial p}{\partial v}\right)_T + \left(\frac{\partial p}{\partial T}\right)_v\right] \frac{dv}{dT} \tag{5.173}$$

The equation analogous to Eq. (5.172)

$$\frac{dv}{dT} = \left(\frac{\partial v}{\partial T}\right)_p + \left(\frac{\partial v}{\partial p}\right)_T \frac{dp}{dT} \tag{5.174}$$

is inserted in the right-hand side of Eq. (5.172):

$$\frac{dp}{dT} = \left(\frac{\partial p}{\partial T}\right)_v + \left(\frac{\partial p}{\partial v}\right)_T \left[\left(\frac{\partial v}{\partial T}\right)_p + \left(\frac{\partial v}{\partial p}\right)_T \frac{dp}{dT}\right]$$

$$\left[1 - \left(\frac{\partial p}{\partial v}\right)_T \left(\frac{\partial v}{\partial p}\right)_T\right] \frac{dp}{dT} = \left(\frac{\partial p}{\partial T}\right)_v + \left(\frac{\partial v}{\partial T}\right)_p \left(\frac{\partial p}{\partial v}\right)_T \tag{5.175}$$

With the relation

$$\left(\frac{\partial p}{\partial v}\right)_T \left(\frac{\partial v}{\partial p}\right)_T = 1 \tag{5.176}$$

it follows from Eq. (5.175)

$$\left(\frac{\partial v}{\partial T}\right)_p \left(\frac{\partial p}{\partial v}\right)_T + \left(\frac{\partial p}{\partial T}\right)_v = 0 \tag{5.177}$$

and thus from Eq. (5.173)

$$c_p - c_v = T \left(\frac{\partial v}{\partial T}\right)_p \left(\frac{\partial p}{\partial T}\right)_v . \tag{5.178}$$

Eq. (5.178) can also be written with Eq. (5.177) as follows:

$$c_p - c_v = -T \left(\frac{\partial v}{\partial T}\right)_p^2 \left(\frac{\partial p}{\partial v}\right)_T = -T \frac{\left(\frac{\partial v}{\partial T}\right)_p^2}{\left(\frac{\partial v}{\partial p}\right)_T} \tag{5.179}$$

$$c_p - c_v = -T \left(\frac{\partial p}{\partial T}\right)_v^2 \left(\frac{\partial v}{\partial p}\right)_T = -T \frac{\left(\frac{\partial p}{\partial T}\right)_v^2}{\left(\frac{\partial p}{\partial v}\right)_T} \tag{5.180}$$

Example 5.12 For the refrigerant R 134 a with the thermal equation of state (5.83) and the constants according to example 5.11, the equations for the specific heat capacities at constant pressure and at constant volume are to be determined.
One needs the following derivations

$$\left(\frac{\partial p}{\partial T}\right)_v = \frac{R}{v-b} + \frac{a}{v(v+b)}\beta\sqrt{\frac{\alpha(T)}{T_k T}}$$

$$\left(\frac{\partial^2 p}{\partial T^2}\right)_v = \frac{a}{v(v+b)}\frac{\beta}{2T}\left(\frac{\beta}{T_k} + \sqrt{\frac{\alpha(T)}{T_k T}}\right)$$

$$\left(\frac{\partial p}{\partial v}\right)_T = -\frac{RT}{(v-b)^2} + \frac{a\,\alpha(T)(2v+b)}{v^2(v+b)^2}$$

and the integral

$$T\int_{v_1}^{v}\left(\frac{\partial^2 p}{\partial T^2}\right)_v dv = \frac{a}{b}\frac{\beta}{2}\left(\frac{\beta}{T_k} + \sqrt{\frac{\alpha(T)}{T_k T}}\right)\ln\left(1+\frac{b}{v}\right).$$

Since at the pressure $p_1 \to 0$ (ideal gas) the specific volume goes to $v_1 \to \infty$ and also $\ln 1 = 0$ has to be taken into account, the expression for the lower limit of the last equation disappears. Using c_v^0 according to Example 5.11, Eq. (5.169) gives:

$$c_v = c_0 \frac{T_k}{T} + c_1 - R + c_2 \frac{T}{T_k} + c_3 \left(\frac{T}{T_k}\right)^2 + \frac{a}{b}\frac{\beta}{2}\left(\frac{\beta}{T_k} + \sqrt{\frac{\alpha(T)}{T_k T}}\right)\ln\left(1+\frac{b}{v}\right)$$

This equation can be checked using Eqs. (5.132) and (5.133).
From Eq. (5.180), it follows with the equation for c_v

$$c_p = c_v + T\frac{\left[\frac{R}{v-b} + \frac{a}{v(v+b)}\beta\sqrt{\frac{\alpha(T)}{T_k T}}\right]^2}{\frac{RT}{(v-b)^2} - \frac{a\,\alpha(T)(2v+b)}{v^2(v+b)^2}}.$$

Eqs. (5.138) and (5.139) can be used for checking this result.

5.4.3 The Isentropic Exponent and the Isothermal Exponent

In section 4.3.4 when dealing with ideal gases, Eq. (4.44)

$$\kappa = \frac{c_p}{c_v} \tag{5.181}$$

is introduced. κ is given the name isentropic exponent. For real gases, κ according to Eq. (5.181) is not equal to the isentropic exponent. If the isentropic exponent is now denoted by n_S, $n_S \neq \kappa$ holds. Eq. (4.117) is no longer valid for real gases.
From the equation

$$pv^n = \text{const} \tag{5.182}$$

for a polytropic change of state follows with constant n

$$d(pv^n) = v^n\,dp + pn v^{n-1}dv = 0 \qquad \frac{dp}{p} + n\frac{dv}{v} = 0$$

$$n = -\frac{v\,dp}{p\,dv} = -\frac{v}{p\,\dfrac{dv}{dp}}. \tag{5.183}$$

The total or exact differential of the specific volume $dv = (\partial v/\partial T)_p\,dT + (\partial v/\partial p)_T\,dp$ results after division by the diffenrential of the pressure in

$$\frac{dv}{dp} = \left(\frac{\partial v}{\partial T}\right)_p \frac{dT}{dp} + \left(\frac{\partial v}{\partial p}\right)_T. \tag{5.184}$$

In the case of an isentropic change of state $(ds = 0)$, Eq. (5.154) becomes

$$\frac{dT}{dp} = \frac{T}{c_p}\left(\frac{\partial v}{\partial T}\right)_p. \tag{5.185}$$

By inserting this equation into Eq. (5.184), one finds

$$\frac{dv}{dp} = \frac{T}{c_p}\left(\frac{\partial v}{\partial T}\right)_p^2 + \left(\frac{\partial v}{\partial p}\right)_T. \tag{5.186}$$

Applying Eq. (5.183) to an isentropic change of state, with Eq. (5.186) this results in

$$n_S = -\frac{v}{p\left[\dfrac{T}{c_p}\left(\dfrac{\partial v}{\partial T}\right)_p^2 + \left(\dfrac{\partial v}{\partial p}\right)_T\right]}. \tag{5.187}$$

Eq. (5.187) can be used to calculate the isentropic exponent n_S of a real gas.

The derivation of an equation for the isothermal exponent follows: For Eq. (5.182), which takes the form $pv = \text{const}$ in the case of an isothermal change of state for ideal gases, one has to calculate the following for the real gases

$$pv^{n_T} = \text{const} \tag{5.188}$$

where n_T is the isothermal exponent. For an isothermal change of state $(dT = 0)$, Eq. (5.184) becomes

$$\frac{dv}{dp} = \left(\frac{\partial v}{\partial p}\right)_T. \tag{5.189}$$

and thus from Eq. (5.183)

$$n_T = -\frac{v}{p\left(\dfrac{\partial v}{\partial p}\right)_T} = -\frac{v}{p}\left(\frac{\partial p}{\partial v}\right)_T. \tag{5.190}$$

Eq. (5.190) gives $n_T = 1$ in the case of an ideal gas.

To derive a relation between the exponent of the isentrope n_S and the exponent of the isotherm n_T, Eq. (5.187) can be written as follows:

$$n_S = -\frac{v}{p\left(\dfrac{\partial v}{\partial p}\right)_T\left[\dfrac{T}{c_p}\left(\dfrac{\partial v}{\partial T}\right)_p^2\left(\dfrac{\partial p}{\partial v}\right)_T + 1\right]} \tag{5.191}$$

Inserting Eq. (5.179) into Eq. (5.191) gives

$$n_S = -\frac{v}{p\left(\dfrac{\partial v}{\partial p}\right)_T \left(1 - \dfrac{c_p - c_v}{c_p}\right)}$$

$$n_S = -\frac{v}{p\left(\dfrac{\partial v}{\partial p}\right)_T} \kappa = -\frac{v}{p}\left(\frac{\partial p}{\partial v}\right)_T \kappa . \qquad (5.192)$$

Eqs. (5.190) and (5.192) yield

$$n_S = n_T \kappa . \qquad (5.193)$$

If one knows the specific heat capacities c_p and c_v and thus κ according to Eq. (5.181), then for the calculation of the isentropic exponent n_S, one can do this without Eq. (5.187). It is easier to determine the isothermal exponent n_T according to Eq. (5.190) and thus, via Eq. (5.193), to calculate the isentropic exponent n_S.

Example 5.13 Using the equations of the specific heat capacities and the partial derivative $(\partial p/\partial v)_T$ for the refrigerant R 134 a according to example 5.12, the values of κ, n_T and n_S of this refrigerant at 1 bar are to be calculated.

The results are given in Table 5.4:

Table 5.4 Specific heat capacity at constant pressure and specific heat capacity at constant volume, isothermal exponent and isentropic exponent of superheated vapor of the refrigerant R 134 a at 1 bar

t	c_p^0	c_p	c_v^0	c_v	κ	n_T	n_S
°C	kJ/(kg K)	kJ/(kg K)	kJ/(kg K)	kJ/(kg K)			
-20	0.76249	0.77354	0.68100	0.68252	1.1333	0.96997	1.0993
0	0.80261	0.81178	0.72112	0.72247	1.1236	0.97621	1.0969
20	0.84174	0.84946	0.76025	0.76146	1.1156	0.98092	1.0943
40	0.87976	0.88635	0.79827	0.79937	1.1088	0.98454	1.0917
60	0.91658	0.92227	0.83509	0.83609	1.1031	0.98737	1.0891
80	0.95213	0.95694	0.87065	0.87153	1.0980	0.96245	1.0568
100	0.98636	0.99058	0.90487	0.90568	1.0937	0.96409	1.0545
120	1.01920	1.02290	0.93773	0.93848	1.0900	0.96541	1.0523

5.4.4 The *Clausius-Clapeyron* Equation

The *Clausius-Clapeyron* equation provides the relationship between the specific enthalpy of vaporisation (specific heat of vaporisation) and the thermal state variables. In the two-phase liquid-vapor region, the pressure is a function of the temperature alone, i.e. it does not depend on the specific volume. Therefore, here is $(\partial p/\partial v)_T = 0$, and Eq. (5.172) gives with Eq. (5.149)

$$\frac{dp}{dT} = \left(\frac{\partial p}{\partial T}\right)_v = \left(\frac{\partial s}{\partial v}\right)_T . \qquad (5.194)$$

5 Real Gases and Vapors

From Eqs. (5.27) and (5.33), by eliminating x one obtains

$$s_N(v_N) = \frac{s'' - s'}{v'' - v'} v_N + s' - \frac{s'' - s'}{v'' - v'} v' \ . \tag{5.195}$$

This function between the variables s_N and v_N is the equation of a straight line at constant temperature. So with Eq. (5.40) there is

$$\left(\frac{\partial s}{\partial v}\right)_T = \frac{s'' - s'}{v'' - v'} = \frac{r}{(v'' - v')T} \ . \tag{5.196}$$

r is the specific enthalpy of vaporisation. Eqs. (5.194) and (5.196) give the *Clausius-Clapeyron*[3]

$$r = (v'' - v')T \frac{dp}{dT} \ . \tag{5.197}$$

The *Clausius-Clapeyron* equation is valid not only for the two-phase liquid-vapor region, but also for the two-phase solid-liquid region and the two-phase solid-vapor region.

Thus, for example, analogous to Eq. (5.197) with σ as the specific enthalpy of melting, v_{fl} and v_{fest} as the specific volumes at the melting line and at the solidification line and T_{Sch} as the melting temperature, it is possible to write for the two-phase solid-liquid region:

$$\sigma = (v_{fl} - v_{fest})T_{Sch}\frac{dp}{dT} \ . \tag{5.198}$$

To calculate the specific heat of vaporisation r, one needs the specific volume of the saturated liquid v', the specific volume of the saturated vapor v'', the vaporisation temperature T and the slope dp/dT of the liquid-vapor line (vapor pressure curve).

At low pressures v' can be neglected with respect to v'':

$$r = v'' T \frac{dp}{dT} \tag{5.199}$$

v'' can be expressed approximately using the thermal equation of state of ideal gases (4.26). Then one can write

$$r = \frac{R}{p} T^2 \frac{dp}{dT} \ . \tag{5.200}$$

With a suitable approach to the specific enthalpy of vaporization, with the simplified *Clausius-Clapeyron* equation (5.200) a relation for the vapor pressure curve can be obtained. From Eq. (5.200) it follows

$$\frac{dp}{p} = \frac{r}{R} \frac{dT}{T^2} \ . \tag{5.201}$$

If the heat of evaporation r is constant

$$r = \beta_0 \ , \tag{5.202}$$

then the integration of Eq. (5.201) yields

$$\ln \frac{p}{p_1} = \frac{\beta_0}{R}\left(\frac{1}{T_1} - \frac{1}{T}\right) \ . \tag{5.203}$$

---

[3] *Rudolf Clausius*: see footnote in section 3.2.2
Emile Clapeyron, 1799 to 1864, French physicist, brought the findings of *Carnot* in thermodynamics before a broader scientific public

5.4 Calculation of State Variables; Property Tables

If one sets
$$\ln \frac{p}{p_1} = \ln \frac{p}{1 \text{ bar}} - \ln \frac{p_1}{1 \text{ bar}} \tag{5.204}$$

and define the constants
$$b_0 = \ln \frac{p_1}{1 \text{ bar}} + \frac{B_0}{R}\frac{1}{T_1} \tag{5.205}$$

$$b_1 = \frac{B_0}{R}, \tag{5.206}$$

so it follows from Eq. (5.203)
$$\ln \frac{p}{1 \text{ bar}} = b_0 - \frac{b_1}{T}. \tag{5.207}$$

Example 5.14 According to the *Maxwell* considerations (section 5.3.2), the vapor pressure, the specific volume of the saturated liquid and the specific volume of the saturated vapor are to be calculated with the *Redlich-Kwong-Soave* thermal equation of state for the refrigerant R 134 a. In addition, the liquid-vapor line (vapor pressure curve) and the integration constants in the equations for the specific enthalpy and for the specific entropy according to example 5.11 are to be determined.

Via the *Maxwell* relation Eq. (5.71), which in this case takes the form
$$RT \ln \frac{v''-b}{v'-b} + \frac{a\alpha(T)}{b} \ln \frac{v'(v''+b)}{v''(v'+b)} - p(v''-v') = 0,$$

and the transformed thermal equation of state
$$v^3 - \frac{RT}{p}v^2 + \left(\frac{a\alpha(T) - RTb}{p} - b^2\right)v - \frac{a\alpha(T)b}{p} = 0$$

one obtains for p, v' and v'' the values of Table 5.5.

With a chosen temperature and an estimated pressure, from the transformed thermal equation of state v' and v'' are calculated. Then it has to be checked whether the *Maxwell* relation is fulfilled, and if not, the calculation is repeated with an improved pressure. The liquid-vapor line as a function of saturation pressure from temperature is obtained as a balancing polynomial using the least squares method:

$$\ln \frac{p}{p_1} = b_0 + b_1 \frac{T}{T_1} + b_2 \left(\frac{T}{T_1}\right)^2 + b_3 \left(\frac{T}{T_1}\right)^3 + b_4 \left(\frac{T}{T_1}\right)^4 + b_5 \left(\frac{T}{T_1}\right)^5 + b_6 \left(\frac{T}{T_1}\right)^6$$

$$p_1 = 1 \text{ bar} \quad T_1 = 1 \text{ K}$$

$b_0 = -73.934387$ $\qquad b_4 = -4.0666504 \cdot 10^{-8}$
$b_1 = 0.97604252$ $\qquad b_5 = 4.5450782 \cdot 10^{-11}$
$b_2 = -5.8070359 \cdot 10^{-3}$ $\qquad b_6 = -2.1627798 \cdot 10^{-14}$
$b_3 = 2.0016140 \cdot 10^{-5}$

With the derivation of the eqution for the liquid-vapor line, according to the temperature and the equation of *Clausius-Clapeyron* one obtains the specific enthalpy of vaporisation r:

$$r = (v'' - v')p\frac{T}{T_1}\left[b_1 + 2b_2\frac{T}{T_1} + 3b_3\left(\frac{T}{T_1}\right)^2 + 4b_4\left(\frac{T}{T_1}\right)^3 + 5b_5\left(\frac{T}{T_1}\right)^4 + 6b_6\left(\frac{T}{T_1}\right)^5\right]$$

For the liquid refrigerant in the saturation state at $0\,°C$, let the following apply

$$h' = 200.00 \text{ kJ/kg} \qquad s' = 1.0000 \text{ kJ/(kg K)}.$$

At 0 °C the vapor pressure is 2.9386 bar, the specific volumes at saturation are $v' = 0.00088884$ m^3/kg and $v'' = 0.070329$ m^3/kg. This gives a specific enthalpy of vaporisation at 0 °C of $r = 203.7379$ kJ/kg and according to Eqs. (5.34) and (5.40)

$$h'' - h' = 203.7379 \text{ kJ/kg} \qquad s'' - s' = 0.74588 \text{ kJ/(kg K)}.$$

Using the equations for the specific enthalpy and the specific entropy in example 5.11, at saturation pressure at 0 °C are

$$h'' - h^* = 98.1379 \text{ kJ/kg} \qquad s'' - s^* = 0.57274 \text{ kJ/(kg K)}.$$

Table 5.5 Table of properties for the refrigerant R 134 a in the saturation state

t °C	p bar	v' m^3/kg	v'' m^3/kg	h' kJ/kg	h'' kJ/kg	r kJ/kg	s' kJ/(kg K)	s'' kJ/(kg K)
−20	1.323	0.0008440	0.14986	173.01	390.49	217.48	0.8980	1.7570
0	2.939	0.0008888	0.07033	200.00	403.74	203.74	1.0000	1.7459
20	5.771	0.0009478	0.03649	228.78	416.23	187.45	1.1007	1.7402
40	10.305	0.0010293	0.02025	259.93	427.28	167.35	1.2020	1.7364
60	17.081	0.0011517	0.01164	294.44	435.76	141.32	1.3065	1.7307
80	26.685	0.0013680	0.00663	334.75	439.10	104.35	1.4204	1.7158

Table 5.6 Table of properties for the refrigerant R 134 a in the superheated vapor state at the pressure of 1 bar

t °C	v m^3/kg	h kJ/kg	u kJ/kg	s kJ/(kg K)	z
−20	0.20025	391.07	371.04	1.7814	0.97073
0	0.21740	406.92	385.18	1.8416	0.97668
20	0.23440	423.54	400.10	1.9003	0.98121
40	0.25128	440.90	415.77	1.9576	0.98472
60	0.26808	458.98	432.18	2.0136	0.98749
80	0.29302	477.81	448.51	2.0707	1.0182
100	0.31018	497.28	466.27	2.1243	1.0200
120	0.32729	517.42	484.69	2.1769	1.0216

The values of the integration constants are determined in the following:

$$h^* = (h'' - h') + h' - (h'' - h^*) = (203.7379 + 200 - 98.1379) \text{ kJ/kg} = 305.6000 \text{ kJ/kg}$$

$$s^* = (s'' - s') + s' - (s'' - s^*) = (0.74588 + 1 - 0.57274) \text{ kJ/(kg K)} = 1.17314 \text{ kJ/(kg K)}$$

After calculating the integration constants, all specific caloric state variables can be determined for the saturation state and the superheated vapor state.

One can also calculate the specific internal energy in the saturation state as follows:

$$u' = h' - pv' \qquad u'' = h'' - pv''$$

5.4.5 Free Energy and Free Enthalpy

5.4.5.1 General

The question of work in frictionless isothermal changes of state leads to free energy and free enthalpy. In a closed system, the volume change work is

$$W_{V12} = U_2 - U_1 - (Q_{12})_{rev} \ . \tag{5.208}$$

With the heat transferred during a frictionless isothermal change of state

$$(Q_{12})_{rev} = T(S_2 - S_1) \tag{5.209}$$

one gets

$$W_{V12} = (U_2 - TS_2) - (U_1 - TS_1) \ . \tag{5.210}$$

The expression

$$F = U - TS \tag{5.211}$$

is called free energy. Thus Eq. (5.210) can be written as $W_{V12} = F_2 - F_1$. Dividing by the mass gives the specific free energy

$$f = u - Ts \ . \tag{5.212}$$

Considering an open system, the pressure change work is

$$W_{p12} = H_2 - H_1 - (Q_{12})_{rev} \ . \tag{5.213}$$

Again, in the case of frictionless isothermal change of state, Eq. (5.209) holds

$$W_{p12} = (H_2 - TS_2) - (H_1 - TS_1) \ . \tag{5.214}$$

The expression

$$G = H - TS \tag{5.215}$$

is called free enthalpy. Thus, Eq. (5.214) can be written as $W_{p12} = G_2 - G_1$. The specific free enthalpy is

$$g = h - Ts \ . \tag{5.216}$$

Free energy and free enthalpy are state variables of universal importance. F is also called *Helmholtz* function (*Helmholtz* free energy), G as *Gibbs*[4] function, *Gibbs* thermodynamic potential or *Gibbs* free enthalpy. Specific free energy and specific free enthalpy are state variables with the total differentials (exact differentials)

$$df = \left(\frac{\partial f}{\partial T}\right)_v dT + \left(\frac{\partial f}{\partial v}\right)_T dv \tag{5.217}$$

$$dg = \left(\frac{\partial g}{\partial T}\right)_p dT + \left(\frac{\partial g}{\partial p}\right)_T dp \ . \tag{5.218}$$

With the relation

$$d(Ts) = s \, dT + T \, ds \tag{5.219}$$

follow from Eqs. (5.212) and (5.216) the equations

$$df = du - d(Ts) = du - s \, dT - T \, ds \tag{5.220}$$

$$dg = dh - d(Ts) = dh - s \, dT - T \, ds \tag{5.221}$$

---

[4] *Josiah Willard Gibbs*, 1839 to 1903, North American scientist, made important contributions to the thermodynamics of equilibria

and with Eqs. (5.126) and (5.127)

$$df = -s\,dT - p\,dv \tag{5.222}$$

$$dg = -s\,dT + v\,dp . \tag{5.223}$$

Comparing Eqs. (5.217) and (5.222) and likewise (5.218) and (5.223), the partial derivatives have the following meanings:

$$\left(\frac{\partial f}{\partial T}\right)_v = -s \tag{5.224}$$

$$\left(\frac{\partial f}{\partial v}\right)_T = -p \tag{5.225}$$

$$\left(\frac{\partial g}{\partial T}\right)_p = -s \tag{5.226}$$

$$\left(\frac{\partial g}{\partial p}\right)_T = v \tag{5.227}$$

Furthermore, according to Eqs. (5.133) and (5.224) as well as (5.139) and (5.226) it follows

$$T\left(\frac{\partial^2 f}{\partial T^2}\right)_v = -c_v \tag{5.228}$$

$$T\left(\frac{\partial^2 g}{\partial T^2}\right)_p = -c_p . \tag{5.229}$$

With the relations

$$\frac{\partial^2 f}{\partial v\,\partial T} = -\left(\frac{\partial p}{\partial T}\right)_v \tag{5.230}$$

$$\frac{\partial^2 g}{\partial p\,\partial T} = \left(\frac{\partial v}{\partial T}\right)_p \tag{5.231}$$

$$\left(\frac{\partial^2 f}{\partial v^2}\right)_T = -\left(\frac{\partial p}{\partial v}\right)_T \tag{5.232}$$

$$\left(\frac{\partial^2 g}{\partial p^2}\right)_T = \left(\frac{\partial v}{\partial p}\right)_T , \tag{5.233}$$

which are obtained from Eqs. (5.227) and (5.225), follows from Eqs. (5.179) and (5.180)

$$c_p - c_v = -T\frac{\left(\dfrac{\partial^2 g}{\partial p\,\partial T}\right)^2}{\left(\dfrac{\partial^2 g}{\partial p^2}\right)_T} \tag{5.234}$$

$$c_p - c_v = T\frac{\left(\dfrac{\partial^2 f}{\partial v\,\partial T}\right)^2}{\left(\dfrac{\partial^2 f}{\partial v^2}\right)_T} \tag{5.235}$$

By inserting Eq. (5.228) in Eq. (5.235) and Eq. (5.229) in Eq. (5.234) one gets:

$$c_p = -T\left(\frac{\partial^2 f}{\partial T^2}\right)_v + T\frac{\left(\frac{\partial^2 f}{\partial v \partial T}\right)^2}{\left(\frac{\partial^2 f}{\partial v^2}\right)_T} \qquad (5.236)$$

$$c_v = -T\left(\frac{\partial^2 g}{\partial T^2}\right)_p + T\frac{\left(\frac{\partial^2 g}{\partial p \partial T}\right)^2}{\left(\frac{\partial^2 g}{\partial p^2}\right)_T} \qquad (5.237)$$

Expressions for κ according to Eq. (5.181) are obtained from Eqs. (5.228) and (5.236) as well as Eqs. (5.229) and (5.237):

$$\kappa = 1 - \frac{\left(\frac{\partial^2 f}{\partial v \partial T}\right)^2}{\left(\frac{\partial^2 f}{\partial v^2}\right)_T \left(\frac{\partial^2 f}{\partial T^2}\right)_v} = \frac{\left(\frac{\partial^2 f}{\partial v^2}\right)_T \left(\frac{\partial^2 f}{\partial T^2}\right)_v - \left(\frac{\partial^2 f}{\partial v \partial T}\right)^2}{\left(\frac{\partial^2 f}{\partial v^2}\right)_T \left(\frac{\partial^2 f}{\partial T^2}\right)_v} \qquad (5.238)$$

$$\kappa = \frac{\left(\frac{\partial^2 g}{\partial T^2}\right)_p \left(\frac{\partial^2 g}{\partial p^2}\right)_T}{\left(\frac{\partial^2 g}{\partial T^2}\right)_p \left(\frac{\partial^2 g}{\partial p^2}\right)_T - \left(\frac{\partial^2 g}{\partial p \partial T}\right)^2} \qquad (5.239)$$

The isothermal exponent given by Eq. (5.190) is expressed with the help of Eqs. (5.225) and (5.232) as well as Eqs. (5.227) and (5.233):

$$n_T = -v\frac{\left(\frac{\partial^2 f}{\partial v^2}\right)_T}{\left(\frac{\partial f}{\partial v}\right)_T} \qquad (5.240)$$

$$n_T = -\frac{1}{p}\frac{\left(\frac{\partial g}{\partial p}\right)_T}{\left(\frac{\partial^2 g}{\partial p^2}\right)_T} \qquad (5.241)$$

The isentropic exponent according to Eq. (5.193) is obtained from Eqs. (5.238) and (5.240) as well as Eqs. (5.239) and (5.241):

$$n_S = v\frac{\left(\frac{\partial^2 f}{\partial v \partial T}\right)^2 - \left(\frac{\partial^2 f}{\partial v^2}\right)_T \left(\frac{\partial^2 f}{\partial T^2}\right)_v}{\left(\frac{\partial f}{\partial v}\right)_T \left(\frac{\partial^2 f}{\partial T^2}\right)_v} \qquad (5.242)$$

$$n_S = \frac{1}{p}\frac{\left(\frac{\partial g}{\partial p}\right)_T \left(\frac{\partial^2 g}{\partial T^2}\right)_p}{\left(\frac{\partial^2 g}{\partial p \partial T}\right)^2 - \left(\frac{\partial^2 g}{\partial T^2}\right)_p \left(\frac{\partial^2 g}{\partial p^2}\right)_T} \qquad (5.243)$$

Functions of the type $f = f(T,v)$ and $g = g(T,p)$ with specific free energy and specific free enthalpy belong to the thermodynamic potentials. Thermodynamic potentials are state functions from which all arbitrary state variables can be obtained by differentiations because they contain all thermodynamic information implicitly. The name potential is based on the analogy to other potentials, e.g. the streaming potential, from which the velocity and acceleration are obtained by derivations. With the considerations so far, the initial point for the calculation of a table of thermodynamic propereties was a thermal equation of state. From this, one obtained the caloric state variables via integrations. This still required additional data, e.g. on the specific heat capacity in the state of the ideal gas. From now on, in addition a new way is offered: Starting from the thermodynamic potentials

$$f = f(T,v)$$

or

$$g = g(T,p),$$

a table of thermodynamic properties can be obtained via differentiations. Since differentiation is simpler than integration and the determination of integration constants is omitted, the second way is more advantageous than the first.

The representation of a thermodynamic potential as a function of the associated independent variables is called a canonical equation of state or fundamental equation. It contains all information about the thermodynamic properties of a substance and is of great importance for technical applications [6], [26]. The thermal equation of state $F(p,v,T) = 0$ and the caloric equations of state $u = u(T,v)$ and $h = h(T,p)$ are not canonical equations of state. Besides the equations $f = f(T,v)$ and $g = g(T,p)$, it is of interest to mention the following canonical equations of state [26]:

$$u = u(s,v) \quad s = s(u,v) \quad v = v(s,u) \quad \text{and} \quad h = h(s,p) \quad s = s(h,p) \quad p = p(s,h)$$

For ideal gases with constant specific heat capacities $c_p^0 = c_{pm}^0 = \text{const}$ and $c_v^0 = c_{vm}^0 = \text{const}$ (e.g., noble gases) and the specific gas constant R, the mentioned canonical equations of state, starting from a reference state 0 with values p_0, v_0, T_0, u_0, h_0, and s_0, take the following forms:

$$f = f(T,v) = u_0 + c_v^0 (T - T_0) - s_0 T - c_v^0 T \ln \frac{T}{T_0} - RT \ln \frac{v}{v_0} \tag{5.244}$$

$$g = g(T,p) = h_0 + c_p^0 (T - T_0) - s_0 T - c_p^0 T \ln \frac{T}{T_0} + RT \ln \frac{p}{p_0} \tag{5.245}$$

$$u = u(s,v) = u_0 \left(\frac{v}{v_0}\right)^{-\frac{R}{c_v^0}} \cdot e^{\frac{s-s_0}{c_v^0}} \tag{5.246}$$

$$s = s(u,v) = s_0 + c_v^0 \ln \frac{u}{u_0} + R \ln \frac{v}{v_0} \tag{5.247}$$

$$v = v(s,u) = v_0 \left(\frac{u}{u_0}\right)^{-\frac{c_v^0}{R}} \cdot e^{\frac{s-s_0}{R}} \tag{5.248}$$

$$h = h(s,p) = h_0 \left(\frac{p}{p_0}\right)^{\frac{R}{c_p^0}} \cdot e^{\frac{s-s_0}{c_p^0}} \tag{5.249}$$

$$s = s(h,p) = s_0 + c_p^0 \ln \frac{h}{h_0} - R \ln \frac{p}{p_0} \tag{5,250}$$

$$p = p(s,h) = p_0 \left(\frac{h}{h_0}\right)^{\frac{c_p^0}{R}} \cdot e^{-\frac{s-s_0}{R}} \tag{5.251}$$

5.4 Calculation of State Variables; Property Tables

Equations of the type $u = u(s,v)$, $s = s(u,v)$ and $v = v(s,u)$ are particularly suitable for describing changes of state in closed systems, equations of the type $h = h(s,p)$, $s = s(h,p)$ and $p = p(s,h)$ are particularly suitable for describing changes of state in open systems. In [26] canonical equations of the forms $h = h(s,p)$ and $s = s(h,p)$ are given for water and water vapor, air and helium, respectively; there are also given algorithms for the calculation of all important state variables from canonical equations of the types $f = f(T,v)$, $g = g(T,p)$, $h = h(s,p)$, $s = s(h,p)$, $p = p(s,h)$, $u = u(s,v)$, $s = s(u,v)$, and $v = v(s,u)$.

Is a canonical equation of state available, in principle no additional information about the saturation state is necessary. According to section 5.3.2, the liquid-vapor line, the density of the saturated liquid and the density of the saturated vapor can be determined with the aid of the *Maxwell* relationship. In the case of additional data on the saturation state, care must be taken to ensure that the *Maxwell* relation does not give reasons to contradictions [104].

Example 5.15 From the van der Waals equation of state (5.57)

$$p(v,T) = \frac{RT}{v-b} - \frac{a}{v^2}$$

and the specific heat capacity at constant volume in the state of the ideal gas

$$c_v^0(T) = c_0 + c_1 T + c_2 T^2$$

the thermodynamic properties of a real gas are to be derived. Then, it shall be shown that the same information can be obtained from the canonical equation of state with the specific free energy

$$f(v,T) = \left[c_0\left(1 - \ln\frac{T}{T_1}\right) - R \ln\frac{v-b}{v_1-b} - s^*\right]T - \frac{c_1}{2}T^2 - \frac{c_2}{6}T^3 - \frac{a}{v} + u^*.$$

With the partial derivative

$$\left(\frac{\partial p}{\partial T}\right)_v = \frac{R}{v-b}$$

one obtains the specific internal energy from Eq. (5.159):

$$u = u_1 + c_0(T - T_1) + \frac{c_1}{2}(T^2 - T_1^2) + \frac{c_2}{3}(T^3 - T_1^3) - a\left(\frac{1}{v} - \frac{1}{v_1}\right)$$

$$u = c_0 T + \frac{c_1}{2}T^2 + \frac{c_2}{3}T^3 - \frac{a}{v} + u^*$$

u^* is the sum of all constants. Eq. (5.161) gives the specific entropy:

$$s = c_0 \ln\frac{T}{T_1} + c_1 T + \frac{c_2}{2}T^2 + R \ln\frac{v-b}{v_1-b} + s^*$$

Eq. (2.41) $h = u + pv$ gives the specific enthalpy:

$$h = c_0 T + \frac{c_1}{2}T^2 + \frac{c_2}{3}T^3 + \frac{RTv}{v-b} - 2\frac{a}{v} + h^*$$

According to Eq. (5.169), because of

$$\left(\frac{\partial^2 p}{\partial T^2}\right)_v = 0$$

the specific heat capacity c_v is a function of temperature only:

$$c_v = c_v^0 = c_0 + c_1 T + c_2 T^2$$

The partial derivatives $(\partial p/\partial T)_v$ and

$$\left(\frac{\partial p}{\partial v}\right)_T = -\frac{RT}{(v-b)^2} + \frac{2a}{v^3}$$

and Eq. (5.180) give the specific heat capacity at constant pressure:

$$c_p = c_0 + c_1 T + c_2 T^2 + \frac{R^2 T}{RT - \frac{2a}{v^3}(v-b)^2} = c_v^0 + \frac{R^2 T}{RT - \frac{2a}{v^3}(v-b)^2}$$

According to Eq. (5.181), κ is

$$\kappa = 1 + \frac{R^2 T}{(c_0 + c_1 T + c_2 T^2)[RT - \frac{2a}{v^3}(v-b)^2]}.$$

According to Eq. (5.190), the isothermal exponent is

$$n_T = \frac{v}{v-b} \frac{RT - \frac{2a}{v^3}(v-b)^2}{RT - \frac{a}{v^2}(v-b)}.$$

Eq. (5.193) gives the isentropic exponent:

$$n_S = \frac{v}{v-b} \frac{(c_0 + c_1 T + c_2 T^2)[RT - \frac{2a}{v^3}(v-b)^2] + R^2 T}{(c_0 + c_1 T + c_2 T^2)[RT - \frac{a}{v^2}(v-b)]}$$

With the given equation of the specific free energy $f(T, v)$, the following derivatives can be found:

$$\left(\frac{\partial f}{\partial T}\right)_v = -c_0 \ln \frac{T}{T_1} - R \ln \frac{v-b}{v_1-b} - s^* - c_1 T - \frac{c_2}{2}T^2$$

$$\left(\frac{\partial^2 f}{\partial T^2}\right)_v = -\frac{c_0}{T} - c_1 - c_2 T$$

$$\left(\frac{\partial f}{\partial v}\right)_T = -\frac{RT}{v-b} + \frac{a}{v^2}$$

$$\left(\frac{\partial^2 f}{\partial v^2}\right)_T = \frac{RT}{(v-b)^2} - \frac{2a}{v^3}$$

$$\frac{\partial^2 f}{\partial v \partial T} = -\frac{R}{v-b}$$

It can be seen immediately that Eqs. (5.224), (5.225), and (5.228) are satisfied:

$$s = c_0 \ln \frac{T}{T_1} + R \ln \frac{v-b}{v_1-b} + s^* + c_1 T + \frac{c_2}{2}T^2$$

$$p = \frac{RT}{v-b} - \frac{a}{v^2}$$

$$c_v = c_0 + c_1 T + c_2 T^2$$

According to Eqs. (5.236) and (5.238) is

$$c_p = c_0 + c_1 T + c_2 T^2 + \frac{R^2 T}{RT - \frac{2a}{v^3}(v-b)^2}$$

$$\kappa = \cfrac{[RT - \cfrac{2a}{v^3}(v-b)^2](-\cfrac{c_0}{T} - c_1 - c_2 T) - R^2}{[RT - \cfrac{2a}{v^3}(v-b)^2](-\cfrac{c_0}{T} - c_1 - c_2 T)}.$$

Eqs. (5.240) and (5.242) yield

$$n_T = \frac{v}{v-b} \frac{RT - \frac{2a}{v^3}(v-b)^2}{RT - \frac{a}{v^2}(v-b)}$$

$$n_S = \frac{v}{v-b} \frac{R^2 + [RT - \frac{2a}{v^3}(v-b)^2](\frac{c_0}{T} + c_1 + c_2 T)}{[RT - \frac{a}{v^2}(v-b)](\frac{c_0}{T} + c_1 + c_2 T)}.$$

The specific internal energy is obtained from the defining equation of the specific free energy:

$$u = f + Ts = c_0 T + \frac{c_1}{2}T^2 + \frac{c_2}{3}T^3 - \frac{a}{v} + u^*$$

The specific enthalpy is obtained from Eq. (2.41) $h = u + pv$ as before.

5.4.5.2 A g,s Diagram for Water and Steam

Structure of the g,s Diagram

In the investigation of technical systems, the specific free enthalpy g can be used to obtain additional statements These can be determined and illustrated by means of state diagrams. Therefore, a g, s diagram for water and steam is presented in Fig. 5.19. Here g is the ordinate, s the abscissa. The pressure p and the *Celsius* temperature t are chosen as parameters.

The g, s diagram for water and steam is based on the 1997 IAPWS formulation [128] and on the h, s diagram with absolute specific enthalpies and entropies according to [18]. Two variants are available [32], where in the first variant (Figure 5.19) the upper limit of the temperature is 800 °C, and in the second variant, also published in [32], it is 2000 °C. g is represented as an absolute specific state variable; only then can g be used in a thermodynamically meaningful way. To calculate g, absolute specific values for h and s were used according to [18] (properties at the triple point: $p_{Tr} = 0.006112$ bar, $T_{Tr} = 273.16$ K ($t_{Tr} = 0.01$ °C), $h_{Tr} = 633.0$ kJ/kg, $s_{Tr} = 3.5214$ kJ/(kg K); $g_{Tr} = -328.9$ kJ/kg for water in the saturated liquid state).

The specific free enthalpy g is — except in the solid region and in the two-phase solid-liquid region at high pressures — a negative state variable; therefore, the diagram coordinate is plotted downward. If g were positive, the diagram shape for the two-phase liquid-vapor region and the superheated vapor region would be similar to that of the T, s diagram.

At the coordinate origin, the absolute zero is $T = 0$ K ($t = -273.15$ °C), $h = 0$ kJ/kg and $s = 0$ kJ/(kg K) and thus $g = 0$ kJ/kg. To the right of the solid region (reproduced here only with the boundary line for low pressures), the two-phase solid-liquid region joins, which is overlaid by the sublimation region in the part of low pressures. The two-phase solid-liquid region contains isobars and isotherms as horizontal lines, for which g is constant.

The narrow — and therefore graphically difficult to reproduce — liquid region is shown from the triple point up to the pressure $p = 1000$ bar; it superimposes the two-phase solid-liquid region below about 250 °C (in regions of very low vapor content).

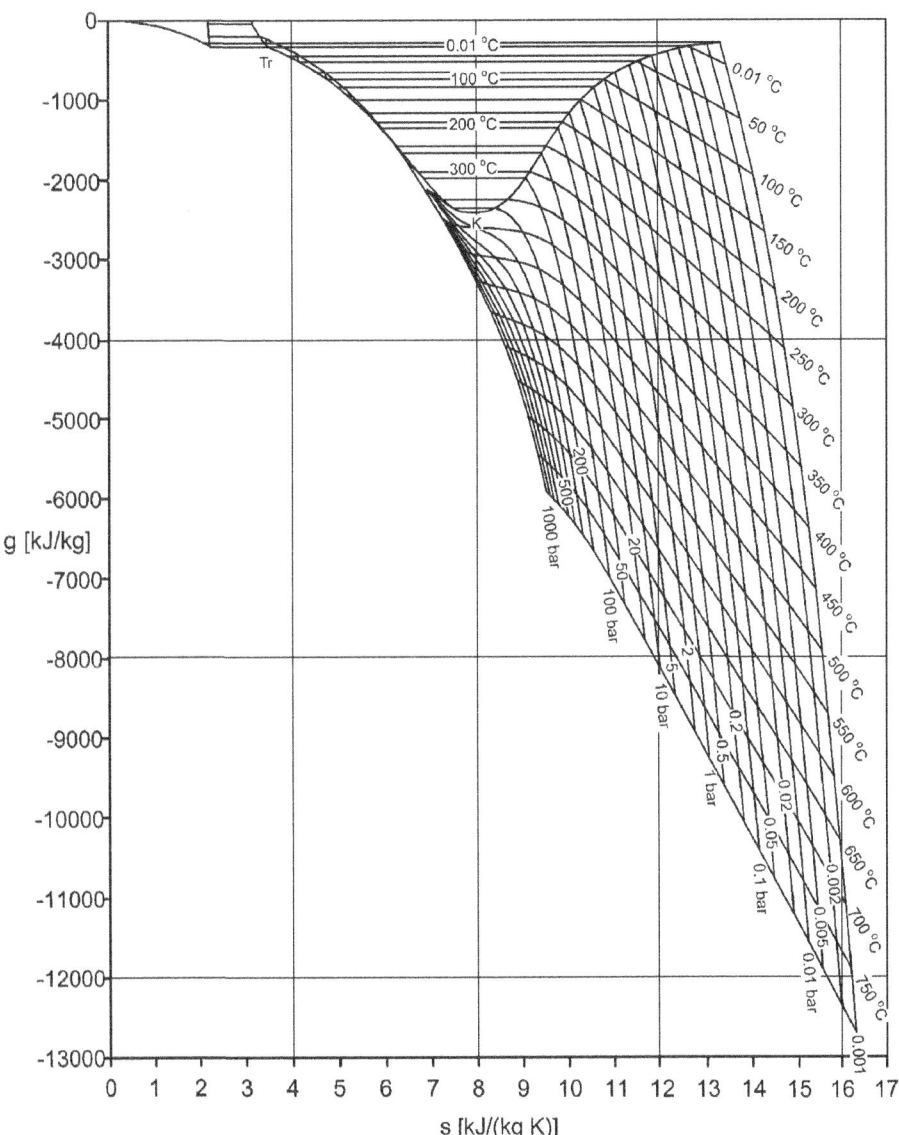

Figure 5.19 g, s diagram for water and steam

The sublimation region lies above the horizontal triple line ($t_{Tr} = 0.01\,°\text{C}$). The two-phase liquid-vapor region follows below; also here isotherms, isobars and lines of constant specific free enthalpy g run horizontally. Downward, the two-phase liquid-vapor region ends in the critical point K ($p_K = 220.64$ bar, $T_K = 647.10$ K ($t_K = 373.95\,°\text{C}$), $h_K = 2720.55$ kJ/kg, $s_K = 7.9334$ kJ/(kg K), $g_K = -2413.15$ kJ/kg), which represents

the low point of the boundary curve (saturated liquid line and saturated vapor line). In the superheated vapor region, the isotherms between $t_{Tr} = 0.01\,°C$ ($t_{Tr} \approx 0\,°C$) and $800\,°C$ in interval steps of $50\ °C$ each are shown. As the temperature increases, the inclination of the isotherms also increases. The isobars — scaled logarithmically — are shown from 0.001 bar to 1000 bar.

Some Useful Applications of the g, s Diagram

For the isothermal change of state of an open system at the absolute temperature T, the difference of the specific free enthalpies $g_2 - g_1$ gives the specific pressure change work w_{p12}:

$$w_{p12} = h_2 - h_1 - (q_{12})_{rev} = h_2 - h_1 - T(s_2 - s_1) = g_2 - g_1 = \Delta g \tag{5.252}$$

The g, s diagram illustrates, for example, that for right-hand cyclic processes with an isothermal expansion in the upper temperature range (e.g., in *Carnot*, *Ericsson*, and *Stirling* processes) the work received for this isothermal change of state is the greater, the higher the temperature and the larger the pressure ratio p_2/p_1 are: Then $w_{p12} = g_2 - g_1 = \Delta g$ takes large values.

The diagram shows that the specific free enthalpy g has a potential character with respect to the isothermal change of state (see section 8.1.13). It is also visible that for right-hand cyclic processes with an isothermal change of state in the lower temperature range (e.g. in the *Clausius-Rankine* process), the amount of work required for isothermal compression is zero, if this is an isothermal condensation: Then $w_{p12} = \Delta g = g_2 - g_1 = 0$ can be seen. Cyclic processes with isothermal condensation at the 'cold end' are thus thermodynamically advantageous.

Another useful application of the g, s diagram is the consideration of changes of state with $g_2 - g_1 = \Delta g = 0$ in the region of the ideal gas.

Relations for Curves $g = $ const and for Other Curves in the Region of the Ideal Gas

As Figure 5.19 shows, during the melting or solidification process, during the sublimation or desublimation process and in the vaporisation or liquefaction process of a pure substance, the specific free enthalpy g remains the same. Therefore, it is obvious to additionally investigate a change of state at $g = $ const in the region of the ideal gas. For this the relations apply:

$$T = T_0 \frac{s_0 - c_p}{s - c_p} \quad \text{with the transformation} \quad s = \frac{T_0}{T}(s_0 - c_p) + c_p \tag{5.253}$$

$$p = p_0 \left(\frac{T}{T_0}\right)^{\frac{\kappa}{\kappa-1}} e^{\frac{T-T_0}{T}\left(\frac{s_0}{R} - \frac{\kappa}{\kappa-1}\right)} \tag{5.254}$$

$$v = v_0 \left(\frac{T}{T_0}\right)^{\frac{1}{1-\kappa}} e^{-\frac{T-T_0}{T}\left(\frac{s_0}{R} - \frac{\kappa}{\kappa-1}\right)} \tag{5.255}$$

The prerequisite here is $c_p = $ const or a constant mean value $c_p = c_{pm}$ in the considered interval. s_0 and T_0 are constant reference values (initial values of the change of state), κ is the isentropic exponent and R is the specific gas constant.

The mentioned relation $p = p(T)$ for curves of constant specific free enthalpy ($g =$ const) is a special form of the general relation for ideal gases

$$p = p_0 \left(\frac{T}{T_0}\right)^{n_1} e^{n_2}, \qquad (5.256)$$

which holds for certain constant state variables:

		n_1	n_2
Isobaric curves	$p = p_0 = $ const	0	0
Isochoric curves	$v = v_0 = $ const	1	0
Isothermal curves	$T = T_0 = $ const	∞	0
Isenthalpic curves	$h = h_0 = $ const	∞	0
Isentropic curves	$s = s_0 = $ const	$\dfrac{\kappa}{\kappa - 1}$	0
Liquid-vapor line		0	$\dfrac{T - T_0}{T} \dfrac{s_0'' - s_0'}{R}$
Curves of equal specific free enthalpy	$g = g_0 = $ const	$\dfrac{\kappa}{\kappa - 1}$	$\dfrac{T - T_0}{T}\left(\dfrac{s_0}{R} - \dfrac{\kappa}{\kappa - 1}\right)$
Curves of equal specific exergy of the specific enthalpy	$e = e_0 = $ const	$\dfrac{\kappa}{\kappa - 1}$	$-\dfrac{T - T_0}{T_u} \dfrac{\kappa}{\kappa - 1}$

In the relation for $e = e_0 = $ const, T_u is the ambient temperature. In the approximation relation for the liquid-vapor line (vapor pressure curve) (as the boundary line of the region of the ideal gas), the state variables p_0, T_0, s_0' and s_0'' denote a chosen reference point 0; the specific enthalpy of vaporisation is assumed to be $r_0 = T_0(s_0'' - s_0') = $ const.

The changes of state on curves with $g = $ const in the region of the ideal gas between the initial state 0 and a general final state, for the process variables specific pressure change work w_p, specific volume change work w_v, specific heat $q_{rev} = q_s$ (also named specific substitute heat or specific entropy change heat) and specific temperature change heat q_T (cf. section 7) the following relations are valid:

$$w_p = q_T = (s_0 - c_p) T_0 \ln \frac{T}{T_0} + c_p (T - T_0) \qquad (5.257)$$

$$w_v = (s_0 - c_p) T_0 \ln \frac{T}{T_0} + c_p (T - T_0) + R T_0 \left[1 - \left(\frac{T}{T_0}\right)^\kappa\right] \qquad (5.258)$$

$$q_s = -(s_0 - c_p) T_0 \ln \frac{T}{T_0} \qquad (5.259)$$

An interpretation of the change of state at constant specific free enthalpy g in the ideal gas region as a continuation of the isobaric-isothermal vaporisation of a pure substance in the two-phase liquid-vapor region is given in [32].

5.4.6 The *Joule-Thomson* Effect

Between the change in pressure and the change in temperature during an isenthalpic throttling (isenthalpic expansion) of a real gas there is the relationship

$$dT = \delta_h \, dp, \quad \text{resp.} \quad \delta_h = \left(\frac{\partial T}{\partial p}\right)_h. \tag{5.260}$$

δ_h is the *Joule-Thomson* coefficient, also named isenthalpic throttling coefficient. In the case of throttling, $dp < 0$ is always given. If δ_h is positive, there is a temperature decrease associated with throttling. The *Joule-Thomson* effect is then called positive. If δ_h is negative, then a temperature increase is associated with throttling. In this case, the *Joule-Thomson* effect is called negative. Since dT is a differential temperature change that occurs as a result of a differential pressure change, δ_h is the differential *Joule-Thomson* coefficient. At the transition to *Joule-Thomson* coefficients with changed sign, $\delta_h = 0$ holds, and no temperature change takes place. By $\delta_h = 0$ the inversion curve of the differential *Joule-Thomson* coefficient or in short the differential inversion curve is determined.

In the case of throttling, the specific enthalpy h is constant, which implies $dh = 0$. Thus it follows from Eq. (5.152)

$$dT = \frac{1}{c_p}\left[T\left(\frac{\partial v}{\partial T}\right)_p - v\right] dp, \tag{5.261}$$

and with $T = T(p,h)$, $dT = (\partial T/\partial p)_h \, dp + (\partial T/\partial h)_p \, dh = (\partial T/\partial p)_h \, dp$ (for $h = $ const) as well as Eq. (5.260), the *Joule-Thomson* coefficient becomes

$$\delta_h = \left(\frac{\partial T}{\partial p}\right)_h = \frac{1}{c_p}\left[T\left(\frac{\partial v}{\partial T}\right)_p - v\right]. \tag{5.262}$$

The *Joule-Thomson* coefficient vanishes when is

$$\left(\frac{\partial T}{\partial v}\right)_p = \frac{T}{v}. \tag{5.263}$$

In the case of ideal gases, Eq. (5.263) is always satisfied, so it becomes $\delta_h = 0$. For an illustrative representation of the *Joule-Thomson* effect, a p, h diagram is suitable with the pressure p as the ordinate, the specific enthalpy h as the abscissa, and isotherms as parameters. The shape of the differential inversion curve is obtained by Eq. (5.263).

The temperature change when the pressure is lowered over a larger range (called the integral *Joule-Thomson* effect), can be described by the following equation:

$$T_1 - T_2 = \delta_{int}(p_1 - p_2) \tag{5.264}$$

p_1, T_1 are the initial values and p_2, T_2 are the final values of the isenthalpic throttling process with δ_{int} as the proportionality factor. When throttling to ambient pressure, the cooling of the gas is greatest when the initial state lies on the differential inversion curve. At higher initial pressure, a temperature rise occurs first, so that the overall cooling is less. Also, the integral *Joule-Thomson* effect can be positive or negative. By

the initial states giving the value $\delta_{int} = 0$ for an isenthalpic throttling to $p \to 0$, the integral inversion curve is defined. It is obtained by connecting special points of the isotherms: For each isotherm such a special point is determined by the condition that it must have the same specific enthalpy as a point of the isotherm in the state of $p \to 0$.

Example 5.16 For a gas represented by the *van der Waals* equation of state and the caloric equations of state given in example 5.15, the differential and integral inversion curves are to be determined.

From Eq. (5.57) follow the two equations

$$\frac{T}{v} = \frac{1}{R}\left(p - p\frac{b}{v} + \frac{a}{v^2} - \frac{ab}{v^3}\right)$$

$$\left(\frac{\partial T}{\partial v}\right)_p = \frac{1}{R}\left(p - \frac{a}{v^2} + \frac{2ab}{v^3}\right),$$

which can be equated according to Eq. (5.263). It follows

$$p = \frac{a}{v}\left(\frac{2}{b} - \frac{3}{v}\right)$$

or, inserting this expression into Eq. (5.56), this leads to

$$T = \frac{2a}{R}\left[\frac{1}{b} - \frac{1}{v}\left(2 - \frac{b}{v}\right)\right].$$

The last two equations, together with the equation for the specific enthalpy given in example 5.15

$$h = c_0 T + \frac{c_1}{2}T^2 + \frac{c_2}{3}T^3 + \frac{RTv}{v-b} - 2\frac{a}{v} + h^*,$$

give the differential inversion curve in a p, h diagram. The independent variable is v, which is used to first calculate p and T. Then v and T give h. The isotherms in the p, h diagram are also obtained with the equation for the specific enthalpy and with the *van der Waals* equation (5.57)

$$p = \frac{RT}{v-b} - \frac{a}{v^2}.$$

v is the independent variable used to calculate the two coordinates p and h at constant T. To determine the integral inversion curve one starts from the enthalpy equation and sets for the final state approximately $v \to \infty$ which is equal to the approximation $p \to 0$. The specific enthalpy in the final state is equal to the specific enthalpy on the integral inversion curve:

$$c_0 T + \frac{c_1}{2}T^2 + \frac{c_2}{3}T^3 + RT + h^* = c_0 T + \frac{c_1}{2}T^2 + \frac{c_2}{3}T^3 + \frac{RTv}{v-b} - 2\frac{a}{v} + h^*$$

This eqution can be simplified to

$$RT = \frac{RTv}{v-b} - 2\frac{a}{v}$$

From this follows

$$T = \frac{2a}{R}\left(\frac{1}{b} - \frac{1}{v}\right).$$

This equation, together with the equation for the specific enthalpy and the *van der Waals* equation Eq. (5.57), gives the integral inversion curve.

The transition to the reduced quantities according to Eqs. (5.78) implies a generalization. With the reduced specific enthalpy

$$h_r = \frac{h}{p_k v_k},$$

the reduced specific heat capacity at constant volume

$$c_{vr} = \frac{c_v T_k}{p_k v_k}$$

$$c_{vr} = c_{0r} + c_{1r} T_r + c_{2r} T_r^2 ,$$

the Eqs. (5.65) to (5.67) and the constants

$$c_{0r} = \frac{c_0 T_k}{p_k v_k} \qquad c_{1r} = \frac{c_1 T_k}{p_k v_k} \qquad c_{2r} = \frac{c_2 T_k}{p_k v_k}$$

$$h_r^* = \frac{h^*}{p_k v_k}$$

are leading to the equations for the differential inversion curve in the p_r, h_r diagram:

$$p_r = \frac{9}{v_r}\left(2 - \frac{1}{v_r}\right)$$

$$T_r = \frac{9}{4}\left[3 - \frac{1}{v_r}\left(2 - \frac{1}{3 v_r}\right)\right]$$

$$h_r = c_{0r} T_r + \frac{c_{1r}}{2}T_r^2 + \frac{c_{2r}}{3}T_r^3 + \frac{8}{3}\frac{T_r v_r}{v_r - \frac{1}{3}} - \frac{6}{v_r} + h_r^*$$

The equations of the isotherms in the p_r, h_r-diagram are

$$p_r = \frac{8}{3}\frac{T_r}{v_r - \frac{1}{3}} - \frac{3}{v_r^2}$$

$$h_r = c_{0r} T_r + \frac{c_{1r}}{2}T_r^2 + \frac{c_{2r}}{3}T_r^3 + \frac{8}{3}\frac{T_r v_r}{v_r - \frac{1}{3}} - \frac{6}{v_r} + h_r^* .$$

The integral inversion curve is described in the p_r, h_r diagram by the following equations:

$$T_r = \frac{9}{4}\left(3 - \frac{1}{v_r}\right)$$

$$p_r = \frac{8}{3}\frac{T_r}{v_r - \frac{1}{3}} - \frac{3}{v_r^2}$$

$$h_r = c_{0r} T_r + \frac{c_{1r}}{2}T_r^2 + \frac{c_{2r}}{3}T_r^3 + \frac{8}{3}\frac{T_r v_r}{v_r - \frac{1}{3}} - \frac{6}{v_r} + h_r^*$$

An evaluation of the given equations with the numerical values

$c_{0r} = 2.5 \qquad c_{1r} = 0.15 \qquad c_{2r} = -0.01 \qquad h_r^* = 5.763676874$

results in Fig. 5.20.

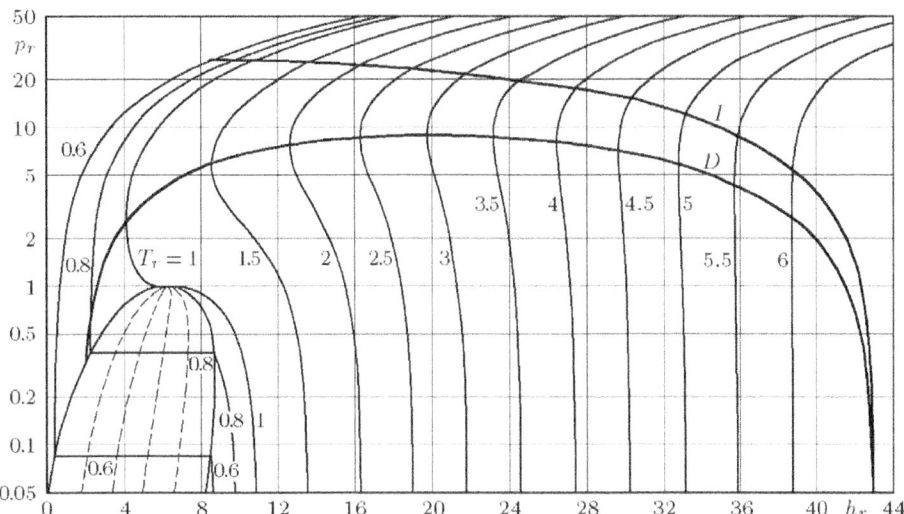

Figure 5.20 p_r, h_r diagram according to example 5.16
D Differential inversion curve
I Integral inversion curve

Tasks for Section 5

1. Boiling water with a gage pressure of $p_1 = 1.2$ bar is withdrawn from a condensation vessel by isenthalpic throttling to $p_2 = 0.05$ bar absolute pressure. The ambient pressure is 990 mbar. What state does the two-phase liquid-vapor reach after throttling?

2. What amount of fuel must be burned hourly in an older steam generating plant if the plant produces the liquid-vapor quantity $\dot{m} = 2.3$ t per hour at a gage pressure of 18 bar and with a steam content of $x = 0.95$? The ambient pressure is 1000 mbar. The incoming feed water has the temperature 190 °C. The calorific value of the fuel is $H_i = 28\,500$ kJ/kg, the plant efficiency (the quotient of used heat and generated thermal energy of the fuel) is 0.76.

3. A steam boiler, into which the condensate returns at saturated liquid temperature, supplies 1.8 t of superheated steam per hour at 11 bar and 220 °C. The steam demand suddenly drops to 1.6 t/h. How high does the temperature of the superheated steam rise with constant pressure and unchanged heat supply from the combustion chamber?

4. After the firing is switched off, in a small steam generator superheated steam of 300 °C and 70 bar is in a closed volume of 28 liters (state 1).

a) The residual heat of the firebox (afterglow of the brick lining) raises the pressure of this steam volume to 120 bar (state 2). What heat was added during this process?

b) When the safety valve is opened, 0.133 kg of steam escapes. After the valve is closed again, the residual mass has filled the volume completely without heat transfer. In which steam state (state 3) is this residual mass then?

5. Flowing superheated steam of $p_1 = 16$ bar and $t_1 = 350$ °C is to be cooled down to $t_2 = 280$ °C by an adiabatic and isobaric injection of water of $t_W = 40$ °C. What amount of water

must be injected per kg of superheated steam? What entropy change per kg of superheated steam occurs during this process?

6. Poperties of the refrigerant R 12 (difluorodichloromethane CF_2Cl_2) can be calculated, according to Baehr and Hicken [3], by the thermal equation of state

$$p = \frac{RT}{v} + \frac{B_0 + \frac{B_1}{T}}{v^2}$$

with the specific gas constant $R = 68.7563$ J/(kg K) and the constants $B_0 = 240$ bar·$(dm^3/kg)^2$, $B_1 = -298 \cdot 10^3$ K bar·$(dm^3/kg)^2$. The specific heat capacities at constant pressure and at constant volume in the state of ideal gas are

$$\frac{c_p^0}{R} = \frac{c_v^0}{R} + 1 = 2.055 + 3.03492 \cdot 10^{-2} \frac{T}{1\,K} - 2.63583 \cdot 10^{-5} \left(\frac{T}{1\,K}\right)^2.$$

The equations for specific volume v, specific enthalpy h, and specific entropy s are to be determined.

7. A vessel with the volume $V_1 = 50$ m^3 contains water vapor in the saturated vapor state ($x_1 = 1$) with $t_1 = 54\,°C$ and $v_1 = 10$ m^3/kg.

a) What is the mass m_1? The properties of p_1, h_1, and s_1 are to be taken from the T, s diagram for H_2O.

b) To the vessel content heat is supplied reversibly by an isochoric change of state (i.e., at $v_1 = v_2 = 10$ m^3/kg) from state 1 to the temperature $t_2 = 262\,°C$ (state 2). The properties of p_2, h_2, and s_2 are to be taken from the T, s diagram. How much heat Q_{12} has been supplied? What value is calculated for the pressure p_2 if the water vapor is simplified as an ideal gas with the specific gas constant $R = 0.4615$ kJ/(kg K)? Is this simplification permissible as a first approximation?

c) What state 3 is reached when the saturated steam is cooled reversibly from state 1 by an isochoric change of state (i.e., at $v_1 = v_3 = 10$ m^3/kg) to the temperature $t_3 = 46\,°C$ (state 3)? The vapor content x_3 as well as p_3, h_3, and s_3 are to be approximated using the T,s diagram. How much heat Q_{13} has to be removed?

d) The entropy differences $S_2 - S_1$ and $S_3 - S_1$ shall be calculated approximately.

8. A pressure vessel has the volume $V_1 = 2.0$ m^3. It contains water and water vapor with the temperature $t_1 = 257\,°C$ and the specific volume $v_1 = 0.01$ m^3/kg.

a) Using the T, s diagram for H_2O, the vapor content x_1, the masses of the two-phase liquid-vapor mixture m_{N1}, of the liquid water m_{F1} and of the saturated vapor m_{D1} as well as the pressure p_1, the specific enthalpy h_1 and the specific entropy s_1 are to be taken from the T, s diagram. By a reversible isochoric heating the state 2 with the pressure $p_2 = 300$ bar (path A) is achieved. The same state 2 can also be achieved by a reversible isentropic compression followed by a reversible isobaric heating (path B).

b) For both process paths A and B, the temperature t_2, the entropy change $S_2 - S_1$, the enthalpy change $H_2 - H_1$, the increase of the internal energy $U_2 - U_1$, the heat $Q_{12\,A}$ respective the heat $Q_{12\,B}$ to be supplied, shall be determined approximately.

9. For the refrigerant R 12 according to task 6, the liquid-vapor line (vapor pressure curve) and the density of the saturated liquid are described by the following equations:

$$\ln \frac{p}{p_k} = \frac{T_k}{T}(-6.64555\,\Theta + 3.4589\,\Theta^2 - 13.774\,\Theta^3 + 25.476\,\Theta^4 - 21.92\,\Theta^5)$$

$$\frac{\varrho'}{\varrho_k} = 1.07244 + 3.30596\,\sqrt{\Theta} - 1.95871\,\Theta + 1.36942\,\Theta\sqrt{\Theta} \qquad \Theta = 1 - \frac{T}{T_k}$$

$$T_k = 385.15 \text{ K} \qquad p_k = 41.482 \text{ bar} \qquad \varrho_k = 0.55809 \text{ kg/dm}^3$$

For the liquid refrigerant in the saturation state at $-40\,^\circ$C shall apply

$$h' = 100.00 \text{ kJ/kg} \qquad\qquad s' = 1.6455 \text{ kJ/(kg K)}\,.$$

The integration constants in the equations for the specific enthalpy h and for the specific entropy s are to be calculated.

6 Thermal Machines

In thermodynamics, the study of machines and apparatuses in which processes with gases (including vapors) and liquids take place is important. In thermal machines, fluids undergo temperature changes during the process. In these machines, energy conversions are necessary between mechanical and thermal energy.

Hereby one has to distinguish between ideal and real machines. The ideal machine would allow the optimal energy transformation, in the real machine this optimum cannot be achieved. Reversible processes take place in ideal machines. Therefore, the main conditions that must be required for ideal machines are:

1) sequence of quasistatic changes of state

2) reversible heat transfer processes

3) absence of friction work

4) combustion processes in the machine are replaced by external heat supply to the working gas

The real machines sometimes considerably deviate from these requirements. Nevertheless, important results can be obtained on the basis of the ideal machine models.

6.1 Classification and Types of Machines

Machines may be subdivided according to various thermodynamic and engineering considerations:

6.1.1 Classification According to the Direction of Energy Conversion

> mechanical energy ⇒ thermal energy: working machine
> thermal energy ⇒ mechanical energy: power machine (prime mover, engine)

Task of the working machine:
Mechanical energy is applied to the drive shaft or piston rod. The enthalpy of a gas flow is increased (compressor, pump), or heat is raised to a higher temperature level (refrigerator, refrigerating machine, heat pump).

Task of the power machine:
Through the expansion of a gas or liquid flow, in which the enthalpy is reduced (turbine, gas engine), or by the conversion of thermal energy resulting from the combustion of fuels (internal combustion engines), mechanical energy is obtained, which can be used at the shaft of the machine or at the piston rod.

6.1.2 Classification According to the Construction of the Machines

> continuous operation: fluid flow machine
> periodic operation: displacement machine

Description of fluid flow machines: The common characteristic of all fluid flow machines is the bladed impeller through which the working gas flows continuously (Fig. 2.11 a to d).

Description of displacement machines: Among the displacement machines, the most important group are reciprocating machines, In addition, there are many types of rotary piston machines, rolling piston machines and screw machines. Their common design characteristic is the sealed space provided inside the machine for the working gas. This space is periodically changed during the cycle (Fig. 2.11 e).

6.1.3 Classification According to the Type of Process Taking Place

> open system: open process
> closed system: cyclic process

Characteristics of an open process: The gas flow entering the machine undergoes one or a series of changes of state in the machine and leaves the machine in a state that differs from the initial state of the entering gas. The working space of the machine represents an open system. The process taking place in the machine is called an open process. If the gas flow through a cross-section (e.g. the inlet cross-section) is constant in time, the open process is called stationary.

Displacement machines generally generate a pulsating gas flow, but its pulsation decays at some distance from the machine, so that the process can also be considered stationary if the frequency and amplitude do not change. In special cases, the gas flow can also be made uniform by means of special devices (e.g. air vessels).

Characteristics of a cyclic process: In a cyclic process, a quantity of substance undergoes a series of changes of state, which follow each other in such a way that the initial state is reached again. This sequence of changes of state is constantly repeated. Cyclic processes can either take place in displacement machines or through the interconnection of stationary flow machines and heat exchangers. Since the circulating mass is always the same, it is a closed system.

6.2 Ideal Machines

In the following discussion of section 6, we will restrict ourselves to open processes; cyclic processes will be discussed in section 7. For the machines to be considered here, the energy equation for open systems can be applied to the gas flow. The change of the potential energy of the gas flow between the inlet and outlet cross section of the machine can be neglected.

6.2 Ideal Machines

For ideal machines, a reversible process can be described according to Eqs. (2.67) and (3.78) as

$$(Q_{12})_{rev} + (W_e)_{id} = H_2 - H_1 + \frac{m}{2}(c_2^2 - c_1^2) \tag{6.1}$$

and according to Eq. (3.75) as

$$(Q_{12})_{rev} + W_{p12} = H_2 - H_1 . \tag{6.2}$$

6.2.1 Compression and Expansion in Ideal Machines

In the following considerations about the change of state to be aimed at during compression or expansion of a gas in ideal machines it shall be assumed that the change of the kinetic energy can be neglected. Then, according to Eqs. (6.1) and (6.2) one can write

$$(W_e)_{id} = W_{p12} = H_2 - H_1 - (Q_{12})_{rev} . \tag{6.3}$$

For different changes of state, the same initial state is assumed. It follows for an isothermal change of state (index a)

$$(W_e)_{id} = W_{p12a} = H_{2a} - H_1 - (Q_{12a})_{rev} , \tag{6.4}$$

for an isentropic change of state (index b)

$$(W_e)_{id} = W_{p12b} = H_{2b} - H_1 , \tag{6.5}$$

for a polytropic change of state (index c)

$$(W_e)_{id} = W_{p12c} = H_{2c} - H_1 - (Q_{12c})_{rev} . \tag{6.6}$$

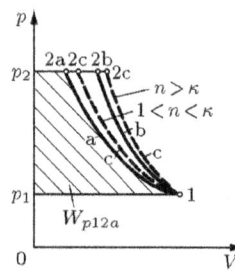

Figure 6.1 Changes of state during a compression in a p, V diagram: a isothermal b isentropic c polytropic

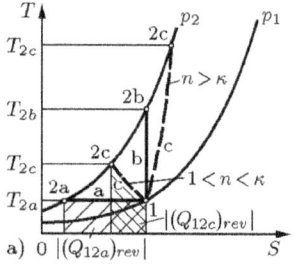

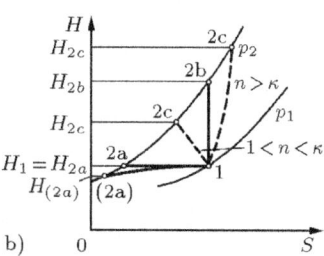

Figure 6.2 Changes of state during a compression in a a) T, S diagram   b) H, S diagram: a isothermal b isentropic c polytropic

Fig. 6.1 and Fig. 6.2 show the changes of state for the compression in the p, V diagram, T, S diagram and H, S diagram from a common initial state 1. From the large number of possible polytropes, those with an exponent $1 < n < \kappa$ and $n > \kappa$ are selected, which are also used for comparison purposes in real machines. In a H, S diagram, the final state 2a is reached with an ideal gas, and the final state (2a) with a real gas.

From the p, V diagram of Fig. 6.1 it can be seen that the pressure change work is smallest with isothermal compression of the gas. Thus, according to Eq. (6.3), the

smallest coupling work is also required to supply it. Fig. 6.2 a) shows that isothermal compression requires the greatest gas cooling to compress the gas. Technically, however, the possibilities for gas cooling during compression are limited, so, in the extreme case, only polytropes with an exponent $1 < n < \kappa$ can be achieved.

Let us consider the expansion of a gas flow in a p, V diagram (Fig. 6.3), where an isothermal, an isentropic and a polytropic change of state with $1 < n < \kappa$ are chosen, we find that the pressure change work along of the isothermal change of state is the largest.

So, the isothermal expansion of the gas provides the largest coupling work. In the T, S diagram according to Fig. 6.4 a), it is clear that isothermal expansion also requires the greatest heat input. Technically, however, the possibilities for such an intensive heat transfer during expansion are limited, so, in the extreme case, only polytropes with an exponent $1 < n < \kappa$ can be achieved.

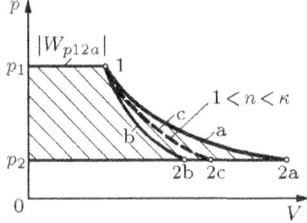

Figure 6.3 Changes of state during an expansion in a p, V diagram:
a isothermal b isentropic c polytropic

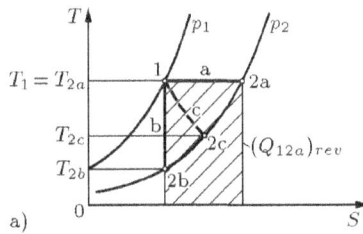

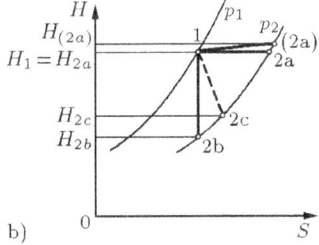

Figure 6.4 Changes of state during expansion in a T, S diagram (left) and in a H, S diagram (right): a isothermal b isentropic c polytropic

6.2.2 Multi-Stage Compression and Expansion

When a gas is compressed, the final temperature that occurs can become so high that there is a risk of combustion of the the lubricating, cooling or sealing oil. In this case, a multi-stage compression is carried out and, ideally, the compressed gas is cooled down to the initial temperature after each stage. In the following is to be investigated the question of the most favorable pressures for recooling in polytropic compression of an ideal gas in an ideal machine.

The p, V diagram of a compression with recooling to the initial temperature according to Fig. 6.5 a) shows that, as a further advantage, a reduction in the pressure change work by ΔW_p and thus, according to Eq. (6.3), in the coupling work is achieved. The energy input approaches the input for isothermal compression W_{pAB}.

The intermediate pressures p_m can be determined in such a way that the pressure change work for multistage compression becomes a minimum. In the case of the two-

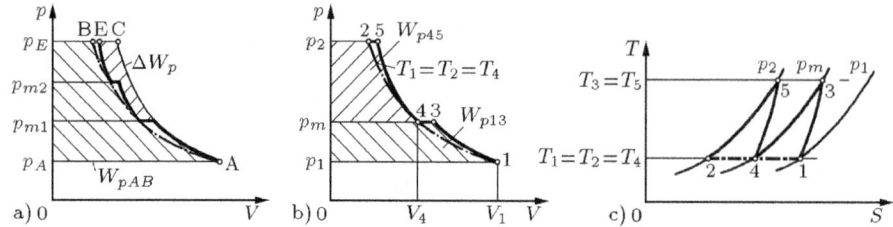

Figure 6.5 a) Three-stage compression with recooling to the initial temperature in a p, V diagram; A–B isothermal curve   b) Two-stage compression with recooling to the initial temperature in a p, V diagram; 1–4–2 isothermal curve   c) Two-stage compression with recooling to the initial temperature in a T, S diagram; 1–4–2 isothermal line

stage compression of an ideal gas according to Fig. 6.5 b) and Fig. 6.5 c) it can be found, if in both stages polytropes of the same exponent n are assumed, according to Table 4.2 Eq. (2):

$$W_{p13} = \frac{n}{n-1} p_1 V_1 \left[\left(\frac{p_m}{p_1}\right)^{\frac{n-1}{n}} - 1 \right] \tag{6.7}$$

$$W_{p45} = \frac{n}{n-1} p_m V_4 \left[\left(\frac{p_5}{p_m}\right)^{\frac{n-1}{n}} - 1 \right] \tag{6.8}$$

As a result of isobaric recooling from state 3 to state 4 with initial temperature $T_4 = T_1$ holds:

$$p_1 V_1 = p_m V_4 = m R T_1 \tag{6.9}$$

The following abbreviation is chosen:

$$\frac{n-1}{n} = a \tag{6.10}$$

The sum of the required compression work should be as low as possible. This minimum depends on the value of the intermediate pressure p_m. As a minimum condition, the derivative of the sum of the compression works as a function of p_m shall be set equal to zero:

$$\frac{d(W_{p13} + W_{p45})}{dp_m} = \frac{m R T_1}{a} \left[a\left(\frac{p_m}{p_1}\right)^{a-1} \frac{1}{p_1} - a\left(\frac{p_5}{p_m}\right)^{a-1} \frac{p_5}{p_m^2} \right] = 0$$

$$\frac{p_m^{a-1}}{p_1^a} = \frac{p_5^a}{p_m^{a+1}} \tag{6.11}$$

This results in

$$p_m^{2a} = p_5^a \, p_1^a \tag{6.12 a}$$

and

$$p_m^2 = p_5 \, p_1 \tag{6.12 b}$$

or also

$$\frac{p_m}{p_1} = \frac{p_5}{p_m} \, . \tag{6.13}$$

The intermediate pressure sought must be chosen so that both stages have the same pressure ratio. Then the pressure change work of both stages will also be the same:

$$W_{p13} = W_{p45} \tag{6.14}$$

In the case of several compression stages (i compression stages), the intermediate pressures are to be selected in such a way that the pressure ratio of each compression stage is the same [27]. With the initial pressure p_a and the final pressure p_e one obtains

$$\frac{p_{m1}}{p_a} = \frac{p_{m2}}{p_{m1}} = \cdots = \frac{p_e}{p_{m(i-1)}}. \tag{6.15}$$

Multiplication gives

$$\frac{p_{m1}}{p_a} \frac{p_{m2}}{p_{m1}} \cdots \frac{p_e}{p_{m(i-1)}} = \left(\frac{p_{m1}}{p_a}\right)^i = \frac{p_e}{p_a}$$

$$\frac{p_{m1}}{p_a} = \left(\frac{p_e}{p_a}\right)^{\frac{1}{i}}. \tag{6.16}$$

To approximate the isothermal change of state during the expansion of a gas flow, a multistage expansion with intermediate superheating, in which heat is added by an isobaric change of state, can be applied. The pressure change work becomes a maximum if the intermediate pressures are chosen in such a way that the pressure ratio of each expansion stage is the same (Fig. 6.6).

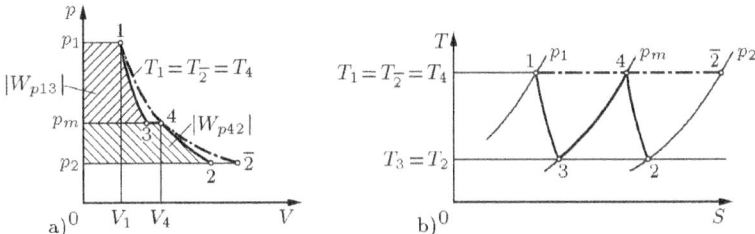

Figure 6.6 Two-stage expansion with intermediate superheating to the initial temperature: 1–4–$\bar{2}$ isothermal curve a) p, V diagram b) T, S diagram

Example 6.1 In a process, helium is heated from the initial pressure $p_a = 10.56$ bar to the final pressure $p_e = 28.76$ bar. The compression is to take place in two stages. After the low-pressure compression, the gas is cooled back to the initial temperature of 25 °C by an isobaric change of state. Helium can be assumed as an ideal gas. Which intermediate pressure is to be selected?

From Eq. (6.12 b) $p_m^2 = p_5\, p_1$ follows: $p_m = \sqrt{p_a\, p_e} = 17{,}43$ bar

Equation (6.16) can also be used:

$$\frac{p_{m1}}{p_a} = \left(\frac{p_e}{p_a}\right)^{\frac{1}{i}} \qquad p_m = p_a \left(\frac{p_e}{p_a}\right)^{\frac{1}{2}} = \left(p_a^2 \frac{p_e}{p_a}\right)^{\frac{1}{2}} = \left(p_a\, p_e\right)^{\frac{1}{2}} = 17{,}43 \text{ bar}$$

6.2.3 The Energy Balance for Flow Machines

The laws discussed with Eqs. 6.1 to 6.16 apply to all cases of the design of the machines for both flow and displacement machines (Fig. 6.7 and Fig. 6.8). The following sections 6.2.3 and 6.2.4 deal with the special conditions of these machine types.

Flow machines include the turbo compressors which are built as axial and radial compressors, and turbines (gas and steam turbines). The change of kinetic energy between the inlet and outlet cross-sections can no longer be neglected in some applications.

From Eqs. (6.1) and (6.2) the following results for the coupling work of ideal machines:

$$(W_e)_{id} = W_{p12} + \frac{m}{2}(c_2^2 - c_1^2) \quad . \tag{6.17}$$

Example 6.2 The compressor of a micro gas turbine unit compresses air as an ideal gas ($p_1 = 0.988$ bar, $t_1 = 25\,°C$) by an isentropic change of state to $p_2 = 4.38$ bar. The compressor is to be considered as an ideal machine, for the air $\kappa = 1.40$ is to be set. The change in kinetic energy is negligible.

a) What is the power required by the compressor if an air flow of 1.244 kg/s is to be delivered?
b) What outlet temperature does the airflow reach?
c) The same machine now is cooled. This succeeds in reducing the outlet temperature of the airflow to $123\,°C$. By which polytropic change of state can the compression now be described?
d) What is the power now required for c), what heat flow has to be cooled away?

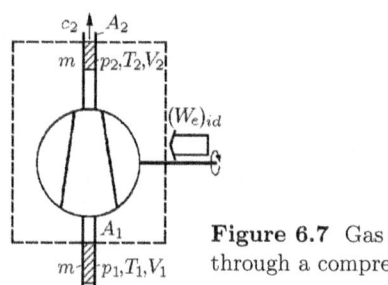

Figure 6.7 Gas flow through a compressor

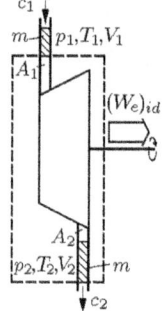

Figure 6.8 Gas flow through a turbine

(a) According to Eqs. (4.105) and (4.111), with $R = 287.06$ J/(kg K) one gains

$$(P_e)_{id} = P_{p12} = \frac{\kappa}{\kappa - 1}\dot{m} R T_1 \left[\left(\frac{p_2}{p_1}\right)^{\frac{\kappa-1}{\kappa}} - 1\right] = 197.614 \text{ kW} \ .$$

b) According to Eq. (4.100) is

$$T_2 = \left(\frac{p_2}{p_1}\right)^{\frac{\kappa-1}{\kappa}} T_1 = 456.26 \text{ K} \qquad t_2 = 183.11\,°C$$

c) According to Eq. (4.144) is

$$n = \frac{\ln\frac{p_2}{p_1}}{\ln\frac{p_2}{p_1} - \ln\frac{T_2}{T_1}} = 1.2359 \ .$$

(d) According to Table 4.2 Eq. (2) it follows

$$(P_e)_{id} = P_{p12} = \frac{n}{n-1}\dot{m} R T_1\left[\left(\frac{p_2}{p_1}\right)^{\frac{n-1}{n}} - 1\right] = 183.374 \text{ kW} \ .$$

According to Table 4.3 Eq. (4) is

$$(\dot{Q}_{12})_{rev} = \frac{n - \kappa}{(\kappa - 1)(n - 1)}\dot{m} R (T_2 - T_1) = -60.862 \text{ kW} \ .$$

6.2.4 The Energy Balance for Displacement Machines

The most common type of displacement machines are reciprocating machines. Therefore, piston compressors (reciprocating compressors) and piston engines will be considered in the following.

To describe the energy conversion, Eqs. (6.1) and (6.2) can be used. For displacement machines, the change in kinetic energy can always be neglected.

$$(Q_{12})_{rev} + (W_e)_{id} = H_2 - H_1 \tag{6.18}$$

$$(Q_{12})_{rev} + W_{p12} = H_2 - H_1 \tag{6.19}$$

$$(W_e)_{id} = W_{p12} \tag{6.20}$$

The gas states 1 and 2 refer to states of the equalised gas flow before and after the machine, respectively.

However, it can be shown that Eqs. (6.18) to (6.20) also describe the energy balance for the quantity of gas passed through the machine.

In reciprocating compressors, a quantity of gas m flows into the cylinder through the open inlet valve (E) (Fig. 6.9). The gas quantity entering the cylinder causes a piston motion until the position s_1 is reached. The gas performs the shift work

$$W_1 = p_1 \, A_K \, s_1 = p_1 \, V_1 \ . \tag{6.21}$$

After closing the inlet valve and piston reversal, the force F acting on the piston rod compresses the gas to the pressure p_2. The work of the piston force F is equal to the volume change work W_{V12}:

$$\int_1^2 F \, \mathrm{d}s = W_{V12} = -\int_1^2 p \, \mathrm{d}V \tag{6.22}$$

Finally, with the exhaust valve (A) open, the piston must push the gas compressed to pressure p_2 out of the cylinder with the shift work W_2 out of the cylinder:

$$W_2 = p_2 \, A_K \, s_2 = p_2 \, V_2 \tag{6.23}$$

(shift work = charge exchange work)

For the machine, the coupling work has to be equal to the algebraic sum of the amounts of work added and removed during a work cycle. Temporal differences in the occurrence of the work amounts are bridged by the flywheel mass of the machine.

In the piston compressor, the gas volume delivers the shift work W_1 to the machine, the volume change work W_{V12} and the shift work W_2 are transferred from the machine to the gas volume:

$$(W_e)_{id} = -|W_1| + |W_{V12}| + |W_2| \tag{6.24}$$

For the amounts of the displacement operations hold

$$|W_1| = W_1 \qquad\qquad |W_2| = W_2 \ . \tag{6.25}$$

The volume change work is positive when the volume of the gas is compressed:

$$|W_{V12}| = W_{V12} = -\int_1^2 p \, \mathrm{d}V \tag{6.26}$$

For the coupling work this gives

$$(W_e)_{id} = p_2 V_2 - p_1 V_1 - \int_1^2 p\,dV = \int_1^2 V\,dp = W_{p12}\;. \tag{6.27}$$

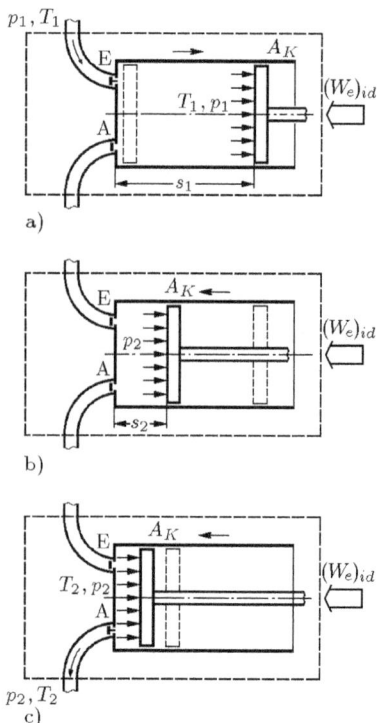

Figure 6.9 Piston compressor
a) Inlet gas flow
b) Compression
c) Displacement of the gas

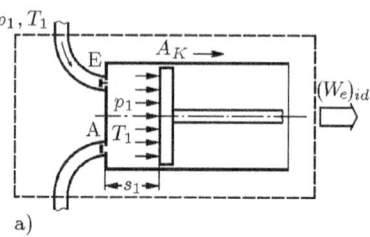
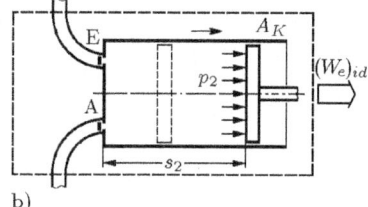
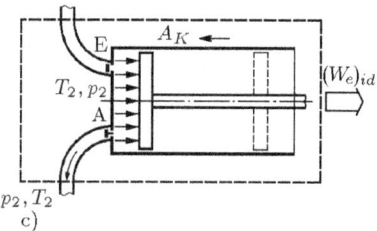

Figure 6.10 Piston engine
a) Inlet gas flow
b) Expansion
c) Displacement of the gas

Using Eq. (2.29), Eq. (6.20) is obtained. By inserting Eq. (3.75), also Eqs. (6.18) and (6.19) can be found.

If the atmospheric pressure $p_L \leq p_1$ acts on the piston according to Fig. 6.9 a) and counteracts the pressure p_1, then when the gas flows in, by the gas the shift work

$$W_1 = (p_1 - p_L) V_1 \tag{6.28}$$

is performed at the piston. The work done by the gas on the outside air is $p_L V_1$. During the compression process from state 1 to state 2, the work to be done via the piston force F given by

$$\int_1^2 F\,ds = W_{V12} - p_L(V_1 - V_2) = -\int_1^2 p\,dV - p_L(V_1 - V_2)\;, \tag{6.29}$$

is lower because the outside air contributes the work $p_L(V_1 - V_2)$. To push out the gas, the shift work of the piston

$$W_2 = (p_2 - p_L) V_2 \tag{6.30}$$

222   6 Thermal Machines

is required; the outside air does the work $p_L V_2$. For the coupling work, the following results as in Eq. (6.27):

$$(W_e)_{id} = (p_2 - p_L)V_2 - (p_1 - p_L)V_1 - \int_1^2 p\,dV - p_L(V_1 - V_2)$$

$$= p_2 V_2 - p_1 V_1 - \int_1^2 p\,dV = \int_1^2 V\,dp = W_{p12}. \qquad (6.31)$$

So the effects of the outside air cancel each other out.

In the following a piston engine is considered. The mode of action is based on the expansion of the gas flowing into the cylinder under the pressure p_1. The gas delivers the work $(W_e)_{id}$ to the machine (Fig. 6.10), thus $(W_e)_{id}$ becomes negative:

$$(W_e)_{id} = -|W_1| - |W_{V12}| + |W_2| \qquad (6.32)$$

For the shift operations, Eqs. (6.25) are to be used. The volume change work is negative when the gas volume expands:

$$|W_{V12}| = -W_{V12} = \int_1^2 p\,dV \qquad (6.33)$$

Eq. (6.20) is also obtained for the piston engine:

$$(W_e)_{id} = p_2 V_2 - p_1 V_1 - \int_1^2 p\,dV = \int_1^2 V\,dp = W_{p12}. \qquad (6.34)$$

Eq. (6.27), Eq. (6.31) and Eq. (6.34) are identical.

Eq. (6.20) gives positive values for the coupling work in the case of working machines, it is therefore supplied to the machine and used to compress the gas flow. Negative values result for prime movers, which therefore gain coupling work from the expansion of the gas flow and deliver it to the shaft of the machine.

Example 6.3 A compressor is to deliver an air flow of $\dot{V}_0 = 100$ m^3/h in the physical norm state with a pressure of $p_2 = 9$ bar. In the intake state, the air as an ideal gas has a pressure of $p_1 = 0.981$ bar and a temperature of $t_1 = 27\,°C$. For an ideal machine, by an isothermal, by an isentropic and by a polytropic change of state are to be calculated

a) the volume of air flowing through the inlet cross section $\dot{V}_1$,

b) the volume of air flowing through the outlet cross section $\dot{V}_2$, the motive power $(P_e)_{id}$ and the heat flow by cooling $(\dot{Q}_{12})_{rev}$,

c) the operating costs $\dot{L}_G$ of the electrically driven compressor with electricity costs of 0.15 Euro/kWh and cooling water costs of 2 Euro/m^3 if a temperature rise of 10 K in the cooling water is assumed.

In the case of polytropic compression, a polytropic change of state with $n = 1.2$ is to be taken as a basis and, as an alternative to water cooling, air cooling should also be calculated ($\kappa = 1{,}40$).

a) With $\quad \dot{V}_1 = \dot{m}R\dfrac{T_1}{p_1}\quad$ and $\quad \dot{m} = \dfrac{\dot{V}_0\, p_0}{R T_0}\quad$ becomes $\quad \dot{V}_1 = \dot{V}_0 \dfrac{p_0}{p_1}\dfrac{T_1}{T_0}$.

b) Isothermal: $\quad p_2 \dot{V}_2 = p_1 \dot{V}_1 \quad \dot{V}_2 = \dot{V}_1 \dfrac{p_1}{p_2} \quad (P_e)_{id} = p_1 \dot{V}_1 \ln \dfrac{p_2}{p_1} \quad (\dot{Q}_{12})_{rev} = -(P_e)_{id}$

Isentropic: $\quad p_2 \dot{V}_2^\kappa = p_1 \dot{V}_1^\kappa \quad \dot{V}_2 = \dot{V}_1\left(\dfrac{p_1}{p_2}\right)^{\frac{1}{\kappa}} \quad (P_e)_{id} = \dfrac{\kappa}{\kappa-1} p_1 \dot{V}_1 \left[\left(\dfrac{p_2}{p_1}\right)^{\frac{\kappa-1}{\kappa}} - 1\right]$

$(\dot{Q}_{12})_{rev} = 0$

Polytropic: $\quad p_2 \dot{V}_2^n = p_1 \dot{V}_1^n \quad \dot{V}_2 = \dot{V}_1 \left(\frac{p_1}{p_2}\right)^{\frac{1}{n}} \quad (P_e)_{id} = \frac{n}{n-1} p_1 \dot{V}_1 \left[\left(\frac{p_2}{p_1}\right)^{\frac{n-1}{n}} - 1\right]$

$(\dot{Q}_{12})_{rev} = \frac{n-\kappa}{(\kappa-1)(n-1)} p_1 \dot{V}_1 \left[\left(\frac{p_2}{p_1}\right)^{\frac{n-1}{n}} - 1\right]$

c) Cooling water flow: $\quad \dot{m}_W = \frac{(\dot{Q}_{12})_{rev}}{c \Delta t} \quad \dot{V}_W = \frac{(\dot{Q}_{12})_{rev}}{\rho c \Delta t}$

In cost accounting mean
$\dot{L}_E$ = electricity costs $\quad \dot{L}_W$ = cooling water costs $\quad \dot{L}_G$ = operating costs.
The calculations yield the results summarised in the following table:

change of state		isothermal	isentropic	polytropic water cooling	polytropic air cooling
$\dot{V}_1$	in m³/h	113.50	113.50	113.50	113.50
$\dot{V}_2$	in m³/h	12.37	23.30	17.90	17.90
$(P_e)_{id}$	in kW	6.85	9.57	8.29	8.29
$(\dot{Q}_{12})_{rev}$	in kW	−6.85	0	−3.46	−3.46
$\dot{L}_E$	in Euro/h	1.03	1.43	1.24	1.24
$\dot{L}_W$	in Euro/h	1.18	0	0.59	0
$\dot{L}_G$	in Euro/h	2.21	1.43	1.83	1.24

The comparison of the different changes of state shows that the air-cooled, polytropic compressor achieves the lowest operating costs. If cooling water were available free of charge, isothermal compression would be the most favourable. However, this would require about twice as much heat to be cooled as with polytropic compression.

6.3 Energy Balances for Real Machines

Friction work occurs in real machines. It arises from several causes:

Displacement machines: Friction of the piston rings (or a corresponding machine part) against the cylinder wall, acceleration, deceleration and turbulence of the gas in the cylinder, throttle loss at the inlet and outlet valves, bearing friction.

Fluid flow machines: Flow pressure losses in the channels of the impeller vanes and guide vanes due to flow boundary layers and vortex formation, shock losses, bearing friction

The distinction between internal and external friction work is made at the system boundary. It is defined as follows:

The external friction work heats up a coolant (cooling air, lubricant), so that this part of the friction work does not reach the actual working chamber of the machine where the gas is compressed or expanded.

The internal friction work influences the change of state of the gas by increasing the internal energy of the gas (section 2.4.5).

For the energy balance in real machines one obtains according to Eqs. (2.67) and (2.70) with negligible change of the potential energy

$$Q_{12} + W_e - |W_{RA}| = H_2 - H_1 + \frac{m}{2}(c_2^2 - c_1^2) \tag{6.35}$$

$$Q_{12} + W_{p12} + |W_{RI}| = H_2 - H_1 . \tag{6.36}$$

6.3.1 Internal or Indexed Work

In an ideal working machine, the coupling work $(W_e)_{id}$ supplied to the machine shaft is completely supplied to the gas flow. In an ideal power machine, work taken from the gas flow is completely available as coupling work at the machine shaft. In real machines, the amount of the transmitted work is reduced by the external friction work. In the following considerations, it makes sense to look at first at the relationship of the amounts of work to each other and then select the sign of the quantities according to the agreement in section 2.1.

The real working machine is supplied with the coupling work W_e (Fig. 6.11). The gas flow is reached by the difference of the work $|W_e|-|W_{RA}|$, which in the case of fluid flow machines is called internal work and in the case of displacement machines as indexed work.

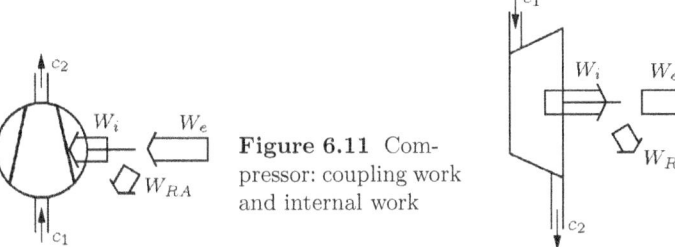

Figure 6.11 Compressor: coupling work and internal work

Figure 6.12 Power machine: coupling work and internal work

$$|W_i| = |W_e| - |W_{RA}| \tag{6.37}$$

As a result of supplying the work amounts to the system, the work quantities are positive:

$$W_i = W_e - |W_{RA}| \tag{6.38}$$

If, in a power machine (prime mover, engine), the gas flow delivers the internal work to the machine, then only a part of this work is gained as coupling work (Fig. 6.12):

$$|W_i| = |W_e| + |W_{RA}| \tag{6.39}$$

Since the system outputs work, internal work and coupling work are negative:

$$-W_i = -W_e + |W_{RA}| \tag{6.40}$$

Eq. (6.40) agrees with Eq. (6.38).

For ideal machines, Eq. (6.37) to (6.40) become simplified:

$$(W_i)_{id} = (W_e)_{id} \,. \tag{6.41}$$

In fluid flow machines, the internal work is transferred between the rotor and the gas flow. In reciprocating machines, the indexed work can be determined from the indicator diagram. It represents the work transferred between piston and gas. If Eq. (6.38) is inserted into Eq. (6.35), — including Eq. (6.36) — new relations between the energy quantities can be obtained (Table 6.1).

6.3.2 Total Work

If the change of the kinetic energy is not negligible (section 6.2.3), it can be combined with the internal work or with the enthalpy (section 6.3.3) to new quantities.

Table 6.1 Relationships between the different types of energy

$$Q_{12} + W_i = H_2 - H_1 + \frac{m}{2}(c_2^2 - c_1^2) = H_{t2} - H_{t1} \quad (1)$$

$$Q_{12} + W_{p12} + W_{RI} = H_2 - H_1 \quad (2)$$

$$W_i = W_{p12} + \frac{m}{2}(c_2^2 - c_1^2) + W_{RI} \quad (3)$$

adiabatic

$$W_i = H_2 - H_1 + \frac{m}{2}(c_2^2 - c_1^2) = H_{t2} - H_{t1} \quad [4]$$

$$W_{p12} + W_{RI} = H_2 - H_1 \quad (5)$$

$$W_i = W_{p12} + \frac{m}{2}(c_2^2 - c_1^2) + W_{RI} \quad (6)$$

no change of kinetic energy

$$Q_{12} + W_i = H_2 - H_1 \quad (10)$$

$$Q_{12} + W_{p12} + W_{RI} = H_2 - H_1 \quad (11)$$

$$W_i = W_{p12} + W_{RI} \quad (12)$$

frictionless and reversible

$$(Q_{12})_{rev} + (W_i)_{id} = H_2 - H_1 + \frac{m}{2}(c_2^2 - c_1^2) = H_{t2} - H_{t1} \quad (7)$$

$$(Q_{12})_{rev} + W_{p12} = H_2 - H_1 \quad (8)$$

$$(W_i)_{id} = W_{p12} + \frac{m}{2}(c_2^2 - c_1^2) \quad (9)$$

reversible adiabatic (isentropic)

$$(W_i)_{id} = H_2 - H_1 + \frac{m}{2}(c_2^2 - c_1^2) = H_{t2} - H_{t1} \quad (13)$$

$$W_{p12} = H_2 - H_1 \quad (14)$$

$$(W_i)_{id} = W_{p12} + \frac{m}{2}(c_2^2 - c_1^2) \quad (15)$$

$$(Q_{12})_{rev} + (W_i)_{id} = H_2 - H_1 \quad (16)$$

$$(Q_{12})_{rev} + W_{p12} = H_2 - H_1 \quad (17)$$

$$(W_i)_{id} = W_{p12} \quad (18)$$

$$(W_i)_{id} = H_2 - H_1 \quad (19)$$

$$W_{p12} = H_2 - H_1 \quad (20)$$

$$(W_i)_{id} = W_{p12} \quad (21)$$

$$W_i = H_2 - H_1 \quad (22)$$

$$W_{p12} + W_{RI} = H_2 - H_1 \quad (23)$$

$$W_i = W_{p12} + W_{RI} \quad (24)$$

The total work W_t in Eq. (6.42) is obtained by transforming Eq. (1) of Table 6.1 in such a way that the mechanical energy is on the left side of the equal sign and the thermal energy is on the right side of the equal sign.

$$W_t = W_i - \frac{m}{2}(c_2^2 - c_1^2) = H_2 - H_1 - Q_{12} \quad (6.42)$$

$$W_t = W_i - \frac{m}{2}(c_2^2 - c_1^2) \quad (6.43)$$

The term mechanical energy is used here in the sense of section 2.4 as a generic term for work and kinetic energy.

In a working machine, the internal work W_i and, with the inflowing gas, the kinetic energy $\frac{m}{2}c_1^2$ are supplied to the system, while the kinetic energy $\frac{m}{2}c_2^2$ leaves the system with the outflowing gas. The total work W_t is therefore the mechanical energy actually available for the compression of a gas, when the change in kinetic energy is no longer negligible. Together with the internal work W_i, it is converted into thermal energy:

$$W_t = W_i + \frac{m}{2}c_1^2 - \frac{m}{2}c_2^2 = H_2 - H_1 - Q_{12} \tag{6.44}$$

In power machines, the total work $-W_t$ consisting of the internal work $-W_i$ and the increase of kinetic energy between the inlet and outlet cross sections is calculated from the thermal energies:

$$H_1 - H_2 + Q_{12} = -W_i + \frac{m}{2}\left(c_2^2 - c_1^2\right) = -W_t \tag{6.45}$$

It should be noted that the internal work W_i and the total work W_t are negative due to the fact that they are released from the system. The total work represents the mechanical energy recoverable from the thermal energy. From the equations of Table 6.1, with the help of Eq. (6.43), corresponding relations for the total work can be derived.

6.3.3 Total Enthalpy

The sum of enthalpy and kinetic energy is called total enthalpy H_t:

$$H_t = H + \frac{m}{2}c^2 \tag{6.46}$$

The change in total enthalpy is

$$H_{t2} - H_{t1} = H_2 - H_1 + \frac{m}{2}\left(c_2^2 - c_1^2\right). \tag{6.47}$$

New relations arise for the first law of thermodynamics for open systems, which are given in Table 6.1. Eq. (4) of Table 6.1 has an important meaning in steam turbine design. It states that for adiabatic machines the internal work is equal to the change in total enthalpy.

6.4 Real Machines

6.4.1 The Uncooled Compressor

In compressors without forced cooling, the heat transfer from the gas flowing through the machine to the generally colder surroundings is so small that uncooled compressors can be considered adiabatic compressors.

The change of state in an ideal machine is isentropic (Fig. 6.13), but in the real machine the internal friction work has an effect on the change of state. For the illustration in a T,S diagram (Fig. 6.14), the internal friction work $|W_{RI}|$ according to Eq. (3.78) is replaced by the reversibly supplied substitute heat $(Q_{12})_{rev}$.

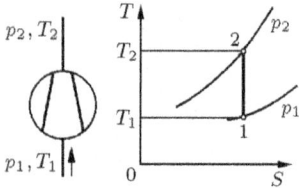
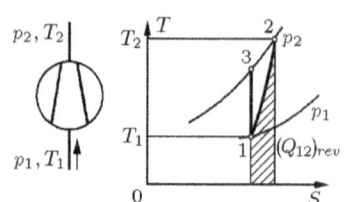
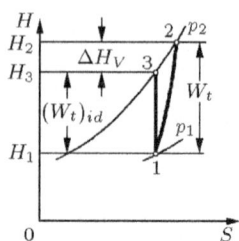

Figure 6.13 Adiabatic compression in an ideal compressor in a T, S diagram

Figure 6.14 Adiabatic compression 1 2 in a real compressor in a T, S diagram

Figure 6.15 Additional work at adiabatic compression ΔH_V in a H, S diagram

For the same initial state (T_1, p_1) and the same final pressure p_2, the final state 2 of the gas compressed in a real machine differs from the final state 3 of the gas compressed in an ideal machine. The change of state from 1 to 2 can be understood as a polytrope with an exponent $n > n_S$ or $n > \kappa$ (for ideal gases).

On the basis of a T, S diagram and a H, S diagram, with the internal friction work a comparison of the additional work required for a real machine in relation to an ideal machine can be made. According to Eq. (13) of Table 6.1 and Eq. (6.43), it follows for the total work of an ideal machine:

$$(W_t)_{id} = H_3 - H_1 \tag{6.48}$$

The total work of a real machine is according to Eq. (4) of Table 6.1 and Eq. (6.43)

$$W_t = H_2 - H_1 . \tag{6.49}$$

$$W_t - (W_t)_{id} = H_2 - H_3 = \Delta H_V \tag{6.50}$$

Fig. 6.15 shows the amount of the additional work ΔH_V as a distance in a H, S diagram.

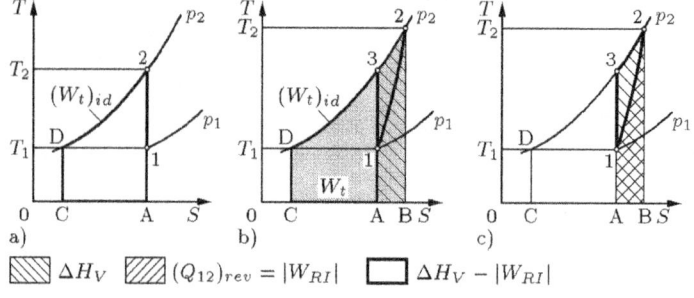

Figure 6.16 a) Total work at ideal adiabatic compression of ideal gases in a T, S diagram b) additional work for adiabatic compression in a T, S diagram c) additional work, internal friction work and heating loss in a T, S diagram

As can be seen from Fig. 4.27 and Fig. 5.15, enthalpy changes can be represented as areas in a T, S diagram. In a T, S diagram for ideal gases according to Fig. 6.16, the additional work ΔH_V appears as area A32BA; it is larger than the internal friction work W_{RI}, which is also shown, represented by the area A12BA. The difference

$$\Delta H_V - |W_{RI}| = \Delta E_V , \tag{6.51}$$

represented by area 1231, is called the heating loss. According to Eq. (6) of Table 6.1 as well as Eq. (6.43), the area ACD21A represents the pressure change work W_{p12}. Fig. 6.17 gives the same considerations for a real gas or a superheated steam.

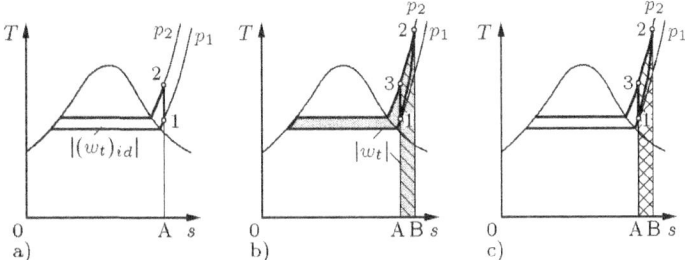

Figure 6.17 Total work, additional work and heating loss for a real gas or a superheated steam in a T, s diagram (hatchings as in Figure 6.16, the specific quantities are shown)

Example 6.4 An uncooled real turbo compressor compresses 3000 m³/h of a natural gas (ideal gas) in the physical norm state ($R = 507.06$ J/(kg K), $\kappa = 1.33$) from 1 bar and 38 °C to 1.3 bar and 66 °C. The change in kinetic energy between the inlet cross section and the outlet cross section is negligible.

a) What mass flow is to be compressed?
b) To which polytrope does the compressor compress?
c) What would be the internal power of an ideal machine?
d) What is the internal power of the real machine?
e) What is the excess power?
f) What is the internal friction power?
g) What is the heating loss?

a) $\dot{m} = 0{,}610$ kg/s

b) $n_u = 1.49$

c) According to Eqs. (19) and (20) of Table 6.1 and Eq. (6.43) is

$$(P_i)_{id} = (P_t)_{id} = \dot{m}(h_3 - h_1) = \dot{m}\, c_p(t_3 - t_1)$$
$$= 0.610 \text{ kg/s} \cdot 2.044 \text{ kJ/(kg K)} \cdot (59\,°C - 38\,°C) = 26.18 \text{ kW}$$

d) According to Eqs. (22) and (23) of Table 6.1 and Eq. (6.43) is

$$P_i = P_t = \dot{m}(h_2 - h_1) = \dot{m}\, c_p(t_2 - t_1)$$
$$= 0.610 \text{ kg/s} \cdot 2.044 \text{ kJ/(kg K)} \cdot (66\,°C - 38\,°C) = 34.91 \text{ kW}$$

e) $\Delta \dot{H}_V = P_t - (P_t)_{id} = 34.91 \text{ kW} - 26.18 \text{ kW} = 8.73 \text{ kW}$

f) $|P_{RI}| = \dot{m} c_n (t_2 - t_1) = 0.610 \text{ kg/s} \cdot 0.502 \text{ kJ/(kg K)} \cdot (66\,°C - 38\,°C) = 8.57 \text{ kW}$

g) $\Delta \dot{E}_V = \Delta \dot{H}_V - |P_{RI}| = 8.73 \text{ kW} - 8.57 \text{ kW} = 0.16 \text{ kW}$

6.4.2 The Cooled Compressor

By cooling the gas during compression, a reduction of the coupling work needed is achieved. In the case of single-stage machines, not only casing, blade or piston cooling is possible, but also cooling with injected oil or water, which is separated again after compression.

In the case of multi-stage machines, the possibility of isobaric intercooling arises (section 6.2.2). In this section, only single-stage machines will be considered.

It shall be assumed that the compressor with cooling has a final state 2, without cooling with the same internal frictional energy $|W_{RI}|$ reaches end state 3, then the cases shown in Fig. 6.18 can be distinguished.

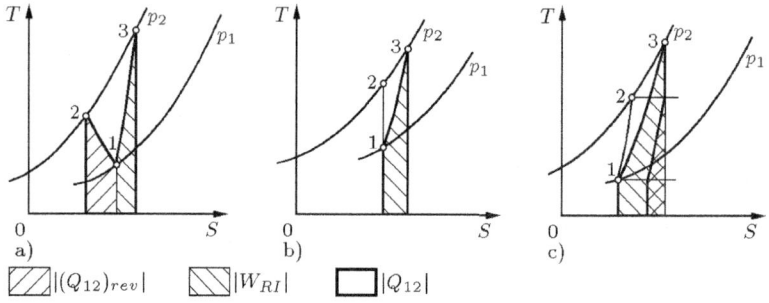

Figure 6.18 T, S diagram for the cooled compressor
a) with large heat release $|Q_{12}| > |W_{RI}|$
b) with heat release equal to the internal friction work $|Q_{12}| = |W_{RI}|$
c) with small heat release $|Q_{12}| < |W_{RI}|$

The relationship between heat release, internal friction work and reversible substitute heat is given by Eq. (3.78):

$$(Q_{12})_{rev} = Q_{12} + |W_{RI}| \tag{6.52}$$

The negative heat Q_{12} can be greater (Fig. 6.18 a), equal (Fig. 6.18 b) or less (Fig. 6.18 c) than the internal friction work $|W_{RI}|$.

For the compressor with cooling to end state 2, the total work according to Eq. (1) of Table 6.1 and Eq. (6.43) is

$$W_t = H_2 - H_1 - Q_{12} . \tag{6.53}$$

For the uncooled compressor, according to Eq. (4) of Table 6.1 and Eq. (6.43), holds

$$W_{tu} = H_3 - H_1 . \tag{6.54}$$

The total work saved by cooling is

$$W_{tu} - W_t = H_3 - H_2 + Q_{12} = (H_3 - H_2) - |Q_{12}| \ . \tag{6.55}$$

It is equal to the area between the state points 1231 which is shown in the T, S diagrams of Fig. 6.18.

For a cooled ideal compressor reaching the same final state 2 as a cooled real compressor, the total work according to Eq. (7) of Table 6.1 as well as Eq. (6.43) is given by

$$(W_t)_{id} = H_2 - H_1 - (Q_{12})_{rev} \ . \tag{6.56}$$

The additional work ΔH_V of a real compressor compared to an ideal compressor is with the same final state 2 according to Eqs. (6.52), (6.53) and 6.56)

$$\Delta H_V = W_t - (W_t)_{id} = (Q_{12})_{rev} - Q_{12} = |W_{RI}| \ . \tag{6.57}$$

Additional work ΔH_V and internal friction work W_{RI} are equal, thus the heating loss according to Eq. (6.51) is

$$\Delta E_V = 0 \ . \tag{6.58}$$

Example 6.5 In a cooled real compressor, $0.12 \text{ m}^3/\text{s}$ CO (ideal gas, $R = 0.2968$ kJ/(kg K); $\kappa = 1.40$) is sucked in at a pressure of 0.96 bar and a temperature of 16 °C and compressed to 4.24 bar. The final temperature is 95 °C. Without cooling the compressor, the gas flow would have an outlet temperature of 208 °C for adiabatic compression with friction at the same final pressure.

a) What is the internal friction power, if one can assume that it is independent of whether cooling is or not?
b) What heat flow the gas gives off by being cooled?
c) What is the internal power if the change in kinetic energy is negligible?
d) What amount of internal power is saved by cooling?

a) $\dot{m} = 0.1342$ kg/s $n_u = 1.5217$

$$|P_{RI}| = (\dot{Q}_{13})_{rev} = \frac{n_u - \kappa}{(\kappa - 1)(n_u - 1)} \dot{m} R (t_3 - t_1) = 4.460 \text{ kW}$$

b) $n_k = 1.1942$

$$(\dot{Q}_{12})_{rev} = \frac{n_k - \kappa}{(\kappa - 1)(n_k - 1)} \dot{m} R (t_2 - t_1) = -8.336 \text{ kW}$$

$$\dot{Q}_{12} = (\dot{Q}_{12})_{rev} - |P_{RI}| = -12.796 \text{ kW}$$

c) According to Eqs. (10) and (11) of Table 6.1 and Eq. (6.43) is

$$P_i = P_t = \dot{m} (h_2 - h_1) - \dot{Q}_{12} = \dot{m} c_p (t_2 - t_1) - \dot{Q}_{12} = 23.809 \text{ kW}$$

d) $(P_i)_u = (P_t)_u = \dot{m} (h_3 - h_1) = \dot{m} c_p (t_3 - t_1) = 26.766 \text{ kW}$

$(P_t)_u - P_t = 26.766 \text{ kW} - 23.809 \text{ kW} = 2.957 \text{ kW}$

6.4.3 Piston Compressor

In displacement machines, e.g. piston compressors, the change in kinetic energy of the gas flow between inlet and outlet at the machine is negligible. For safety reasons, in reciprocating machines the piston cannot reach the bottom of the cylinder. A dead space remains. For the piston compressor this has the consequence that the compressed gas cannot be pushed out completely but expands to its initial pressure when the piston goes back. For an ideal machine with dead space, the coupling work would result in

$$(W_e)_{id} = (W_i)_{id} = W_{p12} + W_{p34} \tag{6.59}$$

For the real machine, the indexed work W_i is larger by the internal friction work (Fig. 6.19):

$$W_i = W_{p12} + W_{p34} + |W_{RI}| \tag{6.60}$$

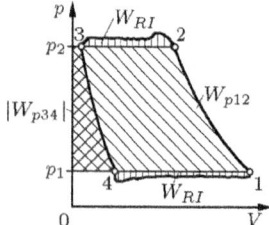
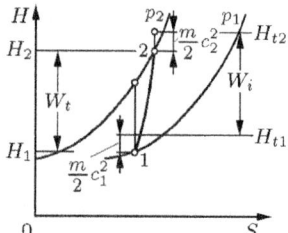

Figure 6.19 Indicator diagram of a piston compressor (schematic)

Figure 6.20 Total work and internal work at a compressor in a H, S diagram

The amount of the internal friction work can be taken from the indicator diagram, it is also called gas friction. The external friction work is called engine friction.

6.4.4 Turbo Compressor

For machines of smaller size, the change in kinetic energy between the inlet state and the outlet state can be neglected. In this case, the internal work is used to assess the energy conversion. For adiabatic machines, Eqs. (22) to (24) of Table 6.1 apply. The calculation of larger machine units is carried out with the total work. Like the total enthalpy, it can be represented in a H, S diagram. For an adiabatic machine (Fig. 6.20) Eqs. (4) to (6) of Table 6.1 as well as Eq. (6.43) apply.

6.4.5 Gas and Steam Turbines

In ideal adiabatic machines, the expansion from the initial state (T_1, p_1) to the final pressure p_2 occurs isentropically according to Fig. 6.21. In the real adiabatic machine the internal friction work modifies the change of state in such a way that, when the pressure is released to the same final pressure p_2, the final state 2 is reached according to Fig. 6.22.

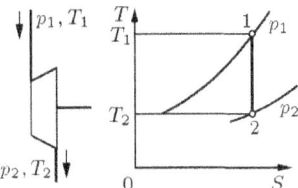
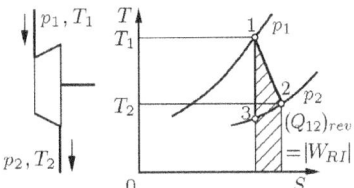

Figure 6.21 Adiabatic expansion in an ideal turbine shown in a T, S diagram

Figure 6.22 Adiabatic expansion in a real turbine shown in a T, S diagram

In a T, S diagram the internal friction work $|W_{RI}|$ is replaced by the reversible substitute heat input $(Q_{12})_{rev}$. The change of state from 1 to 2 is described as a polytrope with an exponent in the range $1 < n < n_S$. For ideal gases, according to Eq. (4.117), follows $n_S = \kappa$. For real gases and vapors according to Eq. (5.193), follows $n_S = n_T \kappa$.

The total work of the ideal adiabatic turbine is according to Eq. (13) of Table 6.1 as well as Eq. (6.43)

$$(W_t)_{id} = H_3 - H_1 \;. \tag{6.61}$$

The total work of the real adiabatic turbine is according to Eq. (4) of Table 6.1 and Eq. (6.43)

$$W_t = H_2 - H_1 \;. \tag{6.62}$$

The representation in a H, S diagram shows that the absolute amount of the total work gained by the real machine $|W_t|$ is smaller than that of the ideal machine $|(W_t)_{id}|$ (Figure 6.23).

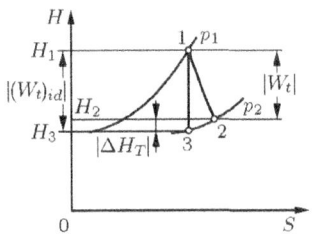

Figure 6.23 Reduced work during adiabatic expansion in a H, S diagram

The reduced work is

$$(W_t)_{id} - W_t = H_3 - H_2 = \Delta H_T \;. \tag{6.63}$$

The reduced work ΔH_T is negative. Therefore it is given by

$$|\Delta H_T| = -\Delta H_T \;. \tag{6.64}$$

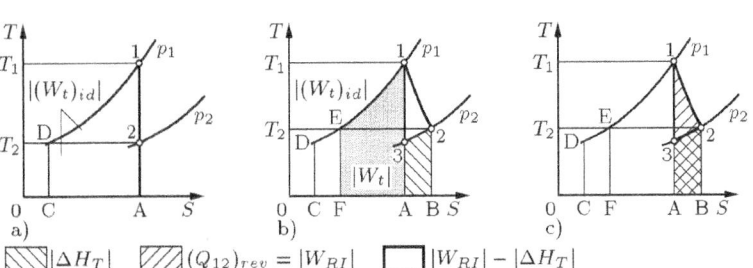

Figure 6.24 Ideal gas: a) Total work at ideal adiabatic expansion in a T, S diagram
b) Reduced work at adiabatic expansion in a T, S diagramm
c) Reduced work, internal friction work and energy recovery in a T, S diagram

The amount $|\Delta H_T|$ appears in the T, S diagram for an ideal gas according to Fig. 6.24 as the area A32BA, which is equal to the area CDEFC.

The reversible substitute heat $(Q_{12})_{rev}$, on the other hand, appears as the area A12BA. The absolute amount of the loss of total work (= work reduction) $|\Delta H_T|$ is therefore smaller than the friction work $|W_{RI}|$ causing it. The difference

$$\Delta E_T = |W_{RI}| - |\Delta H_T| \tag{6.65}$$

(area 3123) is called energy recovery.

Gas turbines are used to expand a hot gas flow. This hot gas flow can be obtained by indirect heating the gas via a heat exchanger or by direct combustion in a combustion chamber. The mechanical energy obtained is calculated as total work, as larger machine units are generally used, (Fig. 6.25).

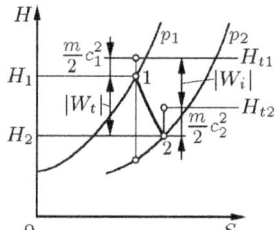

Figure 6.25 Total work and internal work for an adiabatic turbine in a H, S diagram

In steam turbines, steam (superheated steam, saturated steam, superheated vapor, saturated vapor) is expanded. The representation of the working variables in a T, s diagram for steam is shown in Fig. 6.26. The final state of expansion for the real turbine (2) is often in the superheated region, but it can also be in the wet steam region (two phase liquid-vapor region) with values of the quality x usually between about 0.8 and 1. Fig. 6.27 and Fig. 6.28 show the change of state in a T, s digram and in a h, s diagram.

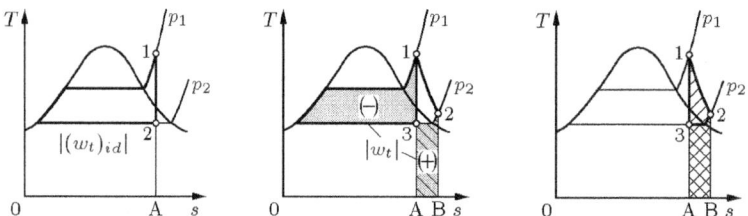

Fig. 6.26 Total work, reduced work and energy recovery for a real gas expansion or a superheated steam expansion (superheated vapor expansion) in a T, s diagram (hatchings as in Fig. 6.24, the specific quantities are shown)

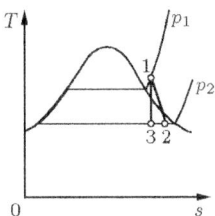

Fig. 6.27 Expansion in a turbine in a T, s diagram

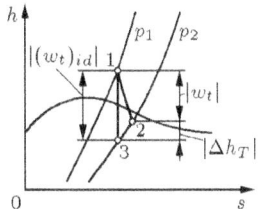

Fig. 6.28 Expansion in a turbine in a h, s diagram

Example 6.6 In an older industrial steam turbine, 14 m³/min of superheated steam of 52 bar and 490 °C are flowing into the turbine with a velocity of 12 m/s and then are irreversibly-adiabatically expanded to 0.1 bar. The velocity in the exhaust cross section is 73 m/s, the exhaust steam state is determined by the quality $x_2 = 0.948$.

a) What is the internal power of the turbine?

b) What is the total power?

a) The mass flow is $\quad \dot{m} = \dfrac{\dot{V}}{v} = \dfrac{14 \text{ m}^3/\text{min}}{0.065 \text{ m}^3/\text{kg}} = 3.59 \text{ kg/s}$

According to Eq. (4) of Table 6.1, using the *Mollier h, s* diagram, one obtains

$$P_i = \dot{m}(h_2 - h_1) + \frac{\dot{m}}{2}(c_2^2 - c_1^2) = 3.59 \text{ kg/s} \cdot (2460 - 3410) \text{ kJ/kg} +$$

$$+ \frac{3.59 \text{ kg/s}}{2} \cdot (73^2 - 12^2) \text{ m}^2/\text{s}^2 = -3411 \text{ kW} + 9 \text{ kW} = -3402 \text{ kW}$$

b) According to Eq. (4) of Table 6.1 and Eq. (6.43)

$$P_t = \dot{m}(h_2 - h_1) = 3.59 \text{ kg/s} \cdot (2460 - 3410) \text{ kJ/kg} = -3411 \text{ kW}$$

Turbines are also cooled in order to reduce the stress on the material, although this leads to a loss of performance. For a cooled turbine is, according to Eq. (1) of Table 6.1 and Eq. (6.43),

$$W_t = H_2 - H_1 - Q_{12} . \tag{6.66}$$

For the uncooled turbine, Eq. (4) of Table 6.1 and Eq. (6.43) apply:

$$W_{tu} = H_3 - H_1 \tag{6.67}$$

Cooling reduces the gained absolute amount of total work, compared to the uncooled machine. Since the total work is negative in the turbine and heat is given off, it is useful to form the amounts of total work and the amounts of the heat transfer:

$$|W_t| = H_1 - H_2 + Q_{12} = H_1 - H_2 - |Q_{12}| \tag{6.68}$$

$$|W_{tu}| = H_1 - H_3 \tag{6.69}$$

Then, the difference is obtained:

$$|W_{tu}| - |W_t| = |Q_{12}| - (H_3 - H_2) \qquad (6.70)$$

This difference corresponds to area 1231 in a T, S diagram shown in Fig. 6.29.

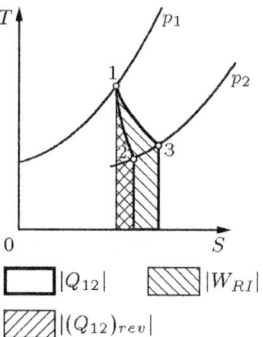

Fig. 6.29 T, S diagram for a cooled turbine

6.5 Efficiencies

Efficiencies are used to describe the quality of an energy conversion. Either according to

case a) the energy conversion at a built real machine can be considered,

or it can be considered according to

case b) a comparison between a real and an ideal machine.

In case a) one should proceed according to the following rule:

$$\text{efficiency} = \frac{\text{benefit}}{\text{effort}}$$

One then always gets a number that is less than one. In order to always obtain a number smaller than one in case b) as well, the comparison must be made in distinguishing the case of expansion machines (e.g. turbines) and the case of compressors:

$$\text{compressor efficiency} = \frac{\text{work of the ideal machine}}{\text{work of the real machine}}$$

$$\text{expansion machine efficiency (turbine efficiency)} = \frac{\text{work of the real machine}}{\text{work of the ideal machine}}$$

In case b) one needs well-defined ideal machines. The most important ideal machines are listed in the following section.

6.5.1 Comparison Processes

The processes that take place in ideal machines are called comparison processes. Some simple comparison processes are defined according to the following single-stage ideal machines:

1) An adiabatic machine with isentropic change of state (final state 3 in Fig. 6.30). According to Eq. (13) of Table 6.1 as well as Eq. (6.43), the following applies to it:

$$(W_t)_{id} = H_3 - H_1 \tag{6.71}$$

$$(W_i)_{id} = H_{t3} - H_{t1} \tag{6.72}$$

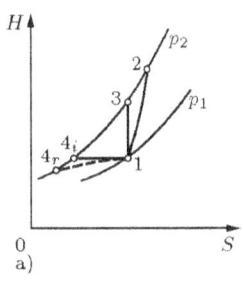
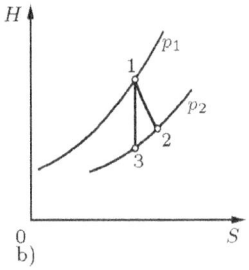

Figure 6.30 Comparison processes for single-stage machines

a) Compression

b) Expansion

2) A compressor with ideal cooling. The change of state is isothermal (final state for ideal gas 4_i, for real gas 4_r in Fig. 6.30). It is according to Eq. (7) of Table 6.1 as well as Eq. (6.43)

$$(W_t)_{id} = H_4 - H_1 - (Q_{14})_{rev} \tag{6.73}$$

$$(W_i)_{id} = H_{t4} - H_{t1} - (Q_{14})_{rev} . \tag{6.74}$$

3) A machine with polytropic change of state. The end point of the change of state coincides with that of the real machine (end state 2 in Fig. 6.30). One obtains according to Eq. (7) of Table 6.1 as well as Eq. (6.43)

$$(W_t)_{id} = H_2 - H_1 - (Q_{12})_{rev} \tag{6.75}$$

$$(W_i)_{id} = H_{t2} - H_{t1} - (Q_{12})_{rev} . \tag{6.76}$$

6.5.2 The Internal Efficiency

The internal efficiency is an efficiency according to case b) and is used to compare the internal or indexed work of a real machine with the work output or work input of an ideal machine.

For an ideal machine, either the internal work $(W_i)_{id}$ or the total work $(W_t)_{id}$ can be used. The internal work $(W_i)_{id}$ is used if the change in kinetic energy cannot be utilised profitably or can be neglected. In special cases, if the change of the kinetic energy is used directly (thrust of an aeroplane gas turbine engine) or is used after conversion into another form of energy (conversion into static pressure in connected pipelines), the total work $(W_t)_{id}$ must be considered.

6.5 Efficiencies

Table 6.2 The internal efficiency

	adiabatic machine ⇒ inner isentropic efficiency	
	real and ideal gas	ideal gas, constant specific isobaric heat capacity
compressor	$\eta_{isV} = \dfrac{(W_i)_{id}}{W_i} = \dfrac{H_{t3} - H_{t1}}{H_{t2} - H_{t1}}$ (1)	$\eta_{isV} = \dfrac{c_p(t_3 - t_1) + \frac{1}{2}(c_3^2 - c_1^2)}{c_p(t_2 - t_1) + \frac{1}{2}(c_2^2 - c_1^2)}$ (2)
turbine	$\eta_{isT} = \dfrac{W_i}{(W_i)_{id}} = \dfrac{H_{t2} - H_{t1}}{H_{t3} - H_{t1}}$ (3)	$\eta_{isT} = \dfrac{c_p(t_2 - t_1) + \frac{1}{2}(c_2^2 - c_1^2)}{c_p(t_3 - t_1) + \frac{1}{2}(c_3^2 - c_1^2)}$ (4)
	cooled machine ⇒ inner isothermal efficiency	
	real and ideal gas	ideal gas, constant specific isobaric heat capacity
compressor	$\eta_{itV} = \dfrac{H_{t4} - H_{t1} - (Q_{14})_{rev}}{H_{t2} - H_{t1} - Q_{12}}$ (5)	$\eta_{itV} = \dfrac{\frac{1}{2}(c_4^2 - c_1^2) + RT_1 \ln \frac{p_2}{p_1}}{c_p(t_2 - t_1) + \frac{1}{2}(c_2^2 - c_1^2) - q_{12}}$ (6)

If one restricts himself to the use of the inner work in the comparison process, then a distinction is to be made according to Table 6.2 between

– the inner isentropic efficiency for uncooled machines with the comparison process 1, and

– the inner isothermal efficiency for cooled machines with the comparison process 2.

The final states of the real machines and the ideal machines (final state 2, 3, 4i, 4r) are different for the listed comparison processes. If one imagines an ideal machine for the comparison process, which has the same constructional design of the flow channels as the real machine, then the velocities in the outlet cross section of the ideal machine c_3 and of the real machine c_2 will differ. However, one can also imagine the outlet cross section of the ideal machine changed in such a way that the velocities c_3 and c_2 become equal.

Example 6.7 What is the internal power of the comparison process 1 of section 6.5.1 and the internal isentropic efficiency for the older industrial steam turbine of example 6.6?

The internal power of the comparison process 1, with the condition $c_3 = c_2$, is given by

$$(P_i)_{id} = \dot{m}(h_3 - h_1) + \frac{\dot{m}}{2}(c_2^2 - c_1^2) = 3.59 \text{ kg/s} \cdot (2195 - 3410) \text{ kJ/kg} + \frac{3.59 \text{ kg/s}}{2} \cdot (73^2 - 12^2) \text{ m}^2/\text{s}^2$$

$$= -4362 \text{ kW} + 9 \text{ kW} = -4353 \text{ kW} \qquad \eta_{isT} = \frac{P_i}{(P_i)_{id}} = 0.7815$$

6.5.3 The Mechanical Efficiency

The mechanical efficiency is an efficiency according to case a). It incorporates the amount of the external frictional work into a comparison of the coupling work with the internal or indexed work.

Table 6.3 The mechanical efficiency

compressor	$\eta_{mV} = \dfrac{W_i}{W_e}$	$= \dfrac{W_i}{W_i + |W_{RA}|}$	$= \dfrac{W_e - |W_{RA}|}{W_e}$	(1)
turbine	$\eta_{mT} = \dfrac{W_e}{W_i}$	$= \dfrac{W_i + |W_{RA}|}{W_i}$	$= \dfrac{W_e}{W_e - |W_{RA}|}$	(2)

6.5.4 The Total Efficiency

The product of internal and mechanical efficiency gives the total efficiency. It shows to what extent the available energy is converted into the desired form of energy. The internal isentropic efficiency or the internal isothermal efficiency can be used as the internal efficiency.

Table 6.4 The total efficiency

adiabatic machine $\Rightarrow$ isentropic total efficiency		
compressor	$\eta_{sgV} = \eta_{isV}\,\eta_{mV}$	(1)
turbine	$\eta_{sgT} = \eta_{isT}\,\eta_{mT}$	(2)
cooled machine $\Rightarrow$ isothermal total efficiency		
compressor	$\eta_{tgV} = \eta_{itV}\,\eta_{mV}$	(3)
turbine	$\eta_{tgT} = \eta_{itT}\,\eta_{mT}$	(4)

Example 6.8 What drive power is required by the turbo compressor in example 6.4 if the mechanical efficiency is $\eta_{mV} = 0.96$? What is the total efficiency?

$$P_e = \frac{P_i}{\eta_{mV}} = \frac{34.91 \text{ kW}}{0.96} = 36.36 \text{ kW}$$

$$\eta_{isV} = \frac{(P_i)_{id}}{P_i} = \frac{26.18 \text{ kW}}{34.91 \text{ kW}} = 0.75 \qquad \eta_{sgV} = \eta_{isV}\,\eta_{mV} = 0.75 \cdot 0.96 = 0.72$$

6.5.5 The Isentropic Efficiency

The isentropic efficiency is an efficiency according to case b). It compares the total work of the real machine W_t with the total work according to the comparison process 1. If the change in kinetic energy of the real machine is negligibly small compared to the change in enthalpy, this will also be assumed for the ideal machine of the comparison process. In this case, isentropic and internal efficiency are equal.

Example 6.9 What is the total power of the comparison process 1 and the isentropic efficiency for the older industrial steam turbine of examples 6.6 and 6.7?

$h_3 = 2195$ kJ/kg $\qquad (P_t)_{id} = \dot{m}(h_3 - h_1) = 3.59$ kg/s $\cdot\ (2195 - 3410)$ kJ/kg $= -4362$ kW

According to Table 6.5 Eq. (3), it follows: $\eta_{sT} = \dfrac{h_2 - h_1}{h_3 - h_1} = \dfrac{(2460 - 3410)\text{ kJ/kg}}{(2195 - 3410)\text{ kJ/kg}} = 0.78189$

Table 6.5 The isentropic efficiency

		adiabatic machines		
	real and ideal gas		ideal gas, constant specific isobaric heat capacity	
compressor	$\eta_{sV} = \dfrac{(W_t)_{id}}{W_t} = \dfrac{W_{p13}}{W_{p12} + |W_{RI}|}$		$\eta_{sV} = \dfrac{t_3 - t_1}{t_2 - t_1}$	(2)
	$= \dfrac{H_3 - H_1}{H_2 - H_1}$	(1)		
turbine	$\eta_{sT} = \dfrac{W_t}{(W_t)_{id}} = \dfrac{W_{p12} + |W_{RI}|}{W_{p13}}$		$\eta_{sT} = \dfrac{t_2 - t_1}{t_3 - t_1}$	(4)
	$= \dfrac{H_2 - H_1}{H_3 - H_1}$	(3)		

6.5.6 The Isothermal Efficiency

For cooled compressors the isothermal efficiency is useful. One compares the total work of the real machine W_t with the total work accordimg to comparison process 2.

Table 6.6 The isothermal efficiency

		cooled machines		
	real and ideal gas		ideal gas, constant specific isobaric heat capacity	
compressor	$\eta_{tV} = \dfrac{(W_t)_{id}}{W_t} = \dfrac{W_{p14}}{W_{p12} + |W_{RI}|}$		$\eta_{tV} = \dfrac{RT_1 \ln \dfrac{p_2}{p_1}}{c_p(t_2 - t_1) - q_{12}}$	(2)
	$= \dfrac{H_4 - H_1 - (Q_{14})_{rev}}{H_2 - H_1 - Q_{12}}$	(1)		

Example 6.10 What is the isothermal efficiency of the cooled compressor of example 6.5?

$(\dot{Q}_{14})_{rev} = -\dot{m} R T \ln \dfrac{p_2}{p_1} = -0.1342 \text{ kg/s} \cdot 0.2968 \text{ kJ/(kg K)} \cdot 289.15 \text{ K} \cdot \ln \dfrac{4.24 \text{ bar}}{0.96 \text{ bar}}$

$= -17.107 \text{ kW}$

According to Table 6.6 Eq. (1), it follows: $\eta_{tV} = \dfrac{-(\dot{Q}_{14})_{rev}}{P_t} = \dfrac{17.107 \text{ kW}}{23.809 \text{ kW}} = 0.7185$

6.5.7 The Polytropic Efficiency

The polytropic efficiency compares the total work of the real machine W_t and the total work of the comparison process 3. The meaning of the polytropic efficiency can be shown by the notation which contains the pressure change work. According to Eq. (9) of Table 6.1 as well as Eq. (6.43), the pressure change work W_{p12} represents that work, which in an ideal machine between the gas states 1 and 2 either for compression is needed or by expansion can be gained. In the polytropic efficiency this work is compared with that

work which in the real machine in the presence of internal friction work $|W_{RI}|$ either for compression is needed or by expansion can be gained.

The polytropic efficiency thus satisfies both meanings listed in section 6.5: For given initial and final states 1 and 2, it compares, on the one hand, the benefit with the effort for the task of compressing or expanding a gas flow, and compares, on the other hand, also the total work of the ideal machine with the total work of the real machine.

Table 6.7 The polytropic efficiency

	real and ideal gas		ideal gas, constant specific isobaric heat capacity	
compressor	$\eta_{pV} = \dfrac{(W_t)_{id}}{W_t} = \dfrac{W_{p12}}{W_{p12} + |W_{RI}|}$		$\eta_{pV} = \dfrac{1}{\dfrac{\kappa}{\kappa-1}\dfrac{n-1}{n} - \dfrac{Q_{12}}{W_{p12}}}$	(2)
	$= \dfrac{H_2 - H_1 - (Q_{12})_{rev}}{H_2 - H_1 - Q_{12}}$	(1)		
turbine	$\eta_{pT} = \dfrac{W_t}{(W_t)_{id}} = \dfrac{W_{p12} + |W_{RI}|}{W_{p12}}$		$\eta_{pT} = \dfrac{\kappa}{\kappa-1}\dfrac{n-1}{n} - \dfrac{Q_{12}}{W_{p12}}$	(4)
	$= \dfrac{H_2 - H_1 - Q_{12}}{H_2 - H_1 - (Q_{12})_{rev}}$	(3)		
	real machine adiabatic, ideal machine with heat turnover		ideal gas, constant specific isobaric heat capacity	
compressor	$\eta_{pV} = \dfrac{H_2 - H_1 - (Q_{12})_{rev}}{H_2 - H_1}$	(5)	$\eta_{pV} = \dfrac{\kappa-1}{\kappa}\dfrac{n}{n-1}$	(6)
turbine	$\eta_{pT} = \dfrac{H_2 - H_1}{H_2 - H_1 - (Q_{12})_{rev}}$	(7)	$\eta_{pT} = \dfrac{\kappa}{\kappa-1}\dfrac{n-1}{n}$	(8)

For ideal gases, a polytropic change of state according to Eqs. (4.43), (4.130) and (4.136) is described by

$$H_2 - H_1 = m\,c_p(t_2 - t_1) \qquad (6.77)$$

$$(Q_{12})_{rev} = m\,c_n(t_2 - t_1) = m\,c_v\frac{n-\kappa}{n-1}(t_2 - t_1)\,. \qquad (6.78)$$

With theses equations, combined with Eqs. (4.49) and (4.50), equations are obtained, which contain the polytropic exponent n and the isentropic exponent κ.

Example 6.11 Steam at 120 bar and 500 °C is polytropically expanded in an adiabatic turbine to the saturated steam state at 4 bar.

a) What are the total work, pressure change work and internal friction work per kilogram of steam?

b) What are the isentropic and polytropic efficiencies?

a) The 'Industrial Standard IAPWS IF97' [128] gives for superheated steam at 120 bar and 500 °C

$$v_1 = 0.026830 \text{ m}^3/\text{kg} \qquad h_1 = 3349.97 \text{ kJ/kg} \qquad s_1 = 6.490191 \text{ kJ/(kg K)}$$

6.5 Efficiencies

and for saturated steam at 4 bar and 143.61 °C [128]
$$v_2 = v'' = 0.462392 \text{ m}^3/\text{kg} \qquad h_2 = h'' = 2738.06 \text{ kJ/kg} \qquad h' = 604.72 \text{ kJ/kg}$$
$$s_2 = s'' = 6.895418 \text{ kJ/(kg K)} \qquad s' = 1.776598 \text{ kJ/(kg K)} .$$

In the final state of the ideal isentropic turbine, one finds with $s_3 = s_1$ according to Eq. (5.33)
$$x_3 = \frac{s_3 - s'}{s'' - s'} = \frac{(6.4902 - 1.7766) \text{ kJ/(kg K)}}{(6.8954 - 1.7766) \text{ kJ/(kg K)}} = 0.92084$$

and according to Eq. (5.29)
$$h_3 = 604.72 \text{ kJ/kg} + 0.92084 \cdot (2738.06 - 604.72) \text{ kJ/kg} = 2569.17 \text{ kJ/kg} .$$

The polytropic exponent n according to Eq. (4.142) is

$$n = \frac{\ln \frac{p_2}{p_1}}{\ln \frac{v_1}{v_2}} = \frac{\ln \frac{4 \text{ bar}}{120 \text{ bar}}}{\ln \frac{0.026830 \text{ m}^3/\text{kg}}{0.462392 \text{ m}^3/\text{kg}}} = 1.194702 .$$

Note that the polytropic exponent is defined by Eq. (4.116) and Eq. (5.182) as $pv^n = \text{const}$. It follows that the polytropic exponent must be calculated according to Eq. (4.142) used above. Eqs. (4.143) and (4.144) are only applicable for ideal gases and lead here to the deviating results $n = 1.217063$ and $n = 1.222028$.

The specific total work is according to Eq. (4) of Table 6.1 as well as Eq. (6.43)
$$w_t = h_2 - h_1 = (2738.06 - 3349.97) \text{ kJ/kg} = -611.908 \text{ kJ/kg} .$$

The specific pressure change work is according to Table 4.2 Eq. (3)
$$w_{p12} = \frac{n}{n-1}(p_2 v_2 - p_1 v_1) = \frac{1.194702}{0.194702} \cdot (4 \text{ bar} \cdot 0.462392 \text{ m}^3/\text{kg} - 120 \text{ bar} \cdot 0.026830 \text{ m}^3/\text{kg})$$
$$= -840.643 \text{ kJ/kg} .$$

The specific pressure change work would also have to be calculated according to the equations

$$w_{p12} = \frac{n}{n-1} p_1 v_1 \left[\left(\frac{v_1}{v_2}\right)^{n-1} - 1 \right] = \frac{n}{n-1} p_1 v_1 \left[\left(\frac{p_2}{p_1}\right)^{\frac{n-1}{n}} - 1 \right] .$$

The other equations in Table 4.2, which contain the specific gas constant and temperatures, are not applicable.

According to Eq. (6) of Table 6.1 as well as Eq. (6.43), the specific internal friction work is
$$|w_{RI}| = w_t - w_{p12} = (-611.908 + 840.643) \text{ kJ/kg} = 228.735 \text{ kJ/kg} .$$

Another way to determine the specific internal friction work is according to Eqs. (3.78) and (3.92)
$$|w_{RI}| = (q_{12})_{rev} = \int_1^2 T \, ds .$$

The following table contains, according to the equations of the 'Industrial Standard IAPWS IF97' [128], the values of the specific entropy s as well as the values of the thermal state variables T, p and v at the boundaries of 20 subdivisions of the state curve in the T, s diagram with a width of $\Delta s = 0.020261$ kJ/(kg K) in each case.

The application of the *Simpson* formula according to example 5.2 for the integration results in
$$|w_{RI}| = (q_{12})_{rev} = 228.734 \text{ kJ/kg} .$$

This result agrees with the previously calculated value.

s	T	p	v	s	T	p	v
kJ/(kg K)	K	bar	m³/kg	kJ/(kg K)	K	bar	m³/kg
6.490191	773.1500	120.0000	0.0268298	6.692804	549.4182	18.15124	0.1246567
6.510452	744.8765	98.56853	0.0316326	6.713066	533.1876	16.22202	0.1432382
6.530714	717.8752	81.08253	0.0372498	6.733327	517.7647	13.77667	0.1642310
6.550975	692.2879	66.88111	0.0437643	6.753589	503.0837	11.72804	0.1879227
6.571237	668.1339	55.35278	0.0512736	6.773850	489.0877	10.00634	0.2146310
6.591498	645.3708	45.97837	0.0598889	6.794111	475.7228	8.374544	0.2447169
6.611759	623.9197	38.33120	0.0697385	6.814373	462.9395	7.327404	0.2785886
6.632020	603.6900	32.06953	0.0809671	6.834634	450.6940	6.286560	0.3167064
6.652282	584.5879	26.92154	0.0937382	6.854895	438.9432	5.401330	0.3596064
6.672543	566.5256	22.67244	0.1082332	6.875156	427.6463	4.646209	0.4079160
				6.895418	416.7625	4.000000	0.4623918

b) The isentropic efficiency is according to Table 6.5 Eq. (3)

$$\eta_{sT} = \frac{(2738.06 - 3349.97) \text{ kJ/kg}}{(2569.17 - 3349.97) \text{ kJ/kg}} = 0.78370 .$$

According to Table 6.7 Eq. (7), the polytropic efficiency is

$$\eta_{pT} = \frac{(2738.06 - 3349.97) \text{ kJ/kg}}{-840.643 \text{ kJ/kg}} = 0.72790 .$$

Additions to a):
According to Eqs. (3.78) and (4.130) one could also proceed as follows for the determination of the specific internal friction work:

$$|w_{RI}| = (q_{12})_{rev} = c_n(T_2 - T_1)$$

The determination of c_n is burdened with difficulties. First, it seems that the mean specific polytropic heat capacity c_n would be calculable from Eq. (4.141):

$$c_n = \frac{s_2 - s_1}{\ln \frac{T_2}{T_1}} = -0.65575 \text{ kJ/(kg K)}$$

This gives

$$c_n(T_2 - T_1) = -0.65575 \text{ kJ/(kg K)} \cdot (416.76 - 773.15) \text{ K} = 233.70 \text{ kJ/kg} .$$

This result is not correct because, according to Example 5.2, the mean specific heat capacities in the calculations of the heat and of the entropy change are different. The two correct ways described before lead to the numerical value $c_n = -0.64181$ kJ/(kg K).

Tasks for Section 6

In the following tasks, the change in kinetic energy is negligible.

1. An ideal gas compressor compresses 600 kg/h of carbon dioxide from 25 °C and 0.95 bar isentropically to 3 bar. Subsequently, the compressed gas is isobarically cooled back to 25 °C by a heat exchanger. Tap water with 8 °C is available for cooling. The outlet temperature of the water shall be 25 °C.

a) What coupling power does the compressor require and what volume of water is needed for cooling?

b) What coupling power would be required for isothermal compression, and what would be the water outlet temperature?

c) The two changes of state of the gas are to be plotted to scale in a T, s diagram.

d) What is the total entropy change in both processes?

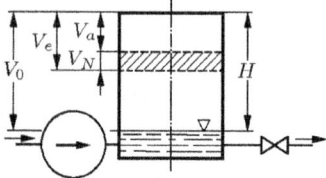

Figure 6.31 For task 2

2. A storage tank with an internal diameter $d_i = 1380$ mm is installed in a water supply system. At start-up, the air pressure in the storage tank is $p_0 = 1$ bar, the air volume is V_0, and the height of the air space above the water level is $H = 1.8$ m. The pump delivers a water flow of 5 ℓ/s into the tank. When the pressure in the tank has risen to $p_a = 9$ bar due to compression of the air cushion, the pump is stopped by a pressure switch device (corresponding air volume V_a). If the pressure drops to $p_e = 7$ bar (associated air volume V_e) due to water withdrawal from the reservoir, the pump is put into operation again (Fig. 6.31).

a) What volume of useful water V_N can be taken from the tank between the switch-off and switch-on points if the described compression and expansion of the air volume take place isothermally?

b) What quantity of useful water V_N can be withdrawn if the processes run isentropically?

c) How long does the pump run during isothermal or isentropic compression of the air volume?

d) How large is the useful water volume V_N for case a), if at start-up the air volume V_0 is not under a pressure $p_0 = 1$ bar, but already under a pressure equal to the switch-on pressure $p_e = 7$ bar?

3. An ideal piston compressor without dead space is operated to fill an air vessel with a capacity of 4.5 m³, which is used for the compressed air supply of an industrial plant. The air in the vessel should have a gage pressure of 7 bar, the variables of state of the ambient air to be compressed are 933 mbar and 12 °C. During the filling process, the compressor passes through the entire range of gage pressure from 0 to 7 bar. The cylinder of the compressor has an inside diameter of 100 mm and a piston stroke of 50 mm, the speed is 600 min⁻¹. The compression is polytropic with $n = 1.35$.

a) What volume of air is compressed per one rotation?

b) The heat release of the air vessel should be just large enough so that the heat supplied to the vessel with the air quantity Δm is is given off until the next air supply Δm takes place. The temperature of the vessel thus remains constant at 12 °C. What is the total amount of air to be compressed?

c) How long does the compressor run if it switches off automatically after reaching the required pressure in the vessel?

4. An ideal two-stage compressor with double-acting piston (Fig. 6.32), which is air-cooled and equipped with isobaric intermediate cooling and final cooling by water, compresses air of 976 mbar and 19 °C. The change of state is polytropic with $n = 1.14$. The cylinder diameter is $D = 80$ mm, the piston stroke $\Delta h = 60$ mm. There are no dead spaces. The final pressure in each stage is four times the initial pressure. The intermediate cooling and the final cooling bring the initial temperature back in each case. The speed is 1450 min⁻¹.

a) What is the piston diameter d_K?
b) What power does the compressor require?
c) What power would be required for isothermal compression to the same end pressure?
d) What is the power saved by the isobaric intermediate cooling?
e) The change of state of the air is to be plotted in a T, s diagram to show the specific heat given off as two areas in the T, s diagram.

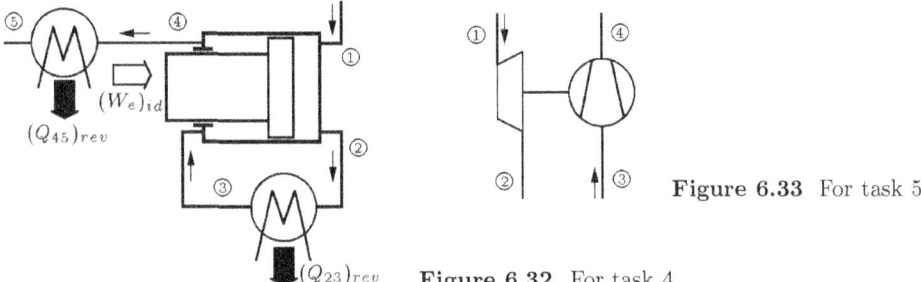

Figure 6.32 For task 4

Figure 6.33 For task 5

5. The ideal turbine of an exhaust gas turbocharger expands exhaust gas ($\varrho_0 = 1.215$ kg/m³ in the physical norm state; $\kappa = 1.38$) from 455 °C to 362 °C and 963 mbar. The ideal compressor operates at an end gage pressure of 0.506 bar and compresses 18.6 kg/s of air from 963 mbar and 18 °C. Expansion and compression are both isentropic (Fig. 6.33).

a) What is the initial gage pressure of the exhaust gas before entering the turbine?
b) What temperature does the compressed air reach?
c) What is the power output of the turbine?
d) What exhaust gas flow rate is required?

6. The compressor of the turbocharger in task 5 should produce the same air flow at the same initial condition and end pressure with the internal isentropic efficiency $\eta_{isV} = 0.86$ and the mechanical efficiency $\eta_{mV} = 0.97$.

a) What final state does the compressed air reach?
b) Which polytropic exponent n characterises the compression?
c) What internal power and coupling power does the compressor require?
d) The turbine is to go from the same initial state as in task 5 to the same end pressure with the same values of the internal isentropic efficiency $\eta_{isT} = \eta_{isV} = 0.86$ and the same mechanical efficiency $\eta_{mT} = \eta_{mV} = 0.97$ as the compressor. What exhaust gas flow rate is required?

7. The turbine of a combined heat and power plant expands 6000 kg/h of superheated steam (water vapor) from 500 °C and 74 bar to 2.2 bar and 180 °C.

a) What is the internal isentropic efficiency of the turbine?
b) What power does the turbine deliver to the shaft if the mechanical efficiency is $\eta_{mT} = 0.96$?
c) What power could be gained in an ideal turbine?

7 Cyclic Processes

7.1 Cyclic Process Work, Heat Input and Heat Output

Thermodynamic changes of state in simple systems can generally be described by the following – usually occurring simultaneously and correlated with each other – four process variables (cf. section 3, Eqs. (3.104) and (3.105)):

- volume change work $W_{V12} = -\int_1^2 p \, dV$ (7.1)

- pressure change work $W_{p12} = \int_1^2 V \, dp$ (7.2)

- reversible substitute heat (here also named entropy change heat)

$(Q_{12})_{rev} = Q_{s12} = \int_1^2 T \, dS$ (7.3)

- temperature change heat $Q_{T12} = \int_1^2 S \, dT$ (7.4)

This becomes clear as follows: If a substance is transferred from state 1 to state 2, then the state variables shift work pV and bound energy TS change from $p_1 V_1$ and $T_1 S_1$ to $p_2 V_2$ and $T_2 S_2$ respectively. Hereby, the above four process variables generally occur simultaneously (Fig. 7.1 and Fig. 7.2).

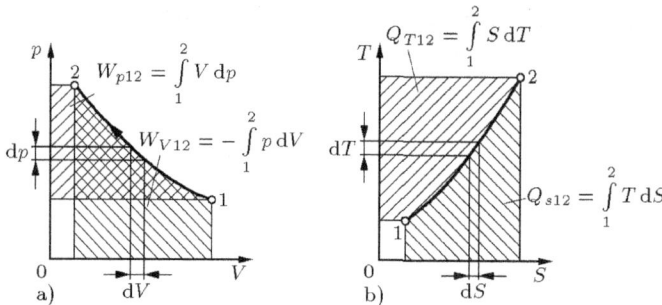

Figure 7.1 a) Volume change work W_{V12} and pressure change work W_{p12} in a p, V diagram. b) entropy change heat Q_{s12} and temperature change heat Q_{T12} in a T, S diagram

The temperature change heat Q_{T12} was referred to in particular in section 5 when introducing the differential forms of the specific free energy f and of the specific free enthalpy g (Eq. (5.222) and Eq. (5.223)); there the differential form of the specific temperature change heat $s \, dT$ occurs, where s is the absolute specific entropy. W_{p12}, W_{V12} as well as the part Q_{12} of $Q_{s12} = (Q_{12})_{rev} = Q_{12} + |W_{RI}|$ are process variables which either can exceed the system boundary or act within the system boundary; $|W_{RI}|$ acts only within the system boundary. Q_{T12} as a process variable related to W_{p12}, W_{V12} and $(Q_{12})_{rev}$ obviously also acts only within the system boundary and describes a change of state only within the system or substance, respectively, and thus does not exceed the system boundary. The following relations hold:

$$\int_1^2 d(pV) = \int_1^2 p \, dV + \int_1^2 V \, dp = p_2 V_2 - p_1 V_1 = -W_{V12} + W_{p12}$$

$$\int_1^2 d(TS) = \int_1^2 T \, dS + \int_1^2 S \, dT = T_2 S_2 - T_1 S_1 = (Q_{12})_{rev} + Q_{T12} = Q_{s12} + Q_{T12}$$

© The Author(s), under exclusive license to
Springer Fachmedien Wiesbaden GmbH, part of Springer Nature 2023
M. Dehli et al., *Fundamentals of Technical Thermodynamics*,
https://doi.org/10.1007/978-3-658-38910-9_7

246   7 Cyclic Processes

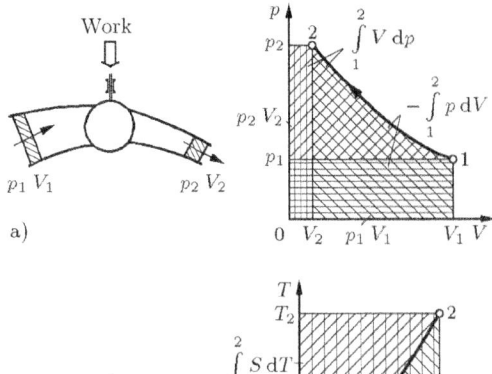

Figure 7.2
a) Areal representation of the difference of the shift work $p_2 V_2 - p_1 V_1$ in relation to the volume change work W_{V12} and to the pressure change work W_{p12} in a p, V diagram
b) Areal representation of the difference of the bound energy $T_2 S_2 - T_1 S_1$ in relation to the entropy change heat Q_{s12} and to the temperature change heat Q_{T12}

In the following, in addition to section 4.3.5, equations for the temperature change heat Q_{T12} are given which for polytropic changes of ideal gases describe the correlation with the process variables mentioned:

$$\frac{Q_{T12}}{W_{V12}} = (n-1)\frac{s_1}{R} + \frac{n-\kappa}{\kappa-1}\left(\frac{T_2}{T_2-T_1}\ln\frac{T_2}{T_1} - 1\right) \tag{7.5}$$

$$\frac{Q_{T12}}{W_{p12}} = \frac{n-1}{n}\frac{s_1}{R} + \frac{n-\kappa}{n(\kappa-1)}\left(\frac{T_2}{T_2-T_1}\ln\frac{T_2}{T_1} - 1\right) \tag{7.6}$$

$$\frac{Q_{T12}}{Q_{s12}} = \frac{(\kappa-1)(n-1)}{n-\kappa}\frac{s_1}{R} + \frac{T_2}{T_2-T_1}\ln\frac{T_2}{T_1} - 1 = \frac{s_1}{c_n} + \frac{T_2}{T_2-T_1}\ln\frac{T_2}{T_1} - 1 \tag{7.7}$$

Table 7.1 Equations for the temperature change heat Q_{T12} for a polytropic change of state of an ideal gas

$$Q_{T12} = \left[s_1 + c_n\left(\frac{T_2}{T_2-T_1}\ln\frac{T_2}{T_1} - 1\right)\right]\frac{p_1 V_1}{R}\left[\left(\frac{V_1}{V_2}\right)^{n-1} - 1\right] \tag{1}$$

$$Q_{T12} = \left[s_1 + c_n\left(\frac{T_2}{T_2-T_1}\ln\frac{T_2}{T_1} - 1\right)\right]\frac{p_1 V_1}{R}\left[\left(\frac{p_2}{p_1}\right)^{\frac{n-1}{n}} - 1\right] \tag{2}$$

$$Q_{T12} = \left[s_1 + c_n\left(\frac{T_2}{T_2-T_1}\ln\frac{T_2}{T_1} - 1\right)\right]m T_1\left[\left(\frac{V_1}{V_2}\right)^{n-1} - 1\right] \tag{3}$$

$$Q_{T12} = \left[s_1 + c_n\left(\frac{T_2}{T_2-T_1}\ln\frac{T_2}{T_1} - 1\right)\right]m T_1\left[\left(\frac{p_2}{p_1}\right)^{\frac{n-1}{n}} - 1\right] \tag{4}$$

$$Q_{T12} = \left[s_1 + c_n\left(\frac{T_2}{T_2-T_1}\ln\frac{T_2}{T_1} - 1\right)\right]m T_1\left(\frac{T_2}{T_1} - 1\right) \tag{5}$$

$$Q_{T12} = \left[s_1 + c_n\left(\frac{T_2}{T_2-T_1}\ln\frac{T_2}{T_1} - 1\right)\right]m(T_2 - T_1) \tag{6}$$

$$Q_{T12} = \left[s_1 + c_n\left(\frac{T_2}{T_2-T_1}\ln\frac{T_2}{T_1} - 1\right)\right]\frac{1}{R}(p_2 V_2 - p_1 V_1) \tag{7}$$

In Table 7.1, the equations for the temperature change heat Q_{T12} are calculated using Eq. (7.7).

Because of the path dependence of these four process variables, it is possible to convert thermal energy into mechanical energy and vice versa in a cyclic process. The mode of action of the energy conversion becomes clear in the following example:

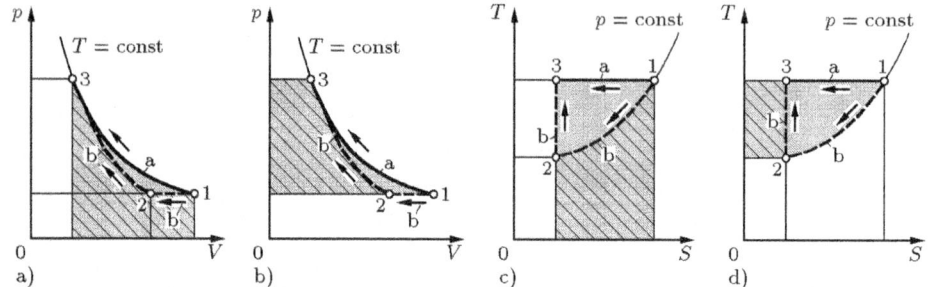

Figure 7.3 Path dependency of the process variables
a) volume change work in a p, V diagram
b) pressure change work in a p, V diagram
c) substitute heat (entropy change heat) in a T, S diagram
d) temperature change heat in a T, S diagram

A gas of state 1 (p_1, T_1) is compressed (Fig. 7.3)

- on path a isothermally to state 3 $(p_3, T_3 = T_1)$,
- on path b first isobaric to state 2 $(p_2 = p_1, T_2)$ and then isentropic to state 3 (p_3, T_3).

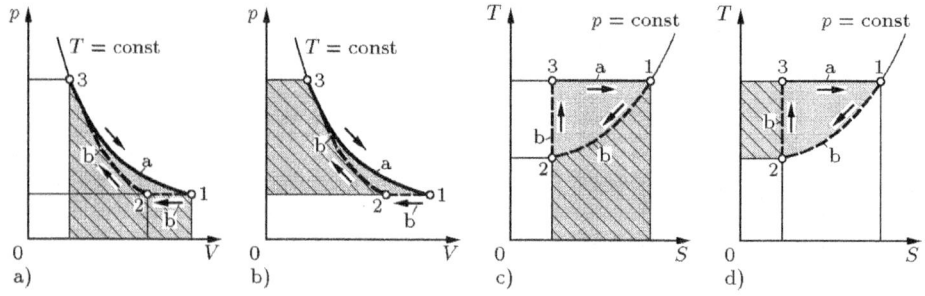

Figure 7.4
Cyclic process isobaric − isentropic − isothermal
a) p, V diagram, enclosed area $-\oint p\,\mathrm{d}V$
b) p, V diagram, enclosed area $\oint V\,\mathrm{d}p$
c) T, S diagram, enclosed area $\oint T\,\mathrm{d}S$
d) T, S diagram, enclosed area $\oint S\,\mathrm{d}T$

The illustrations of the changes of state in a p, V diagram and in a T, S diagram show, as already partly discussed in sections 1.4 and 3.5.2, that the process variables volume

change work, pressure change work, reversible substitute heat (i.e. entropy change heat) as well as temperature change heat lead to different values on both paths a and b, respectively.

If one allows the gas, compressed via path b, to expand again via path a (Fig. 7.4), the gas undergoes a cyclic process. In a p, V diagram and in a T, S diagram, closed curves 1 – 2 – 3 – 1 appear, each enclosing an area. In a cyclic process, this area is circumnavigated.

The realisation of the cyclic process can be imagined in a displacement machine (e.g. a piston engine) (Fig. 7.5); in this case the heat transfer must be controlled accordingly. However, it is also possible to combine fluid flow machines with heat exchangers. In this case, the fluid flow machines can be arranged on a common shaft, for example (Fig. 7.6).

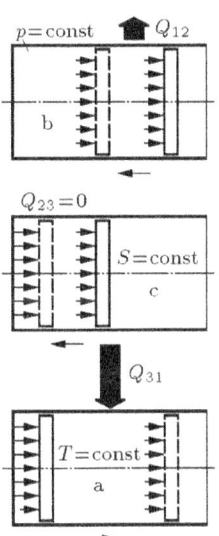

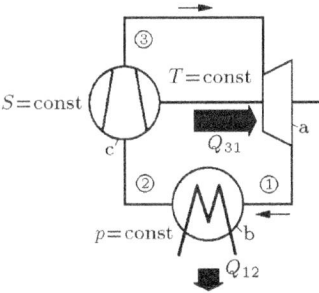

Figure 7.5 Realisation of the cyclic process according to Fig. 7.4 in a displacement machine (e.g. a piston engine)

Figure 7.6 Realisation of the cyclic process according to Fig. 7.4 through series connection of fluid flow machines and heat exchangers

The distinction made in section 6 between ideal and real machines can also be applied to cyclic processes since these machines form the main part of the cyclic processes. In an ideal cyclic process, the same conditions must therefore be fulfilled as for ideal machines. The real cycle must be distinguished from the ideal cycle.

In practice, the conditions of a cyclic process can often not be fulfilled. For example, the necessary supply or release of large heat flows causes considerable difficulties in the design of displacement machines. They can be avoided if the external heat supply is replaced by the internal combustion of a fuel-air mixture and, instead of the heat output, the exhaust gas with a temperature higher than the ambient temperature is released to the ambient air, and a change of the working gas is managed. This technique is used in particular for internal combustion engines, which include combustion engines, but also gas turbine plants with combustion chambers. Although this results in an open process, the processes on which these machines are based are treated as cyclic processes (sections 7.4.1 and 7.4.2).

7.1 Cyclic Process Work, Heat Input and Heat Output

The notion of a circular integral, which is needed in the following, had been introduced in section 3.5.3. According to this, a circular integral becomes equal to zero if the expression to be integrated is a total differential (exact differential). A total differential exists if it is the differential of a state variable (section 3.3.3). For example, the circular integral over the differential of entropy according to Eq. (3.103) is zero. If the integration is to be carried out over a process variable, the circular integral is not equal to zero as it is shown in the case of the Eqs. (3.106) and (3.107).

If one calculates the total volume change work occurring for the cyclic process illustrated in Fig. 7.4, then according to section 3.5.3 it follows

$$W_{V12} + W_{V23} + W_{V31} = \oint dW_V = -\oint p\,dV . \qquad (7.8)$$

Correspondingly, one obtains for the pressure change work

$$W_{p12} + W_{p23} + W_{p31} = \oint dW_p = \oint V\,dp . \qquad (7.9)$$

The absolute value of each of the two circular integrals in Eqs. (7.8) and (7.9) corresponds to the area enclosed by the state curves in a p, V diagram (Fig. 7.4 a and b). Accordingly, it must hold:

$$-\oint p\,dV = \oint V\,dp = W_{Kreis} \qquad (7.10)$$

The same relationship is also shown in Eq. (3.104). W_{Kreis} is called the cyclic process work.

For ideal cyclic processes, the transferred reversible substitute heat is shown in a T, S diagram. According to Fig. 7.4 c one obtains

$$(Q_{12})_{rev} + (Q_{23})_{rev} + (Q_{31})_{rev} = \oint dQ_{rev} = Q_{s12} + Q_{s23} + Q_{s31} = \oint dQ_s . \qquad (7.11)$$

Similarly it becomes (cf. Fig. 7.4 d)

$$Q_{T12} + Q_{T23} + Q_{T31} = \oint dQ_T . \qquad (7.12)$$

The absolute value of each of the two circular integrals in Eqs. (7.11) and (7.12) gives the value obtained from the enclosed area by the state curves in a T, S diagram. Because, according to Fig. 7.4, the sum of the reversible substitute heat (entropy change heat) supplied has a different sign, compared to the sum of the volume change work given off, to the sum of the pressure change work given off and to the sum of the negative temperature change heat, the following applies

$$\oint dW_V = \oint dW_p = W_{Kreis} = -\oint dQ_{rev} = -\oint dQ_s = \oint dQ_T \qquad (7.13)$$

$$-\oint p\,dV = \oint V\,dp = W_{Kreis} = -\oint T\,dS = \oint S\,dT . \qquad (7.14)$$

A distinction can be made with respect to the heat reversibly transferred between the ideal cycle and the surroundings according to Eq. (3.105):

total reversible heat supplied to the ideal cycle $(Q_{zu})_{rev}$
total reversible heat released by the ideal cycle $(Q_{ab})_{rev}$

For ideal cyclic processes one can write

$$\oint dQ_{rev} = (Q_{zu})_{rev} - |(Q_{ab})_{rev}| \ . \tag{7.15}$$

Cyclic processes form closed systems. For ideal cyclic processes, one obtains by applying Eq. (3.76)

$$\oint dQ_{rev} - \oint p\,dV = \oint dU \ . \tag{7.16}$$

If a gas quantity has gone through a cycle process, in the end all state variables are back in the initial state. Therefore, the following applies to the internal energy, related to the cycle process shown in Figure 7.4:

$$U_2 - U_1 + U_3 - U_2 + U_1 - U_3 = 0 \tag{7.17}$$

In general this can be formulated as follows:

$$\oint dU = 0 \tag{7.18}$$

Thus Eqs. (7.16), (7.13) and (7.14), respectively, yield the same statement.

In real cyclic processes, too, the transferred heat can be divided into supplied and released heat, as can be seen at the examples shown in Fig. 7.5 and Fig. 7.6. One denotes the

total heat supplied to the real cycle as Q_{zu}
total heat released by the real cycle as Q_{ab}.

Hereby, for real cycle processes one can write

$$\oint dQ = Q_{zu} - |Q_{ab}| \ . \tag{7.19}$$

If the formulations of the first law of thermodynamics for closed systems according to Eqs. (2.47) and (2.55) will be applied to real cyclic processes, the following equations are obtained:

$$\oint dQ + \oint dW_V + \oint |dW_{RI}| = 0 \tag{7.20}$$

and

$$\oint dQ + \oint dW_e - \oint |dW_{RA}| = 0 \tag{7.21}$$

If one writes the circular integrals of the process variables in simplified eypressions

$$\oint dW_e = W_e \qquad \oint |dW_{RI}| = |W_{RI}| \qquad \oint |dW_{RA}| = |W_{RA}|, \qquad (7.22)$$

then the equations for cyclic processes can be formulated in a way as they are given in Table 7.2:

Tabelle 7.2 Energy equations for cyclic processes

ideal cyclic process	
$(Q_{zu})_{rev} - |(Q_{ab})_{rev}| + W_{Kreis} = 0$	(1)
$(Q_{zu})_{rev} - |(Q_{ab})_{rev}| + (W_e)_{id} = 0$	(2)
$(W_e)_{id} = W_{Kreis}$	(3)
real cyclic process	
$Q_{zu} - |Q_{ab}| + W_{Kreis} + |W_{RI}| = 0$	(4)
$Q_{zu} - |Q_{ab}| + W_e - |W_{RA}| = 0$	(5)
$W_e = W_{Kreis} + |W_{RI}| + |W_{RA}|$	(6)

In an ideal cyclic process, the coupling work is equal to the cycle work:

$$(W_e)_{id} = W_{Kreis}$$

If the fluid flow machines of a cyclic process act on a common shaft, it is possible to calculate the coupling work of the cyclic process $(W_e)_{id}$ or W_e, respectively, from the coupling work of the compressors $(W_{eV})_{id}$ or W_{eV}, respectively, and the turbines $(W_{eT})_{id}$ or (W_{eT}) respectively:

$$(W_e)_{id} = (W_{eV})_{id} + (W_{eT})_{id} \qquad (7.23)$$

$$W_e = W_{eV} + W_{eT} \qquad (7.24)$$

7.2 Right-Hand and Left-Hand Cyclic Processes

The sequence of state changes chosen in Fig. 7.4 results in a right-hand direction of circulation for the cyclic process. If the process takes place in a displacement machine, the compression work being an effort is less than the expansion work being a benefit. The cyclic process therfore supplies mechanical energy. Using a T, S diagram, one can see that more heat at higher temperatures must be added to the cyclic process than the heat removed at lower temperatures. Accordingly, right-hand cyclic processes are called power machine processes.

If one changes the direction of circulation in the machine by compressing via path a and expanding via path b, the result is a left-hand cyclic process. Since the compression work being an effort is now larger than the expansion work being a benefit, the cyclic process needs mechanical energy. Thus, left-hand cyclic processes are called working

machine processes. Here, the heat output is characterised by higher temperatures and is larger than the heat input at lower temperatures.

Table 7.3 Energy conversion of the cyclic processes

right-hand cyclic processes (clockwise cyclic processes)	left-hand cyclic processes (counterclockwise cyclic processes)
power machine processes delivery of mechanical energy heat input greater than heat output	working machine processes supply of mechanical energy heat output greater than heat input

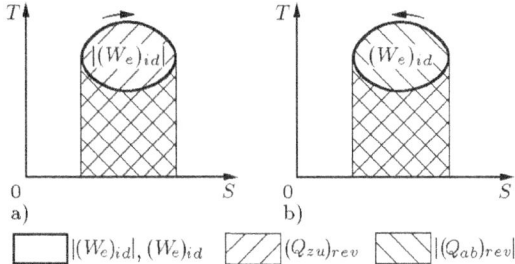

Figure 7.7 Cyclic process in a T, S diagram
(changes of state schematically)
a) right-hand   b) left-hand
(clockwise)   (counterclockwise)

Figure 7.7 shows the energy quantities for ideal cyclic processes in a T, S diagram. The changes of state are assumed schematically.

7.3 The Theory of Right-Hand Cyclic Processes

Right-hand cyclic processes (also named clockwise cyclic processes) form the theoretical basis of calculation for thermal power machines (prime movers, engines) and thermal power plants. In these processes, thermal energy, which is produced e.g. by the combustion of fuels, is converted into mechanical energy.

The required heat is supplied to the process by heat exchangers (e.g. by steam generators of a steam power plant) or by internal combustion (e.g. in internal combustion engines). Generally, heat generation is referred to as heat source. The heat to be released in each clockwise cycle is transferred to a cooling fluid (e.g. in a condenser of a steam power plant) or to the environment (e.g. by a combustion engine). The given off heat of the cycle process and the external friction work $|W_{RA}|$ are given off to a heat sink. According to Table 7.2 Eq. (5) gives

$$Q_{zu} = -W_e + |Q_{ab}| + |W_{RA}| \,. \tag{7.25}$$

Since mechanical energy is generated in a right-hand cycle process, it becomes

$$-W_e = |W_e| \,. \tag{7.26}$$

Only part of the supplied heat Q_{zu} is converted into mechanical energy $|W_e|$, the rest $|Q_{ab}|$ and $|W_{RA}|$ is given off as heat to a heat sink:

$$Q_{zu} = |W_e| + |Q_{ab}| + |W_{RA}| \tag{7.27}$$

A general right-hand cycle process has to be considered, whose reversible substitute process is shown in a T, S diagram according to Fig. 7.8 a) . Let the temperature of the heat source be T_{Qu}, in the cycle process the heat supply occurs in the upper part of the curve from 1 to 2. In order to ensure sufficient heat transfer, the temperature of

the heat source T_{Qu} must be higher than the temperature of the working fluid which obtains the heat. The mean value of this temperature is denoted by T_{zu} in Fig. 7.8 a). Conversely, the temperature of the heat sink T_{Se} must be lower than the temperature of the heat-emitting working fluid (lower part of the curve from 2 to 1). The mean value of the temperature of the heat release is denoted by T_{ab}.

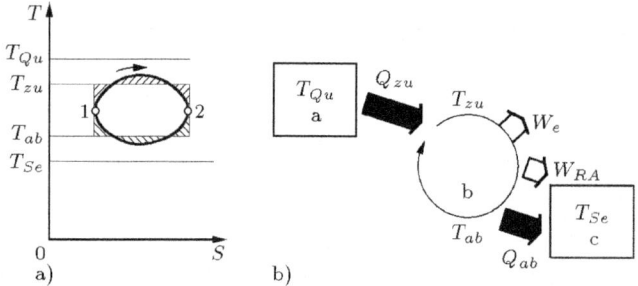

Figure 7.8 Right-hand cyclic process   a) T, S diagram (changes of state schematic)
b) model representation   a heat source   b cyclic process   c heat sink

In Fig. 7.8 b) these considerations are translated into a model representation.

7.3.1 Conversion of Thermal to Mechanical Energy

In a right-hand cycle process, the heat Q_{zu} supplied can never be completely converted into mechanical energy. A cycle process always requires a heat release Q_{ab}. Only then, if one would change the cycle process according to Fig. 7.8 in such a way that the lower curve would run from 2 to 1 on the abscissa axis, the heat release would be zero (Fig. 7.9). However, then the absolute temperature T_{ab} would also have to be zero, which, as is well known, is not possible. It can be shown that the second law of thermodynamics would be violated if the heat transferred from the heat source to the cycle Q_{zu} could be completely transformed into mechanical energy.

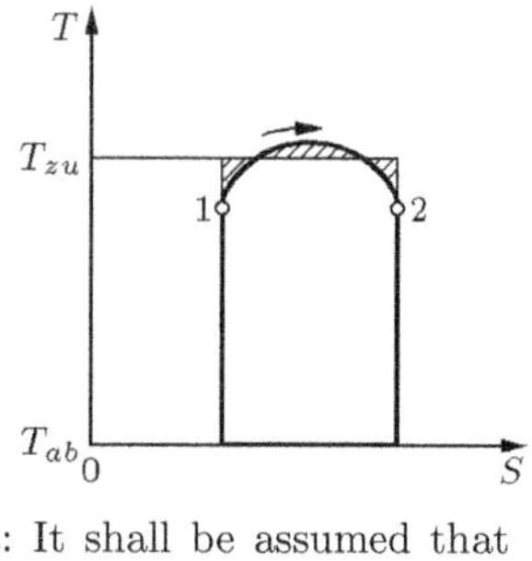

Figure 7.9 Right-hand cycle process without releasing heat

According to *Baehr*[1] [7] this proof can be carried out as follows: It shall be assumed that there exists a machine that does not need a heat sink and would thus perform the complete conversion of the heat into mechanical energy. One can then calculate the total entropy change for the substitute process. For the heat source, since heat is always given off by the heat source, the entropy change is

$$S_{2Qu} - S_{1Qu} = -\int_1^2 \frac{dQ_{rev}}{T} \,. \tag{7.28}$$

---

[1] *Hans Dieter Baehr*, 1928 to 2015, German professor of thermodynamics in Hannover, made numerous contributions to the field of thermodynamics.

For the cycle process, since the entropy S is a state variable,

$$\oint dS = 0 \tag{7.29}$$

must be. By adding the two entropy changes according to Eqs. (7.28) and (7.29,) one obtains the total entropy change

$$S_{2Qu} - S_{1Qu} + \oint dS = -\int_1^2 \frac{dQ_{rev}}{T} . \tag{7.30}$$

The total entropy change is negative. According to the second law of thermodynamics, a process for whose substitute process the total entropy change is negative, is not possible.

Could a machine be constructed in which the heat transferred from the heat source to the cyclic process could be completely changed into mechanical energy, then it would be conceivable to use the inexhaustible supplies of thermal energy in the air or in sea water for this machine and to obtain mechanical energy from it. However, according to the knowledge gained, such a conversion of energy can only take place if a part of the thermal energy can be transferred to a heat sink with a lower temperature. Without this heat sink, no "periodically operating thermal machine" is possible that can perform the planned energy conversion.

Such a machine without a heat sink by *Ostwald*[2] was called perpetuum mobile of the second kind. His version of the second law of thermodynamics reads:

> **A perpetuum mobile of the second kind is impossible.**

7.3.2 Thermal Efficiency

In order to evaluate the various cycle processes, one needs a standard of evaluation. One possibility of a comparative yardstick is to form the quotient of benefit and effort. In the case of a machine to generate work that is realised as a right-hand cyclic process, the benefit is the work that is available at the coupling. The effort is the heat supplied.

The ratio of the coupling work W_e to the heat Q_{zu} is called thermal efficiency η_{th}. Since in such a process the coupling work W_e is negative, in order to obtain positive values for the thermal efficiency, the absolute value of W_e must be used:

$$\eta_{th} = \frac{-W_e}{Q_{zu}} = \frac{|W_e|}{Q_{zu}} \tag{7.31}$$

Using Table 7.2 Eq. (6) this leads to

$$\eta_{th} = \frac{-W_{Kreis} - |W_{RI}| - |W_{RA}|}{Q_{zu}} \tag{7.32}$$

---

[2] *Wilhelm Ostwald*, 1853 to 1932, German professor of physical chemistry in Leipzig, 1909 Nobel Prize for Chemistry

7.3 The Theory of Right-Hand Cyclic Processes

and with Table 7.2 Eq. (5) to

$$\eta_{th} = \frac{Q_{zu} - |Q_{ab}| - |W_{RA}|}{Q_{zu}} \tag{7.33}$$

and to

$$\eta_{th} = 1 - \frac{|Q_{ab}| + |W_{RA}|}{Q_{zu}}. \tag{7.34}$$

For an ideal cyclic process, Eq. (3) of Table 7.2 gives

$$\eta_{th} = \frac{|(W_e)_{id}|}{(Q_{zu})_{rev}} = \frac{-W_{Kreis}}{(Q_{zu})_{rev}}, \tag{7.35}$$

and Eq. (2) of Table 7.2 gives

$$\eta_{th} = \frac{(Q_{zu})_{rev} - |(Q_{ab})_{rev}|}{(Q_{zu})_{rev}} = 1 - \frac{|(Q_{ab})_{rev}|}{(Q_{zu})_{rev}}. \tag{7.36}$$

7.3.3 Right-Hand *Carnot* Process

It is of great practical interest what maximum value the thermal efficiency can reach or — in other words — how much mechanical energy can be obtained from an available heat Q_{zu} in a cycle process.

It is to be assumed as an additional condition that the heat source supplies the heat Q_{zu} at the constant temperature T_{Qu} and the heat sink Q_{ab} takes up the heat at the constant temperature T_{Se}. The heat source (e.g. a hot water boiler) and the heat sink (e.g. the free atmosphere) are therefore so large that their temperatures change only negligibly due to the heat transfer (Fig. 7.10 a).

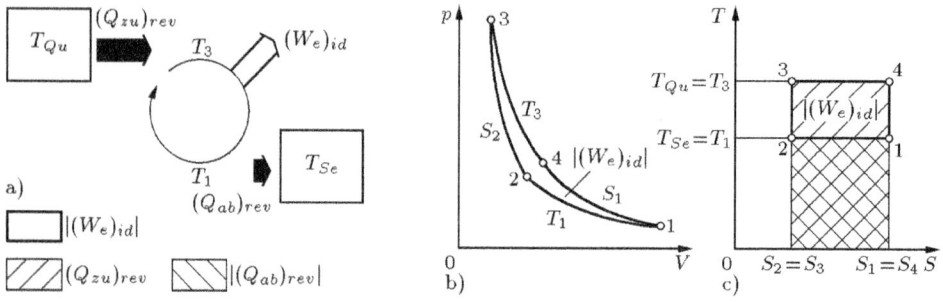

Figure 7.10 Right-hand *Carnot* process
a) model representation   b) p, V diagram   c) T, S diagram

As will be proved in the next section, a first condition for reaching the maximum thermal efficiency is the reversibility of the whole process. This condition is fulfilled if all subprocesses are reversible:

1) The heat transfer from the heat source to the cyclic process: Heat source and the substance flow carrying out this change of state of the cycle must have the same temperature. The heat absorption $Q_{zu} = (Q_{zu})_{rev}$ in the cycle must on the isotherm

$$T_{Qu} = T_{zu} = T_3 \tag{7.37}$$

take place.

2) The operation of the machines in the cycle: All machines used must be ideal machines without friction.

3) The heat transfer from the cycle to the heat sink: The substance flow of the cycle must transfer the heat $Q_{ab} = (Q_{ab})_{rev}$ at the temperature of the heat sink and thus on the isotherm

$$T_{Se} = T_{ab} = T_1 \ . \tag{7.38}$$

The connection of both isotherms of the cycle T_1 and T_3 can only be caused by changes of state, where there is no heat transfer between the heat source or heat sink and the cyclic process; these changes of state can be e.g. isentropic changes of state (cf. also sections 7.4.3 to 7.4.5). Thus one obtains the course of the cyclic process shown in Fig. 7.10 b) and c). It was first described by *Carnot*[3] and is therefore referred to as the *Carnot* process. To find the thermal efficiency, one calculates the heat transfer on the isotherms:

$$(Q_{zu})_{rev} = (Q_{34})_{rev} = T_3(S_4 - S_3) = T_{Qu}(S_4 - S_3) \tag{7.39}$$

$$|(Q_{ab})_{rev}| = |(Q_{12})_{rev}| = T_1(S_1 - S_2) = T_{Se}(S_4 - S_3) \tag{7.40}$$

According to Eq. (7.36) it follows

$$\eta_{th\ C} = 1 - \frac{|(Q_{ab})_{rev}|}{(Q_{zu})_{rev}} = 1 - \frac{T_1}{T_3} = 1 - \frac{T_{Se}}{T_{Qu}} \ . \tag{7.41}$$

> **The thermal efficiency of the *Carnot* process only depends on the temperatures of the heat source and of the heat sink.**

7.3.4 Effect of Irreversible Processes

It can be shown that irreversible processes reduce thermal efficiency. Among the possible irreversible processes, we will single out the two most important ones: heat transfer with temperature gradient and friction work. The investigation is based on a right-hand *Carnot* process. This process is to be changed by the irreversible processes mentioned. Both irreversible processes should not occur simultaneously. First of all, the effect of the irreversible heat transfer shall be investigated, while the machines in which the changes of state occur are operated without friction.

---
[3] *Nicolas Leonard Sadi Carnot*, 1796 to 1832, joined the French Corps of Genius in 1814, but took his leave in 1828 as a Captain in 1828 and lived as a private scientist.

7.3 The Theory of Right-Hand Cyclic Processes

In the T, S diagram of Fig. 7.11 the equivalent process of this irreversible cycle is shown. The heat supplied Q_{zu}, corresponding to the area AB6CA, is supposed to be as large as in the *Carnot* process (section 7.3.3) and to be available to the cycle at the same temperature T_{Qu} as in the *Carnot* process.

However, the cycle absorbs this heat at the lower temperature

$$T_3 = T_{Qu} - \Delta T_{Qu} \, . \tag{7.42}$$

Thus, the area 34DC3 is equal to the area AB6CA:

$$Q_{zu} = T_{Qu}(S_6 - S_2) = T_3(S_1 - S_2) \tag{7.43}$$

The heat release of the cycle (area 21DC2) occurs at the temperature

$$T_1 = T_{Se} + \Delta T_{Se} \, . \tag{7.44}$$

Figure 7.11 Irreversible process: heat transfer under temperature gradient
a) model representation
b) T, S diagram

☐ $|W_e|$ ⊘ Q_{zu}

⊠ $|Q_{ab}|$

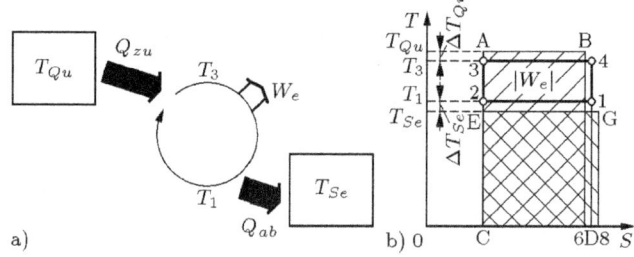

The sink receives the same heat at the temperature T_{Se} that is the temperature of the sink of the *Carnot* process: Thus, the area 21DC2 is equal to the area EG8CE.

$$|Q_{ab}| = T_1(S_1 - S_2) = T_{Se}(S_8 - S_2) \tag{7.45}$$

According to Eq. (7.31), it follows for the thermal efficiency of the cycle

$$\eta_{th} = 1 - \frac{|Q_{ab}|}{Q_{zu}} = 1 - \frac{T_1(S_1 - S_2)}{T_3(S_1 - S_2)} = 1 - \frac{T_1}{T_3} = 1 - \frac{T_{Se} + \Delta T_{Se}}{T_{Qu} - \Delta T_{Qu}} \, . \tag{7.46}$$

Eq. (7.46) gives a smaller thermal efficiency than Eq. (7.41).

Figure 7.12 Irreversible process with real machinery
a) model representation
b) T, S diagram

☐ $|W_e| + |W_{RI}| + |W_{RA}|$

⊘ Q_{zu} ⊠ $|Q_{ab}|$

☐ $|W_{RI}|$

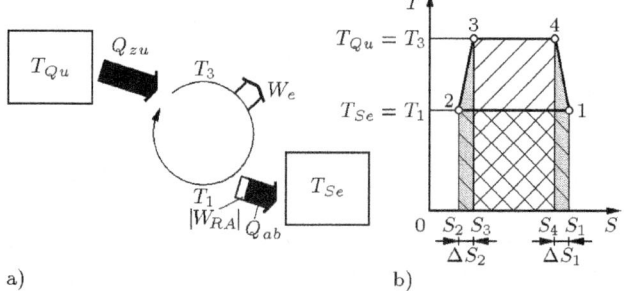

The effect of the friction work on the thermal efficiency can be illustrated by substituting the ideal machines used in the *Carnot* process by real machines.

The compression 2 – 3 is to be effected by an irreversible adiabatic compressor, the expansion 4 – 1 by an irreversible adiabatic turbine. The result is a cyclic process, the

equivalent of which is shown in Figure 7.12. The heat input to the cycle is reversible at the temperature of the heat source $T_{Qu} = T_3$:

$$Q_{zu} = T_{Qu}(S_4 - S_3) = T_3(S_4 - S_3) \qquad (7.47)$$

The heat release is also reversible at the temperature $T_1 = T_{Se}$:

$$|Q_{ab}| = T_1(S_1 - S_2) = T_{Se}(S_1 - S_2) = T_{Se}(S_4 - S_3 + \Delta S_1 + \Delta S_2) \qquad (7.48)$$

For the thermal efficiency one obtains according to Eq. (7.34)

$$\eta_{th} = 1 - \frac{|Q_{ab}| + |W_{RA}|}{Q_{zu}} = 1 - \frac{T_{Se}}{T_{Qu}} - \frac{T_{Se}(\Delta S_1 + \Delta S_2) + |W_{RA}|}{T_{Qu}(S_4 - S_3)} . \qquad (7.49)$$

It can be seen that at the same temperature of heat source T_{Qu} and heat sink T_{Se} as in Eq. (7.41), the thermal efficiency becomes smaller due to the irreversibility of the friction work.

In practice, heat transfer under a temperature gradient often provides the largest irreversibilities of a cyclic process. For example, the thermal energy in the steam generator (boiler) of a steam power plant is generated by the combustion of the fuels at approximately 1200 °C, but in the best case it is transferred to the steam in the steam boiler below approximately 540 °C to 650 °C.

In internal combustion engines and gas turbine plants in which the exhaust gases leave the engine at temperatures of about 350 °C to 850 °C (gas turbine plants: 450 °C to 620 °C; naturally aspirated *Otto* engines: 700 °C to 850 °C; naturally aspirated *Diesel* engines: 500 °C to 700 °C; turbocharged *Diesel* engines: 350 °C to 400 °C), the temperature gradient is on the side of the heat sink, which is represented by the ambient air with the constant temperature of e.g. approximately 20 °C.

Combined processes can be used to reduce these irreversibilities: In combined cycle power plants (CCPP), the steam power process is preceded by a gas turbine process, with which a temperature gradient between about 1250 °C and about 600 °C is used to generate work; the downstream steam power process uses a temperature gradient between about 540 °C and about 30 °C.

Internal combustion engines can be combined with an exhaust gas turbine, which in turn drives a compressor to compress the combustion air: The internal combustion engine, which operates according to the *Diesel* or *Otto* process, uses a temperature gradient, e.g. between about 2000 °C and about 550 °C, while the exhaust gas turbine uses the temperature range between about 550 °C and about 370 °C.

The list of irreversibilities is not complete. In internal combustion engines, the combustion of the fuel is one of the main sources of irreversibility. To avoid it, one tries to replace the combustion process by a reversible oxidation process. This goal is pursued by energy conversion plants that work according to the principle of fuel cells. Different types of fuel cells are discussed in section 12.12.

7.3.5 Carnot Factor

The special importance of the *Carnot* process arises from a comparison with other cyclic processes. Here one can confine oneself to the consideration of ideal cyclic processes, since the irreversibility in each case reduces the thermal efficiency.

For the *Carnot* process and an arbitrary cyclic process, the maximum temperature of the heat source T_{Qu} and the minimum temperature of the heat sink T_{Se} shall be the same. Heat input and heat output correspond to the areas in the T,S diagram according to Fig. 7.13:

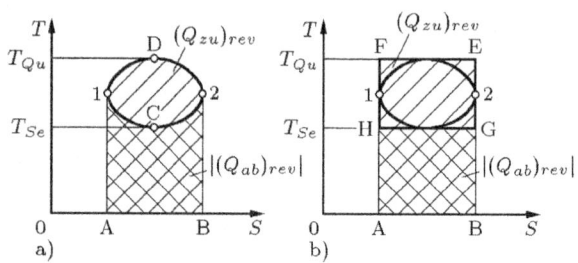

Figure 7.13 Comparison of an arbitrary ideal cyclic process (a) with the circumscribed *Carnot* process (b) at same temperatures T_{Qu} and T_{Se}

Carnot process:

$$\eta_{th\ C} = 1 - \frac{|(Q_{ab})_{rev}|}{(Q_{zu})_{rev}} = 1 - \frac{\text{area ABGHA}}{\text{area ABEFA}} \qquad (7.50)$$

Any circular process:

$$\eta_{th} = 1 - \frac{|(Q_{ab})_{rev}|}{(Q_{zu})_{rev}} = 1 - \frac{\text{area AB2C1A}}{\text{area AB2D1A}} \qquad (7.51)$$

The area comparison shows that the *Carnot* process provides the highest possible thermal efficiency at given temperature limits of heat source and heat sink.

The *Carnot* process is therefore of fundamental importance for the conversion of thermal energy into mechanical energy. Solving Eq. (7.35) for the coupling work $(W_e)_{id}$ and setting as the thermal efficiency that of the *Carnot* process according to Eq. (7.41), one obtains

$$|(W_e)_{id}| = \left(1 - \frac{T_{Se}}{T_{Qu}}\right)(Q_{zu})_{rev} = \eta_C (Q_{zu})_{rev}\ . \qquad (7.52)$$

η_C is called the *Carnot* factor. It is equal to the thermal efficiency of the ideal *Carnot* process.

The product of the *Carnot* factor η_C and the heat $(Q_{zu})_{rev}$ available at the temperature T_{Qu} gives that part of the heat transferred by an ideal right-hand *Carnot* process between the temperatures of the heat source T_{Qu} and the heat sink T_{Se} which can be converted into mechanical energy $(W_e)_{id}$.

From a technical point of view, it makes sense to ask for that part of the heat that can be transformed into mechanical energy between two temperature limits. The maximum temperature of a cycle is often given by the technological properties of the materials used to construct the machine. The minimum temperature is determined by the need for heat release into a heat sink.

In internal combustion engines, the temperature of the heat source is predetermined by the combustion of the fuel: Depending on the process design, the temperature of the resulting gas mixture can rise for a short time, e.g. to about 2000 °C ... 2200 °C . This maximum temperature occurs during such a short period that the cylinder walls and the upper piston surface are not damaged.

The blades of the turbines in gas turbines and steam power plants, on the other hand, have the temperature of the hot gases entering. Therefore, the gas inlet temperature in small gas turbines is about 1000 °C. In large gas turbines of combined gas and steam turbine power plants (combined cycle power plants) the temperature is approximately 1250 °C (blade cooling). For steam turbines, the steam inlet temperature is generally approximately 540 °C to 600 °C, in special cases approximately 650 °C.

Due to the replacement of the working gas, the lowest temperature of the heat sink in internal combustion engines is the same as the temperature of the ambient air. This lowest temperature of the heat sink is also decisive for the share of heat given off, which is caused by heat transfer to the cooling water or to the cooling air.

In gas turbine plants and steam power plants, either the same heat sink (cooling water or cooling air) is used, or the heat output of the cycle takes place at a higher temperature, which allows the utilisation of thermal energy for heat supply (process heat, heat for district heating, etc.).

Example 7.1 What is the thermal efficiency of a right-hand *Carnot* process, operating between a heat source of 700 °C and a heat sink of 25 °C ? What part of the supplied heat is converted into mechanical energy? How does the *Carnot* factor increase when the temperature of the heat source is raised to 1100 °C ?

According to Eq. (7.41), the thermal efficiency is

$$\eta_C = 1 - \frac{T_1}{T_3} = 1 - \frac{298.15 \text{ K}}{973.15 \text{ K}} = 0.694 \ .$$

So 69.4 % of the supplied heat is convertible into mechanical energy. When the temperature of the heat source is increased to $T_3 = 1373.15$ K, the thermal efficiency is

$$\eta_C = 1 - \frac{298.15 \text{ K}}{1373.15 \text{ K}} = 0.783 \ .$$

So 78.3 % of the supplied heat is convertible into mechanical energy..

7.4 Technically Used Right-Hand Cyclic Processes

The *Carnot* process achieves the highest thermal efficiency at given temperatures of heat source and heat sink, its realisation, however, encounters considerable technical difficulties (cf. section 7.5). Therefore, one concentrates on cyclic processes, which are technically more favourable to realise. Internal combustion engines (*Otto* and *Diesel* engines) as well as gas turbines, gas turbine power plants and steam power plants have achieved great importance.

Internal combustion engines are positive displacement machines that use an open process to take in and compress a mixture of fuel and air (carburettor *Otto* engine) or only air (stratified fuel injection *Otto* engine, *Diesel* engine), then burn the fuel with oxygen of the air, then expand the resulting exhaust gases and finally expel them. The air used is not only subjected to a series of changes of state, but is also chemically altered to a limited extent by the combustion process. With good reason, the machine process is approximated by a comparative process in which air, as a gas mixture of constant composition, runs through an ideal cycle.

Turbo machines are used in gas turbine plants. In this process, a gas is compressed by a turbo compressor and then heated. The gas then flows through a turbine. Gas turbine plants have developed in two directions:

In the seldom used closed design for stationary plants, the gas flows through two heat exchangers: one in which heat is supplied to the gas and another in which heat is given off by the gas (Fig. 7.14 a). These systems are primarily used to drive generators and thus to produce electrical energy. The gas runs through a closed cycle in these plants.

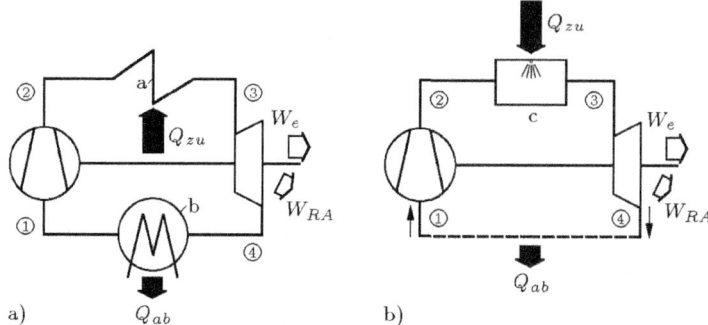

Figure 7.14 Gas turbine processes   a) closed design   b) open design
a gas heater   b gas cooler   c combustion chamber

In the open design, the gas turbine — e.g. as a unit for aircrafts or in stationary plants — the gas turbine is designed for low weight and small space requirements. Air flows through the compressor; kerosine, light fuel oil or natural gas is then burned with the compressed air in a combustion chamber. The resulting exhaust gases are expanded in the turbine. This eliminates the need for heavy heat exchangers (Fig. 7.14 b).

Instead of the open process that actually takes place, for comparison the *Joule* process as an ideal cycle is used with air as a gas of constant composition.

In order to calculate the comparative processes of combustion engines and gas turbine plants in an easier way, constant specific heat capacities are used.

Steam power plants, like stationary gas turbine plants, are used to generate electrical energy. The steam produced in the steam generator is expanded in a turbine and liquefied in an condenser. The condensation temperature can be selected so high that the heat that has to be given off can be used for heating purposes or for process heat supply — e.g. as part of a district heating system (combined heat and power plant).

7.4.1 *Seiliger* Process, *Otto* Process, *Diesel* Process, Generalised *Diesel* Process

Reversible comparison processes of internal combustion engines are described by the *Seiliger*[4] process or mixed comparison process. *Otto*[5] process and *Diesel*[6] process are included as special cases.

---

[4] *Myron Seiliger*, born 1874, German engineer, proposed in 1922 the process named after him for combustion engines.

[5] *Nikolaus August Otto*, 1832 to 1891, German engineer, invented the four-stroke gas engine named after him with a compressed charge.

[6] *Rudolf Diesel*, 1858 to 1913, German engineer, developed the *Diesel* engine according to thermodynamic principles.

In the beginning of the process, the compression of the gas quantity brought into the cylinder takes place isentropically. Hereby, the volume V_1 is reduced to the volume V_2. The following terms are used to describe the compression (Fig. 7.15):

stroke volume
$$V_H = V_1 - V_2 \tag{7.53}$$

compression volume
$$V_K = V_2 = V_3 \tag{7.54}$$

compression ratio
$$\varepsilon = \frac{V_1}{V_2} = \frac{V_H + V_K}{V_K} \tag{7.55}$$

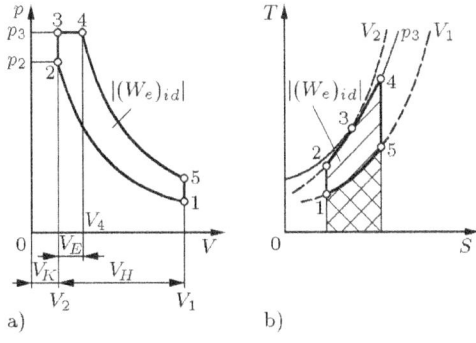

Figure 7.15 Seiliger process (hatchings like at Figure 7.10)
a) p, V diagram   b) T, S diagram

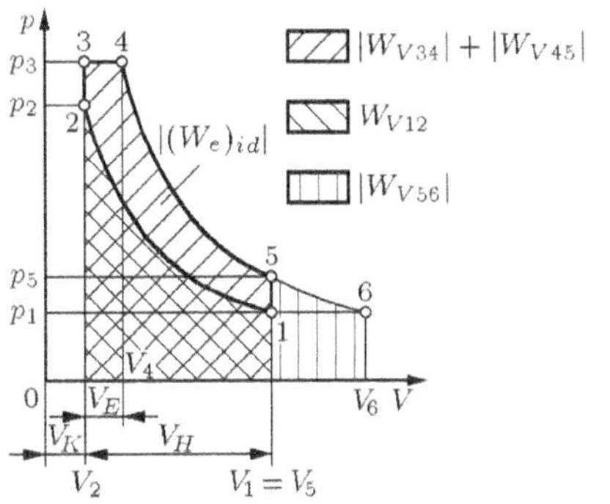

Figure 7.16 Seiliger process as an open process in the p, V diagram

Instead of the combustion taking place in the engine, one calculates with an isochoric heat input from 2 to 3 and a subsequent isobaric heat input from 3 to 4 to the gas. The following terms describe this process:

pressure increase ratio
$$\psi = \frac{p_3}{p_2} \tag{7.56}$$

injection volume
$$V_E = V_4 - V_2 = V_4 - V_3 \tag{7.57}$$

injection ratio
$$\varphi = \frac{V_4}{V_2} = \frac{V_4}{V_3} = \frac{V_K + V_E}{V_K} \tag{7.58}$$

The expansion of the combustion gas (exhaust gas) is associated with an isentropic change of state. After reaching the volume $V_5 = V_1$, an isochoric heat release to the environment replaces the change of exhaust gas and fresh gas. In the following, an ideal gas with constant specific heat capacities is assumed as the working fluid. The reversibly supplied heat is

$$(Q_{zu})_{rev} = (Q_{23})_{rev} + (Q_{34})_{rev} = m\, c_v(T_3 - T_2) + m\, c_p(T_4 - T_3) \,. \tag{7.59}$$

The reversibly released heat is

$$|(Q_{ab})_{rev}| = |(Q_{51})_{rev}| = m\, c_v(T_5 - T_1) \,. \tag{7.60}$$

From this, according to Table 7.2 Eqs. (1) and (3) the coupling work becomes

$$(W_e)_{id} = W_{Kreis} = -m\left[c_v(T_3 + T_1 - T_2 - T_5) + c_p(T_4 - T_3)\right]. \tag{7.61}$$

With Eqs. (7.59), (7.60) and (4.44) one obtains the thermal efficiency according to Eq. (7.36)

$$\eta_{th} = 1 - \frac{T_5 - T_1}{T_3 - T_2 + \kappa(T_4 - T_3)}. \tag{7.62}$$

The following relationships hold between the temperatures at the respective states:

$$\frac{T_5}{T_4} = \left(\frac{V_4}{V_1}\right)^{\kappa-1} = \left(\frac{V_4}{V_2}\frac{V_2}{V_1}\right)^{\kappa-1} = \left(\frac{\varphi}{\varepsilon}\right)^{\kappa-1} \qquad T_5 = T_4\left(\frac{\varphi}{\varepsilon}\right)^{\kappa-1} \tag{7.63}$$

$$\frac{T_4}{T_3} = \frac{V_4}{V_3} = \frac{V_4}{V_2} = \varphi \qquad T_4 = T_3\,\varphi \tag{7.64}$$

$$\frac{T_3}{T_2} = \frac{p_3}{p_2} = \psi \qquad T_3 = T_2\,\psi \tag{7.65}$$

$$\frac{T_2}{T_1} = \left(\frac{V_1}{V_2}\right)^{\kappa-1} = \varepsilon^{\kappa-1} \qquad T_2 = T_1\,\varepsilon^{\kappa-1} \tag{7.66}$$

From these relations result the following equations:

$$T_3 = T_1\,\psi\,\varepsilon^{\kappa-1} \qquad T_4 = T_1\,\varphi\,\psi\,\varepsilon^{\kappa-1} \qquad T_5 = T_1\,\varphi^\kappa\,\psi \tag{7.67}$$

Eqs. (7.66) and (7.67) are inserted into Eq. (7.62):

$$\eta_{th} = 1 - \frac{\varphi^\kappa\,\psi - 1}{\varepsilon^{\kappa-1}\left[\psi - 1 + \kappa\,\psi\,(\varphi - 1)\right]} \tag{7.68}$$

In a *Otto* process, the isobaric heat input from 3 to 4 is omitted; this gives $T_3 = T_4$ as well as $V_4 = V_3 = V_2$ and thus $\varphi = 1$. From Eqs. (7.62) resp. (7.68) result the following equations:

$$\eta_{th} = 1 - \frac{T_5 - T_1}{T_3 - T_2} \quad \text{and} \quad \eta_{th} = 1 - \frac{1}{\varepsilon^{\kappa-1}} \tag{7.69}$$

In a *Diesel* process, there is no isochoric heat input from 2 after 3; this gives $T_2 = T_3$ as well as $p_2 = p_3 = p_4$ and thus $\psi = 1$. From Eqs. (7.62) resp. (7.68) result the following equations:

$$\eta_{th} = 1 - \frac{T_5 - T_1}{\kappa(T_4 - T_3)} \quad \text{and} \quad \eta_{th} = 1 - \frac{\varphi^\kappa - 1}{\varepsilon^{\kappa-1}\,\kappa\,(\varphi - 1)} \tag{7.70}$$

An improvement of the thermal efficiency η_{th} of the *Diesel* process is possible with the help of a multistage injection of the fuel and thus a modified combustion process. Thus, the isobaric heat input from 3 to 4 can be changed to a polytropic heat input, whereby the polytropic exponent can advantageously lie between $n = 0$ (isobaric change of state) and $n = 1$ (isothermal change of state). This generalised *Diesel* process is discussed in more detail in section 7.5.

The coupling work of the ideal *Seiliger* process can also be calculated from the energy balance for the gas quantity transferred by the machine during a working cycle through the machine (cf. section 6.2.4). When flowing into the cylinder of the internal combustion engine, the gas performs the shift work (Figure 7.16)

$$W_{ein} = p_1 V_H .\qquad(7.71)$$

The isentropic compression of the gas from V_1 to V_2 gives the volume change work

$$W_{V12} = U_2 - U_1 = m\, c_v(T_2 - T_1) .\qquad(7.72)$$

Moreover, the isochoric change of state from 2 to 3 is described by

$$W_{V23} = 0 .$$

The volume change work for isobaric expansion from V_3 to V_4 and the volume change work for isentropic expansion from V_4 to V_5 are

$$W_{V34} = p_3(V_3 - V_4) = m\, R\,(T_3 - T_4) = m\,(c_p - c_v)(T_3 - T_4)\qquad(7.73)$$

$$W_{V45} = U_5 - U_4 = m\, c_v(T_5 - T_4) .\qquad(7.74)$$

In state 5 the outlet valve opens although the pressure p_5 is still higher than the pressure p_1 in the outlet pipe. The gas expands under pressure equalisation (section 3.2.3) from V_5 to V_6. The occurring volume change work is transformed into internal friction work. Thus, for the changes of state from 5 to 6 and from 6 to 1, $W_{V51} = 0$ applies as a substitute:

$$-W_{V56} = |W_{RI}|\qquad(7.75)$$

The gas volume $V_6 - V_5$ escapes from the cylinder by itself. Only the gas in the displacement volume under pressure p_1 has to be expelled:

$$W_{aus} = p_1 V_H\qquad(7.76)$$

If one forms the balance of the amounts of work supplied or removed during a working cycle supplemented by the coupling work, one obtains for the prime mover

$$(W_e)_{id} + |W_{ein}| - |W_{V12}| + |W_{V34}| + |W_{V45}| - |W_{aus}| = 0\qquad(7.77)$$

$$(W_e)_{id} = W_{aus} - W_{ein} + W_{V12} + W_{V34} + W_{V45}\qquad(7.78)$$

$$(W_e)_{id} = -m\,[c_v(T_1 - T_2) + c_p(T_4 - T_3) - c_v(T_4 - T_3) + c_v(T_4 - T_5)] .\qquad(7.79)$$

Example 7.2 For an ideal four-stroke engine, operating at a speed of 5500 rpm according to the *Otto* process, the influence of the compression ratio shall be investigated. The intake air has a temperature of $t_1 = 20\,°C$ at a pressure of $p_1 = 1$ bar. The heat supplied is 2.5 kJ. The specific heat capacities are assumed to be constant ($\kappa = 1{,}40$).

ε		6	7	8	9	10	11	12	13
V_1	litre	2.330	2.190	2.073	1.980	1.900	1.825	1.770	1.708
V_2	litre	0.388	0.313	0.259	0.220	0.190	0.166	0.148	0.131
T_2	K	600.3	638.5	673.5	706.0	736.4	765.0	792.1	817.9
t_2	°C	327.1	365.3	400.3	432.8	463.2	491.8	518.9	544.7
p_2	bar	12.29	15.25	18.38	21.67	25.12	28.70	32.42	36.27
m	g	2.763	2.597	2.463	2.349	2.254	2.168	2.094	2.028
T_3	K	1863	1981	2088	2190	2285	2371	2459	2535
t_3	°C	1590	1708	1815	1917	2012	2098	2185	2262
p_3	bar	38.06	47.31	56.98	67.17	77.94	88.99	100.2	112.5
η_{th}	%	51.16	54.08	56.47	58.48	60.19	61.68	62.99	64.16
$(W_e)_{id}$	kJ	1.281	1.354	1.412	1.463	1.508	1.542	1.578	1.604
$(P_e)_{id}$	kW	58.72	62.07	64.70	67.08	69.09	70.68	72.31	73.51

7.4.2 Joule Process

In the reversible *Joule* process according to Fig. 7.17, compression and expansion take place in ideal fluid flow machines, for which isentropic changes of state are assumed. In the following, an ideal gas with constant specific heat capacities is assumed as the working medium. With isobaric changes of state in the heat exchangers one obtains

$$(Q_{zu})_{rev} = (Q_{23})_{rev} = m\, c_p(T_3 - T_2) \tag{7.80}$$

$$|(Q_{ab})_{rev}| = |(Q_{41})_{rev}| = m\, c_p(T_4 - T_1)\,. \tag{7.81}$$

The coupling work is given by Table 7.2 Eqs. (1) and (3)

$$(W_e)_{id} = W_{Kreis} = -m\, c_p(T_1 + T_3 - T_2 - T_4)\,. \tag{7.82}$$

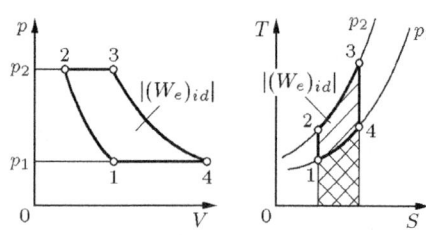

Figure 7.17 Joule process (hatchings as for Figure 7.10)
a) p, V diagram
b) T, S diagram

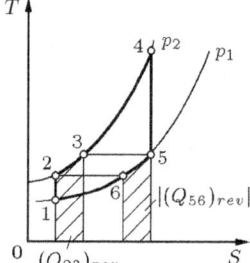

Figure 7.18 Joule process with in-process ideal heat transfer

The thermal efficiency according to Eq. (7.36) is

$$\eta_{th} = 1 - \frac{T_4 - T_1}{T_3 - T_2} = 1 - \frac{T_1\left(\frac{T_4}{T_1} - 1\right)}{T_2\left(\frac{T_3}{T_2} - 1\right)}\,. \tag{7.83}$$

For isentropic compression and expansion, the following applies in each case

$$\frac{T_2}{T_1} = \left(\frac{p_2}{p_1}\right)^{\frac{\kappa-1}{\kappa}} \tag{7.84}$$

$$\frac{T_3}{T_4} = \left(\frac{p_3}{p_4}\right)^{\frac{\kappa-1}{\kappa}} = \left(\frac{p_2}{p_1}\right)^{\frac{\kappa-1}{\kappa}}\,. \tag{7.85}$$

From these relations it follows

$$\frac{T_2}{T_1} = \frac{T_3}{T_4} \qquad \frac{T_4}{T_1} = \frac{T_3}{T_2}\,. \tag{7.86}$$

Eq. (7.83) for the thermal efficiency, with Eqs. (7.84) and (7.86), takes the following form:

$$\eta_{th} = 1 - \frac{T_1}{T_2} = 1 - \frac{T_4}{T_3} = 1 - \left(\frac{p_1}{p_2}\right)^{\frac{\kappa-1}{\kappa}} \tag{7.87}$$

The thermal efficiency of the *Joule* process can be improved if a part of the heat — that has to be reversibly removed — is not released to the environment, but is used in the process to supply heat.

This in-process heat transfer is carried out in an ideal counterflow heat exchanger (Fig. 7.18). Since the expressions $(Q_{zu})_{rev}$ and $(Q_{ab})_{rev}$ only represent the heat reversibly transferred externally to the process in connection with the heat source respectively transferred outside the process in connection with the heat sink, $(Q_{23})_{rev}$ and $(Q_{56})_{rev}$ are omitted in the formulation of $(Q_{zu})_{rev}$ and $(Q_{ab})_{rev}$:

$$(Q_{zu})_{rev} = (Q_{34})_{rev} = m\, c_p (T_4 - T_3) = m\, c_p\, T_3 \left(\frac{T_4}{T_3} - 1\right) \tag{7.88}$$

$$|(Q_{ab})_{rev}| = |(Q_{61})_{rev}| = m\, c_p (T_6 - T_1) = m\, c_p (T_2 - T_1) = m\, c_p\, T_1 \left(\frac{T_2}{T_1} - 1\right) \tag{7.89}$$

The coupling work is

$$(W_e)_{id} = W_{Kreis} = -[(Q_{zu})_{rev} - |(Q_{ab})_{rev}|] = -m\, c_p (T_1 + T_4 - T_3 - T_6) . \tag{7.90}$$

According to Eq. (7.36) one gets

$$\eta_{th} = 1 - \frac{T_1 \left(\frac{T_2}{T_1} - 1\right)}{T_3 \left(\frac{T_4}{T_3} - 1\right)} . \tag{7.91}$$

Analogous to Eqs. (7.86), the following applies in the case of ideal in-process heat transfer:

$$\frac{T_2}{T_1} = \frac{T_4}{T_5} = \frac{T_4}{T_3} . \tag{7.92}$$

By using the Eqs. (7.91), (7.92) and (7.84), the thermal efficiency of a *Joule* process with ideal in-process heat exchanger (without temperature gradient) is given by

$$\eta_{th} = 1 - \frac{T_1}{T_3} = 1 - \frac{T_1 T_2}{T_2 T_3} = 1 - \frac{T_2}{T_3}\left(\frac{p_1}{p_2}\right)^{\frac{\kappa-1}{\kappa}} . \tag{7.93}$$

The improvement of the efficiency compared to the *Joule* process without in-process heat transfer depends on the temperature ratio T_2/T_3.

Example 7.3 A *Joule* process for the generation of electrical energy and the provision of useful heat (cogeneration) with internal heat transfer under a temperature gradient (Δt) is operated with helium ($R = 2077.3$ J/(kg K)) (Fig. 7.19). The irreversibly adiabatic compression is carried out in two stages (1–2, 4–5) combined with isobaric intermediate cooling (2–H_Z–4) to the initial temperature of 25 °C. The irreversibly adiabatic expansion (8–9) takes place in one step to 460 °C.

initial pressure: 10.56 bar   intermediate pressure: 17.38 bar   final pressure: 28.76 bar

The isentropic efficiency for compression is 0.84 in both stages. The isobaric heating is achieved by the in-process heat transfer (5–7) and the subsequent heating to 753 °C (7–8). The isobaric cooling is performed by the in-process heat transfer (9–11) to 168 °C and the subsequent cooling (11–H_K–1) to the initial temperature of 25 °C.

The process serves by providing a coupling power of 50 MW to drive an electric generator, and in addition to supply a useful heat output of 53.5 MW. To provide the useful heat output, heat is given off by the cycle during intercooling (2–H_Z) and cooling (11-H_K).

The external friction work is assumed to be zero, the specific isobaric heat capacity of helium is kept constant at $c_p = 5.24$ kJ/(kg K).

a) What temperature is reached after low-pressure compression, which after high pressure compression?

b) To what temperature does the gas heat up through heat transfer within the process?

c) What is the isentropic efficiency of the turbine?

d) What is the mass flow rate in the plant?

e) What power does the turbine deliver, what do the compressors require?

f) To what equal temperature t_H does the gas cool down in the intercooler and cooler?

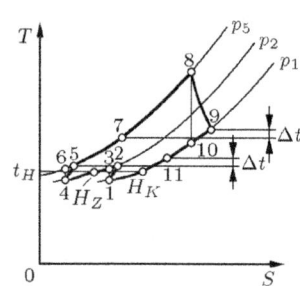

Figure 7.19 T, S diagram for example 7.3

a) $\kappa = 1.6568 \qquad T_3 = T_1 \left(\dfrac{p_2}{p_1}\right)^{\frac{\kappa-1}{\kappa}} = 298.15 \text{ K}\left(\dfrac{17.38}{10.56}\right)^{0.39641} = 363.26 \text{ K} \qquad t_3 = 90.11\,°C$

According to Table 6.5 Eq. (2) is $\quad t_2 = t_1 + \dfrac{t_3 - t_1}{\eta_{sV}} = 25\,°C + \dfrac{90.11\,°C - 25\,°C}{0.84} = 102.51\,°C$

$T_6 = T_4 \left(\dfrac{p_5}{p_2}\right)^{\frac{\kappa-1}{\kappa}} = 298.15 \text{ K}\left(\dfrac{28.76}{17.38}\right)^{0.39641} = 364.04 \text{ K} \qquad t_6 = 90.89\,°C$

$t_5 = t_4 + \dfrac{t_6 - t_4}{\eta_{sV}} = 25\,°C + \dfrac{90.89\,°C - 25\,°C}{0.84} = 103.44\,°C$

b) In-process heat transfer: $\quad \dot{Q}_{57} = \dot{m}\,c_p(t_7 - t_5) \qquad \dot{Q}_{9\,11} = \dot{m}\,c_p(t_{11} - t_9) \qquad \dot{Q}_{57} = -\dot{Q}_{9\,11}$

$t_7 = t_5 + t_9 - t_{11} = 103.44\,°C + 460\,°C - 168\,°C = 395.44\,°C$

c) $T_{10} = T_8\left(\dfrac{p_1}{p_5}\right)^{\frac{\kappa-1}{\kappa}} = 1026.15 \text{ K} \cdot \left(\dfrac{10.56}{28.76}\right)^{0.39641} = 689.80 \text{ K} \qquad t_{10} = 416.65\,°C$

According to Table 6.5 Eq. (4) is

$\eta_{sT} = \dfrac{t_8 - t_9}{t_8 - t_{10}} = \dfrac{753\,°C - 460\,°C}{753\,°C - 416.65\,°C} = 0.8711$

d) $\dot{Q}_{78} = \dot{m}\,c_p(t_8 - t_7) = \dot{Q}_{zu} \qquad -(\dot{Q}_{11\,1} + \dot{Q}_{24}) = \dot{m}\,c_p(t_{11} - t_1 + t_2 - t_4) = |\dot{Q}_{ab}|$

According to Table 7.2 Eq. (5) is with $|W_{RA}| = 0 \qquad P_e = |\dot{Q}_{ab}| - \dot{Q}_{zu}$

$P_e = \dot{m}\,c_p(t_{11} - t_1 + t_2 - t_4 - t_8 + t_7)$

$\dot{m} = \dfrac{-50\,000 \text{ kW}}{5.24 \text{ kJ/(kg K)} \cdot (168 - 25 + 102.51 - 25 - 753 + 395.44)\,°C} = 69.62 \text{ kg/s}$

e) According to Eq. (6.40) and Table 6.1 Eq. (4) is, with negligible change of the kinetic energy,

$P_{eT} = \dot{m}(h_9 - h_8) = \dot{m}\,c_p(t_9 - t_8) = 69.62 \text{ kg/s} \cdot 5.24 \text{ kJ/(kg K)} \cdot (-293\,°C) = -106.9 \text{ MW}$

According to Eq. (7.24) is $\qquad P_{eV} = P_e - P_{eT} = -50 \text{ MW} + 106.9 \text{ MW} = 56.9 \text{ MW}$

f) $\dot{Q}_H = \dot{m}\, c_p (t_H - t_{11} + t_H - t_2)$

$$2t_H = t_{11} + t_2 + \frac{\dot{Q}_H}{\dot{m}\, c_p} = 168\,°\mathrm{C} + 102.51\,°\mathrm{C} - \frac{53\,500\ \mathrm{kW}}{69.62\ \mathrm{kg/s} \cdot 5.24\ \mathrm{kJ/(kg\,K)}} = 2 \cdot 61.93\,°\mathrm{C}$$

7.4.3 Ericsson Process

An improvement of the *Joule* process with ideal heat exchanger leads to the *Ericsson*[7] process. A higher thermal efficiency can be achieved if, according to Fig. 7.20, the compression is carried out in stages with intermediate cooling and the expansion in stages with intermediate heating.

If one again assumes an ideal gas with constant isobaric heat capacity c_p and chooses for the compression and for the expansion i stages, then according to section 6.2.2 the

$$\text{pressure ratio per stage is} \quad \left(\frac{p_2}{p_1}\right)^{\frac{1}{i}}. \tag{7.94}$$

From this it follows for the temperatures T_2 and T_4

$$T_2 = T_1 \left(\frac{p_2}{p_1}\right)^{\frac{\kappa-1}{i\kappa}} \tag{7.95}$$

$$T_4 = \frac{T_3}{\left(\dfrac{p_2}{p_1}\right)^{\frac{\kappa-1}{i\kappa}}}. \tag{7.96}$$

The reversibly released heat $|(Q_{46})_{rev}|$ is used internally in the cycle as heat $(Q_{25})_{rev}$ for the reversible heating of the gas. This internally used heat is neither with the heat source nor with the heat sink in external connection. The heat, reversibly extracted from the heat source, is

$$(Q_{zu})_{rev} = i\, m\, c_p (T_3 - T_4)$$

$$= i\, m\, c_p\, T_3 \left[1 - \frac{1}{\left(\dfrac{p_2}{p_1}\right)^{\frac{\kappa-1}{i\kappa}}}\right]. \tag{7.97}$$

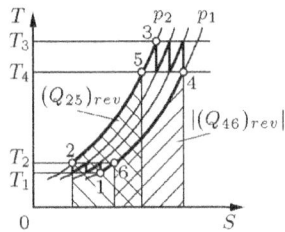

Figure 7.20 *Joule* process with in-process heat transfer and three-stage expansion and compression

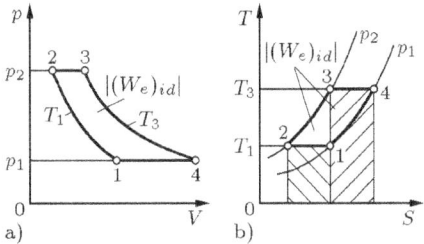

Figure 7.21 *Ericsson* process (hatching as in Fig. 7.10)

---

[7] *Johan Ericsson*, 1803 to 1889, Swedish engineer, was engaged in England and North America in development of locomotives, hot-air engines and shipbuilding.

To the heat sink the heat

$$|(Q_{ab})_{rev}| = i\,m\,c_p(T_2 - T_1) = i\,m\,c_p T_1\left[\left(\frac{p_2}{p_1}\right)^{\frac{\kappa-1}{i\kappa}} - 1\right] \tag{7.98}$$

is reversibly given off.

The coupling work is

$$(W_e)_{id} = W_{Kreis} = -i\,m\,c_p(T_1 + T_3 - T_2 - T_4). \tag{7.99}$$

Eqs. (7.97) and (7.98) with Eq. (7.36) give the thermal efficiency

$$\eta_{th} = 1 - \frac{T_1}{T_3}\left(\frac{p_2}{p_1}\right)^{\frac{\kappa-1}{i\kappa}}. \tag{7,100}$$

The thermal efficiency η_{th} increases with an increasing number of stages i.
With the boundary transition $i \to \infty$, the *Ericsson* process is obtained (Fig. 7.21). For the *Ericsson* process, Eqs. (7.95) to (7.100) are transformed into Eqs. (7.101) to (7.104), because one gets $T_2 = T_1$ and $T_4 = T_3$:

$$(Q_{zu})_{rev} = m\,R\,T_3 \ln\frac{p_2}{p_1} \tag{7.101}$$

$$|(Q_{ab})_{rev}| = m\,R\,T_1 \ln\frac{p_2}{p_1} \tag{7.102}$$

$$(W_e)_{id} = W_{Kreis} = -m\,R(T_3 - T_1)\ln\frac{p_2}{p_1} \tag{7.103}$$

$$\eta_{th} = 1 - \frac{T_1}{T_3} \tag{7.104}$$

The *Ericsson* process, like the *Carnot* process, achieves the highest possible thermal efficiency.

If a finite temperature gradient Δt is required for the in-process heat transfer according to Fig. 7.22, then the counterflow heat exchanger is not ideal, then the thermal efficiency becomes smaller. Therefore, Eqs. (7.101) and (7.102) are replaced by the equations

$$(Q_{zu})_{rev} = m\,R\,T_3 \ln\frac{p_2}{p_1} + m\,c_p\Delta t \tag{7.105}$$

$$|(Q_{ab})_{rev}| = m\,R\,T_1 \ln\frac{p_2}{p_1} + m\,c_p\Delta t. \tag{7.106}$$

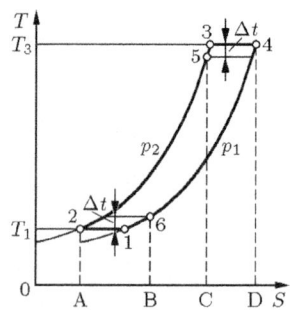

Figure 7.22 *Ericsson* process with non-ideal in-process heat transfer

$(Q_{zu})_{rev}$ is the area C534DC and $|(Q_{ab})_{rev}|$ is the area A216BA. Eq. (7.100) for the cyclic process work remains valid. Eq. (7.104) for the thermal efficiency is replaced by the following equations:

$$\eta_{th} = 1 - \frac{RT_1 \ln \frac{p_2}{p_1} + c_p \Delta t}{RT_3 \ln \frac{p_2}{p_1} + c_p \Delta t} = 1 - \frac{T_1 \ln \frac{p_2}{p_1} + \frac{\kappa}{\kappa-1} \Delta t}{T_3 \ln \frac{p_2}{p_1} + \frac{\kappa}{\kappa-1} \Delta t} \qquad (7.107)$$

Example 7.4 What improvement in efficiency is reached if a *Joule* process, a *Joule* process with ideal in-process heat transfer and an *Ericsson* process with ideal in-process heat transfer are compared with each other? For the processes, the pressures 1 bar and 4 bar, the extreme temperatures $20\,°C$ and $400\,°C$ as well as air as the working fluid ($\kappa = 1.4$) are to be taken as a basis.

a) *Joule* process: $\quad \eta_{th} = 1 - \left(\frac{p_1}{p_2}\right)^{\frac{\kappa-1}{\kappa}} = 0.3270 \qquad \kappa = 1.4$

b) *Joule* process with ideal in-process heat transfer: $\quad t_1 = 20\,°C \quad t_4 = 400\,°C$

$$T_2 = T_1 \left(\frac{p_2}{p_1}\right)^{\frac{\kappa-1}{\kappa}} = 435.62 \text{ K} = T_6 \qquad T_5 = T_4 \left(\frac{p_1}{p_2}\right)^{\frac{\kappa-1}{\kappa}} = 453.00 \text{ K} = T_3$$

$$\eta_{th} = 0.3529$$

c) *Ericsson* process with ideal in-process heat transfer: $\quad \eta_{th} = 0.5645$

7.4.4 Stirling Process

In the following, it is assumed that the right-hand cycle processes to be considered use an ideal gas with constant specific heat capacities as the working fluid. Both the reversible *Carnot* process discussed in section 7.3.3 and the reversible *Ericsson* process studied in section 7.4.3 are characterized by an isothermal change of state at a low temperature level ($T_1 = T_2$) and at a high temperature level ($T_3 = T_4$). For both isothermal changes of state, no temperature change heat appears: $Q_{T12} = Q_{T34} = 0$.

Furthermore, it is characteristic for the *Carnot* process that between the two — the isotherms connecting — isentropic changes of state, a complete in-process transfer of the process variables pressure change work $W_{p23} = -W_{p41}$ or volume change work $W_{V23} = -W_{V41}$ takes place, whereby the simultaneously transferred reversible substitute heat (entropy change heat) is zero: $(Q_{23})_{rev} = -(Q_{41})_{rev} = Q_{s23} = -Q_{s41} = 0$.

In contrast, the *Ericsson* process is characterised by the fact that between the two — the isotherms connecting — isobaric changes of state a complete in-process transfer of the process variables reversible substitute heat (entropy change heat) $(Q_{23})_{rev} = -(Q_{41})_{rev} = Q_{s23} = -Q_{s41}$ and volume change work $W_{V23} = -W_{V41}$ takes place, where the simultaneously transferred pressure change work is zero: $W_{p23} = -W_{p41} = 0$.

Another process — the reversible *Stirling*[8] process —, like the reversible *Carnot* process and the reversible *Ericsson* process, is characterised by an isothermal change of state at a low temperature level ($T_1 = T_2$) as well as at a high temperature level ($T_3 = T_4$). In

---

[8] *Robert Stirling*, 1790 to 1878, Scottish clergyman, was granted a patent in 1816 for the piston engine he had developed, using hot air as the working fluid.

this case, the two isotherms are connected by two isochoric changes of state, in which a complete in-process transfer of the process variables reversible substitute heat (entropy change heat) $Q_{23})_{rev} = -(Q_{41})_{rev} = Q_{s23} = -Q_{s41}$ and at the same time pressure change work $W_{p23} = -W_{p41}$ takes place, where the simultaneously transferred volume change work is zero: $W_{V23} = -W_{V41} = 0$, since $V_2 = V_3$ and $V_4 = V_1$. The *Stirling* process is shown in Fig. 7.23.

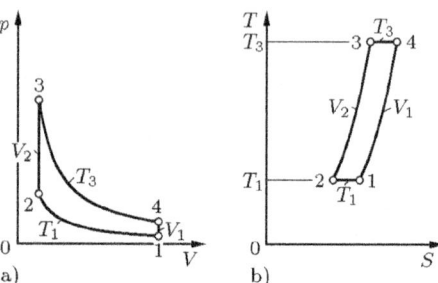

Figure 7.23 *Stirling* process
a) p, V diagram   b) T, S diagram

As in the *Carnot* process and in the *Ericsson* process, also in the *Stirling* process between the two — the isotherms connecting — similar changes of state a complete in-process transfer of the process variables reversible substitute heat (entropy change heat) as well as pressure change work or volume change work is assumed and the occuring temperature change heat is only an internal process variable.

Thus, it follows immediately that in the case of the two isochoric changes of state, no transmission of process variables in connection with devices outside of the *Stirling* process are possible.

Thus, only in the case of the two isothermal changes of state, process variables occur that are associated with equipment outside the Stirling process. These are a heat source, a heat sink and devices for work transfer.

With Eqs. (4.86) and (4.87) and Eq. (1) from Table 7.2, the following relationships hold:

$$(Q_{zu})_{rev} = (Q_{34})_{rev} = m\,R\,T_3 \ln \frac{V_4}{V_3} = m\,R\,T_3 \ln \frac{V_1}{V_2} \qquad (7.108)$$

$$|(Q_{ab})_{rev}| = |(Q_{12})_{rev}| = m\,R\,T_1 \ln \frac{V_1}{V_2} \qquad (7,109)$$

$$(W_e)_{id} = W_{Kreis} = -(Q_{zu})_{rev} + |(Q_{ab})_{rev}| = -m\,R\,(T_3 - T_1) \ln \frac{V_1}{V_2} \qquad (7.110)$$

The thermal efficiency is given by Eq. (7.36):

$$\eta_{th} = 1 - \frac{|(Q_{ab})_{rev}|}{(Q_{zu})_{rev}} = 1 - \frac{T_1}{T_3} \qquad (7.111)$$

The reversible *Stirling* process thus, like the reversible *Carnot* process and the reversible *Ericsson* process has the highest possible thermal efficiency.

7.4.5 Single-Polytropic *Carnot* Process

As a generalisation of these three processes, the simple-polytropic *Carnot* process is introduced here: Its characteristics are an isothermal change of state at the low temperature level T_1 and at the high temperature level T_3 as well as two polytropic changes of state between the low and the high temperature level.

The two polytropes are linked with each other in such a way that, with an ideal gas as the working fluid, the polytropic exponents n characterising them are equal equal in each case. Moreover, between the polytropes a complete in-process transfer of the process variables reversible substitute heat (entropy change heat), pressure change work and volume change work occur (Fig. 7.24) [33].

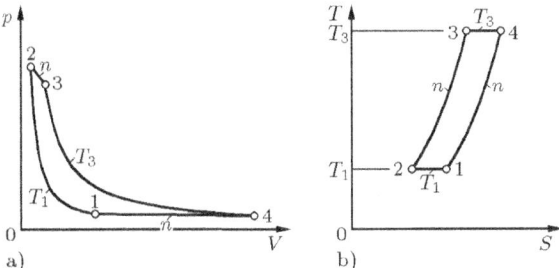

Figure 7.24 Single-polytropic *Carnot* process   a) p, V diagram   b) T, S diagram

The polytropic exponent is a variable with values $\infty \leq n \leq \infty$.
Special cases of the single-polytropic *Carnot* process result in:

- the *Carnot* process with the polytropic exponent $n = \kappa$
- the *Ericsson* process with the polytropic exponent $n = 0$
- the *Stirling* process with the polytropic exponent $n = \infty$

Under the preconditions mentioned, the thermal efficiency is

$$\eta_{th} = 1 - \frac{|(Q_{ab})_{rev}|}{(Q_{zu})_{rev}} = 1 - \frac{T_1}{T_3} \ . \tag{7.112}$$

A multi-polytropic *Carnot* process is also conceivable: From the single-polytropic *Carnot* process it differs in a way that between the low and the high temperature level not only one pair of polytropic changes of state, opposing each other, with equal polytropic exponents n are arranged, but several pairs of polytropes, opposing each other, are assumed between the low and the high temperature level.

The different pairs of polytropes are described by different polytropic exponents n. As each pair of polytropes opposing each other has polytropic exponents of equal size n within the same ranges between two temperature levels, between them a complete transfer of the sums of the process variables reversible substitute heat (entropy change heat) $\sum Q_{23i})_{rev} = -\sum (Q_{41i})_{rev} = \sum Q_{s23i} = -\sum Q_{s41i}$, pressure change work $\sum W_{p23i} = -\sum W_{p41i}$ and volume change work $\sum W_{V23i} = -\sum W_{V41i}$ takes place. This multi-polytropic *Carnot* process is not considered in more detail here.

7.4.6 Gas Expansion Process

The Gas expansion process as another reversible cyclic process consists of an isothermal compression from the initial pressure p_1 to the final pressure p_2 at the temperature level $T_1 = T_2$, an isobaric change of state from T_2 to the temperature T_3 at the pressure level $p_2 = p_3$, an isentropic expansion from the final pressure p_2 to the initial pressure p_1 with simultaneous decrease of the temperature T_3 to T_4 and an isobaric change of state to the temperature T_1 at the pressure level $p_4 = p_1$ (Fig. 7.25).

Part of the reversible supply of substitute heat (entropy change heat) during the isobaric change of state from the temperature T_2 to the temperature T_3, namely from T_2 to the intermediate temperature $T_Z = T_4$, is realised by the completely in-process and reversibly transferred substitute heat (entropy change heat)

$$(Q_{2Z})_{rev} = -(Q_{41})_{rev} = Q_{s2Z} = -Q_{s41}. \tag{7.113}$$

Therefore, for the isobaric change of state from temperature T_2 to temperature T_3, only from the intermediate temperature $T_Z = T_4$ to the temperature T_3 an additional reversible substitute heat (entropy change heat) $(Q_{Z3})_{rev} = m\, c_p\, (T_3 - T_4)$ is reversibly transferred from the outside. Thus, in the lower temperature range between T_1 and $T_Z = T_4$, the gas expansion process has the characteristics of the *Ericsson* process and in the upper temperature range between $T_Z = T_4$ and T_3 the characteristics of the *Joule* process [28], [29], [30].

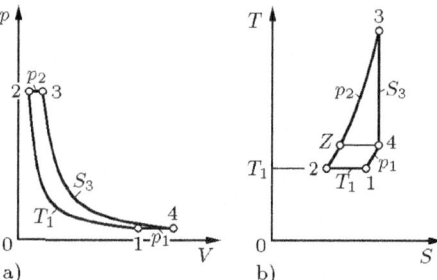

Figure 7.25 Gas expansion process
a) p, V diagram   b) T, S diagram

For the thermal efficiency, Eq. (7.35) holds:

$$\eta_{th} = -\frac{W_{Kreis}}{(Q_{zu})_{rev}} = -\frac{W_{p12} + W_{p34}}{(Q_{Z3})_{rev}} \tag{7.114}$$

Using Eqs. (4.86), (4.87), (4.108), (4.49) and (4.111), it follows:

$$\eta_{th} = -\frac{m\,RT_1 \ln \frac{p_2}{p_1} + m\,c_p\,(T_4 - T_3)}{m\,c_p\,(T_3 - T_4)} = 1 - \frac{m\,RT_1 \ln \frac{p_2}{p_1}}{m\,c_p\,(T_3 - T_4)} \tag{7.115}$$

From the total balance of all works and reversible substitute heats supplied from outside and released to outside, it follows:

$$W_{p12} + (Q_{12})_{rev} + (Q_{Z3})_{rev} + W_{p34} = 0 \tag{7.116}$$

As a consequence of Eq. (4.86) one gets:

$$W_{p34} = -(Q_{Z3})_{rev} = -m\,c_p\,(T_3 - T_4) \tag{7.117}$$

From Eqs. (4.105) and (4.111) it follows:

$$W_{p34} = -\frac{\kappa}{\kappa-1} m R T_3 \left(1 - \left(\frac{p_1}{p_2}\right)^{\frac{\kappa-1}{\kappa}}\right) \qquad (7.118)$$

Thus, Eq. (7.115) becomes

$$\eta_{th} = 1 - \frac{m R T_1 \ln \frac{p_2}{p_1}}{\frac{\kappa}{\kappa-1} m R T_3 \left(1 - \left(\frac{p_1}{p_2}\right)^{\frac{\kappa-1}{\kappa}}\right)} = 1 - \frac{\kappa-1}{\kappa} \frac{T_1 \ln \frac{p_2}{p_1}}{T_3 \left(1 - \left(\frac{p_1}{p_2}\right)^{\frac{\kappa-1}{\kappa}}\right)} \qquad (7.119)$$

7.4.7 Clausius-Rankine Process

Steam power plants are generally based on the reversible *Clausius-Rankine* process.[9] In this right-hand cycle, as in the *Joule* process, power and working machines as well as heat exchangers are connected with each other. Water in liquid state and vapor state (steam) is usually used as working fluid. Water in the state of boiling liquid with the pressure p_1 and the temperature T_1, in the feed water pump (a) (Fig. 7.26 a) is brought to the steam generator pressure p_2 by an isentropic compression, which is at the same time an approximately isochoric compression.

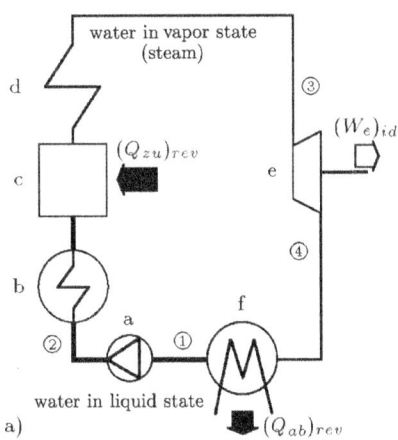

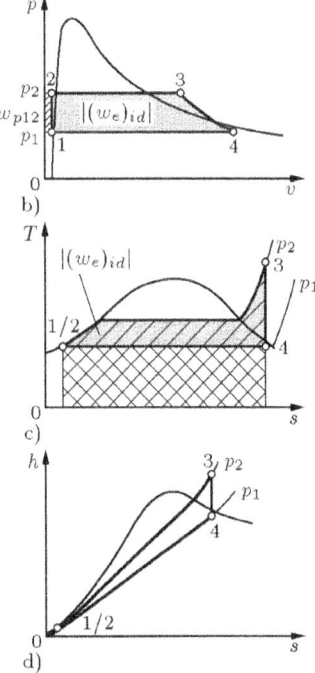

Figure 7.26 *Clausius-Rankine* process

a) circuit diagram b) p, v diagram
c) T, s diagram d) h, s diagram

In the steam generator, the working fluid is isobarically heated, vaporised and transformed into superheated steam. In most cases, the steam generator consists of the

---
[9] *William John Macquorn Rankine*, 1820 to 1872, Scottish engineer, made important contributions to the theory of the steam engine.

preheater (b), which heats the condensate to boiling temperature, the vaporiser (c) and the superheater (d) for superheating the saturated steam formed in the vaporiser. The superheated steam with the pressure $p_2 = p_3$ and the temperature T_3 is fed to the steam turbine (e) in which the steam is expanded isentropically to the condenser pressure $p_4 = p_1$. The steam can leave the turbine as superheated steam, saturated steam or wet steam (where $x > 0{,}85$). In the condenser (f), the expanded steam is isobarically liquefied again to the initial state by cooling water.

Like the *Joule* process, the comparison process consists of isobars and isentropes. Fig. 7.26 b to d shows the mapping of the process in a p, v diagram, a T, s diagram and a h, s diagram. For the heat input on the isobar from point 2 to point 3, the following applies

$$(Q_{zu})_{rev} = m(h_3 - h_2) \ , \tag{7.120}$$

for the isobaric heat release in the condenser

$$|(Q_{ab})_{rev}| = m(h_4 - h_1) \ . \tag{7.121}$$

The coupling work is according to Table 7.2 Eqs. (1) and (3)

$$(W_e)_{id} = W_{Kreis} = -m(h_3 + h_1 - h_4 - h_2) \ . \tag{7.122}$$

As thermal efficiency one gets

$$\eta_{th} = 1 - \frac{h_4 - h_1}{h_3 - h_2} \ . \tag{7.123}$$

According to Table 6.1 Eq. (19) as well as Eq. (6.41), the coupling work of the turbine becomes, when the change in kinetic energy is neglected,

$$(W_{eT})_{id} = m(h_4 - h_3) \ . \tag{7.124}$$

The coupling work of the feed water pump is

$$(W_{eV})_{id} = m(h_2 - h_1) \ . \tag{7.125}$$

The enthalpy change during the pressure increase of the water in the pump is small compared to all other enthalpy changes. If one sets $h_2 = h_1$ approximately, this results in

$$(W_e)_{id} = W_{Kreis} = -m(h_3 - h_4) \ . \tag{7.126}$$

According to Eq. (7.124), this is the coupling work of the ideal turbine

$$(W_e)_{id} = W_{Kreis} = (W_{eT})_{id} \ . \tag{7.127}$$

The coupling work of the compressor (in this case the feed water pump) is neglected. In the p, v diagram of Figure 7.26 b it can be seen that the pressure change work of the pump is very small. For an ideal machine it is equal to the coupling work and can be calculated according to Table 6.1 Eqs. (19) and (20).

For the thermal efficiency, one obtains with $h_2 = h_1$

$$\eta_{th} = \frac{h_3 - h_4}{h_3 - h_1} \ . \tag{7.128}$$

If one neglects in the thermal efficiency additionally the specific enthalpy h_1 which is small compared to h_3, then the following remains as an approximation

$$\eta_{th} = 1 - \frac{h_4}{h_3} \ . \tag{7.129}$$

Example 7.5 In a waste-to-energy plant capable of providing electricity and district heating, 40 t/h of steam of 100 bar and 400 °C is produced and expanded to 6 bar in a turbine. After complete condensation in a heat exchanger, the condensate is pumped back into the steam generator by the feed water pump (Fig. 7.27).

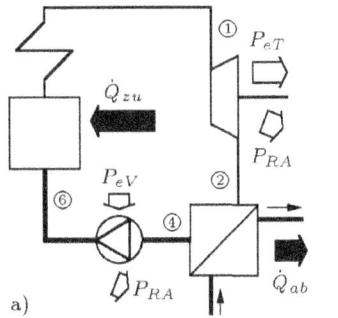
a)

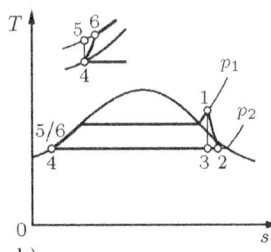
b)

Figure 7.27 a) and b) To Example 7.5

The heat released during condensation is used to generate 230 t/h of hot water with an inlet temperature of 60 °C and an outlet temperature of 140 °C . The change in kinetic energy in the turbine and in the feed water pump is negligible.

a) What is the thermal efficiency of the *Clausius-Rankine* process?

b) What is the heat output of the process in the heat exchanger?

c) What is the internal isentropic efficiency of the turbine?

d) What is the power output at a mechanical efficiency of 0.96?

e) What drive power does the pump require with an internal isentropic efficiency of 0.86 and a mechanical efficiency of 0.94?

f) What heat is to be supplied to the process?

g) What is the thermal efficiency of the real process?

a) In the *Clausius-Rankine* process, expansion and compression occur isentropically.

$h_1 = 3097.37 \text{ kJ/kg}$[10] $s_1 = 6.2139 \text{ kJ/(kg K)} = s_3$ Eq. (5.41) : $s_3 = s_3' + x_3 \dfrac{r_3}{T_3}$

$x_3 = \dfrac{s_3 - s_3'}{r_3} T_3 = \dfrac{6.2139 \text{ kJ/(kg K)} - 1.9311 \text{ kJ/(kg K)}}{2085.64 \text{ kJ/kg}} \cdot 431.98 \text{ K} = 0.8871$

$h_3 = h_3' + x_3 \, r_3 = 670.50 \text{ kJ/kg} + 0.8871 \cdot 2085.64 \text{ kJ/kg} = 2520.67 \text{ kJ/kg}$

$h_4 = h_4' = 670.50 \text{ kJ/kg}$ $s_4 = s_5 = 1.9311 \text{ kJ/(kg K)}$ $h_5 = 680.83 \text{ kJ/kg}$

$\eta_{th} = 1 - \dfrac{h_3 - h_4}{h_1 - h_5} = 1 - \dfrac{(2520.67 - 670.50) \text{ kJ/kg}}{(3097.37 - 680.83) \text{ kJ/kg}} = 0.2344$

---

[10] For numerical values see [128] or Tab. A.9b and Tab. A.10 in the appendix.

b) $|\dot{Q}_{ab}| = \dot{m}_W\, c_W\, \Delta t_W = 230 \cdot 10^3$ kg/h $\cdot$ 4.215 kJ/(kg K) $\cdot$ 80 K = 21.543 MW

c) $|\dot{Q}_{ab}| = \dot{m}(h_2 - h_4)$ $h_2 = \dfrac{|\dot{Q}_{ab}|}{\dot{m}} + h_4 = \dfrac{21.543\ \text{MW}}{40 \cdot 10^3\ \text{kg/h}} + 670.50$ kJ/kg $= 2609.40$ kJ/kg

Table 6.5 Eq. (3): $\eta_{isT} = \dfrac{h_2 - h_1}{h_3 - h_1} = \dfrac{(2609.40 - 3097.37)\ \text{kJ/kg}}{(2520.67 - 3097.37)\text{kJ/kg}} = 0.8461$

d) $P_{eT} = \eta_{mT}\, \dot{m}(h_2 - h_1) = 0{,}96 \cdot 40 \cdot 10^3$ kg/h $\cdot (2609.40 - 3097.37)$ kJ/kg $= -5.205$ MW

e) $P_{eV} = \dfrac{\dot{m}(h_5 - h_4)}{\eta_{isV}\, \eta_{mV}} = \dfrac{40 \cdot 10^3\ \text{kg/h} \cdot (680.83 - 670.50)\ \text{kJ/kg}}{0.86 \cdot 0.94} = 142$ kW

f) Table 6.5 Eq. (1): $\eta_{isV} = \dfrac{h_5 - h_4}{h_6 - h_4}$ $h_6 = h_4 + \dfrac{h_5 - h_4}{\eta_{isV}}$

$h_6 = 670.50\ \text{kJ/kg} + \dfrac{(680.83 - 670.50)\ \text{kJ/kg}}{0.86} = 682.51$ kJ/kg

$\dot{Q}_{zu} = \dot{m}(h_1 - h_6) = 40 \cdot 10^3$ kg/h $\cdot (3097.37 - 682.51)$ kJ/kg $= 26.832$ MW

g) Table 6.3 Eq. (2): $|P_{RA\,T}| = P_{eT} - \dfrac{P_{eT}}{\eta_{mT}} = 0.217$ MW

Table 6.3 Eq. (1): $|P_{RA\,V}| = P_{eV} - \eta_{mV}\, P_{eV} = 0.009$ MW

$|P_{RA}| = |P_{RA\,T}| + |P_{RA\,V}| = 0.217$ MW $+ 0.009$ MW $= 0.226$ MW

Eq. (7.34): $\eta_{th} = 1 - \dfrac{|\dot{Q}_{ab}| + |P_{RA}|}{\dot{Q}_{zu}} = 1 - \dfrac{(21.543 + 0.226)\ \text{MW}}{26.832\ \text{MW}} = 0.1887$

Eq. (7.128): $\eta_{th} = \dfrac{h_1 - h_2}{h_1 - h_4} = \dfrac{(3097.37 - 2609.40)\ \text{kJ/kg}}{(3097.37 - 670.50)\ \text{kJ/kg}} = 0.2011$

Eq. (7.129): $\eta_{th} = 1 - \dfrac{h_2}{h_1} = 1 - \dfrac{2609.40\ \text{kJ/kg}}{3097.37\ \text{kJ/kg}} = 0.1575$

7.5 Comparative Evaluation of Right-Hand Cyclic Processes

Considerable success has been achieved in the further development of thermal machines in recent decades. In order to increase thermal efficiencies, pressures and temperatures have been raised, the possibilities of different types of thermodynamic cycles and the combination of different cycles have been optimised.

Examples of this are

- increased live steam temperatures and pressures in steam turbine processes,

- the application of sequential combustion in gas turbine processes,

- the realisation of controlled high-pressure injection of fuel in combustion engine processes,

- the optimisation of the combination of gas turbine and steam turbine (combined cycle) processes, of combustion engine processes with steam turbine processes, and combustion engine processes with expansion turbines and compressors (e.g. exhaust gas turbochargers).

With regard to further improvement possibilities, a comparative evaluation of the most important ideal thermodynamic cycles, underlying the real processes, seems of special interest.

For various types of real machine processes, the machine-specific effort is very different. In addition, mechanical and thermal irreversibilities, which lead to a reduction in the intended technical benefit, have different degrees of importance.

In the following, with the help of five new defined thermodynamic evaluation variables — two mechanical effort ratios, two thermal effort ratios and their sum — a comparative evaluation of the most important reversible cyclic processes will be possible.

7.5.1 Process Variables and Cyclic Processes

At the beginning of section 7, the usefulness of the four process variables, used in the following as specific variables, was pointed out:

specific volume change work

$$w_{V12} = -\int_1^2 p\,dv \tag{7.130}$$

specific pressure change work

$$w_{p12} = \int_1^2 v\,dp \tag{7.131}$$

specific reversible substitute heat (specific entropy change heat)

$$(q_{12})_{rev} = q_{s12} = \int_1^2 T\,ds \tag{7.132}$$

specific temperature change heat

$$q_{T12} = \int_1^2 s\,dT \tag{7.133}$$

These can be represented as areas in a p, v diagram and in a T, s diagram respectively (cf. Figs. 7.1 and 7.2).

In the following, statements are made for right-hand cyclic processes — i.e. processes for the production of work. Equivalent statements can also be made for left-hand cyclic processes — i.e. refrigeration or heat pump processes.

In order to assess the efficiency of right-hand ideal cyclic processes, the thermal efficiency as the quotient of the absolute value of the specific cycle work obtained

$$|w_{Kreis}| = |\sum w_{pij}| \tag{7.134}$$

and the sum of the specific entropy change heats supplied from outside to the cyclic process at higher temperatures

$$\sum q_{zuij} = \sum q_{szuij} \qquad (7.135)$$

is of interest:

$$\eta_{th} = \frac{|\sum w_{pij}|}{\sum q_{zuij}} = \frac{|w_{Kreis}|}{\sum q_{szuij}} \qquad (7.136)$$

In the case of ideal gases and real substances as working fluids, η_{th} depends on the pressures p or pressure ratios reached as well as on the temperatures T or temperature ratios reached (cf. sections 7.3.3 and 7.4).

Here, the position of the process in a p, v diagram is particularly important with regard to the ordinate p, or in a T, s diagram with respect to the ordinate T. With regard to achieving the highest possible thermal efficiency, for cyclic processes the following should be aimed for:

- starting from a low initial pressure (such as the ambient pressure) in the course of the cyclic process, a high pressure p and / or

- starting from a low initial temperature (e.g. the ambient temperature) in the course of the cyclic process a high temperature T

For this purpose, a

- supply of specific volume change work w_{Vij} corresponding with the

- supply of specific pressure change work w_{pij}

as well as a

- supply of specific entropy change heat q_{sij} corresponding with an

- altering of specific temperature change heat q_{Tij}

is required.

7.5.2 Mechanical Effort Ratios and Thermal Effort Ratios

The displacement of a reversible cyclic process with an ideal gas as the working fluid in a p, v diagram and in a T, s diagram along lines of constant temperature does not affect the thermal efficiency. Such a shift is shown in the example of the *Stirling* process in Fig. 7.28. Its thermal efficiency depends — like that of the *Carnot*-, the *Ericsson*- and the single-polytropic *Carnot* process — only on the respective characteristic temperature ratio T_3/T_1 (cf. Eqs. (7.111), (7.41), (7.104) and (7.112)).

Here, the area enclosed by the cyclic process corresponding to the absolute value of the specific cyclic process work $|w_{Kreis}|$, is the same in the considered cases 1234 and 1'2'3'4'. On the other hand, it can be seen from the p, v diagram of Fig. 7.28 that in the case of the cyclic process shifted to the left and using a higher compressed working fluid 1'2'3'4', in comparison to the original process 1234, which is operated with a lower

compressed working fluid, a significantly smaller machine is required. As far as e.g. a piston engine is used, in the chosen example for the cyclic process 1234 more than ten times the cylinder volume is necessary.

If the cylinder volume V_a and the initial pressure p_a are given, the utility of the machine can be improved, if the working fluid is compressed much more than initially specified, i.e. to greater specific volume ratios v_a/v_b or corresponding pressure ratios p_b/p_a: e.g. starting from the cycle 1234 to such an extent that the cyclic process $1'2'3'4'$ is also enclosed and an extended cyclic process $12'3'4$ with a clearly enlarged working area and thus a clearly enlarged specific cyclic process work $|w_{Kreis}|$ is achieved. However, the thermal efficiency does not change.

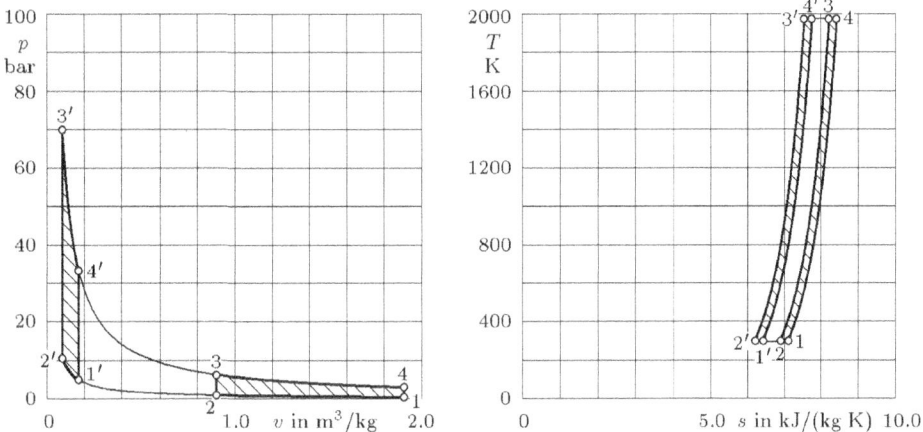

Figure 7.28 Displacement of a *Stirling* process 1234 to $1'2'3'4'$ with respectively equal absolute value of the specific cyclic process work $|w_{Kreis}|$ (enclosed areas) in a p, v diagram and in a T, s diagram along isotherms

Also for processes of heat transfer, e.g. in the *Stirling* process as a closed process with the aid of heat-transferring surfaces, the equipment required is considerably lower, if one can work with a higher compressed instead of a lower compressed working fluid. In addition, in the case of a real, irreversible cyclic process, it is possible to realise a process with a more highly compressed working fluid and with higher pressure ratios p_b/p_a in order to achieve a desired specific cyclic process work, significantly lower irreversibilities are to be expected than in the case of a cyclic process with less compressed working fluid and with lower pressure ratios p_b/p_a.

The designation "lower compressed working fluid" describes a lower molar density, i.e. a lower molar quantity per volume unit; it corresponds to a larger volume and a higher entropy per molar quantity, specified as absolute entropy, or a higher specific entropy, specified as absolute entropy.

Obviously, the efficiency alone is not sufficient to evaluate the thermodynamic quality of a cyclic process. If, therefore, in addition to the thermal efficiency, additional parameters are to be obtained for the evaluation of cyclic processes, it is obvious that with ideal gases as the working medium, in addition to a suitable temperature ratio T_d/T_c, also an absolute entropy and a characteristic specific volume ratio v_a/v_b or a corresponding pressure ratio p_b/p_a are to be included in the consideration.

The above facts can also be expressed in the following way: For given volumetric dimensions of plant components such as piston and fluid flow machines as well as heat-transferring surfaces, the gain in cyclic process work can be all the greater, the more highly compressed the working fluid is and the higher the characteristic pressure ratios are.

The friction work, e.g. between piston and cylinder, depends strongly on the volumetric dimensions of a piston machine. The higher the working fluid can be compressed, the less the friction work tends to have a reducing effect on the gain of cyclic process work.

For given dimensions of a heat-transferring surface and a given temperature difference the following can be assumed: If the density of the working fluid is high, this has a positive effect on the amount of transferred heat.

A reversible cyclic process with given thermal efficiency and given initial state T_1, v_1, p_1 and s_1 needs in its practical implementation — related to the gain of specific cyclic process work — the less machine effort and is the less sensitive to irreversibilities, the more it can be extended in a p,v diagram along isotherms in the direction of smaller specific volumes and thus in the direction of high pressures towards the ordinate axis, and the more it can be extended in a T,s diagram along isotherms to the left in the direction of smaller specific entropies towards the ordinate axis T (Fig. 7.28).

In the example of Fig. 7.28, the extended process $12'3'4$ has, compared to the original process 1234, a significantly larger specific working area and thus a significantly larger absolute amount of specific cyclic process work $|w_{Kreis}|$. This increased benefit is to be seen against the background of the effort expended — in this case the absolute amounts of the specific pressure change works $|w_{pij}|$ required (cf. the p,v diagram of Fig. 7.28).

Therefore it is obvious to calculate for each individual change of state of the process the value of the specific pressure change work w_{pij}, to form the absolute value and to sum up the respective absolute values.

The increased benefit of the extended process $12'3'4$ compared to the original process 1234 can also be determined on the basis of the effort expended with regard to the absolute values of the specific temperature change heats $|q_{Tij}|$ (cf. the T,s diagram of Figure 7.28).

Therefore, it is also obvious to calculate for each individual change of state of the process the value of the specific temperature change heat q_{Tij}, to form the absolute value of this and then to sum up the respective absolute values.

Thus, the absolute value of the specific cyclic process work $|w_{Kreis}|$ of a reversible cyclic process is to set

- in relation to the sum of all absolute amounts of the specific pressure change works of the cyclic process $\sum |w_{pij}|$ and

- in relation to the sum of all absolute amounts of the specific temperature change heats of the cyclic process $\sum |q_{Tij}|$.

Thus, taking into account Eq. (7.13), the mechanical effort ratio

$$a_{wp} = \frac{\sum |w_{pij}|}{|w_{Kreis}|} = \frac{\sum |w_{pij}|}{|\oint dw_p|} = \frac{\sum |w_{pij}|}{|\sum w_{pij}|} \quad (7.137)$$

can be defined. Also the thermal effort ratio

$$a_{qT} = \frac{\sum |q_{Tij}|}{|w_{Kreis}|} = \frac{\sum |q_{Tij}|}{|\oint dq_T|} = \frac{\sum |q_{Tij}|}{|\sum q_{Tij}|} \quad (7.138)$$

can be defined [33], [34].

The mechanical effort ratio a_{wp} indicates how large the multiple is that has to be converted in a cyclic process in total in terms of the absolute values of the specific pressure change work $\sum |w_{pij}|$ in order to obtain the desired specific cycle process work $|w_{Kreis}|$. The thermal effort ratio a_{qT} indicates how large the multiple is that is required in a cyclic process in total in terms of the absolute values of the specific temperature change heat $\sum |q_{Tij}|$ in order to obtain the desired specific cycle work $|w_{Kreis}|$.

a_{wp} and a_{qT} can take values between the limits 1 and ∞. The better corresponding values for a given reversible cyclic process approximate the numerical value 1, the more limited is the machine effort, and the less sensitive this cyclic process is to irreversibilities in its technical implementation.

High values, on the other hand, characterise the respective cyclic process as less favourable. If the reciprocals $1/a_{wp}$ and $1/a_{qT}$ are formed, the achievable values lie between 0 and 1. Values close to 1 indicate favourable conditions.

A first approach to capture the problem addressed here is the definition of the "mean pressure" p_m. This is the ratio of the area of the cyclic process in the p, V diagram to the diagram base line, i.e. to the maximum volume difference:

$$p_m = \frac{\text{area of the cyclic process in a } p, V \text{ diagram}}{\Delta V \text{ line in a } p, V \text{ diagram}} \quad (7.139)$$

However, the mean pressure p_m used, for example, in combustion engines, does not fully represent the thermodynamic nature of the respective cyclic process.

Accordingly, two further effort ratios can be defined [34], for which the absolute value of the specific cyclic process work $|w_{Kreis}|$ of a reversible cycle process

- is related to the sum of all absolute values of the specific volume change works of the cycle process $\sum |w_{Vij}|$ and

- in relation to the sum of all absolute values of the specific entropy change heats $\sum |q_{sij}|$.

Thus, taking into account Eq. (7.13), the following take place:

$$a_{wv} = \frac{\sum |w_{Vij}|}{|w_{Kreis}|} = \frac{\sum |w_{Vij}|}{|\oint dw_V|} = \frac{\sum |w_{Vij}|}{|\sum w_{Vij}|} \quad (7.140)$$

7.5 Comparative Evaluation of Right-Hand Cyclic Processes

$$a_{qs} = \frac{\sum |q_{sij}|}{|w_{Kreis}|} = \frac{\sum |q_{sij}|}{|\oint dq_s|} = \frac{\sum |q_{sij}|}{|w_{Kreis}|} \qquad (7.141)$$

According to the defining equation (7.141) for a_{qs}, the reciprocal $1/a_{qs}$ has a proximity to the thermal efficiency $\eta_{th} = |w_{Kreis}|/(q_{zu})_{rev} = |w_{Kreis}|/q_{s\,zu}$ (cf. Eq. (7.35)). A merit of the definitions made is that, for example, according to Eq. (5.223)

$$\int_i^j dg = \int_i^j v\,dp - \int_i^j s\,dT = w_{pij} - q_{Tij}, \qquad (7.142)$$

numerical values of a_{wp} and a_{qT} can be directly compared with each other, since in their numerator only the absolute values of the directly comparable quantities w_{pij} and q_{Tij} are contained and their denominator in each case is equal [33], [34].

In general, a direct comparability of the numerical values of all effort ratios a_{wp}, a_{wv}, a_{qT} and a_{qs} is given; this can be recognised analogous to Eq. (5.222) and the two Eqs.(3.77) from

$$\int_i^j df = -\int_i^j p\,dv - \int_i^j s\,dT \qquad (7.143)$$

for a_{wv} and a_{qT},

$$\int_i^j dh = \int_i^j v\,dp + \int_i^j T\,ds \qquad (7.144)$$

for a_{wp} and a_{qs}, and

$$\int_i^j du = -\int_i^j p\,dv + \int_i^j T\,ds \qquad (7.145)$$

for a_{wv} and a_{qs}.

Through this, the total evaluation variable of a cyclic process as the sum of

$$a_g = a_{wp} + a_{wv} + a_{qT} + a_{qs} \qquad (7.146)$$

is suggested. a_g is named total effort ratio. In this way, the position of the respective cycle in a p, v diagram and in a T, s diagram is clearly recorded via the effort ratios.

7.5.3 Evaluation Criteria For Important Thermodynamic Cyclic Processes

7.5.3.1 General Thermodynamic Relations

In order to gain a comparative evaluation of technically important cyclic processes, the single-polytropic *Carnot* process, the *Carnot* process, the *Ericsson* process, the *Stirling* process, the *Joule* process, the *Otto* process, the *Diesel* process, the generalised *Diesel* process, the Gas expansion process and the *Clausius-Rankine* process are investigated. Except for the latter process, ideal gases are assumed as working fluids in all processes. Table 7.4 lists the individual changes of state which characterise the mentioned

reversible cyclic processes. Nothing is said about the type of machine technology to be selected.

Table 7.4 Changes of state of reversible right-hand cyclic processes
a) with an ideal gas as working fluid
b) with a real fluid as working fluid

a)

Change of state	Single-polytropic Carnot process	Carnot process	Ericsson process	Stirling process	Joule process	Otto process	Diesel process	Generalised Diesel process	Gas expansion process
12	Isothermal compression	Isothermal compression	Isothermal compression	Isothermal compression	Isentropic compression	Isentropic compression	Isentropic compression	Isentropic compression	Isothermal compression
23	Polytropic change of state	Isentropic compression	Isobaric heat input	Isochoric heat input	Isobaric heat input	Isochoric heat input	Isobaric heat input	Polytropic heat input	Isobaric heat input
34	Isothermal expansion	Isothermal expansion	Isothermal expansion	Isothermal expansion	Isentropic expansion	Isentropic expansion	Isentropic expansion	Isentropic expansion	Isentropic expansion
41	Polytropic change of state	Isentropic expansion	Isobaric heat output	Isochoric heat output	Isobaric heat output	Isochoric heat output	Isochoric heat output	Isochoric heat output	Isobaric heat output

b)

change of state	Clausius-Rankine process
12	Isentropic compression
23	Isobaric heat input
34	Isentropic expansion
41	Isobaric heat output

Tables 7.5 to 7.14 (cf. [33]) give the respective thermodynamic relationships for the individual processes, which apply for the thermal efficiency η_{th}, the exergetic efficiency ζ (cf. section 8 Eq. (8.135)), the mechanical effort ratios a_{wp} and a_{wv} as well as the thermal effort ratios a_{qT} and a_{qs}. The derivation of these relations for η_{th}, ζ, a_{wp}, a_{wv}, a_{qT} and a_{qs} is omitted here due to lack of space.

In the case of ideal gases as working fluids, constant specific isobaric and isochoric heat capacities c_p and c_v and thus also a constant isentropic exponent κ are assumed. To be considered as independent variables are the pressure p_1 and the temperature T_1 of the initial state 1, the pressure p_2 of the state 2 as well as the highest cycle temperature T_3 of state 3. If instead of the pressure p_2, the highest cyclic process pressure p_3 is chosen as an independent variable, in Tables 7.5 to 7.14 also corresponding auxiliary equations for the calculation of p_2 from p_3 are listed.

The combination of several reversible cyclic processes — such as the *Joule* process with a downstream *Clausius Rankine* process and its practical implementation as a combined gas and steam turbine (CCGT) process — can also be evaluated according to the evaluation criteria η_{th}, ζ, a_{wp}, a_{wv}, a_{qT}, a_{qs} and a_g; however, this will not be discussed further here.

7.5.3.2 Examples

In order to demonstrate the usefulness of the evaluation criteria η_{th}, ζ, a_{wp}, a_{wv}, a_{qT}, a_{qs} and a_g, in the following (Figs. 7.29 to 7.36 and Table 7.15), all the above mentioned cyclic processes which use ideal gases as the working fluid, are compared with each other [33]. In each case, the same working fluid will be assumed[11]. In addition, the same initial state 1 is taken as a basis, and initially the same state 3 is fictitiously selected in each case, at which the highest cyclic process pressure p_3 occurs at the highest cyclic process temperature T_3. From a materials engineering point of view, p_3 and T_3 are important limiting factors for the design of the respective cyclic process.

Table 7.5 Single-polytropic *Carnot* process: Thermodynamic relations for the thermal efficiency η_{th}, the exergetic efficiency ζ, the mechanical effort ratios a_{wp} and a_{wv} as well as the thermal effort ratios a_{qT} and a_{qs}

$$p_2 = p_3 \left(\frac{T_1}{T_3}\right)^{\frac{n}{n-1}} \quad (1) \qquad \eta_{th} = \frac{T_3 - T_1}{T_3} \quad (2) \qquad \zeta = \frac{T_3 - T_1}{T_3 - T_b} \quad (3)$$

$$a_{wp} = \frac{T_3 + T_1}{T_3 - T_1} + \frac{2|n|}{|n-1|\ln\frac{p_2}{p_1}} \quad (4) \qquad a_{wv} = \frac{T_3 + T_1}{T_3 - T_1} + \frac{2}{|n-1|\ln\frac{p_2}{p_1}} \quad (5)$$

$$a_{qT} = \frac{2s_1}{R \ln\frac{p_2}{p_1}} - 1 + \frac{2|\kappa - n|}{|n-1|(\kappa-1)\ln\frac{p_2}{p_1}} \left(\frac{T_3}{T_3 - T_1} \ln\frac{T_3}{T_1} - 1\right) \quad (6)$$

$$a_{qs} = \frac{T_3 + T_1}{T_3 - T_1} + \frac{2|\kappa - n|}{|n-1|(\kappa-1)\ln\frac{p_2}{p_1}} \quad (7) \qquad a_{wp} = \kappa\, a_{wv} - (\kappa - 1)\, a_{qs} \quad \text{(for } \kappa - n \geq 0\text{)} \quad (8)$$

Table 7.6 *Carnot* process: Thermodynamic relations for the thermal efficiency η_{th}, the exergetic efficiency ζ, the mechanical effort ratios a_{wp} and a_{wv} as well as the thermal effort ratios a_{qT} and a_{qs}

$$p_2 = p_3 \left(\frac{T_1}{T_3}\right)^{\frac{\kappa}{\kappa-1}} \quad (1) \qquad \eta_{th} = \frac{T_3 - T_1}{T_3} \quad (2) \qquad \zeta = \frac{T_3 - T_1}{T_3 - T_b} \quad (3)$$

$$a_{wp} = \frac{T_3 + T_1}{T_3 - T_1} + \frac{2\kappa}{(\kappa - 1)\ln\frac{p_2}{p_1}} \quad (4) \qquad a_{wv} = \frac{T_3 + T_1}{T_3 - T_1} + \frac{2}{(\kappa - 1)\ln\frac{p_2}{p_1}} \quad (5)$$

$$a_{qT} = \frac{2s_1}{R \ln\frac{p_2}{p_1}} - 1 \quad (6) \qquad a_{qs} = \frac{T_3 + T_1}{T_3 - T_1} \quad (7) \qquad a_{wp} = \kappa\, a_{wv} - (\kappa - 1)\, a_{qs} \quad (8)$$

For the comparison of all cyclic processes with ideal gases as working fluids, air is used as the working fluid in the following ($R = 0.2872$ kJ/(kg K), $s_1 = 6.867$ kJ/(kg K),

---

[11] In order to achieve favourable values for the seven evaluation criteria mentioned, κ should normally be large. This requirement is best fulfilled by monoatomic gases such as helium and to a reasonable extent also by diatomic gases such as nitrogen and oxygen or gas mixtures such as e.g. air. With regard to low values for a_{qT}, it is also important to choose working fluids with the smallest possible quotient $s_1/R = (S_{m1}/M)/(R_m/M) = S_{m1}/R_m$, i.e. a molar standard entropy S_{m1} makes sense, which is as low as possible. This applies to hydrogen, for example, although safety considerations may make such a working fluid of less interest.

$c_p = 1.005$ kJ/(kg K) $=$ const, $\kappa = 1.4 =$ const; s_1 is the absolute specific entropy). Here, state 1 with $p_1 = 1.0$ bar and $T_1 = 298.15$ K (25 °C) is assumed; as the ambient temperature also $T_b = 298.15$ K (25 °C) is selected as the ambient temperature. For state 3, $p_3 = 70$ bar and $T_3 = 1973.15$ K (1700 °C) are initially taken as a basis for illustration.

Table 7.7 *Ericsson* process: Thermodynamic relations for the thermal efficiency η_{th}, the exergetic efficiency ζ, the mechanical effort ratios a_{wp} and a_{wv} as well as the thermal effort ratios a_{qT} and a_{qs}

$$p_2 = p_3 \quad (1) \qquad \eta_{th} = \frac{T_3 - T_1}{T_3} \quad (2) \qquad \zeta = \frac{T_3 - T_1}{T_3 - T_b} \quad (3)$$

$$a_{wp} = \frac{T_3 + T_1}{T_3 - T_1} \quad (4) \qquad a_{wv} = \frac{T_3 + T_1}{T_3 - T_1} + \frac{2}{\ln \frac{p_2}{p_1}} \quad (5)$$

$$a_{qT} = \frac{2 s_1}{R \ln \frac{p_2}{p_1}} - 1 + \frac{2\kappa}{(\kappa - 1) \ln \frac{p_2}{p_1}} \left(\frac{T_3}{T_3 - T_1} \ln \frac{T_3}{T_1} - 1 \right) \quad (6)$$

$$a_{qs} = \frac{T_3 + T_1}{T_3 - T_1} + \frac{2\kappa}{(\kappa - 1) \ln \frac{p_2}{p_1}} \quad (7) \qquad a_{wp} = \kappa \, a_{wv} - (\kappa - 1) \, a_{qs} \quad (8)$$

Table 7.8 *Stirling* process: Thermodynamic relations for the thermal efficiency η_{th}, the exergetic efficiency ζ, the mechanical effort ratios a_{wp} and a_{wv} as well as the thermal effort ratios a_{qT} and a_{qs}

$$p_2 = p_3 \frac{T_1}{T_3} \quad (1) \qquad \eta_{th} = \frac{T_3 - T_1}{T_3} \quad (2) \qquad \zeta = \frac{T_3 - T_1}{T_3 - T_b} \quad (3)$$

$$a_{wp} = \frac{T_3 + T_1}{T_3 - T_1} + \frac{2}{\ln \frac{p_2}{p_1}} \quad (4) \qquad a_{wv} = \frac{T_3 + T_1}{T_3 - T_1} \quad (5)$$

$$a_{qT} = \frac{2 s_1}{R \ln \frac{p_2}{p_1}} - 1 + \frac{2}{(\kappa - 1) \ln \frac{p_2}{p_1}} \left(\frac{T_3}{T_3 - T_1} \ln \frac{T_3}{T_1} - 1 \right) \quad (6)$$

$$a_{qs} = \frac{T_3 + T_1}{T_3 - T_1} + \frac{2}{(\kappa - 1) \ln \frac{p_2}{p_1}} \quad (7) \qquad a_{wp} = (\kappa - 1) \, a_{qs} - (\kappa - 2) \, a_{wv} \quad (8)$$

Figs. 7.29 to 7.36 show the respective cyclic processes under these conditions in the p, v diagram and in the T, s diagram for air. It can be seen that *Stirling* process, *Otto* process, *Diesel* process and generalised *Diesel* process because of their comparatively limited volume changes with the help of displacement machines — i.e. reciprocating machines and screw machines respectively — can be realised. In contrast, for the realisation of the chosen single-polytropic *Carnot* process as well as the *Ericsson* process, the *Joule* process and the Gas expansion process, the use of thermal fluid flow machinery makes sense because of their large volume changes. The scales of the diagrams are chosen in such a way that the areas representing the absolute value of the specific cyclic process work $|w_{Kreis}|$ for the respective processes in the p, v diagram and in the T, s diagram have the same numerical size.

7.5 Comparative Evaluation of Right-Hand Cyclic Processes

With respect to the technical implementation of the *Otto* process as well as the *Diesel* process, in the case of non-supercharged engines, the selected state 3, for example, is realistic. The generalised *Diesel* process included here differs from the *Diesel* process only in a way that, with regard to the material stress, for heat input between state 2 and state 3, in the medium temperature range higher pressures are assumed than in the higher temperature range: At pressures $p_2 = 86.15$ bar and $p_3 = 70$ bar, reaching the temperature $T_3 = 1973.15$ K, the polytropic exponent results in $n = 0.2519$ (cf. also Table 7.15).

Table 7.9 *Joule* **process:** Thermodynamic relations for the thermal efficiency η_{th}, the exergetic efficiency ζ, the mechanical effort ratios a_{wp} and a_{wv} as well as the thermal effort ratios a_{qT} and a_{qs}

$$p_2 = p_3 \quad (1) \qquad \eta_{th} = 1 - \left(\frac{p_1}{p_2}\right)^{\frac{\kappa-1}{\kappa}} \quad (2)$$

$$\zeta = \frac{1 - \left(\frac{p_1}{p_2}\right)^{\frac{\kappa-1}{\kappa}}}{1 - \dfrac{T_b}{T_3 - T_1 \left(\frac{p_2}{p_1}\right)^{\frac{\kappa-1}{\kappa}}} \ln\left(\frac{T_3}{T_1}\left(\frac{p_1}{p_2}\right)^{\frac{\kappa-1}{\kappa}}\right)} \quad (3) \qquad a_{wp} = \frac{T_3 + T_1 \left(\frac{p_2}{p_1}\right)^{\frac{\kappa-1}{\kappa}}}{T_3 - T_1 \left(\frac{p_2}{p_1}\right)^{\frac{\kappa-1}{\kappa}}} \quad (4)$$

$$a_{wv} = \frac{\left(\frac{p_2}{p_1}\right)^{\frac{\kappa-1}{\kappa}} + 1 - \dfrac{2}{\kappa} \dfrac{T_3 \left(\frac{p_1}{p_2}\right)^{\frac{\kappa-1}{\kappa}} - T_1 \left(\frac{p_2}{p_1}\right)^{\frac{\kappa-1}{\kappa}}}{T_3 \left(\frac{p_1}{p_2}\right)^{\frac{\kappa-1}{\kappa}} - T_1}}{\left(\frac{p_2}{p_1}\right)^{\frac{\kappa-1}{\kappa}} - 1} \quad (5)$$

$$a_{qT} = \left(\frac{s_1}{R}\frac{\kappa-1}{\kappa}(T_3 - T_1) + T_3 \ln \frac{T_3}{T_1 \left(\frac{p_2}{p_1}\right)^{\frac{\kappa-1}{\kappa}}}\right) \frac{2\left(\frac{p_2}{p_1}\right)^{\frac{\kappa-1}{\kappa}}}{\left(T_3 - T_1 \left(\frac{p_2}{p_1}\right)^{\frac{\kappa-1}{\kappa}}\right)\left(\left(\frac{p_2}{p_1}\right)^{\frac{\kappa-1}{\kappa}} - 1\right)} - \frac{\left(\frac{p_2}{p_1}\right)^{\frac{\kappa-1}{\kappa}} + 1}{\left(\frac{p_2}{p_1}\right)^{\frac{\kappa-1}{\kappa}} - 1} \quad (6)$$

$$a_{qs} = \frac{\left(\frac{p_2}{p_1}\right)^{\frac{\kappa-1}{\kappa}} + 1}{\left(\frac{p_2}{p_1}\right)^{\frac{\kappa-1}{\kappa}} - 1} \quad (7) \qquad a_{wp} = \kappa\, a_{wv} - (\kappa - 1)\, a_{qs} \quad (8)$$

For the *Joule* process in stationary plants, state 3a with $p_{3a} = 20$ bar and $t_{3a} = 1523.15$ K (1250 °C) currently represents a realisable upper value (cf. Table 7.15). As far as the *Joule* process is approximated to the *Ericsson* process (section 7.4.3) by multistage isentropic compression combined with isobaric intercooling in the low temperature range

and by multistage isentropic expansion combined with isobaric intermediate heating in the high temperature range, the chosen state 3a could also be used for the *Ericsson* process.

Table 7.10 *Otto* **process:** Thermodynamic relations for the thermal efficiency η_{th}, the exergetic efficiency ζ, the mechanical effort ratios a_{wp} and a_{wv} as well as the thermal effort ratios a_{qT} and a_{qs}

$$p_2 = \frac{p_3^\kappa}{p_1^{\kappa-1}} \left(\frac{T_1}{T_3}\right)^\kappa \quad (1) \qquad \eta_{th} = 1 - \left(\frac{p_1}{p_2}\right)^{\frac{\kappa-1}{\kappa}} \quad (2)$$

$$\zeta = \frac{1 - \left(\frac{p_1}{p_2}\right)^{\frac{\kappa-1}{\kappa}}}{1 - \frac{T_b}{T_3 - T_1 \left(\frac{p_2}{p_1}\right)^{\frac{\kappa-1}{\kappa}}} \ln\left(\frac{T_3}{T_1}\left(\frac{p_1}{p_2}\right)^{\frac{\kappa-1}{\kappa}}\right)} \quad (3)$$

$$a_{wp} = \kappa \frac{T_3 + T_1 \left(\frac{p_2}{p_1}\right)^{\frac{\kappa-1}{\kappa}}}{T_3 - T_1 \left(\frac{p_2}{p_1}\right)^{\frac{\kappa-1}{\kappa}}} + (\kappa - 1)\frac{\left(\frac{p_2}{p_1}\right)^{\frac{\kappa-1}{\kappa}} + 1}{\left(\frac{p_2}{p_1}\right)^{\frac{\kappa-1}{\kappa}} - 1} \quad (4) \qquad a_{wv} = \frac{T_3 + T_1 \left(\frac{p_2}{p_1}\right)^{\frac{\kappa-1}{\kappa}}}{T_3 - T_1 \left(\frac{p_2}{p_1}\right)^{\frac{\kappa-1}{\kappa}}} \quad (5)$$

$$a_{qT} = \left(\frac{s_1}{R}(\kappa-1)(T_3-T_1) + T_3 \ln \frac{T_3}{T_1\left(\frac{p_2}{p_1}\right)^{\frac{\kappa-1}{\kappa}}}\right)\left(T_3 - T_1\left(\frac{p_2}{p_1}\right)^{\frac{\kappa-1}{\kappa}}\right)\left(\left(\frac{p_2}{p_1}\right)^{\frac{\kappa-1}{\kappa}} - 1\right) - \frac{2\left(\frac{p_2}{p_1}\right)^{\frac{\kappa-1}{\kappa}}}{\left(\frac{p_2}{p_1}\right)^{\frac{\kappa-1}{\kappa}} - 1} \quad (6)$$

$$a_{qs} = \frac{\left(\frac{p_2}{p_1}\right)^{\frac{\kappa-1}{\kappa}} + 1}{\left(\frac{p_2}{p_1}\right)^{\frac{\kappa-1}{\kappa}} - 1} \quad (7) \qquad a_{wp} = \kappa\, a_{wv} + (\kappa - 1)\, a_{qs} \quad (8)$$

The same applies to the Gas expansion process with air as the working fluid, which corresponds to the *Ericsson* process in the lower temperature range and to the *Joule* process in the upper temperature range. The special feature of the single-polytropic *Carnot* process considered here is that it is aproximated to the *Ericsson* process. It differs from the latter only in that, with regard to the material stress, in the low temperature range, higher pressures are permitted than in the high temperature range. In addition, at the high temperature T_3, the isothermal expansion extends into a range of pressure slight lower than ambient pressure. If one wants to use the single-polytropic *Carnot* process at the selected state 3 ($p_3 = 70$ bar, $T_3 = 1973.15$ K) analogous to the

Table 7.11 *Diesel* process: Thermodynamic relations for the thermal efficiency η_{th}, the exergetic efficiency ζ, the mechanical effort ratios a_{wp} and a_{wv} as well as the thermal effort ratios a_{qT} and a_{qs}

$$p_2 = p_3 \quad (1) \qquad \eta_{th} = 1 - \frac{\left(\frac{T_3}{T_1}\right)^\kappa \left(\frac{p_1}{p_2}\right)^{\kappa-1} - 1}{\kappa\left(\frac{T_3}{T_1} - \left(\frac{p_2}{p_1}\right)^{\frac{\kappa-1}{\kappa}}\right)} \quad (2)$$

$$\zeta = \frac{\frac{T_3}{T_1} - \left(\frac{p_2}{p_1}\right)^{\frac{\kappa-1}{\kappa}} - \frac{1}{\kappa}\left(\left(\frac{T_3}{T_1}\right)^\kappa \left(\frac{p_1}{p_2}\right)^{\kappa-1} - 1\right)}{\frac{T_3}{T_1} - \left(\frac{p_2}{p_1}\right)^{\frac{\kappa-1}{\kappa}} - \frac{T_b}{T_1}\ln\left(\frac{T_3}{T_1}\left(\frac{p_1}{p_2}\right)^{\frac{\kappa-1}{\kappa}}\right)} \quad (3)$$

$$a_{wp} = \frac{\kappa\left(\frac{T_3}{T_1} + \left(\frac{p_2}{p_1}\right)^{\frac{\kappa-1}{\kappa}} - 2\right) - \left(\left(\frac{T_3}{T_1}\right)^\kappa \left(\frac{p_1}{p_2}\right)^{\kappa-1} - 1\right)}{\kappa\left(\frac{T_3}{T_1} - \left(\frac{p_2}{p_1}\right)^{\frac{\kappa-1}{\kappa}}\right) - \left(\left(\frac{T_3}{T_1}\right)^\kappa \left(\frac{p_1}{p_2}\right)^{\kappa-1} - 1\right)} \quad (4)$$

$$a_{wv} = \frac{\kappa\left(\frac{T_3}{T_1} - \left(\frac{p_2}{p_1}\right)^{\frac{\kappa-1}{\kappa}}\right) + 2\left(\frac{p_2}{p_1}\right)^{\frac{\kappa-1}{\kappa}} - \left(\frac{T_3}{T_1}\right)^\kappa \left(\frac{p_1}{p_2}\right)^{\kappa-1} - 1}{\kappa\left(\frac{T_3}{T_1} - \left(\frac{p_2}{p_1}\right)^{\frac{\kappa-1}{\kappa}}\right) - \left(\frac{T_3}{T_1}\right)^\kappa \left(\frac{p_1}{p_2}\right)^{\kappa-1} + 1} \quad (5)$$

$$a_{qT} = \frac{\frac{2s_1}{R}\frac{\kappa-1}{\kappa}(T_3 - T_1) + 2T_3\ln\left(\frac{T_3}{T_1}\left(\frac{p_1}{p_2}\right)^{\frac{\kappa-1}{\kappa}}\right) - T_3 + T_1\left(\frac{p_2}{p_1}\right)^{\frac{\kappa-1}{\kappa}}}{T_3 - T_1\left(\frac{p_2}{p_1}\right)^{\frac{\kappa-1}{\kappa}} - \frac{T_1}{\kappa}\left(\left(\frac{T_3}{T_1}\right)^\kappa \left(\frac{p_1}{p_2}\right)^{\kappa-1} - 1\right)}$$

$$\phantom{a_{qT}} - \frac{\frac{T_1}{\kappa}\left(\left(\frac{T_3}{T_1}\right)^\kappa \left(\frac{p_1}{p_2}\right)^{\kappa-1} - 1\right)}{T_3 - T_1\left(\frac{p_2}{p_1}\right)^{\frac{\kappa-1}{\kappa}} - \frac{T_1}{\kappa}\left(\left(\frac{T_3}{T_1}\right)^\kappa \left(\frac{p_1}{p_2}\right)^{\kappa-1} - 1\right)} \quad (6)$$

$$a_{qs} = \frac{\kappa\left(\frac{T_3}{T_1} - \left(\frac{p_2}{p_1}\right)^{\frac{\kappa-1}{\kappa}}\right) + \left(\frac{T_3}{T_1}\right)^\kappa \left(\frac{p_1}{p_2}\right)^{\kappa-1} - 1}{\kappa\left(\frac{T_3}{T_1} - \left(\frac{p_2}{p_1}\right)^{\frac{\kappa-1}{\kappa}}\right) - \left(\frac{T_3}{T_1}\right)^\kappa \left(\frac{p_1}{p_2}\right)^{\kappa-1} + 1} \quad (7)$$

generalised *Diesel* process, and to realise the pressure $p_2 = 86.36$ bar by isothermal compression at the temperature T_1, this needs a polytropic exponent of $n = 0.1$. If e.g. at $p_{3a} = 20$ bar and $T_{3a} = 1523.15$ K a pressure $p_{2a} = 40$ bar is to be reached, for this a polytropic exponent of $n = 0.2983$ is needed (cf. Table 7.15). For the *Stirling* process, nowadays e.g. the state 3b characterised by $p_{3b} = 20$ bar and $T_{3b} = 973.15$ K (700 °C) seems to be realistic.

Table 7.12 Generalised *Diesel* process: Thermodynamic relations for the thermal efficiency η_{th}, the exergetic efficiency ζ, the mechanical effort ratios a_{wp} and a_{wv} as well as the thermal effort ratios a_{qT} and a_{qs}

$$p_2 = \left(\frac{T_3\, p_1^{\frac{\kappa-1}{\kappa}}}{T_1\, p_3^{\frac{n-1}{n}}}\right)^{\frac{n\kappa}{\kappa-n}} \quad (1)$$

$$\eta_{th} = 1 - \frac{n-1}{n-\kappa}\, \frac{\left(\frac{T_3}{T_1}\right)^{\frac{\kappa-n}{1-n}} \left(\frac{p_2}{p_1}\right)^{\frac{(\kappa-n)(\kappa-1)}{\kappa(n-1)}} - 1}{\frac{T_3}{T_1} - \left(\frac{p_2}{p_1}\right)^{\frac{\kappa-1}{\kappa}}} \quad (2)$$

$$\zeta = \frac{\left(\frac{T_3}{T_1} - \left(\frac{p_2}{p_1}\right)^{\frac{\kappa-1}{\kappa}}\right) - \frac{n-1}{n-\kappa}\left(\left(\frac{T_3}{T_1}\right)^{\frac{\kappa-n}{1-n}}\left(\frac{p_2}{p_1}\right)^{\frac{(\kappa-n)(\kappa-1)}{\kappa(n-1)}} - 1\right)}{\left(\frac{T_3}{T_1} - \left(\frac{p_2}{p_1}\right)^{\frac{\kappa-1}{\kappa}}\right) - \frac{T_b}{T_1}\ln\left(\frac{T_3}{T_1}\left(\frac{p_1}{p_2}\right)^{\frac{\kappa-1}{\kappa}}\right)} \quad (3)$$

$$a_{wp} = \frac{\kappa\left(\left(\frac{p_2}{p_1}\right)^{\frac{\kappa-1}{\kappa}} - 1\right) + \left|\frac{n(\kappa-1)}{n-1}\right|\left(\frac{T_3}{T_1} - \left(\frac{p_2}{p_1}\right)^{\frac{\kappa-1}{\kappa}}\right)}{\frac{n-\kappa}{n-1}\left(\frac{T_3}{T_1} - \left(\frac{p_2}{p_1}\right)^{\frac{\kappa-1}{\kappa}}\right) - \left(\left(\frac{T_3}{T_1}\right)^{\frac{\kappa-n}{1-n}}\left(\frac{p_2}{p_1}\right)^{\frac{(\kappa-n)(\kappa-1)}{\kappa(n-1)}} - 1\right)} +$$

$$+ \frac{\kappa\left(\frac{T_3}{T_1} - 1\right) - \left(\left(\frac{T_3}{T_1}\right)^{\frac{\kappa-n}{1-n}}\left(\frac{p_2}{p_1}\right)^{\frac{(\kappa-n)(\kappa-1)}{\kappa(n-1)}} - 1\right)}{\frac{n-\kappa}{n-1}\left(\frac{T_3}{T_1} - \left(\frac{p_2}{p_1}\right)^{\frac{\kappa-1}{\kappa}}\right) - \left(\left(\frac{T_3}{T_1}\right)^{\frac{\kappa-n}{1-n}}\left(\frac{p_2}{p_1}\right)^{\frac{(\kappa-n)(\kappa-1)}{\kappa(n-1)}} - 1\right)} \quad (4)$$

$$a_{wv} = \frac{\left(\frac{T_3}{T_1} + \left(\frac{p_2}{p_1}\right)^{\frac{\kappa-1}{\kappa}}\right) + \frac{\kappa-1}{|n-1|}\left(\frac{T_3}{T_1} - \left(\frac{p_2}{p_1}\right)^{\frac{\kappa-1}{\kappa}}\right)}{\frac{n-\kappa}{n-1}\left(\frac{T_3}{T_1} - \left(\frac{p_2}{p_1}\right)^{\frac{\kappa-1}{\kappa}}\right) - \left(\left(\frac{T_3}{T_1}\right)^{\frac{\kappa-n}{1-n}}\left(\frac{p_2}{p_1}\right)^{\frac{(\kappa-n)(\kappa-1)}{\kappa(n-1)}} - 1\right)} -$$

$$- \frac{\left(\left(\frac{T_3}{T_1}\right)^{\frac{\kappa-n}{1-n}}\left(\frac{p_2}{p_1}\right)^{\frac{(\kappa-n)(\kappa-1)}{\kappa(n-1)}} + 1\right)}{\frac{n-\kappa}{n-1}\left(\frac{T_3}{T_1} - \left(\frac{p_2}{p_1}\right)^{\frac{\kappa-1}{\kappa}}\right) - \left(\left(\frac{T_3}{T_1}\right)^{\frac{\kappa-n}{1-n}}\left(\frac{p_2}{p_1}\right)^{\frac{(\kappa-n)(\kappa-1)}{\kappa(n-1)}} - 1\right)} \quad (5)$$

In a right-hand *Carnot* process — starting from state 1 — the states 3, 3a and 3b cannot be realised; therefore a *Carnot* process with state 3c with $p_{3c} = 40$ bar and $T_{3c} = 673.15$ K (400 °C) is chosen here (cf. Table 7.15).

The Gas expansion process, with natural gas as the working fluid, is realisable e.g. with a state 3c characterised by $p_{3c} = 40$ bar and $T_{3c} = 673.15$ K (400 °C). In the result of calculations shown here, air is used as an alternative.

Finally, the *Clausius-Rankine* process with water as working medium and the states 1 ($p_1 = 0.032$ bar, $T_1 = 298.15$ K (25 °C), $s_1 = 3.888$ kJ/(kg K) and optionally 3d (p_{3d}

7.5 Comparative Evaluation of Right-Hand Cyclic Processes

$= 350$ bar, $T_{3d} = 873.15$ K ($600\,°C$)) and 3e ($p_{3e} = 200$ bar, $T_{3e} = 823.15$ K ($550\,°C$)) is included in the comparison (cf. Table 7.15).

Table 7.15 shows the obtained values for the thermal efficiency η_{th}, the exergetic efficiency ζ, the mechanical effort ratios a_{wp} and a_{wv} as well as the thermal effort ratios a_{qT} and a_{qs}; the total effort ratio a_g is also listed.

Table 7.12 (continued) Generalised *Diesel* process: Thermodynamic relations for the thermal efficiency η_{th}, the exergetic efficiency ζ, the mechanical effort ratios a_{wp} and a_{wv} as well as the thermal effort ratios a_{qT} and a_{qs}

$$a_{qT} = \frac{\frac{2s_1}{R}(\kappa-1)(T_3 - T_1) + T_3 \ln\left(\left(\frac{T_3}{T_1}\right)^{\frac{\kappa-n}{1-n}} \left(\frac{p_2}{p_1}\right)^{\frac{(\kappa-n)(\kappa-1)}{\kappa(n-1)}}\right)}{\frac{n-\kappa}{n-1}\left(T_3 - T_1\left(\frac{p_2}{p_1}\right)^{\frac{\kappa-1}{\kappa}}\right) - T_1\left(\left(\frac{T_3}{T_1}\right)^{\frac{\kappa-n}{1-n}}\left(\frac{p_2}{p_1}\right)^{\frac{(\kappa-n)(\kappa-1)}{\kappa(n-1)}} - 1\right)} +$$

$$+ \frac{\left|\frac{n-\kappa}{n-1}\right|\left(T_3 \ln\left(\frac{T_3}{T_1}\left(\frac{p_1}{p_2}\right)^{\frac{\kappa-1}{\kappa}}\right) - \left(T_3 - T_1\left(\frac{p_2}{p_1}\right)^{\frac{\kappa-1}{\kappa}}\right)\right)}{\frac{n-\kappa}{n-1}\left(T_3 - T_1\left(\frac{p_2}{p_1}\right)^{\frac{\kappa-1}{\kappa}}\right) - T_1\left(\left(\frac{T_3}{T_1}\right)^{\frac{\kappa-n}{1-n}}\left(\frac{p_2}{p_1}\right)^{\frac{(\kappa-n)(\kappa-1)}{\kappa(n-1)}} - 1\right)}$$

$$- \frac{T_1\left(\left(\frac{T_3}{T_1}\right)^{\frac{\kappa-n}{1-n}}\left(\frac{p_2}{p_1}\right)^{\frac{(\kappa-n)(\kappa-1)}{\kappa(n-1)}} - 1\right)}{\frac{n-\kappa}{n-1}\left(T_3 - T_1\left(\frac{p_2}{p_1}\right)^{\frac{\kappa-1}{\kappa}}\right) - T_1\left(\left(\frac{T_3}{T_1}\right)^{\frac{\kappa-n}{1-n}}\left(\frac{p_2}{p_1}\right)^{\frac{(\kappa-n)(\kappa-1)}{\kappa(n-1)}} - 1\right)} \quad (6)$$

$$a_{qs} = \frac{\left|\frac{n-\kappa}{n-1}\right|\left(\frac{T_3}{T_1} - \left(\frac{p_2}{p_1}\right)^{\frac{\kappa-1}{\kappa}}\right) + \left(\left(\frac{T_3}{T_1}\right)^{\frac{\kappa-n}{1-n}}\left(\frac{p_2}{p_1}\right)^{\frac{(\kappa-n)(\kappa-1)}{\kappa(n-1)}} - 1\right)}{\frac{n-\kappa}{n-1}\left(\frac{T_3}{T_1} - \left(\frac{p_2}{p_1}\right)^{\frac{\kappa-1}{\kappa}}\right) - \left(\left(\frac{T_3}{T_1}\right)^{\frac{\kappa-n}{1-n}}\left(\frac{p_2}{p_1}\right)^{\frac{(\kappa-n)(\kappa-1)}{\kappa(n-1)}} - 1\right)} \quad (7)$$

It can be seen that at the selected highest state 3, which is currently only realisable for the *Otto* process, the *Diesel* process and the generalised *Diesel* process, the single-polytropic *Carnot* process, the *Ericsson* process and the *Stirling* process have the most favourable values with regard to thermal and exergetic efficiency; these processes are each equivalent to each other.

In the case of the effort ratios, the single-polytropic *Carnot* process and the *Ericsson* process are very favourable; the *Stirling* process also performs well — apart from a higher value for the thermal effort ratio a_{qT}.

Table 7.13 Gas expansion process: Thermodynamic relations for the thermal efficiency η_{th}, the exergetic efficiency ζ, the mechanical effort ratios a_{wp} and a_{wv} as well as the thermal effort ratios a_{qT} and a_{qs}

$$p_2 = p_3 \quad (1) \qquad \eta_{th} = 1 - \frac{\kappa-1}{\kappa} \frac{T_1 \ln \frac{p_2}{p_1}}{T_3 \left(1 - \left(\frac{p_1}{p_2}\right)^{\frac{\kappa-1}{\kappa}}\right)} \quad (2)$$

$$\zeta = \frac{T_3 \left(1 - \left(\frac{p_1}{p_2}\right)^{\frac{\kappa-1}{\kappa}}\right) - \frac{\kappa-1}{\kappa} T_1 \ln \frac{p_2}{p_1}}{T_3 \left(1 - \left(\frac{p_1}{p_2}\right)^{\frac{\kappa-1}{\kappa}}\right) - \frac{\kappa-1}{\kappa} T_b \ln \frac{p_2}{p_1}} \quad (3)$$

$$a_{wp} = \frac{\frac{\kappa}{\kappa-1} T_3 \left(1 - \left(\frac{p_1}{p_2}\right)^{\frac{\kappa-1}{\kappa}}\right) + T_1 \ln \frac{p_2}{p_1}}{\frac{\kappa}{\kappa-1} T_3 \left(1 - \left(\frac{p_1}{p_2}\right)^{\frac{\kappa-1}{\kappa}}\right) - T_1 \ln \frac{p_2}{p_1}} \quad (4)$$

$$a_{wv} = \frac{\frac{\kappa}{\kappa-1} T_3 \left(1 + \frac{\kappa-2}{\kappa}\left(\frac{p_1}{p_2}\right)^{\frac{\kappa-1}{\kappa}}\right) + T_1 \left(\ln\left(\frac{p_2}{p_1}\right) - 2\right)}{\frac{\kappa}{\kappa-1} T_3 \left(1 - \left(\frac{p_1}{p_2}\right)^{\frac{\kappa-1}{\kappa}}\right) - T_1 \ln\left(\frac{p_2}{p_1}\right)} \quad (5)$$

$$a_{qT} = \frac{\left(\frac{s_1}{R} - \ln \frac{p_2}{p_1} - \frac{\kappa}{\kappa-1}\right) 2(T_3 - T_1) + 2 T_3 \frac{\kappa}{\kappa-1} \ln \frac{T_3}{T_1}}{T_3 \frac{\kappa}{\kappa-1}\left(1 - \left(\frac{p_1}{p_2}\right)^{\frac{\kappa-1}{\kappa}}\right) - T_1 \ln \frac{p_2}{p_1}} + 1 \quad (6)$$

$$a_{qs} = \frac{\frac{\kappa}{\kappa-1} T_3 \left(1 + \left(\frac{p_1}{p_2}\right)^{\frac{\kappa-1}{\kappa}}\right) + T_1 \left(\ln\left(\frac{p_2}{p_1}\right) - \frac{2\kappa}{\kappa-1}\right)}{\frac{\kappa}{\kappa-1} T_3 \left(1 - \left(\frac{p_1}{p_2}\right)^{\frac{\kappa-1}{\kappa}}\right) - T_1 \ln\left(\frac{p_2}{p_1}\right)} \quad (7)$$

$$a_{wp} = \kappa\, a_{wv} - (\kappa - 1) a_{qs} \quad (8)$$

This also tends to apply to the Gas expansion process. In contrary, *Joule* process, *Otto* process, *Diesel* process and the generalised *Diesel* process in a comparison to that have a lower thermal efficiency η_{th}, a lower exergetic efficiency ζ and higher mechanical effort ratios a_{wp} and a_{wv} as well as the thermal effort ratio a_{qT} and the total effort ratio a_g; they are only more advantageous with respect to the thermal effort ratio a_{qT}. In tendency, the same evaluation results considering the conditions 3a and 3b.

If one compares the different cyclic processes from the point of view of the current technical feasibility, it can be seen that the *Joule* process with the highest achievable

state 3a has a lower thermal efficiency η_{th} compared to the *Diesel* process and the *Otto* process at state 3, and the exergetic efficiency ζ is about the same. But on the other hand the *Joule* process is advanageous with respect to the mechanical effort ratio a_{wp}, the thermal effort ratio a_{qT} and the total effort ratio a_g compared to the *Diesel* process and the *Otto* process; slight disadvantages are to be found in the mechanical effort ratio a_{wv} and the thermal effort ratio a_{qs}.

The comparison also shows that the *Joule* process, insofar as it is further developed and in future will move more in the direction to the *Ericsson* process, to the single-polytropic *Carnot* process or to the Gas expansion process, could open up a very considerable potential for improvement.

Tabelle 7.14 Clausius-Rankine process: Thermodynamic relations for the thermal efficiency η_{th}, the exergetic efficiency ζ, the mechanical effort ratios a_{wp} and a_{wv} as well as the thermal effort ratios a_{qT} and a_{qs}

$$p_2 = p_3 \quad (1) \qquad \eta_{th} = 1 - \frac{h_4 - h_1}{h_3 - h_2} \quad (2) \qquad \zeta = \frac{(h_3 - h_4) - (h_2 - h_1)}{(h_3 - h_2) - (s_3 - s_1)T_b} \quad (3)$$

$$a_{wp} = \frac{(h_3 - h_4) + (h_2 - h_1)}{(h_3 - h_4) - (h_2 - h_1)} \quad (4) \qquad a_{wv} = \frac{(h_3 - h_2) + (h_4 - h_1) - 2(u_4 - u_2)}{(h_3 - h_2) - (h_4 - h_1)} \quad (5)$$

$$a_{qT} = \frac{2s_3(T_3 - T_1) + (s_3 - s_1)T_1 - (h_3 - h_2)}{(h_3 - h_4) - (h_2 - h_1)} \quad (6) \qquad a_{qs} = \frac{(h_3 - h_2) + (h_4 - h_1)}{(h_3 - h_2) - (h_4 - h_1)} \quad (7)$$

A direct comparison of the *Otto* and *Diesel* processes at condition 3 shows that the *Diesel* process offers advantages over the *Otto* process for almost all evaluation criteria; this applies to the evaluation criteria η_{th}, ζ, a_{wp}, a_{qT}, a_{qs} and a_g; only a_{wv} shows a disadvantage. The generalised *Diesel* process, under the above assumptions, represents an improvement of the *Diesel* process. The intensive efforts in the further development of *Diesel* engines thus prove to be — measured by the above-mentioned evaluation criteria — as purposeful.

For the *Stirling* process, state 3b, for example, currently appears to be the highest achievable state that is technically well controllable. It turns out that the *Stirling* process, already being operated at this state with comparatively moderate pressure p_{3b} and moderate temperature T_{3b}, has advantages in comparison to the generalised *Diesel* process, the *Diesel* process and the *Otto* process when evaluated according to the criteria η_{th}, ζ, a_{wp}, a_{wv}, a_{qT} and a_g; only a_{qs} shows a disadvantage. This illustrates the — so far little used — theoretical potential of the *Stirling* process.

The fact that the *Carnot* process does not offer any significant scope for effective technical implementation is shown by the comparisons in Table 7.15: Since it could only be realised at comparatively low highest process temperatures — for example at temperature T_{3c} —, very considerable disadvantages would have to be accepted with regard to η_{th}, ζ, a_{wp}, a_{wv}, a_{qT} and a_g.

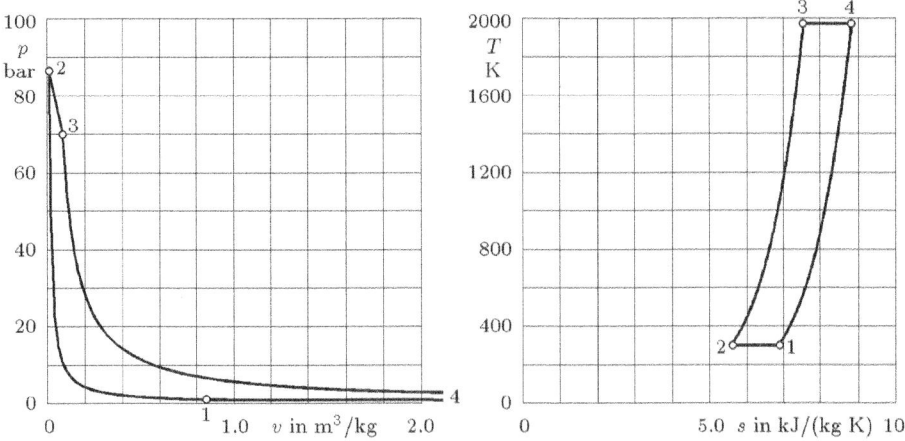

Figure 7.29 Single-polytropic *Carnot* process in a p,v diagram and a T,s diagram of air; $n = 0{,}1$

$p_1 = 1.000$ bar	$v_1 = 0.85629$ m³/kg	$T_1 = 298.15$ K	$s_1 = 6.8670$ kJ/(kg K)
$p_2 = 86.355$ bar	$v_2 = 0.00992$ m³/kg	$T_2 = 298.15$ K	$s_2 = 5.5865$ kJ/(kg K)
$p_3 = 70.000$ bar	$v_3 = 0.08096$ m³/kg	$T_3 = 1973.15$ K	$s_3 = 7.5464$ kJ/(kg K)
$p_4 = 0.811$ bar	$v_4 = 6.99095$ m³/kg	$T_4 = 1973.15$ K	$s_4 = 8.8269$ kJ/(kg K)
$w_{V12} = 381.77$ kJ/kg	$w_{p12} = 381.77$ kJ/kg	$q_{s12} = -381.77$ kJ/kg	$q_{T12} = 0$ kJ/kg
$w_{V23} = -534.51$ kJ/kg	$w_{p23} = -53.451$ kJ/kg	$q_{s23} = 1737.2$ kJ/kg	$q_{T23} = 11488$ kJ/kg
$w_{V34} = -2526.6$ kJ/kg	$w_{p34} = -2526.6$ kJ/kg	$q_{s34} = 2526.6$ kJ/kg	$q_{T34} = 0$ kJ/kg
$w_{V41} = 534.51$ kJ/kg	$w_{p41} = 53.451$ kJ/kg	$q_{s41} = -1737.2$ kJ/kg	$q_{T41} = -13632$ kJ/kg
$w_{Kreis} = -2144.8$ kJ/kg		$\eta_{th} = 0.8489$	$\zeta = 1.0000$
$a_{wp} = 1.4058$	$a_{wv} = 1.8544$	$a_{qT} = 11.712 \qquad a_{qs} = 2.9759$	$a_g = 17.948$

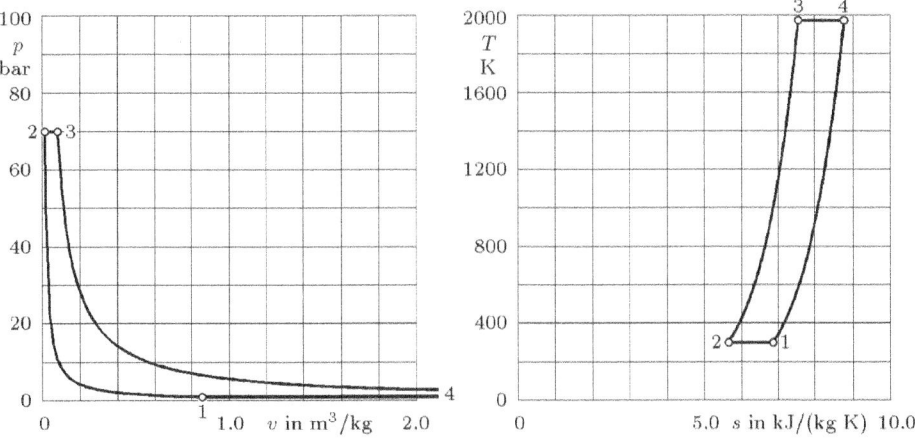

Figure 7.30 *Ericsson* process in a p,v diagram und in a T,s diagram of air

$p_1 = 1.000$ bar	$v_1 = 0.85629$ m³/kg	$T_1 = 298.15$ K	$s_1 = 6.8670$ kJ/(kg K)
$p_2 = 70.000$ bar	$v_2 = 0.01223$ m³/kg	$T_2 = 298.15$ K	$s_2 = 5.6468$ kJ/(kg K)
$p_3 = 70.000$ bar	$v_3 = 0.08096$ m³/kg	$T_3 = 1973.15$ K	$s_3 = 7.5464$ kJ/(kg K)
$p_4 = 1.000$ bar	$v_4 = 5.66689$ m³/kg	$T_4 = 1973.15$ K	$s_4 = 8.7666$ kJ/(kg K)
$w_{V12} = 363.79$ kJ/kg	$w_{p12} = 363.79$ kJ/kg	$q_{s12} = -363.79$ kJ/kg	$q_{T12} = 0$ kJ/kg
$w_{V23} = -481.06$ kJ/kg	$w_{p23} = 0$ kJ/kg	$q_{s23} = 1683.7$ kJ/kg	$q_{T23} = 11523$ kJ/kg
$w_{V34} = -2407.6$ kJ/kg	$w_{p34} = -2407.6$ kJ/kg	$q_{s34} = 2407.6$ kJ/kg	$q_{T34} = 0$ kJ/kg
$w_{V41} = 481.06$ kJ/kg	$w_{p41} = 0$ kJ/kg	$q_{s41} = -1683.7$ kJ/kg	$q_{T41} = -13567$ kJ/kg
$w_{Kreis} = -2043.8$ kJ/kg		$\eta_{th} = 0.8489$	$\zeta = 1.0000$
$a_{wp} = 1.3560$	$a_{wv} = 1.8268$	$a_{qT} = 12.276 \qquad a_{qs} = 3.0036$	$a_g = 18.463$

7.5 Comparative Evaluation of Right-Hand Cyclic Processes

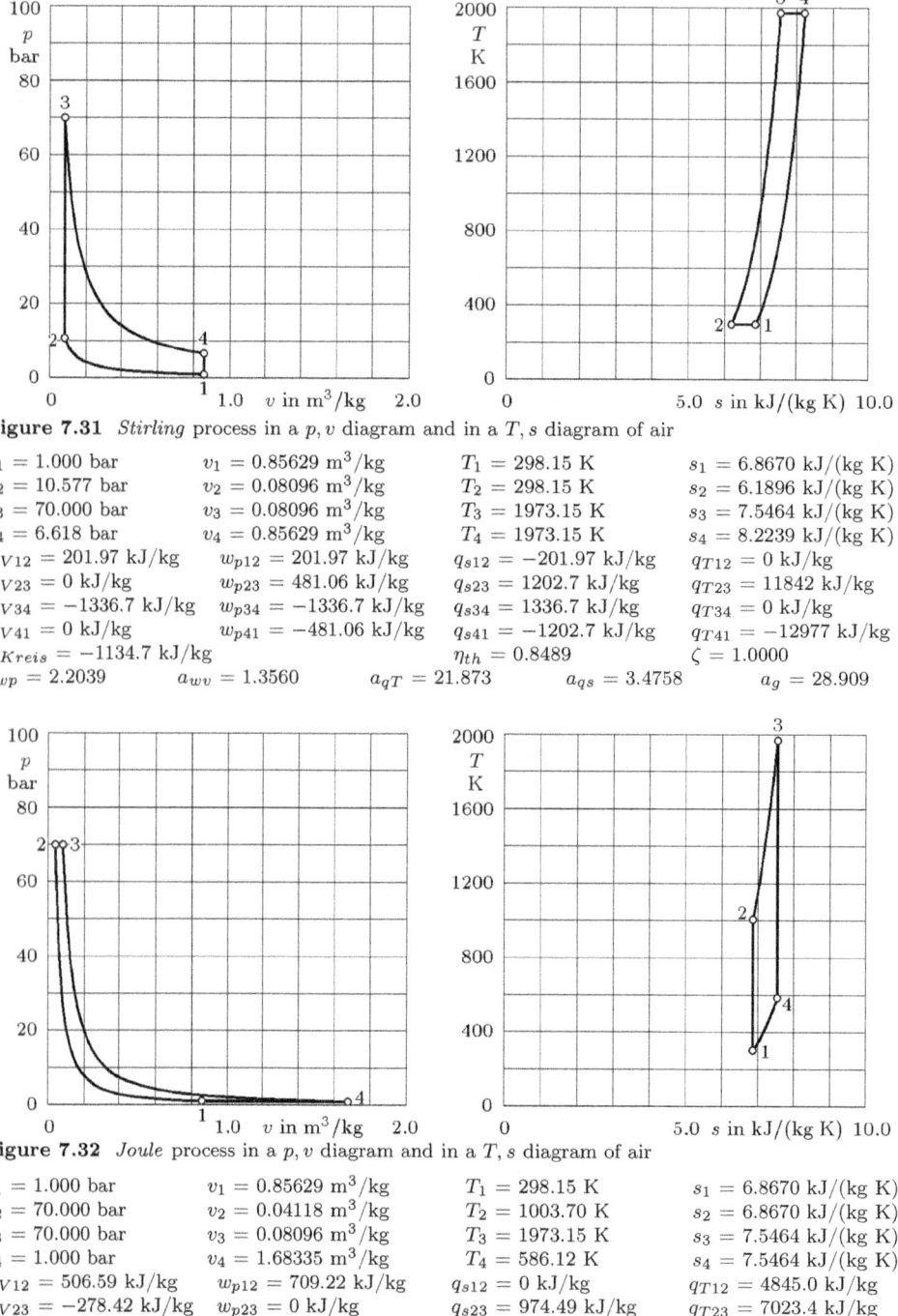

Figure 7.31 *Stirling* process in a p, v diagram and in a T, s diagram of air

$p_1 = 1.000$ bar	$v_1 = 0.85629$ m^3/kg	$T_1 = 298.15$ K	$s_1 = 6.8670$ kJ/(kg K)
$p_2 = 10.577$ bar	$v_2 = 0.08096$ m^3/kg	$T_2 = 298.15$ K	$s_2 = 6.1896$ kJ/(kg K)
$p_3 = 70.000$ bar	$v_3 = 0.08096$ m^3/kg	$T_3 = 1973.15$ K	$s_3 = 7.5464$ kJ/(kg K)
$p_4 = 6.618$ bar	$v_4 = 0.85629$ m^3/kg	$T_4 = 1973.15$ K	$s_4 = 8.2239$ kJ/(kg K)
$w_{V12} = 201.97$ kJ/kg	$w_{p12} = 201.97$ kJ/kg	$q_{s12} = -201.97$ kJ/kg	$q_{T12} = 0$ kJ/kg
$w_{V23} = 0$ kJ/kg	$w_{p23} = 481.06$ kJ/kg	$q_{s23} = 1202.7$ kJ/kg	$q_{T23} = 11842$ kJ/kg
$w_{V34} = -1336.7$ kJ/kg	$w_{p34} = -1336.7$ kJ/kg	$q_{s34} = 1336.7$ kJ/kg	$q_{T34} = 0$ kJ/kg
$w_{V41} = 0$ kJ/kg	$w_{p41} = -481.06$ kJ/kg	$q_{s41} = -1202.7$ kJ/kg	$q_{T41} = -12977$ kJ/kg
$w_{Kreis} = -1134.7$ kJ/kg		$\eta_{th} = 0.8489$	$\zeta = 1.0000$
$a_{wp} = 2.2039$	$a_{wv} = 1.3560$	$a_{qT} = 21.873$ $\quad$ $a_{qs} = 3.4758$	$a_g = 28.909$

Figure 7.32 *Joule* process in a p, v diagram and in a T, s diagram of air

$p_1 = 1.000$ bar	$v_1 = 0.85629$ m^3/kg	$T_1 = 298.15$ K	$s_1 = 6.8670$ kJ/(kg K)
$p_2 = 70.000$ bar	$v_2 = 0.04118$ m^3/kg	$T_2 = 1003.70$ K	$s_2 = 6.8670$ kJ/(kg K)
$p_3 = 70.000$ bar	$v_3 = 0.08096$ m^3/kg	$T_3 = 1973.15$ K	$s_3 = 7.5464$ kJ/(kg K)
$p_4 = 1.000$ bar	$v_4 = 1.68335$ m^3/kg	$T_4 = 586.12$ K	$s_4 = 7.5464$ kJ/(kg K)
$w_{V12} = 506.59$ kJ/kg	$w_{p12} = 709.22$ kJ/kg	$q_{s12} = 0$ kJ/kg	$q_{T12} = 4845.0$ kJ/kg
$w_{V23} = -278.42$ kJ/kg	$w_{p23} = 0$ kJ/kg	$q_{s23} = 974.49$ kJ/kg	$q_{T23} = 7023.4$ kJ/kg
$w_{V34} = -995.89$ kJ/kg	$w_{p34} = -1394.2$ kJ/kg	$q_{s34} = 0$ kJ/kg	$q_{T34} = -10467$ kJ/kg
$w_{V41} = 82.706$ kJ/kg	$w_{p41} = 0$ kJ/kg	$q_{s41} = -289.47$ kJ/kg	$q_{T41} = -2086.3$ kJ/kg
$w_{Kreis} = -685.02$ kJ/kg		$\eta_{th} = 0.7030$	$\zeta = 0.8874$
$a_{wp} = 3.0707$	$a_{wv} = 2.7205$	$a_{qT} = 35.651$ $\quad$ $a_{qs} = 1.8452$	$a_g = 43.2878$

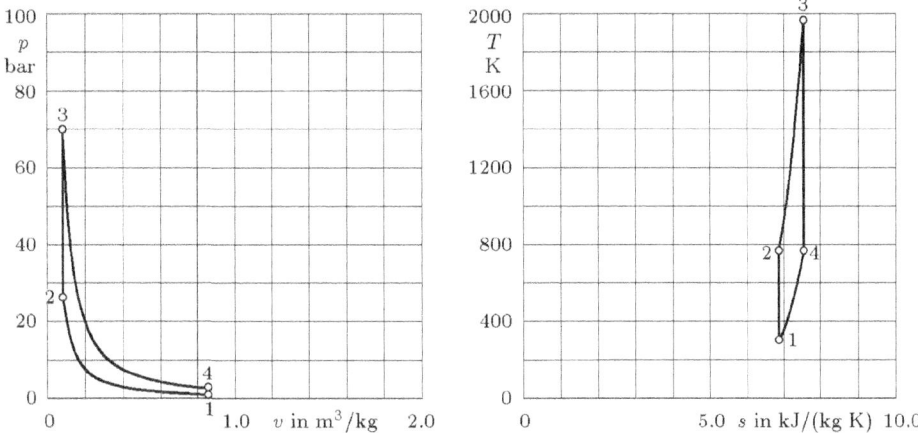

Figure 7.33 *Otto* process in a p, v diagram and in a T, s diagram of air

$p_1 = 1.000$ bar	$v_1 = 0.85629$ m³/kg	$T_1 = 298.15$ K	$s_1 = 6.8670$ kJ/(kg K)
$p_2 = 27.172$ bar	$v_2 = 0.08096$ m³/kg	$T_2 = 765.92$ K	$s_2 = 6.8670$ kJ/(kg K)
$p_3 = 70.000$ bar	$v_3 = 0.08096$ m³/kg	$T_3 = 1973.15$ K	$s_3 = 7.5464$ kJ/(kg K)
$p_4 = 2.576$ bar	$v_4 = 0.85629$ m³/kg	$T_4 = 768.09$ K	$s_4 = 7.5464$ kJ/(kg K)
$w_{V12} = 335.86$ kJ/kg	$w_{p12} = 470.20$ kJ/kg	$q_{s12} = 0$ kJ/kg	$q_{T12} = 3212.2$ kJ/kg
$w_{V23} = 0$ kJ/kg	$w_{p23} = 346.72$ kJ/kg	$q_{s23} = 866.79$ kJ/kg	$q_{T23} = 8763.9$ kJ/kg
$w_{V34} = -865.23$ kJ/kg	$w_{p34} = -1211.3$ kJ/kg	$q_{s34} = 0$ kJ/kg	$q_{T34} = -9093.9$ kJ/kg
$w_{V41} = 0$ kJ/kg	$w_{p41} = -134.97$ kJ/kg	$q_{s41} = -337.42$ kJ/kg	$q_{T41} = -3411.5$ kJ/kg
$w_{Kreis} = -529.38$ kJ/kg		$\eta_{th} = 0.6107$	$\zeta = 0.7970$
$a_{wp} = 4.0864$	$a_{wv} = 2.2689$	$a_{qT} = 46.2461 \quad a_{qs} = 2.2748$	$a_g = 54.8762$

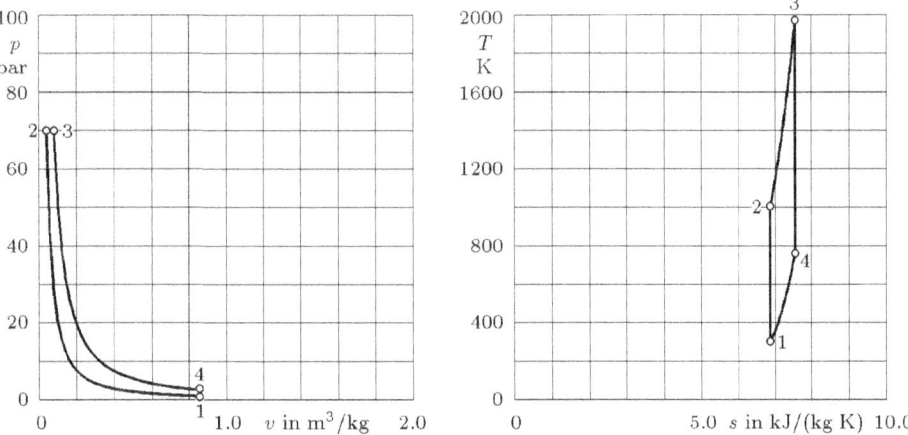

Figure 7.34 *Diesel* process in a p, v diagram and in a T, s diagram of air

$p_1 = 1.000$ bar	$v_1 = 0.85629$ m³/kg	$T_1 = 298.15$ K	$s_1 = 6.8670$ kJ/(kg K)
$p_2 = 70.000$ bar	$v_2 = 0.04118$ m³/kg	$T_2 = 1003.70$ K	$s_2 = 6.8670$ kJ/(kg K)
$p_3 = 70.000$ bar	$v_3 = 0.08096$ m³/kg	$T_3 = 1973.15$ K	$s_3 = 7.5464$ kJ/(kg K)
$p_4 = 2.576$ bar	$v_4 = 0.85629$ m³/kg	$T_4 = 768.09$ K	$s_4 = 7.5464$ kJ/(kg K)
$w_{V12} = 506.59$ kJ/kg	$w_{p12} = 709.22$ kJ/kg	$q_{s12} = 0$ kJ/kg	$q_{T12} = 4845.0$ kJ/kg
$w_{V23} = -278.42$ kJ/kg	$w_{p23} = 0$ kJ/kg	$q_{s23} = 974.49$ kJ/kg	$q_{T23} = 7023.4$ kJ/kg
$w_{V34} = -865.23$ kJ/kg	$w_{p34} = -1211.3$ kJ/kg	$q_{s34} = 0$ kJ/kg	$q_{T34} = -9093.9$ kJ/kg
$w_{V41} = 0$ kJ/kg	$w_{p41} = -134.97$ kJ/kg	$q_{s41} = -337.42$ kJ/kg	$q_{T41} = -3411.5$ kJ/kg
$w_{Kreis} = -63707$ kJ/kg		$\eta_{th} = 0.6538$	$\zeta = 0.8253$
$a_{wp} = 3.2265$	$a_{wv} = 2.5904$	$a_{qT} = 38.259 \quad a_{qs} = 2.0593$	$a_g = 46.135$

7.5 Comparative Evaluation of Right-Hand Cyclic Processes

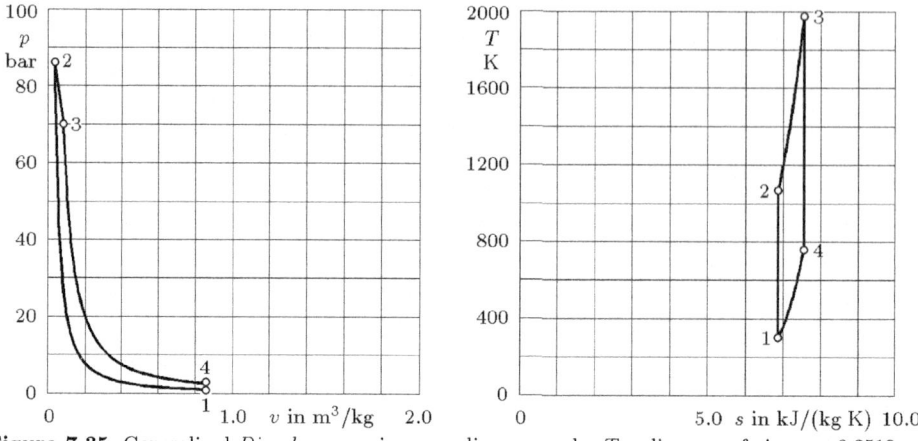

Figure 7.35 Generalised *Diesel* process in a p, v diagram and a T, s diagram of air; $n = 0.2519$

$p_1 = 1.000$ bar	$v_1 = 0.85629$ m³/kg	$T_1 = 298.15$ K	$s_1 = 6.8670$ kJ/(kg K)
$p_2 = 86.150$ bar	$v_2 = 0.03551$ m³/kg	$T_2 = 1065.04$ K	$s_2 = 6.8670$ kJ/(kg K)
$p_3 = 70.000$ bar	$v_3 = 0.08096$ m³/kg	$T_3 = 1973.15$ K	$s_3 = 7.5464$ kJ/(kg K)
$p_4 = 2.576$ bar	$v_4 = 0.85629$ m³/kg	$T_4 = 768.09$ K	$s_4 = 7.5464$ kJ/(kg K)
$w_{V12} = 550.63$ kJ/kg	$w_{p12} = 770.88$ kJ/kg	$q_{s12} = 0$ kJ/kg	$q_{T12} = 4845.0$ kJ/kg
$w_{V23} = -348.62$ kJ/kg	$w_{p23} = -87.806$ kJ/kg	$q_{s23} = 1000.6$ kJ/kg	$q_{T23} = 7023.4$ kJ/kg
$w_{V34} = -865.23$ kJ/kg	$w_{p34} = -1211.3$ kJ/kg	$q_{s34} = 0$ kJ/kg	$q_{T34} = -9093.9$ kJ/kg
$w_{V41} = 0$ kJ/kg	$w_{p41} = -134.97$ kJ/kg	$q_{s41} = -337.42$ kJ/kg	$q_{T41} = -3411.5$ kJ/kg
$w_{Kreis} = -663.23$ kJ/kg		$\eta_{th} = 0.6628$	$\zeta = 0.8310$
$a_{wp} = 3.3246$	$a_{wv} = 2.6604$	$a_{qT} = 36.711$ $a_{qs} = 2.0175$	$a_g = 44.714$

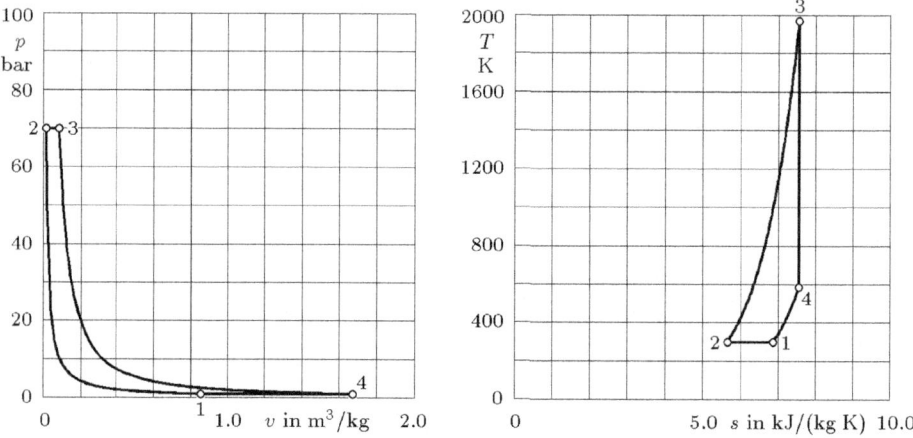

Figure 7.36 Gas expansion process in a p, v diagram and in a T, s diagram of air

$p_1 = 1.000$ bar	$v_1 = 0.85629$ m³/kg	$T_1 = 298.15$ K	$s_1 = 6.8670$ kJ/(kg K)
$p_2 = 70.000$ bar	$v_2 = 0.01223$ m³/kg	$T_2 = 298.15$ K	$s_2 = 5.6468$ kJ/(kg K)
$p_3 = 70.000$ bar	$v_3 = 0.08096$ m³/kg	$T_3 = 1973.15$ K	$s_3 = 7.5464$ kJ/(kg K)
$p_4 = 1.000$ bar	$v_4 = 1.68335$ m³/kg	$T_4 = 586.12$ K	$s_4 = 7.5464$ kJ/(kg K)
$w_{V12} = 363.79$ kJ/kg	$w_{p12} = 363.79$ kJ/kg	$q_{s12} = -363.79$ kJ/kg	$q_{T12} = 0$ kJ/kg
$w_{V23} = -481.06$ kJ/kg	$w_{p23} = 0$ kJ/kg	$q_{s23} = 1683.7$ kJ/kg	$q_{T23} = 11523$ kJ/kg
$w_{V34} = -995.89$ kJ/kg	$w_{p34} = -1394.2$ kJ/kg	$q_{s34} = 0$ kJ/kg	$q_{T34} = -10467$ kJ/kg
$w_{V41} = 82.706$ kJ/kg	$w_{p41} = 0$ kJ/kg	$q_{s41} = -289.47$ kJ/kg	$q_{T41} = -2086.3$ kJ/kg
$w_{Kreis} = -1030.5$ kJ/kg		$\eta_{th} = 0{,}7391$	$\zeta = 1.0000$
$a_{wp} = 1.7061$	$a_{wv} = 1.8666$	$a_{qT} = 23.365$ $a_{qs} = 2.2679$	$a_g = 29.2056$

Table 7.15 Thermal efficiency, exergetic efficiency, mechanical effort ratios and thermal effort ratios as well as the total effort ratio of reversible cyclic processes [33]
a) with air (ideal gas) as working fluid ($p_1 = 1.00$ bar; $T_1 = T_b = 298.15$ K)
b) with water and steam as working fluid ($p_1 = 0.032$ bar; $T_1 = T_b = 298.15$ K)

a)

	Single-polytropic Carnot process	Carnot process	Ericsson process	Stirling process	Joule process	Otto process	Diesel process	Generalised Diesel process; n=0.2519	Gas expansion process
state 3: $p_3 = 70$ bar; $T_3 = 1973.15$ K								
η_{th}	0.849	As a right-hand cycle process not possible	0.849	0.849	0.703	0.611	0.654	0.663	0.739
ζ	1		1	1	0.887	0.797	0.825	0.831	1
a_{wp}	1.406		1.356	2.204	3.071	4.086	3.227	3.325	1.706
a_{wv}	1.854		1.827	1.356	2.721	2.269	2.590	2.660	1.867
a_{qT}	11.712		12.276	21.873	35.651	46.246	38.259	36.711	23.365
a_{qs}	2.976		3.004	3.476	1.845	2.275	2.059	2.018	2.268
a_g	17.948		18.463	28.909	43.288	54.876	46.135	44.714	29.206
staate 3a: $p_{3a} = 20$ bar; $T_{3a} = 1523.15$ K								
η_{th}	0.804	As a right-hand cycle process not possible	0.804	0.804	0.575	0.421	0.492	0.507	0.709
ζ	1		1	1	0.800	0.619	0.678	0.697	1
a_{wp}	1.557		1.487	2.952	2.708	4.331	2.997	3.111	1.822
a_{wv}	2.186		2.029	1.487	2.643	2.021	2.427	2.508	2.143
a_{qT}	16.389		17.365	37.804	37.949	59.267	44.191	41.570	28.640
a_{qs}	3.760		3.823	5.150	2.478	3.754	3.065	2.945	2.947
a_g	23.892		24.704	47.393	45.778	69.373	52.680	50.134	35.552
state 3b: $p_{3b} = 20$ bar; $T_{3b} = 973.15$ K								
η_{th}	0.694	As a right-hand cycle process not possible	0.694	0.694	0.575	0.484	0.544	0.551	0.545
ζ	1		1	1	0.897	0.834	0.850	0.853	1
a_{wp}	1.954		1.883	2.987	6.170	7.374	6.461	6.597	2.676
a_{wv}	2.594		2.551	1.883	5.115	4.445	4.901	4.998	2.893
a_{qT}	15.921		16.611	27.325	60.677	71.988	64.038	62.437	30.623
a_{qs}	4.193		4.220	4.642	2.478	2.878	2.673	2.632	3.434
a_g	24.662		25.265	36.837	74.440	86.685	78.073	76.664	39.626
state 3c: $p_{3c} = 40$ bar; $T_{3c} = 673.15$ K								
η_{th}	0.557	0.557	0.557	0.557	As a right-hand cycle process not possible	As a right-hand cycle process not possible	As a right-hand cycle process not possible	As a right hand cycle process not possible	0.285
ζ	1	1	1	1					1
a_{wp}	2.649	10.938	2.590	3.286					6.057
a_{wv}	3.178	8.553	3.132	2.590					5.765
a_{qT}	12.536	56.025	12.840	16.439					38.652
a_{qs}	4.501	2.590	4.488	4.330					5.035
a_g	22.864	78.106	23.050	26.645					55.509

b)

Clausius-Rankine process	
state 3d:	
$p_{3d} = 350$ bar	
$T_{3d} = 873.15$ K	
η_{th}	0.473
ζ	1
a_{wp}	1.045
a_{wv}	1.125
a_{qT}	6.191
a_{qs}	3.226
a_g	11.587
state 3e:	
$p_{3e} = 200$ bar	
$T_{3e} = 823.15$ K	
η_{th}	0.455
ζ	1
a_{wp}	1.027
a_{wv}	1.135
a_{qT}	5.948
a_{qs}	3.391
a_g	11.501

Practical difficulties in the realisation of a *Carnot* process arise with an ideal gas as a working fluid if the upper process temperature is to assume a desired high value; then a technically unrealisable very high process pressure is required. In contrast, with a pure substance in the two-phase region and in its vicinity, approximations to the *Carnot* process can be realised. This shows, as an example, the Cold vapor compression process, which is a modification of the left-hand *Carnot* process and can be realised at low specific values of volume and entropy in the liquid region, in the two-phase liquid-vapor region and in the superheated vapor region in the vicinity of the saturated vapor line (cf. section 7.6.5).

The Gas expansion process is positioned between the *Ericsson* process and the *Joule* process with regard to the seven evaluation criteria. As far as — e.g. in the case of electrical power recovery in natural gas supply systems at the interfaces of transport networks to distribution networks — the process is designed for safety rea-

sons with a comparatively low upper process temperature T_{3c}, only moderate values for the thermal efficiency η_{th}, for the mechanical effort ratios a_{wp} and a_{wv}, as well as for the thermal effort ratio a_{qs} result.

In this case, however, this is hardly of importance, because the compression with regard to the natural gas transport has to be carried out anyway and therefore, strictly speaking, should not be counted as part of the power recovery process ([28] to [31]). Compared to the *Carnot* process with the state T_{3c}, the Gas expansion process has a lower thermal efficiency η_{th}, but significantly better values for the effort ratios a_{wp}, a_{wv}, a_{qT} and a_g.

The *Clausius-Rankine* process, as the comparison shows, cannot achieve the favourable theoretical thermal efficiencies η_{th} of the generalised *Diesel* process, the *Diesel* process, the *Otto* process and the *Joule* process. On the other hand, the *Clausius-Rankine* process has, with respect to the exergetic efficiency ζ as well as to all effort ratios a_{wp}, a_{wv}, a_{qT}, a_{qs} and a_g, special advantages: The low values for these illustrate the long-known fact that the *Clausius-Rankine* process is comparatively less sensitive to irreversibilities; thus, the thermal efficiency of the irreversible steam power process can reach that of irreversible combustion engine processes and irreversible gas turbine processes in practice or even exceed them significantly.

Possible improvements of the *Clausius-Rankine* process such as a multi-stage isentropic expansion combined with respective multi-stage isobaric intermediate superheating in the high-temperature range as well as multi-stage feedwater preheating are not included in the comparison made here, but they would illustrate the advantages of the *Clausius-Rankine* process.

The favourable values for the mechanical as well as the thermal effort ratios further imply that the process gets by with comparatively limited mechanical effort (in particular with a low mechanical effort for the feedwater compression); the same applies, with restrictions, to the equipment required for heat transfer. An important reason for this is that the *Clausius-Rankine* process is not limited to the gas phase, but is to be realised also in the two phase liquid-vapor region as well as in the liquid region — i.e. also in regions of low specific volumes and entropies. Furthermore, the process can be carried out with extremely large volume ratios v_4/v_1 and pressure ratios p_3/p_1. Overall, this leads to very favourable values of a_{wp}, a_{wv}, a_{qT}, a_{qs} and a_g. This opens up favourable prospects for the *Clausius-Rankine* process also in the future for possibilities for improvement.

7.5.3.3 Graphical Representation of the Thermodynamic Relations

The equations in Tables 7.5 to 7.14 gain clarity through graphical representations. These graphical representations can be understood as surfaces in space; this is exemplified by Fig. 7.37 for the thermal effort ratio a_{qT} of the *Otto* process. The relationships captured in the equations of Tables 7.5 to 7.14 are shown in Figs. 7.38 to 7.46 for the single-polytropic *Carnot* process for $n = 0.1$, the *Carnot* process, the *Ericsson* process, the *Stirling* process, the *Joule* process, the *Otto* process, the *Diesel* process, the generalised *Diesel* process for $n = 0.3$ and the Gas expansion process; for this purpose air as working fluid shall be chosen. In addition, state 1 according to section 7.5.3.2 is chosen as the initial state.

Here, for the respective reversible cyclic processes the thermal efficiency η_{th}, the exergetic efficiency ζ, the mechanical effort ratios a_{wp} and a_{wv}, the thermal effort ratios a_{qT} and a_{qs}, as well as the total effort ratio a_g as a function of temperature T_3 and pressure p_3 are considered. T_3 and p_3 are — with the exception of the selected single-polytropic *Carnot* process and the generalised *Diesel* process — the respectively highest temperatures and pressures of the cyclic processes considered. Temperatures and pressures have values of 373.15 K to 2273.15 K (100 °C to 2000 °C) and of 10 bar to 150 bar. The representations are suitable e.g. for a comparison of thermodynamic advantages or disadvantages of certain selected reversible cyclic processes as well as for an illustration of their possibilities for further improvements.

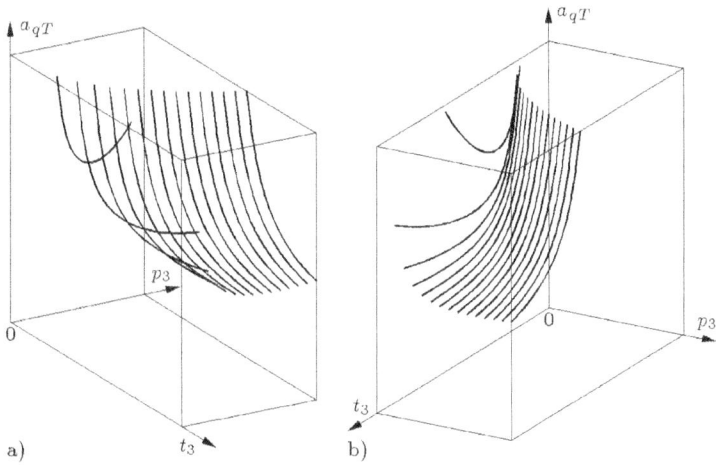

Figure 7.37 Spatial representation of the thermal effort ratio a_{qT} of the *Otto* process

The graphs for the thermal efficiency η_{th} and the thermal effort ratio a_{qs} show within the chosen coordinate ranges for the *Joule* process, the *Otto* process, the *Diesel* process and the generalised *Diesel* process curves with defined initial states for the respective isobars; this also applies to the thermal efficiency η_{th} of the Gas expansion process. These are characterised by the condition that a right-hand cycle process must be realisable.

If the curves were extended to the left towards lower temperatures T_3, left-hand cyclic processes or transition cyclic processes would be recorded; the representation of which is omitted here. The respective initial states of the isobars are also shown in the diagrams for the exergetic efficiency ζ at $\zeta = 1$. The limits for the *Carnot* process are shown in the diagram for its thermal efficiency η_{th} with an additional scale.

A consideration of the *Otto* process, the *Diesel* process and the generalised *Diesel* process (Figs. 7.43 to 7.45) leads in particular to the following observations: Their thermal efficiencies η_{th} and exergetic efficiencies ζ increase with increasing pressure p_3 and decrease with increasing temperature T_3. (The temperature-dependent decrease of η_{th} and ζ is, however, smaller at high pressures p_3 than at low pressures p_3.) From this, without knowledge of the courses of a_{wp}, a_{wv}, a_{qT}, a_{qs} and a_g, it could be concluded that a high pressure p_3 is advantageous, but a high temperature T_3 is disadvantageous. The curves of a_{wp}, a_{wv}, a_{qT} and a_g show, however, that with regard to the technical effort not only the highest possible pressure p_3, but also the highest possible tempera-

ture T_3 is useful, because all effort ratios except a_{qs} at the desired higher pressures p_3 decrease with increasing temperature T_3.

This suggests a further increase in p_3 and T_3 as a development goal. However, the pictures also illustrate that the increase in improvements with increasing p_3 and T_3 at already high p_3 and T_3 is no longer as significant as in the region of low p_3 and T_3. The diagrams further show that the *Diesel* process has significant advantages over the *Otto* process. However, these advantages are not as pronounced at high values of p_3 and T_3 as at lower values of p_3 and T_3.

The *Stirling* process as a cyclic process realisable with displacement machines (piston machines or screw machines) is technically comparable with the *Otto* process and the *Diesel* process. As Fig. 7.41 shows, a transition to higher pressures p_3 has a positive effect on the thermal effort ratios a_{qT} and a_{qs}, but has no influence on the thermal efficiency η_{th}, the exergetic efficiency ζ and the mechanical effort ratio a_{wv}, and only a limited influence on the mechanical effort ratio a_{wp}.

An increase of the highest process temperature T_3 leads to an improvement of η_{th}, a_{wv} and partly a_{wp}, but to a deterioration of a_{qT} and a_g. Compared to *Otto* and *Diesel* processes at high pressures p_3 and high temperatures T_3, the *Stirling* process already shows advantages at moderate values of p_3 and T_3. The technical realisation of the *Stirling* process, from a thermodynamic point of view, appears to make sense.

The *Carnot* process opens — despite its in principle outstanding thermal and exergetic efficiency η_{th} and ζ — only a very limited range for technical implementation, as Fig. 7.39 shows: The effort ratios a_{wp}, a_{wv}, a_{qT} and a_g reach acceptable values only at very moderate temperatures T_3; at higher temperatures T_3 the *Carnot* process cannot be represented as a right-hand cyclic process. Thus, with the chosen initial state 1, high values for the thermal efficiency η_{th} are not realisable.

A transition to higher pressures p_3 has a positive effect on the effort ratios a_{wp}, a_{wv}, a_{qT} and a_g, but has no influence on the thermal efficiency η_{th}, the exergetic efficiency ζ and the thermal effort ratio a_{qs}. A technical implementation of the *Carnot* process therefore appears to be reasonable only within limits.

This is shown in particular by a comparison with the *Stirling* process (Fig. 7.41), the *Ericsson* process (Fig. 7.40) and the single-polytropic *Carnot* process with $n = 0.1$ investigated here (Fig. 7.38); these achieve the same good values of the thermal and exergetic efficiencies η_{th} and ζ, but avoid the disadvantages of the *Carnot* process in the evaluation criteria a_{wp}, a_{wv}, a_{qT} and a_g.

302 7 Cyclic Processes

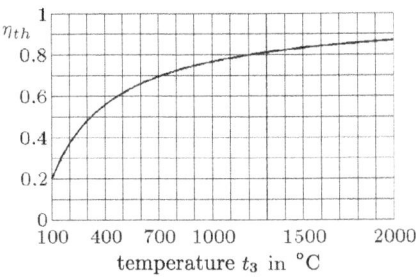

a) Thermal efficiency η_{th}

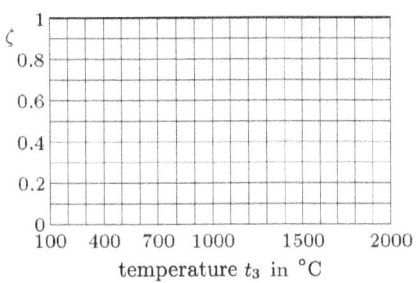

b) Exergetic efficiency ζ

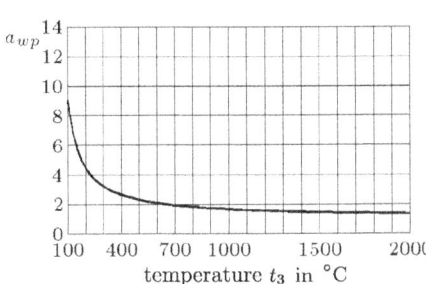

c) Mechanical effort ratio a_{wp}

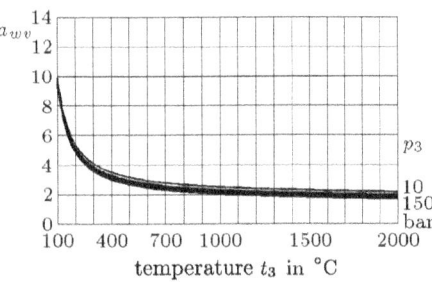

d) Mechanical effort ratio a_{wv}

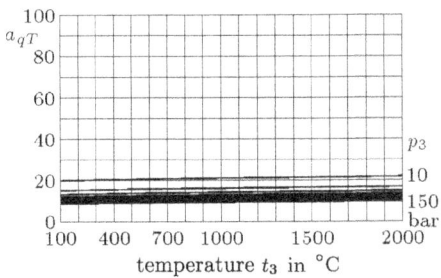

e) Thermal effort ratio a_{qT}

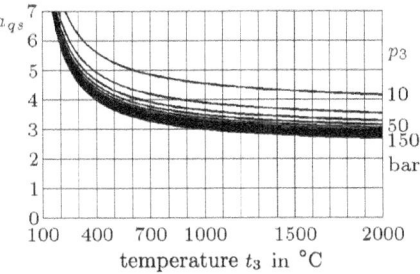

f) Thermal effort ratio a_{qs}

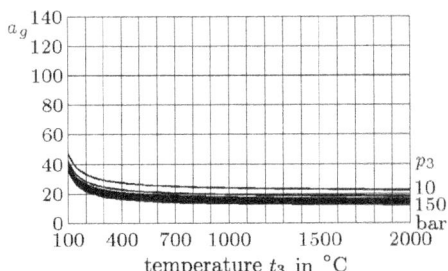

g) Total effort ratio a_g

Figure 7.38 Single-polytropic *Carnot* process with polytropic exponent $n = 0.1$
a) thermal efficiency η_{th}
b) exergetic efficiency ζ
c) mechanical effort ratio a_{wp}
d) mechanical effort ratio a_{wv}
e) thermal effort ratio a_{qT}
f) thermal effort ratio a_{qs}
g) total effort ratio a_g

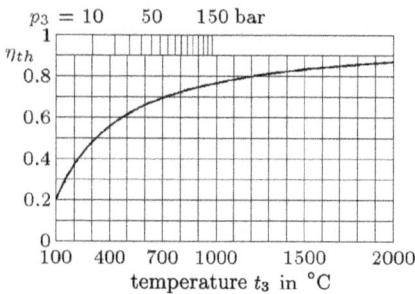
a) Thermal efficiency η_{th}

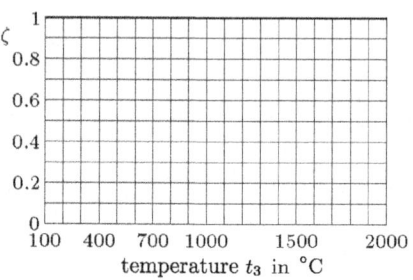
b) Exergetic efficiency ζ

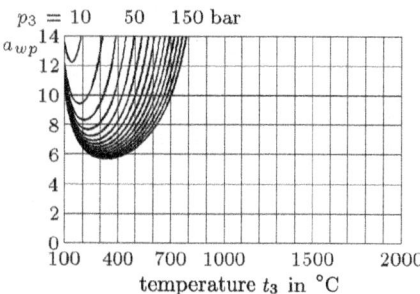
c) Mechanical effort ratio a_{wp}

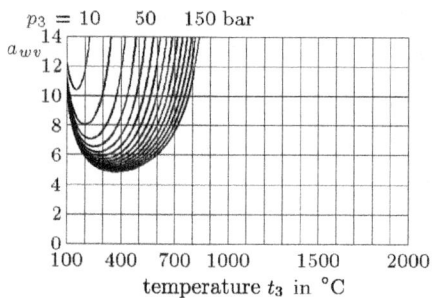
d) Mechanical effort ratio a_{wv}

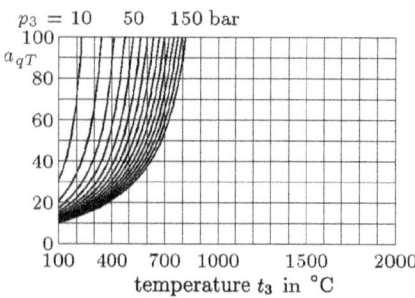
e) Thermal effort ratio a_{qT}

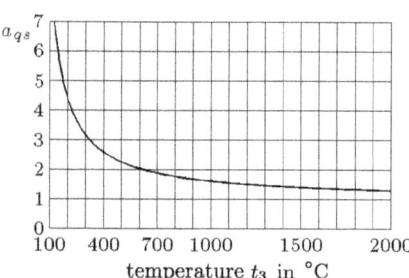
f) Thermal effort ratio a_{qs}

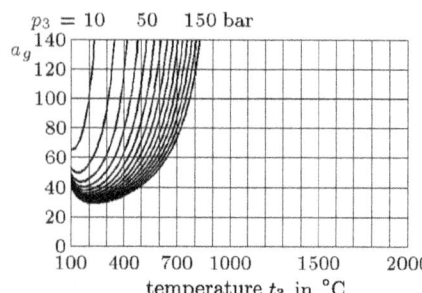
g) Total effort ratio a_g

Figure 7.39 *Carnot* process
a) thermal efficiency η_{th}
b) exergetic efficiency ζ
c) mechanical effort ratio a_{wp}
d) mechanical effort ratio a_{wv}
e) thermal effort ratio a_{qT}
f) thermal effort ratio a_{qs}
g) total effort ratio a_g

304 7 Cyclic Processes

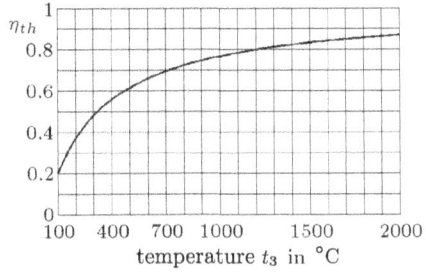

a) Thermal efficiency η_{th}

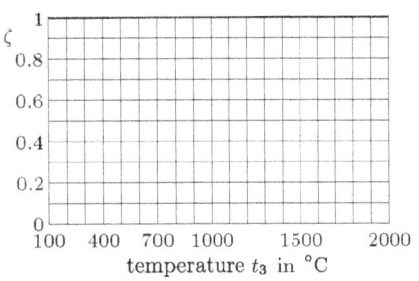

b) Exergetic efficiency ζ

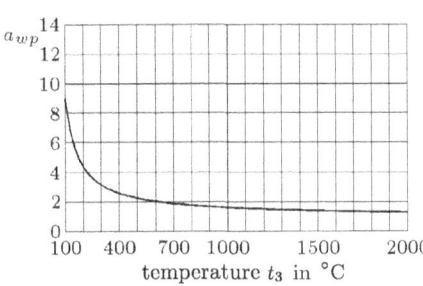

c) Mechanical effort ratio a_{wp}

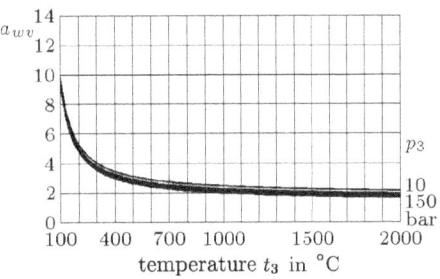

d) Mechanical effort ratio a_{wv}

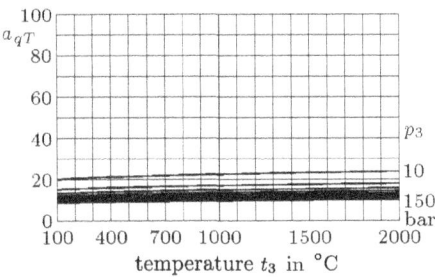

e) Thermal effort ratio a_{qT}

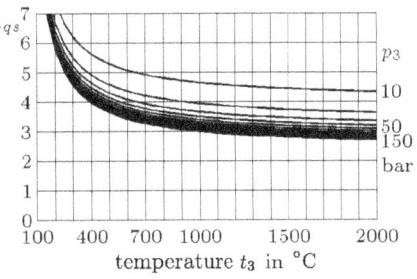

f) Thermal effort ratio a_{qs}

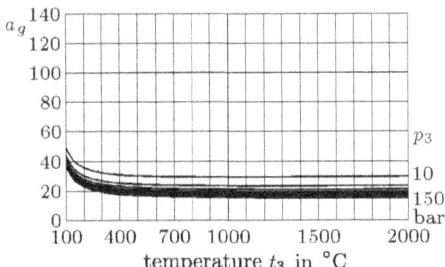

g) Total effort ratio a_g

Figure 7.40 *Ericsson* process
a) thermal efficiency η_{th}
b) exergetic efficiency ζ
c) mechanical effort ratio a_{wp}
d) mechanical effort ratio a_{wv}
e) thermal effort ratio a_{qT}
f) thermal effort ratio a_{qs}
g) total effort ratio a_g

7.5 Comparative Evaluation of Right-Hand Cyclic Processes

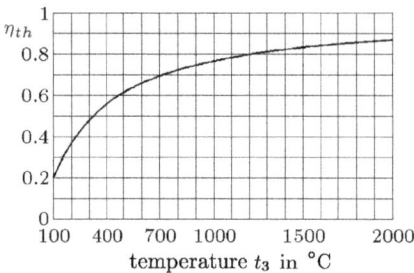
a) Thermal efficiency η_{th}

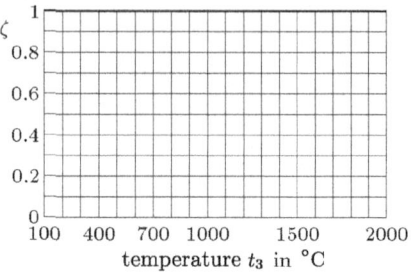
b) Exergetic efficiency ζ

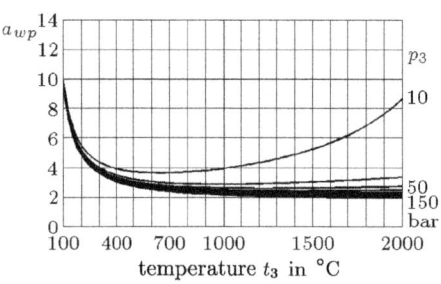
c) Mechanical effort ratio a_{wp}

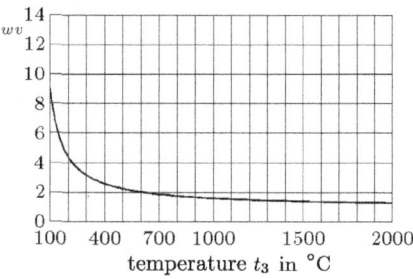
d) Mechanical effort ratio a_{wv}

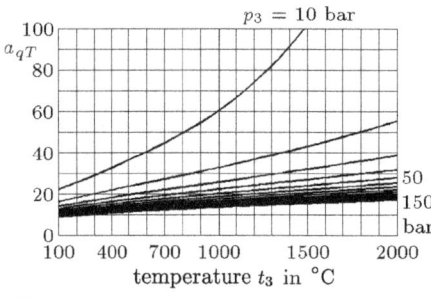
e) Thermal effort ratio a_{qT}

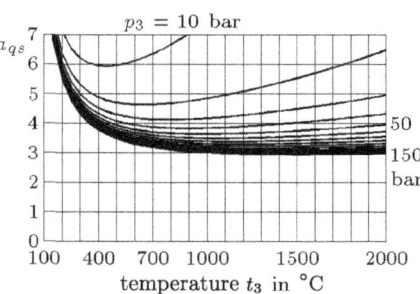
f) Thermal effort ratio a_{qs}

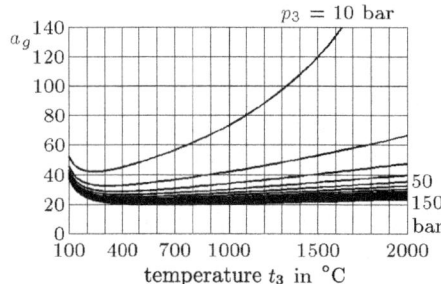
g) Total effort ratio a_g

Figure 7.41 *Stirling* process
a) thermal efficiency η_{th}
b) exergetic efficiency ζ
c) mechanical effort ratio a_{wp}
d) mechanical effort ratio a_{wv}
e) thermal effort ratio a_{qT}
f) thermal effort ratio a_{qs}
g) total effort ratio a_g

306   7 Cyclic Processes

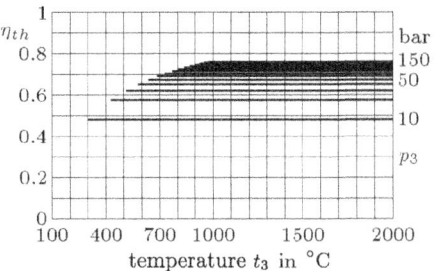

a) Thermal efficiency η_{th}

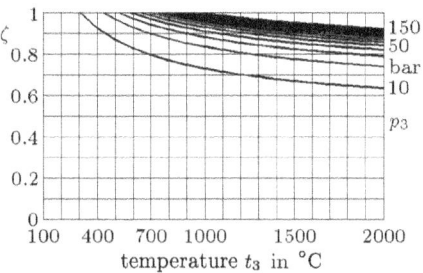

b) Exergetic efficiency ζ

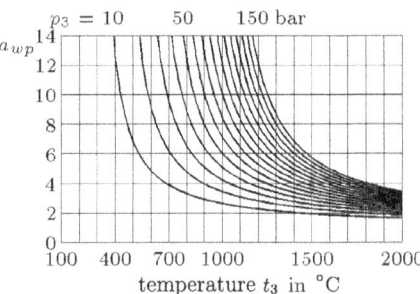

c) Mechanical effort ratio a_{wp}

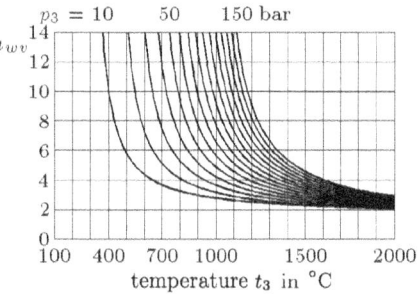

d) Mechanical effort ratio a_{wv}

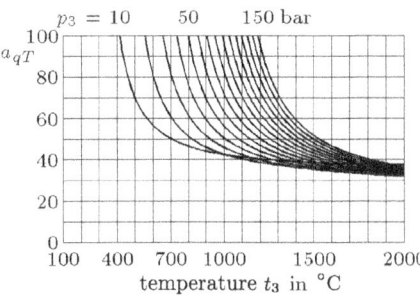

e) Thermal effort ratio a_{qT}

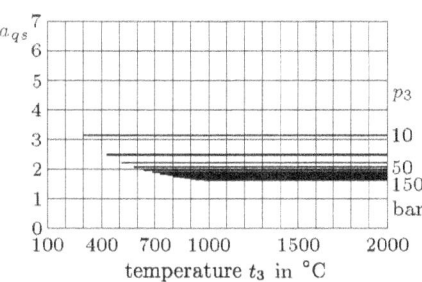

f) Thermal effort ratio a_{qs}

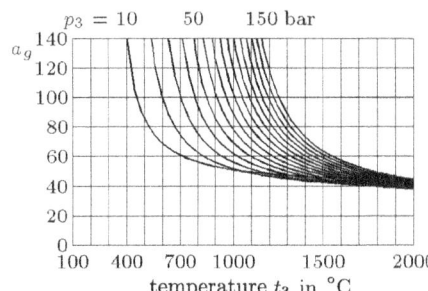

g) Total effort ratio a_g

Figure 7.42 *Joule* process
a) thermal efficiency η_{th}
b) exergetic efficiency ζ
c) mechanical effort ratio a_{wp}
d) mechanical effort ratio a_{wv}
e) thermal effort ratio a_{qT}
f) thermal effort ratio a_{qs}
g) total effort ratio a_g

7.5 Comparative Evaluation of Right-Hand Cyclic Processes   307

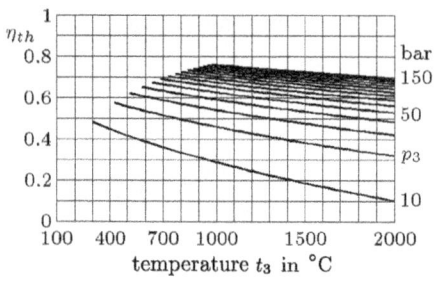

a) Thermal efficiency η_{th}

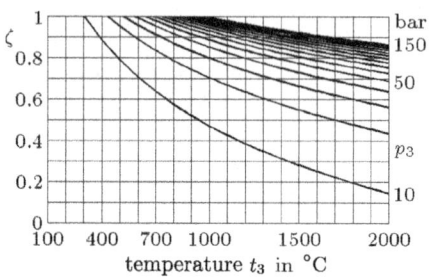

b) Exergetic efficiency ζ

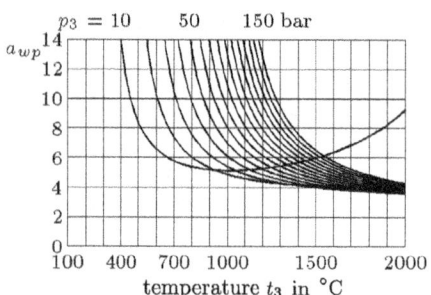

c) Mechanical effort ratio a_{wp}

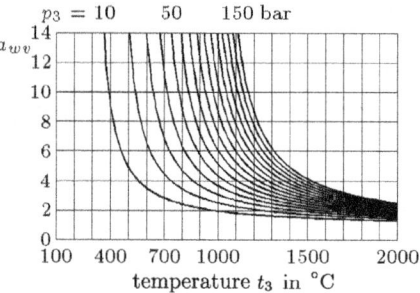

d) Mechanical effort ratio a_{wv}

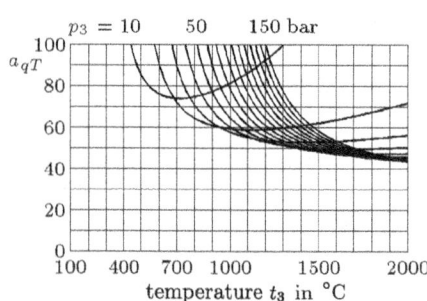

e) Thermal effort ratio a_{qT}

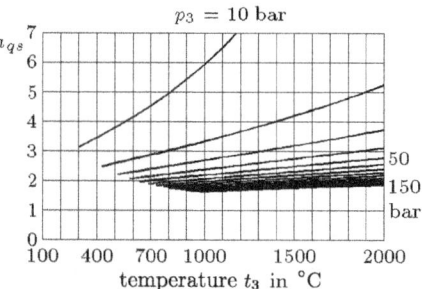

f) Thermal effort ratio a_{qs}

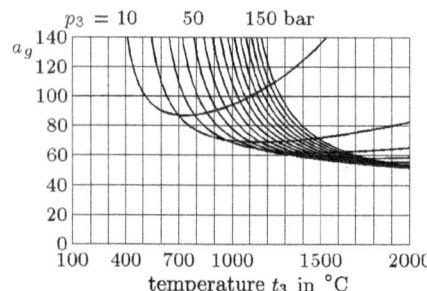

g) Total effort ratio a_g

Figure 7.43 *Otto* process
a) thermal efficiency η_{th}
b) exergetic efficiency ζ
c) mechanical effort ratio a_{wp}
d) mechanical effort ratio a_{wv}
e) thermal effort ratio a_{qT}
f) thermal effort ratio a_{qs}
g) total effort ratio a_g

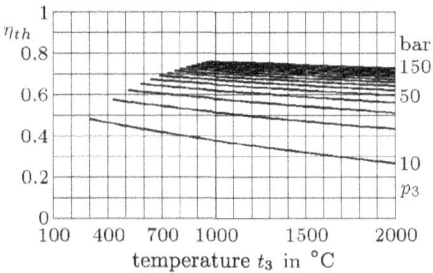
a) Thermal efficiency η_{th}

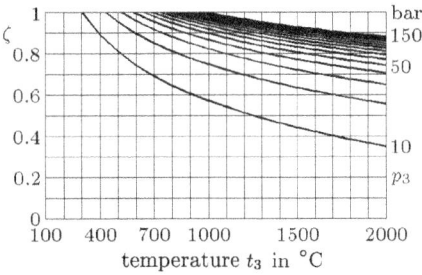
b) Exergetic efficiency ζ

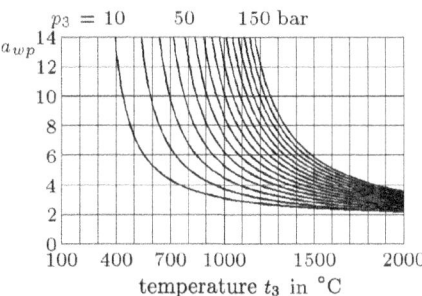
c) Mechanical effort ratio a_{wp}

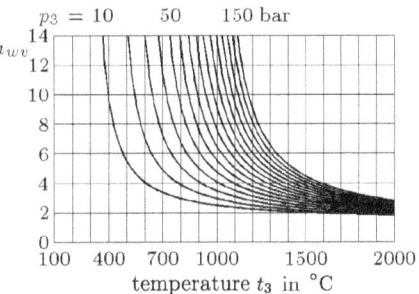
d) Mechanical effort ratio a_{wv}

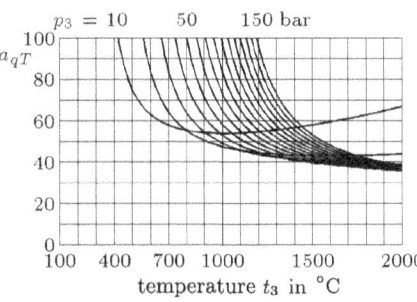
e) Thermal effort ratio a_{qT}

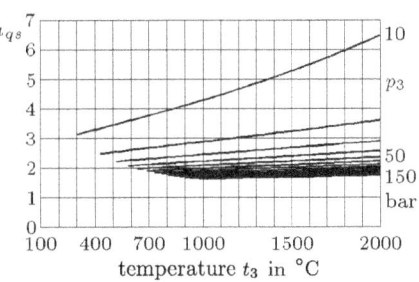
f) Thermal effort ratio a_{qs}

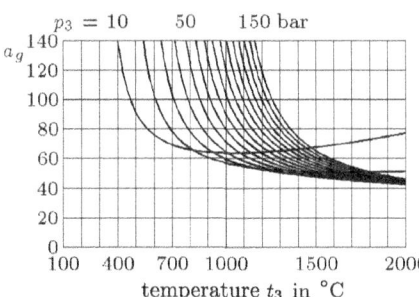
g) Total effort ratio a_g

Figure 7.44 *Diesel* process
a) thermal efficiency η_{th}
b) exergetic efficiency ζ
c) mechanical effort ratio a_{wp}
d) mechanical effort ratio a_{wv}
e) thermal effort ratio a_{qT}
f) thermal effort ratio a_{qs}
g) total effort ratio a_g

7.5 Comparative Evaluation of Right-Hand Cyclic Processes

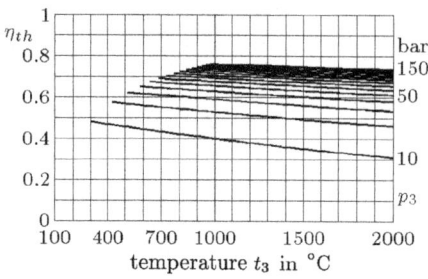

a) Thermal efficiency η_{th}

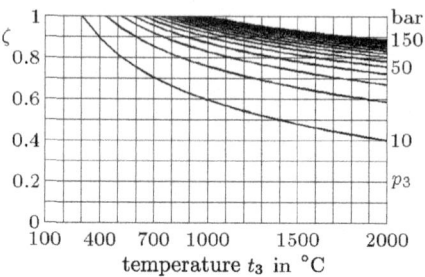

b) Exergetic efficiency ζ

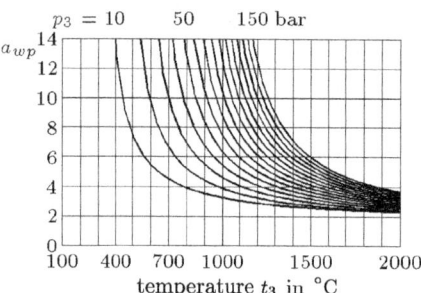

c) Mechanical effort ratio a_{wp}

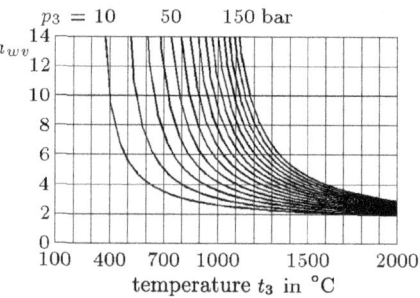

d) Mechanical effort ratio a_{wv}

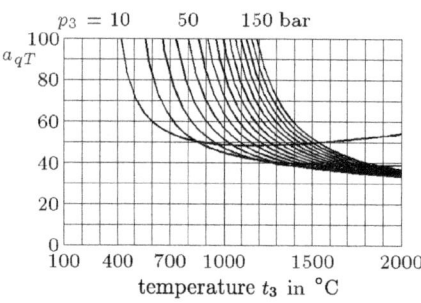

e) Thermal effort ratio a_{qT}

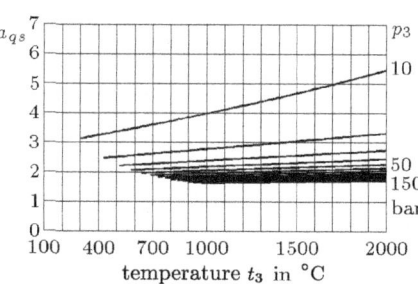

f) Thermal effort ratio a_{qs}

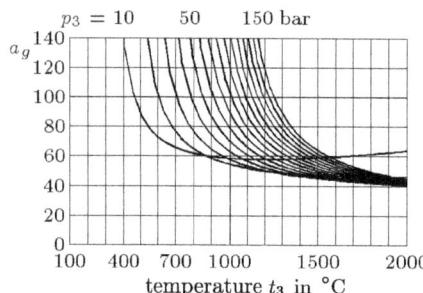

g) Total effort ratio a_g

Figure 7.45 Generalised *Diesel* process with polytropic exponent $n = 0.3$
a) thermal efficiency η_{th}
b) exergetic efficiency ζ
c) mechanical effort ratio a_{wp}
d) mechanical effort ratio a_{wv}
e) thermal effort ratio a_{qT}
f) thermal effort ratio a_{qs}
g) total effort ratio a_g

310 7 Cyclic Processes

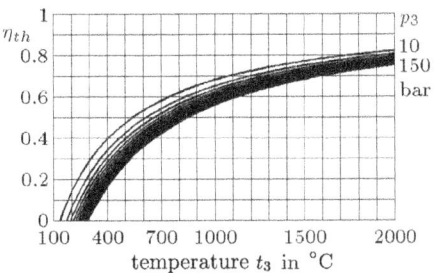

a) Thermal efficiency η_{th}

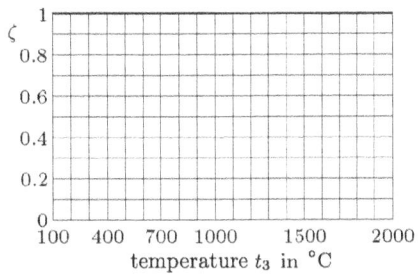

b) Exergetic efficiency ζ

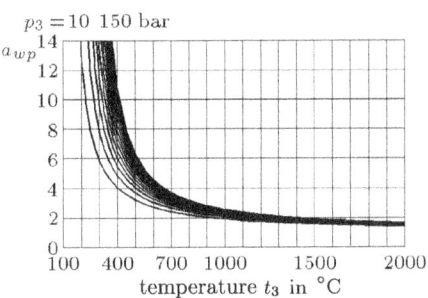

c) Mechanical effort ratio a_{wp}

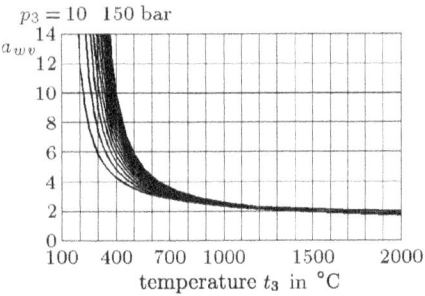

d) Mechanical effort ratio a_{wv}

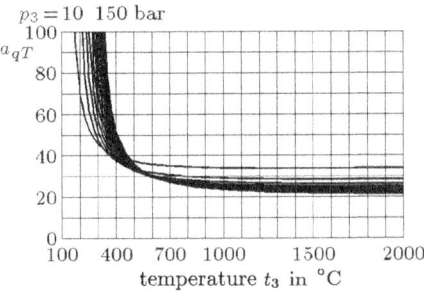

e) Thermal effort ratio a_{qT}

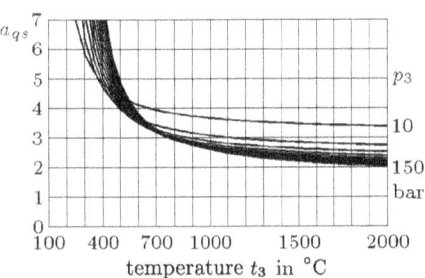

f) Thermal effort ratio a_{qs}

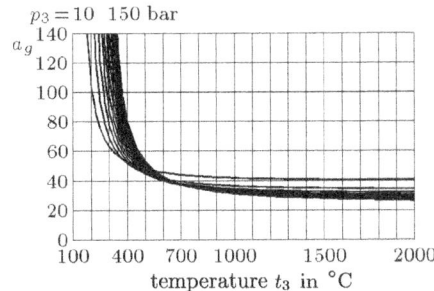

g) Total effort ratio a_g

Figure 7.46 Gas expansion process
a) thermal efficiency η_{th}
b) exergetic efficiency ζ
c) mechanical effort ratio a_{wp}
d) mechanical effort ratio a_{wv}
e) thermal effort ratio a_{qT}
f) thermal effort ratio a_{qs}
g) total effort ratio a_g

Joule process (Fig. 7.42) and *Ericsson* process (Fig. 7.40) are best realised in thermal fluid flow machinery. Single-polytropic *Carnot* process (Fig. 7.38) and Gas expansion process (Fig. 7.46) can be used — depending on the chosen values for the polytropic exponent n or for the working pressures and temperatures — either in displacement machines (piston or screw machines) or in thermal fluid flow machines. The choice of $n = 0.1$ for the single-polytropic *Carnot* process makes it similar to the *Ericsson* process; therefore, Fig. 7.38 and Fig. 7.40 show that the curves for the evaluation criteria η_{th}, ζ, a_{wp}, a_{wv}, a_{qT}, a_{qs} and a_g are very similar for both processes. This is also true to a first approximation for the Gas expansion process in the range of higher temperatures T_3 (Fig. 7.46).

For the selected single-polytropic *Carnot* process as well as for the *Ericsson* process, the thermal and exergetic efficiencies η_{th} and ζ as well as the mechanical effort ratios a_{wp} and a_{wv} are not or barely pressure-dependent; for a_{qT}, a_{qs} and a_g higher pressures p_3 lead to more favourable values. Similarly, these tendencies are also evident for the Gas expansion process in the range of higher temperatures T_3. The implementation of high temperatures T_3 leads to noticeable improvements for all three processes considered for η_{th}, a_{wp}, a_{wv}, a_{qs} and a_g, while high temperatures T_3 have a stronger effect on a_{qT} only in the case of the Gas expansion process.

In comparison, the *Joule* process (Fig. 7.42) deviates significantly in some cases: The thermal efficiency η_{th} and the exergetic efficiency ζ increase with increasing pressure p_3, whereas T_3 has no influence on η_{th}. The effort ratios a_{wp}, a_{wv}, a_{qT} and a_g can be significantly improved by high values of the temperature T_3 while higher pressures p_3 lead to a deterioration. For a_{qs}, higher pressures p_3 lead to more favourable values, while here an influence of the temperature T_3 is not given. The illustrations show that for the *Joule* process a further increase of p_3 and T_3 seems to be reasonable. In addition, a further development of the *Joule* process in the direction of the Gas expansion process opens up possibilities as well as — in the sense of additional improvements — towards the *Ericsson* process and towards the single-polytropic *Carnot* process [34].

7.5.3.4 Cyclic Process Calculations for Real Fluids

In the calculation of cyclic processes with ideal gases as working fluids, where $c_p = \text{const}$ is assumed, the relations for the thermal and the exergetic efficiency as well as for the effort ratios are shown in Tables 7.5 to 7.13 in section 7.5.3.1. The only exception to this are the relations in Table 7.14, which apply to the *Clausius-Rankine* process with a real fluid.

In the following, corresponding relations for real fluids are presented for all cyclic processes discussed in section 7.5.3.1. Knowledge of suitable equations of state for the respective working fluid is presupposed, as they are available in particular according to [128] and [130]. Thus, with the software [130], the most important thermodynamic state variables of numerous pure substances can be calculated with the help of thermodynamic equations of state for real fluids.

When calculating cyclic processes with ideal gases as working fluids, it can be directly illustrated how e.g. the highest process pressure $p_3 = p_{max}$ and the highest process temperature $T_3 = T_{max}$ as well as further state variables have an effect on important evaluation parameters.

This is no longer directly possible when using real fluids. However, the relevant calculation variables specific enthalpy h, specific internal energy u, specific free enthalpy g and specific free energy f can be represented as functions of p and T, so that — e.g. by diagrams — the influence of $p_3 = p_{max}$ and $T_3 = T_{max}$ as well as other state variables can be shown.

In Table 7.16 — starting from the thermodynamic basic equations suitable for this purpose — corresponding relations are compiled for this aim. With them for simple systems for the important cases of the

- isothermal change of state,
- isobaric change of state,
- isochoric change of state,
- isentropic change of state

the differentials dw_p, dw_v, dq_T and $dq_s = dq_{rev}$ of the four process variables specific pressure change work, specific volume change work, specific temperature change heat and specific entropy change heat (i.e. the specific reversible substitute heat) respectively, with the help of the differentials dh, du, dg and df of the four state variables specific enthalpy, specific internal energy, specific free enthalpy and specific free energy can be calculated.

Accordingly, between the states i and j, the four process variables specific pressure change work $w_{p\,ij}$, specific volume change work $w_{v\,ij}$, specific temperature change heat $q_{T\,ij}$ and specific entropy change heat (i.e. the specific reversible substitute heat) $q_{s\,ij} = (q_{ij})_{rev}$ can be calculated, using the difference of the specific enthalpy $h_j - h_i$, of the specific internal energy $u_j - u_i$, of the specific free enthalpy $g_j - h_i$, and of the specific free energy $f_j - f_i$.

These four process variables are necessary for the calculation of the thermal efficiency η_{th}, the exergetic efficiency ζ, the four effort ratios a_{wp}, a_{wv}, a_{qT} and a_{qs} and the total effort ratio a_g, with which the efficiency of reversible thermodynamic cycles can be characterised.

The corresponding relations for the *Carnot* process, *Ericsson* process, *Stirling* process, *Joule* process, *Otto* process, *Diesel* process, Gas expansion process and the *Clausius-Rankine* process are compiled in the following pages. The same counting of the respective state variables 1, 2, 3 and 4 is used as in sections 7.5.3.1 to 7.5.3.3: From state 1 with the lowest process temperature, the working fluid is compressed; from state 3 with the highest process temperature and the highest process pressure, the working fluid is expanded to a lower pressure.

From these relations it can be seen that here the four state variables specific enthalpy h, specific internal energy u, specific free enthalpy g and specific free energy f have the same importance for the calculations and thus are equivalent to each other.

In contrast to this, in engineering thermodynamics, the specific enthalpy h has been given a special significance. This is due to the fact that in right-hand and left-hand cyclic processes with real fluids as a working medium, which are composed of isobaric

and isentropic changes of state (in particular seen in the *Joule* process, in the *Clausius-Rankine* process and in the Cold vapor compression cooling machine process (cf. section 7.6.5) as well as in their further developments), for the calculation of the thermal efficiency η_{th} and of the exergetic efficiency ζ only the knowledge of the respective differences $h_j - h_i$ between the individual states i and j is required. In a generalisation of the procedure for cyclic process calculations, however, there is no reason to prefer h and not to treat h, u, g and f as equivalent to each other.

Table 7.16 Calculation of process variables using state variables

Basic equation	Transformation	Result
Isothermic change of state		
$dg = -sdT + vdp$	$vdp = dg$	$dw_p = dg$
$df = -sdT - pdv$	$-pdv = df$	$dw_v = df$
$sdT = 0$		$dq_T = 0$
$dh = Tds + vdp = Tds + dg$	$Tds = dh - dg$	$dq_s = dh - dg$
$du = Tds - pdv = Tds + df$	$Tds = du - df$	$dq_s = du - df$
Isobaric change of state		
$vdp = 0$		$dw_p = 0$
$df = -sdT - pdv = dg - pdv$	$-pdv = df - dg$	$dw_v = df - dg$
$du = Tds - pdv = dh - pdv$	$-pdv = du - dh$	$dw_v = du - dh$
$dg = -sdT + vdp$	$-sdT = dg$	$dq_T = -dg$
$dh = Tds + vdp$	$Tds = dh$	$dq_s = dh$
Isochoric change of state		
$dg = -sdT + vdp = df + vdp$	$vdp = dg - df$	$dw_p = dg - df$
$dh = Tds + vdp = du + vdp$	$vdp = dh - du$	$dw_p = dh - du$
$-pdv = 0$		$dw_v = 0$
$df = -sdT - pdv$	$-sdT = df$	$dq_T = -df$
$du = Tds - pdv$	$Tds = du$	$dq_s = du$
Isentropic change of state		
$dh = Tds + vdp$	$vdp = dh$	$dw_p = dh$
$du = Tds - pdv$	$-pdv = du$	$dw_v = du$
$dg = -sdT + vdp = -sdT + dh$	$sdT = dh - dg$	$dq_T = dh - dg$
$df = -sdT - pdv = -sdT + du$	$sdT = du - df$	$dq_T = du - df$
$Tds = 0$		$dq_s = 0$

The necessity to use h, u, g and f in an equivalent way accordimg to Table 7.16 is no barrier for calculations, as with fundamental equations and computational programs such as [128] and [130] for important real fluids, the specific state variables h, u, g and f — e.g. as functions of p and T — can be calculated. However, when using them, care must be taken that g and f are each available as absolute specific state variables — and not merely as relative specific state variables.

For the processes mentioned, as far as they are characterised by an internal heat transfer or work transfer, calculating these reversible processes questions arise that need to be answered more comprehensively than for processes with ideal gases. Since here, for example, the specific isobaric heat capacity c_p cannot be assumed as constant but depends on temperature and also on pressure, an internal heat transfer alone can no longer take place at temperature gradients approaching zero. To ensure reversible

heat transfer within a differential temperature interval — in addition to the original differential heat transfer — a further supplementary differential heat supply or heat removal within a differential temperature interval — in addition to the original differential heat transfer — is required in order to make a differential total heat transfer possible with a temperature gradient approaching zero in each case. The sum of the additional differential heat quantities required in each case is taken into account in the thermodynamic equations of Table 7.16 for η_{th}, ζ, a_{wp}, a_{wv}, a_{qT} and a_{qs}. The same considerations are necessary for the calculation of an internal work transfer.

Carnot process

$$T_2 = T_1 \quad s_3 = s_2 \quad T_4 = T_3 \quad s_4 = s_1$$

$$\eta_{th\,car} = 1 - \frac{(g_2 - g_1) + (h_1 - h_2)}{(g_3 - g_4) + (h_4 - h_3)}$$

$$\zeta_{car} = 1 - \frac{(g_2 - g_1) + (h_1 - h_2) - T_b(s_1 - s_2)}{(g_3 - g_4) + (h_4 - h_3) - T_b(s_4 - s_3)} \quad \text{(If } T_b = T_1, \text{ then } \zeta_{car} = 1.\text{)}$$

$$a_{wp\,car} = \frac{(g_3 - g_4) + (g_2 - g_1) + (h_3 - h_2) + (h_4 - h_1)}{(g_3 - g_4) - (g_2 - g_1) + (h_4 - h_3) - (h_1 - h_2)}$$

$$a_{wv\,car} = \frac{(f_3 - f_4) + (f_2 - f_1) + (u_3 - u_2) + (u_4 - u_1)}{(g_3 - g_4) - (g_2 - g_1) + (h_4 - h_3) - (h_1 - h_2)}$$

$$a_{qT\,car} = \frac{(g_2 - g_3) + (g_1 - g_4) + (h_3 - h_2) + (h_4 - h_1)}{(g_3 - g_4) - (g_2 - g_1) + (h_4 - h_3) - (h_1 - h_2)}$$

$$a_{qs\,car} = \frac{(g_3 - g_4) + (g_2 - g_1) + (h_4 - h_3) + (h_1 - h_2)}{(g_3 - g_4) - (g_2 - g_1) + (h_4 - h_3) - (h_1 - h_2)}$$

Ericsson process

$$T_2 = T_1 \quad p_3 = p_2 \quad T_4 = T_3 \quad p_4 = p_1$$

$$\eta_{th\,er} = 1 - \frac{(g_2 - g_1) + (h_1 - h_2)}{(g_3 - g_4) + (h_1 - h_2)}$$

$$\zeta_{er} = 1 - \frac{(g_2 - g_1) + (h_1 - h_2) - T_b(s_1 - s_2)}{(g_3 - g_4) + (h_1 - h_2) - T_b(s_1 - s_2)} \quad \text{(If } T_b = T_1, \text{ then } \zeta_{er} = 1.\text{)}$$

$$a_{wp\,er} = \frac{(g_3 - g_4) + (g_2 - g_1)}{(g_3 - g_4) - (g_2 - g_1)}$$

$$a_{wv\,er} = \frac{(f_3 - f_4) + (f_2 - f_1) + (h_3 - h_2) - (u_3 - u_2) + (h_4 - h_1) - (u_4 - u_1)}{(g_3 - g_4) - (g_2 - g_1)}$$

$$a_{qT\,er} = \frac{(g_2 - g_3) + (g_1 - g_4)}{(g_3 - g_4) - (g_2 - g_1)}$$

$$a_{qs\,er} = \frac{(g_3 - g_4) + (g_2 - g_1) + 2(h_4 - h_2)}{(g_3 - g_4) - (g_2 - g_1)}$$

7.5 Comparative Evaluation of Right-Hand Cyclic Processes

Stirling process

$$T_2 = T_1 \quad v_3 = v_2 \quad T_4 = T_3 \quad v_4 = v_1$$

$$\eta_{th\,st} = 1 - \frac{(f_2 - f_1) + (u_1 - u_2)}{(f_3 - f_4) + (u_1 - u_2)}$$

$$\zeta_{st} = 1 - \frac{(f_2 - f_1) + (u_1 - u_2) - T_b(s_1 - s_2)}{(f_3 - f_4) + (u_1 - u_2) - T_b(s_1 - s_2)} \quad \text{(If } T_b = T_1, \text{ then } \zeta_{st} = 1.)$$

$$a_{wp\,st} = \frac{(g_3 - g_4) + (g_2 - g_1) + (h_3 - h_2) - (u_3 - u_2) + (h_4 - h_1) - (u_4 - u_1)}{(f_3 - f_4) - (f_2 - f_1)}$$

$$a_{wv\,st} = \frac{(f_3 - f_4) + (f_2 - f_1)}{(f_3 - f_4) - (f_2 - f_1)}$$

$$a_{qT\,st} = \frac{(f_2 - f_3) + (f_1 - f_4)}{(f_3 - f_4) - (f_2 - f_1)}$$

$$a_{qs\,st} = \frac{(f_3 - f_4) + (f_2 - f_1) + 2(u_4 - u_2)}{(f_3 - f_4) - (f_2 - f_1)}$$

Joule process

$$s_2 = s_1 \quad p_3 = p_2 \quad s_4 = s_3 \quad p_4 = p_1$$

$$\eta_{th\,j} = 1 - \frac{(h_4 - h_1)}{(h_3 - h_2)}$$

$$\zeta_j = 1 - \frac{(h_4 - h_1) - T_b(s_4 - s_1)}{(h_3 - h_2) - T_b(s_4 - s_1)}$$

$$a_{wp\,j} = \frac{(h_3 - h_4) + (h_2 - h_1)}{(h_3 - h_2) - (h_4 - h_1)}$$

$$a_{wv\,j} = \frac{(h_3 - h_2) + (h_4 - h_1) - 2(u_4 - u_2)}{(h_3 - h_2) - (h_4 - h_1)}$$

$$a_{qT\,j} = \frac{(h_3 - h_4) + (h_2 - h_1) + 2(g_1 - g_3)}{(h_3 - h_2) - (h_4 - h_1)}$$

$$a_{qs\,j} = \frac{(h_3 - h_2) + (h_4 - h_1)}{(h_3 - h_2) - (h_4 - h_1)}$$

Otto process

$$s_2 = s_1 \quad v_3 = v_2 \quad s_4 = s_3 \quad v_4 = v_1$$

$$\eta_{th\,o} = 1 - \frac{(u_4 - u_1)}{(u_3 - u_2)}$$

$$\zeta_o = 1 - \frac{(u_4 - u_1) - T_b(s_4 - s_1)}{(u_3 - u_2) - T_b(s_4 - s_1)}$$

$$a_{wp\,o} = \frac{2(h_3 - h_1) - (u_3 - u_2) - (u_4 - u_1)}{(u_3 - u_2) - (u_4 - u_1)}$$

$$a_{wv\,o} = \frac{(u_3 - u_4) + (u_2 - u_1)}{(u_3 - u_4) - (u_2 - u_1)}$$

$$a_{qT\,o} = \frac{(u_3 - u_4) + (u_2 - u_1) + 2(f_1 - f_3)}{(u_3 - u_4) - (u_2 - u_1)}$$

$$a_{qs\,o} = \frac{(u_3 - u_2) + (u_4 - u_1)}{(u_3 - u_2) - (u_4 - u_1)}$$

Diesel process

$$s_2 = s_1 \quad p_3 = p_2 \quad s_4 = s_3 \quad v_4 = v_1$$

$$\eta_{th\,d} = 1 - \frac{(u_4 - u_1)}{(h_3 - h_2)}$$

$$\zeta_d = 1 - \frac{(u_4 - u_1) - T_b(s_4 - s_1)}{(h_3 - h_2) - T_b(s_4 - s_1)}$$

$$a_{wp\,d} = \frac{(h_3 - h_1) + (h_2 - h_1) - (u_4 - u_1)}{(h_3 - h_2) - (u_4 - u_1)}$$

$$a_{wv\,d} = \frac{(h_3 - h_2) + (u_2 - u_1) - (u_4 - u_2)}{(h_3 - h_2) - (u_4 - u_1)}$$

$$a_{qT\,d} = \frac{2(f_1 - f_3) - (h_3 - h_2) + (u_3 - u_4) + (u_3 - u_1)}{(h_3 - h_2) - (u_4 - u_1)}$$

$$a_{qs\,d} = \frac{(h_3 - h_2) + (u_4 - u_1)}{(h_3 - h_2) - (u_4 - u_1)}$$

Gas expansion process

$$T_2 = T_1 \quad p_3 = p_2 \quad s_4 = s_3 \quad p_4 = p_1$$

$$\eta_{th\,gex} = 1 - \frac{(g_2 - g_1) + (h_1 - h_2)}{(h_3 - h_4) + (h_1 - h_2)}$$

$$\zeta_{gex} = 1 - \frac{(g_2 - g_1) + (h_1 - h_2) - T_b(s_1 - s_2)}{(h_3 - h_4) + (h_1 - h_2) - T_b(s_1 - s_2)} \qquad \text{(If } T_b = T_1, \text{ then } \zeta_{gex} = 1.\text{)}$$

$$a_{wp\,gex} = \frac{(h_3 - h_4) + (g_2 - g_1)}{(h_3 - h_4) - (g_2 - g_1)}$$

$$a_{wv\,gex} = \frac{(f_2 - f_1) + (h_3 - h_2) - (u_4 - u_2) + (h_4 - h_1) - (u_4 - u_1)}{(h_3 - h_4) - (g_2 - g_1)}$$

$$a_{qT\,gex} = \frac{(h_3 - h_4) + (g_2 - g_3) + (g_1 - g_3)}{(h_3 - h_4) - (g_2 - g_1)}$$

$$a_{qs\,gex} = \frac{(h_3 - h_2) + (h_4 - h_2) + (g_2 - g_1)}{(h_3 - h_4) - (g_2 - g_1)}$$

Clausius-Rankine process

$$s_2 = s_1 = s_{1'} \quad p_3 = p_2 \quad s_4 = s_{4''} = s_3 \quad p_4 = p_1$$

$$\eta_{th\,cr} = 1 - \frac{h_4 - h_1}{h_3 - h_2}$$

$$\zeta_{cr} = 1 - \frac{(h_4 - h_1) - T_b(s_4 - s_1)}{(h_3 - h_2) - T_b(s_4 - s_1)} \quad \text{(If } T_b = T_1, \text{ then } \zeta_{cr} = 1.\text{)}$$

$$a_{wp\,cr} = \frac{(h_3 - h_4) + (h_2 - h_1)}{(h_3 - h_4) - (h_2 - h_1)}$$

$$a_{wv\,cr} = \frac{(h_3 - h_2) + (h_4 - h_1) - 2(u_4 - u_2)}{(h_3 - h_2) - (h_4 - h_1)}$$

$$a_{qT\,cr} = \frac{(f_1 - f_2) + (f_4 - f_3) + (g_2 - g_3) + (u_2 - u_1) + (u_3 - u_4)}{(h_3 - h_4) - (h_2 - h_1)}$$

$$a_{qs\,cr} = \frac{(h_3 - h_2) + (h_4 - h_1)}{(h_3 - h_2) - (h_4 - h_1)}$$

7.6 Left-Hand Cyclic Processes

Left-hand cyclic processes are used in vapor compression refrigeration systems and vapor compression heat pumps as well as in absorption refrigeration systems and absorption heat pumps. Since the direction of circulation is different in left-hand cyclic processes compared to right-hand cyclic processes, the signs of the process variables work and heat occurring between the cycle and the environment also change. At a low temperature of the heat source T_{Qu}, the heat Q_{zu} is supplied to the cycle (Fig. 7.47). The cycle absorbs this heat at the averaged temperature T_{zu} and releases the heat Q_{ab} at the averaged temperature T_{ab}, which is higher than the temperature T_{zu}, to the heat sink with the temperature T_{Se}.

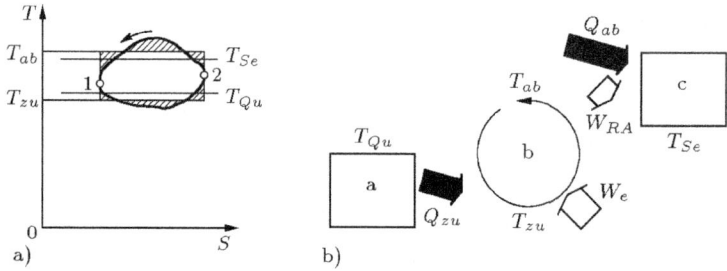

Figure 7.47 Left-hand cyclic process
a) T, S diagram (changes of state schematic)
b) model representation   a heat source   b cyclic process   c heat sink

In vapor compression refrigeration systems and vapor compression heat pumps considered in the following, the cycle is supplied with the coupling work W_e; therefore, it is positive for left-hand cyclic processes. According to Table 7.2 Eq. (5), the energy balance is given as

$$|Q_{ab}| = Q_{zu} + W_e - |W_{RA}| \,. \tag{7.147}$$

Refrigeration systems can serve various purposes: preservation and storage of food, cooling in a process of manufacturing products, drying of substances by moisture separation, cooling of air in air-conditioning systems, separation of substances by liquefaction at low temperatures, use of physical effects at low temperatures (e.g. superconductivity), etc. The basic task of all refrigeration systems is to cool a room, substance or substance flow below the ambient temperature. For this purpose, heat must be extracted from the room, substance or substance flow, which represents the heat source, at a lower temperature than the ambient temperature. In general, the environment serves as the heat sink.

In the case of heat pumps, the focus is on the heat release of the left-hand cyclic process. They can also be used in different ways. In building heat supply, for example, they have the task of supplying heat to a room that represents the heat sink. The heat source is, for example, the ambient air, geothermal heat, water, the exhaust air of a ventilation system or the waste heat of a production process, whose temperature is too low for direct heat utilisation.

From an energy point of view, the application of left-hand cyclic processes is particularly favourable when both the heat source (e.g. artificial ice rink, household freezer) as well as the heat sink (e.g. indoor swimming pool, drinking water reservoir) can be used in practise.

Vapor compression refrigeration systems and vapor compression heat pumps can be realised by the same machine process, but generally the temperature range of the refrigeration process is lower than that of the heat pump.

7.6.1 Performance Number

The quotient of benefit and effort is suitable for evaluating a machine or system. This results in the thermal efficiency of right-hand cyclic processes, which is always less than one or, in the best case, equal to one. In many cases of left-hand cyclic processes, however, the quotient of benefit and effort is greater than one. Therefore, in this case, one speaks of a performance number.

The benefit of a refrigeration system is to extract heat Q_{zu} from a room, substance or substance flow. The effort required is the coupling work W_e to drive the vapor compression refrigeration system. Therefore, the performance number of a vapor compression refrigeration system is[12]

$$\varepsilon_K = \frac{Q_{zu}}{W_e} \ . \qquad (7.148)$$

For a vapor compression heat pump, the benefit is the heat release Q_{ab}. The effort is, equal to a vapor compression refrigeration system, the coupling work W_e. The performance number of a vapor compression heat pump is therefore

$$\varepsilon_W = \frac{|Q_{ab}|}{W_e} \ . \qquad (7.149)$$

With Eq. (7.147) one obtains

---

[12] Another often used efficiency ratio ist the coefficient of performance (COP) in which the effort contains not only the coupling work W_e but also additional work necessary for auxiliary techniques of the system.

$$\varepsilon_W = 1 + \frac{Q_{zu} - |W_{RA}|}{W_e} \, . \tag{7.150}$$

The performance number of a vapor compression heat pump is greater than one. For an ideal cyclic process, according to Table 7.2 Eq. (1), the equations follow

$$\varepsilon_K = \frac{(Q_{zu})_{rev}}{(W_e)_{id}} = \frac{(Q_{zu})_{rev}}{W_{Kreis}} = \frac{(Q_{zu})_{rev}}{|(Q_{ab})_{rev}| - (Q_{zu})_{rev}} \tag{7.151}$$

$$\varepsilon_W = 1 + \frac{(Q_{zu})_{rev}}{(W_e)_{id}} = 1 + \frac{(Q_{zu})_{rev}}{W_{Kreis}} = 1 + \frac{(Q_{zu})_{rev}}{|(Q_{ab})_{rev}| - (Q_{zu})_{rev}} = 1 + \varepsilon_K. \tag{7.152}$$

7.6.2 Left-Hand *Carnot* Process

The reversible *Carnot* process also has great theoretical significance for left-hand cyclic processes. For given temperatures of heat source T_{Qu} and heat sink T_{Se}, the left-hand *Carnot* process achieves the highest numbers of performance. For the heat transfer on the isotherms, the following holds (Fig. 7.48):

$$(Q_{zu})_{rev} = (Q_{41})_{rev} = T_1(S_1 - S_4) = T_{Qu}(S_1 - S_4) \tag{7.153}$$

$$|(Q_{ab})_{rev}| = |(Q_{23})_{rev}| = T_3(S_2 - S_3) = T_{Se}(S_1 - S_4) \tag{7.154}$$

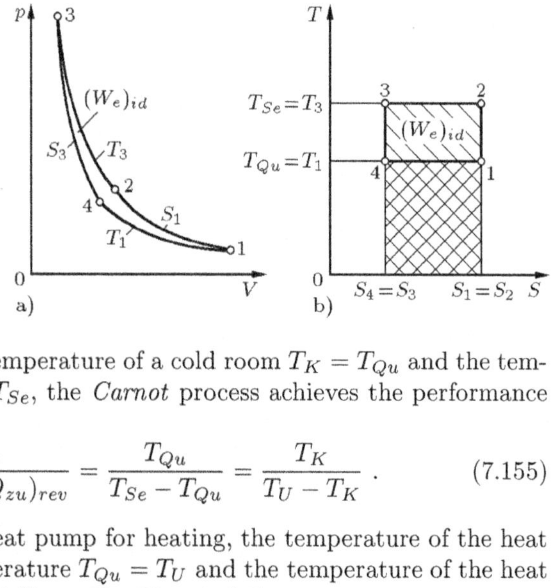

Figure 7.48 Left-hand *Carnot* process
a) p, V diagram
b) T, S diagram
(hatchings as in Fig. 7.10)

As a refrigeration system with the temperature of a cold room $T_K = T_{Qu}$ and the temperature of the environment $T_U = T_{Se}$, the *Carnot* process achieves the performance number

$$\varepsilon_K = \frac{(Q_{zu})_{rev}}{|(Q_{ab})_{rev}| - (Q_{zu})_{rev}} = \frac{T_{Qu}}{T_{Se} - T_{Qu}} = \frac{T_K}{T_U - T_K} \, . \tag{7.155}$$

If the *Carnot* process is used as a heat pump for heating, the temperature of the heat source is equal to the ambient temperature $T_{Qu} = T_U$ and the temperature of the heat sink is that of the room to be heated $T_{Se} = T_H$. Thus it follows:

$$\varepsilon_W = \frac{T_{Se}}{T_{Se} - T_{Qu}} = 1 + \frac{T_{Qu}}{T_{Se} - T_{Qu}} = \frac{T_H}{T_H - T_U} = 1 + \frac{T_U}{T_H - T_U} \tag{7.156}$$

Example 7.6 What are the performance numbers of a left-hand *Carnot* process as a refrigerating process and as a heat pump for the seasonally most unfavourable case, when the extreme ambient temperatures are $t_U = 35\,°C$ in summer and $t_U = -20\,°C$ in winter? The

required cold room temperature is $t_K = -10\,°C$ and the required temperature of the room to be heated is $t_H = 25\,°C$.

Refrigeration system: $T_K = 263.15\,\text{K}$ $T_U = 308.15\,\text{K}$

$$\varepsilon_K = \frac{263.15\,\text{K}}{45\,\text{K}} = 5.8478$$

Heat pump: $T_U = 253.15\,\text{K}$ $T_H = 298.15\,\text{K}$

$$\varepsilon_W = 1 + \frac{253.15\,\text{K}}{45\,\text{K}} = 6.6256\ .$$

7.6.3 Left-Hand *Joule* Process

The realisation of a left-hand cyclic process with ideal gases is possible in the left-hand *Joule* process. The following applies to the heat input and heat output (Fig. 7.49):

$$(Q_{zu})_{rev} = (Q_{41})_{rev} = m\,c_p(T_1 - T_4) \tag{7.157}$$

$$|(Q_{ab})_{rev}| = |(Q_{23})_{rev}| = m\,c_p(T_2 - T_3) \tag{7.158}$$

The coupling work is given by Table 7.2 Eqs. (1) and (3)

$$(W_e)_{id} = W_{Kreis} = m\,c_p(T_4 + T_2 - T_1 - T_3)\ . \tag{7.159}$$

For the performance numbers, one obtains according to Eqs. (7.151) and (7.152)

$$\varepsilon_K = \frac{T_1 - T_4}{T_4 + T_2 - T_1 - T_3} \tag{7.160}$$

$$\varepsilon_W = \frac{T_2 - T_3}{T_4 + T_2 - T_1 - T_3}\ . \tag{7.161}$$

Both equations can be expressed using Eqs. (7.84) to (7.86):

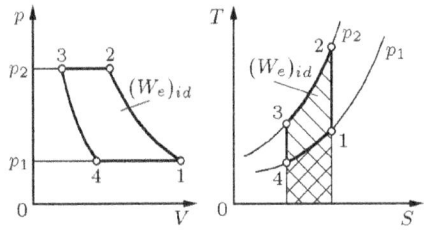

$$\varepsilon_K = \frac{1}{\left(\dfrac{p_2}{p_1}\right)^{\frac{\kappa-1}{\kappa}} - 1} \tag{7.162}$$

$$\varepsilon_W = \frac{1}{1 - \left(\dfrac{p_1}{p_2}\right)^{\frac{\kappa-1}{\kappa}}} \tag{7.163}$$

Figure 7.49 Left-hand *Joule* process
a) p, V diagram   b) T, S diagram
(hatchings as in Fig. 7.10)

Example 7.7 A cold gas system supplies cold air at an ambient pressure of 0.96 bar to a room to be air-conditioned at $25\,°C$. For this purpose, $12\,\text{m}^3/\text{s}$ of outside air at $25\,°C$ and ambient pressure is drawn in and then isentropically compressed to 1.5 bar. The compressed air is isobarically cooled to $51\,°C$. This is managed by a heat exchanger that is flowed through with the room's exhaust air flow, which is then let out to the surroundings. By isentropic expansion in a turbine to ambient pressure the previously compressed air is cooled.

a) At what temperature does the cooled air flow out of the cold gas cooling system?

b) What temperature does the air reach after compression'
c) What exhaust air flow must pass through the heat exchanger if the exhaust air is to heat up by 13 K?
d) What cooling power does the refrigeration system reach?
e) What drive power must be applied?
f) What is the performance number of the refrigeration system?
g) What performance number would be achieved, if the same system were to operate as a heat pump?
h) What heating power (e.g. for drinking water heating) would the system achieve?

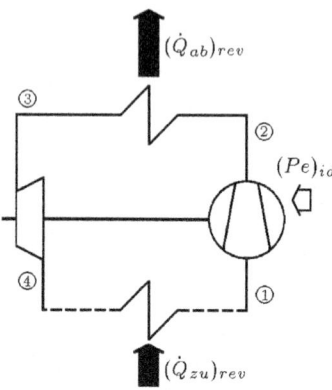

Figure 7.50 For example 7.7

The plant operates on a left-hand reversible *Joule* process (Fig. 7.50). The open process can be thought of as a cycle process completed by the dashed isobaric change of state.

a) $T_4 = T_3 \left(\dfrac{p_1}{p_2}\right)^{\frac{\kappa-1}{\kappa}} = 324.15 \cdot \left(\dfrac{0.96 \text{ bar}}{1.5 \text{ bar}}\right)^{\frac{0.4}{1.4}} = 285.34 \text{ K} \qquad t_4 = 12.19\,°C$

b) $T_2 = T_1 \left(\dfrac{p_2}{p_1}\right)^{\frac{\kappa-1}{\kappa}} = 298.15 \text{ K} \cdot \left(\dfrac{1.5 \text{ bar}}{0.96 \text{ bar}}\right)^{\frac{0.4}{1.4}} = 338.70 \text{ K} \qquad t_2 = 65.55\,°C$

c) $|(\dot{Q}_{ab})_{rev}| = \dot{m}_A\, c_p (T_2 - T_3) = \dfrac{p_1 \dot{V}_{A1}}{R_L T_1} c_p (T_2 - T_3) = 13.46 \text{ kg/s} \cdot 1.0062 \text{ kJ/(kg K)} \cdot (338.70 - 324.15)\text{ K} = 197.06 \text{ kW}$

$\dot{m}_F = \dfrac{|(\dot{Q}_{ab})_{rev}|}{c_p \Delta t} = \dfrac{197.06 \text{ kW}}{1.0046 \text{ kJ/(kg K)} \cdot 13 \text{ K}} = 15.09 \text{ kg/s}$

d) $(\dot{Q}_{zu})_{rev} = \dot{m}_A\, c_p (T_1 - T_4) = 13.46 \text{ kg/s} \cdot 1.0040 \text{ kJ/(kg K)} \cdot (298.15 - 285.34) \text{ K} = 173.11 \text{ kW}$

e) $(P_e)_{id} = -[(\dot{Q}_{zu})_{rev} - |(\dot{Q}_{ab})_{rev}|] = 23.95 \text{ kW}$

f) $\varepsilon_K = \dfrac{(\dot{Q}_{zu})_{rev}}{|(\dot{Q}_{ab})_{rev}| - (\dot{Q}_{zu})_{rev}} = \dfrac{173.11 \text{ kW}}{197.06 \text{ kW} - 173.11 \text{ kW}} = 7.23$

g) $\varepsilon_W = 1 + \varepsilon_K = 8.23$

h) $|(\dot{Q}_{ab})_{rev}| = 197.06 \text{ kW}$

7.6.4 Gas Expansion Process as a Left-Hand Cycle Process

The right-hand Gas expansion process with an ideal gas as the working fluid (section 7.4.6; cf. the T, S diagram of Fig. 7.51 a) can be changed by lowering the highest process temperature T_3 into a left-hand cyclic process:

In terms of a first process transformation for this (Fig. 7.51 b), the isothermal compression from initial pressure p_1 to pressure p_2 at the temperature level $T_1 = T_2$ (ambient temperature) is maintained, but the highest process temperature T_3 is reduced to such an extent that the isobaric recuperative heat transfer of the completely in-process and reversibly transferred heat (entropy change heat) becomes zero:

$(Q_{2Z})_{rev} = -(Q_{41})_{rev} = Q_{s2Z} = -Q_{s41} = 0$ The temperature T_3 is chosen in such a way that the isentropic expansion from the final pressure p_2 to the initial pressure p_1 (with simultaneous decrease of temperature T_3 to T_4) and the subsequent isobaric change of state to the temperature T_1 at the pressure level $p_4 = p_1$ leads to an equality of the enclosed area 23M2 and the enclosed area 41M4.

The area 23M2 corresponds to the absolute value of the obtained cyclic process work of the right-hand reversible partial cyclic process 23M2, where the entropy change heat supplied above the ambient temperature $T_1 = T_2$ is $(Q_{23})_{rev}$ as effort and the sum of the work $W_{pM2} + W_{p3M}$ (W_{pM2}: supplied; W_{p3M}: released) are to be accounted for as benefit for this. The area 41M4 corresponds to the required cyclic process work of the left-hand reversible partial cyclic process 41M4, with which below the ambient temperature $T_1 = T_2$ the entropy change heat $(Q_{41})_{rev}$ for the intended purpose of cooling can be supplied to the process as benefit; the sum of the works $W_{p1M} + W_{pM4}$ (W_{p1M}: supplied; W_{pM4}: released) is to be accounted for as the effort for this. Fig. 7.51 b shows that with the right-hand reversible partial cyclic process 23M2 just enough cyclic process work is gained so that the left-hand partial cyclic process 41M4 can be used as a cooling process. The thermal efficiency η_{th} of the total process 12341 thus results in zero (abscissa $\eta_{th} = 0$ in Fig. 7.46 a).

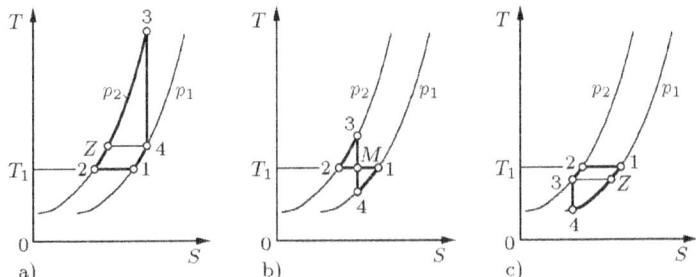

Bild 7.51 Gas expansion process in a T, S diagram

a) right-hand cyclic process
b) combination of a right-hand cyclic process und a left-hand cyclic process
c) left-hand cyclic process

In the sense of a second process transformation (Fig. 7.51 c) — with unchanged isothermal compression from the pressure p_1 to the final pressure p_2 at the temperature level $T_1 = T_2$ (ambient temperature) — the process temperature T_3 is lowered below the ambient temperature; this is associated with a Gas expansion process that has an inner isobaric recuperative heat transfer opposite to the right-hand gas expansion process. The heat $-(Q_{23})_{rev}$ (entropy change heat) on the high pressure level p_2 is completely and reversibly transferred to the working fluid at low pressure level p_1: $-(Q_{23})_{rev} = (Q_{Z1})_{rev}$.

This is followed by an isentropic expansion from the pressure p_2 to the pressure p_1 with a simultaneous decrease of the temperature T_3 to the temperature T_4, followed by an isobaric transfer of entropy change heat $(Q_{4Z})_{rev}$ delivering the intended purpose of cooling as a benefit. This heat supply rises the temperature T_4 of the working fluid

7.6 Left-Hand Cyclic Processes

to the temperature T_Z at the pressure level $p_4 = p_1$. Finally, the already mentioned isobaric recuperative heat transfer with $-(Q_{23})_{rev} = (Q_{Z1})_{rev}$ takes place.

This process is related to the *Claude* process (there with a real fluid as working fluid); it has in the temperature range between T_1 and $T_Z = T_3$ the characteristics of the *Ericsson* process and in the temperature range between $T_Z = T_3$ and T_4 the characteristics of the *Joule* process.

The area 12341 corresponds to the required cyclic process work of the left-hand reversible cyclic process 12341, with which below the temperature $T_Z = T_3 < T_1 = T_2$ the entropy change heat $(Q_{4Z})_{rev}$ can be extracted from a heat source as benefit for the cooling process; the sum of the work $W_{p12} + W_{p34}$ (W_{p12}: supplied; W_{p34}: given off) is to be accounted for as an effort for this purpose.

If one defines a thermal efficiency η_{th} of the ideal left-hand cyclic process 12341 in accordance with the thermal efficiency of an ideal right-hand cyclic process, one has to take into account the fact that in section 7.3.2, Eq. (7.35) was determined with the intention to gain positive values for the thermal efficiency, despite the negative values of the output of the cycle process work W_{Kreis}:

$$\eta_{th} = \frac{-W_{Kreis}}{(Q_{zu})_{rev}} \qquad (7.164)$$

In the sense of continuity, this definition is also adopted for the ideal left-hand Gas expansion process.[13] Thus, η_{th} here is the negative quotient of effort to benefit. Then, Eqs. (7.115) and (7.119) for the thermal efficiency of the right-hand Gas expansion process can be retained unchanged, as the following considerations show:

$$\eta_{th} = -\frac{W_{Kreis}}{(Q_{zu})_{rev}} = -\frac{w_{Kreis}}{(q_{zu})_{rev}} = -\frac{W_{p12} + W_{p34}}{(Q_{4Z})_{rev}} \qquad (7.165)$$

With Eqs. (4.86), (4.87), (4.108), (4.49) and (4.111) for ideal gases follows:

$$\eta_{th} = -\frac{m R T_1 \ln \frac{p_2}{p_1} + m c_p (T_4 - T_3)}{m c_p (T_3 - T_4)} = 1 - \frac{m R T_1 \ln \frac{p_2}{p_1}}{m c_p (T_3 - T_4)} \qquad (7.166)$$

The total energy balance of all works and reversible substitute heats, from outside supplied respectively to outside released, leads to

$$W_{p12} + (Q_{12})_{rev} + (Q_{4Z})_{rev} + W_{p34} = 0 . \qquad (7.167)$$

As a consequence of Eq. (4.86) follows

$$W_{p34} = -(Q_{4Z})_{rev} = -m c_p (T_3 - T_4) . \qquad (7.168)$$

Eqs. (4.105) and (4.111) also give

$$W_{p34} = -\frac{\kappa}{\kappa - 1} m R T_3 \left(1 - \left(\frac{p_1}{p_2}\right)^{\frac{\kappa-1}{\kappa}}\right) . \qquad (7.169)$$

---

[13] This means that — with respect to the history of the efficiency definition for ideal right-hand processes — a certain logical inconsistency must be accepted for the efficiency definition of ideal left-hand cyclic processes.

Eq. (7.166), hereby leads to

$$\eta_{th} = 1 - \frac{m R T_1 \ln \frac{p_2}{p_1}}{\frac{\kappa}{\kappa - 1} m R T_3 \left(1 - \left(\frac{p_1}{p_2}\right)^{\frac{\kappa-1}{\kappa}}\right)} = 1 - \frac{\kappa - 1}{\kappa} \frac{T_1 \ln \frac{p_2}{p_1}}{T_3 \left(1 - \left(\frac{p_1}{p_2}\right)^{\frac{\kappa-1}{\kappa}}\right)} \quad (7.170)$$

These relationships reflect the fact that the left-hand Gas expansion process emerges steadily from the right-hand Gas expansion process with the help of the lowering of the temperature T_3. In Fig. 7.46 a), at η_{th}, there is then a continuation of the efficiency curves into the negative value range, but these curves are not shown there.

The thermal efficiency with negative sign $-\eta_{th}$ according to Eq. (7.170) can be used for by multiplication with the equations for a_{wp}, a_{wv}, a_{qT} and a_{qs} of Table 7.13 for the representation of corresponding effort ratios a^*_{wp}, a^*_{wv}, a^*_{qT} and a^*_{qs} as the appropriate ratios for the left-hand Gas expansion process. Thus, in the equations of these effort ratios, in the mathematical denominator appears as benefit instead of $|w_{Kreis}|$ now $(q_{zu})_{rev}$.

This possibility opens up analogously, among other things, for the left-hand single-polytropic *Carnot* process, the left-hand *Carnot* process, the left-hand *Ericsson* process, the left-hand *Stirling* process and the left-hand *Joule* process with in-process ideal heat transfer.[14]

If in Eq. (7.170) T_3 is replaced by the lowest cycle process temperature

$$T_4 = T_3 \left(\frac{p_1}{p_2}\right)^{\frac{\kappa-1}{\kappa}}, \quad (7.171)$$

the result is

$$\eta_{th} = 1 - \frac{\kappa - 1}{\kappa} \frac{T_1 \ln \frac{p_2}{p_1}}{T_4 \left(\left(\frac{p_2}{p_1}\right)^{\frac{\kappa-1}{\kappa}} - 1\right)}. \quad (7.172)$$

The performance number ε_K of the refrigeration process as a ratio of benefit to effort is calculated accordingly:

$$\varepsilon_K = -\frac{1}{\eta_{th}} = \frac{1}{\frac{\kappa-1}{\kappa} \frac{T_1 \ln \frac{p_2}{p_1}}{T_3 \left(1 - \left(\frac{p_1}{p_2}\right)^{\frac{\kappa-1}{\kappa}}\right)} - 1} = \frac{1}{\frac{\kappa-1}{\kappa} \frac{T_1 \ln \frac{p_2}{p_1}}{T_4 \left(\left(\frac{p_2}{p_1}\right)^{\frac{\kappa-1}{\kappa}} - 1\right)} - 1} \quad (7.173)$$

---

[14]It also makes sense from a thermodynamic point of view to define modified performance numbers and effort ratios in such a way that instead of the reversibly supplied specific heat $(q_{zu})_{rev}$, the specific exergy $(eq_{zu})_{rev}$ appears in the mathematical denominator.

7.6 Left-Hand Cyclic Processes

For the Gas expansion process as a heat pump process this gives:

$$\varepsilon_W = 1 + \varepsilon_K = 1 - \frac{1}{\eta_{th}} \tag{1}$$

$$= 1 + \frac{1}{\frac{\kappa-1}{\kappa} \frac{T_1 \ln \frac{p_2}{p_1}}{T_3 \left(1 - \left(\frac{p_1}{p_2}\right)^{\frac{\kappa-1}{\kappa}}\right)} - 1} = 1 + \frac{1}{\frac{\kappa-1}{\kappa} \frac{T_1 \ln \frac{p_2}{p_1}}{T_4 \left(\left(\frac{p_2}{p_1}\right)^{\frac{\kappa-1}{\kappa}} - 1\right)} - 1} \tag{2}$$

(7.174)

7.6.5 Cold Vapor Compression Process

For the most part, the Cold vapor compression process is used in refrigeration systems and in heat pumps as the reference process. It is operated with refrigerants that flow through the cyclic process partly in the liquid state and partly in the vapor state. The left-hand cycle process is created by a series connection of open systems (Fig. 7.52).

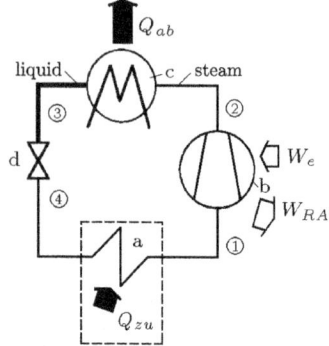

Figure 7.52 Cold vapor compression refrigeration system or cold vapor compression heat pump (real cycle)

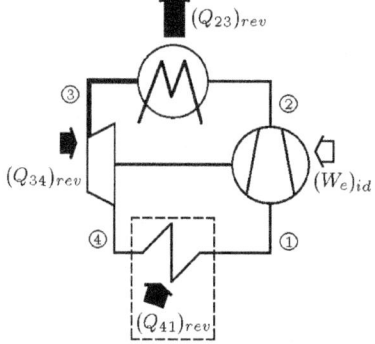

Figure 7.53 Cold vapor compression refrigeration process (ideal cycle)

In a vaporiser (a), the refrigerant at the low pressure p_1 is vaporated by heat supply from e.g. the cold room (refrigeration systems) or from e.g. the environment (heat pumps). The saturated steam is compressed to the pressure p_2 in a displacement machine or a fluid flow machine (b).

In a heat exchanger (condenser) (c) the refrigerant is cooled isobarically; in the process it passes successively from the state of superheated vapor to saturated vapor, by condensation to boiling liquid and then, if necessary, to the cooled liquid state. Subsequently, expansion takes place by isenthalpic throttling in an expansion valve (d) to the pressure p_1. The resulting two-phase liquid-vapor mixture flows into the vaporiser.

The choice of the refrigerant is made — in addition to environmental aspects — primarily according to the vaporisation temperature and the condensation temperature, which in turn depends on the cooling task (for refrigeration systems) or e.g. the ambient temperature (for heat pumps) and the temperature of the heat release in the condenser.

The pressure in the vaporiser p_1 should be technically well controllable and the pressure ratio p_2/p_1 should not be too high. In addition, the specific heat of vaporisation should be as high as possible and the specific volume of the saturated vapor to be compressed should be as small as possible in order to achieve a small compressor size.

When setting up the comparison process, it should be noted that throttling is an irreversible process (section 3.4.4). As a reversible substitute process for throttling, one can imagine a turbine which, with the addition of heat, converts the refrigerant to an isenthalp $h = $ const (section 5.2.2). With this reversible substitute process, an ideal cycle can be assembled according to Fig. 7.53, in which the pressure change work W_{p34h} is released by the working fluid and the heat $(Q_{34})_{rev}$ is supplied to the working fluid. In a T, s diagram of this ideal cycle (Fig. 7.54 a) the specific heat supply appears as area $(q_{34})_{rev}$.

Another way to define an ideal cycle is to replace the isenthalpic throttling by a reversible isentropic expansion with gain of pressure change work W_{p34s}; this gives the left-hand *Clausius-Rankine* process.

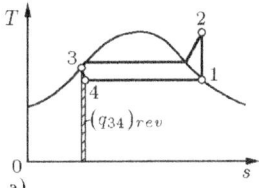
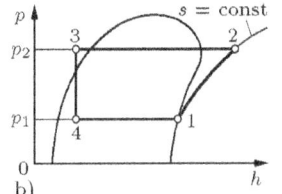

Figure 7.54 Cold vapor compression process in a
a) T, s diagram
b) p, h diagram

However, the ideal cycle is not used as the comparison process. The specific heat input $(q_{34})_{rev}$ according to Fig. 7.54 a in terms of a specific cold gain as well as the specific work gain w_{p34} would improve the performance number considerably, but would not be in accordance with the achievable values of conventional technical systems. For the comparison process, therefore, frictionless isentropic compression of the refrigerant and isenthalpic throttle expansion are taken as a basis. This means that for the comparison process according to Fig. 7.54 b is

$$Q_{zu} = Q_{41} = H_1 - H_4 \tag{7.175}$$

$$|Q_{ab}| = |Q_{23}| = H_2 - H_3 \ . \tag{7.176}$$

For the coupling work according to Table 7.2 Gl. (5) holds

$$W_e = |Q_{ab}| - Q_{zu} + |W_{RA}| \ . \tag{7.177}$$

The isentropic compression in an ideal compressor with $W_{RA} = 0$ results in

$$W_e = H_2 + H_4 - H_3 - H_1 \ . \tag{7.178}$$

As the throttle expansion is an isenthalpic change of state, this means

$$H_3 = H_4 \ . \tag{7.179}$$

Therefore the coupling work is

$$W_e = H_2 - H_1 \ . \tag{7.180}$$

7.6 Left-Hand Cyclic Processes

The performance numbers ε_K and ε_W can be expressed according to Eqs. (7.148), (7.149) and (7.150) as

$$\varepsilon_K = \frac{H_1 - H_4}{H_2 - H_1} \qquad (7.181)$$

$$\varepsilon_W = 1 + \varepsilon_K = 1 + \frac{H_1 - H_4}{H_2 - H_1} = \frac{H_2 - H_4}{H_2 - H_1} = \frac{H_2 - H_3}{H_2 - H_1}. \qquad (7.182)$$

Numerator and denominator of the performance numbers can be represented as areas in a T, S diagram (Figure 7.55).

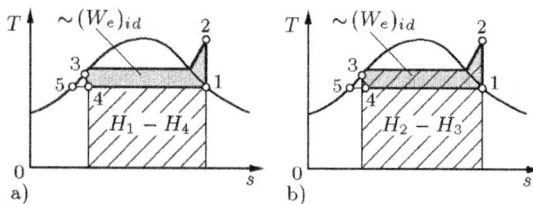

Figure 7.55 Cold vapor compression process in a T, S diagram
a) refrigeration process
b) heat pump process

Example 7.8 A heat pump system with solar absorber and ammonia (R 717) as refrigerant is to be operated at temperatures above $0\,°C$ for the heating of a house (Fig. 7.56). The heat output requirement of the house is 10.4 kW for a living area of 130 m². The vaporisation temperature is to be $-10\,°C$, the pressure $p_2 = 20.00$ bar. The compressor operates with an internal isentropic efficiency of 0.65 and a mechanical efficiency of 0.94. Complete condensation takes place in the condenser.

a) What is the specific enthalpy after the compressor?
b) What is the mass flow through the unit?
c) What heat flow is extracted from the environment?
d) What absorber area is required if the specific heat flow input from the environment is 200 W/m²?
e) What coupling power does the compressor require?
f) What performance number does the unit achieve?

According to [36], the extended *Benedict-Webb-Rubin* equation (5.92) applies to ammonia with the constants

$A_2 = -5.0420969 \quad D_2 = -1.0331007 \quad C_3 = 1.9041669 \quad A_6 = 0.013864372$
$B_2 = 2.1518441 \quad A_3 = -0.98753174 \quad A_4 = -1.1754653 \quad \beta = 1.1$
$C_2 = -2.0263398 \quad B_3 = 1.2179486 \quad A_5 = 2.207015$

The specific gas constant and the critical properties are

$R = 0.48818$ kJ/(kg K) $\quad p_k = 113.53$ bar $\quad v_k = 0.0042735$ m³/kg $\quad T_k = 405.5$ K

The *van der Waals* number according to Eq. (5.94) is $Wa = 4.080153$. The liquid-vapor line is given by the equation

$$\ln \pi = A + B\vartheta + \frac{C}{\vartheta} + D\vartheta^2 + E\vartheta^3 + \frac{F}{\vartheta}(1 - \vartheta)^{\frac{3}{2}}$$

$A = 19.667984 \qquad C = -11.079219 \qquad E = -2.152941$
$B = -15.54993 \qquad D = 9.114107 \qquad F = 1.81269$

328 7 Cyclic Processes

The reduced density of the saturated liquid is

$$\delta' = 1 + c_1(1-\vartheta)^{\frac{1}{3}} + c_2(1-\vartheta)^{\frac{2}{3}} + c_3(1-\vartheta) + c_4(1-\vartheta)^{\frac{4}{3}} + c_5(1-\vartheta)^{\frac{5}{3}}$$

The specific vaporisation heat is

$$r = (v'' - v')\vartheta\,\pi\,p_k\left\{B + 2D\vartheta + 3E\vartheta^2 - \frac{1}{\vartheta^2}\left[C + F\sqrt{1-\vartheta}\left(1+\frac{\vartheta}{2}\right)\right]\right\}.$$

The specific isobaric heat capacity in the state of ideal gas is

$$c_p^0 = a_1 + a_2\,T + a_3\,T^2$$

$a_1 = 1.86926$ kJ/(kg K) $a_2 = 1.15645 \cdot 10^{-4}$ kJ/(kg K^2) $a_3 = 2.19428 \cdot 10^{-6}$ kJ/(kg K^3).

The equations representng the specific enthalpy and the specific entropy are

$$h = a_1 T + \frac{a_2}{2}T^2 + \frac{a_3}{3}T^3 + p_k v_k\left\{(2A_2 + B_2\vartheta + \frac{4C_2}{\vartheta^2} + \frac{6D_2}{\vartheta^4})\delta + (\frac{3}{2}A_3 + B_3\vartheta + \frac{5C_3}{2\vartheta^2})\delta^2+\right.$$

$$+\left(\frac{A_4}{\vartheta^2} + \frac{A_5}{\vartheta^4}\right)\delta^2(1+\beta\delta^2)e^{-\beta\delta^2} + \left(\frac{3A_4}{\vartheta^2} + \frac{5A_5}{\vartheta^4}\right)\frac{1}{\beta}\left[1-\left(1+\frac{\beta\delta^2}{2}\right)e^{-\beta\delta^2}\right] + \frac{6}{5}A_6\delta^5\right\} + h^*$$

$$s = (a_1 - R)\ln\vartheta + a_2 T + \frac{a_3}{2}T^2 - R\ln\delta - \frac{R}{Wa}\left\{(B_2 - \frac{2C_2}{\vartheta^3} - \frac{4D_2}{\vartheta^5})\delta + (\frac{1}{2}B_3 - \frac{C_3}{\vartheta^3})\delta^2 - \right.$$

$$\left. - (\frac{A_4}{\vartheta^3} + \frac{2A_5}{\vartheta^5})\frac{2}{\beta}\left[1-\left(1+\frac{\beta\delta^2}{2}\right)e^{-\beta\delta^2}\right]\right\} + s^*$$

with the integration constants $h^* = 972.50$ kJ/kg and $s^* = 4.0747$ kJ/(kg K).

a) state 1: $t_1 = -10\,°C$ $p_1 = 2.907$ bar $h_1'' = 1451.15$ kJ/kg $s_1'' = 5.7582$ kJ/(kg K)

state 3 (Fig. 7.56): $p_2 = 20.00$ bar $t_S = t_4 = 49.37\,°C$ $h_4 = h_5 = 433.25$ kJ/kg

$t_3 = 133.90\,°C$ $h_3 = 1744.37$ kJ/kg $s_3 = s_1''$

According to Table 6.2 Eq.(1), neglecting the change of kinetic energy, is

$$h_2 = h_1 + \frac{h_3 - h_1}{\eta_{isV}} = 1902.26 \text{ kJ/kg}.$$

b) $|q_{ab}| = h_2 - h_4 = 1469.01$ kJ/kg $\dot{m} = \frac{|\dot{Q}_{ab}|}{|q_{ab}|} = \frac{10.4 \text{ kW}}{1469.01 \text{ kJ/kg}} = 7.0796 \cdot 10^{-3}$ kg/s

c) $\dot{Q}_{zu} = \dot{m}\,q_{zu} = \dot{m}(h_1 - h_5) = 7.2063$ kW

d) $A = \dfrac{7.2063 \text{ kW}}{0.2 \text{ kW/m}^2} = 36.03 \text{ m}^2$

e) $P_e = \dot{m}\dfrac{h_2 - h_1}{\eta_{mV}} = 3.40$ kW

f) $\varepsilon_W = \dfrac{|\dot{Q}_{ab}|}{P_e} = \dfrac{10.4 \text{ kW}}{3.40 \text{ kW}} = 3.06$

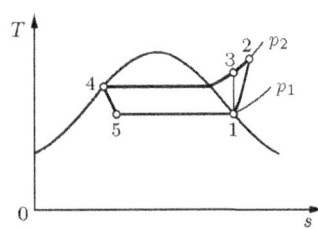

Figure 7.56 To example 7.8

Tasks for Section 7

1. A reversible *Otto* process and a reversible *Diesel* process are to be compared with each other. The suction volume for both processes is $V_1 = 0.5$ ℓ, the suction pressure is $p_1 = 0.96$ bar. The isentropic compression in the *Otto* process occurs with the compression ratio $\varepsilon_O = 10$, in the *Diesel* process the compression ratio is $\varepsilon_D = 22$ (Fig. 7.57). After the heat supply, being realised isochoric in the *Otto* process and isobaric in the *Diesel* process, for both processes pressure and temperature ($t = 1827\,°C$) are equal. The expansion following is isentropic in both processes. Air as an ideal gas can be assumed with $c_p = 1.050$ kJ/(kg K) and $\kappa = 1.4$.

a) Pressures and temperatures in both cyclic processes are to be calculated.

b) Which values of the thermal efficiency in both processes shall be reached?

c) How large is the performance gain of the *Diesel* process compared to the *Otto* process, if a four-stroke engine with a speed of 4000 min^{-1} is used?

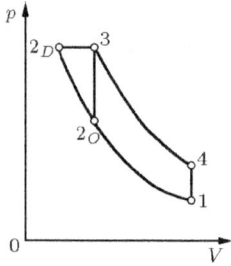

Figure 7.57 To task 1

Figure 7.58 To task 2

2. In a reversible *Joule* process with internal heat recovery (Fig. 7.58), 0.88 kg/s air of 21 °C and 1.04 bar is isentropically compressed to 5.82 bar. The combustion gas temperature at the inlet of the turbine is 650 °C, the turbine expands isentropically to the initial pressure. In the heat exchanger, the heat is transferred without temperature difference (i.e. reversibly). The process shall be calculated wit air as an ideal gas ($c_p = 1.040$ kJ/(kg K); $\kappa = 1.4$)).

a) What heat flow is transferred in the heat exchanger?

b) What coupling power can be achieved in the system?

c) What is the thermal efficiency of the plant?

d) What would be the thermal efficiency of a process without internal heat transfer at the same pressure ratio?

e) What is the thermal efficiency of an *Ericsson* process between the same temperatures of 21 °C and 650 °C?

f) The process shall be plotted to scale in a T, s diagram.

Figure 7.59 To task 3

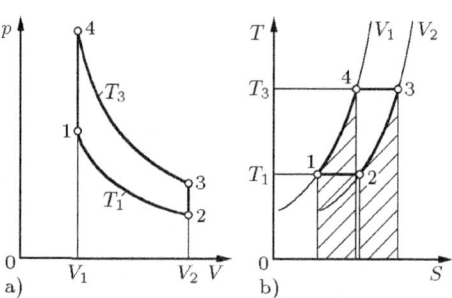

330   7 Cyclic Processes

3. The *Stirling* process consists of two isochores and two isotherms. Reversible internal heat transfer takes place on the isochores. The left-hand *Stirling* process forms the comparative process for the *Philips* gas refrigeration machine. A left-hand *Stirling* process (Fig. 7.59) is operated with hydrogen and is characterised by the following changes of state:

1. 1–2 isothermal expansion at $T_1 = 130$ K from $V_1 = 0.15$ ℓ and $p_1 = 18.2$ bar to $V_2 = 0.3$ ℓ.
2. 2–3 isochoric heating to $T_3 = 300$ K
3. 3–4 isothermal compression
4. 4–1 isochoric cooling

a) What cooling power can the process deliver at speed 1440 min^{-1} (number of process cycles per minute)?

b) What is the performance number of the process?

c) What is the coupling power required?

4. In an irreversibly operating combined heat and power plant (Fig. 7.60) 300 t/h of superheated steam (state 1) of 200 bar and 530 °C is produced. The steam is first expanded irreversibly and adiabatic in a steam turbine to 4 bar and a quantity of $x_2 = 95$ % in the two-phase liquid-vapor region (state 2). After 40 t/h of wet steam as a heating fluid have been extracted for heating purposes, the expansion in the turbine is continued with the remaining steam to 0.10 bar and a quantity of $x_3 = 85$ % in the two-phase liquid-vapor region (state 3). In the heating system, the quantity x is reduced at approximately the same pressure to such an extent that, by means of a subsequent throttling at $h =$ const, the state (3) which the remaining steam has reached, is also reached by the heating fluid.

a) The changes of state are to be plotted in a T, s diagram for water. The heat given off in the heating system shall be shown in a T, s diagram for water. I a first approximation, e.g. Diagram 7 for H$_2$O in the appendix can be used.

b) What are the coupling powers obtained by the two expansions in the turbine at a mechanical efficiency of 0.96?

c) What is the heat flow output?

d) What are the internal isentropic efficiencies of the two turbine stages?

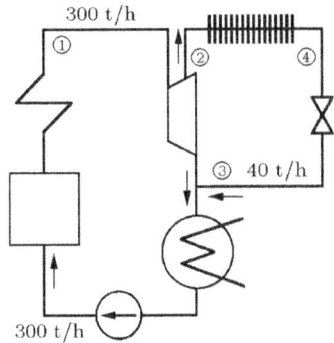

Figure 7.60 To task 4

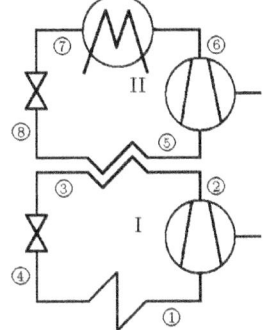

Figure 7.61 To task 5

5. A two-stage Cold vapor compression refrigeration process with ammonia (NH$_3$; R 717) as refrigerant (Fig. 7.61) with ideal compressors is to produce a refrigerating capacity of 80 kW.

It consists of two completely separate cycles I and II. Cycle I takes heat in the vaporiser at a refrigerant temperature of $-35\,°C$ from the cold room, whose temperature is $-25\,°C$, and transfers heat at a refrigerant pressure of 5 bar in a heat exchanger (condenser of cycle I) to cycle II, whose refrigerant in the heat exchanger (vaporiser of cycle II) has a pressure of 4 bar. Cycle II transfers heat to the environment in the condenser at a pressure of 18 bar, the temperature of which is maximum $+35\,°C$. Before compression, the refrigerant is in both cycles in the saturated vapor state; condensation takes place in both cycles to the saturated liquid state. It is to be calculated with the equation of example 7.8 or by using Diagram 3 and Diagram 4 for NH_3 in the appendix:

a) the properties of states 1 to 8 with respect to p, t, x and h,

b) the flow rates of the refrigerant in both cycles,

c) the heat flow transferred in the heat exchanger,

d) the heat flow output to the environment,

e) the coupling power of both ideal compressors,

f) the performance number of the entire system.

g) What is the performance number of a reversible *Carnot* process operating between a heat source of $-25\,°C$ and a heat sink of $+35\,°C$?

6. For operating a smaller combined heat and power unit (CHP), an internal combustion engine is provided according to the Gas-*Otto* method. (Fig. 7.15). Its ideal comparison process is the *Otto* process. Compared to the *Seiliger* process, this process is characterised by the fact that the injection ratio has the value $\varphi = 1$: After the isentropic compression 12 and the isochoric pressure increase 23, the isobaric heat supply 34 is therefore omitted; thus the states 3 and 4 are identical. Air is used as the working fluid with m = 2 g (air as an ideal gas with $R = 0.2872$ kJ/(kg K); $c_p = 1.005$ kJ/(kg K); $c_v = 0.718$ kJ/(kg K); $\kappa = 1.4$). The intake air is characterised by the state $t_1 = 25\,°C$ and $p_1 = 1.0$ bar; the compression ratio is $\varepsilon = 12$, the pressure increase ratio is $\psi = 2.5$.

The state variables V_1, V_2, T_2, p_2, T_3, p_3, V_3, V_5, T_5 and p_5 are to be calculated. Furthermore, the supplied heat Q_{23}, the heat output Q_{51}, the gained work $(W_e)_{id}$, the thermal efficiency η_{th} and the power $(P_e)_{id}$ gained at a speed of 6000 revolutions per minute are to be calculated.

7. For the operation of a combined heat and power plant (CHP) with the energy source Bio-*Diesel* oil (rapeseed oil methyl ester RME), the *Diesel* process shall be investigated (Fig. 7.15). This process is characterised in comparison to the *Seiliger* process by the fact that the pressure increase ratio has the value $\psi = 1$: Thus, after the isentropic compression 12, the isochoric pressure increase 23 is omitted; thus states 2 and 3 are identical. Air with m = 10 g is used as the working fluid (properties of air: cf. task 6). The intake air is characterised by the state $t_1 = 25\,°C$ and $p_1 = 1.0$ bar; the compression ratio is $\varepsilon = 20$, the injection ratio has the value $\varphi = 2.2$.

The state variables $V_2 = V_3$, $T_2 = T_3$, $p_2 = p_3$, T_4, p_4, V_4, V_5, T_5 and p_5 are to be calculated. Furthermore, the supplied heat Q_{34}, the heat output Q_{51}, the gained work $(W_e)_{id}$, the thermal efficiency η_{th} and the power $(P_e)_{id}$ gained at a speed of 3000 revolutions per minute are to be calculated.

8. For the *Stirling* process, the mechanical effort ratios a_{wp} and a_{wv} as well as the thermal effort ratios a_{qT} and a_{qs} are to be derived in general.

9. A *Otto* process is to be calculated with air as working fluid (properties of air: cf. task 6). The initial state 1 is given by the following properties: $p_1 = 1.0$ bar, $t_1 = 25\,°C$, $s_1 = 6.867$ kJ/(kg K). The highest process state is characterised by $p_3 = 70.0$ bar and $T_3 = 1973.15$ K. The following numerical values are to be calculated: the thermal and the exergetic efficiency

η_{th} and ζ, the mechanical effort ratios a_{wp} and a_{wv}, the thermal effort ratios a_{qT} and a_{qs} and the total effort ratio a_g.

10. A *Diesel* process is to be calculated with air as working fluid (properties of air: cf. task 6). The initial state 1 is given by $p_1 = 1.0$ bar, $t_1 = 25\,°\text{C}$, $s_1 = 6.867$ kJ/(kg K). The highest process state is characterised by $p_3 = 70.0$ bar and $T_3 = 1973.15$ K. The following numerical values are to be calculated: the thermal and the exergetic efficiency η_{th} and ζ, the mechanical effort ratios a_{wp} and a_{wv}, the thermal effort ratios a_{qT} and a_{qs} and the total effort ratio a_g.

11. A gas turbine is provided to cover the electricity peak load and as a back-up power plant. The reversible comparison process is the *Joule* process, which is to be calculated with air as the working fluid (properties of air: cf. task 6). The initial state 1 is characterised by $p_1 = 1.0$ bar, $t_1 = 25\,°\text{C}$ and $s_1 = 6.867$ kJ/(kg K), the highest process state is defined by $p_3 = 20.0$ bar and $T_3 = 1523.15$ K. The following numerical values are to be calculated: the thermal and the exergetic efficiency η_{th} and ζ, the mechanical effort ratios a_{wp} and a_{wv}, the thermal effort ratios a_{qT} and a_{qs} and the total effort ratio a_g.

12. Three different *Stirling* processes (subtasks a) to c)) are to be calculated with air and then with helium as working fluid. For subtask a), air is to be calculated as an ideal gas (properties: cf. task 6). For subtasks b) and c), helium is to be assumed as an ideal gas ($R = 2.0773$ kJ/(kg K); $c_p = 5.1931$ kJ/(kg K); $c_v = 3.1158$ kJ/(kg K); $\kappa = 1.6667$).

a) The initial state 1 of the process with air is defined by $p_1 = 1.0$ bar, $t_1 = 25\,°\text{C}$ and $s_1 = 6.867$ kJ/(kg K), the highest process state is defined by $p_3 = 20.0$ bar and $T_3 = 973.15$ K.

b) The initial state 1 of the process with helium is defined by $p_1 = 1.0$ bar, $t_1 = 25\,°\text{C}$ and $s_1 = 31.5170$ kJ/(kg K), the highest process state is given by $p_3 = 20.0$ bar and $T_3 = 973.15$ K.

c) The initial state 1 of the process with helium in the higher pressure range is defined by $p_1 = 7.5$ bar, $t_1 = 25\,°\text{C}$ and $s_1 = 27.3314$ kJ/(kg K), the highest process state by $p_3 = 150.0$ bar and $T_3 = 973.15$ K.

For the subtasks a) to c) the following numerical values are to be determined: the thermal and the exergetic efficiency η_{th} and ζ, the mechanical effort ratios a_{wp} and a_{wv}, the thermal effort ratios a_{qT} and a_{qs}, and the total effort ratio a_g.

13. A conventional steam power plant is designed for use as an electricity base load power plant. The reversible comparison process is the simple *Clausius-Rankine* process (without feedwater preheating, without reheating), which is to be calculated with water (H_2O) as liquid and as vapor (steam). H_2O is to be treated as a real fluid. In a first approximation, the T,s diagram for H_2O (e.g. Diagram 7 in the appendix) can be used. Then is to be considered that there no absolute specific entropies are contained. Thus, in it for the initial state 1 ($p_1 = 0.032$ bar, $t_1 = 25\,°\text{C}$) is $s_1 = 0.3672$ kJ/(kg K)) instead of the absolute specific entropy $s_1 = 3.888$ kJ/(kg K). The highest process state is characterised by $p_3 = 350.0$ bar and $T_3 = 873.15$ K.

The following numerical values are to be calculated: the thermal and the exergetic efficiency η_{th} and ζ, the mechanical effort ratios a_{wp} and a_{wv}, the thermal effort ratios a_{qT} and a_{qs} and the total effort ratio a_g.

8 Exergy

Energy conversion is restricted according to the second law of thermodynamics: For example, coupling work can be completely converted into internal energy; a complete conversion of internal energy into coupling work, on the other hand, is impossible. In this section, the conditions under which energy conversions take place are examined in more detail. This section is thus a continuation of the treatment of the second law. Here, the term energy is used as a generic term that includes work.

8.1 Energy and Exergy

Energy is referred to as stored work or work capacity in section 2.4. This thermodynamic description of energy is unsatisfactory to the engineer, because it does not reveal the extent to which the energy is capable of work and can be converted into work. Therefore, that part of energy which can be completely converted into any other kind of energy should be emphasised and given a special name. The part of energy that can be converted into work and thus, in the most favourable case, be used with the cooperation of the environment, is to be given the name exergy.

From mechanics, in the frictionless pendulum motion, the constantly repeating, complete conversion of potential energy into kinetic energy and vice versa is known. Both types of energy thus consist entirely of exergy. A lake with a water temperature equal to the ambient temperature represents a large energy store in the form of the internal energy of the water. However, this internal energy cannot be harnessed for conversion into electrical energy: The internal energy of the water in the lake has no exergy.

Convertibility and transferability of energy are closely related. The internal energy of warm springs on Iceland, for example, is used to heat greenhouse crops, while the numerically much larger internal energy of seawater in the vicinity of the island is not usable. Between the water of a warm spring and the surrounding area, an almost constant temperature gradient causes a heat flow. That portion of the internal energy of the warm water that is available above the ambient temperature can be converted into heat as a form of energy transfer. A heat flow from the seawater to the surroundings that could be technically used to the same extent cannot be produced, because there is not a sufficient temperature gradient. Decisive for the evaluation of an energy storage as an energy source is the environmental condition. It is not only the temperature gradient that is important; a gradient due to an altitude, a pressure gradient, a negative entropy difference and a concentration gradient in relation to the ambient state can also be decisive.

The exergy of an energy is obtained as the useful work in the case of reversible changes of state with the participation of the environment. To determine the useful work, one assumes ideal processes in closed or open systems. The exergy of the heat (more precisely: the exergy of the reversible substitute heat, also called the exergy of the entropy change heat), the exergy of the bound energy, the exergy of the volume change work, the exergy of the pressure change work and the exergy of the shift work can be determined by ideal cyclic processes:

8 Exergy

In these ideal cyclic processes, the sum of the occurring volume change works results in the cyclic process work; The same applies to the sum of the pressure change works, entropy change heats and temperature change heats that occurs (cf. temperature change heat see Eqs. (7.12) and (7.13)). The obtained cycle work — i.e. the useful work — thus results in the exergy of the sum of the respective process variable. Furthermore, the sum of the respective exergies of each of the four process variables mentioned also results in the cyclic process work. The exergy of a process variable is itself a process variable.

The exergies of the internal energy, enthalpy, free energy and free enthalpy result from reversible changes of state to the ambient state. In ideal processes, the environment can act as a source or sink of heat (more precisely: entropy change heat) and can release or absorb bound energy. The environment can give off shift work by ambient air pressure, or shift work is given off to the environment. Since the air pressure is assumed to be constant, the difference in shift work of the environment between beginning and end of a change of state according to Eq. (4.74) also means a volume change work. The environment is not capable of performing pressure change work.

A consideration of the path-dependent energies heat (more precisely: entropy change heat), temperature change heat, volume change work and pressure change work lead to exergies that are process variables. Since bound energy, shift work, internal energy, enthalpy, free energy and free enthalpy are path-independent, the associated exergies are state variables.

The exergy part of an energy is identified by the letter E in front of it. For the specific exergy, e is used.

If an ideal cycle is used to determine the exergy of the heat (more precisely: the entropy change heat) $(Q_{12})_{rev} = Q_{s12}$ or the bound energy TS, then it is chosen in a T, S diagram in such a way that one of the changes of state at the ambient temperature T_0 and two further changes of state at in each case constant entropy — i.e. without heat transfer — take place. If the exergies of the heat (entropy change heat) and of the bound energy are known, then the exergy of the temperature change heat Q_{T12} can be determined.

In order to obtain the exergy of the heat, one considers the heat transfer in the ideal cycle in a T, S diagram. The useful work appears as the cycle work given off. According to Table 7.2, Eq. (1) is

$$-W_{Kreis} = (Q_{zu})_{rev} - |(Q_{ab})_{rev}| = (Q_{12})_{rev} + (Q_0)_{rev} . \qquad (8.1)$$

$(Q_{12})_{rev}$ is the heat to be investigated for its exergy part, $(Q_0)_{rev}$ describes the heat given off to the environment or extracted from the environment.

If an ideal cyclic process is assumed for the determination of the exergy of the volume change work W_{V12} or of the shift work pV, then it is chosen in the p, V diagram in such a way that one of the changes of state at the ambient pressure p_0 and two further changes of state each take place at constant volume — i.e. without the transfer of volume change work. If the exergies of the volume change work and of the shift work are known, the exergy of the pressure change work W_{p12} can be determined.

In order to determine the exergy of a work, the representation of the ideal cyclic process in the p, V diagram is of interest. With W_{12} as the work to be investigated and W_0 as the work of the environment, according to Eq. (7.10) one can write

$$W_{Kreis} = \oint dW = W_{12} + W_0 = W_{12} - (-W_0) . \tag{8.2}$$

If the work W_{12} is compression work, the work of the environment W_0 must be expansion work and vice versa.

8.1.1 Exergy of Heat

If into a system above ambient temperature the heat (entropy change heat) $(Q_{12})_{rev} = Q_{s12}$ is reversibly supplied, then the exergy of heat $(EQ_{12})_{rev} = EQ_{s12}$ is that part of the heat which is reversible changeable into useful work in a right-hand cycle with the environment as heat sink. At the ambient temperature T_0, the absolute value of the reversible isothermal heat release to the environment according to Eq. (3.107) is $T_0 (S_2 - S_1)$. In Eq. (8.1), $-T_0 (S_2 - S_1)$ takes the place of $(Q_0)_{rev}$ according to Fig. 8.1 a). The part of the supplied heat that can be converted into useful work is equal to the negative cyclic process work:

$$(EQ_{12})_{rev} = EQ_{s12} = (Q_{12})_{rev} - T_0 (S_2 - S_1) = Q_{s12} - T_0 (S_2 - S_1) \tag{8.3}$$

The heat $(Q_{12})_{rev}$ supplied to the system and the exergy of this heat $(EQ_{12})_{rev}$ are to be specified positively with respect to the system.

When heat is given off above the ambient temperature, the environment is regarded as the heat source of a left-hand cycle process. Useful work must be supplied to the system as cyclic process work. The heat $(Q_{12})_{rev}$ given off by the system and the exergy of this heat $(EQ_{12})_{rev}$ have a negative sign (Fig. 8.1 b).

If the heat supply is below the ambient temperature, a left-hand cycle results with the environment as the heat sink. The heat $(Q_{12})_{rev}$ supplied to the system is positive, the exergy of this heat $(EQ_{12})_{rev}$ is negative (Fig. 8.1 c).

If the ambient temperature is higher than the temperature of the heat given off, then in a right-hand cyclic process useful work is gained. The heat $(Q_{12})_{rev}$ given off by the system is negative, the exergy of this heat $(EQ_{12})_{rev}$ is positive (Fig. 8.1 d).

The sign of the exergy of the heat $(EQ_{12})_{rev}$ depends on the direction of circulation of the reversible cyclic process, which is imagined for utilisation of the heat. Imaging a right-hand cyclic process, the exergy of the heat is positive, in a left-hand cyclic process, the exergy of the heat is negative. In either case, Eq. (8.3) applies to the exergy of heat. Using Eq. (3.28), Eq. (8.3) reads

$$(EQ_{12})_{rev} = (Q_{12})_{rev} - T_0 \int_1^2 \frac{dQ_{rev}}{T} . \tag{8.4}$$

The integral can be transformed — with the introduction of an average temperature T_m — as follows:

$$\int_1^2 \frac{dQ_{rev}}{T} = \frac{1}{T_m} \int_1^2 dQ_{rev} = \frac{(Q_{12})_{rev}}{T_m} = S_2 - S_1 \tag{8.5}$$

From Eqs. (8.4) and (8.5) follows

$$(EQ_{12})_{rev} = \left(1 - \frac{T_0}{T_m}\right)(Q_{12})_{rev}. \tag{8.6}$$

The average temperature T_m is according to Eq. (8.5)

$$T_m = \frac{(Q_{12})_{rev}}{S_2 - S_1}. \tag{8.7}$$

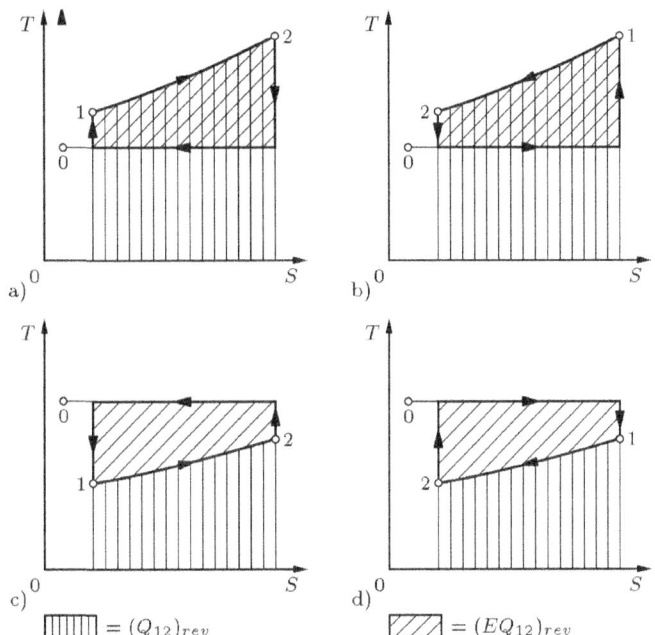

Bild 8.1 Representation of the cyclic processes for determining the exergy of the heat $(EQ_{12})_{rev}$ 0 ambient state 1 initial state 2 end state

	a)	b)	c)	d)
$(Q_{12})_{rev}$	+	−	+	−
$(EQ_{12})_{rev}$	+	−	−	+
direction of circulation	right	left	left	right
example	power machine	heat pump	refrigerator	power machine

Example 8.1 2.5 kg of air is isobarically heated from 100 °C to 300 °C at a pressure of 2 bar. The specific gas constant is $R = 0.2872$ kJ/(kg K), the specific isobaric heat capacity is $c_p = 1.0045$ kJ/(kg K). The heat (entropy change heat) and its exergy with respect to the ambient state of 1 bar and 15 °C are to be calculated.

Accordimg to Eqs. (4.77), (4.78), (4.79) and (8.3) are
$(Q_{12})_{rev} = H_2 - H_1 = m\, c_p\, (t_2 - t_1) = 2.5\,\text{kg} \cdot 1.0045\,\text{kJ/(kg K)} \cdot (300 - 100)\,\text{K} = 502.25\,\text{kJ}$
$S_2 - S_1 = m\, c_p\, \ln\frac{T_2}{T_1} = 2.5\,\text{kg} \cdot 1.0045\,\text{kJ/(kg K)} \cdot \ln\frac{573.15}{373.15} = 1.07775\,\text{kJ/K}$
$(EQ_{12})_{rev} = (Q_{12})_{rev} - T_0\,(S_2 - S_1) = 502.25\,\text{kJ} - 288.15\,\text{K} \cdot 1.07775\,\text{kJ/K} = 191.70\,\text{kJ}$
The heat and the exergy of the heat are represented in the T, S diagram of Fig. 8.1 a).

8.1.2 Exergy of Bound Energy

The state variable bound energy TS can only be positive, since T and S are always absolute values and thus are never less than zero. TS can only be partially converted into any other energy, since its ability to be converted is limited by the ambient temperature T_0 and by the entropy of the substance at ambient state S_0. The exergy of the bound energy ETS is obtained for the case $T > T_0$ and $S > S_0$ from TS, if in a T, S diagram a right-hand reversible *Carnot* cycle process between the state 1 of the working medium and its ambient state 0 is carried out according to Fig. 8.2 a), there to the bound energy TS a reversibly heat at constant temperature T (entropy change heat) $T(S - S_0)$ is assigned. This shall be interpreted as an area and finally from this area the area of the heat $T_0(S - S_0)$ reversibly released to the environment at constant ambient temperature T_0 is subtracted. According to Eq. (8.1), for the given off work, i.e. negative cycle work, which is equal to the positive useful work and thus equal to the exergy of the bound energy ETS, the following results:

$$ETS = -W_{Kreis} = (Q_{zu})_{rev} - |(Q_{ab})_{rev}| = T(S - S_0) - T_0(S - S_0) \qquad (8.8)$$

$$ETS = (T - T_0)(S - S_0) = TS - TS_0 - T_0(S - S_0) \qquad (8.9)$$

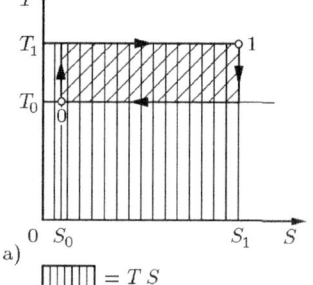
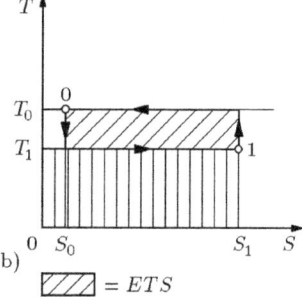

Figure 8.2 The bound energy TS and the exergy of the bound energy ETS for $S > S_0$ in a T, S diagram

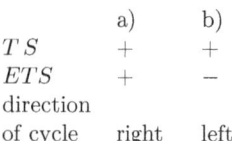

This procedure is consistent with the derivation for the exergy of the free enthalpy presented in section 8.1.10.

If are $T < T_0$ and $S > S_0$, then with the left-hand cyclic process according to Fig. 8.2 b) and Eqs. (8.8) and (8.9), the exergy of the bound energy ETS is negative. If is $S < S_0$, the signs reverse.

Figure 8.3 Exergy difference of the difference of the bound energies $ET_2S_2 - ET_1S_1$

The exergy difference of the bound energies at the entry and exit of an open system or during the change of state in a closed system is (cf. Figure 8.3)

$$ET_2S_2 - ET_1S_1 = (T_2 - T_0)(S_2 - S_0) - (T_1 - T_0)(S_1 - S_0)$$
$$= T_2 S_2 - T_1 S_1 - T_0(S_2 - S_1) - (T_2 - T_1) S_0 \quad . \qquad (8.10)$$

Example 8.2 For example 8.1, the difference of the bound energies and the corresponding difference of the exergies of the bound energies of air are to be determined.
It should be noted that absolute entropies are to be calculated. Thus, s_0, s_1 and s_2 are to be determined with i. e. Eq. (4.58) from the value given in section 7.5.3.2 page 286/287 for $s = 6.867$ kJ/(kg K) at $25\,°C$ and 1.0 bar.

$$T_2\, S_2 - T_1\, S_1 = m\,(T_2\, s_2 - T_1\, s_1)$$
$$= 2.5 \text{ kg}\,(573.15 \text{ K} \cdot 7.325 \text{ kJ/(kg K)} - 373.15 \text{ K} \cdot 6.894 \text{ kJ/(kg K)}) = 4064.57 \text{ kJ}$$
$$E T_2\, S_2 - E T_1\, S_1 = T_2\, S_2 - T_1\, S_1 - T_0\, m\,(s_2 - s_1) - (T_2 - T_1)\,m\,s_0$$
$$= 4064.57 \text{ kJ} - 288.15 \text{ K} \cdot 2.5 \text{ kg}\,(7.325 - 6.894) \text{ kJ/(kg K)}$$
$$- (573.15 \text{ K} - 373.15 \text{ K})\,2.5 \text{ kg} \cdot 6.833 \text{ kJ/(kg K)} = 337.59 \text{ kJ}$$

8.1.3 Exergy of Temperature Change Heat

According to Eqs. (7.3) and (7.4) and the relation $d(T\,S) = T\,dS + S\,dT$ the temperature change heat Q_{T12} results from the reversible heat (entropy change heat) $(Q_{12})_{rev} = Q_{s12}$ and the difference of the bound energies $T_2\, S_2 - T_1\, S_1$:

$$\int_1^2 S\,dT = -\int_1^2 T\,dS + T_2\, S_2 - T_1\, S_1 \tag{8.11}$$

$$Q_{T12} = -(Q_{12})_{rev} + T_2\, S_2 - T_1\, S_1 = -Q_{s12} + T_2\, S_2 - T_1\, S_1 \tag{8.12}$$

Here, T und S are absolute values.

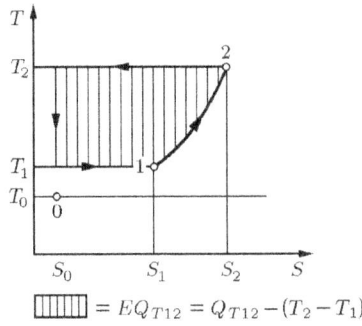

Figure 8.4 Cycle process to determine the exergy of the temperature change heat $EQ_{T12} = Q_{T12} - (T_2 - T_1)\,S_0$

Accordingly, the exergy of the temperature change heat is composed of the exergy of the reversible substitute heat (entropy change heat) and the exergy difference of the difference of the bound energies:

$$EQ_{T12} = -(EQ_{12})_{rev} + ET_2 S_2 - ET_1 S_1 = -EQ_{s12} + ET_2 S_2 - ET_1 S_1 \tag{8.13}$$

Eqs. (8.3) and (8.10) are inserted in Eq. (8.13):

$$EQ_{T12} = -(Q_{12})_{rev} + T_0\,(S_2 - S_1) + T_2\, S_2 - T_1\, S_1 - T_0\,(S_2 - S_1) - (T_2 - T_1)\,S_0$$
$$= -Q_{s12} + T_0\,(S_2 - S_1) + T_2\, S_2 - T_1\, S_1 - T_0\,(S_2 - S_1) - (T_2 - T_1)\,S_0 \tag{8.14}$$
$$EQ_{T12} = -(Q_{12})_{rev} + T_2\, S_2 - T_1\, S_1 - (T_2 - T_1)\,S_0$$
$$= -Q_{s12} + T_2\, S_2 - T_1\, S_1 - (T_2 - T_1)\,S_0 \tag{8.15}$$

From Eqs. (8.12) and (8.15) for the exergy of the temperature change heat EQ_{T12} follows:

$$EQ_{T12} = Q_{T12} - (T_2 - T_1)\,S_0 \tag{8.16}$$

8.1 Energy and Exergy

The same result is obtained with the help of Eq. (7.13), if with a reversible cycle process according to Fig. 8.4 $W_{Kreis} = \oint dQ_T$ is determined, where S_0 is to be considered as the limit.

Combining the exergy areas in Figs. 8.1a) and 8.3 to form Fig. 8.4 confirms Eqs. (8.13) and (8.16).

Example 8.3 The exergy of the temperature change heat EQ_{T12} for the reversible polytropic change of state of an ideal gas between states 1 and 2 as a function of m, c_n, T_1 and T_2
a) according to Eq. (8.16) is to be calculated and
b) using the exergy difference of the bound energies $ET_2 S_2 - ET_1 S_1$ and the exergy of the reversible substitute heat $(EQ_{12})_{rev} = (Q_{s12}$ is to be determined.

a) The temperature change heat can be determined as follows for the reversible polytropic change of state of an ideal gas according to Table 7.1 Eq. (6):

$$Q_{T12} = \left[s_1 + c_n \left(\frac{T_2}{T_2 - T_1} \ln \frac{T_2}{T_1} - 1\right)\right] m (T_2 - T_1)$$

$$= m \left[s_1 (T_2 - T_1) + c_n T_2 \ln \frac{T_2}{T_1} - c_n (T_2 - T_1)\right]$$

$$Q_{T12} = S_1 (T_2 - T_1) + m c_n T_2 \ln \frac{T_2}{T_1} - m c_n (T_2 - T_1)$$

According to Eq. (8.16), the exergy of the temperature change heat is

$$EQ_{T12} = Q_{T12} - (T_2 - T_1) S_0$$

$$EQ_{T12} = S_1 (T_2 - T_1) + m c_n T_2 \ln \frac{T_2}{T_1} - m c_n (T_2 - T_1) - (T_2 - T_1) S_0$$

$$EQ_{T12} = T_2 (S_1 - S_0) - T_1 (S_1 - S_0) + m c_n T_2 \ln \frac{T_2}{T_1} - m c_n (T_2 - T_1) \ .$$

Is $\quad S_1 - S_0 = m c_n \ln \frac{T_1}{T_0} \quad$ inserted according to section 4.3.5 Eq. (4.141), this results in

$$EQ_{T12} = m c_n T_2 \ln \frac{T_1}{T_0} - m c_n T_1 \ln \frac{T_1}{T_0} + m c_n T_2 \ln \frac{T_2}{T_1} - m c_n (T_2 - T_1) \ .$$

It follows:

$$EQ_{T12} = m c_n T_2 \ln \frac{T_2}{T_0} - m c_n T_1 \ln \frac{T_1}{T_0} - m c_n (T_2 - T_1)$$

$$= m c_n \left[T_2 \ln \frac{T_2}{T_0} - T_1 \ln \frac{T_1}{T_0} - (T_2 - T_1)\right]$$

b) $EQ_{T12} = ET_2 S_2 - ET_1 S_1 - (EQ_{12})_{rev}$

$EQ_{T12} = (T_2 - T_0)(S_2 - S_0) - (T_1 - T_0)(S_1 - S_0) - (Q_{12})_{rev} + T_0 (S_2 - S_1)$

From section 4.3.5, with Eqs. (4.130) and (4.141), one obtains:

$$EQ_{T12} = (T_2 - T_0) m c_n \ln \frac{T_2}{T_0} - (T_1 - T_0) m c_n \ln \frac{T_1}{T_0} - m c_n (T_2 - T_1) + T_0 m c_n \ln \frac{T_2}{T_1}$$

$$EQ_{T12} = T_2 m c_n \ln \frac{T_2}{T_0} - T_0 m c_n \ln \frac{T_2}{T_0} - T_1 m c_n \ln \frac{T_1}{T_0} + T_0 m c_n \ln \frac{T_1}{T_0}$$

$$- m c_n (T_2 - T_1) + T_0 m c_n \ln \left[\frac{T_2}{T_0} \frac{T_0}{T_1}\right]$$

$$EQ_{T12} = m\,c_n\,T_2 \ln\frac{T_2}{T_0} - m\,c_n\,T_0 \ln\frac{T_2}{T_0} + m\,c_n\,T_0 \ln\frac{T_2}{T_0} - m\,c_n\,T_1 \ln\frac{T_1}{T_0} + m\,c_n\,T_0 \ln\frac{T_1}{T_0}$$
$$- m\,c_n\,T_0 \ln\frac{T_1}{T_0} - m\,c_n\,(T_2 - T_1)$$
$$EQ_{T12} = m\,c_n\,T_2 \ln\frac{T_2}{T_0} - m\,c_n\,T_1 \ln\frac{T_1}{T_0} - m\,c_n\,(T_2 - T_1)$$
$$= m\,c_n\left[T_2 \ln\frac{T_2}{T_0} - T_1 \ln\frac{T_1}{T_0} - (T_2 - T_1)\right]$$

The examples 8.3 a) and 8.3 b) have the same result.

8.1.4 Exergy of Volume Change Work

If the volume change work W_{V12} is reversibly supplied to a closed system according to Fig. 2.17, it is composed of the work done by the air at pressure p_0 $p_0\,(V_1 - V_2)$ and the work $\int_1^2 F\,ds$ which is done by the piston force F acting on the piston rod. Because of the reversibility of the change of state, $|W_{RI}| = 0$ and $|W_{RA}| = 0$ are given.

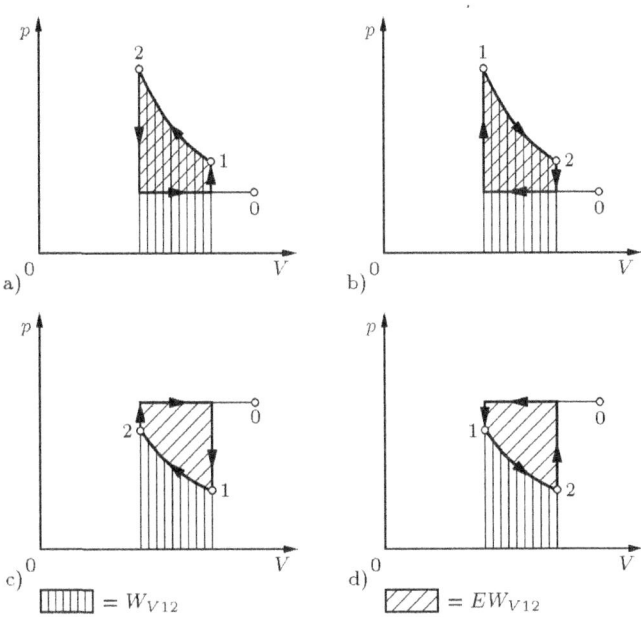

Figure 8.5 Illustration of the cyclic processes used to determine the exergy of the volume change work EW_{V12}

	a)	b)	c)	d)
W_{V12}	+	−	+	−
EW_{V12}	+	−	−	+
direction of cycle	left	right	right	left

The exergy of the volume change work EW_{V12} is equal to the work done by the piston force F acting on the piston rod during the piston movement. In reversible operation, according to the summarized Eqs. (2.57) and (2.58) follows

$$W_{12} = \int_1^2 F\,\mathrm{d}s + p_0\,(V_1 - V_2) = W_{V12}\,. \tag{8.17}$$

Thus, to calculate EW_{V12}, from the volume change work W_{V12} the work of the ambient air at ambient pressure $p_0(V_1 - V_2)$ is to be subtracted because the integral describes the work done on the piston rod and thus describes the exergy EW_{V12}:

$$EW_{V12} = W_{V12} - p_0\,(V_1 - V_2) \tag{8.18}$$

If the change of state with the volume change work W_{V12} is supplemented to form a cyclic process according to Fig. 8.5 a), then in Eq. (8.2) W_{Kreis} is to be replaced by EW_{V12}, W_{12} by W_{V12} and W_0 by $p_0\,(V_2 - V_1)$ according to Eq. (2.21). One obtains

$$EW_{V12} = W_{V12} + p_0\,(V_2 - V_1)\,. \tag{8.19}$$

Eqs. (8.18) and (8.19) are equal to each other. The same result is obtained by integrating Eq. (2.21) and replacing for the absolute pressure p the gage pressure $p - p_0$:

$$EW_{V12} = -\int_1^2 (p - p_0)\,\mathrm{d}V \tag{8.20}$$

The cycle process shown in Figure 8.5 a) is a left-hand cycle process, the volume change work W_{V12} and the exergy of the volume change work EW_{V12} are positive. From the four cases shown in Fig. 8.5 it can be seen that the direction of the cyclic process assumed for the calculation of the exergy determines the sign of the exergy of the volume change work. The calculation is always performed according to Eq. (8.19).

Example 8.4 3 kg of oxygen (specific gas constant $R = 0.25984$ kJ/(kg K); isentropic exponent $\kappa = 1.4$) are compressed in a closed system from 1.5 bar and 80 °C to 3.2 bar. The change of state can be described by the polytropic exponent $n = 1.6$ (supply of volume change work and heat). For the ambient state, 1 bar and 15 °C apply. What are the volume change work and the exergy of the volume change work?

According to Table 4.1 Eq. (4) and Eqs. (4.25), (4.126) and (8.19) are

$T_2 = 469.20$ K $\qquad W_{V12} = 150.77$ kJ $\qquad EW_{V12} = 81.54$ kJ .

The volume change work W_{V12} and the exergy of the volume change work EW_{V12} are shown in Figure 8.5 a).

8.1.5 Exergy of Shift Work

The exergy of the shift work EW is obtained from the shift work according to Eq. (2.25) $W = pV$, if the shift work of the ambient pressure $p_0 V$ (Fig. 8.6 a) is subtracted:

$$EW = (p - p_0)\,V \tag{8.21}$$

In Eq. (8.2), W_{Kreis} is to be replaced by EW and W_0 by $-p_0 V$. The shift work W is always supposed to be positive according to section 2.4.3; then W_0 must have a negative sign. If $p > p_0$, then the exergy of the shift work EW (or W_{Kreis} respectively) is positive. The cyclic process in which one can imagine the difference formation according to Eq. (8.21), is a left-hand cyclic process (Fig. 8.6 a). If $p < p_0$, EW (or W_{Kreis}

respectively) is negative. The corresponding cycle process is a right-hand cycle process (Fig. 8.6 b). The exergy difference of the shift works at the entry and exit of an open system is

$$EW_2 - EW_1 = W_2 - W_1 - p_0\,(V_2 - V_1)\,. \tag{8.22}$$

In the p, V diagram, positive and negative areas result (Figure 8.7).

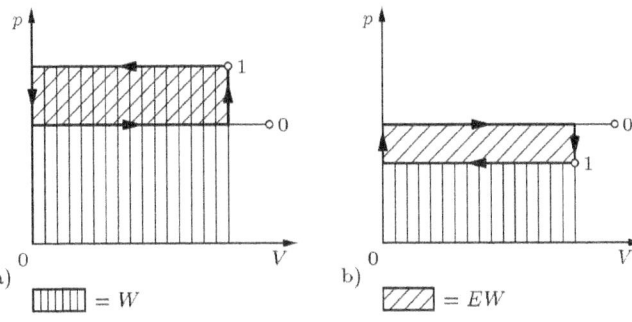

Figure 8.6 The shift work W and the exergy of the shift work EW in a p, V diagram W in a) and b) positive, EW in a) positive, in b) negative, direction of cycle in a) left, in b) right

In the isothermal change of state of an ideal gas, the difference of the shift works becomes zero, the exergy difference of the shift works, however, does not.

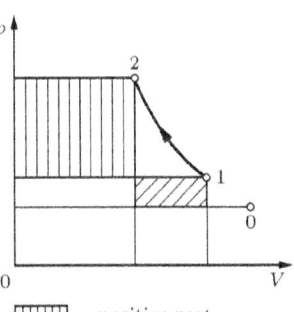

Figure 8.7 Exergy difference of the shift works $EW_2 - EW_1$ according to example 8.5

▦ = positive part
▨ = negative part

Example 8.5 From the data of example 8.4, according to Eqs. (2.27) and (8.22) one gets the following values for the difference of the shift works and the corresponding exergy difference:

$$W_2 - W_1 = 90.46 \text{ kJ} \qquad EW_2 - EW_1 = 159.69 \text{ kJ}$$

According to Figure 8.7, the positive part of the exergy difference is 194.30 kJ, and the negative part is 34.61 kJ.

8.1.6 Exergy of Pressure Change Work

According to Eq. (2.28), the pressure change work W_{p12} is equal to the sum of the volume change work W_{V12} and the difference of the shift works $W_2 - W_1$:

$$W_{p12} = W_{V12} + W_2 - W_1 \tag{8.23}$$

Accordingly, the exergy of the pressure change work is composed of the exergy of the volume change work and the exergy difference of the difference of the shift works:

$$EW_{p12} = EW_{V12} + EW_2 - EW_1 \tag{8.24}$$

Eqs. (8.19) and (8.22) are inserted into Eq. (8.24):

$$EW_{p12} = W_{V12} + p_0(V_2 - V_1) + W_2 - W_1 - p_0(V_2 - V_1) = W_{V12} + W_2 - W_1 \quad (8.25)$$

From Eqs. (8.23) and (8.25), it follows that the exergy of the pressure change work EW_{p12} coincides with the pressure change work W_{p12}:

$$EW_{p12} = W_{p12} \quad (8.26)$$

The same result is obtained with the help of Eq. (8.2) if W_{Kreis} is replaced by EW_{p12} and W_{12} by W_{p12}. Hereby, it is to be noted that $W_0 = 0$ because the environment is not able to do any pressure change work (Fig. 8.8).

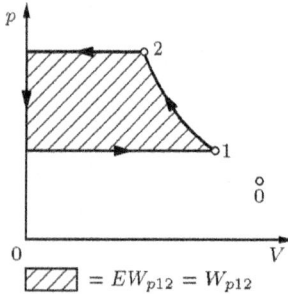

Figure 8.8 Cyclic process for determining the exergy of the pressure change work $EW_{p12} = W_{p12}$

According to Fig. 8.8, a left-hand cyclic process can also be considered with inclusion of the change of state from state 1 to state 2 as well as with two additional isobaric changes of state at p_1 and p_2; in this case this also results in $W_{Kreis} = \oint V \, dp = EW_{p12} = W_{p12}$ according to Eq. (7.14). Combining the exergy areas in Figs. 8.5 a) and 8.7 to form Fig. 8.8 confirms Eqs. (8.24) and (8.26).

8.1.7 Exergy of Internal Energy

The exergy of the internal energy EU of a substance is equal to that part of the internal energy which is in a closed system, e.g. in a cylinder with a movable wall, during an imaginary reversible change of state into the state of equilibrium with the environment, converted into useful work. In the thought model, according to Fig. 8.9, the change of state to the ambient state occurs in two steps (shown as expansion in Fig. 8.9). The work done in the reversible change of state in a closed system is volume change work.

The first step leads from the initial state 1 with the internal energy U in a reversible-adiabatic, i.e. isentropic change of state to an intermediate state Z with the internal energy U_Z and the ambient temperature T_0. According to Eq. (2.53), the volume change work done in this process is

$$W_{V1Z} = U_Z - U \, . \quad (8.27)$$

For an ideal gas, the internal energy in the intermediate state U_Z is equal to the internal energy in the ambient state U_0. Thus

$$W_{V1Z} = U_0 - U \quad (8.28)$$

is obtained.

The second step leads in an isothermal change of state from the intermediate state Z with entropy S_Z equal to the entropy in the initial state S, as a result of the reversible

transfer of heat (entropy change heat) between the System and the environment to the ambient state 0 with the entropy S_0. For an ideal gas, the volume change work in the second step is given by Eq. (4.86)

$$W_{VZ0} = -T_0 (S_0 - S) . \tag{8.29}$$

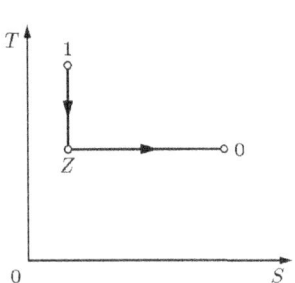

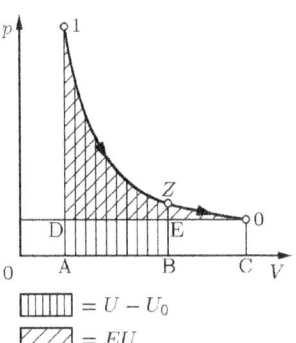

Figure 8.9 Changes of state in the determination of the exergy of the internal energy EU and the exergy of the enthalpy EH in a T, S diagram   1 initial state   Z intermediate state   0 ambient state

Figure 8.10 The change of the internal energy $U - U_0$ and the exergy of the internal energy EU according to example 8.6 in a p, V diagram

The useful work is obtained by subtracting from the negative volume change work the difference of the shift work of the ambient air with the pressure p_0. With V for the volume in the initial state and V_0 for the volume in the ambient state, the difference of the shift work is $p_0 (V_0 - V)$. Summarizing the useful work of both steps, one obtains the exergy of the internal energy according to the following equation:

$$EU = U - U_0 - T_0 (S - S_0) + p_0 (V - V_0) \tag{8.30}$$

For a real gas, in addition to Eq. (8.27), instead of Eq. (8.29), the following equation applies

$$W_{VZ0} = -T_0 (S_0 - S) + U_0 - U_Z . \tag{8.31}$$

The sum of the negative volume change work of both steps and the subtraction of the difference of the shift works caused by the ambient air also gives Eq. (8.30).

The three components of the right-hand side of Eq. (8.30) can be illustrated in a p, V diagram shown in Fig. 8.10. The area 1ZBA1 represents the change in internal energy according to Eq. (8.28). The area Z0CBZ is the heat (entropy change heat) transferred with the participation of the environment according to Eq. (8.29). The area 0CAD0 corresponds to the difference of the shift works of the external atmospheric pressure. The exergy of the internal energy EU appears in the p, V diagram according to Fig. 8.10 as area 1Z0D1.

Example 8.6 The exergy of the internal energy EU for a quantity of air of 4.1 kg with the pressure of 8 bar and the temperature $200\,°C$ is to be determined. The specific gas constant is $R = 0.287$ kJ/(kg K), and the isentropic exponent is $\kappa = 1.4$. The ambient state is described by $p_0 = 1$ bar and $t_0 = 20\,°C$.

The assumed reversible state changes are shown in Fig. 8.9.

The change in internal energy is given by Table 4.4 Eq. (4)

$$U - U_0 = 529.515 \text{ kJ} .$$

According to Eq. (4.58), the entropy change is
$$S - S_0 = -0.47526 \text{ kJ/K} .$$

The volume change is given by Eq. (4.25)
$$V - V_0 = 0.69594 \text{ m}^3 - 3.44950 \text{ m}^3 = -2.75356 \text{ m}^3 .$$

This gives the exergy of the internal energy according to Eq. (8.30)
$$EU = 529.515 \text{ kJ} + 293.15 \text{ K} \cdot 0.47526 \text{ kJ/K} - 1 \text{ bar} \cdot 2.75356 \text{ m}^3 \cdot 100 \text{ kJ/(bar m}^3) = 393.48 \text{ kJ} .$$

The change in internal energy $U - U_0$ and the exergy of the internal energy EU are shown in Fig. 8.10.

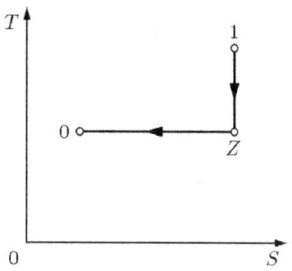

Figure 8.11 Changes of state in the determination of the exergy of the internal energy EU and the exergy of the enthalpy EH in a T, S diagram
1 initial state   Z intermediate state
0 ambient state

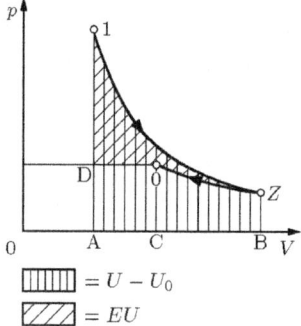

Figure 8.12 Change of the internal energy $U - U_0$ and the exergy of the internal energy EU according to example 8.7 in a p, V diagram

In technical processes gas states occur, from which in a thought model by reversible expansion alone, the ambient state cannot be reached. The isentropic and the isothermal state changes can also be reversible compressions. For example, according to Fig. 8.11, an isentropic expansion from 1 to Z must be followed by an isothermal compression from Z to 0. Fig. 8.12 shows the changes of state in a p, V diagram. The interpretation and designation of the areas can be taken from the explanation given in connection with Fig. 8.10.

Example 8.7 2.7 kg of air ($R = 0.287$ kJ/(kg K); $\kappa = 1.4$) with a pressure of 3 bar and the temperature 200 °C are reversibly transferred to the ambient state $p_0 = 1$ bar, $t_0 = 20\,°$C. The changes of state are shown in the T, S diagram of Fig. 8.11 and in the p, V diagram of Fig. 8.12. The change of the internal energy and the exergy of the internal energy are to be calculated.

The change of the internal energy is according to table 4.4 Eq. (4)
$$U - U_0 = 348.705 \text{ kJ} .$$
According to Eq. (4.58), the entropy change is
$$S - S_0 = 0.44707 \text{ kJ/K} .$$
The volume change is given by Eq. (4.25)
$$V - V_0 = 1.2221 \text{ m}^3 - 2.2716 \text{ m}^3 = -1.0495 \text{ m}^3 .$$

According to Eq. (8.30), the exergy of the internal energy is
$EU = 348.705 \text{ kJ} - 293.15 \text{ K} \cdot 0.44707 \text{ kJ/K} - 1 \text{ bar} \cdot 1.0495 \text{ m}^3 \cdot 100 \text{ kJ/(bar m}^3) = 112.70 \text{ kJ}$.
The change of the internal energy $U - U_0$ and the exergy of the internal energy EU are shown in Fig. 8.12.

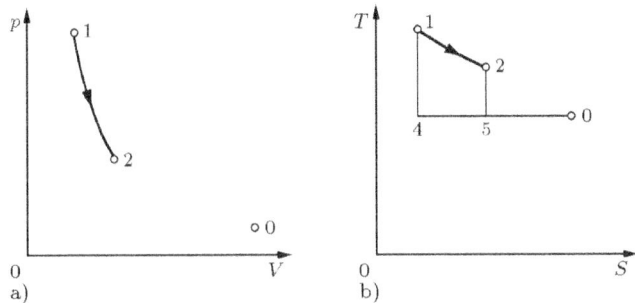

Figure 8.13 Polytropic expansion of a volume of air from state 1 to state 2 according to example 8.8 in a   a) p, V diagram   b) T, S diagram
0 is the ambient state of the air volume
1 − 4 − 0 reversible expansion to determine EU_1
2 − 5 − 0 reversible expansion to determine EU_2

The change in exergy of the internal energy in the case of an actually executed change of state according to Fig. 8.13 a) is given by the difference of the exergies in the initial state 1 and in the final state 2:
$$EU_2 - EU_1 = U_2 - U_1 - T_0 (S_2 - S_1) + p_0 (V_2 - V_1) \qquad (8.32)$$
In the isothermal change of state of an ideal gas, the difference of the internal energies becomes zero, the exergy difference of the internal energies does not.

Fig. 8.13 b) describes the thought model of the reversible change of state for determining EU_1 and EU_2. Assuming that the exergy of the internal energy in the initial state EU_1 has already been determined in example 8.6 and is shown in Fig. 8.10, this still has to be done for the exergy of the internal energy in the final state EU_2 in example 8.8 and Fig. 8.14.

Example 8.8 The volume of air in example 8.6 is expanded polytropically to 3.5 bar and 120 °C. What is the change in exergy of the internal energy?

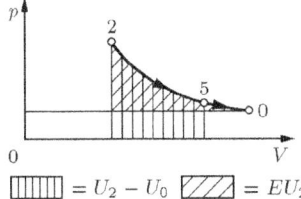

Figure 8.14 The change in internal energy $U_2 - U_0$ and the exergy of the internal energy EU_2 according to example 8.8 in a p, V diagram

▥ $= U_2 - U_0$ ▨ $= EU_2$

The exergy of the internal energy in the final state EU_2 is (cf. Fig. 8.13 and Fig. 8.14):
$EU_2 = U_2 - U_0 - T_0 (S_2 - S_0) + p_0 (V_2 - V_0)$
$= 294.175 \text{ kJ} + 293.15 \text{ K} \cdot 0.26533 \text{ kJ/K} - 1 \text{ bar} \cdot 2.12773 \text{ m}^3 \cdot 100 \text{ kJ/(bar m}^3)$
$= 159.18 \text{ kJ}$

With $U_1 - U_0$ and EU_1 according to example 8.6 one obtains for the change of state according to Fig. 8.13
$$U_2 - U_1 = -235.34 \text{ kJ} \quad \text{and} \quad EU_2 - EU_1 = -234.30 \text{ kJ}.$$

8.1.8 Exergy of Enthalpy

The exergy of the enthalpy EH of a substance is equal to that part of the enthalpy that is given off in an open system, e.g. by a turbine, in an assumed reversible change of state to the state of equilibrium with the environment into useful work. Here, too, in the thought model the transition to the ambient state usually takes place in two steps with an isentropic as well as an isothermal change of state; these are similar to the determination of the exergy of the internal energy (cf. Fig. 8.9 and 8.11).

The work that occurs is pressure change work. With the enthalpy H and the entropy S in the initial state 1, the enthalpy H_0 and the entropy S_0 in the ambient state 0 as well as the ambient temperature T_0, the pressure change works in the two steps for an ideal gas are

$$W_{p1Z} = H_0 - H \tag{8.33}$$

$$W_{pZ0} = -T_0 (S_0 - S) . \tag{8.34}$$

The useful work is equal to the sum of the negative pressure change work. The exergy of enthalpy is therefore

$$EH = H - H_0 - T_0 (S - S_0) . \tag{8.35}$$

In the isothermal change of state of an ideal gas, the difference in enthalpies becomes zero, but the exergy difference of enthalpies does not.

If reversible expansions take place in both steps, the change in enthalpy $H - H_0$ and the exergy of the enthalpy EH in a T, S diagram and in a p, V diagram are shown in Fig. 8.15.

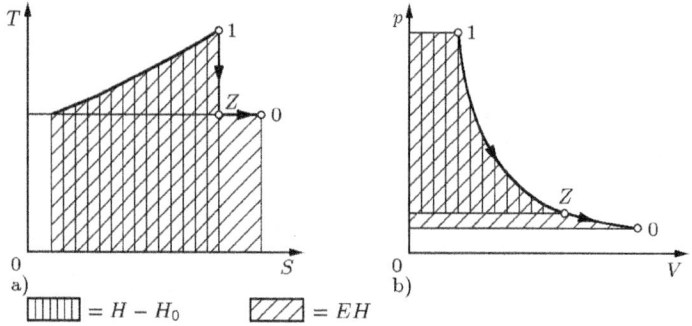

Figure 8.15 The change in enthalpy $H - H_0$ and the exergy of the enthalpy EH according to example 8.9 in a   a) T, S diagram (ideal gas)   b) p, V diagram

Example 8.9 The exergy of the enthalpy for the amount of air in example 8.6 is to be determined.
The change in enthalpy with the assumed reversible expansion to the ambient state is given by Table 4.5 Eq. (4)

$$H - H_0 = 741.321 \text{ kJ} .$$

With the numerical values of example 8.6 one obtains according to Eq. (8.35) (cf. Fig. 8.15)

$$EH = 741.321 \text{ kJ} - 293.15 \text{ K} \cdot (-0.47526 \text{ kJ/K}) = 880.64 \text{ kJ} .$$

If the change of state in the second step is a compression, representations as in Fig. 8.16 result.

Example 8.10 Using the data from example 8.7, what is the exergy of the enthalpy?

$$EH = 488.187 \text{ kJ} - 293.15 \text{ K} \cdot 0.44707 \text{ kJ/K} = 357.13 \text{ kJ}$$

Fig. 8.16 shows the enthalpy difference $H - H_0$ and the exergy of the enthalpy EH.

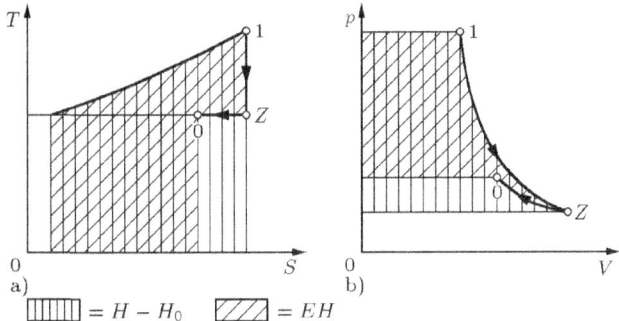

Figure 8.16 The change in enthalpy $H - H_0$ and the exergy of the enthalpy EH according to example 8.10   a) in a T, S diagram (ideal gas)   b) in a p, V diagram

If one subtracts from the exergy of the enthalpy according to Eq. (8.35) the exergy of the internal energy according to Eq. (8.30), one obtains with Eq. (2.41) the difference of the shift works $pV - p_0 V_0 - p_0(V - V_0)$ called the replenishment work [17]:

$$EH - EU = (p - p_0) V \tag{8.36}$$

It can be found as a rectangular area in the p, V diagram according to Fig. 8.17. $EH - EU$ is equal to the exergy of the shift work EW according to Eq. (8.21).

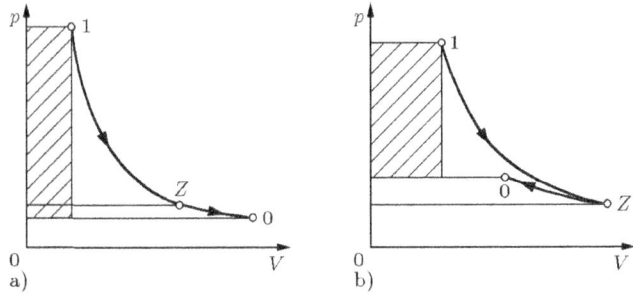

Figure 8.17 The difference of the shift works (replenishment work) $(p - p_0) V$ in a p, V diagram according to
a) example 8.11   b) example 8.12

Example 8.11 From examples 8.9 and 8.6 the following replenishment work is obtained:

$$EH - EU = 880.64 \text{ kJ} - 393.48 \text{ kJ} = 487.16 \text{ kJ}$$

Example 8.12 Using the exergy of the enthalpy and the exergy of the internal energy according to the examples 8.10 and 8.7, one obtains the replenishment work

$$EH - EU = 357.13 \text{ kJ} - 112.70 \text{ kJ} = 244.43 \text{ kJ} .$$

For a change of state between the states 1 and 2, the change in exergy of the enthalpy is
$$EH_2 - EH_1 = H_2 - H_1 - T_0(S_2 - S_1). \tag{8.37}$$

Fig. 8.18 shows for ideal gases a representation of the exergy difference of the enthalpy $EH_2 - EH_1$ as an area in a T, S diagram.

According to section 4.4, the enthalpy differences $H_2 - H_0$ and $H_1 - H_0$ are equal to the areas 2FA42 and 1EB51 in a T, S diagram according to Fig. 8.18. The area 1EB51 can be shifted to the left so that it coincides with the area 3DA43. Then $H_2 - H_1$ is equal to the area 2FD32. The area 67FE6 corresponding to the expression $T_0(S_2 - S_1)$ is subtracted from this area. Thus the area 276ED32 represents the exergy difference $EH_2 - EH_1$ of the enthalpy difference $H_2 - H_1$.

Example 8.13 2 kg of air ($R = 0.287$ kJ/(kg K); $\kappa = 1.4$) are compressed from 3 bar and 240 °C to 7 bar and 720 °C. What is the exergy increase of the enthalpy if the ambient state is described by 1 bar and 0 °C ?

According to Table 4.5 Eq. (4) is
$$H_2 - H_1 = 964.32 \text{ kJ}.$$

Eq. (4.58) gives the entropy difference
$$S_2 - S_1 = 0.84022 \text{ kJ/K}.$$

With Eq. (8.37) one obtains (cf. Fig. 8.18)
$$EH_2 - EH_1 = 964.32 \text{ kJ} - 273.15 \text{ K} \cdot 0.84022 \text{ kJ/K} = 734.81 \text{ kJ}.$$

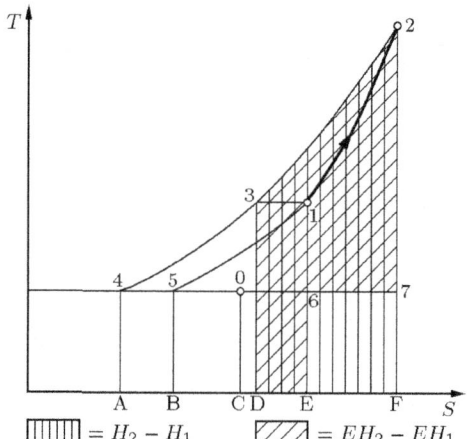

Figure 8.18 The change in enthalpy $H_2 - H_1$ and the change in exergy of enthalpy $EH_2 - EH_1$ according to example 8.13 in a T, S diagram (ideal gas)

In the ambient state is $EH = 0$. For this case it follows from Eq. (8.35)
$$H = T_0 S + (H_0 - T_0 S_0). \tag{8.38}$$

Eq. (8.38) represents a straight line in a H, S diagram. It connects all states with the exergy of enthalpy $EH = 0$. It is called the ambient straight line. For $EH = \text{const}$, the straight lines parallel to the ambient straight line in a H, S diagram result according to the following equation:
$$H = T_0 S + (H_0 - T_0 S_0 + EH). \tag{8.39}$$

The slope of the straight line of constant exergy of enthalpy is given by Eqs. (8.38) and (8.39)

350   8 Exergy

$$\left(\frac{\partial H}{\partial S}\right)_{EH} = \left(\frac{\partial h}{\partial s}\right)_{eh} = T_0 \,. \tag{8.40}$$

According to Eq. (5.55), this is the slope of the isotherm and isobar in the two-phase liquid-vapor region at ambient temperature.

Example 8.14 For ammonia (NH_3, R 717) a h, s diagram with the ambient line $eh = 0$ and the lines of constant specific exergy of the enthalpy $eh = $ const is to be drawn. The ambient state is assumed to be the saturated vapor state of ammonia at $T_0 = 290.00$ K.

The reference point for the calculation of the exergy at $T_0 = 290.00$ K or $t_0 = 16.85\,°C$ is described by the following values: $p_0 = 7.745$ bar, $v_0 = 0.1647$ m^3/kg, $h_0 = 1477.83$ kJ/kg, $s_0 = 5.4126$ kJ/(kg K)

According to Eq. (8.35), the specific exergy of the enthalpy is

$$eh = h - h_0 - T_0\,(s - s_0)\,.$$

As a state of equilibrium for ammonia with the environment, $p_0 = 1$ bar and $T_0 = 290.00$ K (superheated steam) can be assumed - deviating from the state chosen here. Then the values of the specific exergy eh calculated with this differ from those of Figure 8.19 by the constant value $\Delta eh = -278.13$ kJ/kg. The fact of negative exergy values in Figure 8.19 is related to the choice of the equilibrium state with the environment.

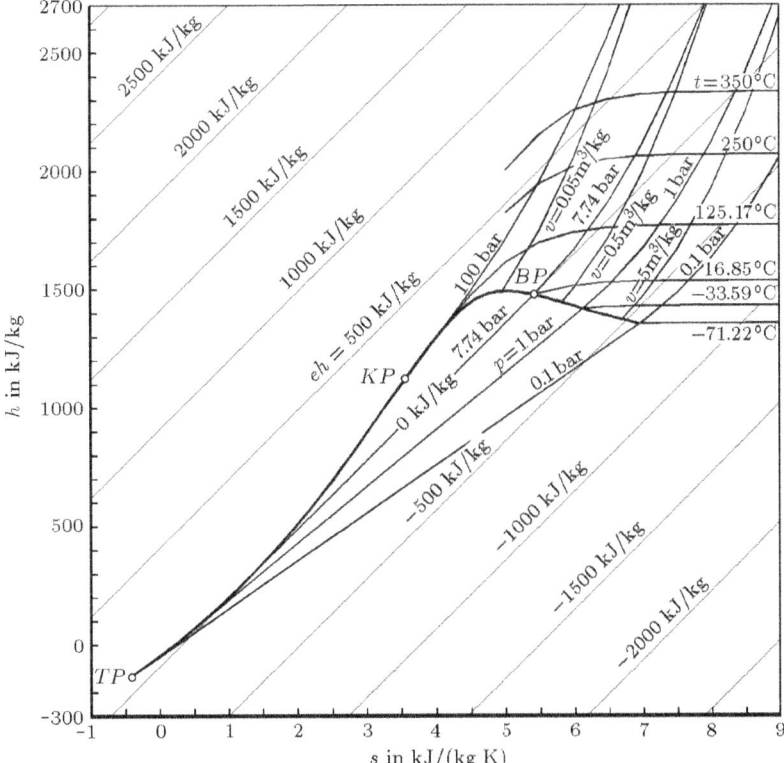

Figure 8.19 h, s diagram of ammonia according to example 8.14
TP triple point KP critical point BP reference point for the calculation of exergy
The lines of constant exergy have in parts only a formal meaning.

8.1.9 Exergy of Free Energy

The free energy with which e.g. states of a substance in a closed system can be described, according to Eq. (5.211) is given by its internal energy minus its bound energy:

$$F = U - TS \tag{8.41}$$

Thus, the exergy of the free energy can be calculated as the difference of the exergy of the internal energy and the exergy of the bound energy of the substance under consideration:

$$EF = EU - ETS \tag{8.42}$$

Using Eqs. (8.30) and (8.9), one obtains from Eq. (8.42):

$$EF = U - U_0 - T_0(S - S_0) + p_0(V - V_0) - (T - T_0)(S - S_0)$$
$$= U - U_0 - T_0 S + T_0 S_0 + p_0(V - V_0) - TS - T_0 S_0 + T_0 S + T S_0 \tag{8.43}$$

$$EF = U - TS + p_0(V - V_0) + (T - T_0)S_0 - (U_0 - T_0 S_0) \tag{8.44}$$

Considering Eq. (8.41) for F and F_0, one obtains

$$EF = F - F_0 + (T - T_0)S_0 + p_0(V - V_0). \tag{8.45}$$

In the case of a change of state between states 1 and 2, the change of exergy of the free energy according to Eq. (8.45) is given as follows:

$$EF_2 - EF_1 = F_2 - F_0 + (T_2 - T_0)S_0 + p_0(V_2 - V_0) - (F_1 - F_0 + (T_1 - T_0)S_0 + p_0(V_1 - V_0)) \tag{8.46}$$

$$EF_2 - EF_1 = F_2 - F_1 + (T_2 - T_1)S_0 + p_0(V_2 - V_1) \tag{8.47}$$

8.1.10 Exergy of Free Enthalpy

The free enthalpy, which can be used to describe e.g. states of a substance in an open system, is given by according to Eq. (5.215) from its enthalpy minus its bound energy:

$$G = H - TS \tag{8.48}$$

Thus, the exergy of the free enthalpy can be calculated as the difference between the exergy of the enthalpy and the exergy of the bound energy of the substance under consideration:

$$EG = EH - ETS \tag{8.49}$$

Using Eqs. (8.35) and (8.9), one obtains from Eq. (8.48):

$$EG = H - H_0 - T_0(S - S_0) - (T - T_0)(S - S_0)$$
$$= H - H_0 - T_0 S + T_0 S_0 - TS - T_0 S_0 + T_0 S + T S_0 \tag{8.50}$$

$$EG = H - TS - (H_0 - T_0 S_0) + (T - T_0)S_0 \tag{8.51}$$

Considering Eq. (8.48) for G and G_0, one gets:

$$EG = G - G_0 + (T - T_0)S_0 \tag{8.52}$$

Example 8.15 Equation (8.52) for the calculation of the exergy of the free enthalpy is to be derived with the aid of two reversible changes of state in an open system between an assumed initial state 1 and the state 0 of the substance in equilibrium with the environment. The total reversible change of state shall consist of an isothermal change of state from the initial state 1 to the intermediate state Z and an isentropic change of state from the intermediate state Z to the final state 0 (Fig. 8.20). For the change of state between the states 1 and Z, the ability of the free enthalpy to interact with a heat source shall be taken into account. The amount of

the sum of the respective recoverable pressure change works represents the exergy of the free enthalpy.

Isothermal change of state from 1 to Z with $T_1 = T_Z = T$:

$$W_{p1Z} = G_Z - G_1 = \int_1^Z V\,dp - \int_1^Z S\,dT = \int_1^Z V\,dp \qquad -W_{p1Z} = G - G_Z$$

Isentropic change of state from Z to 0:

$W_{pZ0} = H_0 - H_Z = G_0 + T_0 S_0 - (G_Z + T_Z S_Z) = G_0 - G_Z - T_Z S_Z + T_0 S_0$

With $\quad T_Z = T_1 = T \quad$ and $\quad S_Z = S_0 \quad$ becomes:

$-W_{pZ0} = G_Z - G_0 + T S_0 - T_0 S_0$

$EG = -W_{p1Z} - W_{pZ0} = G - G_Z + G_Z - G_0 + T S_0 - T_0 S_0$

$EG = G - G_0 + (T - T_0) S_0$

Thus, Eq. (8.52) is confirmed.

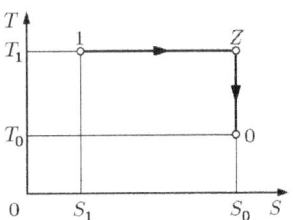

Figure 8.20 Changes of state in the determination of the exergy of the free enthalpy EG in a T, S diagram
1 initial state   Z intermediate state
0 ambient state

For a thermodynamic change of state from 1 to 2, the associated exergy difference $EG_2 - EG_1$ can be determined using Eq. (8.52):

$$EG_2 - EG_1 = G_2 - G_0 + (T_2 - T_0) S_0 - \bigl(G_1 - G_0 + (T_1 - T_0) S_0\bigr) \qquad (8.53)$$

$$EG_2 - EG_1 = G_2 - G_1 + (T_2 - T_1) S_0 \qquad (8.54)$$

Example 8.16 In example 8.1, a volume of air of 2.5 kg is heated isobarically from 100 °C to 300 °C at a pressure of 2 bar.

a) The difference of zhe enthalpies $H_2 - H_1$, the difference of the exergies of the enthalpies $EH_2 - EH_1$, the difference of the bound energies $T_2 S_2 - T_1 S_1$, the difference of the exergies of the bound energies $ET_2 S_2 - ET_1 S_1$, the difference of the free enthalpies $G_2 - G_1$, the exergies of the free enthalpies EG_1 and EG_2 and the difference of the exergies of the free enthalpies $EG_2 - EG_1$ are to be determined.

b) For the isobaric change of state, an energy balance and an exergy balance are to be prepared. The enthalpies and the reversible substitute heat on the one hand and the free enthalpies, the bound energies and the reversible substitute heat on the other hand are to be used as energetic quantities.

a) $H_2 - H_1 = m\,c_p\,(T_2 - T_1) = 2.5\,\text{kg} \cdot 1.0045\,\text{kJ/(kg K)}\,(573.15 - 373.15)\,\text{K} = 502.25\,\text{kJ}$

s_0 as the absolute value of the specific entropy is calculated with Eq. (4.58), using the value s_0^* given in Section 7.5.3.2 (page 286/287) at $t_0^* = 25\,°\text{C}$ and $p_0^* = 1$ bar, as $s_0 = 6.833$ kJ/(kg K) at 15 °C and 1 bar. From this, s_1 and s_2 can be determined with Eq. (4.58) to $s_1 = 6.894$ kJ/(kg K) and $s_2 = 7.325$ kJ/(kg K).

$EH_2 - EH_1 = H_2 - H_1 - T_0\,m\,(s_2 - s_1) = 502.25\,\text{kJ} - 288.15\,\text{K} \cdot 2.5\,\text{kg}\,(7.325 - 6.894)\,\text{kJ/(kg K)}$
$= 191.77\,\text{kJ}$

$T_2 S_2 - T_1 S_1 = m\,(T_2 s_2 - T_1 s_1)$

$$= 2.5 \,\text{kg} \cdot (573.15 \,\text{K} \cdot 7.325 \,\text{kJ/(kg K)} - 373.15 \,\text{K} \cdot 6.894 \,\text{kJ/(kg K)}) = 4064.57 \,\text{kJ}$$

$$ET_2 S_2 - ET_1 S_1 = T_2 S_2 - T_1 S_1 - T_0 (S_2 - S_1) - (T_2 - T_1) S_0$$
$$= T_2 S_2 - T_1 S_1 - T_0 m (s_2 - s_1) - (T_2 - T_1) m s_0$$
$$= 4064.57 \,\text{kJ} - 288.15 \,\text{K} \cdot 2.5 \,\text{kg} \cdot (7.325 - 6.894) \,\text{kJ/(kg K)}$$
$$-(573.15 - 373.15) \,\text{K} \cdot 2.5 \,\text{kg} \cdot 6.833 \,\text{kJ/(kg K)}$$
$$= 4064.57 \,\text{kJ} - 310.48 \,\text{kJ} - 3416.50 \,\text{kJ} = 337.59 \,\text{kJ}$$

$$G_2 - G_1 = H_2 - H_1 - (T_2 S_2 - T_1 S_1)$$
$$G_2 - G_1 = 502.25 \,\text{kJ} - 4064.57 \,\text{kJ} = -3562.32 \,\text{kJ}$$

$$EG_1 = G_1 - G_0 + (T_1 - T_0) S_0 = H_1 - H_0 - (T_1 S_1 - T_0 S_0) + (T_1 - T_0) S_0$$
$$= m \left[c_p (T_1 - T_0) - (T_1 s_1 - T_0 s_0) + (T_1 - T_0) s_0 \right]$$
$$= 2.5 \,\text{kg} \left[1.0045 \,\text{kJ/(kg K)} \cdot (373.15 - 288.15) \,\text{K} - 373.15 \,\text{K} \cdot 6.894 \,\text{kJ/(kg K)} \right.$$
$$\left. + 288.15 \,\text{K} \cdot 6.833 \,\text{kJ/(kg K)} + (373.15 - 288.15) \,\text{K} \cdot 6.833 \,\text{kJ/(kg K)} \right] = 156.55 \,\text{kJ}$$

$$EG_2 = G_2 - G_0 + (T_2 - T_0) S_0 = H_2 - H_0 - (T_2 S_2 - T_0 S_0) + (T_2 - T_0) S_0$$
$$= m \left[c_p (T_2 - T_0) - (T_2 s_2 - T_0 s_0) + (T_2 - T_0) s_0 \right]$$
$$= 2.5 \,\text{kg} \left[1.0045 \,\text{kJ/(kg K)} \cdot (573.15 - 288.15) \,\text{K} - 573.15 \,\text{K} \cdot 7.325 \,\text{kJ/(kg K)} \right.$$
$$\left. + 288.15 \,\text{K} \cdot 6.833 \,\text{kJ/(kg K)} + (573.15 - 288.15) \,\text{K} \cdot 6.833 \,\text{kJ/(kg K)} \right] = 10.73 \,\text{kJ}$$

$$EG_2 - EG_1 = 10.73 \,\text{kJ} - 156.55 \,\text{kJ} = -145.82 \,\text{kJ}$$

b) On the one hand, the energy balance can be established directly with the help of the enthalpies H_1 and H_2 as well as the reversible substitute heat $(Q_{12})_{rev}$ (Fig. 8.21 a); on the other hand, it can be formulated indirectly with the help of the free enthalpies G_1 and G_2, the bound energies $T_1 S_1$ and $T_2 S_2$ as well as the reversible substitute heat $(Q_{12})_{rev}$ (Fig. 8.21 b):

$$H_2 - H_1 = (Q_{12})_{rev} = m c_p (T_2 - T_1) = 2.5 \,\text{kg} \cdot 1.0045 \,\text{kJ/(kg K)} \,(573.15 - 373.15) \,\text{K} = 502.25 \,\text{kJ}$$

$$G_2 - G_1 + (T_2 S_2 - T_1 S_1) = (Q_{12})_{rev} = -3562.32 \,\text{kJ} + 4064.57 \,\text{kJ} = 502.25 \,\text{kJ}$$

On the one hand, the exergy balance can be established directly with the help of the exergies of the enthalpies EH_1 and EH_2 as well as the exergy of the reversible substitute heat $(EQ_{12})_{rev}$ (Fig. 8.21 c); on the other hand, it can be formulated indirectly with the help of the exergies of the free enthalpies EG_1 and EG_2, the exergies of the bound energies $ET_1 S_1$ and $ET_2 S_2$ as well as the exergy of the reversible substitute heat $(EQ_{12})_{rev}$ (Figure 8.21 d):

$$EH_2 - EH_1 = (EQ_{12})_{rev} = 191.70 \,\text{kJ}$$

$$EG_2 - EG_1 + ET_2 S_2 - ET_1 S_1 = (EQ_{12})_{rev} = -145.82 \,\text{kJ} + 337.59 \,\text{kJ} = 191.77 \,\text{kJ}$$

It can be seen that here the direct balancing with the help of the enthalpies and with the help of the exergies of the enthalpies respectively, according to the Figs. 8.21 a and 8.21 c, is easier.

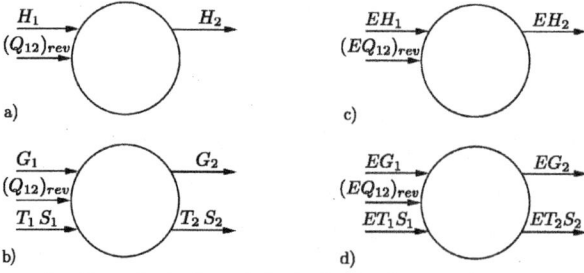

Figure 8.21 Balances for the calculation of the isobaric change of state according to example 8.16

8.1.11 Difference between EU and EF

The difference between the exergy of the internal energy EU and the exergy of the free energy EF results from Eqs. (8.30) and (8.43):

$$EU - EF = U - U_0 - T_0(S - S_0) + p_0(V - V_0) -$$
$$- (U - U_0 - T_0(S - S_0) + p_0(V - V_0) - (T - T_0)(S - S_0)) \quad (8.55)$$
$$EU - EF = ETS = (T - T_0)(S - S_0) \quad (8.56)$$

This exergy difference is shown as an area in Figure 8.22. This corresponds to the area of the exergy of the bound energy ETS in Figure 8.2.

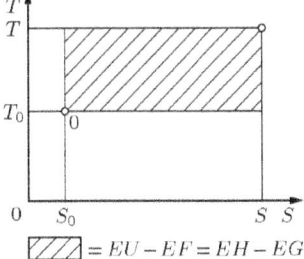

Figure 8.22 Difference of the exergy of the internal energy EU and the exergy of the free energy EF; difference of the exergy of the enthalpy EH and the exergy of the free enthalpy EG

For a change of state between states 1 and 2, the difference between the change of the exergy of the internal energy $EU_2 - EU_1$ and the change of the exergy of the free energy $EF_2 - EF_1$ results from Eqs. (8.32), (8.47) as well as Eq. (8.41):

$$(EU_2 - EU_1) - (EF_2 - EF_1) = U_2 - U_1 - T_0(S_2 - S_1) + p_0(V_2 - V_1) -$$
$$- (U_2 - U_1 - (T_2 S_2 - T_1 S_1) + (T_2 - T_1) S_0 + p_0(V_2 - V_1)) \quad (8.57)$$
$$(EU_2 - EU_1) - (EF_2 - EF_1) = T_2 S_2 - T_1 S_1 - T_0(S_2 - S_1) - (T_2 - T_1) S_0 \quad (8.58)$$

This exergy difference is shown as an area in Figure 8.23. This corresponds to the area for $ET_2 S_2 - ET_1 S_1$ in Figure 8.3.

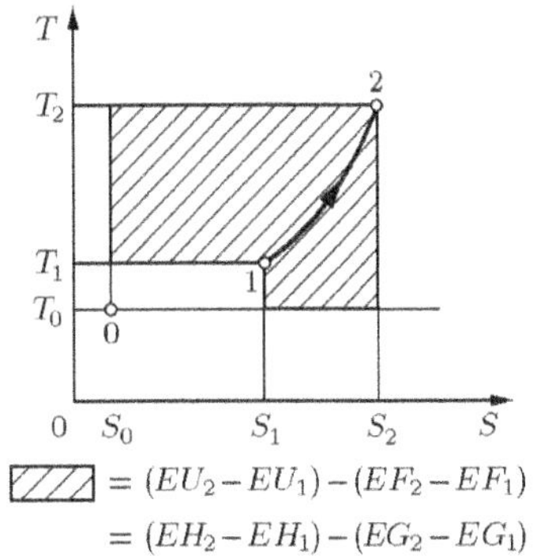

Figure 8.23 Difference between the exergy difference of the internal energies $EU_2 - EU_1$ and the exergy difference of the free energies $EF_2 - EF_1$; difference between the exergy difference of enthalpies $EH_2 - EH_1$ and the exergy difference of the free enthalpies $EG_2 - EG_1$

The right side of Eq. (8.58) is also the right side of Eq. (8.10). From this follows

$$(EU_2 - EU_1) - (EF_2 - EF_1) = ET_2 S_2 - ET_1 S_1 . \quad (8.59)$$

The right side of Eq. (8.59) can also be represented by the transformed Eq. (8.13). From this follows

$$(EU_2 - EU_1) - (EF_2 - EF_1) = EQ_{T12} + (EQ_{12})_{rev} = EQ_{T12} + EQ_{s12} . \quad (8.60)$$

The difference between the exergy difference of the internal energies $EU_2 - EU_1$ and the exergy difference of the free energies $EF_2 - EF_1$ is the sum of the exergy of the entropy change heat $(EQ_{s12})_{rev}$ and the exergy of the temperature change heat EQ_{T12}.

8.1.13 Difference between EH and EG

The difference between the exergy of the enthalpy EH and the exergy of the free enthalpy EG is given by Eqs. (8.35) and (8.50):

$$EH - EG = H - H_0 - T_0(S - S_0) - \big(H - H_0 - T_0(S - S_0) - (T - T_0)(S - S_0)\big) \quad (8.61)$$

$$EH - EG = (T - T_0)(S - S_0) \quad (8.62)$$

This exergy difference is shown as an area in Figure 8.22. This corresponds to the area of the exergy of the bound energy ETS in Figure 8.2.

In the case of a change of state between states 1 and 2, the difference between the change of the exergy of the enthalpy $EH_2 - EH_1$ and the change of the exergy of the free enthalpy $EG_2 - EG_1$ follows from Eqs. (8.37), (8.54) as well as Eq. (8.48):

$$(EH_2 - EH_1) - (EG_2 - EG_1) = H_2 - H_1 - T_0(S_2 - S_1) -$$
$$- \big(H_2 - H_1 - (T_2 S_2 - T_1 S_1) + (T_2 - T_1) S_0\big) \quad (8.63)$$

$$(EH_2 - EH_1) - (EG_2 - EG_1) = T_2 S_2 - T_1 S_1 - T_0(S_2 - S_1) - (T_2 - T_1) S_0 \quad (8.64)$$

This exergy difference is shown as an area in Figure 8.23. This corresponds to the area for $ET_2 S_2 - ET_1 S_1$ in Figure 8.3.

The right side of Eq. (8.64) is also the right side of Eq. (8.10). Thus follows

$$(EH_2 - EH_1) - (EG_2 - EG_1) = ET_2 S_2 - ET_1 S_1 . \quad (8.65)$$

The right side of Eq. (8.65) can also be represented by the transformed Eq. (8.13). From this follows

$$(EH_2 - EH_1) - (EG_2 - EG_1) = EQ_{T12} + (EQ_{12})_{rev} = EQ_{T12} + EQ_{s12} . \quad (8.66)$$

The difference between the exergy difference of the enthalpies $EH_2 - EH_1$ and the exergy difference of the free enthalpies $EG_2 - EG_1$ is the sum of the exergy of the entropy change heat EQ_{s12} and the exergy of the temperature change heat EQ_{T12}.

8.1.13 Free Energy and Free Enthalpy as Thermodynamic Potentials

In the case of an isothermal change of state between states 1 and 2 ($T_1 = T_2 = T$), the exergy differences can be calculated according to Eqs. (8.58) and (8.64) according to Figure 8.24 as a rectangular area. With $T_2 = T_1$ follows:

$$(EU_2 - EU_1) - (EF_2 - EF_1) = (EH_2 - EH_1) - (EG_2 - EG_1)$$
$$= T_1 S_2 - T_1 S_1 - T_0(S_2 - S_1) - (T_1 - T_1) S_0$$
$$= T_1(S_2 - S_1) - T_0(S_2 - S_1)$$

$$(EU_2 - EU_1) - (EF_2 - EF_1) = (EH_2 - EH_1) - (EG_2 - EG_1) = (T_1 - T_0)(S_2 - S_1) \quad (8.67)$$

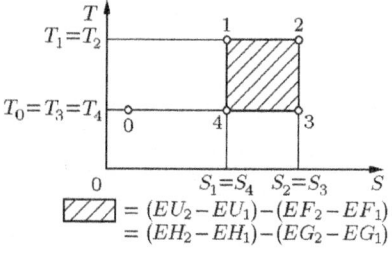

Figure 8.24 Isothermal change of state of a substance: Difference between the exergy difference of enthalpies $EH_2 - EH_1$ and the exergy difference of free enthalpies $EG_2 - EG_1$; right-hand *Carnot* cyclic process

If one interprets the changes of state $1 \to 2 \to 3 \to 4 \to 1$ in Figure 8.24 as a right-hand reversible *Carnot* cyclic process with the ambient temperature T_0 as the lower temperature limit, then the absolute value of the work done in the cycle is obtained by applying Eqs. (7.35), (7.36), (7.39) and (7.40) to

$$|W_{Kreis}| = (T_1 - T_0)(S_2 - S_1) \ . \tag{8.68}$$

With Eq. (8.67) and Eq. (8.68) one obtains

$$(EG_2 - EG_1) - (EH_2 - EH_1) = -(T_1 - T_0)(S_2 - S_1) = -|W_{Kreis}| \tag{8.69}$$

and

$$(EF_2 - EF_1) - (EU_2 - EU_1) = -(T_1 - T_0)(S_2 - S_1) = -|W_{Kreis}| \ . \tag{8.70}$$

For a reversible isothermal change of state from 1 to 2, the difference between the exergy difference of the enthalpies $EH_2 - EH_1$ and the exergy difference of the free enthalpies $EG_2 - EG_1$ differ according to Eq. (8.69) by the exergy of the reversibly and isothermally provided heat (entropy change heat):

$$(EQ_{12})_{rev} = EQ_{s12} = (T_1 - T_0)(S_2 - S_1) = |W_{Kreis}| \tag{8.71}$$

According to Eq. (8.70), this also applies to the difference between the exergy difference of the internal energies $EU_2 - EU_1$ and the exergy difference of the free energies $EF_2 - EF_1$.

From this, an essential feature of the free enthalpy and the free energy as thermodynamic potentials becomes clear. This will be explained below with reference to the free enthalpy: According to Eq. (8.54), with $T_2 = T_1$ becomes $EG_1 - EG_2 = G_1 - G_2$ If a substance with state 1 is reversibly and isothermally transformed into state 2, it has the ability, i.e. the potential

$$G_1 - G_2 = EG_1 - EG_2 = EH_1 - EH_2 + (T_1 - T_0)(S_2 - S_1) \tag{8.72}$$

not only to perform useful work due to the difference of the exergy of the enthalpies $EH_1 - EH_2$, but it also has the ability to activate reversibly the heat (entropy change heat) $(Q_{12})_{rev} = Q_{s12}$ with respect to a heat reservoir with constant temperature $T_2 = T_1 = T$.

Thus — depending on the position of states 1 and 2 — the exergy of this heat $(EQ_{12})_{rev} = EQ_{s12}$ can be used either for providing additional useful work, or it must be given off. However, the exergy of this heat is not included in the difference of the exergy of the enthalpies $EH_1 - EH_2$ but is either to be supplied to the substance from the outside or is to be given off by the substance to the outside during the change of state from 1 to 2 across the system boundary.

In the case of an ideal gas, the following applies with regard to the right-hand reversible *Carnot* cyclic process $1 \to 2 \to 3 \to 4 \to 1$ considered in Fig. 8.24 for $|W_{Kreis}|$ according to the transformed Eqs. (5.215) and (8.69) with $H_2 = H_1$ and $T_2 = T_1$:

$$G_1 - G_2 = H_1 - H_2 - T_1 S_1 + T_2 S_2 = T_1 (S_2 - S_1) \tag{8.73}$$

$$EH_1 - EH_2 = H_1 - H_2 - T_0 (S_1 - S_2) = T_0 (S_2 - S_1) \tag{8.74}$$

This becomes

$$G_1 - G_2 - (EH_1 - EH_2) = T_1 (S_2 - S_1) - T_0 (S_2 - S_1) = |W_{Kreis}| \ . \tag{8.75}$$

Eq. (8.75) makes visible that, to the difference of the free enthalpy $G_1 - G_2$, the reversibly added heat (entropy change heat) $(Q_{zu})_{rev} = (Q_{12})_{rev} = T_1(S_2 - S_1)$ is to be assigned, while to the negative difference of the exergy of the enthalpy $-(EH_1 - EH_2)$, the waste heat reversibly given off to the environment $(Q_{ab})_{rev} = -T_0(S_2 - S_1)$ is to be assigned.

Example 8.17 3.0 kg of air (ideal gas; $\kappa = 1.4$; $R = 0.2872$ kJ/(kg K)) is expanded reversibly and isothermally at $t_1 = 400\,°C$ from $p_1 = 40.0$ bar to $p_2 = 2.313$ bar. The ambient temperature is $t_0 = 25\,°C$, the ambient pressure is $p_0 = 1.0$ bar.

a) What are the absolute values of the specific entropy s_1 and s_2, if for the absolute value of the specific entropy of the air $s_0 = 6.867$ kJ/(kg K) at $t_0^* = 25\,°C$ and $p_0^* = 1$ bar is to be utilized (cf. section 7.5.3.2, page 286/287)?

b) The change of the free enthalpy $G_1 - G_2$ as well as the change of the exergy of the free enthalpy $EG_1 - EG_2$ is to be calculated.

c) What is the value of the change of enthalpy $H_1 - H_2$ and what is the value of the change of exergy of the enthalpy $EH_1 - EH_2$?

d) What heat (entropy change heat) $(Q_{12})_{rev}$ is to be supplied for the change of state from 1 to 2? What is the exergy $(EQ_{12})_{rev}$ of this heat?

e) The change of state from 1 to 2 is to be understood as part of a reversible Carnot cycle process, in which the heat $(Q_{ab})_{rev}$ is given off to the environment at $t_0 = 25\,°C$. What are the ideal cyclic process work $|W_{Kreis}|$, the heat $|(Q_{ab})_{rev}|$ given off and the thermal efficiency η_{th}?

a) $s_1 = s_0 + c_p \ln \dfrac{T_1}{T_0} - R \ln \dfrac{p_1}{p_0} = (6.867 + 0.8186 - 1.0594)$ kJ/(kg K) $= 6.626$ kJ/(kg K)

$s_2 = s_0 + c_p \ln \dfrac{T_2}{T_0} - R \ln \dfrac{p_2}{p_0} = (6.867 + 0.8186 - 0.2408)$ kJ/(kg K) $= 7.445$ kJ/(kg K)

b) $G_1 - G_2 = H_1 - H_2 - (T_1 S_1 - T_2 S_2) = T_1(S_2 - S_1) = T_1 m (s_2 - s_1)$

$ = 673.15$ K $\cdot$ 3.0 kg $(7.445 - 6.626)$ kJ/(kg K) $= 1653.93$ kJ

$T_2 = T_1$ leads to $\quad EG_1 - EG_2 = G_1 - G_2 - (T_1 - T_2) S_0 = G_1 - G_2 = 1653.93$ kJ .

c) $H_1 - H_2 = m c_p (T_1 - T_2) = 0$ kJ

$EH_1 - EH_2 = H_1 - H_2 - T_0(S_1 - S_2) = T_0(S_2 - S_1) = T_0 m (s_2 - s_1)$

$ = 298.15$ K $\cdot$ 3.0 kg $(7.445 - 6.626)$ kJ/(kg K) $= 732.55$ kJ

d) $(Q_{12})_{rev} = -W_{p12} = m R T_1 \ln \dfrac{p_1}{p_2}$

$\phantom{(Q_{12})_{rev}} = 3$ kg $\cdot$ 0.2872 kJ/(kg K) $\cdot$ 673.15 K $\cdot \ln \dfrac{40\ \text{bar}}{2.313\ \text{bar}} = 1653.15$ kJ $\approx G_1 - G_2$

$(EQ_{12})_{rev} = \dfrac{T_1 - T_0}{T_1} (Q_{12})_{rev} = \dfrac{673.15\ \text{K} - 298.15\ \text{K}}{673.15\ \text{K}} \cdot 1653.93$ kJ $= 921.38$ kJ

e) $|W_{Kreis}| = (T_1 - T_0)(S_2 - S_1) = (T_1 - T_0) m (s_2 - s_1)$

$\phantom{|W_{Kreis}|} = (673.15 - 298.15)$ K $\cdot$ 3.0 kg $(7.445 - 6.626)$ kJ/(kg K) $= 921.38$ kJ

$|W_{Kreis}| = EG_1 - EG_2 - (EH_1 - EH_2) = G_1 - G_2 - (EH_1 - EH_2)$

$\phantom{|W_{Kreis}|} = (1653.93 - 732.55)$ kJ $= 921.38$ kJ

$|(Q_{ab})_{rev}| = (Q_{12})_{rev} - |W_{Kreis}| = (Q_{12})_{rev} - (EQ_{12})_{rev} = (1653.93 - 921.38)\,\text{kJ} = 732.55\,\text{kJ}$

$(Q_{ab})_{rev} = -|(Q_{ab})_{rev}| = -732.55\,\text{kJ}$

$\eta_{th} = 1 - \dfrac{|(Q_{ab})_{rev}|}{(Q_{zu})_{rev}} = 1 - \dfrac{732.55\,\text{kJ}}{1653.93\,\text{kJ}} = 0.557 \qquad \eta_{th} = \dfrac{T_1 - T_0}{T_1} = 0.557$

This result agrees with the value of η_{th} given in Table 7.15 for the *Carnot* cycle process at state 3c.

8.2 Exergy and Anergy

The determination of the part of energy that can be converted as exergy into any other form of energy and thus be used, also suggests the determination of the energy share that cannot be made usable. The part of the energy that cannot be converted as exergy is called anergy. Every energy can thus be disassembled into exergy and anergy:

$$\text{energy} = \text{exergy} + \text{anergy}$$

The application of this equation to the reversible substitute heat (entropy change heat), the bound energy, the temperature change heat, the volume change work, the shift work, the pressure change work, the internal energy, the enthalpy, the free energy and the free enthalpy results in the way that the anergy part of an energy is indicated by the letter A in front of it (or a for specific energies):

heat (entropy change heat)	$(Q_{12})_{rev} = (EQ_{12})_{rev} + (AQ_{12})_{rev}$	(8.76)
bound energy	$TS = ETS + ATS$	(8.77)
	$T_2 S_2 - T_1 S_1 = ET_2 S_2 - ET_1 S_1 + AT_2 S_2 - AT_1 S_1$	(8.78)
temperature change heat	$Q_{T12} = EQ_{T12} + AQ_{T12}$	(8.79)
volume change work	$W_{V12} = EW_{V12} + AW_{V12}$	(8.80)
shift work	$W = EW + AW$	(8.81)
	$W_2 - W_1 = EW_2 - EW_1 + AW_2 - AW_1$	(8.82)
pressure change work	$W_{p12} = EW_{p12}$	(8.83)
internal energy	$U = EU + AU$	(8.84)
	$U_2 - U_1 = EU_2 - EU_1 + AU_2 - AU_1$	(8.85)
enthalpy	$H = EH + AH$	(8.86)
	$H_2 - H_1 = EH_2 - EH_1 + AH_2 - AH_1$	(8.87)
free energy	$F = EF + AF$	(8.88)
	$F_2 - F_1 = EF_2 - EF_1 + AF_2 - AF_1$	(8.89)
free enthalpy	$G = EG + AG$	(8.90)
	$G_2 - G_1 = EG_2 - EG_1 + AG_2 - AG_1$	(8.91)

8.2 Exergy and Anergy

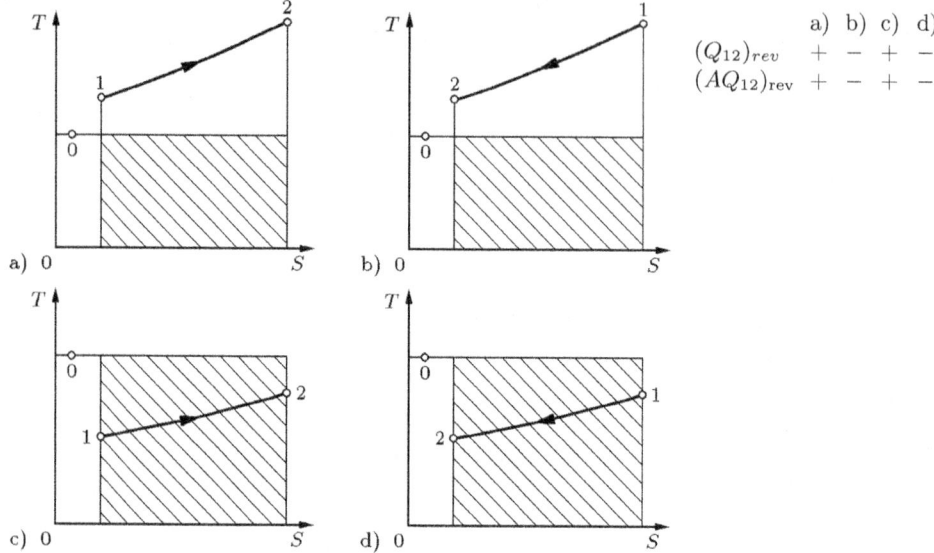

Figure 8.25 The anergy of the heat $(AQ_{12})_{rev} = AQ_{s12}$ in a T, S diagram

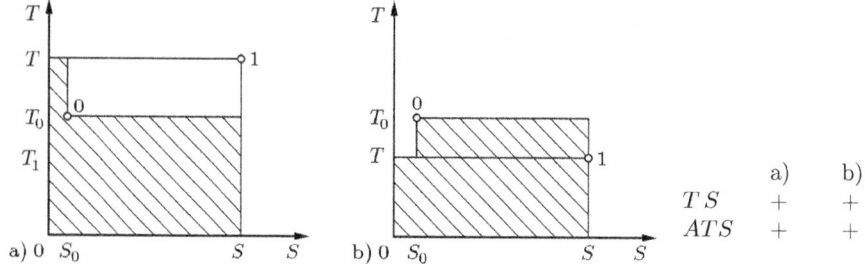

Figure 8.26 The anergy of the bound energy ATS in a T, S diagram for $S > S_0$.

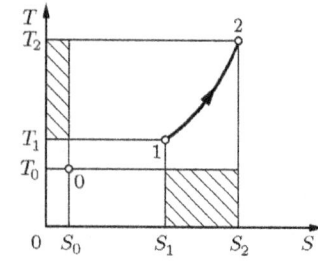

Figure 8.27 The anergy of the difference of the bound energies $AT_2S_2 - AT_1S_1$ in a T, S diagram

From Eqs. (8.3), (8.9), (8.10), (8.16), (8.19), (8.21), (8.22), (8.26), (8.30), (8.32), (8.35), (8.37), (8.45), (8.47), (8.52) and (8.54) for the exergies, with Eqs. (8.76) to (8.91) the following equations are obtained for the anergies:

heat (entropy change heat) $(AQ_{12})_{rev} = T_0 (S_2 - S_1)$ (8.92)
bound energy $ATS = T S_0 + T_0 (S - S_0)$ (8.93)
                                   $AT_2 S_2 - AT_1 S_1 = T_0 (S_2 - S_1) + (T_2 - T_1) S_0$ (8.94)
temperature change heat $AQ_{T12} = (T_2 - T_1) S_0$ (8.95)
volume change work $AW_{V12} = -p_0 (V_2 - V_1)$ (8.96)
shift work $AW = p_0 V$ (8.97)
                                   $AW_2 - AW_1 = p_0 (V_2 - V_1)$ (8.98)

pressure change work	$AW_{p12} = 0$	(8.99)
internal energy	$AU = U_0 + T_0(S - S_0) - p_0(V - V_0)$	(8.100)
	$AU_2 - AU_1 = T_0(S_2 - S_1) - p_0(V_2 - V_1)$	(8.101)
enthalpy	$AH = H_0 + T_0(S - S_0)$	(8.102)
	$AH_2 - AH_1 = T_0(S_2 - S_1)$	(8.103)
free energy	$AF = F_0 - p_0(V - V_0) - (T - T_0)S_0$	(8.104)
	$AF_2 - AF_1 = -p_0(V_2 - V_1) - (T_2 - T_1)S_0$	(8.105)
free enthalpy	$AG = G_0 - (T - T_0)S_0$	(8.106)
	$AG_2 - AG_1 = -(T_2 - T_1)S_0$	(8.107)

8.2.1 Anergy in a p, V Diagram and in a T, S Diagram

According to Figs. 8.25 and 8.28 to 8.29, for heat (entropy change heat), volume change work and shift work, the sign of the anergy coincides with the sign of the energy. This also follows from Eqs. (8.92), (8.96) and (8.97). According to Eq. (3.24), the heat is positive when the entropy increases. Eq. (8.92) shows that the anergy of the heat is positive at $S_2 > S_1$.

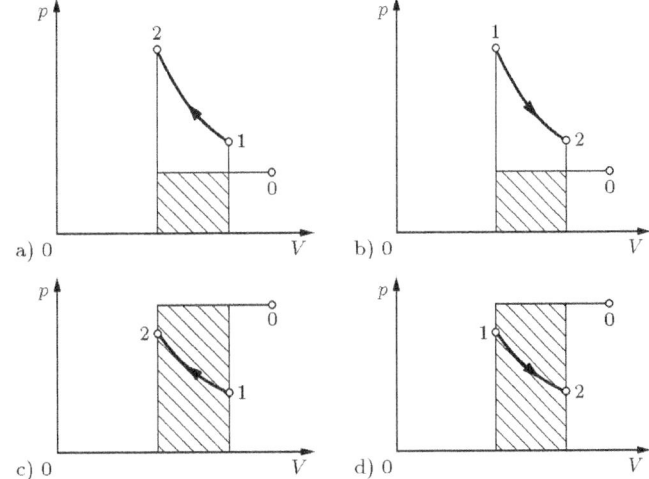

Figure 8.28 The anergy of the volume change work AW_{V12} in a p, V diagram

	a)	b)	c)	d)
W_{V12}	+	−	+	−
AW_{V12}	+	−	+	−

The anergy of the bound energy can be positive or negative according to Eq. (8.93), depending on the values of T, T_0, S and S_0 (Fig. 8.26). According to Eq. (8.94), the anergy of the difference of the bound energies is positive for $S_2 > S_1$ and $T_2 > T_1$ and negative for $S_2 < S_1$ as well as $T_2 < T_1$ (cf. Fig. 8.27). At the occurrence of $S_2 > S_1$ and $T_2 < T_1$ as well as at the occurrence of $S_2 < S_1$ and $T_2 > T_1$ one partial area is positive and one partial area is negative. Depending on which partial area predominates, the anergy of the difference of the bound energy can be positive, negative or equal to zero.

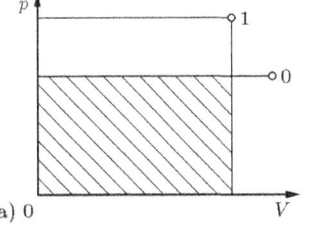

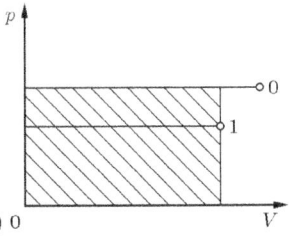

Figure 8.29 The anergy of the shift work AW in a p, V diagram

	a)	b)
W	+	+
AW	+	+

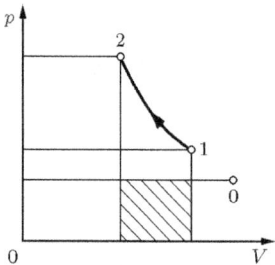

Figure 8.30 The anergy of the difference of the shift work $AW_2 - AW_1$ according to example 8.5 in a p, V diagram

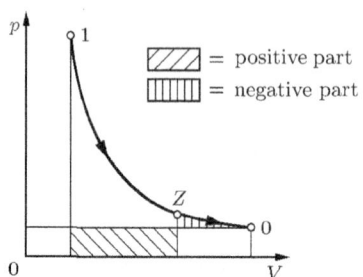

Figure 8.31 The anergy difference of the internal energy $AU - U_0$ according to example 8.6 in a p, V diagram

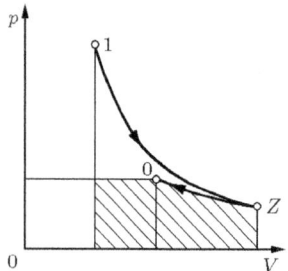

Figure 8.32 The anergy difference of the internal energy $AU - U_0$ according to example 8.7 in a p, V diagram

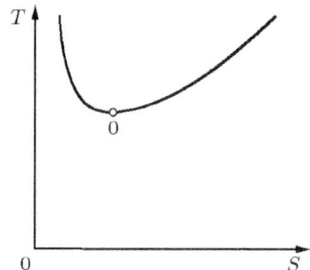

Figure 8.33 Curve of constant anergy of the internal energy $AU - U_0 = 0$ according to example 8.6

In the case of compression, the volume change work is positive according to Eq. (2.21). Eq. (8.96) gives a positive value for the anergy of the volume change work at $V_2 < V_1$. In the case of an expansion, the volume change work and the corresponding anergy are negative. The anergy of the shift work according to Eq. (8.97) is always positive like the shift work. The anergy of the difference of the shift work according to Eq. (8.98) is positive in case of an expansion and negative in case of a compression (Fig. 8.30).

Figs. 8.31 and 8.32 show the anergy difference of the internal energy $AU - U_0$ as an area in the p, V diagram. In Fig. 8.31, one partial area is positive and one partial area is negative. Depending on which partial area is predominant, the anergy difference of the internal energy can be positive, negative or equal to zero. Fig. 8.33 contains the curve $AU - U_0 = 0$ in a T, S diagram for the data of example 8.6. It results from

$$p_0 (V - V_0) = T_0 (S - S_0) \tag{8.108}$$

or with the thermal equation of state of ideal gases (4.25)

$$\frac{V}{V_0} = 1 + \frac{S - S_0}{m R} . \tag{8.109}$$

Inserting Eq. (8.109) into Eq. (4.57), one obtains for $AU - U_0 = 0$ the equation

$$T = T_0 \frac{e^{\frac{S - S_0}{m R}(\kappa - 1)}}{\left(1 + \frac{S - S_0}{m R}\right)^{\kappa - 1}} . \tag{8.110}$$

Above the boundary curve according to Eq. (8.110) $AU - U_0$ is positive, below the boundary curve negative.

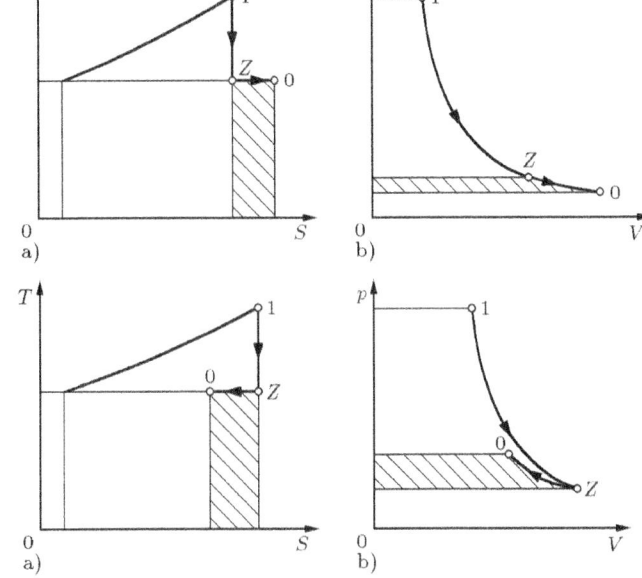

Figure 8.34 The anergy difference of the enthalpy $AH - H_0$ in example 8.9
a) in a T, S diagram
b) in a p, V diagram

Figure 8.35 The anergy difference of the enthalpy $AH - H_0$ in example 8.10
a) in a T, S diagram
b) in a p, V diagram

According to Eq. (8.102), the anergy difference of the enthalpy $AH - H_0$ is negative if the point representing the state in the T, S diagram lies to the left of the ambient state (Fig. 8.34). For a state with greater entropy than that of the ambient state, $AH - H_0$ is positive (Fig. 8.35). For an isentropic change of state, according to Eq. (8.103), there is no change in the anergy of the enthalpy.

Since the change in the anergy of the enthalpy is equal to the anergy of the reversible substitute heat (entropy change heat), one obtains in the case of a change of state with heat input, a positive value for the anergy difference of the enthalpy. The value becomes negative when heat is released. The isentropes are curves of constant anergy of enthalpy. Example 8.13 deals with a change of state where the change in anergy of enthalpy is positive (Fig. 8.36).

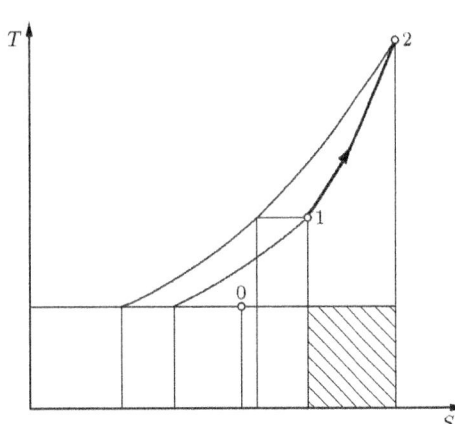

Figure 8.36 The change of the anergy of the enthalpy $AH_2 - AH_1$ according to example 8.13 in a T, S diagram

The h, s diagram for ammonia according to Fig. 8.19 contains lines of constant exergy of enthalpy. The lines of constant anergy of enthalpy must be thought of as vertical lines, where the straight line for $ah = 0$ is at $s = - 0.7129$ kJ/(kg K). To the right, the amount of the specific anergy of the enthalpy ah increases. In the reference point for the calculation of the exergy, it becomes $ah = h_0$.

8.2.2 Anergy-Free Energies

All types of energy that are completely convertible into any other type of energy consist only of exergy. These are especially the following types of energy:

- potential energy
- kinetic energy
- pressure change work
- coupling work
- internal work
- total work
- cyclic process work

In a special position,

- friction work

must be seen. Although it consists only of exergy, its ability to transform is limited to one type of energy: Friction work always leads directly to an increase in internal energy. Internal friction work causes an increase of the internal energy of the system or of a substance, external friction work increases the internal energy of the environment.

Because of this limitation, it is better not to say: Friction work consists only of exergy, because then one would assign to friction work an unrestricted transformability into any kind of energy. On the contrary, friction work 'consumes' exergy as a process variable. In an exergy balance, however, the friction work must be treated as an anergy-free energy.

8.3 Exergy Loss

In contrast to the law of conservation of energy, there is no generally valid law of conservation of exergy. Conservation of exergy only exists for reversible processes. Irreversible processes are associated with a loss of exergy. Since, according to the second law of thermodynamics, all natural and technical processes are irreversible, the second law can also be called the principle of exergy destruction: In an isolated system, energy conversion processes are always accompanied by a reduction of exergy.

8.3.1 Irreversibility and Exergy Loss

The causes of irreversibility are friction and the equalisation processes of temperature equalisation, pressure equalisation, concentration equalisation nad entropy equalisation respectively, heat transfer, throttling and substance mixing. The internal friction work W_{RI} of a real process appears in the reversible substitute process as reversible substitute heat (entropy change heat) $(Q_{12})_{rev}$, if the real process is adiabatic, or according to Eq. (3.78) as a part of this heat, if the real process is non-adiabatic.

In any case, that part of the heat which corresponds to the internal friction work in the real process, in the reversible substitution process consists of exergy and anergy. In the real process, however, the internal friction work cannot contain an anergy part, because friction work is a mechanical work and, as such, requires a full mechanical work and as such requires a full exergy supply.

The exergy of mechanical work to deliver the internal friction work W_{RI} in the real process is by the anergy portion of the reversible substitute heat higher. This additional effort of exergy is the exergy loss E_{VI} of the internal friction work W_{RI}:

$$E_{VI} = (AQ_{12})_{rev} \tag{8.111}$$

Correspondingly, the external friction work W_{RA} also consists only of exergy and causes the exergy loss E_{VA}.

The reversible substitute process in a closed system is described by Eq. (3.73):

$$(Q_{12})_{rev} + W_{V12} = U_2 - U_1 \tag{8.112}$$

The exergy balance belonging to this equation is

$$(EQ_{12})_{rev} + EW_{V12} = EU_2 - EU_1 \,. \tag{8.113}$$

For the real process, Eq. (3.72) applies

$$Q_{12} + |W_{RI}| + W_{V12} = U_2 - U_1 \,. \tag{8.114}$$

Since the internal friction work is fully used in the exergy balance, as a compensation for it, the exergy loss E_{VI} must appear on the right side:

$$EQ_{12} + |W_{RI}| + EW_{V12} = EU_2 - EU_1 + E_{VI} \tag{8.115}$$

Example 8.18 The change of state described in example 8.8 is to be understood as an adiabatic expansion with friction. What is the exergy loss caused by the internal friction work?

The energy balance of the reversible substitute process according to Eq. (8.112) gives

$$(Q_{12})_{rev} + W_{V12} \quad = U_2 - U_1$$
$$90.67 \text{ kJ} - 326.01 \text{ kJ} = -235.34 \text{ kJ} \,.$$

The following is the split of the respective energies into exergy and anergy according to Eqs. (8.76), (8.80) and (8.85):

$$(Q_{12})_{rev} = 29.13 \text{ kJ} + 61.54 \text{ kJ}$$
$$W_{V12} = -263.43 \text{ kJ} - 62.58 \text{ kJ}$$
$$U_2 - U_1 = -234.30 \text{ kJ} - 1.04 \text{ kJ}$$

Thus, the exergy balance according to Eq. (8.113) is:

$$(EQ_{12})_{rev} + EW_{V12} \quad = EU_2 - EU_1$$
$$29.13 \text{ kJ} \quad - 263.43 \text{ kJ} = -234.30 \text{ kJ}$$

Since the change of state is supposed to be adiabatic, in Eq. (8.115) is $EQ_{12} = 0$. Since according to Eq. (3.78) is

$$|W_{RI}| = (Q_{12})_{rev} \,,$$

the exergy loss is

$$E_{VI} = (AQ_{12})_{rev} = 61.54 \text{ kJ} \,.$$

The exergy balance according to Eq. (8.115) is

$$|W_{RI}| \quad + EW_{V12} \quad = EU_2 - EU_1 + E_{VI}$$
$$90.67 \text{ kJ} - 263.43 \text{ kJ} \quad = -234.30 \text{ kJ} + 61.54 \text{ kJ}$$

The first law for the reversible substitution process in an open system can be written according to Eq. (3.75) as follows:

$$(Q_{12})_{rev} + W_{p12} = H_2 - H_1 \qquad (8.116)$$

In the exergy balance

$$(EQ_{12})_{rev} + W_{p12} = EH_2 - EH_1 \qquad (8.117)$$

appears the pressure change work, because it does not contain any anergy. For the real process Eq. (3.74) applies

$$Q_{12} + |W_{RI}| + W_{p12} = H_2 - H_1 . \qquad (8.118)$$

$|W_{RI}|$ also occurs in the exergy balance. On the right side one finds the exergy loss E_{VI}:

$$EQ_{12} + |W_{RI}| + W_{p12} = EH_2 - EH_1 + E_{VI} \qquad (8.119)$$

Example 8.19 What is the exergy loss due to internal friction work if the compression in example 8.13 is to be considered adiabatic and irreversible?

According to Eq. (4.144), the polytropic exponent is $n = 4.5314$. The heat and the pressure change work are calculated according to Table 4.3 Eq. (4) and Table 4.2 Eq. (4).

For the reversible substitute process according to Eqs. (8.116) and (8.117) one obtailns:

Energy: $(Q_{12})_{rev} + W_{p12} = H_2 - H_1$ $\quad$ 610.78 kJ + 353.54 kJ = 964.32 kJ

Exergy: $(EQ_{12})_{rev} + W_{p12} = EH_2 - EH_1$ $\quad$ 381.27 kJ + 353.54 kJ = 734.81 kJ

The real process according to Eq. (8.119) is

Exergy: $|W_{RI}| + W_{p12} = EH_2 - EH_1 + E_{VI}$ $\quad$ 610.78 kJ + 353.54 kJ = 734.81 kJ + 229.51 kJ.

The exergy loss is

$$E_{VI} = (AQ_{12})_{rev} = 229,51 \text{ kJ} .$$

There are two systems to consider in heat transfer. The fluid with the higher temperature releases heat Q_{12} between the states 1 and 2. Since the entropy decreases during the heat release, Q_{12} is negative.

To the fluid with the lower temperature the heat $Q_{1'2'}$ is supplied which causes the change of state from 1' to 2'. To simplify the notation, it will be denoted as Q'_{12} in the following. It is positive. Both systems together form an adiabatic total system (section 3.2.2), so that

$$Q'_{12} = -Q_{12} \qquad (8.120)$$

becomes. For the corresponding exergies one can write

$$EQ'_{12} < -EQ_{12} . \qquad (8.121)$$

Subtracting the smaller exergy of the colder fluid EQ'_{12} from the larger exergy of the warmer fluid $-EQ_{12}$, one obtains the exergy loss of the heat transfer E_{VW}

$$E_{VW} = -EQ_{12} - EQ'_{12} . \qquad (8.122)$$

If one considers heat flows and thus the exergy flow loss, the following applies:

$$\dot{E}_{VW} = -E\dot{Q}_{12} - E\dot{Q}'_{12} \qquad (8.123)$$

The following example shows that the lower the temperatures are, the greater becomes the exergy loss during the heat transfer.

Example 8.20 In a counterflow heat exchanger, 0.5 kg/s of helium ($R = 2.0773$ kJ/(kg K); $\kappa = 1.6667$) are cooled by 1.8 kg/s air ($R = 0.287$ kJ/(kg K); $\kappa = 1.4$). The temperature decrease of the helium is 56 K, the difference between the inlet temperature of helium and the inlet temperature of air is 150 K. This results in a heat flow of 145.411 kW for isobaric changes of state and constant mass flows. What is the exergy flow loss of the heat transfer $\dot{E}_{VW}$ (ambient state 1 bar, 20 °C) in the four cases with the inlet temperatures of the helium flow assumed to 200 °C, 125 °C, 83 °C and -10 °C?

The exergy flows $E\dot{Q}_{12}$ and $E\dot{Q}'_{12}$ are calculated using Eq. (8.3). The results are shown in the following table and in Fig. 8.37.

t_1	t'_1	$E\dot{Q}_{12}$	$E\dot{Q}'_{12}$	$\dot{E}_{VW}$
°C	°C	kW	kW	kW
200	50	-49.53	27.61	21.92
125	-25	-30.03	-3.38	33.41
83	-67	-15.19	-29.18	44.37
-10	-160	36.73	-139.19	102.46

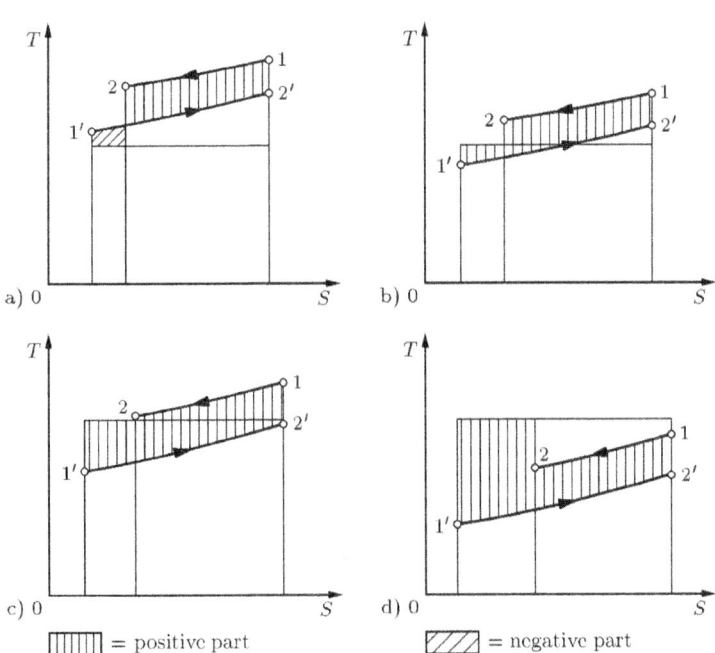

▦ = positive part   ▨ = negative part

Figure 8.37 The exergy loss of the heat transfer according to example 8.20

Fig. 8.37 shows energies obtained from the heat flows by multiplying them by the time duration τ. Here the areas for $\dot{Q}_{12}\tau$ and $\dot{Q}'_{12}\tau$ are drawn in such a way that the points 1 and 2' lie vertically above of each other. For the representation of the ambient state in the T, S diagram, this results in a different point for each fluid. Therefore, the marking of the ambient state is omitted.

According to section 3.2.4, internal friction work is converted by throttling. The exergy loss of the throttling E_{VD} can therefore be determined like the exergy loss of the

internal friction work E_{VI}. Analogous to Eq. (8.111), the exergy loss of the throttling is

$$E_{VD} = (AQ_{12})_{rev} . \qquad (8.124)$$

Example 8.21 Out of a vessel at a pressure of 10 bar and at a temperature of 450 K, 2.5 kg of an ideal gas ($R = 0.287$ kJ/(kg K); $\kappa = 1.4$) are taken out and are throttled to 2 bar. How large is the exergy loss? What would be the exergy loss of the throttling at a temperature of 150 K, if one may calculate with the same substance values? The ambient temperature is 20 °C.

According to Eq. (8.124) and Eq. (8.92), in both cases the exergy loss of the throttling is

$$E_{VD} = 338.52 \text{ kJ} .$$

In Fig. 8.38, the exergy loss of the throttling E_{VD} is shown as a hatched area.

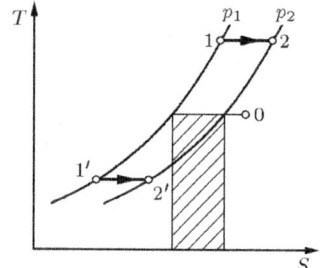

Figure 8.38 The exergy loss by throttling of an ideal gas according to example 8.21. The curves $1 - 1'$ and $2 - 2'$ are the isobars for $p_1 = 10$ bar and $p_2 = 2$ bar, respectively.

8.3.2 Exergy Loss and Anergy Gain

Because according to section 8.2 energy is a sum of exergy and anergy, there must be a decrease in exergy in irreversible processes associated with an increase in anergy. Such an increase of anergy is often called — not very smart — as an anergy gain.

A disappearance of exergy while the anergy remains the same would be a disappearance of energy, which, according to the first law of thermodynamics, is just as impossible as a new creation of energy. This would be a perpetuum mobile of the first kind, because it would violate the first law of thermodynamics.

The production of exergy from anergy would be a perpetual motion machine of the second kind, because it would violate the second law of thermodynamics.

With E_V for the exergy loss and A_G for the anergy gain, the following applies

$$A_G = E_V . \qquad (8.125)$$

Both quantities shall always be positive.

Analogous to the loss of exergy, the gain in anergy due to internal friction work is denoted with A_{GI}, the gain in anergy due to external friction work is denoted with A_{GA}, the gain in anergy due to heat transfer is denoted with A_{GW} and the gain in anergy due to throttling is denoted with A_{GD}.

To the energy and exergy balances of a closed system according to the Eqs. (8.112) to (8.115), the anergy balances for the reversible substitute process are given by

$$(AQ_{12})_{rev} + AW_{V12} = AU_2 - AU_1 \qquad (8.126)$$

and for the real process

368 8 Exergy

$$AQ_{12} + AW_{V12} = AU_2 - AU_1 - A_{GI} . \qquad (8.127)$$

Since the part of the anergy $(AQ_{12})_{rev}$, which is formally due to the internal friction work, is missing on the left side of Eq. (8.127), the same amount must be subtracted as anergy gain A_{GI} on the right side of Eq. (8.127).

Example 8.22 The anergy balance to example 8.18 (adiabatic irreversible expansion) is to be added.

For the reversible substitute process, according to Eqs. (8.12), (8.113) and (8.126), one obtains

Energy: $(Q_{12})_{rev} + W_{V12} = U_2 - U_1$ 90.67 kJ $-$ 326.01 kJ $= -$ 235.34 kJ

Exergy: $(EQ_{12})_{rev} + EW_{V12} = EU_2 - EU_1$ 29.13 kJ $-$ 263.43 kJ $= -$ 234.30 kJ

Anergy: $(AQ_{12})_{rev} + AW_{V12} = AU_2 - AU_1$ 61.54 kJ $-$ 62.58 kJ $\ = -$ 1.04 kJ.

The real process, according to the Eqs. (8.114), (8.115) and (8.127), is given by

Energy: $|W_{RI}| + W_{V12} \ = U_2 - U_1$ 90.67 kJ $-$ 326.01 kJ $\ = -$ 235.34 kJ

Exergy: $|W_{RI}| + EW_{V12} = EU_2 - EU_1 + E_{VI}$ 90.67 kJ $-$ 263.43 kJ $= -$ 234.30 kJ $+$ 61.54 kJ

Anergy: $AQ_{12} + AW_{V12} = AU_2 - AU_1 - A_{GI} 0$ $-$ 62.58 kJ $\ = -$ 1.04 kJ $-$ 61.54 kJ.

The anergy balances leading to the energy and exergy balances of an open system according to Eqs. (8.116) to (8.119) are for the reversible substitute process

$$(AQ_{12})_{rev} = AH_2 - AH_1 \qquad (8.128)$$

and for the real process

$$AQ_{12} = AH_2 - AH_1 - A_{GI} . \qquad (8.129)$$

In the expression for the anergy of the reversible substitute heat $(AQ_{12})_{rev}$, according to Eq. (3.78), the parts AQ_{12} and $|AW_{RI}|$ are included.

According to section 8.2.2, in the anergy balance of the real process is $|AW_{RI}| = 0$. To compensate, one has to introduce the anergy gain $A_{GI} = |AW_{RI}|$ of the reversible substitute process.

If one brings the anergy gain to the right side of Eq. (8.129), one must add a minus sign to it. This is the same consideration that applies to Eq. (8.127).

Example 8.23 Adding the energy, exergy and anergy balances to example 8.19, this gives the followinmg results for the reversible substitute process according to Eqs. (8.116), (8.117) and (8.128):

$(Q_{12})_{rev} + W_{p12} \ = H_2 - H_1$

Energy: 610.78 kJ $+$ 353.54 kJ $=$ 964.32 kJ

Exergy: 381.27 kJ $+$ 353.54 kJ $=$ 734.81 kJ

Anergy: 229.51 kJ $+$ 0 kJ $=$ 229.51 kJ

The real process according to Eqs. (8.118), (8.119) and (8.129) is given by

	$\lvert W_{RI}\rvert$	$+\ W_{p12}$	$=H_2-H_1$
Energy:	610.78 kJ	+ 353.54 kJ	= 964.32 kJ
Exergy:	610.78 kJ	+ 353.54 kJ	= 734.81 kJ + 229.51 kJ
Anergy:	0 kJ	+ 0 kJ	= 229.51 kJ − 229.51 kJ

For the heat transfer follows from Eqs. (8.76) and (8.120)

$$EQ'_{12} + AQ'_{12} = -EQ_{12} - AQ_{12} \tag{8.130}$$

or

$$-EQ_{12} - EQ'_{12} = -AQ_{12} + AQ'_{12} \tag{8.131}$$

Eqs. (8.122) and (8.125) with Eq. (8.131) give the anergy gain of the heat transfer

$$A_{GW} = AQ_{12} + AQ'_{12}\ . \tag{8.132}$$

Accordingly, the anergy gain flow is

$$\dot{A}_{GW} = A\dot{Q}_{12} + A\dot{Q}'_{12}\ . \tag{8.133}$$

A representation of the anergy gain of the heat transfer in a T,S diagram according to example 8.20 is shown in Fig. 8.39.

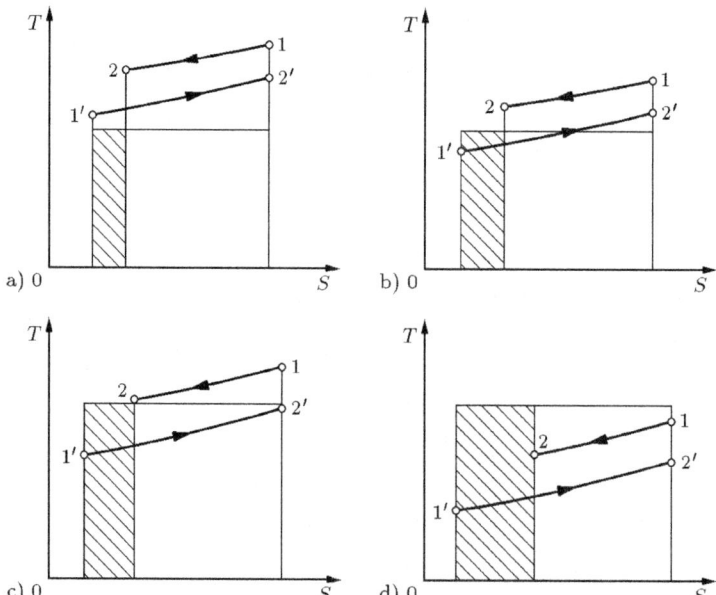

Figure 8.39 The anergy gain of the heat transfer according to example 8.20

The anergy gain by throttling A_{GD} is given by Eqs. (8.124) and (8.125):

$$A_{GD} = (AQ_{12})_{rev} \tag{8.134}$$

The hatched area in Fig. 8.38 represents both the exergy loss E_{VD} and the anergy gain A_{GD}.

8.3.3 Exergetic Efficiencies

An efficiency can be, for example, the quotient of benefit and effort. If one expresses the benefit and the effort by exergies, one obtains an exergetic efficiency. While the thermal efficiency η_{th} can never reach the maximum value of 1, even theoretically, an exergetic efficiency of $\zeta = 1$ is conceivable. The upper limit for the thermal efficiency $\eta_{th} < 1$ results from the fact that in a right-handed cycle process, heat must not only be supplied but also given off. An exergetic efficiency $\zeta = 1$ can be achieved for a right-hand reversible cycle process. If one sets in analogy to equation (7.36)

$$\zeta = \frac{(EQ_{zu})_{rev} - |(EQ_{ab})_{rev}|}{(EQ_{zu})_{rev}}, \tag{8.135}$$

so, in the case of reversible heat transfer to the surroundings with a temperature gradient reaching the limit value zero, the exergy of the given off heat is $(EQ_{ab})_{rev} = 0$ and thus the exergetic efficiency becomes $\zeta = 1$.

Exergetic efficiencies are obtained from exergy balances according to the first law of thermodynamics, which are expressed in the following form:

$$\text{exergy benefit} = \text{exergy effort} - \text{exergy loss}$$

Each expression of the first law must be classified as either a benefit, an effort or a loss. The exergetic efficiency is then obtained as

$$\zeta = \frac{\text{exergy benefit}}{\text{exergy effort}} = 1 - \frac{\text{exergy loss}}{\text{exergy effort}}.$$

In the following, an application of this consideration to an open system is given for the case of a compressor, where the changes of the potential energy and the kinetic energy are neglected. Since there are different ways of formulating the energy use and energy effort, there are also different exergetic efficiencies possible [4].

Assuming Table 6.1 Eq. (10) for the first law, the exergy balance can be written as follows:

$$EH_2 - EH_1 - EQ_{12} = W_i - E_{VI}. \tag{8.136}$$

The exergy increase of the enthalpy and the exergy of the heat given off during cooling are shown as benefits. EQ_{12} is negative. The internal work W_i represents the effort. To Eq. (8.136) belongs the exergetic efficiency

$$\zeta = \frac{EH_2 - EH_1 - EQ_{12}}{W_i} = 1 - \frac{E_{VI}}{W_i}. \tag{8.137}$$

If the exergy of the heat is not to be included in the exergy benefit, it appears on the right side as a reduction of the effort:

$$EH_2 - EH_1 = W_i + EQ_{12} - E_{VI} \tag{8.138}$$

The exergetic efficiency in this case is

$$\zeta = \frac{EH_2 - EH_1}{W_i + EQ_{12}} = 1 - \frac{E_{VI}}{W_i + EQ_{12}}. \tag{8.139}$$

If the heat is cooled by cooling air or cooling water of the ambient state, then its exergy is lost according to Eq. (8.122) with $EQ'_{12} = 0$:

$$E_{VW} = -EQ_{12} \tag{8.140}$$

$$EH_2 - EH_1 = W_i - E_{VI} - E_{VW} \tag{8.141}$$

The exergetic efficiency is then

$$\zeta = \frac{EH_2 - EH_1}{W_i} = 1 - \frac{E_{VI} + E_{VW}}{W_i}. \tag{8.142}$$

Inserting Eq. (6.38) into Eq. (10) of Table 6.1, one can obtain with

$$|W_{RA}| = E_{VA} \tag{8.143}$$

the following exergy balance:

$$EH_2 - EH_1 - EQ_{12} = W_e - E_{VI} - E_{VA} \tag{8.144}$$

Here the exergy of the heat is again part of the exergy benefit. The following exergetic efficiency takes into account the exergy loss of the internal and external friction work:

$$\zeta = \frac{EH_2 - EH_1 - EQ_{12}}{W_e} = 1 - \frac{E_{VI} + E_{VA}}{W_e} \tag{8.145}$$

If one brings the exergy of heat to the right side of Eq. (8.144), it appears there as a reduction in the effort, and the exergy balance becomes

$$EH_2 - EH_1 = W_e + EQ_{12} - E_{VI} - E_{VA}. \tag{8.146}$$

Thus the exergetic efficiency becomes

$$\zeta = \frac{EH_2 - EH_1}{W_e + EQ_{12}} = 1 - \frac{E_{VI} + E_{VA}}{W_e + EQ_{12}}. \tag{8.147}$$

If the exergy of the heat becomes a component of the exergy loss according to Eq. (8.140), thus the exergy balance is

$$EH_2 - EH_1 = W_e - E_{VI} - E_{VA} - E_{VW}. \tag{8.148}$$

Then the exergetic efficiency takes the following form:

$$\zeta = \frac{EH_2 - EH_1}{W_e} = 1 - \frac{E_{VI} + E_{VA} + E_{VW}}{W_e} \tag{8.149}$$

Example 8.24 A cooled compressor takes in a volume flow of $0.15 \text{ m}^3/\text{s}$ of an ideal gas ($c_p = 1.05$ kJ/(kg K); $\kappa = 1.4$) at the ambient state 1 bar and 18 °C. In the final state, the gas has a pressure of 3.5 bar and a density of 3.34 kg/m^3. A heat flow of 14 kW is given off to the environment by cooling. The average temperature of the heat given off can be considered to be the final temperature of compression. The mechanical efficiency is 0.96.

The exergetic efficiencies according to Eqs. (8.137), (8.139), (8.142), (8.145), (8.147) and (8.149) are to be calculated.

According to Eq. (4.49), the gas has the specific gas constant $R = 0.3$ kJ/(kg K) and according to Eq. (4.25) for the state 1, the mass flow is $\dot{m} = 0.17173$ kg/s.

According to Eq. (4.27), after the compression the final temperature of state 2 is $T_2 = 349.30$ K. The polytropic exponent of the compression according to Eq. (4.144) is $n = 1.17008$.

With Eq. (4) from Table 4.3 one obtains $(\dot{Q}_{12})_{rev} = 10.125$ kW. The exergy of the heat flow given off by cooling is, following Eq. (8.6)

$$E\dot{Q}_{12} = \left(1 - \frac{T_0}{T_m}\right)\dot{Q}_{12} = \left(1 - \frac{291.15 \text{ K}}{349.30 \text{ K}}\right) \cdot (-14 \text{ kW}) = -2.331 \text{ kW}.$$

The reversible substitute heat can be splitted according to Eq. (3.78) as follows:

	$(\dot{Q}_{12})_{rev}$	$= \dot{Q}_{12}$	$+ |W_{RI}|$	
Energy:	-10.125 kW	$= -14$ kW	$+3.875$ kW	
Exergy:	-0.894 kW	$= -2.331$ kW	$+3.875$ kW -2.438 kW	
Anergy:	-9.231 kW	$= -11.669$ kW $+0$ kW	$+2.438$ kW	

For the calculation of $(EQ_{12})_{rev} = -0.894$ kW Eq. (8.6) can be used; hereby, T_m can be determined with Eq. (8.7) and herein the difference of the entropy flow with Eq. (4.58).

Neglecting the potential energy and the kinetic energy, Eq. (4.43), Eq. (1) of Table 6.1, Eq. (2.77) and Eq. (1) of Table 6.3 lead to the following balances:

	$\dot{Q}_{12}$	$+ \dot{W}_i$	$= \dot{H}_2 - \dot{H}_1$	
Energy:	-14 kW	$+24.486$ kW $= 10.486$ kW		
Exergy:	-2.331 kW	$+24.486$ kW $= 19.717$ kW	$+2.438$ kW	
Anergy:	-11.669 kW $+0$ kW		$= -9.231$ kW -2.438 kW	

	$\dot{Q}_{12}$	$+ \dot{W}_e$	$- |\dot{W}_{RA}|$	$= \dot{H}_2 - \dot{H}_1$
Energy:	-14 kW	$+25.506$ kW	-1.020 kW	$= 10.486$ kW
Exergy:	-2.331 kW	$+25.506$ kW	-1.020 kW	$= 19.717$ kW $+2.438$ kW
Anergy:	-11.669 kW $+0$ kW		-0 kW	$= -9.231$ kW -2.438 kW

The exergetic efficiencies are

Eq. (8.137): $\zeta = 0.9004$ Gl. (8.145): $\zeta = 0.8644$

Eq. (8.139): $\zeta = 0.8900$ Gl. (8.147): $\zeta = 0.8508$

Eq. (8.142): $\zeta = 0.8052$ Gl. (8.149): $\zeta = 0.7730$.

For a right-hand ideal cycle process, the exergetic efficiency is according to Eq. (8.135) or as follows:

$$\zeta = \frac{|(W_e)_{id}|}{(EQ_{zu})_{rev}} = \frac{-W_{Kreis}}{(EQ_{zu})_{rev}} \qquad (8,150)$$

For a right-hand real cycle process, the exergetic efficiency is given by Eqs. (7.31) to (7.34)

$$\zeta = \frac{-W_e}{EQ_{zu}} = \frac{|W_e|}{EQ_{zu}} = \frac{-W_{Kreis} - |W_{RI}| - |W_{RA}|}{EQ_{zu}} \qquad (8.151)$$

$$\zeta = \frac{EQ_{zu} - |EQ_{ab}| - |W_{RA}|}{EQ_{zu}} = 1 - \frac{|EQ_{ab}| + |W_{RA}|}{EQ_{zu}}. \qquad (8.152)$$

Example 8.25 For the steam power plant according to example 7.5, the exergetic efficiencies of the reversible *Clausius-Rankine* process and the real process shall be determined. Ambient pressure and ambient temperature are 1 bar and 20 °C.

Punkt	h kJ/kg	s kJ/(kg K)
1	3097.37	6.2139
2	2609.40	6.4195
3	2520.67	6.2139
4	670.50	1.9311
5	680.83	1.9311
6	682.51	1.9352

The designations of example 7.5 apply. For the reversible *Clausius-Rankine* process, Eq. (8.150) states:

$$\zeta = \frac{-\dot{W}_{Kreis}}{(E\dot{Q}_{zu})_{rev}}$$

According to Table 7.2 Eq. (1) is

$$-\dot{W}_{Kreis} = (\dot{Q}_{zu})_{rev} - |(\dot{Q}_{ab})_{rev}|.$$

$$(\dot{Q}_{zu})_{rev} = \dot{m}(h_1 - h_5) = 40 \cdot 10^3 \text{ kg/h} \cdot (3097.37 - 680.83) \text{ kJ/kg} = 26.850 \text{ MW}$$

$$|(\dot{Q}_{ab})_{rev}| = \dot{m}(h_3 - h_4) = 40 \cdot 10^3 \text{ kg/h} \cdot (2520.67 - 670.50) \text{ kJ/kg} = 20.557 \text{ MW}$$

$$-\dot{W}_{Kreis} = 6.293 \text{ MW}$$

$$\begin{aligned}(E\dot{Q}_{zu})_{rev} &= (\dot{Q}_{zu})_{rev} - T_0 \dot{m}(s_1 - s_5) \\ &= 26.850 \text{ MW} - 293.15 \text{ K} \cdot 40 \cdot 10^3 \text{ kg/h} \cdot (6.2139 - 1.9311) \text{ kJ/(kg K)} \\ &= 12.900 \text{ MW}\end{aligned}$$

$$\zeta = \frac{6.293 \text{ MW}}{12.900 \text{ MW}} = 0.4878$$

For the real process, Eq. (8.152) states:

$$\zeta = 1 - \frac{|E\dot{Q}_{ab}| + |\dot{W}_{RA}|}{E\dot{Q}_{zu}}$$

$$\begin{aligned}|E\dot{Q}_{ab}| &= |\dot{Q}_{ab}| - T_0 \dot{m}(s_2 - s_4) \\ &= 21.543 \text{ MW} - 293.15 \text{ K} \cdot 40 \cdot 10^3 \text{ kg/h} \cdot (6.4195 - 1.9311) \text{ kJ/(kg K)} = 6.923 \text{ MW}\end{aligned}$$

$$\begin{aligned}E\dot{Q}_{zu} &= \dot{Q}_{zu} - T_0 \dot{m}(s_1 - s_6) \\ &= 26.832 \text{ MW} - 293.15 \text{ K} \cdot 40 \cdot 10^3 \text{ kg/h} \cdot (6.2139 - 1.9352) \text{ kJ/(kg K)} = 12.895 \text{ MW}\end{aligned}$$

According to Example 7.5, the total external friction power is

$$|\dot{W}_{RA}| = 0.226 \text{ MW}.$$

Thus, one obtains

$$\zeta = 1 - \frac{6.923 \text{ MW} + 0.226 \text{ MW}}{12.895 \text{ MW}} = 0.4456.$$

Tasks for Section 8

1. What is the exergy of internal energy and the exergy of enthalpy of 4.6 kg of oxygen ($R = 259.84$ J/(kg K); $n = 1.4$) at a pressure of 3.2 bar and at a temperature of 53 °C, if the ambient pressure is 1 bar and the ambient temperature is 20 °C ?

2. The gas according to task 1 is to be heated isobarically and frictionless in a way, that the exergy of the internal energy becomes twice as large.

a) What is the final temperature?

b) By what amount does the internal energy increase?

c) How much does the anergy of the internal energy increase?

d) How much heat must be supplied?

e) What is the exergy and the anergy of the heat supplied?

3. The gas according to task 1 is polytropically compressed from the ambient state to the final state according to task 2. The process takes place in an open adiabatic system with friction. The exergy loss, the anergy gain and, for the first law of thermodynamics in the form of Eq. (2.70), the balances for the energy, for the exergy and for the anergy are to be determined.

4. What are the exergetic efficiencies of the two *Carnot* cycle processes according to example 7.1, if the ambient temperature is 15 °C ?

9 Heat Transfer

Heat can be transferred from a warmer to a colder substance by radiation, conduction and convection. Heat transfer by radiation does not require a transfer medium. Convection in liquids or gases occurs when heat conduction is superimposed by a flow.

9.1 Heat Radiation

The flow of heat that reaches the earth from the sun through radiation is a prerequisite for life on the earth. From the extensive wavelength range of electromagnetic radiation (in order of increasing wavelength λ: Cosmic rays, γ rays, X-rays (Röntgen rays), ultraviolet rays, visible light rays, infrared rays (also called heat rays), radio and hertzian waves), ultraviolet rays, visible light rays and infrared rays are of particular interest with regard to the topic under discussion.

9.1.1 Stefan-Boltzmann Law

The fundamental law of thermal radiation is the law of *Stefan*[1]-*Boltzmann*[2]

$$\mathrm{d}\dot{Q} = C\, T^4\, \mathrm{d}A = E\, \mathrm{d}A \,. \tag{9.1}$$

According to this, the heat flow $\dot{Q}$ (heat flux $\dot{Q}$) is proportional to the fourth power number of the absolute temperature T and the surface area A of the body. The proportionality factor C is the radiation coefficient. E is the emission (also named emittance, total emissive power, radiant flux density, radiance, total hemispherical intensity); it represents a heat flow related to the surface of the radiating body. The radiation coefficient C depends on the type of body. The black body has the largest radiation coefficient.

The black body (Fig. 9.1) is a model using a hollow enclosure provided with only a very small hole. The area of the opening can be considered a surface with complete absorptiion. The black body has the largest radiation coefficient. The *Stefan-Boltzmann* law applies exactly to black bodies, for other bodies only approximately [46].

The radiation coefficient for black bodies is $C_S = 5.67028 \cdot 10^{-8}$ W/(m² K⁴) [64].

The emission ratio (also known as emissivity) ε of any body is the ratio of the emission E of that body $E = C\,T^4$ to the emission E_S of a black body $E_S = C_S\,T^4$ with the same temperature. The absorption ratio (also known as absorptivity) a of any body is the ratio of the absorbed radiation to the incident radiation. Since a black body absorbs all of the incident radiation, the absorption ratio of any body is also the ratio of the absorbed radiation to the radiation that a black body would absorb.

9.1.2 Kirchhoff's Law

According to the law of *Kirchhoff*[3], the absorption ratio of any body being in thermal equilibrium is equal to its emission ratio:

$$a = \varepsilon = \frac{E}{E_S} = \frac{C}{C_S} \tag{9.2}$$

---

[1] *Josef Stefan*, Austrian physicist, 1835 to 1893, dealt among other things with questions of optics.
[2] *Ludwig Boltzmann*, 1844 to 1906, Austrian physicist, gave thermodynamics a new form, founded statistical mechanics and was a pioneer of quantum mechanics.
[3] *Gustav Robert Kirchhoff*, 1824 to 1887, German professor for theoretical physics in Berlin, was mainly concerned with the study of electricity and thermal radiation.

© The Author(s), under exclusive license to
Springer Fachmedien Wiesbaden GmbH, part of Springer Nature 2023
M. Dehli et al., *Fundamentals of Technical Thermodynamics*,
https://doi.org/10.1007/978-3-658-38910-9_9

The incident radiation that is not absorbed by a body can be reflected or be transmitted. The reflection ratio (also known as reflectivity) r is the ratio of the reflected radiation to the incident radiation. The transmittance ratio (also known as transmissivity) d is the ratio of the transmitted radiation to the incident radiation. For the sum of the absorption ratio, the reflection ratio and the transmission ratio,

$$a + r + d = 1$$

applies. For a black body is $a = 1$. A body for which holds $r = 1$, is called a white body. If the absorption ratio $0 < a < 1$ is independent of the wavelength, it is called a grey body. A selective radiation behaviour exists if the absorption ratio of a body depends on the wavelength or disappears completely at some wavelengths.

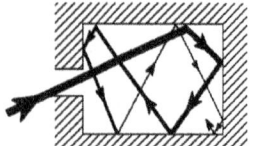

Figure 9.1 Absorption of the black body

9.1.3 *Planck*'s Radiation Law

The emission E is obtained by integrating the intensity I over the wavelength λ:

$$E = \int_0^\infty I \, d\lambda \qquad (9.3)$$

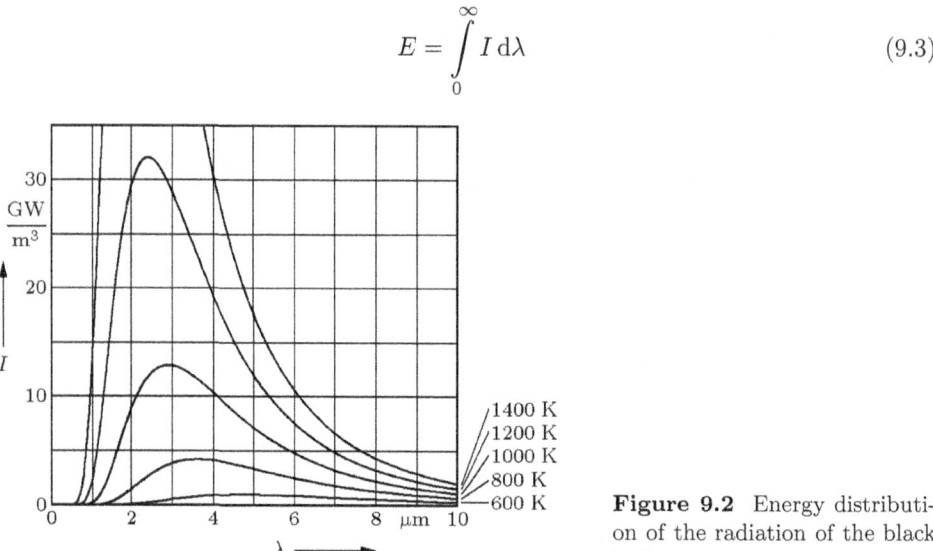

Figure 9.2 Energy distribution of the radiation of the black body

According to *Max Planck* [94], the intensity distribution of the radiation of the black body is:

$$I = \frac{c_1}{\lambda^5} \frac{1}{e^{\frac{c_2}{\lambda T}} - 1} \qquad (9.4)$$

Here, according to [64] c_1 is the first radiation constant

$$c_1 = 2\pi h c^2 = 3.7418024 \cdot 10^{-16} \text{ W m}^2,$$

the second radiation constant
$$c_2 = h\,c/k = 1.438786 \cdot 10^{-2} \text{ m K},$$
the *Planck* constant (elementary quantum of action)[4]
$$h = 6.626124 \cdot 10^{-34} \text{ J s},$$
the speed of light
$$c = 2.99792458 \cdot 10^8 \text{ m/s},$$
the *Boltzmann* constant
$$k = 1.380652 \cdot 10^{-23} \text{ J/K}.$$

Fig. 9.2 presents a graphical evaluation of Eq. (9.4). For the emission of the black body one obtains by integration

$$E_S = \int_0^\infty I \, \mathrm{d}\lambda = c_1 \left(\frac{T}{c_2}\right)^4 \frac{6\,\pi^4}{90} = C_S\,T^4 \,. \tag{9.5}$$

The equation
$$C_S = \frac{c_1}{c_2^4} \frac{6\,\pi^4}{90} \tag{9.6}$$

represents the given numerical value for the radiation coefficient of the black body C_S [44].

9.1.4 *Wien*'s Displacement Law

Zeroing the derivative of intensity with respect to wavelength gives

$$\frac{c_2}{\lambda T} e^{\frac{c_2}{\lambda T}} - 5\left(e^{\frac{c_2}{\lambda T}} - 1\right) = 0\,. \tag{9.7}$$

This equation describes a zero problem with the solution
$$c_2/(\lambda T) = 4.965114.$$

From this follows *Wien*'s displacement law[5] for the maxima of intensity:

$$\lambda T = 0.00289779 \text{ m K} = \text{const} \tag{9.8}$$

The maximum of the radiation intensity is therefore in the range of smaller wavelengths for bodies of higher temperatures, for bodies of lower temperatures in the range of longer wavelengths.

This is important, for example, for the so-called greenhouse effect: Water vapor and carbon dioxide are selective radiators, nitrogen and oxygen are radiolucent. Because the

---

[4]Instead of the quantum of action h, also the reduced *Planck* quantum of action $\hbar = h/2\pi$ is used. Thus, any harmonic oscillation with the frequency f and angular frequency $\omega = 2\pi f$ can only absorb or release energy in discrete amounts which are integer multiples of the oscillation quantum $\Delta E = h\,f = \hbar\,\omega$. *Planck*'s reduced quantum of action is also (rarely) called *Dirac*'s constant after Paul *Dirac*; its value is $\hbar = 1.054\,571\,817 \cdot 10^{-34}$ J s.

[5] *Wilhelm Carl Wien*, 1864 to 1928, German professor in Munich, published work on thermal radiation, received the Nobel Prize for Physics in 1911.

radiation intensity of the sun, with its high surface temperature, lies more strongly in the range of smaller wavelengths solar radiation is predominantly transmitted through the atmosphere, which contains water vapor and carbon dioxide. In contrast, radiation from Earth surface with its lower temperature, is more strongly absorbed — and thus also emitted — by water vapor and carbon dioxide [39].

9.1.5 Lambert's Cosine Law

E is the emission into half-space (radiation emanating from a point on a plane surface into space, cf. Fig. 9.3). E_n is the emission in the normal direction. According to the cosine law of *Lambert*[6] for the emission E_β that deviates from the normal direction by the angle β, is $E_\beta = E_n \cos \beta$. The *Lambert* cosine law applies exactly for black bodies, for other bodies only approximately. Integration over the half-space gives $E = \pi E_n$.

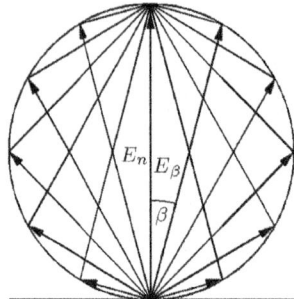

Figure 9.3 Emission into the half-space: illustration of the *Lambert*'s cosine law

For radiation into a solid angle Ω,

$$E_\Omega = \int_\Omega E_\beta \, d\Omega. \tag{9.9}$$

A solid angle that cuts out part of a spherical surface is given by the ratio of the cut out spherical surface to the square of the radius of the sphere. According to Eq. (9.1), the heat flow emanating from the surface element dA into the solid angle Ω is,

$$d\dot{Q} = E_\Omega \, dA = \int_\Omega E_\beta \, d\Omega \, dA \,. \tag{9.10}$$

Integration over the area A gives the double integral

$$\dot{Q}_\Omega = \int_A \int_\Omega E_\beta \, d\Omega \, dA = E \, A \, \varphi \,. \tag{9.11}$$

$\dot{Q}_\Omega$ is the radiation of a body with surface A into solid angle Ω. E is the emission into the half-space, φ is the irradiance number (also named irradiance, angular ratio, shape factor, configuration factor).

9.1.6 Irradiance Number

The irradiance number φ is in the range $0 < \varphi \leq 1$. φ_{12} is the fraction of radiation emanating from the surface A_1 that hits the surface A_2. If all the radiation emanating from the surface A_1 also hits the surface A_2, then results $\varphi_{12} = 1$.

---

[6]*Johann Heinrich Lambert*, Alsatian mathematician and logician, 1728 to 1777, founded the methods of the measurement of the intensity of light as a science.

If there is an exchange of radiation between the surfaces A_1 and A_2, then according to Eq. (9.11) with the equations for E_β and E follow

$$E_1 A_1 \varphi_{12} = \int_{A_1} \int_{\Omega_1} \frac{E_1}{\pi} \cos\beta_1 \, d\Omega_1 \, dA_1 \tag{9.12}$$

$$E_2 A_2 \varphi_{21} = \int_{A_2} \int_{\Omega_2} \frac{E_2}{\pi} \cos\beta_2 \, d\Omega_2 \, dA_2 \,. \tag{9.13}$$

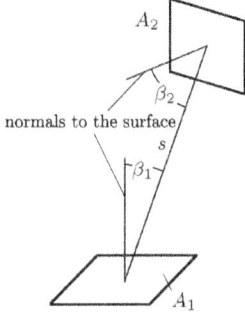

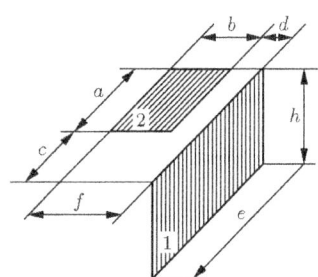

Figure 9.4 Radiation between two surfaces in space

Figure 9.5 To equation (9.17)

Figure 9.6 To equation (9.18)

According to Fig. 9.4 and the definition of the solid angle is

$$d\Omega_1 = \frac{\cos\beta_2}{s^2} dA_2 \qquad d\Omega_2 = \frac{\cos\beta_1}{s^2} dA_1 \,. \tag{9.14}$$

Eqs. (9.12) and (9.13) give for the irradiance numbers

$$\varphi_{12} = \frac{1}{\pi A_1} \int_{A_1} \int_{A_2} \frac{1}{s^2} \cos\beta_1 \cos\beta_2 \, dA_2 \, dA_1 \tag{9.15}$$

$$\varphi_{21} = \frac{1}{\pi A_2} \int_{A_2} \int_{A_1} \frac{1}{s^2} \cos\beta_2 \cos\beta_1 \, dA_1 \, dA_2 \,. \tag{9.16}$$

From Eqs. (9.15) and (9.16) it follows $A_1 \varphi_{12} = A_2 \varphi_{21}$. A series of solutions of the double integrals in Eqs. (9.15) and (9.16) can be found in [132]. Two very versatile applicable equations, which are taken from [42], are communicated here: The irradiance number of two parallel rectangular surfaces (Fig. 9.5) is described by Eq. (9.17). The irradiance number of two rectangular surfaces lying at right angle to each other (Fig. 9.6) can be calculated with Eq. (9.18).

$$\varphi_{12} = \frac{1}{2\pi \left(\frac{e}{h}\right)\left(\frac{f}{h}\right)} \left\{ \frac{f}{h}\sqrt{\left(\frac{e}{h}\right)^2 + 1} \arctan \frac{\frac{f}{h}}{\sqrt{\left(\frac{e}{h}\right)^2 + 1}} + \frac{f}{h}\sqrt{\left(\frac{a}{h}\right)^2 + 1} \arctan \frac{\frac{f}{h}}{\sqrt{\left(\frac{a}{h}\right)^2 + 1}} - \right.$$

$$-\frac{f}{h}\sqrt{1+\left(\frac{c}{h}\right)^2}\arctan\frac{\frac{f}{h}}{\sqrt{1+\left(\frac{c}{h}\right)^2}}-\frac{f}{h}\arctan\frac{f}{h}+\frac{e}{h}\sqrt{\left(\frac{f}{h}\right)^2+1}\arctan\frac{\frac{e}{h}}{\sqrt{\left(\frac{f}{h}\right)^2+1}}-$$

$$-\frac{e}{h}\sqrt{\left(\frac{d}{h}\right)^2+1}\arctan\frac{\frac{e}{h}}{\sqrt{\left(\frac{d}{h}\right)^2+1}}+\frac{e}{h}\sqrt{1+\left(\frac{b}{h}\right)^2}\arctan\frac{\frac{e}{h}}{\sqrt{1+\left(\frac{b}{h}\right)^2}}-\frac{e}{h}\arctan\frac{e}{h}+$$

$$+\frac{a}{h}\sqrt{\left(\frac{f}{h}\right)^2+1}\arctan\frac{\frac{a}{h}}{\sqrt{\left(\frac{f}{h}\right)^2+1}}-\frac{a}{h}\sqrt{\left(\frac{d}{h}\right)^2+1}\arctan\frac{\frac{a}{h}}{\sqrt{\left(\frac{d}{h}\right)^2+1}}+\frac{a}{h}\sqrt{1+\left(\frac{b}{h}\right)^2}\times$$

$$\times\arctan\frac{\frac{a}{h}}{\sqrt{1+\left(\frac{b}{h}\right)^2}}-\frac{a}{h}\arctan\frac{a}{h}-\frac{d}{h}\sqrt{\left(\frac{e}{h}\right)^2+1}\arctan\frac{\frac{d}{h}}{\sqrt{\left(\frac{e}{h}\right)^2+1}}-\frac{d}{h}\sqrt{\left(\frac{a}{h}\right)^2+1}\times$$

$$\times\arctan\frac{\frac{d}{h}}{\sqrt{\left(\frac{a}{h}\right)^2+1}}+\frac{d}{h}\sqrt{1+\left(\frac{c}{h}\right)^2}\arctan\frac{\frac{d}{h}}{\sqrt{1+\left(\frac{c}{h}\right)^2}}+\frac{d}{h}\arctan\frac{d}{h}+\frac{b}{h}\sqrt{\left(\frac{e}{h}\right)^2+1}\times$$

$$\times\arctan\frac{\frac{b}{h}}{\sqrt{\left(\frac{e}{h}\right)^2+1}}+\frac{b}{h}\sqrt{\left(\frac{a}{h}\right)^2+1}\arctan\frac{\frac{b}{h}}{\sqrt{\left(\frac{a}{h}\right)^2+1}}-\frac{b}{h}\sqrt{1+\left(\frac{c}{h}\right)^2}\arctan\frac{\frac{b}{h}}{\sqrt{1+\left(\frac{c}{h}\right)^2}}-$$

$$-\frac{b}{h}\arctan\frac{b}{h}-\frac{c}{h}\sqrt{\left(\frac{f}{h}\right)^2+1}\arctan\frac{\frac{c}{h}}{\sqrt{\left(\frac{f}{h}\right)^2+1}}+\frac{c}{h}\sqrt{\left(\frac{d}{h}\right)^2+1}\arctan\frac{\frac{c}{h}}{\sqrt{\left(\frac{d}{h}\right)^2+1}}-$$

$$-\frac{c}{h}\sqrt{1+\left(\frac{b}{h}\right)^2}\arctan\frac{\frac{c}{h}}{\sqrt{1+\left(\frac{b}{h}\right)^2}}+\frac{c}{h}\arctan\frac{c}{h}+$$

$$+\frac{1}{2}\ln\left\{\frac{\left[\left(\frac{e}{h}\right)^2+\left(\frac{d}{h}\right)^2+1\right]\left[\left(\frac{a}{h}\right)^2+\left(\frac{d}{h}\right)^2+1\right]\left[1+\left(\frac{b}{h}\right)^2+\left(\frac{c}{h}\right)^2\right]}{\left[\left(\frac{e}{h}\right)^2+1+\left(\frac{b}{h}\right)^2\right]\left[\left(\frac{a}{h}\right)^2+1+\left(\frac{b}{h}\right)^2\right]\left[\left(\frac{e}{h}\right)^2+\left(\frac{f}{h}\right)^2+1\right]}\times\right.$$

$$\left.\times\frac{\left[\left(\frac{f}{h}\right)^2+1+\left(\frac{c}{h}\right)^2\right]\left[\left(\frac{e}{h}\right)^2+1\right]\left[\left(\frac{a}{h}\right)^2+1\right]\left[\left(\frac{f}{h}\right)^2+1\right]\left[1+\left(\frac{b}{h}\right)^2\right]}{\left[\left(\frac{a}{h}\right)^2+\left(\frac{f}{h}\right)^2+1\right]\left[\left(\frac{d}{h}\right)^2+1+\left(\frac{c}{h}\right)^2\right]\left[\left(\frac{d}{h}\right)^2+1\right]\left[1+\left(\frac{c}{h}\right)^2\right]}\right\} \qquad (9.17)$$

$$\begin{aligned}
\varphi_{12} = \frac{1}{2\pi} \Bigg\{ & \frac{f}{h}\arctan\frac{e}{f} + \frac{a}{e}\frac{f}{h}\arctan\frac{a}{f} - \frac{c}{e}\frac{f}{h}\arctan\frac{c}{f} - \\
& -\frac{d}{h}\arctan\frac{e}{d} - \frac{a}{e}\frac{d}{h}\arctan\frac{a}{d} + \frac{c}{e}\frac{d}{h}\arctan\frac{c}{d} - \sqrt{1+\left(\frac{f}{h}\right)^2} \times \\
& \times \arctan\frac{\frac{e}{h}}{\sqrt{1+\left(\frac{f}{h}\right)^2}} + \sqrt{1+\left(\frac{d}{h}\right)^2}\arctan\frac{\frac{e}{h}}{\sqrt{1+\left(\frac{d}{h}\right)^2}} - \\
& -\frac{a}{e}\sqrt{1+\left(\frac{f}{h}\right)^2}\arctan\frac{\frac{a}{h}}{\sqrt{1+\left(\frac{f}{h}\right)^2}} + \frac{a}{e}\sqrt{1+\left(\frac{d}{h}\right)^2}\times\arctan\frac{\frac{a}{h}}{\sqrt{1+\left(\frac{d}{h}\right)^2}} + \\
& +\frac{c}{e}\sqrt{1+\left(\frac{f}{h}\right)^2}\arctan\frac{\frac{c}{h}}{\sqrt{1+\left(\frac{f}{h}\right)^2}} - \frac{c}{e}\sqrt{1+\left(\frac{d}{h}\right)^2}\arctan\frac{\frac{c}{h}}{\sqrt{1+\left(\frac{d}{h}\right)^2}} - \\
& -\frac{e}{4h}\ln\frac{\left[1+\left(\frac{f}{h}\right)^2+\left(\frac{e}{h}\right)^2\right]\left[\left(\frac{e}{h}\right)^2+\left(\frac{d}{h}\right)^2\right]}{\left[1+\left(\frac{e}{h}\right)^2+\left(\frac{d}{h}\right)^2\right]\left[\left(\frac{f}{h}\right)^2+\left(\frac{e}{h}\right)^2\right]} - \\
& -\frac{1}{4}\frac{a}{h}\frac{a}{e}\ln\frac{\left[\left(\frac{a}{h}\right)^2+1+\left(\frac{f}{h}\right)^2\right]\left[\left(\frac{a}{h}\right)^2+\left(\frac{d}{h}\right)^2\right]}{\left[\left(\frac{a}{h}\right)^2+\left(\frac{f}{h}\right)^2\right]\left[\left(\frac{a}{h}\right)^2+1+\left(\frac{d}{h}\right)^2\right]} + \\
& +\frac{1}{4}\frac{f}{h}\frac{f}{e}\ln\frac{\left[1+\left(\frac{f}{h}\right)^2+\left(\frac{e}{h}\right)^2\right]\left[\left(\frac{a}{h}\right)^2+1+\left(\frac{f}{h}\right)^2\right]\left[\left(\frac{f}{h}\right)^2+\left(\frac{c}{h}\right)^2\right]\left(\frac{f}{h}\right)^2}{\left[\left(\frac{f}{h}\right)^2+\left(\frac{e}{h}\right)^2\right]\left[\left(\frac{a}{h}\right)^2+\left(\frac{f}{h}\right)^2\right]\left[1+\left(\frac{f}{h}\right)^2\right]\left[1+\left(\frac{f}{h}\right)^2+\left(\frac{c}{h}\right)^2\right]} - \\
& -\frac{1}{4}\frac{d}{h}\frac{d}{e}\ln\frac{\left[1+\left(\frac{e}{h}\right)^2+\left(\frac{d}{h}\right)^2\right]\left[\left(\frac{a}{h}\right)^2+1+\left(\frac{d}{h}\right)^2\right]\left[\left(\frac{d}{h}\right)^2+\left(\frac{c}{h}\right)^2\right]\left(\frac{d}{h}\right)^2}{\left[\left(\frac{e}{h}\right)^2+\left(\frac{d}{h}\right)^2\right]\left[\left(\frac{a}{h}\right)^2+\left(\frac{d}{h}\right)^2\right]\left[1+\left(\frac{d}{h}\right)^2\right]\left[1+\left(\frac{d}{h}\right)^2+\left(\frac{c}{h}\right)^2\right]} + \\
& +\frac{1}{4}\frac{h}{e}\ln\frac{\left[1+\left(\frac{f}{h}\right)^2+\left(\frac{e}{h}\right)^2\right]\left[\left(\frac{a}{h}\right)^2+1+\left(\frac{f}{h}\right)^2\right]\left[1+\left(\frac{d}{h}\right)^2+\left(\frac{c}{h}\right)^2\right]\left[1+\left(\frac{d}{h}\right)^2\right]}{\left[1+\left(\frac{e}{h}\right)^2+\left(\frac{d}{h}\right)^2\right]\left[\left(\frac{a}{h}\right)^2+1+\left(\frac{d}{h}\right)^2\right]\left[1+\left(\frac{f}{h}\right)^2+\left(\frac{c}{h}\right)^2\right]\left[1+\left(\frac{f}{h}\right)^2\right]} + \\
& +\frac{1}{4}\frac{c}{h}\frac{c}{e}\ln\frac{\left[1+\left(\frac{f}{h}\right)^2+\left(\frac{c}{h}\right)^2\right]\left[\left(\frac{d}{h}\right)^2+\left(\frac{c}{h}\right)^2\right]}{\left[1+\left(\frac{d}{h}\right)^2+\left(\frac{c}{h}\right)^2\right]\left[\left(\frac{f}{h}\right)^2+\left(\frac{c}{h}\right)^2\right]} \Bigg\}
\end{aligned} \qquad (9.18)$$

With such extensive and complicated formulas, it is easy for errors to creep in when copying or typesetting. For example, in the quoted original in [42] the expression $\left[\left(\frac{f}{h}\right)^2 + \left(\frac{e}{h}\right)\right]^2$ is surely to be replaced by $\left[\left(\frac{f}{h}\right)^2 + \left(\frac{e}{h}\right)^2\right]$.

It is more difficult, for example, in the case of a sign error, which is not so easy to recognise. Since there is obviously such a sign error in Eq. (9.18) — the equation given here is corrected — the following example shows how to find the error.

Example 9.1 For two surfaces perpendicular to each other according to Fig. 9.7 a) with dimensions $a = 5$, $b = 3$, $c = 2$, $d = 1$ and $h = 6$ — the length units are omitted because they can be chosen arbitrarily — the irradiance number φ is to be determined. The corrected Eq. (9.18) gives the result:

Fig. 9.7 a)	$(a = 5, b = 3, c = 2, d = 1, h = 6)$	$\varphi_{12} = 0.0809049$	(1)

Another way via a difference of areas, namely $\varphi_{12} = \varphi_{13} - \varphi_{14}$ must lead to the same result:

Fig. 9.7 b)	$(a = 5, b = 4, c = 2, d = 0, h = 6)$	$\varphi_{13} = 0.1306251$	(2)
Fig. 9.7 c)	$(a = 5, b = 1, c = 2, d = 0, h = 6)$	$\varphi_{14} = 0.0497202$	(3)
$\varphi_{12} = \varphi_{13} - \varphi_{14}$	$0.1306251 - 0.0497202 = 0.0809049$		(4)

A control possibility is provided by the relation $\varphi_{13} + \varphi_{15} = \varphi_{16}$:

Fig. 9.7 d)	$(a = 2, b = 4, c = 5, d = 0, h = 6)$	$\varphi_{15} = 0.0467231$	(5)
Fig. 9.7 e)	$(a = 7, b = 4, c = 0, d = 0, h = 6)$	$\varphi_{16} = 0.1773572$	(6)
$\varphi_{13} + \varphi_{15} = \varphi_{16}$	$0.1306251 + 0.0467321 = 0.1773572$		(7)

Using the equation given in the original [42], one obtains:

Fig. 9.7 a)	$(a = 5, b = 3, c = 2, d = 1, h = 6)$	$\varphi_{12} = 0.0641232$	(8)

The other way $\varphi_{12} = \varphi_{13} - \varphi_{14}$ leads to the result:

Fig. 9.7 b)	$(a = 5, b = 4, c = 2, d = 0, h = 6)$	$\varphi_{13} = 0.1306251$	(9)
Fig. 9.7 c)	$(a = 5, b = 1, c = 2, d = 0, h = 6)$	$\varphi_{14} = 0.0497202$	(10)
$\varphi_{12} = \varphi_{13} - \varphi_{14}$	$0.1306251 - 0.0497202 = 0.0809049$		(11)

The values of φ_{12} calculated in different ways according to (8) and (11) do not agree. So there must be an error in Eq. (9.18) of the original.

The control calculation according to the relation $\varphi_{13} + \varphi_{15} = \varphi_{16}$ gives:

Fig. 9.7 d)	$(a = 2, b = 4, c = 5, d = 0, h = 6)$	$\varphi_{15} = 0.0467321$	(12)
Fig. 9.7 e)	$(a = 7, b = 4, c = 0, d = 0, h = 6)$	$\varphi_{16} = 0.1773572$	(13)
$\varphi_{13} + \varphi_{15} = \varphi_{16}$	$0.1306251 + 0.0467321 = 0.1773572$		(14)

Obviously, the error occurs only in the calculation of φ_{12}. Since $d = 0$ or $c = 0$ causes some elements of the equation to be omitted in the various calculations in Eq. (9.18), the error is to be supposed in the elements which occur only in φ_{12} in the case of Fig. 9.7 a). The suspicion falls on the element

$$\frac{c}{d} \arctan \frac{c}{d}$$

with a wrong sign. The difference

$$(\varphi_{13} - \varphi_{14}) - \varphi_{12} = 0.0809049 - 0.0641232 = 0.0167817$$

provides the numerical value for the defective element

$$\frac{0.0167817}{2} \, 2\pi = 0.052721;$$

which is exactly the element

$$\frac{c}{d} \arctan \frac{c}{d},$$

whose sign must be changed from minus to plus.

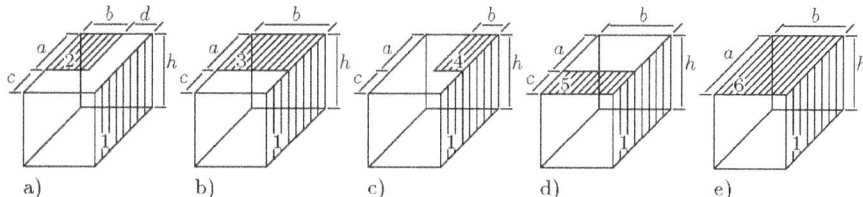

Figure 9.7 To example 9.1

9.2 Radiation Exchange

The calculation of the radiation exchange can be carried out according to the reflection method or according to the cavity method (= method of the enclosed space). The reflection method gives the equation for the exchange of radiation between two surfaces

$$\dot{Q}_{12} = \frac{C_S \, a_1 \, a_2 \, A_1 \, \varphi_{12}}{1 - (1 - a_1)(1 - a_2) \, \varphi_{12} \, \varphi_{21}} (T_1^4 - T_2^4) \,. \tag{9.19}$$

[132]. For example, one could think of the radiation exchange between the floor and the ceiling of a room, where underfloor heating provides an increased temperature of the floor compared to the ceiling. If one asks what heat must be supplied to the floor by the heating and what heat must be released from the ceiling in order to maintain a stationary exchange of radiation, the calculation described does not provide the desired answer.

It does not cover the radiation exchange of the floor and ceiling with the walls. Even if the radiation exchange between the floor and the walls and between the ceiling and the walls is calculated using Eq. (9.19), the sum of the different radiantion exchange calculations would not give the answer to the previous question. It is still not taken into account that, for example, the radiation from the floor to the ceiling, in addition to the direct path, can also reach the ceiling via reflection from the walls.

If one would want to take all this into account with the reflection method, the calculation would become so complicated that it would be practically impossible to manage. The cavity method is better suited for solving problems of the kind mentioned. Here, the radiation exchange between all surfaces enclosing a space is taken into account [110].

9.2.1 Cavity Method

When calculating the radiation exchange between several surfaces, the following terms and designations are of meaning:

H_i is the total incident radiation per square metre on the surface A_i,

B_i is the total radiation per square metre emitted from the surface A_i,

$\dot{Q}_i$ is the heat flow to be supplied to the surface A_i or the heat flow to be released by A_i in order to maintain the stationary state of the surface A_i. Thereby A_i is in radiation exchange with a total of n surfaces A_k.

The following equations [65] apply:

$$H_i A_i = \sum_{k=1}^{n} B_k A_k \varphi_{ki} \tag{9.20}$$

$$B_i = E_i + (1 - a_i) H_i \tag{9.21}$$

$$\dot{Q}_i = (B_i - H_i) A_i \tag{9.22}$$

With the relation $A_i \varphi_{ik} = A_k \varphi_{ki}$, Eq. (9.20) becomes

$$H_i = \sum_{k=1}^{n} B_k \varphi_{ik} . \tag{9.23}$$

Eq. (9.23) is inserted into Eq. (9.21):

$$B_i = E_i + (1 - a_i) \sum_{k=1}^{n} B_k \varphi_{ik} \tag{9.24}$$

Eq. (9.24) describes a linear system of equations by which the B_i are calculated. With H_i according to Eq. (9.23), one can determine from Eq. (9.22) the heat flow $\dot{Q}_i$.

If the surface A_i is in radiative exchange with n surfaces A_k, then $\sum_{k=1}^{n} \varphi_{ik} = 1$ holds.

φ_{ii} then has the value zero if the radiation emanating from the surface A_i does not directly impinge again on the surface A_i. If the surface A_1 is enveloped by the surface A_2, as for example in the case of a complete sphere with the surface A_1 inside a hollow sphere with surface A_2 (Fig. 9.8), then $\varphi_{11} = 0$ is given, while φ_{22} is expressed by a certain numerical value.

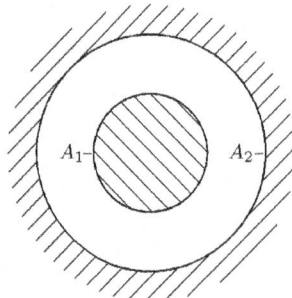

Figure 9.8 Envelopment of a body with surface area A_1 by a body with surface A_2

9.2.2 Envelopment of One Surface by Another

In the special case of the envelopment of the inner surface A_1 by the outer surface A_2, the linear system of equations (9.24) consists of the following two equations:

$$B_1 = E_1 + (1 - a_1)B_2\,\varphi_{12} \tag{9.25}$$

$$B_2 = E_2 + (1 - a_2)(B_1\,\varphi_{21} + B_2\,\varphi_{22}) \tag{9.26}$$

The associated irradiance numbers take the following values:

$$\varphi_{12} = 1 \qquad \varphi_{21} = \frac{A_1}{A_2} \qquad \varphi_{22} = 1 - \frac{A_1}{A_2} \tag{9.27}$$

The Eqs. (9.27) are inserted into Eqs. (9.25) and (9.26):

$$B_1 = E_1 + (1 - a_1)B_2 \tag{9.28}$$

$$B_2 = E_2 + (1 - a_2)[B_1\frac{A_1}{A_2} + B_2(1 - \frac{A_1}{A_2})] \tag{9.29}$$

The two equations can be solved to B_1 and B_2. One obtains

$$B_1 = \frac{E_1[a_2 + (1 - a_2)\frac{A_1}{A_2}] + E_2(1 - a_1)}{a_2 + a_1(1 - a_2)\frac{A_1}{A_2}} \tag{9.30}$$

$$B_2 = \frac{E_1(1 - a_2)\frac{A_1}{A_2} + E_2}{a_2 + a_1(1 - a_2)\frac{A_1}{A_2}} \;. \tag{9.31}$$

According to Eq. (9.23) follow

$$H_1 = B_2\,\varphi_{12} = B_2 \tag{9.32}$$

$$H_2 = B_1\,\varphi_{21} + B_2\,\varphi_{22} = B_1\frac{A_1}{A_2} + B_2(1 - \frac{A_1}{A_2}) \;. \tag{9.33}$$

Eq. (9.22), together with Eqs. (9.32) and (9.33), become

$$\dot{Q}_1 = (B_1 - B_2)A_1 \tag{9.34}$$

$$\dot{Q}_2 = (B_2 - B_1)A_1 \;. \tag{9.35}$$

Eqs. (9.30) and (9.31) are inserted into Eqs. (9.34) and (9.35) [38]:

$$\dot{Q}_1 = -\dot{Q}_2 = \frac{E_1\,a_2 - E_2\,a_1}{a_2 + a_1(1 - a_2)\frac{A_1}{A_2}} A_1 \tag{9.36}$$

9.2.3 Two Parallel Surfaces of Equal Size

In the special case of radiation transfer between two parallel surfaces of equal size $A_1 = A_2 = A$ with a small distance, it becomes $\varphi_{12} = \varphi_{21} = 1$. According to Eq. (9.23), $H_1 = B_2$ and $H_2 = B_1$ result. According to Eq. (9.24) are

$$B_1 = E_1 + (1 - a_1)B_2 \qquad B_2 = E_2 + (1 - a_2)B_1 \ . \tag{9.37}$$

The Eqs. (9.37) can be solved for B_1 and B_2:

$$B_1 = \frac{E_1 + E_2(1 - a_1)}{a_1 + a_2 - a_1 a_2} \qquad B_2 = \frac{E_2 + E_1(1 - a_2)}{a_1 + a_2 - a_1 a_2} \tag{9.38}$$

Eq. (9.22) gives with Eqs. (9.38)

$$\dot{Q}_1 = -\dot{Q}_2 = \frac{E_1 a_2 - E_2 a_1}{a_1 + a_2 - a_1 a_2} A \ . \tag{9.39}$$

The same result is also obtained by Eq. (9.36) if one takes into account the equal size of the two surfaces A_1 and A_2 [38].

9.2.4 Matrix Representation

For the solution of the linear system of equations, it is advantageous to represent Eq. (9.24) in matrix notation:

$$B = E + r\,\Phi\,B \tag{9.40}$$

B and E are vectors. r is a matrix with the reflection ratios $r_i = 1 - a_i$ (for the transmittance ratios $d_i = 0$) in the main diagonal. Φ is the matrix of the irradiance numbers. Eq. (9.40) retains its validity when the left-hand side is multiplied by the unit matrix e:

$$e\,B = E + r\,\Phi\,B \tag{9.41}$$

Both expressions with B are brought to the left-hand side:

$$(e - r\,\Phi)\,B = E \tag{9.42}$$

This is the final form of the representation of the linear system of equations to prepare the solution procedure [65].

Example 9.2 In a room to be treated as a closed system, with the length 4.5 m, the width 3 m and the height 2.7 m, an end wall with the dimensions $3 \cdot 2.7$ m^2 is used as surface A_1 and the opposite wall with the same dimensions as surface A_2. The other two wall areas of $4.5 \cdot 2.7$ m^2 each, together with floor and ceiling of $3 \cdot 4.5$ m^2 shall be considered as area A_3. The surface areas are: $A_1 = A_2 = 8.1\,\text{m}^2$; $A_3 = 51.3\,\text{m}^2$. The absorption ratios and the temperatures of the three surfaces are: $a_1 = 0.8$; $a_2 = 0.2$; $a_3 = 0.9$; $t_1 = 30\,°\text{C}$; $t_2 = 10\,°\text{C}$; $t_3 = 20\,°\text{C}$. Using $E_S = C_S\,T^4$, Eq. (9.2) and the equation for C_S, one obtains the emissions

$$E_1 = 383.1\ \text{W/m}^2 \qquad E_2 = 72.90\ \text{W/m}^2 \qquad E_3 = 376.9\ \text{W/m}^2\ .$$

While the surfaces A_1 and A_2 are in radiation exchange only with the other two surfaces, the surface A_3, in addition to the radiation exchange with the surfaces A_1 and A_2, is is also in a

radiation exchange with its own surface A_3. Eq. (9.18) provides the irradiance number φ_{12}. The other irradiance numbers are obtained from $A_i\,\varphi_{ik} = A_k\varphi_{ki}$ and $\sum\limits_{k=1}^{n} \varphi_{ik} = 1$.

$$\varphi_{12} = \varphi_{21} = 0.10166 \qquad \varphi_{13} = 1 - \varphi_{12} = \varphi_{23} = 0.89834$$

$$\varphi_{31} = \varphi_{13}\frac{A_1}{A_3} = \varphi_{32} = 0.14184 \qquad \varphi_{33} = 1 - \varphi_{31} - \varphi_{32} = 0.71632$$

To explain Eq. (9.42), the individual components of this equation, as they occur in this example, are shown in the following.

E and B are vectors. E is known, B is unknown:

$$E = \begin{pmatrix} E_1 \\ E_2 \\ E_3 \end{pmatrix} \qquad B = \begin{pmatrix} B_1 \\ B_2 \\ B_3 \end{pmatrix}$$

e is the unit matrix:

$$e = \begin{pmatrix} 1 & 0 & 0 \\ 0 & 1 & 0 \\ 0 & 0 & 1 \end{pmatrix}$$

The matrix r contains the reflection ratios:

$$r = \begin{pmatrix} 1-a_1 & 0 & 0 \\ 0 & 1-a_2 & 0 \\ 0 & 0 & 1-a_3 \end{pmatrix}$$

The matrix Φ contains the irradiance numbers:

$$\Phi = \begin{pmatrix} 0 & \varphi_{12} & \varphi_{13} \\ \varphi_{21} & 0 & \varphi_{23} \\ \varphi_{31} & \varphi_{32} & \varphi_{33} \end{pmatrix}$$

Eq. (9.42) represents the following system of linear equations:

$$\begin{array}{rcl} B_1 \quad - \quad (1-a_1)\varphi_{12}B_2 \quad - \quad (1-a_1)\varphi_{13}B_3 & = & E_1 \\ -(1-a_2)\varphi_{21}B_1 \quad + \quad B_2 \quad - \quad (1-a_2)\varphi_{23}B_3 & = & E_2 \\ -(1-a_3)\varphi_{31}B_1 \quad - \quad (1-a_3)\varphi_{32}B_2 \quad + \quad [1-(1-a_3)\varphi_{33}]B_3 & = & E_3 \end{array}$$

The solution of the linear system of equations gives the total radiation emitted by the surfaces:

$$B_1 = 466.8 \text{ W/m}^2 \qquad B_2 = 412.3 \text{ W/m}^2 \qquad B_3 = 419.4 \text{ W/m}^2$$

According to Eq. (9.23) the incident radiation is

$$H_1 = B_2\,\varphi_{12} + B_3\,\varphi_{13} \qquad H_2 = B_1\,\varphi_{21} + B_3\,\varphi_{23} \qquad H_3 = B_1\,\varphi_{31} + B_2\,\varphi_{32} + B_3\,\varphi_{33}$$

$$H_1 = 418.7 \text{ W/m}^2 \qquad H_2 = 424.2 \text{ W/m}^2 \qquad H_3 = 425.1 \text{ W/m}^2$$

According to Eq. (9.22), the heat flows occurring at the surfaces in the stationary state are:

$$\dot{Q}_1 = (B_1 - H_1)A_1 \qquad \dot{Q}_2 = (B_2 - H_2)A_2 \qquad \dot{Q}_3 = (B_3 - H_3)A_3$$

$$\dot{Q}_1 = 390.2 \text{ W} \qquad \dot{Q}_2 = -96.8 \text{ W} \qquad \dot{Q}_3 = -293.4 \text{ W}$$

A positive sign means that a surface emits more energy through radiation than it absorbs through the radiation of the other surfaces. Such a surface must be supplied with a heat flow from the outside in order to maintain a stationary state.

A negative sign means that a surface receives more heat flow through radiation than it emits through radiation. In order for the temperature of such a surface to remain unchanged, a heat flow must be released from the surface to the outside. The total heat flows supplied to all surfaces must be equal to the total heat flows given away from all surfaces.

Example 9.3 A room, which is to be treated as a closed system, with the length 5.5 m, the width 3.5 m and the height 2.5 m has the data given in the following table for each of the six boundary surfaces (area size, absorption ratio, temperature).

The total emitted radiation, the incident radiation and the total heat flow are to be calculated according to Eqs. (9.21), (9.23) and (9.22). In addition, the radiation exchanges of each surface with the other surfaces are to be determined according to the reflection method.

Solution: The sum

$$\dot{Q}_i^{(1)} = \sum_{k=1}^{n} \dot{Q}_{ik}^{(1)}$$

represents a first approximate value for the total heat flow. The difference between the values $\dot{Q}_i$ and $\dot{Q}_1$ is that when calculating $\dot{Q}_{ik}$ according to Eq. (9.19), from which the approximate value $\dot{Q}_1$ is composed, only the direct radiation exchange (self-radiation and reflection) between the surfaces i and k is taken into account, while the reflection from other surfaces is not taken into account. A further simplification is the sum

$$\dot{Q}_i^{(2)} = \sum_{k=1}^{n} \dot{Q}_{ik}^{(2)}$$

because here the total reflection is neglected. This means that Eq. (9.19) has the form

$$\dot{Q}_{ik}^{(2)} = \frac{C_i C_k}{C_S} A_i \, \varphi_{ik} (T_i^4 - T_k^4) \; .$$

The matrix Φ (cf. example 9.2) contains the values given by Eqs. (9.17) and (9.18)

$$\Phi = \begin{pmatrix} 0 & 0.07719 & 0.19439 & 0.19439 & 0.26702 & 0.26702 \\ 0.07719 & 0 & 0.19439 & 0.19439 & 0.26702 & 0.26702 \\ 0.12370 & 0.12370 & 0 & 0.20169 & 0.27545 & 0.27545 \\ 0.12370 & 0.12370 & 0.20169 & 0 & 0.27545 & 0.27545 \\ 0.12137 & 0.12137 & 0.19675 & 0.19675 & 0 & 0.36375 \\ 0.12137 & 0.12137 & 0.19675 & 0.19675 & 0.36375 & 0 \end{pmatrix} .$$

In the following table mean:

Column 1: Number of the area i
Column 2: Area size A_i in m^2
Column 3: Absorption ratio a_i
Column 4: Temperature in °C
Column 5: Emission E_i in W/m^2
Column 6: Total radiation emitted B_i in W/m^2 according to Eq. (9.21)
Column 7: Incident radiation H_i in W/m^2 according to Eq. (9.23)
Column 8: Total heat flow $\dot{Q}_i$ in W (exact value according to Eq. (9.22))
Column 9: Total heat flow (1st approximation) $\dot{Q}_i^{(1)}$ in W
Column 10: Total heat flow (2nd approximation) $\dot{Q}_i^{(2)}$ in W

1	2	3	4	5	6	7	8	9	10
1	8.75	0.95	20	397.8	418.7	417.6	10.0	2.5	2.5
2	8.75	0.90	10	328.0	370.2	421.3	−447.4	−361.5	−361.3
3	13.75	0.85	12	318.7	382.1	422.7	−558.9	−464.1	−463.2
4	13.75	0.80	16	317.1	400.9	418.9	−247.9	−209.6	−208.9
5	19.25	0.75	30	359.2	460.4	404.9	1068.4	900.8	897.8
6	19.25	0.70	22	301.2	426.4	417.3	175.8	131.9	133.1

9.3 Stationary One-Dimensional Heat Conduction

If the temperature gradient between neighbouring particles is unchanged in time, stationary heat conduction (also named steady state heat conduction) occurs. The heat flow $\mathrm{d}\dot{Q}$ in the x-direction perpendicular to the surface element $\mathrm{d}A$ is

$$\mathrm{d}\dot{Q} = -\lambda \frac{\mathrm{d}\vartheta}{\mathrm{d}x}\mathrm{d}A \ . \tag{9.43}$$

The proportionality factor λ is the thermal conductivity. Heat can only flow in x-direction if the temperature gradient $\mathrm{d}\vartheta/\mathrm{d}x$ is negative. If the heat flow is to be positive, then there must be a minus sign on the right-hand side of Eq. (9.43).

9.3.1 Plane Wall

If the same area $\mathrm{d}A$ is always available to the heat flow $\mathrm{d}\dot{Q}$ in x-direction for stationary heat conduction (middle part of Fig. 9.9), then, under the prerequisite of the same thermal conductivity everywhere, the temperature gradient $\mathrm{d}\vartheta/\mathrm{d}x$ is always the same. This means that the temperature gradient is linear. The integration of Eq. (9.43) with the wall thickness δ and the surface temperatures ϑ_0 and ϑ_0' results in

$$\dot{Q} = \frac{\lambda}{\delta}(\vartheta_0 - \vartheta_0')A \ . \tag{9.44}$$

There is an analogy between the conduction of heat and the conduction of electric current. With the voltage u, the current i and the resistance R, Ohm's law $u = iR$ applies to the conduction of the electric current. The analogous notation of Eq. (9.44) is

$$\vartheta_0 - \vartheta_0' = \dot{Q}\frac{\delta}{\lambda A} \tag{9.45}$$

with R_L being the thermal resistance [112], [56]:

$$R_L = \frac{\delta}{\lambda A} \tag{9.46}$$

If a wall consists of n layers, the thermal resistances add up:

$$R_L = \sum_{i=1}^{n} R_{Li} \tag{9.47}$$

If one wants to treat a three-layer wall as a single-layer wall, the equivalent thermal conductivity is

$$\lambda = \frac{\delta}{\frac{\delta_1}{\lambda_1} + \frac{\delta_2}{\lambda_2} + \frac{\delta_3}{\lambda_3}} \ , \tag{9.48}$$

where δ is the sum of the individual layer thicknesses. For equal layer thicknesses, Eq. (9.48) becomes

$$\lambda = \frac{3}{\frac{1}{\lambda_1} + \frac{1}{\lambda_2} + \frac{1}{\lambda_3}} \ . \tag{9.49}$$

According to Eq. (9.49), λ is the harmonic mean of the individual thermal conductivity coefficients λ_i.

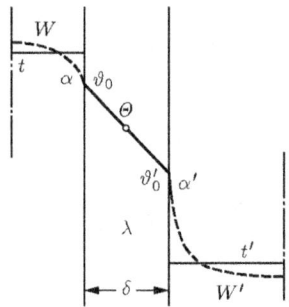

Figure 9.9 Stationary heat transfer through a plane wall. t and t' are the fluid temperatures averaged over the respective flow cross-section. The boundary layer temperatures directly at the wall are equal to the surface temperatures ϑ_0 and ϑ'_0. W and W' are the time-related heat capacities of the fluid flows (heat capacity flows)

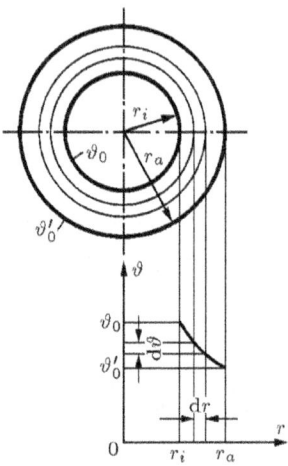

Figure 9.10 Heat conduction through a circular cylindrical tube wall

9.3.2 Pipe Wall

For the wall of a pipe with inner radius r_i, outer radius r_a and length dl (Fig. 9.10), Eq. (9.43) takes the following form:

$$\mathrm{d}\dot{Q} = -\lambda \frac{\mathrm{d}\vartheta}{\mathrm{d}r} 2\pi r\,\mathrm{d}l \tag{9.50}$$

The area element $2\pi r\,\mathrm{d}l$ has a size dependent on the radius r. With a heat flow from the inside to the outside, the cross-section for the heat flow becomes steadily larger, and with the same heat conductivity everywhere, the temperature gradient decreases. With the wall thickness $\delta = r_a - r_i$, the integration results in

$$\dot{Q} = \frac{2\pi l \lambda}{\ln \frac{r_a}{r_i}}(\vartheta_0 - \vartheta'_0) = \frac{\lambda}{\delta}(\vartheta_0 - \vartheta'_0)A_m \ . \tag{9.51}$$

A_m is the mean area with A_a and A_i as the external and internal areas [46]:

$$A_m = \frac{2\pi(r_a - r_i)l}{\ln \frac{r_a}{r_i}} = \frac{A_a - A_i}{\ln \frac{A_a}{A_i}} \tag{9.52}$$

If one uses from the mathematical series

$$\ln x = 2\left[\frac{x-1}{x+1} + \frac{1}{3}\left(\frac{x-1}{x+1}\right)^3 + \frac{1}{5}\left(\frac{x-1}{x+1}\right)^5 + \ldots\right] \tag{9.53}$$

only the first element, then this results in the following approximate expression:

$$\ln \frac{A_a}{A_i} = 2\frac{A_a/A_i - 1}{A_a/A_i + 1} = 2\frac{A_a - A_i}{A_a + A_i} \tag{9.54}$$

From Eqs. (9.52) and (9.54) one obtains the approximation $A_m = \dfrac{A_a + A_i}{2}$. Similar to Eq. (9.46) one obtains from Eq. (9.51) for the thermal resistance of a tube wall

$$R_L = \frac{\ln \dfrac{r_a}{r_i}}{2\pi l \lambda} = \frac{\delta}{\lambda A_m} . \tag{9.55}$$

Eq. (9.46) is also valid for the multilayer tube wall.

9.4 Instationary One-Dimensional Heat Conduction

If the heat flow gradually decreases the temperature gradient at the same location or if external influences cause a temporal change of the temperature gradient, the heat conduction is unsteady (also called instationary or transient).
The heat flow $d\dot{Q}_x$ entering a volume element at the point x and the heat flow $d\dot{Q}_{x+dx}$ leaving the volume element at the point $x + dx$ cause a differential time change of the internal energy $d\dot{U}$:

$$d\dot{Q}_x = -\lambda \left(\frac{\partial \vartheta}{\partial x}\right)_x dA \qquad d\dot{Q}_{x+dx} = -\lambda \left(\frac{\partial \vartheta}{\partial x}\right)_{x+dx} dA \tag{9.56}$$

$$d\dot{U} = dV \varrho c \frac{\partial \vartheta}{\partial \tau} \tag{9.57}$$

In Eq. (9.57), V, ϱ and c are volume, density and specific heat capacity of the volume element. τ is the time. The balance $d\dot{Q}_x = d\dot{U} + d\dot{Q}_{x+dx}$ holds. With

$$\vartheta_{x+dx} = \vartheta_x + \left(\frac{\partial \vartheta}{\partial x}\right)_x dx \qquad \left(\frac{\partial \vartheta}{\partial x}\right)_{x+dx} = \left(\frac{\partial \vartheta}{\partial x}\right)_x + \left(\frac{\partial^2 \vartheta}{\partial x^2}\right)_x dx , \tag{9.58}$$

the second Eq. (9.56) is

$$d\dot{Q}_{x+dx} = -\lambda \left(\frac{\partial \vartheta}{\partial x}\right)_x dA - \lambda \left(\frac{\partial^2 \vartheta}{\partial x^2}\right)_x dx\, dA . \tag{9.59}$$

In Eq. (9.59), one can substitute $dx\, dA$ by dV. The first Eq. (9.56), Eq. (9.57) and Eq. (9.59) are inserted into the above balance $d\dot{Q}_x = d\dot{U} + d\dot{Q}_{x+dx}$:

$$\varrho c \frac{\partial \vartheta}{\partial \tau} = \lambda \frac{\partial^2 \vartheta}{\partial x^2} \tag{9.60}$$

With the thermal diffusivity as a physical property of the material

$$a = \frac{\lambda}{\varrho c} \tag{9.61}$$

one obtains the differential equation of unsteady one-dimensional heat conduction [38], [45]:

$$\frac{\partial \vartheta}{\partial \tau} = a \frac{\partial^2 \vartheta}{\partial x^2} \tag{9.62}$$

9.4.1 Plane Single-Layer Wall

A plane wall through which a steady-state heat flux (stationary heat flux) flows has, in the initial state, left-hand and right-hand surface temperatures ϑ_{l1} and ϑ_{r1}. It shall be assumed that at the time $\tau = 0$, both surface temperatures are abruptly changed to the values on the left ϑ_{l2} and on the right ϑ_{r2} and then kept constant again.

The initially stationary heat conduction through the wall becomes transient (instationary) until, after some time, a stationary state is reached again. The temperature curve as a function of the spatial coordinate x in the wall with the thickness δ, the density ϱ, the specific heat capacity c and the thermal conductivity λ is given by the following solution of the differential equation (9.62) [22]:

$$\vartheta = \vartheta_{l2} + (\vartheta_{r2} - \vartheta_{l2})\frac{x}{\delta} + \sum_{n=1}^{\infty} e^{-a(\frac{n\pi}{\delta})^2 \tau} C_n \sin\left(n\pi\frac{x}{\delta}\right) \tag{9.63}$$

$$C_n = \frac{2}{n\pi}\left[(\vartheta_{l1} - \vartheta_{l2})[1 - \cos(n\pi)] + (\vartheta_{l1} - \vartheta_{r1} - \vartheta_{l2} + \vartheta_{r2})\cos(n\pi)\right] \tag{9.64}$$

a is the thermal diffusivity according to Eq. (9.61). The surface temperatures on the left side and on the right side are obtained for $x = 0$ and $x = \delta$. To calculate the heat flow according to Eq. (9.43) one needs the derivative of the temperature according to the spatial coordinate:

$$\frac{\partial \vartheta}{\partial x} = (\vartheta_{r2} - \vartheta_{l2})\frac{1}{\delta} + \sum_{n=1}^{\infty} e^{-a(\frac{n\pi}{\delta})^2 \tau} \frac{n\pi}{\delta} C_n \cos\left(n\pi\frac{x}{\delta}\right) \tag{9.65}$$

At the left surface, the transient heat flux is

$$\dot{Q}_l = -\frac{\lambda}{\delta}(\vartheta_{r2} - \vartheta_{l2} + \sum_{n=1}^{\infty} e^{-a(\frac{n\pi}{\delta})^2 \tau} n\pi C_n)A \ . \tag{9.66}$$

At the right surface, the transient heat flux is

$$\dot{Q}_r = -\frac{\lambda}{\delta}[\vartheta_{r2} - \vartheta_{l2} + \sum_{n=1}^{\infty} e^{-a(\frac{n\pi}{\delta})^2 \tau} n\pi C_n \cos(n\pi)]A \ . \tag{9.67}$$

If the numerical value is positive, the heat flow is in the direction of the spatial coordinate; if the numerical value is negative, the heat flows in the opposite direction.

Example 9.4 A plane wall with thickness $\delta = 0.2$ m and the substance values density $\varrho = 2000$ kg/m^3, specific heat capacity $c = 1250$ J/(kg K) and thermal conductivity $\lambda = 1.2$ W/(m K) has in the steady state the surface temperatures on the left side $\vartheta_{l1} = 30\,°C$ and on the right side $\vartheta_{r1} = 20\,°C$. After a sudden drop in temperature, the surface temperatures on the left side and on the right side, measured as $\vartheta_{l2} = 24\,°C$ and $\vartheta_{r2} = 4\,°C$, are kept constant.

9.4 Instationary One-Dimensional Heat Conduction

The temperature curve in the wall and the heat flows at the surfaces after 12 minutes, 1 hour and 5 hours are to be calculated.

The solution of the problem is illustrated in Figure 9.11 a. Location- and time-dependent values of temperature and heat flow are shown in the corresponding table.

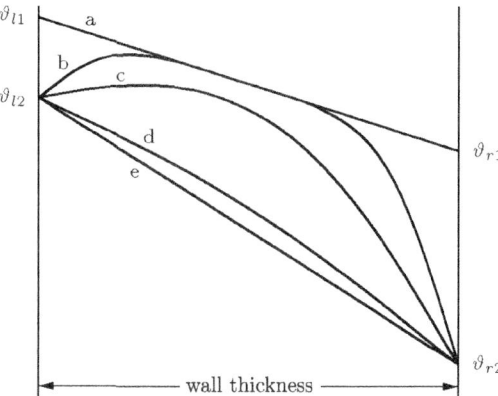

Figure 9.11 a Temperature profile in a plane wall according to example 9.4
a) initial state ($\tau < 0$)
b) $\tau = 12$ minutes
c) $\tau = 1$ hour
d) $\tau = 5$ hours
e) final state (stationary)

In the following table mean:

Column 1: x-coordinate in m
Column 2: temperature in °C at initial steady state
Column 3: heat flow in W/m² at initial steady state
Column 4: temperature in °C after 12 minutes
Column 5: heat flow in W/m² after 12 minutes
Column 6: temperature in °C after 1 hour
Column 7: heat flow in W/m² after 1 hour
Column 8: temperature in °C after 5 hours
Column 9: heat flow in W/m² after 5 hours
Column 10: temperature in °C at steady state
Column 11: heat flow in W/m² at steady state final condition

1	2	3	4	5	6	7	8	9	10	11
0.00	30.00	60.0	24.00	−158.5	24.00	−36.1	24.00	88.7	24.00	120.0
0.02	29.00		26.32		24.57		22.51		22.00	
0.04	28.00		27.23		24.92		20.98		20.00	
0.06	27.00		26.87		24.88		19.34		18.00	
0.08	26.00		25.99		24.30		17.58		16.00	
0.10	25.00		25.00		23.04		15.66		14.00	
0.12	24.00		23.96		20.98		13.58		12.00	
0.14	23.00		22.64		17.98		11.34		10.00	
0.16	22.00		19.95		14.02		8.98		8.00	
0.18	21.00		13.85		9.25		6.51		6.00	
0.20	20.00	60.0	4.00	642.7	4.00	320.0	4.00	151.3	4.00	120.0

9.4.2 Semi-Infinite Body

In a number of technical tasks — e.g. heat conduction processes in the ground in relation to an adjoining basement or in welding — the conceptual model of the semi-infinite body can be helpful. It shall be assumed that a semi-infinite (= infinitely extended on one side) body and the adjoining room have up to the time $\tau = 0$ the temperature ϑ_1.

The spatial coordinate x runs from left to right first through the room and then through the body whose surface A is cut vertically at $x = 0$. At the time $\tau = 0$, the room temperature and the surface temperature are abruptly reduced to the temperature ϑ_2 and kept at this temperature. A transient heat flow towards the surface is caused in the semi-infinite body.

After a long time, the temperature ϑ_2 is established everywhere in the semi-infinite body. If the semi-infinite body has the thermal diffusivity a according to Eq. (9.61), then the solution of the differential equation (9.62) for the temperature course is

$$\frac{\vartheta - \vartheta_2}{\vartheta_1 - \vartheta_2} = \frac{2}{\sqrt{\pi}} \int_0^{\frac{x}{2\sqrt{a\tau}}} e^{-u^2} \, du = \mathrm{erf}\left(\frac{x}{2\sqrt{a\tau}}\right). \tag{9.68}$$

[45], [46]. Here erf(z) is the *Gaussian* error function which can also be represented by an infinite series [22], [61]:

$$\mathrm{erf}(z) = \frac{2}{\sqrt{\pi}} \int_0^z e^{-u^2} \, du = \frac{2}{\sqrt{\pi}}\left(z - \frac{z^3}{1!\,3} + \frac{z^5}{2!\,5} - \frac{z^7}{3!\,7} + - \ldots\right) \tag{9.69}$$

The temperature gradient is

$$\frac{\partial \vartheta}{\partial x} = (\vartheta_1 - \vartheta_2) \frac{2}{\sqrt{\pi}} e^{-\frac{x^2}{4a\tau}} \frac{1}{2\sqrt{a\tau}}. \tag{9.70}$$

The heat flow at the point x is

$$d\dot{Q} = -\lambda \frac{\partial \vartheta}{\partial x} dA = -(\vartheta_1 - \vartheta_2) \sqrt{\frac{\lambda \varrho c}{\pi \tau}} \, e^{-\frac{x^2}{4a\tau}} \, dA. \tag{9.71}$$

The maximum heat flow occurs at the surface ($x = 0$):

$$d\dot{Q}_{x=0} = -(\vartheta_1 - \vartheta_2)\sqrt{\frac{\lambda \varrho c}{\pi \tau}}\, dA \tag{9.72}$$

With the relations

$$\frac{dQ}{d\tau} = \dot{Q} \qquad dQ = \dot{Q}\, d\tau \qquad d^2 Q = d\dot{Q}\, d\tau$$

follows from Eq. (9.72) [45]:

$$d^2 Q_{x=0} = -\frac{\vartheta_1 - \vartheta_2}{\sqrt{\pi}} \sqrt{\lambda \varrho c}\, dA \frac{d\tau}{\sqrt{\tau}} \qquad\qquad dQ_{x=0} = -(\vartheta_1 - \vartheta_2)\frac{2}{\sqrt{\pi}} \sqrt{\lambda \varrho c}\, \sqrt{\tau}\, dA$$

$$Q_{x=0} = -(\vartheta_1 - \vartheta_2)\frac{2}{\sqrt{\pi}}b\sqrt{\tau}\,A \qquad (9.73)$$

In Eq. 9.73, $b = \sqrt{\lambda \varrho c}$ is the heat penetration coefficient. $Q_{x=0}$ indicates the heat that is emitted through the surface of the semi-infinite body since the temperature drop. The negative numerical value for the heat indicates the direction of the heat flow opposite to the direction of the spatial coordinate.

If, with the x-axis in the same position, the room is on the right and the semi-infinite body is on the left with surface at $x = 0$, then the values of the x-coordinate within the body are negative. From Eq. (9.68) results

$$\frac{\vartheta - \vartheta_2}{\vartheta_1 - \vartheta_2} = \frac{2}{\sqrt{\pi}} \int_{\frac{x}{2\sqrt{a\tau}}}^{0} e^{-u^2}\,du = -\frac{2}{\sqrt{\pi}} \int_{0}^{\frac{x}{2\sqrt{a\tau}}} e^{-u^2}\,du = -\mathrm{erf}\!\left(\frac{x}{2\sqrt{a\tau}}\right). \qquad (9.74)$$

When x is negative, the error function also becomes negative and the right-hand side of Eq. (9.74) becomes positive again. In this case, Eq. (9.73) becomes

$$Q_{x=0} = (\vartheta_1 - \vartheta_2)\frac{2}{\sqrt{\pi}}b\sqrt{\tau}\,A \;. \qquad (9.75)$$

Eqs. (9.68) to (9.73) also apply to a sudden increase in temperature at the surface when the body is on the right side. Eqs. (9.74) and (9.75) for the body on the left side describe the cases of a sudden temperature decrease and a sudden temperature increase.

9.4.3 Contact Temperature

A semi-infinite body with temperature ϑ_1 on the right side is brought into contact with a semi-infinite body with lower temperature ϑ_1' on the left side. At the surfaces A and A' the common contact temperature $\vartheta_2 = \vartheta_2' = \vartheta_K$ is established. For the warmer body on the right side, Eq. (9.73) applies:

$$Q_r = -(\vartheta_1 - \vartheta_K)\frac{2}{\sqrt{\pi}}b\sqrt{\tau}\,A \qquad (9.76)$$

For the colder body on the left side, Eq. (9.75) holds:

$$Q_l = (\vartheta_1' - \vartheta_K)\frac{2}{\sqrt{\pi}}b'\sqrt{\tau}\,A' \qquad (9.77)$$

With $A = A'$ and common time τ, $Q_r = Q_l$ must be. This gives the contact temperature

$$\vartheta_K = \frac{\vartheta_1 b + \vartheta_1' b'}{b + b'}\;. \qquad (9.78)$$

In Eq. (9.78), time does not occur. Accordingly, the contact temperature ϑ_K remains unchanged. Eq. (9.78) is also valid for finite bodies as long as the temperature disturbance, emanating from the contact surfaces, has not reached the surfaces facing away from the contact surfaces [45].

Example 9.5 A thick copper plate (thermal conductivity $\lambda = 393$ W/(m K), density $\varrho = 8900$ kg/m^3, specific heat capacity $c = 390$ J/(kg K)) with the temperature $\vartheta_1 = 80\,°$C is brought into contact with a thick iron plate (thermal conductivity $\lambda' = 67$ W/(m K), density $\varrho' = 7860$ kg/m^3, specific heat capacity $c' = 465$ J/(kg K)) with the temperature $\vartheta'_1 = 20\,°$C (Fig. 9.11 b).

a) What are the heat penetration coefficients of copper b and iron b'?
b) What is the contact temperature ϑ_K?
c) What area-related heat q is transferred from the warmer to the colder plate in the first 3 seconds after contact?
d) What is the temperature profile in both plates after 3 seconds?

a) The heat penetration coefficients are:

$$\text{copper}: b = 36.93 \,\frac{\text{kJ}}{\text{m}^2\,\text{K}}\,\frac{1}{\sqrt{\text{s}}} \qquad \text{iron}: b' = 15.65 \,\frac{\text{kJ}}{\text{m}^2\,\text{K}}\,\frac{1}{\sqrt{\text{s}}}$$

b) The contact temperature is $\vartheta_K = 62.14\,°$C.

c) In 3 seconds after contact $q = 1289$ kJ/m^2 are transferred.

d)

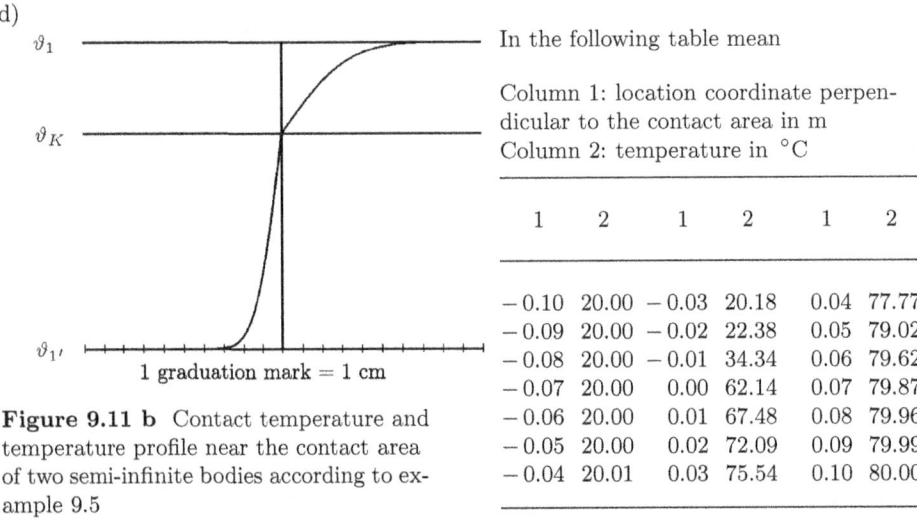

Figure 9.11 b Contact temperature and temperature profile near the contact area of two semi-infinite bodies according to example 9.5

In the following table mean

Column 1: location coordinate perpendicular to the contact area in m
Column 2: temperature in °C

1	2	1	2	1	2
−0.10	20.00	−0.03	20.18	0.04	77.77
−0.09	20.00	−0.02	22.38	0.05	79.02
−0.08	20.00	−0.01	34.34	0.06	79.62
−0.07	20.00	0.00	62.14	0.07	79.87
−0.06	20.00	0.01	67.48	0.08	79.96
−0.05	20.00	0.02	72.09	0.09	79.99
−0.04	20.01	0.03	75.54	0.10	80.00

9.5 Heat Transfer by Convection

In convection, a flow process causes a constantly changing contact between fluid particles of different temperatures, between which energy is transferred by heat conduction. A distinction is made between forced convection and free convection (also named natural convection) according to the cause of the flow. In forced convection, the flow is generated by an externally imposed pressure gradient. With free convection, density differences due to temperature differences cause the formation of a flow.

Near a wall, the mixing effect of the fluid flow is less than at a greater distance from the wall. A boundary layer of stationary and laminar flowing fluid particles forms directly at the wall, through which the heat can only pass by means of a heat conduction process.

9.5.1 Heat Transfer Coefficient

The convective heat transfer from a fluid to a wall (or vice versa) is proportional to the wall area A and the difference of the fluid temperature t and the surface temperature ϑ_0 of the wall. Using the heat transfer coefficient (also called film coefficient and film conductance) α as a proportionality factor, the following results

$$d\dot{Q} = \alpha(t - \vartheta_0)\,dA \ . \tag{9.79}$$

Eq. (9.79) can initially only be written in differential form because the temperatures t and ϑ_0 are locally different in many cases, e.g. in heat exchangers. The heat transfer coefficient α is also generally not the same everywhere. The fluid temperature t, which is decisive for the heat transfer at a certain point of the partition wall of a heat exchanger, results from an averaging over the flow cross-section. Just as a velocity profile is formed in the case of flow through a tube or pipe, a temperature profile is formed in the case of heat transfer. In a circular tube or pipe, velocity w and temperature t in a certain cross-section are a function of the radius r. Since the substance properties of a fluid generally depend on the temperature, the density ϱ and the specific heat capacity c_p are also functions of the radius r. The average fluid temperature in a certain cross-section for a tube or pipe with a circular cross-section is

$$t = \frac{\int_0^{r_i}(t\,\varrho\,c_p\,w)_r\,r\,dr}{\int_0^{r_i}(\varrho\,c_p\,w)_r\,r\,dr} \quad \text{or} \quad t = \frac{\int_0^{r_i}(t\,w)_r\,r\,dr}{\int_0^{r_i}w_r\,r\,dr} \ . \tag{9.80}$$

The second equation is valid if one can neglect the temperature dependence of the substance properties. The index r denotes the dependence on the radius. The physical interpretation of Eq. (9.80) states that t is the fluid temperature which results from perfect mixing in a cross section. This statement is important for measure technology [46].

Assuming that α and ϑ_0 are constant, Eq. (9.79) can be integrated:

$$\dot{Q} = \alpha\,A(t - \vartheta_0)_m \tag{9.81}$$

Here $(t - \vartheta_0)_m$ is the mean value of the temperature difference between the variable fluid temperature t and the constant wall temperature ϑ_0. The heat flow $d\dot{Q}$ transferred to the surface element dA according to Eq. (9.79) causes a temperature change dt of the fluid, which is inversely proportional to the heat capacity flow $W = \dot{V}\varrho\,c_p = \dot{m}\,c_p$ of the fluid flow — it is also denoted by $\dot{C}$. In the case of Eq. (9.79) it is assumed that the fluid is warm and the wall is cold and therefore dt has a negative numerical value:

$$d\dot{Q} = -W\,dt \tag{9.82}$$

In order for $d\dot{Q}$ to become positive, there must be a minus sign in Eq. (9.82) on the right side. From Eqs. (9.79) and (9.82) one obtains a differential equation for the temperature curve:

$$\alpha(t - \vartheta_0)dA = -W\,dt \qquad \frac{dt}{t - \vartheta_0} = -\frac{\alpha}{W}dA \tag{9.83}$$

With the boundary conditions $A = 0 \to t = t_1$ and $A = A \to t = t_2$ one obtains the solution

$$\ln \frac{t_2 - \vartheta_0}{t_1 - \vartheta_0} = -\frac{\alpha A}{W} \qquad t_2 = t_1 - (t_1 - \vartheta_0)(1 - e^{\frac{-\alpha A}{W}}) \; . \tag{9.84}$$

Integrating Eq. (9.82) gives the heat flow equation

$$\dot{Q} = (t_1 - t_2)W \; . \tag{9.85}$$

From Eqs. (9.84) and (9.85) follows

$$\dot{Q} = (t_1 - \vartheta_0)W(1 - e^{\frac{-\alpha A}{W}}) \; . \tag{9.86}$$

Eqs. (9.81) and (9.85), with Eq. (9.84), give an expression for the mean value of the temperature difference $(t - \vartheta_0)_m$

$$(t - \vartheta_0)_m = \frac{t_1 - t_2}{\dfrac{\alpha A}{W}} = \frac{(t_1 - \vartheta_0) - (t_2 - \vartheta_0)}{\ln \dfrac{t_1 - \vartheta_0}{t_2 - \vartheta_0}} \; . \tag{9.87}$$

Knowing the heat transfer area A, the surface temperature ϑ_0, the mass flow $\dot{m}$, the inlet temperature t_1 and the outlet temperature t_2, then $W = \dot{m}\,c_p$ can be used to calculate the heat flux $\dot{Q}$ and with Eqs. (9.87) and (9.81) to determine the heat transfer coefficient α.

9.5.2 Similarity Theory

Heat transfer coefficients can only be determined mathematically in a few cases. As a rule, they are obtained by evaluating measurements. If one wanted to take all dependencies into account, a representation of the heat transfer coefficient in the case of a tube or pipe flow would look as follows:

$$\alpha = \mathrm{f}\,(w, \lambda, \eta, \varrho, c_p, \lambda_{Wd}, \eta_{Wd}, c_{pWd}, d, l) \tag{9.88}$$

Meaning:

- α heat transfer coefficient in W/(m² K)
- w velocity in m/s
- λ thermal conductivity in W/(m K)
- η dynamic viscosity in kg/(m s)
- ϱ density in kg/m³
- c_p specific heat capacity at constant pressure in kJ/(kg K)
- λ_{Wd} thermal conductivity at wall temperature in W/(m K)
- η_{Wd} dynamic viscosity at wall temperature in kg/(m s)
- c_{pWd} specific heat capacity at constant pressure at the wall temperature in kJ/(kg K)
- d tube or pipe diameter in m
- l tube or pipe length in m

The similarity theory allows a considerably simplified representation of the measurement results according to Eq. (9.88). It provides dimensionless key figures. The individual substance properties do not determine the heat transfer coefficient independently of each other, but occur only in certain combinations, so that the number of independent variables is reduced.

As an example, such a key figure is to be derived. The heat flow, which is transferred to a wall element by convection according to Eq. (9.79), passes through the boundary layer by way of heat conduction analogous to Eq. (9.43). From this follows

$$\alpha(t - \vartheta_0) = -\lambda \frac{\mathrm{d}t_x}{\mathrm{d}x} . \qquad (9.89)$$

In this equation t is the fluid temperature averaged over the cross-section with the spatial coordinate x. t_x is the fluid temperature in the boundary layer dependent on the spatial coordinate x. The differential equation (9.89) can be made dimensionless with the aid of a characteristic dimension L, e.g. in the case of the tube or pipe flow with the help of the diameter [112]:

$$\frac{\alpha L}{\lambda} = -\frac{\mathrm{d}\frac{t_x}{t - \vartheta_0}}{\mathrm{d}\frac{x}{l}} \qquad Nu = \frac{\alpha L}{\lambda} \qquad (9.90)$$

On the left-hand side is a dimensionless ratio, the *Nußelt* number[7] Nu, which in the case of pipe flow is written as

$$Nu = \frac{\alpha d}{\lambda} \qquad (9.91)$$

and can be interpreted as a dimensionless heat transfer coefficient. In a similar way, further ratios can be derived which allow the dependence according to Eq. (9.88) as follows:

$$Nu = \mathrm{f}\,(Re, Pr, Pr_{Wd}, \frac{d}{l}) \qquad (9.92)$$

The number of independent variables has reduced from 10 to 4. The ratios Re and Pr are the *Reynolds* number[8] and the *Prandtl* number[9]

$$Re = \frac{w\,\varrho\,d}{\eta} = \frac{w\,d}{\nu} \qquad (9.93)$$

$$Pr = \frac{c_p\,\eta}{\lambda} = \frac{c_p\,\varrho\,\nu}{\lambda} = \frac{\nu}{a} . \qquad (9.94)$$

Pr_{Wd} is the *Prandtl* number calculated with the substance values valid at the wall temperature. ν in Eqs. (9.93) and (9.94) is the kinematic viscosity in m^2/s:

$$\nu = \frac{\eta}{\varrho} \qquad (9.95)$$

---

[7] *Wilhelm Nußelt*, German engineer, 1882 to 1957, professor at the Technical University of Munich, published the similarity theory of heat transfer in 1915.
[8] *Osborne Reynolds*, Northern Irish physicist and engineer, 1842 to 1912, worked in, among other fields, fluid mechanics and turbulence.
[9] *Ludwig Prandtl*, 1875 to 1953, German professor at the University of Göttingen, published numerous fundamental works on fluid mechanics and developed, among other things, the boundary layer theory.

a is the thermal diffusivity in m²/s according to Eq. (9.61). As an example of a relationship in the form of Eq. (9.92), an equation describing the heat transfer in forced turbulent flow of liquids in tubes or pipes is given by *Gnielinski* [132]:

$$Nu = \frac{\frac{\zeta}{8}(Re - 1000)Pr}{1 + 12.7\sqrt{\frac{\zeta}{8}}(Pr^{2/3} - 1)} \left[1 + \left(\frac{d}{l}\right)^{2/3}\right] \left(\frac{Pr}{Pr_{Wd}}\right)^{0.11} \qquad (9.96\,\text{a})$$

$$\zeta = (1.82 \log_{10} Re - 1.64)^{-2} \qquad (9.96\,\text{b})$$

With the expression $(Pr/Pr_{Wd})^{0.11}$ in Eq. (9.96 a), the different influence of heating or cooling on the temperature profile and thus on the heat transfer coefficient is given.

The substance properties are to be formed with T as the mean temperature of the fluid (averaged from the inlet and outlet temperatures) and with T_{Wd} as the mean temperature of the wall, respectively (T and T_{Wd} each in *Kelvin*). Eqs. (9.96 a) and (9.96 b) are also applicable to the heat transfer of gases. However, then in Eq. (9.96 a) the expression $(Pr/Pr_{Wd})^{0.11}$ has to be replaced by the expression $(T/T_{Wd})^n$. In the case of cooling ($T/T_{Wd} > 1$) n has to be calculated as $n = 0$, in the case of heating air at ($T/T_{Wd} = 0.5\ldots 1$) n can be set $n = 0.45$, for gaseous CO_2 at ($T/T_{Wd} = 0.5\ldots 1$) n has to be calculated as $n = 0.12$, for steam at pressures between 21 and 100 bar and at ($T/T_{Wd} = 0.67\ldots 1$), the value of n is $n = -0.18$ (cf. also Eqs. (9.132 e) and (9.132 f) [132]).

Other dimensionless numbers frequently used in forced flow are the *Stanton* number St and the *Peclet* number Pe[10]. In processes with free convection, the *Grashof* number Gr[11]. and the *Raleigh* number Ra[12] is used:

$$St = \frac{Nu}{Re\,Pr} = \frac{\alpha}{w\,\varrho\,c_p} \qquad (9.97)$$

$$Pe = Re\,Pr = \frac{w\,\varrho\,d\,c_p}{\lambda} = \frac{w\,d}{a} \qquad (9.98)$$

$$Gr = \frac{g\,\beta\,\Delta t\,l^3}{\nu^2} \qquad (9.99)$$

$$Ra = Gr\,Pr = \frac{g\,\beta\,\Delta t\,l^3}{\nu\,a} \qquad (9.100)$$

Here g is the acceleration due to gravity (gravity constant), β the thermal expansion coefficient, Δt the temperature difference causing the buoyancy in free convection (natural convection) and l the characteristic length of the body under consideration. The series of dimensionless numbers is not complete; further dimensionless numbers can be formed.

---

[10] *Jean Claude Eugene Peclet*, 1793 to 1857, was a French physicist, professor at the College de Marseille and then at the Ecole Centrale in Paris.

[11] *Franz Grashof*, German engineer, 1826 to 1893, professor at the Königliches Gewerbeinstitut in Berlin and at the Polytechnikum Karlsruhe, taught general mechanical engineering, materials science, hydraulics and thermodynamics.

[12] *Lord Raleigh (John William Strutt)*, English physicist, 1842 to 1919, worked in the fields of acoustics, fluid mechanics and noble gases, among others.

9.5 Heat Transfer by Convection

For the flow through a non-circular duct, the characteristic length L is the hydraulic diameter d_h:

$$d_h = \frac{4\,A_0}{U} \qquad (9{,}101\text{a})$$

The hydraulic diameter is also called equivalent diameter d_{gl} [17], [41], [39], [53], [56]. A_0 is the cross-section of the flowing substance, and U is the circumference of the duct. The velocity is calculated with the actual cross-section of the flowing substance. On that side of the partition wall where the heat transfer coefficient is relatively small and therefore the surface area is increased by fins, ribs or similar, the following applies:

$$d_{gl} = \frac{\left(\dfrac{A_{\min}}{A_S}\right)_L}{\left(\dfrac{A}{V}\right)_L} \qquad (9.101\text{b})$$

[68], [69]. $A_{\min L}$ is the narrowest flow cross-section, and A_{SL} is the face area, so that $(A_{\min}/A_S)_L$ is the cross-sectional constriction on the finned side. $(A/V)_L$ is the total surface area on the finned side related to the block volume. The velocity w is calculated with the narrowest flow cross-section.

The similarity theory not only reduces the number of independent variables and thus allows a compact description of the heat transfer; it also allows the transfer of experimental results to other substances and other experimental conditions, as long as the similarity is maintained.

A characteristic of similarity is the correspondence of the dimensionless numbers. Thus, Eqs. (9.96 a) and (9.96 b) are valid for the heat transfer in turbulent flow of liquids and gases (with corresponding modification of Eqs (9.96 a) and (9.96 b) without restriction to certain substances, provided that it is a question of tube or pipe flows.

The similarity theory only provides the key figures. Thus, the similarity theory deliveres the conditions under which a result that is already known can be transferred to other cases. It says nothing about the form of the equation in which the various dimensionless numbers are linked together. Eqs. (9.96 a) and (9.96 b) are empirical.

Similarity theory is not a substitute for an exact solution. Since the similarity theory does not contribute to the form of the equations for the description of the heat transfer, physically based approaches and analogies are of particular interest. These are often modified by empirical approaches based on measurement results.

9.5.3 *Reynolds* Analogy

A flow of substance $\dot{m}$ with specific heat capacity at constant pressure c_p, velocity w and temperature t is cooled at a tube or pipe wall with surface $A = \pi\,d\,l$ and temperature ϑ_0. Hereby, the heat flow

$$\dot{Q} = \alpha\,A(t - \vartheta_0) \qquad (9.102)$$

is given off. In a model, one can assume that a part of the substance flow $z\,\dot{m}$ comes to rest at the wall and reaches the wall temperature. The heat given off in this process is

$$\dot{Q} = z\,\dot{m}\,c_p(t - \vartheta_0)\ . \qquad (9.103)$$

The associated momentum loss is $z \dot{m} w$. The other part of the substance flow $(1-z)\dot{m}$ retains the temperature t and the velocity w.

After the heat release of the partial flow $z\dot{m}$, the partial flow $z\dot{m}$ mixes with the other partial flow again and reaches its velocity w. To the partial flow $z\dot{m}$ a change in momentum $\Delta p\, A_0$ happens, where $A_0 = \pi d^2/4$ is the flow cross-section:

$$\Delta p\, A_0 = z\, \dot{m}\, w \tag{9.104}$$

This gives the pressure drop

$$\Delta p = \frac{\varrho}{2} w^2 \zeta \frac{l}{d}, \tag{9.105}$$

where ζ is the individual resistance coefficient. Using the equation

$$\dot{m} = \varrho\, w\, A_0 \tag{9.106}$$

it follows from Eqs. (9.104), (9.105) and (9.106)

$$z = \frac{\zeta}{2} \frac{l}{d}. \tag{9.107}$$

According to Eqs. (9.102) and (9.103) is

$$\alpha = z \frac{\dot{m}}{A} c_p. \tag{9.108}$$

Using Eqs. (9.106) and (9.107) and the expressions for A and A_0 one obtains

$$\alpha = w\, \varrho\, c_p \frac{\zeta}{8}. \tag{9.109}$$

The result is as a dimensionless quantity, the *Stanton* number according to Eq. (9.97) and the relation

$$St = \frac{\alpha}{w\, \varrho\, c_p} = \frac{\zeta}{8}, \tag{9.110}$$

which describes a very simple relationship between heat transfer and pressure drop [46], [44].

9.5.4 *Prandtl* Analogy

The forced turbulent flow in a tube or pipe can, in terms of a simplified approach, be subdivided into a laminar boundary layer and into a turbulent core flow. In the boundary layer with the thickness s, the velocity distribution and the temperature distribution are assumed to be linear.

Since the boundary layer thickness s is small in comparison with the pipe radius r, the equilibrium between differential viscosity forces and pressure forces in the direction of flow can be formulated as follows:

$$2\pi r\, dx\, \eta\, \frac{dw}{dy} = -\pi r^2\, dp \quad \text{resp.} \quad 2\pi r\, \eta\, \frac{dw}{dy} = -\pi r^2\, \frac{dp}{dx} \tag{9.111}$$

η is the dynamic viscosity, x is the spatial coordinate in the direction of flow, y is the spatial coordinate perpendicular to the wall.

Due to the linear velocity profile in the boundary layer

$$\frac{dw}{dy} = \frac{w_s}{s} \qquad (9.112)$$

with w_s being the velocity at the transition from the boundary layer to the core flow. Eqs. (9.111) and (9.112) give an expression for the boundary layer thickness s:

$$s = -\frac{2\eta\, w_s}{r\dfrac{dp}{dx}} \qquad (9.113)$$

According to the model, heat transport through the boundary layer occurs only by heat conduction. At the tube or pipe surface $A = 2\pi r l$ the following heat flow is transferred:

$$\dot{Q} = \lambda \frac{t_s - \vartheta_0}{s}\, 2\pi r l \qquad (9.114)$$

λ is the thermal conductivity of the fluid, t_s is the temperature at the transition from the boundary layer to the core flow. ϑ_0 is the temperature of the wall surface. Eq. (9.113) is inserted into Eq. (9.114):

$$\dot{Q} = -\pi r^2 l \frac{\lambda}{\eta}\frac{t_s - \vartheta_0}{w_s}\frac{dp}{dx} \qquad (9.115)$$

Because of the linear distribution of the temperature and of the velocity in the boundary layer, with k as the proportionality factor it follows

$$\frac{t_s - \vartheta_0}{s} = k\frac{w_s}{s} \qquad\qquad t_s - \vartheta_0 = k\, w_s \;. \qquad (9.116)$$

If t and w are the values of temperature and velocity averaged over the cross-section of the tube or pipe — and not over the length — then, analogous to Eq. (9.116), the following simplified relationship can be assumed:

$$t - \vartheta_0 = k\, w \qquad (9.117)$$

From Eqs. (9.116) and (9.117) follows:

$$\vartheta_0 = t_s - k\, w_s = t - k\, w \qquad\qquad t - t_s = k(w - w_s)$$

$$k = \frac{t - t_s}{w - w_s} = \frac{t_s - \vartheta_0}{w_s} \qquad (9.118)$$

To maintain the temperature profile, *Prandtl* made a heat source approach: Between the differential of the heat flow and the differential of the fluid temperature there is the relation

$$d\dot{Q} = W\, dt = \dot{C}\, dt = \dot{V}\, \varrho\, c_p\, dt = \frac{dV}{d\tau}\, \varrho\, c_p\, dt = dV\, \varrho\, c_p\, \frac{dt}{d\tau}\;. \qquad (9.119)$$

$W = \dot{C}$ is the heat capacity flow of the fluid flow. Integration gives

$$\dot{Q} = V\varrho\, c_p\, \frac{dt}{d\tau} \qquad\qquad \varrho\, \frac{dt}{d\tau} = \frac{\dot{Q}}{V c_p} = \frac{\dot{Q}}{\pi r^2 l\, c_p}\;. \qquad (9.120)$$

If one compares the differential equation (9.120) with the equation of motion

$$\varrho \frac{dw}{d\tau} = -\frac{dp}{dx}, \qquad (9.121)$$

where pressure forces but no friction forces are taken into account, one can see the similarity of the two differential equations. From Eq. (9.117) follows $dt = k\,dw$ and with Eqs. (9.120), (9.121), (9.118) and (9.115)

$$\frac{dt}{dw} = k = -\frac{\dot{Q}}{\pi r^2 l\, c_p \dfrac{dp}{dx}} \qquad \dot{Q} = -\pi r^2 l\, c_p \frac{t-t_s}{w-w_s}\frac{dp}{dx} = -\pi r^2 l\, \frac{\lambda}{\eta}\frac{t_s - \vartheta_0}{w_s}\frac{dp}{dx}$$

$$\frac{c_p\,\eta}{\lambda}\frac{t-t_s}{w-w_s} = \frac{t_s - \vartheta_0}{w_s}. \qquad (9.122)$$

Eq. (9.122) contains as a dimensionless ratio the *Prandtl* number given by Eq. (9.94). Eq. (9.122) can be transformed as follows:

$$t_s - \vartheta_0 = (t-\vartheta_0)\frac{Pr}{Pr - 1 + \dfrac{w}{w_s}} \qquad (9.123)$$

Eq. (9.123) and the following equation obtained from Eq. (9.105),

$$-\frac{dp}{dx} = \frac{\varrho}{2}w^2\frac{\zeta}{d} \qquad (9.124)$$

are inserted into Eq. (9.115)

$$\dot{Q} = (t-\vartheta_0)\pi r^2 l\,\frac{\lambda}{\eta}\frac{1}{w_s}\frac{Pr}{Pr - 1 + \dfrac{w}{w_s}}\frac{\varrho}{2}w^2\frac{\zeta}{d}$$

$$\dot{Q} = (t-\vartheta_0)A\frac{\lambda}{d}\frac{w\,\varrho\,d}{\eta}Pr\,\frac{\dfrac{\zeta}{8}}{1 + (Pr-1)\dfrac{w_s}{w}}, \qquad (9.125)$$

where A is the inner partition wall area of the tube or pipe. Eq. (9.125) contains as a dimensionless number the *Reynolds* number given in Eq. (9.93). According to the definition equation of the heat transfer coefficient, with Eqs. (9.125) and (9.93) follows:

$$\alpha = \frac{\dot{Q}}{A(t-\vartheta_0)} = \frac{\lambda}{d}Re\,Pr\,\frac{\dfrac{\zeta}{8}}{1+(Pr-1)\dfrac{w_s}{w}} \qquad (9.126)$$

Eq. (9.126) contains as another dimensionless number the *Nußelt* number given by Eq. (9.91):

$$Nu = Re\,Pr\,\frac{\dfrac{\zeta}{8}}{1+(Pr-1)\dfrac{w_s}{w}} \qquad (9.127)$$

In Eq. (9.127) one could introduce the *Peclet* number according to Eq. (9.98). When $Pr = 1$, Eq. (9.127) merges with Eq. (9.98) into Eq. (9.110). In Eq. (9.127) the expressions for $\zeta/8$ and w_s/w still have to be determined. *Prandtl* used for the individual resistance coefficient the equation of *Blasius*

$$\zeta = 0.3164\, Re^{-1/4} \qquad (9.128)$$

and for the velocity ratio at the transition from the boundary layer to the core flow

$$\frac{w_s}{w} = 1.74\, Re^{-1/8}\ . \qquad (9.129)$$

In [96] one finds the expression $1.60\, Re^{-1/8}$ for the right-hand side of Eq. (9.129), but in the original work the *Reynolds* number is formed with the tube or pipe radius r. The transformation, when forming the *Reynolds* number with diameter d, gives: $1.60/2^{-1/8} = 1.74$. Eqs. (9.128) and (9.129) are inserted into Eq. (9.127) [95], [96], [46]:

$$Nu = \frac{0.03955\, Re^{3/4}\, Pr}{1 + (Pr - 1)\, 1.74\, Re^{-1/8}} \qquad (9.130\,\text{a})$$

Hofmann [57] made a revision of Eq. (9.130 a) and gave the following equation:

$$Nu = \frac{0.03955\, Re^{3/4}\, Pr}{1 + (Pr - 1)\, 1.5\, Re^{-1/8}\, Pr^{-1/6}} \qquad (9.130\,\text{b})$$

This involves the following empirical equation for the reference temperature of the substance values:

$$t^* = t - \frac{0.1\, Pr + 40}{Pr + 72}(t - \vartheta_0) \qquad (9.130\,\text{c})$$

9.5.5 Power Number Approaches for Laminar and Turbulent Flow

In addition to the approaches with a physical background, there are also approaches that use the dimensionless numbers of the similarity theory, but otherwise use approaches with power number expressions, because they seem advantageous for practical applications. Thus, there are equations of the following forms:

$$Nu = C\, Pe^n \qquad (9.131\,\text{a})$$

$$Nu = C\, Re^m\, Pr^n \qquad (9.131\,\text{b})$$

$$Nu = \mathrm{f}\,(Re, Pr, \frac{d}{l}) \qquad (9.131\,\text{c})$$

$$St\, Pr^{2/3} = \mathrm{f}\,(Re) \qquad (9.131\,\text{d})$$

Eq. (9.131 d) follows from Eq. (9.131 b) if one sets n = 1/3:

$$Nu = C\, Re^m\, Pr^{1/3} = St\, Re\, Pr \quad (9.131\,\text{e}) \qquad St\, Pr^{2/3} = C\, Re^{m-1} \quad (9.131\,\text{f})$$

In the following, equations for the calculation of the mean *Nußelt* number Nu or the mean heat transfer coefficient α are given for technically important cases, which are power number approaches based on the similarity theory. To use these, one proceeds in the following manner:

1. Determination of the respective model case to be considered with regard to the heat transfer
2. Selection of the appropriate equation with the mean *Nußelt* number Nu on the left side
3. Determination of the characteristic length l and the reference temperature t or the wall temperature t_{Wd} using the additional information for the respective equation
4. Calculation of the dimensionless numbers required for the respective equation
5. Verification of the range of validity of the equation
6. Calculation of the mean *Nußelt* number Nu
7. Calculation of the mean heat transfer coefficient α from the mean *Nußelt* number Nu

Forced Turbulent Flow Inside a Pipe: *Kraussold* has given separate equations for the heating and cooling of fluids, which have been combined into one equation by *Hausen* [50], [131]:

$$Nu = 0{,}024 \left[1 + \left(\frac{d}{l}\right)^{2/3}\right] Re^{0.8} Pr^{0.33} \left(\frac{\eta}{\eta_{Wd}}\right)^{0.14} \qquad (9.132\,\text{a})$$

It is valid in the range $Re = 7000$ to $1\,000\,000$, $Pr = 1$ to 500, $l/d = 1$ to ∞. The reference temperature for the substance values is the mean liquid temperature t; the dynamic viscosity η_{Wd} shall be related to the wall temperature $\vartheta_0 = t_{Wd}$.

Between 1943 and 1974, *Hausen* gave three equations for the heat transfer in a pipe with turbulent flow of gases and liquids, which are to be listed as examples in the sense of a step-by-step development. The first [49], [50] is

$$Nu = 0.116\,(Re^{2/3} - 125)Pr^{1/3}\left[1 + \left(\frac{d}{l}\right)^{2/3}\right]\left(\frac{\eta}{\eta_{Wd}}\right)^{0.14}. \qquad (9.132\,\text{b})$$

It is valid in the range $Re = 2\,320$ to $1\,000\,000$, $Pr = 0.6$ to 500, $l/d = 1$ to ∞. The equation from 1959 [51], [131] is

$$Nu = 0.037\left[1 + \left(\frac{d}{l}\right)^{2/3}\right](Re^{0.75} - 180)Pr^{0.42}\left(\frac{\eta}{\eta_{Wd}}\right)^{0.14}. \qquad (9.132\,\text{c})$$

In 1974 [52], [53] the following equation was published:

$$Nu = 0.0235\,(Re^{0.8} - 230)(1.8\,Pr^{0.3} - 0.8)\left[1 + \left(\frac{d}{l}\right)^{2/3}\right]\left(\frac{\eta}{\eta_{Wd}}\right)^{0.14} \qquad (9.132\,\text{d})$$

In [132], two equations are given as more developed forms in flows in the transition and turbulence region ($Re = 2\,320$ to $1\,000\,000$):

For $Pr = 0.5$ to 1.5 the following applies

$$Nu = 0.0214\,(Re^{0.8} - 100)\,Pr^{0.4}\left[1 + \left(\frac{d}{l}\right)^{2/3}\right] K\;, \qquad (9.132\,\text{e})$$

for $Pr = 1.5$ to 500 applies

$$Nu = 0.012\,(Re^{0.87} - 280)\,Pr^{0.4}\left[1 + \left(\frac{d}{l}\right)^{2/3}\right] K \; , \qquad (9.132\,\text{f})$$

each with

$$K = \left(\frac{Pr}{Pr_{Wd}}\right)^{0.11} \text{ for liquids and } K = \left(\frac{T}{T_{Wd}}\right)^{n} \text{ for gases and vapors.}$$

The substance values are to be formed with the mean temperatures T of the fluid and T_{Wd} of the wall, respectively. Here, as in equations (9.96 a) and (9. 96 b) T represents the mean temperature of the fluid (averaged from the inlet and outlet temperatures) and T_{Wd} the mean temperature of the wall in *Kelvin*; in the case of cooling ($T/T_{Wd} > 1$) $n = 0$ is valid, in the case of heating air at $T/T_{Wd} = 0.5$ to 1 $n = 0.45$ can be set, for gaseous CO_2 at $T/T_{Wd} = 0.5$ to 1 $n = 0.12$ is valid, for water vapor at pressures between 21 and 100 bar and at $T/T_{Wd} = 0.67$ to 1 $n = -0.18$ can be set. d is the internal diameter of pipes with a circular cross-section; it must be replaced by the hydraulic diameter $d_h = 4\,A_0/\mathrm{U}$ (Eq. (9.100)) for other cross-sectional shapes [132].

Forced Laminar Flow Inside a Pipe: For the heat transfer in forced laminar flow ($Re < 2320$) in a pipe with circular cross-section and for arbitrary values of Pr, an equation of *Gnielinski* is given in [132]:

$$Nu = \left[49.371 + \left(1.615\left(Re\,Pr\,\frac{d}{l}\right)^{1/3} - 0.7\right)^{3}\right]^{1/3} K \qquad (9.133)$$

The respective designations apply in the same way as for Eqs. (9.132 e and f). K is to be calculated as $K = \left(\dfrac{Pr}{Pr_{Wd}}\right)^{0.11}$ for liquids; for gases and vapors $K \approx 1$ applies.

Longitudinally Flowed Plane Plate and Transversely Flowed Cylinder in Forced Laminar Flow: For this, *Pohlhausen* and *Krouzhiline* gave the following equation for the heat transfer [132]:

$$Nu = 0.664\,Re^{1/2}\,Pr^{1/3}\,K \qquad (9.134)$$

Here the *Reynolds* number Re is formed with the plate length l or with the overflow length $l = \pi r$ at the transversely flowed cylinder: $Re = w\,l/\nu$. The equation is valid in the case of the plane plate for $Re < 100\,000$ and for Pr from 0.6 to 2 000; the critical Reynolds number can be assumed to be about $Re = 500\,000$. Eq. (9.134) applies in the case of the transversely flowed cylinder for $Re < 10$ and for Pr from 0.6 to 1 000. The substance values are to be formed with the fluid temperatures T and T_{Wd}, respectively. Here T represents the mean temperature of the fluid (averaged from the values before inflow and after outflow). T_{Wd} represents the mean temperature of the wall. Both temperatures are to be applied in *Kelvin*; w is the inflow velocity. In Eq. (9.134), K is given by

$$K = \left(\frac{Pr}{Pr_{Wd}}\right)^{0.25} \text{ for liquids and } K = \left(\frac{T}{T_{Wd}}\right)^{0.12} \text{ for gases and vapors.}$$

Longitudinally Flowed Plane Plate and Transversely Flowed Cylinder in Forced Turbulent Flow: This case is represented by an equation due to *Petukhov* and *Popov* [132]:

$$Nu = \frac{\zeta/8 \, Re \, Pr}{1 + 12.7 \, (\zeta/8)^{1/2} \, (Pr^{2/3} - 1)} \, K \qquad (9.135\,\text{a})$$

Inserting into this equation the relation for the mean individual resistance coefficient of a plate at turbulent boundary layer $\zeta/8 = 0.037 \, Re^{-0.2}$ given by *Schlichting*, the following equation [132] is obtained:

$$Nu = \frac{0.037 \, Re^{0.8} \, Pr}{1 + 2.443 \, Re^{-0.1} \, (Pr^{2/3} - 1)} \, K \qquad (9.135\,\text{b})$$

Here the *Reynolds* number Re is formed with the plate length l or with the overflow length $l = \pi r$ at the transversely flowed cylinder: $Re = w \, l/\nu$. The equation is valid in the case of the flat plate for Re between 500 000 and 10 000 000 and for Pr from 0.6 to 2000. Eq. (9.135 b) is valid in the case of the cross-flowed cylinder for Re between 10 and 10 000 000 and for Pr from 0.6 to 1000. The substance values are to be formed with the temperatures T and T_{Wd} of the fluid. Here T represents the mean temperature of the fluid (averaged from the values before inflow and after outflow) and T_{Wd} represents the mean temperature of the wall, both in *Kelvin*; w is the inflow velocity. The relations for K correspond to those for Eq. (9.134).

To *Krischer* and *Kast* as well as *Gnielinski* goes back a summarising representation of the heat transfer in the cases of laminar and turbulent flow, in which the equations (9.134) with $Nu = Nu_{lam}$ and (9.135 b) with $Nu = Nu_{turb}$ are connected in the following way [132]: $Nu_{l,0} = (Nu_{lam}^2 + Nu_{turb}^2)^{1/2}$

Shell-and-Tube Heat Exchanger with Baffles in Forced Turbulent Flow: In [24], attention is drawn to an equation by *Donohue* which can be used to roughly calculate the heat transfer in a shell-and-tube heat exchanger with baffles where forced flow prevails across the tube bundles:

$$Nu = C \, Re^{0.6} \, Pr^{0.33} \left(\frac{\eta}{\eta_{Wd}}\right)^{0.14} \qquad (9.136)$$

Here, the outer diameter of the pipe d is to be used as the characteristic length. The equation is valid for Re between 4 and 50 000 and for Pr from 0.5 to 500. In Eq. (9.136) $C = 0.22$ for undrilled casing pipe and $C = 0.25$ for drilled casing pipe. The substance values are to be formed with the fluid temperatures t and t_{Wd}, respectively. Here t represents the mean temperature of the fluid (averaged from the values at the inlet cross-section and at the outlet cross-section) and t_{Wd} represents the mean temperature of the wall, both in *Celsius* degrees. The required calculated velocity w is to be formed with the expression $w = (w_q \, w_l)^{1/2}$, where w_q is the velocity transverse to the pipes in the narrowest cross-section and w_l is the velocity in longitudinal direction at the deflection.

Free Flow at a Vertical Wall and Around a Sphere: In [132] and [24] an equation of *Churchill* and *Chu* is given, with the help of which the heat transfer can be calculated at a vertical wall and at a sphere, which are flowed against and around freely, respectively (free convection); it is valid for the laminar as well as for the turbulent range:

$$Nu = \left[0.825 + 0.387 \left(Ra \, f_1\right)^{1/6}\right]^2 \qquad (9.137\,\text{a})$$

with

$$f_1 = \frac{1}{\left[1 + \left(\frac{0.492}{Pr}\right)^{9/16}\right]^{16/9}} \qquad (9.137\,\text{b})$$

In the case of the vertical wall, the range of validity is characterised by Ra between 0.1 and 10^{12} and in the case of the sphere by Ra between 1000 and 10^{12} and Nu greater than 2; for Pr - from 0.001 upwards to arbitrarily large values - no restrictions apply. The characteristic length l is equal to the wall height h in the case of the vertical wall and equal to the diameter d in the case of the sphere. The substance values are to be calculated with the help of the temperature $(t + t_{Wd})/2$ - obtained by averaging the fluid temperature t and the wall temperature t_{Wd}; the volume expansion coefficient β is determined with the fluid temperature t.

Free Flow Around a Vertical Cylinder: In [132], an equation of $Fujii$ and $Uehara$ is given which can be used to calculate the heat transfer at a vertical cylinder which is freely flowed around (free convection):

$$Nu_{Zyl} = Nu + 0.435 \frac{h}{d} \qquad (9.138)$$

In this h is the height and d the diameter of the cylinder; Nu is to be calculated from equations (9.137 a) and (9.137 b). With regard to the conditions, the same applies as for Eqs. (9.137 a) and (9.137 b).

Free Laminar Flow at a Horizontal Wall With Heat Release at the Top Side or Heat Supply at the Bottom Side: In [132], an equation by $Churchill$ is given which can be used to represent the heat transfer at a horizontal plane wall which has free laminar flow and where the heat is given off at the top side or the heat is supplied at the bottom side (free convection):

$$Nu = 0.766 \left(Ra\ f_2\right)^{1/5} \qquad (9.139\,\text{a})$$

with

$$f_2 = \frac{1}{\left[1 + \left(\frac{0.322}{Pr}\right)^{11/20}\right]^{20/11}} \qquad (9.139\,\text{b})$$

Here the range of validity is characterised by $Ra\ f_2 < 70\,000$ and arbitrarily large values of Pr. The characteristic length is $l = a\,b/(2\,(a+b))$ for rectangular surfaces with side lengths a and b and $l = d/4$ for circular disks.

Free Turbulent Flow at a Horizontal Wall with Heat Release at the Top Side or Heat Supply at the Bottom Side: In the case of a free and turbulent flow at a horizontal plane wall with heat given off at the top side or heat supplied at the bottom side, according to $Churchill$ holds[132]

$$Nu = 0.15 \left(Ra\ f_2\right)^{1/3}. \qquad (9.139\,\text{c})$$

f_2 is determined in the same way as in the case of laminar flow (Eq. (9.139 b)); this also applies to the characteristic length l. The range of validity is given with $Ra\ f_2$ equal to or greater than 70 000 and with arbitrarily large values of Pr - from 0.001 upwards. With regard to the conditions, the same applies as for Eqs. (9.139 a) and (9.139 b).

Free Laminar Flow at a Horizontal Wall With Heat Release at the Bottom Side or with Heat Supply at the Top Side: In [132] an equation by *Churchill* is given which can be used to calculate the heat transfer on a horizontal plane wall which has free laminar flow and where the heat is given off at the bottom side or supplied at the top side (free convection):

$$Nu = 0.6 \left(Ra\ f_1\right)^{1/5} \tag{9.140}$$

f_1 is obtained according to Eq. (9.137 b). The range of validity is described by $Ra\ f_1$ between 1000 and 10^{10} and by arbitrarily large values of Pr - from 0.001 upwards; otherwise the same applies as for Eqs. (9.139 a) and (9.139 b).

Horizontal Cylinder Being Transversely Flowed Freely: In [132] an equation by *Churchill* and *Chu* is given, with the help of which the heat transfer at a horizontal cylinder being transversely flowed by an external flow can be determined (free convection); it is valid for the laminar as well as for the turbulent range:

$$Nu = \left[0.752 + 0.387 \left(Ra\ f_3\right)^{1/6}\right]^2 \tag{9.141 a}$$

with

$$f_3 = \frac{1}{\left[1 + \left(\frac{0.559}{Pr}\right)^{9/16}\right]^{16/9}} \tag{9.141 b}$$

Here the range of validity is characterised by arbitrary values for Ra and Pr. The characteristic length l results as the overflowed length $l = \pi r$. With regard to the conditions, the remarks are valid which are made for Eqs. (9.137 a) and (9.137 b): The substance values are to be calculated with the help of the temperature $(t + t_{Wd})/2$ obtained by averaging from the fluid temperature t and the wall temperature t_{Wd}; the thermal expansion coefficient β is determined with the fluid temperature t.

Free Flow Around a Sphere: In [24] an equation by *Raithby* and *Hollands* is given, with the help of which the heat transfer at a sphere which is externally flowed can be calculated (free convection); it is valid for the laminar as well as for the turbulent range:

$$Nu = 0.56 \left[\frac{Pr\ Ra}{0.846 + Pr}\right]^{1/4} + 2 \tag{9.142}$$

It is valid for arbitrary values of Ra and Pr. The characteristic length is the sphere diameter d. With regard to the conditions, the remarks are valid which are made for Eqs. (9.137 a) and (9.137 b): The substance values are to be calculated with the help of the temperature $(t + t_{Wd})/2$ obtained by averaging from the fluid temperature t and the wall temperature t_{Wd}; the thermal expansion coefficient β is calculated with the fluid temperature t.

Free Laminar Flow on an Inclined Wall: For the heat transfer at an inclined wall where the heat is given off at the top side or the heat is supplied at the bottom side, the same equations (9.137 a) and (9.137 b) apply in the case of free laminar flow when calculating Nu according to *Vliet*, *Fujii* and *Imura* [132] as in the case of a vertical wall, but to be calculated with the modified *Raleigh* number Ra_φ which is derived from the *Raleigh* number Ra by $Ra_\varphi = Ra \cos \varphi$.

Here the range of validity for an angle of inclination φ is in the range smaller than 60° to the perpendicular. The characteristic length is the wall length in the direction of inclination l.

Free Turbulent Flow on an Inclined Wall: The transition region from laminar to turbulent free flow is described by the critical *Raleigh* number Ra_c, for which, depending on the angle of inclination φ, the following applies : for $\varphi = 0°$: $Ra_c = 8 \cdot 10^8$; for $\varphi = 15°$: $Ra_c = 4 \cdot 10^8$; for $\varphi = 30°$: $Ra_c = 10^8$; for $\varphi = 45°$: $Ra_c = 10^7$; for $\varphi = 60°$: $Ra_c = 8 \cdot 10^5$. In the case of free turbulent flow, where the heat is given off at the top side or the heat is supplied at the bottom side, the following equation is then to be calculated [132]:

$$Nu = 0.56 \, (Ra_c \, \cos \varphi)^{1/4} + 0.13 \, (Ra^{1/3} - Ra_c^{1/3}) \tag{9.143}$$

With regard to the conditions, the remarks are valid which are made for Eqs. (9.137 a) and (9.137 b).

Further aspects of heat transfer in free flow (free convection) are not discussed here due to lack of space. For this, reference is made to e.g. [38], [46].

9.5.6 Approaches for Phase Transitions

Laminar Film Condensation: In heat transfer with phase change (phase transition), in the case of vaporisation, the enthalpy of vaporisation is transferred across the surface of a solid to a boiling liquid. Conversely, in the case of condensation, a gas in the saturated vapor state transfers the enthalpy of condensation to the surface of a solid.

(The amount of the specific enthalpy of condensation of a pure substance r corresponds to the specific enthalpy of vaporisation r.) The heat transfer is usually much better than in processes without phase transition.

In condensation, the vapor liquefies at wall temperatures below the saturation temperature of a vapor in contact with the wall; this also applies if the mean vapor temperature is still higher than the saturation temperature.

In the context of technical processes, film condensation is more common than droplet condensation; both laminar and turbulent condensation can occur. Drop condensation occurs when there is a poorly wettable or non-wettable surface. In this case, the heat transfer coefficients are higher than for film condensation.

Nußelt developed far-reaching considerations for film condensation with the water skin theory. For the case of film condensation of stationary or slowly laminar flowing saturated steam (laminar condensate skin) on a vertical wall or on a vertical pipe, where no or only low shear stresses occur in the steam, the equation given by him is [132]:

$$Nu = 0.943 \left[\frac{1 - \rho_D/\rho}{\frac{Ph \, Ga^{1/3}}{Pr}} \right]^{1/4} \tag{9.144 a}$$

In this Ph is the phase transformation number and Ga is the *Galilei* number, which according to *Nußelt* can be represented in the following way:

$$Ph = \frac{c_p \, (t_S - t_{Wd})}{r} \tag{9.144 b} \qquad Ga^{1/3} = \frac{l}{\left(\frac{\nu^2}{g}\right)^{1/3}} \tag{9.144 c}$$

The characteristic length for the *Nußelt* number is the expression $(\nu^2/g)^{1/3}$, so that for the *Nußelt* number one gets:

$$Nu = \alpha \, (\nu^2/g)^{1/3}/\lambda \qquad (9.144\,\text{d})$$

The transformation to α gives:

$$\alpha = 0.943 \left[\frac{\lambda^3 \, (\rho - \rho_D) \, g \, r}{\nu \, l \, (t_S - t_{Wd})} \right]^{1/4} \qquad (9.144\,\text{e})$$

For Eqs. (9.144 a) to (9.144 e), l represents the wall height or the pipe length, t_S the boiling temperature in *Celsius* degrees and r the specific enthalpy of vaporisation or condensation at temperature t_S. The other substance values apply to the liquid phase at the mean temperature t, which is formed from the boiling temperature t_S and the wall temperature t_{Wd}: $t = (t_S + t_{Wd})/2$

g is the acceleration due to gravity (gravity constant), λ the thermal conductivity of the liquid, ν the kinematic viscosity of the liquid, ρ the density of the liquid and ρ_D the density of the saturated vapor. For the *Reynolds* number of the film Re, which is below $Re = 400$ for laminar flow and above $Re = 400$ for tubulent flow, the following applies [24]:

$$Re = 0.943 \left[\frac{\lambda \, g^{1/3} \, l \, (t_S - t_{Wd})}{\nu^{5/3} \, \rho \, r} \right]^{3/4} \qquad (9.144\,\text{f})$$

Turbulent Film Condensation: If saturated vapor condenses in the form of a turbulent condensate film on the outside of a horizontal tube, the heat transfer can be calculated using an equation given by *Grigull*:

$$\alpha = 0.003 \left[\frac{\lambda^3 \, g \, l \, (t_S - t_{Wd})}{\rho \, \nu^3 \, r} \right]^{1/2} \qquad (9.145)$$

The remarks given for laminar film condensation apply in the same way to turbulent film condensation.

Vaporisation: In the vaporisation or boiling process of a pure substance - e.g. water - saturated vapor is formed from boiling liquid. In the following, the vaporisation process inside a vessel is illustrated; however, it can also take place, for example, in pipes with a stationary flow. As only the vaporisation process in a vessel is considered, here the flow velocity has no influence on the heat transfer from a solid surface to the liquid or to the resulting bubbles.

On a technical scale, vaporisation in a vessel is realised in vaporisation boilers with flat heating walls (e.g. the vessel bottom) or with tube or pipe bundles as heating surfaces. If the surface temperature t_{Wd} of a heating element, which is in contact with a liquid with boiling temperature t_S is increased above the value of t_S by supplying the heat flow $\dot{Q}$, the driving temperature difference $t_{Wd} - t_S$ determines the heat flux density (heating surface load) $\dot{q} = \dot{Q}/A$ (with A as the heating surface).

Fig. 9.12 shows the dependence of the heat flux density $\dot{q}$ and the heat transfer coefficient α on the temperature difference $t_{Wd} - t_S$ in the case of boiling water at ambient pressure $p_{amb} = 0.981$ bar. This results in three different ranges: the range of vaporisation due to free convection, the range of bubble vaporisation and the range of film vaporisation.

For small temperature differences of $t_{Wd} - t_S$ up to about 6 K, the heat flux density $\dot{q}$ is less than about 15 kW/m^2. Here, the laws of free convection apply approximately,

although it must be noted that the convective movement in the boundary layer is increased by the formation of very small vapor bubbles on the heated wall and therefore the heat transfer coefficient α is greater than with free convection alone. This can be described according to Eq. (9.102):

$$\dot{q} = \alpha \, (t_{Wd} - t_S)$$

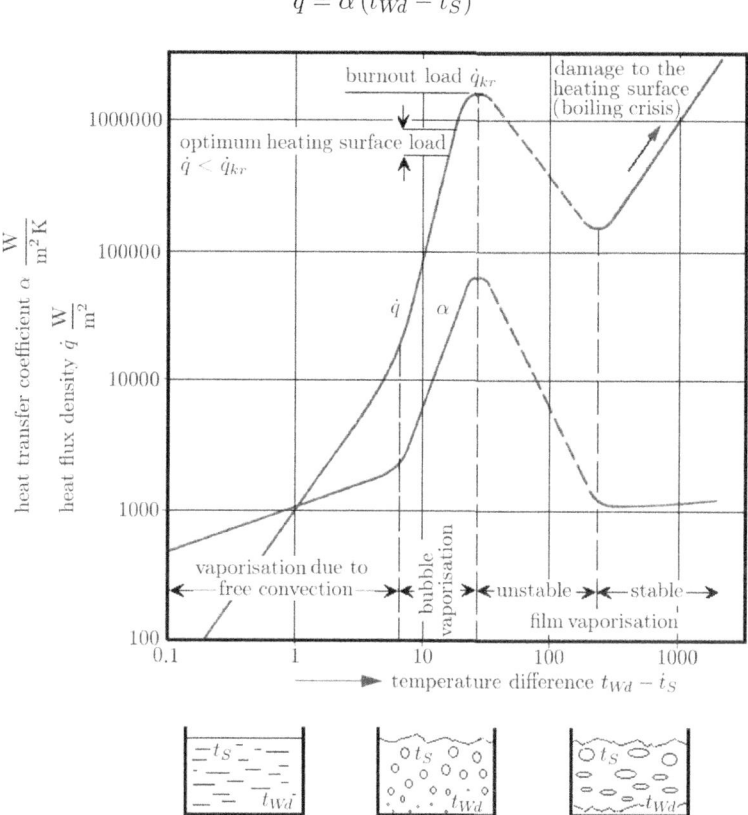

Figure 9.12 Heat flux density and heat transfer coefficient for vaporisation of water at 981 mbar

In [24], two equations according to *Jakob* and *Linke* are shown as examples for water, which are taken from [132] and which apply in the range of the heat flux density $\dot{q}$ below 17 kW/m² and for values of the pressure p between 0.5 and 20 bar for the heat transfer coefficient α in the dimension kW/(m² K):

$$\alpha = 1.026 \, \dot{q}^{0.26} \, p^{0.25} \quad (9.146 \text{ a}) \qquad \alpha = 1.034 \, (t_{Wd} - t_S)^{0.351} \, p^{0.338} \quad (9.146 \text{ b})$$

When the temperature difference is increased, larger vapor bubbles form on the heated surface, which rise rapidly upwards due to buoyancy, thus greatly increasing the heat transfer. Because liquid continuously vaporises at the bubble surfaces, the bubbles increase in size as they rise. The vaporisation process now takes place mainly on the bubble surfaces - in contrast to vaporisation during free convection.

If the temperature difference and thus the heat flux density continues to rise, the bubb-

les grow even more. This process continues until the bubbles forming on the heating surfaces touch each other and thus a maximum is reached in the heat flux density $\dot{q}$ and in the heat transfer coefficient α.

According to [132], there is still no summarising theory to be able to predict the heat transfer coefficient α in bubble vaporisation with the accuracy necessary for technical tasks. Nevertheless, approaches for the determination of α for different tasks are given in [132]; from these, two equations according to *Fritz* are selected in [24] as an example for water, which are valid for the heat transfer coefficient α in the dimension kW/(m² K) in the range of the heat flow density $\dot{q}$ from greater than 17 kW/m² to $\dot{q}_{kr}$ and in the pressure range p between 0.5 and 20 bar:

$$\alpha = 0.274\,\dot{q}^{0.75}\,p^{0.25} \qquad (9.146\,\text{c}) \qquad\qquad \alpha = 5.65\cdot 10^{-3}\,(t_{Wd} - t_S)^3\,p \qquad (9.146\,\text{d})$$

If the surface temperature has risen further due to the heating, the burnout load $\dot{q}_{kr}$ has been reached. Now the bubbles on the heated surface have coalesced into individual larger film-like areas. These are still separated from each other by the liquid; this causes an unstable vaporisation process.

As the temperature difference $t_{Wd} - t_S$ continues to increase, a coherent film of vapor forms between the heated surface and the liquid. Thus, the heat transfer coefficient α decreases very strongly because the heat transport, which now has to take place predominantly with the help of radiation and convection in the vapor space, is considerably hindered. The consequence is a rapidly increasing temperature difference $t_{Wd} - t_S$. This can lead to irreversible damage to the heating surface. Therefore, it is necessary to avoid such a 'boiling crisis' and, according to Fig. 9.12, not to exceed the optimum heating surface load $\dot{q} < \dot{q}_{kr}$, which still remains within the bubble vaporisation range.

Other aspects of heat transfer in phase transitions are not discussed here due to lack of space. For this, reference is made to [38], [46], for example.

9.6 Over-All Heat Transfer

In a partition heat exchanger (also called recuperator), heat is transferred from the warmer fluid with the mean local temperature t to the colder fluid with the mean local temperature t' through the partition wall (Fig. 9.9 for the case of a plane wall).

The heat transfer thus consists of the heat transfer on the warm side, the heat conduction through the partition and the heat transfer on the cold side. For a flat partition wall element, the three processes mentioned are described by the following equations:

$$\mathrm{d}\dot{Q} = \alpha\,(t - \vartheta_0)\,\mathrm{d}A \qquad (9.147\,\text{a})$$

$$\mathrm{d}\dot{Q} = \frac{\lambda}{\delta}\,(\vartheta_0 - \vartheta_0')\,\mathrm{d}A \qquad (9.147\,\text{b})$$

$$\mathrm{d}\dot{Q} = \alpha'\,(\vartheta_0' - t')\,\mathrm{d}A \qquad (9.147\,\text{c})$$

Analogous to Eqs. (9.147 a) to (9.147 c), the following approach can be made for the over-all heat transfer

$$\mathrm{d}\dot{Q} = k\,(t - t')\,\mathrm{d}A\,, \qquad (9.147\,\text{d})$$

where k is the over-all heat transfer coefficient (also named heat transition coefficient).

9.6.1 Over-All Heat Transfer Coefficient

By transforming Eqs. (9.147 a) to (9.147 d) according to the respective temperature differences, adding up the transformed Eqs. (9.147 a) to (9.147 c) and equating with the transformed Eq. (9.147 d) one obtains with equal sized area elements dA for the over-all heat transfer coefficient k for a flat partition wall

$$\frac{1}{k} = \frac{1}{\alpha} + \frac{\delta}{\lambda} + \frac{1}{\alpha'} \ . \tag{9.148 a}$$

Instead of the distinction between the heat transfer on the hot and on the cold side, the distinction between the heat transfer on the inside (index i) and the heat transfer on the outside (index a) can also be used:

$$\frac{1}{k} = \frac{1}{\alpha_i} + \frac{\delta}{\lambda} + \frac{1}{\alpha_a} \tag{9.148 b}$$

Eqs. (9.148 a) and (9.148 b) hold for a plane wall where $A_i = A_m = A_a$. If it is a pipe wall, the outer surface A_a and the inner surface A_i shall be considered:

$$\frac{1}{k\,A} = \frac{1}{(\alpha\,A)_i} + \frac{\delta}{\lambda}\frac{1}{A_m} + \frac{1}{(\alpha\,A)_a} \tag{9.148 c}$$

For A_m the calculation rule according to Eq. (9.52) applies. The area A as a component of the product $k\,A$ cannot - strictly speaking - be physically interpreted.

Any splitting of the product $k\,A$ is arbitrary. Therefore, one should not make use of such a splitting.

In analogy to the thermal resistance R_L according to Eq. (9.55) one can also define heat transfer resistances R_i and R_a and the over-all heat transfer resistance R_D [112], [56]:

$$R_i = \frac{1}{(\alpha\,A)_i} \qquad R_a = \frac{1}{(\alpha\,A)_a} \qquad R_D = \frac{1}{k\,A} \tag{9.149}$$

The over-all heat transfer resistance $R_D = R_i + R_L + R_a$ is the sum of the two transfer resistances and the thermal resistance. If the partition consists of a good heat conductor, then the thermal resistance $R_L = \delta/\lambda\,A_m$ can be neglected in many cases. Eq. (9.148 c) simplifies to

$$\frac{1}{k\,A} = \frac{1}{(\alpha\,A)_i} + \frac{1}{(\alpha\,A)_a} \qquad R_D = R_i + R_a \ . \tag{9.150}$$

In the case of a plane wall, one obtains from Eqs. (9.150)

$$\frac{1}{k} = \frac{1}{\alpha_i} + \frac{1}{\alpha_a} = \frac{1}{\alpha} + \frac{1}{\alpha'} \qquad k = \frac{\alpha_i\,\alpha_a}{\alpha_i + \alpha_a} = \frac{\alpha\,\alpha'}{\alpha + \alpha'} \ . \tag{9.151}$$

Tables 9.1 a and 9.1 b give reference values of heat transfer coefficients α as well as over-all heat transfer coefficients k for different technically relevant cases. Table 9.1 b shows that in many cases the smallest value of the heat transfer coefficient has a determining effect on the size of the heat transfer coefficient.

Table 9.1 a Guide values for heat transfer coefficients α in W/(m² K) [24]

	Heat transfer coefficient W/(m² K)	
	Achievable values	Practical values
Gases and vapors		
- Free flow	5 to 25	8 to 15
- Forced flow	12 to 120	20 to 60
Water		
- Free flow	70 to 700	200 to 400
- Forced flow	600 to 12 000	2 000 to 4 000
- Vaporisation	2 000 to 12 000	about 4 000
- Film condensation	4 000 to 12 000	about 6 000
- Droplet condensation	35 000 to 45 000	-
Viscous liquids		
- Forced flow	60 to 600	300 to 400

Table 9.1 b Guide values of heat transfer coefficients α in W/(m² K) and over-all heat transfer coefficients k in W/(m² K) for heat transfer tubes

Fluid inside of tubes (i)	Fluid outside of tubes (a)	α_i W/(m² K)	α_a W/(m² K)	Typical range for k W/(m² K)
Air	Combustion gas	20	20	10 to 15
Water	Air	4 000	20	15 to 20
Oil	Water	1 000	500	300 to 400
Water	Water	4 000	1 000	800 to 1 000
Water	Vapor. water	2 000	10 000	1 500 to 3 000
Water	Condens. steam	2 000	10 000	1 500 to 3 000

9.6.2 Fin Efficiency and Area Efficiency

The product kA is always smaller than the product $(\alpha A)_i$ and than the product $(\alpha A)_a$. If one wants to increase the performance of a heat exchanger by increasing the product kA, then an increase in the smaller product of heat transfer coefficient and surface area - of $(\alpha A)_i$ or $(\alpha A)_a$ respectively - is particularly effective.

In many cases, an increase in the heat transfer coefficient can be achieved by increasing the flow rate and the power input. In many cases there are narrow limits to increase the heat transfer coefficient α by increasing the flow velocity or other parameters. In this case, it is possible to increase the surface area on the side of the smaller heat transfer coefficient by ribbing or finning respectively, the partition wall and thus achieve the same effect.

Increasing the surface area by ribbing is not equivalent to a corresponding increase in the heat transfer coefficient on the same side. Since the heat conduction process in a fin requires a temperature gradient, the temperature difference between the fin and the surrounding fluid is diminished; this reduces the heat transferred per unit area at the fin compared to the unfinned surface. This is taken into account by defining an efficiency called fin efficiency. If the total area A on one side of the partition wall of a heat exchanger consists of the finless base area A_G and the fin surface A_R with the fin efficiency η_R, then the following applies:

$$\alpha(A_G + A_R \eta_R) = \alpha A \eta \qquad (9.152)$$

Here η is the area efficiency of the total area $A = A_G + A_R$. Thus one obtains for the surface efficiency

$$\eta = 1 - \frac{A_R}{A}(1 - \eta_R) \,. \tag{9.153}$$

[68], [69]. If the outer surface of a heat exchanger is finned, then the first Eq. (9.150) becomes

$$\frac{1}{k\,A} = \frac{1}{(\alpha\,A)_i} + \frac{1}{(\alpha\,A\,\eta)_a} \,. \tag{9.154}$$

Taking into account the heat conduction in the partition wall, one obtains

$$\frac{1}{k\,A} = \frac{1}{(\alpha\,A)_i} + \frac{\delta}{\lambda}\frac{1}{A_m} + \frac{1}{(\alpha\,A\,\eta)_a} \,. \tag{9.155}$$

The fin efficiency depends on the dimensions and the material of the fin as well as on the heat transfer coefficient at the fin surface.

9.6.3 Mean Temperature Difference

On an provisional aspect, the heat transfer equation (9.147 d) is only valid for a partition element, because in this case the local fluid temperatures t and t' can be assumed as constant. Applying the heat transfer equation to the whole heat exchanger it becomes

$$\dot{Q} = k\,A\,(t - t')_m \,, \tag{9.156}$$

where $(t - t')_m$ is the mean temperature difference. $(t - t')_m$ is the integral mean value of the locally different temperature differences.

If one wants to determine the product $k\,A$ from the power measurement of a heat exchanger with the help of Eq. (9.156), one must first determine $\dot{Q}$ and $(t - t')_m$. The heat flow is obtained from the heat flow equation $\dot{Q} = (t_1 - t_2)W$ for the fluid which is warm at the inlet, or from the heat flow equation $\dot{Q} = (t'_2 - t'_1)W'$ for the fluid which is cold at the inlet. Since the mean temperature difference is the result of an integral averaging, an equation for the mean temperature difference must take into account the temperature curve of both fluids in the heat exchanger. The result will depend on the design of the heat exchanger. Since only the inlet temperatures t_1 and t'_1 and the outlet temperatures t_2 and t'_2 can be measured, the mean temperature difference must be calculable from these temperatures:

$$(t - t')_m = \mathrm{f}\,(t_1, t'_1, t_2, t'_2) \tag{9.157}$$

For the parallel flow types of heat exchangers - unidirectional flow and counterflow heat exchangers - such equations can be given. For the technically important case of cross flow heat exchangers there is no such equation. Here, the operating characteristic (also called effectiveness) of a heat exchanger takes on a special significance.

9.6.4 Operating Characteristic (Effectiveness)

According to Eq. (9.156), the mean temperature difference can be regarded as the link between the heat flow $\dot{Q}$ and the product $k\,A$. The operating characteristic (effectiveness) of a heat exchanger has the same function as the mean temperature difference. Eq. (9.156) is replaced by

$$\dot{Q} = (t_1 - t'_1)W_1\,\Phi \tag{9.158}$$

[19], [17], [131]. $W_1 = \dot{C}_1$ belongs to the substance flow with the smaller heat capacity flow; the larger heat capacity flow is denoted by $W_2 = \dot{C}_2$. Further, W denotes the heat capacity flow of the warmer fluid flow and W' that of the colder fluid flow. Φ is the operating characteristic (effectiveness). It can be interpreted as the thermal efficiency

$$\Phi = \frac{\dot{Q}}{\dot{Q}_{\max}} = \frac{\dot{Q}}{(t_1 - t'_1)W_1} \tag{9.159}$$

[11], [68], [69]. $\dot{Q}_{\max}$ is the theoretical limiting capacity of a counterflow heat exchanger when the fluid with the smaller heat capacity flow at the outlet reaches the inlet temperature of the other fluid:

$$\dot{Q}_{\max} = (t_1 - t'_1)W_1 \tag{9.160}$$

Inserting Eq. (9.160) into Eq. (9.159), one gets Eq. (9.158). If the warmer fluid has the smaller heat capacity flow, $\dot{Q} = (t_1 - t_2)W_1$ and Eq. (9.158) become

$$\Phi = \frac{t_1 - t_2}{t_1 - t'_1} \ .$$

If the colder fluid has the smaller heat capacity flow, then from $\dot{Q} = (t'_2 - t'_1)W_1$ and Eq. (9.158) one obtains

$$\Phi = \frac{t'_2 - t'_1}{t_1 - t'_1} \ .$$

The operating characteristic Φ can be expressed solely in terms of temperatures by dividing the temperature change of the fluid with the smaller heat capacity flow by the difference of the inlet temperatures of both fluids.

Since both the mean temperature difference and the operating characteristic describe the relationship between the heat flow $\dot{Q}$ and the product kA, one can also use the two equations (9.156) and (9.159) to find a relationship between the mean temperature difference $(t - t')_m$ and the operating characteristic Φ:

$$(t - t')_m = (t_1 - t'_1) \frac{\Phi}{\frac{kA}{W_1}} \tag{9.161}$$

According to Eq. (9.159), the operating characteristic is a dimensionless quantity whose numerical value lies between 0 and 1. It can be plotted in a diagram as an ordinate above the dimensionless quantity kA/W_1 as an abscissa with the dimensionless parameters a_1, a_2/a_1 or a_2 [19]:

$$\Phi = \mathrm{f}\,(a_1, a_2/a_1) \qquad \Phi = \mathrm{f}\,(a_1, a_2) \tag{9.162}$$

The independent variables are

$$a_1 = \frac{kA}{W_1}, \qquad a_2 = \frac{kA}{W_2}, \qquad \frac{W_1}{W_2} = \frac{a_2}{a_1}.$$

Further possibilities with other parameters characterising certain tasks and yielding implicit equations will be referred to later.

9.7 Finned Heat Transfer Surfaces

The fin efficiency is defined as the quotient of the heat flux $\dot{Q}$ actually transferred by the fin and the heat flux $\dot{Q}_0$ that the fin would transfer if it had the temperature of the base of the fin or the base surface,

$$\eta_R = \frac{\dot{Q}}{\dot{Q}_0} \tag{9.163}$$

[47], [48], [53]. The heat flow transferred by the fin is obtained from the temperature curve in the fin and this in turn is obtained by solving a differential equation. The result of the mathematical treatment is to be shown by way of example in the case of a straight rib or fin with a rectangular cross-section.

9.7.1 Straight Fin with Rectangular Cross-Section

If the heat transfer at the outer edge of the fin is neglected, the fin efficiency of a straight fin with rectangular cross-section (Fig. 9.13) is

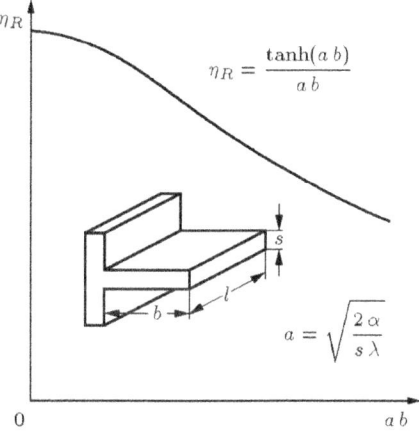

Figure 9.13 Fin efficiency of a straight fin with rectangular cross-section

$$\eta_R = \frac{\tanh(a\,b)}{a\,b}. \tag{9.164}$$

a is an abbreviation for the following expression:

$$a = \sqrt{\frac{2\alpha}{s\lambda}} \tag{9.165}$$

λ is the thermal conductivity of the fin material [47], [111], [62], [53]. If the heat transfer at the outer edge of the fin is considered, the fin efficiency according to [108] is

$$\eta_R = \frac{\tanh(a\,b) + a\frac{s}{2}}{a(b + \frac{s}{2})[1 + a\frac{s}{2}\tanh(a\,b)]}. \tag{9.166}$$

9.7.2 Circular Fin with Rectangular Cross-Section

The heat transfer at the outer edge of the fin is neglected. The fin efficiency is

$$\eta_R = \frac{2\,r_0}{a(R^2 - r_0^2)} \frac{[-iJ_1(iaR)][-H_1(iar_0)] - [-H_1(iaR)][-iJ_1(iar_0)]}{J_0(iar_0)[-H_1(iaR)] + iH_0(iar_0)[-iJ_1(iaR)]} \tag{9.167}$$

[47], [111], [62], [53]. R is the distance of the outer edge of the fin from the tube or pipe axis, r_0 is the distance of the foot of the fin from the tube or pipe axis. $J_0(iz)$ and $-iJ_1(iz)$ are the modified *Bessel* functions of zeroth and first order. $iH_0(iz)$ and $-H_1(iz)$ are the modified *Hankel* functions of zeroth and first order [61], [100].

$$J_0(iz) = 1 + \frac{1}{(1!)^2}\left(\frac{z}{2}\right)^2 + \frac{1}{(2!)^2}\left(\frac{z}{2}\right)^4 + \frac{1}{(3!)^2}\left(\frac{z}{2}\right)^6 + \ldots \tag{9.168}$$

$$-iJ_1(iz) = \frac{z}{2} + \frac{1}{1!\,2!}\left(\frac{z}{2}\right)^3 + \frac{1}{2!\,3!}\left(\frac{z}{2}\right)^5 + \frac{1}{3!\,4!}\left(\frac{z}{2}\right)^7 + \ldots \tag{9.169}$$

$$iH_0(iz) = \frac{2}{\pi}\left[\frac{1}{(1!)^2}\left(\frac{z}{2}\right)^2\left(1-\ln\frac{\gamma z}{2}\right) + \frac{1}{(2!)^2}\left(\frac{z}{2}\right)^4\left(1+\frac{1}{2}-\ln\frac{\gamma z}{2}\right) + \right.$$
$$\left. + \frac{1}{(3!)^2}\left(\frac{z}{2}\right)^6\left(1+\frac{1}{2}+\frac{1}{3}-\ln\frac{\gamma z}{2}\right) + \cdots - \ln\frac{\gamma z}{2}\right] \tag{9.170}$$

$$-H_1(iz) = \frac{2}{\pi}\left\{\frac{1}{z} + \frac{z}{2}\left(\ln\frac{\gamma z}{2} - \frac{1}{2}\right) + \frac{1}{1!\,2!}\left(\frac{z}{2}\right)^3\left(\ln\frac{\gamma z}{2} - 1 - \frac{1}{4}\right) + \right.$$
$$\left. + \frac{1}{2!\,3!}\left(\frac{z}{2}\right)^5\left[\ln\frac{\gamma z}{2} - \left(1+\frac{1}{2}\right) - \frac{1}{6}\right] + \right.$$
$$\left. + \frac{1}{3!\,4!}\left(\frac{z}{2}\right)^7\left[\ln\frac{\gamma z}{2} - \left(1+\frac{1}{2}+\frac{1}{3}\right) - \frac{1}{8}\right] + \ldots \right\} \tag{9.171}$$

Here, $C = \ln\gamma = 0.577\,215\,664\,9$ is the *Euler* constant [25], [61].

Example 9.6 A circular fin of rectangular cross-section has a thickness $s = 0.1$ mm, an outer radius $R = 15$ mm and an inner radius $r_0 = 5$ mm. The fin material has a thermal conductivity $\lambda = 372.2$ W/(m K). What is the fin efficiency with a heat transfer coefficient $\alpha = 70$ W/(m^2 K)?

According to Eq. (9.165)

$$a = \sqrt{\frac{2\alpha}{s\lambda}} = \sqrt{2 \cdot \frac{70\,\text{W}}{\text{m}^2\,\text{K}} \cdot \frac{1}{0.1\,\text{mm}} \cdot \frac{\text{m K}}{372.2\,\text{W}} \cdot \frac{1000\,\text{mm}}{1\,\text{m}}} = 61.3304\,\frac{1}{\text{m}}$$

$$aR = 0.919956 \qquad ar_0 = 0.306652 \qquad \frac{2r_0}{a(R^2 - r_0^2)} = 0.815256$$

$$-iJ_1(i\,a\,R) = 0.510386 \qquad -H_1(i\,a\,r_0) = 1.897596 \qquad -H_1(i\,a\,R) = 0.440237$$

$$-iJ_1(i\,a\,r_0) = 0.155135 \qquad J_0(i\,a\,r_0) = 1.023647 \qquad iH_0(i\,a\,r_0) = 0.860954$$

$$\eta_R = 0.815256 \cdot \frac{0.510386 \cdot 1.897596 - 0.440237 \cdot 0.155135}{1.023647 \cdot 0.440237 + 0.860954 \cdot 0.510386} = 0.824547$$

Other technically important fin or rib shapes are also treated mathematically. Literature references are: [47], [53], [62], [111], [116].

9.8 Partition Wall Heat Exchangers

The design and mode of operation of a partition wall heat exchanger (recuperator) determine the fluid temperature curve, the mean temperature difference and the operating characteristic (Fig. 9.14). Cross flow heat exchangers and parallel flow heat exchangers differ in their design. Unidirectional flow and counterflow describe two operating modes of a parallel flow heat exchanger. The distinguishing marks in different types and modes of operation with respect to the fluid temperature curve, to the mean temperature difference and to the operating characteristic disappear if a phase transition occurs in a fluid.

In addition to the designations a_1 and a_2, as introduced in section 9.6.4, the following abbreviations are used:

$$a = \frac{kA}{W} \qquad b = \frac{kA}{W'} \tag{9.172}$$

In the equations for the temperature profile in a heat exchanger, x and y are related length coordinates, which are chosen so that the numerical values in the range of the heat exchanger are between 0 and 1.

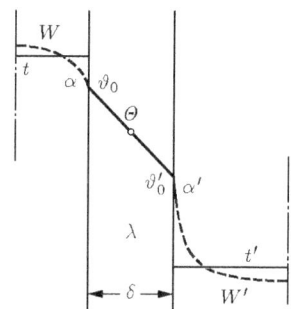

Figure 9.14 Heat transfer designations through a partition wall in unidirectional flow, counterflow and cross flow heat exchangers:

on the side of the heat releasing fluid:
- t mean fluid temperature
- W heat capacity flow of the fluid warm at the inlet
- α heat transfer coefficient
- ϑ_0 wall surface temperature

on the side of the heat absorbing fluid:
- t' averaged fluid temperature
- W' heat capacity flow of the fluid cold at the inlet
- α' heat transfer coefficient
- ϑ'_0 wall surface temperature

λ and δ are thermal conductivity and thickness of the partition wall.
- Θ mean wall temperature

$$\dot{Q} = \dot{m}c_p(t_1 - t_2) = (t_1 - t_2)W \quad \text{and} \quad \dot{Q} = \dot{m}'c'_p(t'_2 - t'_1) = (t'_2 - t'_1)W' \quad \text{hold.}$$

9.8.1 Unidirectional Flow Heat Exchanger

In an unidirectional flow heat exchanger (Fig. 9.15), both fluids have the same direction of flow. By integration, the temperature profiles become

$$t = t_1 - (t_1 - t'_1)\frac{a}{a+b}[1 - e^{-(a+b)x}] \tag{9.173}$$

$$t' = t'_1 + (t_1 - t'_1)\frac{b}{a+b}[1 - e^{-(a+b)x}] . \tag{9.174}$$

The mean temperature difference is given by

$$(t - t')_m = \frac{(t_1 - t'_1) - (t_2 - t'_2)}{\ln \dfrac{t_1 - t'_1}{t_2 - t'_2}} . \tag{9.175}$$

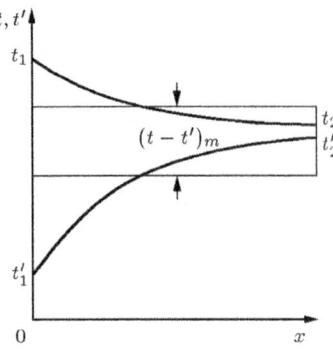

Figure 9.15 Temperature profiles in an unidirectional flow heat exchanger. $(t - t')_m$, as the mean temperature difference, is the integral mean value of the locally different temperature differences of the two fluids:

$$\dot{Q} = k\,A(t - t')_m$$
$$\dot{Q} = (t_1 - t'_1)W_1\,\Phi \qquad (t - t')_m = (t_1 - t'_1)\frac{\Phi}{k\,A/W_1}$$

The operating characteristic (Fig. 9.16) ([19], [17]) is given by

$$\Phi = \frac{1}{1 + a_2/a_1}[1 - \mathrm{e}^{-(1 + a_2/a_1)a_1}]\,. \qquad (9.176)$$

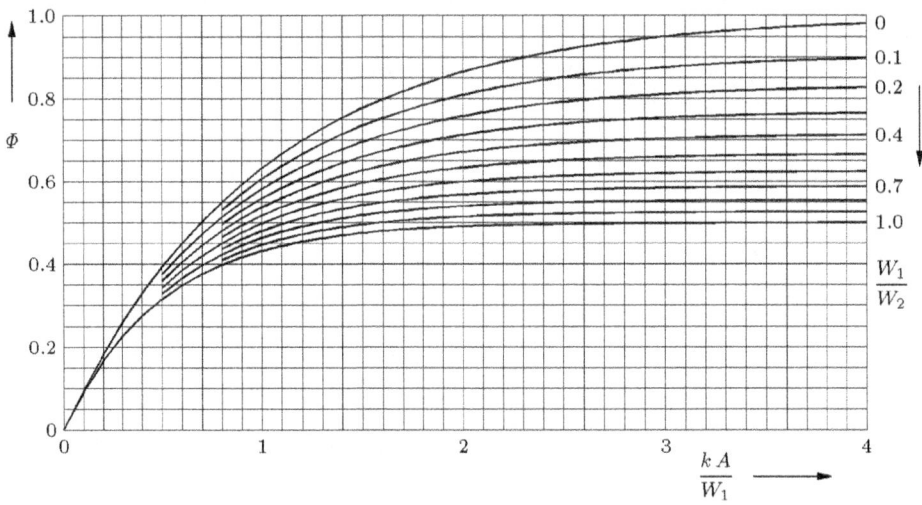

Figure 9.16 Operating characteristic of an unidirectional flow heat exchanger (parameter W_1/W_2)

9.8.2 Counterflow Heat Exchanger

In a counterflow heat exchanger (Fig. 9.17), both fluids have opposite flow directions. By integration, the temperature profiles become

$$t = t_1 - (t_1 - t'_1)\frac{1 - \mathrm{e}^{(b-a)x}}{1 - \frac{b}{a}\mathrm{e}^{b-a}} \qquad (a \neq b) \qquad (9.177)$$

$$t' = t'_1 + (t_1 - t'_1)\frac{b}{a}\frac{\mathrm{e}^{(b-a)x} - \mathrm{e}^{b-a}}{1 - \frac{b}{a}\mathrm{e}^{b-a}} \qquad (a \neq b) \qquad (9.178)$$

The mean temperature difference is given by

$$(t - t')_m = \frac{(t_1 - t'_2) - (t_2 - t'_1)}{\ln\dfrac{t_1 - t'_2}{t_2 - t'_1}} \qquad (a \neq b)\,. \qquad (9.179)$$

The operating characteristic ([19], [17]) (Fig. 9.18) is given by

$$\Phi = \frac{1 - e^{-(1 - a_2/a_1)a_1}}{1 - a_2/a_1\, e^{-(1 - a_2/a_1)a_1}} \quad (a_2/a_1 < 1) . \tag{9.180}$$

The temperature profile depends on the ratio of the heat capacity flows of the two substance flows to each other.

In the special case of $W = W'$ (equivalent to $a = b$), the temperature curves are given by

$$t = t_1 - (t_1 - t'_1)\frac{a}{1+a}x \tag{9.181}$$

$$t' = t'_1 + (t_1 - t'_1)\frac{a}{1+a}(1 - x) . \tag{9.182}$$

The mean temperature difference and the operating characteristic at $W = W'$ are

$$(t - t')_m = t_1 - t'_2 = t_2 - t'_1 \tag{9.183}$$

$$\Phi = \frac{a_1}{1 + a_1} \quad (a_2/a_1 = 1) . \tag{9.184}$$

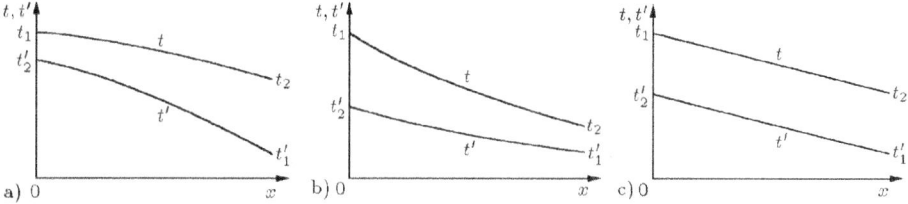

Figure 9.17 Temperature profiles in counterflow heat exchangers
a) $W > W'$ b) $W < W'$ c) $W = W'$

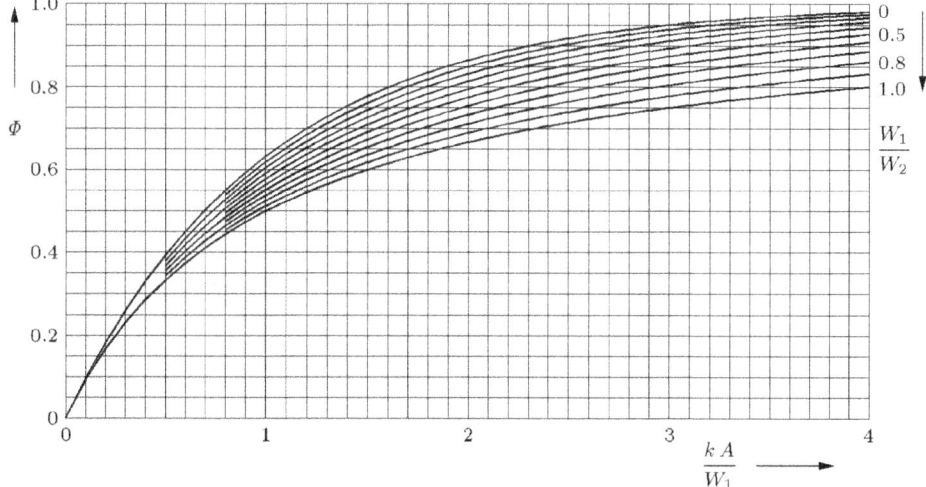

Figure 9.18 Operating characteristic of a counterflow heat exchanger (parameter W_1/W_2)

9.8.3 Crossflow Heat Exchanger

In a crossflow heat exchanger (Fig. 9.19), the flow directions of both fluids are at right angles to each other.
By integration, the temperature profiles become

$$t = t_1 - (t_1 - t_1') e^{-bx - ay} \sum_{n=1}^{\infty} \frac{(ay)^n}{n!} \sum_{p=0}^{n-1} \frac{(bx)^p}{p!} \quad . \tag{9.185}$$

$$t' = t_1' + (t_1 - t_1') e^{-bx - ay} \sum_{n=1}^{\infty} \frac{(bx)^n}{n!} \sum_{p=0}^{n-1} \frac{(ay)^p}{p!} \tag{9.186}$$

Since temperature profiles are obtained in the outlet cross sections, in addition the integral mean values of the outlet temperatures are of interest:

$$t_2 = t_1 - (t_1 - t_1') \frac{1}{b} \sum_{n=0}^{\infty} \left(1 - e^{-b} \sum_{p=0}^{n} \frac{b^p}{p!}\right) \left(1 - e^{-a} \sum_{p=0}^{n} \frac{a^p}{p!}\right) \tag{9.187}$$

$$t_2' = t_1' + (t_1 - t_1') \frac{1}{a} \sum_{n=0}^{\infty} \left(1 - e^{-b} \sum_{p=0}^{n} \frac{b^p}{p!}\right) \left(1 - e^{-a} \sum_{p=0}^{n} \frac{a^p}{p!}\right) \tag{9.188}$$

[93], [109].
For the crossflow heat exchanger, there is no equation with the help of which one can explicitly express the mean temperature difference only by inlet and outlet temperatures. Here, the operating characteristic Φ is helpful [123], [109] (Figs. 9.20, 9.21 and 9.22):

$$\Phi = \frac{1}{a_2} \sum_{n=0}^{\infty} \left(1 - e^{-a_1} \sum_{p=0}^{n} \frac{a_1^p}{p!}\right) \left(1 - e^{-a_2} \sum_{p=0}^{n} \frac{a_2^p}{p!}\right) \tag{9.189}$$

The detailed notation of this equation

$$\Phi = \frac{1}{a_2} \left[1 - e^{-a_1} (1)\right] \left[1 - e^{-a_2} (1)\right] +$$

$$+ \frac{1}{a_2} \left[1 - e^{-a_1} (1 + a_1)\right] \left[1 - e^{-a_2} (1 + a_2)\right] +$$

$$+ \frac{1}{a_2} \left[1 - e^{-a_1} \left(1 + a_1 + \frac{a_1^2}{2!}\right)\right] \left[1 - e^{-a_2} \left(1 + a_2 + \frac{a_2^2}{2!}\right)\right] +$$

$$+ \frac{1}{a_2} \left[1 - e^{-a_1} \left(1 + a_1 + \frac{a_1^2}{2!} + \frac{a_1^3}{3!}\right)\right] \left[1 - e^{-a_2} \left(1 + a_2 + \frac{a_2^2}{2!} + \frac{a_2^3}{3!}\right)\right] +$$

$$+ \frac{1}{a_2} \left[1 - e^{-a_1} \left(1 + a_1 + \frac{a_1^2}{2!} + \frac{a_1^3}{3!} + \frac{a_1^4}{4!}\right)\right] \left[1 - e^{-a_2} \left(1 + a_2 + \frac{a_2^2}{2!} + \frac{a_2^3}{3!} + \frac{a_2^4}{4!}\right)\right] + \ldots$$

provides the prerequisite for the simple function instruction used further below for calculating the operating characteristic of the crossflow heat exchanger. From the comparison with the exponential function

$$e^x = 1 + x + \frac{x^2}{2!} + \frac{x^3}{3!} + \frac{x^4}{4!} + \ldots$$

one can recognise in the round brackets incomplete mathematical series of exponential functions, which can be easily programmed. Since the operating characteristic can be represented as a dimensionless mathematical function dependent on various independent variables, there are a number of diagrams for technical tasks, each with parameters adapted to the task.

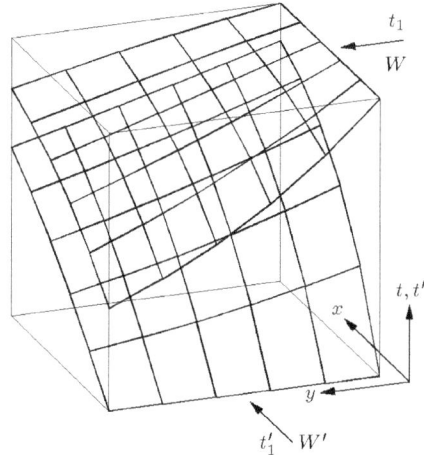

Figure 9.19 Temperature curves in a crossflow heat exchanger

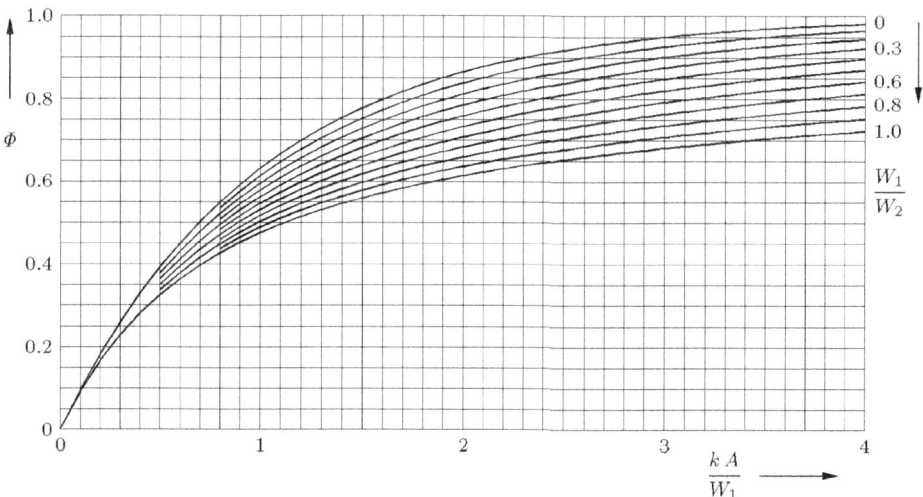

Figure 9.20 Operating characteristic of a crossflow heat exchanger (parameter W_1/W_2)

If in a design $\dfrac{kA}{W_2}$ is given or in case of a change of the operating data $\dfrac{kA}{W_2}$ remains constant, it is advantageous to have a diagram of the operating characteristic with the parameter $\dfrac{kA}{W_2}$:

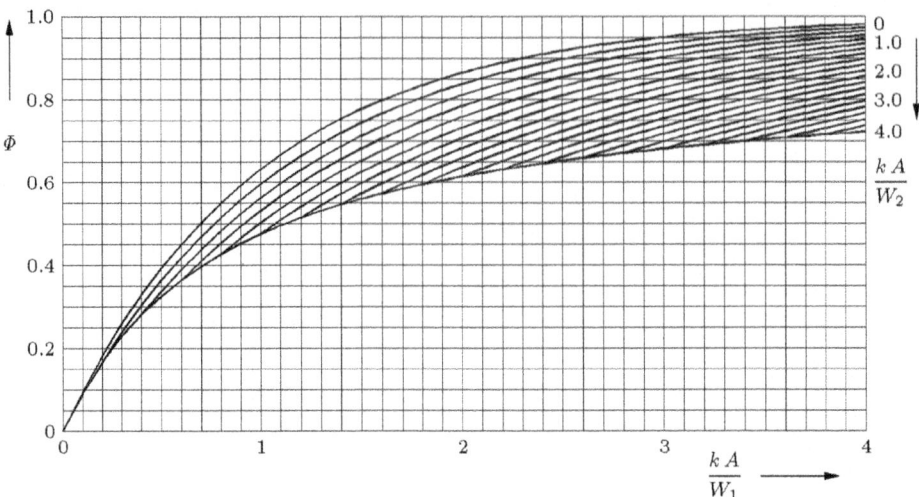

Figure 9.21 Operating characteristic of a crossflow heat exchanger (parameter $a_2 = \dfrac{kA}{W_2}$)

If both inlet temperatures and one outlet temperature are given in the design, this leads to a numerical value for the operating characteristic Φ or the expression $\Phi \dfrac{W_1}{W_2}$. Therefore, it is advantageous to have a diagram of the operating characteristic with the parameter $\Phi \dfrac{W_1}{W_2}$:

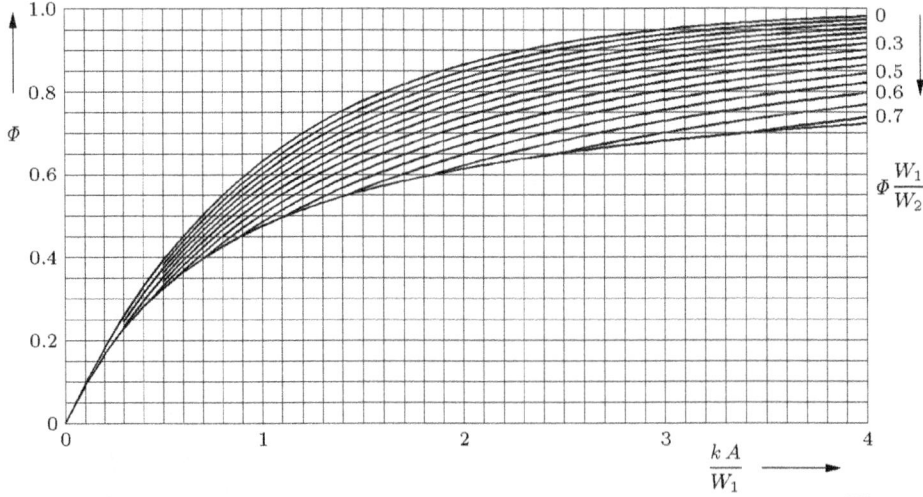

Figure 9.22 Operating characteristic of a crossflow heat exchanger (parameter $\Phi \dfrac{W_1}{W_2}$)

If both outlet temperatures and one inlet temperature are given, the following expressions are thus determined:

$$\frac{t_1-t_2}{t_1-t'_2} = \frac{\Phi\dfrac{W_1}{W}}{1-\Phi\dfrac{W_1}{W'}} \quad \text{and} \quad \frac{t'_2-t'_1}{t_2-t'_1} = \frac{\Phi\dfrac{W_1}{W'}}{1-\Phi\dfrac{W_1}{W}}$$

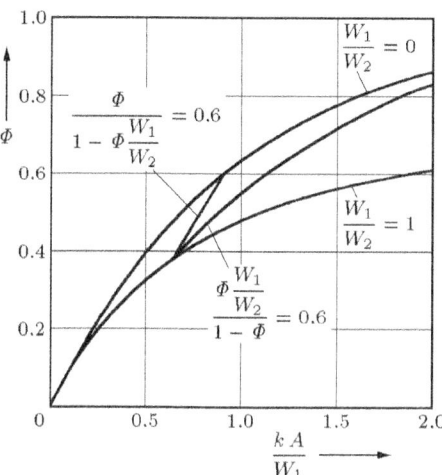

Figure 9.23 "Geometric location" of all solutions for given values according to Eqs. (9.190)

The solutions in a diagram of the operating characteristic lie on the curves

$$\frac{\Phi\dfrac{W_1}{W_2}}{1-\Phi} = \text{const} \quad \text{or} \quad \frac{\Phi}{1-\Phi\dfrac{W_1}{W_2}} = \text{const}, \qquad (9.190)$$

indicating the geometric location of the solutions (Fig. 9.23).

In the operating characteristic of a tube heat exchanger, a distinction must be made whether the fluid with the greater heat capacity flow W_2 or the fluid with the smaller heat capacity flow W_1 flows in the tubes.

Crossflow heat exchanger with 1 row of tubes and fluid with W_2 in the tubes:

$$\Phi = \frac{W_2}{W_1}\left(1 - e^{-\frac{W_1}{W_2}\left(1 - e^{-\frac{kA}{W_1}}\right)}\right) \qquad (9.191)$$

Crossflow heat exchanger with N rows of tubes and fluid W_2 in the tubes with
$c = \dfrac{W_1}{W_2}N\left(1 - e^{-\frac{1}{N}\frac{kA}{W_1}}\right)$:

$$\Phi = \frac{W_2}{W_1}\left\{1 - \frac{1}{N}\left[1 + \sum_{p=1}^{N-1}\sum_{m=0}^{p}\binom{p}{m}\left(1-e^{-\frac{1}{N}\frac{kA}{W_1}}\right)^m e^{-(p-m)\frac{1}{N}\frac{kA}{W_1}}\sum_{r=0}^{m}\frac{c^r}{r!}\right]e^{-c}\right\}$$
$$(9.192)$$

Crossflow heat exchanger with 1 row of tubes and fluid with W_1 in the tubes:

$$\Phi = 1 - e^{-\frac{W_2}{W_1}\left(1 - e^{-\frac{kA}{W_2}}\right)} \tag{9.193}$$

Crossflow heat exchanger with N rows of tubes and fluid with W_1 in the tubes with $c = \frac{W_2}{W_1} N \left(1 - e^{-\frac{1}{N}\frac{kA}{W_2}}\right)$:

$$\Phi = 1 - \frac{1}{N}\left[1 + \sum_{p=1}^{N-1}\sum_{m=0}^{p}\binom{p}{m}\left(1 - e^{-\frac{1}{N}\frac{kA}{W_2}}\right)^m e^{-(p-m)\frac{1}{N}\frac{kA}{W_2}} \sum_{r=0}^{m}\frac{c^r}{r!}\right] e^{-c} \tag{9.194}$$

As the number of tube rows increases, the values of Eqs. (9.192) and (9.194) approach the values of Eq. (9.189).

Example 9.6 For $kA/W_1 = 1$ and $W_1/W_2 = 0.5$, the operating characteristic of a crossflow heat exchanger as a plate heat exchanger and the operating characteristic of a crossflow heat exchanger as a tube heat exchanger with different numbers of rows of tubes N are to be considered.

Φ_P is the operating characteristic of the plate heat exchanger, Φ_{2R} is the operating characteristic of the tube heat exchanger with a fluid with W_2 in the tubes, and Φ_{1R} is the operating characteristic of the tube heat exchanger with a fluid with W_1 in the tubes.

Results:

$\Phi_P = 0.547490$

N	1	2	3	4	5	6	7	8	9
Φ_{2R}	0.541969	0.546128	0.546881	0.547148	0.547271	0.547338	0.547378	0.547404	0.547422
Φ_{1R}	0.544764	0.546806	0.547186	0.547319	0.547380	0.547414	0.547434	0.547447	0.547456
N	10	20	30	40	50	60	80	100	130
Φ_{2R}	0.547435	0.547476	0.547484	0.547486	0.547488	0.547488	0.547489	0.547489	0.547490
Φ_{1R}	0.547462	0.547483	0.547487	0.547488	0.547489	0.547489	0.547489	0.547490	0.547490

9.8.4 Heat Transfer with Phase Transition in a Heat Exchanger

During the phase change of a pure fluid (single-substance system) in a heat exchanger the temperature of the fluid remains constant. Since $\dot{Q} \neq 0$, the heat capacity flow of the fluid in which the phase transition takes place must become infinite. In the heat flow equations, the product of the temperature change $t_1 - t_2$ or $t'_2 - t'_1$ and the heat capacity flow of the condensing or vaporising fluid W or W' is an indeterminate expression $0 \cdot \infty$.

a) Condensation (Fig. 9.24 a)

The temperature profiles become

$$t = t_1 = t_2 = \text{const} \tag{9.195}$$

$$t' = t'_1 + (t_1 - t'_1)(1 - e^{-bx}) \ . \tag{9.196}$$

The mean temperature difference is given by

$$(t - t')_m = \frac{t'_2 - t'_1}{\ln\frac{t_1 - t'_1}{t_1 - t'_2}} \ . \tag{9.197}$$

The operating characteristic [2], [1], [38] is

$$\Phi = 1 - e^{-a_1} . \tag{9.198}$$

b) Vaporisation (Fig. 9.24 b)

The temperature profiles become

$$t = t_1 - (t_1 - t'_1)(1 - e^{-a x}) \tag{9.199}$$

$$t' = t'_1 = t'_2 = \text{const} . \tag{9.200}$$

The mean temperature difference is given by

$$(t - t')_m = \frac{t_1 - t_2}{\ln \dfrac{t_1 - t'_1}{t_2 - t'_1}} . \tag{9.201}$$

The operating characteristic agrees with Eq. (9.198).

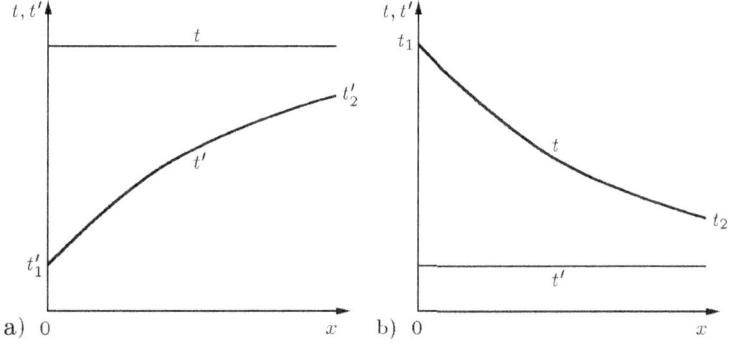

Figure 9.24 Temperature profile in heat exchangers with phase transition of a pure fluid. (single-substance system)
a) condensation   b) vaporisation

9.9 Evaluation and Design

The determination of the dimensions of a heat exchanger requires knowledge of empirical values which describe the behavior of identical or similar apparatus [122]. These empirical values are obtained from measurements and their evaluation. On a partition wall heat exchanger (recuperator), the substance flows of the hot and cold fluids $\dot{m}$ and $\dot{m}'$, the two inlet temperatures t_1 and t'_1 as well as the two outlet temperatures t_2 and t'_2 are measured. The two heat flow equations

$$\dot{Q} = (t_1 - t_2) W \qquad \dot{Q} = (t'_2 - t'_1) W' \tag{9.202}$$

provide with the possibility of control the transferred heat flow. It is obvious to determine $k A$ from this with Eq. (9.156) as the first evaluation step:

$$kA = \frac{\dot{Q}}{(t-t')_m} \tag{9.203}$$

This is possible in the cases of unidirectional flow or counterflow, but not without additional help in the case of a crossflow heat exchanger.

9.9.1 Correction Factor for a Crossflow Heat Exchanger

There is no equation to calculate the mean temperature difference for a crossflow heat exchanger from the inlet and outlet temperatures of the two fluids. In section 9.8.3, it was suggested that in solving this problem, the operating characteristic for a crossflow heat exchanger can be useful.

Either one takes from the operating characteristic for a crossflow heat exchanger the relation between the heat flow $\dot{Q}$ and the product kA or one determines a correction factor, which by multiplication with the mean temperature difference for a counterflow heat exchanger results in the mean temperature difference for a crossflow heat exchanger.

With the designations $(t-t')_{m\,cr}$ and $(t-t')_{m\,co}$ for the mean temperature differences for a crossflow heat exchanger and for a counterflow heat exchanger, the correction factor κ is introduced as follows:

$$\kappa = \frac{(t-t')_{m\,cr}}{(t-t')_{m\,co}} \tag{9.204}$$

Eqs. (9.161) and (9.179) are inserted into Eq. (9.204):

$$\kappa = \frac{t_1 - t_1'}{(t_1-t_2')-(t_2-t_1')} \frac{\Phi}{kA} \cdot \frac{1}{\ln \frac{t_1-t_2'}{t_2-t_1'}} = \ln \frac{t_2-t_1'}{t_1-t_2'} \frac{1}{\frac{t_1-t_2'}{t_1-t_1'} - \frac{t_2-t_1'}{t_1-t_1'}} \frac{\Phi}{kA}{W_1} \tag{9,205}$$

From Eqs. (9.158) and (9.202) follows:

$$\frac{t_1-t_2}{t_1-t_1'} = \Phi \frac{W_1}{W} \quad 1 - \frac{t_1-t_2}{t_1-t_1'} = \frac{t_1-t_1'-t_1+t_2}{t_1-t_1'} = \frac{t_2-t_1'}{t_1-t_1'} = 1 - \Phi \frac{W_1}{W} \tag{9.206}$$

$$\frac{t_2'-t_1'}{t_1-t_1'} = \Phi \frac{W_1}{W'} \quad 1 - \frac{t_2'-t_1'}{t_1-t_1'} = \frac{t_1-t_1'-t_2'+t_1'}{t_1-t_1'} = \frac{t_1-t_2'}{t_1-t_1'} = 1 - \Phi \frac{W_1}{W'} \tag{9.207}$$

Eq. (9.207) is divided by Eq. (9.206):

$$\frac{t_1-t_2'}{t_2-t_1'} = \frac{1 - \Phi \frac{W_1}{W'}}{1 - \Phi \frac{W_1}{W}} \tag{9.208}$$

Eqs. (9.206) to (9.208) are inserted into Eq. (9.205):

$$\kappa = \ln \frac{1 - \Phi \dfrac{W_1}{W'}}{1 - \Phi \dfrac{W_1}{W}} \frac{1}{1 - \Phi \dfrac{W_1}{W'} - \left(1 - \Phi \dfrac{W_1}{W}\right)} \frac{\Phi}{\dfrac{kA}{W_1}} \qquad (9.209)$$

Setting $W = W_1$ and $W' = W_2$, one gets from Eq. (9.209)

$$\kappa = \ln \frac{1 - \Phi \dfrac{W_1}{W_2}}{1 - \Phi} \frac{1}{1 - \dfrac{W_1}{W_2}} \frac{1}{\dfrac{kA}{W_1}} \,. \qquad (9.210)$$

If one assumes $W' = W_1$ and $W = W_2$, one also obtains Eq. (9.210).

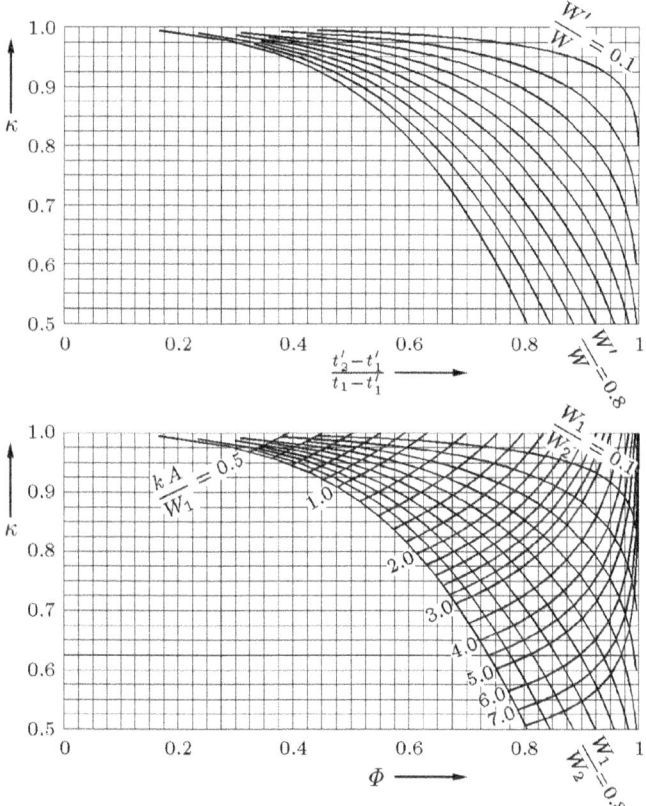

Figure 9.25 Correction factor for for a crossflow heat exchanger.
Into the usual representation of Fig. 9.25 above, the curves $kA/W_1 = $ const can be inserted. This results in the representation of Fig. 9.25 below. Thus one does not need the correction factor any more.

The result of this investigation can be summarised as follows: If, in order to calculate the correction factor for the mean temperature difference for the crossflow heat exchanger, one uses the operating characteristic for the crossflow heat exchanger, then the curves of constant $\dfrac{kA}{W_1}$ values can be plotted directly into the diagram for the correction factor.

Then one does not need the correction factor and the mean temperature difference for the crossflow heat exchanger to find the value for $\dfrac{kA}{W_1}$. It follows that if the operating characteristic is known, the mean temperature difference is an unnecessary quantity (Fig. 9.25).

9.9.2 Representation of the Operating Characteristic

The graphical representation of the operating characteristic is possible in different ways. In the form chosen here, with Φ as ordinate and kA/W_1 as abscissa, it corresponds to the original form [19], which has the great advantage that it clearly shows the non-linear relationship between effort and power.

On the abscissa, the construction effort is represented by the area A, and the energy effort is represented by the over-all heat transfer coefficient k. On the ordinate, the heat flow as heat transfer performance is expressed in the value of the operating characteristic Φ. The in some cases diminished increase in performance with an increase in effort is clearly visible. The corresponding representations in the VDI-Wärmeatlas ([132], [133]) cannot reflect this clarity.

What stands out about the values of heat transfer coefficients, are the uncertainties with which these values are associated with. An explanation is provided by the course of the operating characteristic.

Every power measurement is associated with certain inaccuracies, which affect the numerical values of the operating characteristic and describe a tolerance range there. When evaluating the power measurement — in the diagram of the operating characteristic: transition from the ordinate to the abscissa — the tolerance range for the value of kA/W_1 is considerably larger. If one tries to split values of kA into values of α for the two partition surfaces [115], the tolerance range is increased again.

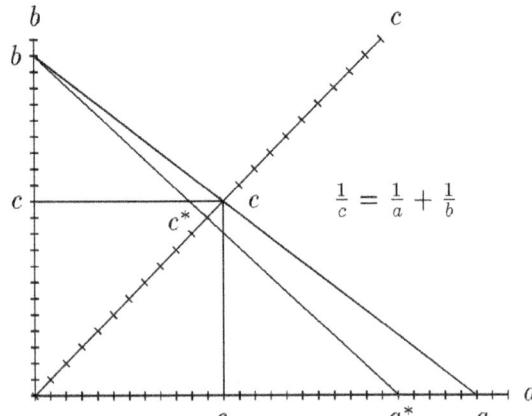

Figure 9.26 Illustration of the equation $\dfrac{1}{c} = \dfrac{1}{a} + \dfrac{1}{b}$: Given a value b and a small tolerance $c - c^*$, the result is a large tolerance $a - a^*$.

For heat exchangers, one can often neglect the heat conduction in the tube or pipe wall and use the approximation of Eq. (9.154). Then Fig. 9.26 also illustrates the relationship between the products αA for the heat transfer on both sides of the tube wall and the product αA for the heat transfer on both sides of the tube or pipe wall and the product kA for the over-all heat transfer. With constant heat transfer on one side of the partition wall, a small tolerance in the determination of the product kA causes a large tolerance in the variable product αA. This is the second reason for the uncertainties in the heat transfer coefficients.

On the other hand, large tolerances in the values of α have only a small effect on the tolerances in the values of Φ and thus on the performance of the heat exchangers.

The graphical representation of the operating characteristic gives a good impression of the basic interrelationships. But the graphical representation of the operating characteristic is only suitable for numerical calculations if a large distance (e.g. 1 m) is available for the ordinate. Such diagram sections were used in the pre-computer age. Today the practical application is done by a computer programme.

Table 9.2 Heat transfer coefficients in $W/(m^2\,K)$ for water with an average temperature of $20\,^\circ C$ flowing through a pipe of 20 mm internal diameter and 1 m length, at a wall temperature of $40\,^\circ C$ according to Eqs. (9.96), (9.130 a), (9.132 a), (9.132 b) (9.132 c) and (9.132 d)

The calculation of the heat transfer coefficients shown here has been carried out with the following
Prandtl number at $15\,^\circ C$: $Pr = 8.07$
Prandtl number at $40\,^\circ C$: $Pr = 4.34$
dynamic viscosity at $15\,^\circ C$:
$\eta = 0.001139\ kg/(m\ s)$
dynamic viscosity at $40\,^\circ C$:
$\eta = 0.0006531\ kg/(m\ s)$
thermal conductivity at $20\,^\circ C$:
$\lambda = 0.5911\ W/(m\ K)$
density at $15\,^\circ C$: $\varrho = 999.2\ kg/m^3$
kinematic viscosity at $15\,^\circ C$:
$\nu = 0.00000114\ m^2/s$

The velocity of water is given by

$$w = Re\,\frac{\nu}{d} = Re \cdot 0.000057\,\frac{m}{s}$$

Re number	9.96	9.130 a	9.132 a	9.132 b	(9.132 c	9.132 d
3000.0	800.5	1062.5	992.0	662.5	687.5	776.0
4000.0	1131.1	1352.9	1248.7	1013.5	985.3	1100.0
5000.0	1442.1	1631.4	1492.7	1336.1	1264.9	1408.1
6000.0	1739.6	1900.8	1727.1	1637.7	1530.7	1703.9
7000.0	2026.9	2162.9	1953.8	1922.9	1785.6	1990.1
8000.0	2306.2	2418.7	2174.1	2194.8	2031.5	2268.1
9000.0	2578.9	2669.3	2388.9	2455.6	2269.8	2539.2
10000.0	2846.1	2915.3	2599.0	2706.9	2501.6	2804.4
11000.0	3108.4	3157.1	2804.9	2949.9	2727.7	3064.3
12000.0	3366.6	3395.3	3007.1	3185.7	2948.7	3319.5
13000.0	3621.1	3630.2	3205.9	3415.0	3165.1	3570.5
14000.0	3872.3	3862.0	3401.8	3638.5	3377.4	3817.7
15000.0	4120.4	4091.1	3594.8	3856.7	3585.9	4061.4
16000.0	4365.9	4317.5	3785.3	4070.1	3791.0	4301.8
17000.0	4608.9	4541.6	3973.4	4279.1	3992.9	4539.2
18000.0	4849.5	4763.5	4159.3	4484.1	4191.8	4773.9
19000.0	5088.0	4983.3	4343.1	4685.3	4388.0	5005.9
20000.0	5324.6	5201.1	4525.1	4883.0	4581.6	5235.6

As a rule, four substance values of a fluid are needed to calculate heat transfer coefficients. It is advantageous to calculate these substance values with the help of a formula, e.g. with the following polynomial

$$y = \sum_{i=0}^{n} a_i\, t^i \ . \tag{9.211}$$

The following Tables 9.3 a and 9.3 b contain the coefficients for the density, the specific heat capacity, the thermal conductivity and the dynamic viscosity of liquid water and gaseous air. This can be used to determine further substance values or key figures, e.g. the kinematic viscosity

$$\nu = \frac{\eta}{\varrho}$$

or the *Prandtl* number

$$Pr = \frac{c_p\,\eta}{\lambda}\ .$$

Table 9.3 a Coefficients of the balancing polynomials for density, specific isobaric heat capacity, thermal conductivity and dynamic viscosity of water as a function of temperature

	ϱ kg/m^3	c_p kJ/(kg K)	λ kJ/(m K)	η kg/(m s)
a_0	999.800	4.21700	$5.62000 \cdot 10^{-1}$	558.095
a_1	$8.08172 \cdot 10^{-2}$	$-3.66156 \cdot 10^{-3}$	$2.14133 \cdot 10^{-3}$	19.3111
a_2	$-8.22201 \cdot 10^{-3}$	$1.50976 \cdot 10^{-4}$	$-1.48687 \cdot 10^{-5}$	$1.37640 \cdot 10^{-1}$
a_3	$-9.54720 \cdot 10^{-6}$	$-3.90834 \cdot 10^{-6}$	$1.15551 \cdot 10^{-7}$	$-1.22947 \cdot 10^{-4}$
a_4	$2.73628 \cdot 10^{-6}$	$7.07011 \cdot 10^{-8}$	$-1.42957 \cdot 10^{-9}$	$-8.94454 \cdot 10^{-6}$
a_5	$-5.51573 \cdot 10^{-8}$	$-8.13119 \cdot 10^{-10}$	$1.26580 \cdot 10^{-11}$	$1.45048 \cdot 10^{-7}$
a_6	$4.65161 \cdot 10^{-10}$	$5.21612 \cdot 10^{-12}$	$-5.94881 \cdot 10^{-14}$	$-1.03550 \cdot 10^{-9}$
a_7	$-1.45203 \cdot 10^{-12}$	$-1.39191 \cdot 10^{-14}$	$1.03161 \cdot 10^{-16}$	$2.80488 \cdot 10^{-12}$

Table 9.3 b Coefficients of the balancing polynomials for density, specific isobaric heat capacity, thermal conductivity and dynamic viscosity of air as a function of temperature

	ϱ kg/m^3	c_p kJ/(kg K)	λ kJ/(m K)	η kg/(m s)
a_0	1.27533	1.00643	$2.417874 \cdot 10^{-2}$	$1.72439 \cdot 10^{-5}$
a_1	$-4.69088 \cdot 10^{-3}$	$2.32876 \cdot 10^{-6}$	$7.642165 \cdot 10^{-5}$	$5.04899 \cdot 10^{-8}$
a_2	$1.73598 \cdot 10^{-5}$	$4.79185 \cdot 10^{-7}$	$-4.894379 \cdot 10^{-8}$	$-3.95680 \cdot 10^{-11}$
a_3	$-6.30112 \cdot 10^{-8}$	$1.97884 \cdot 10^{-9}$	$5.443279 \cdot 10^{-11}$	$2.69570 \cdot 10^{-14}$
a_4	$2.05441 \cdot 10^{-10}$	$-3.01337 \cdot 10^{-11}$	$1.932858 \cdot 10^{-13}$	$2.20529 \cdot 10^{-16}$
a_5	$-5.32666 \cdot 10^{-13}$	$1.89720 \cdot 10^{-13}$	$-2.047763 \cdot 10^{-15}$	$-1.10592 \cdot 10^{-18}$
a_6	$9.85792 \cdot 10^{-16}$	$-6.51331 \cdot 10^{-16}$	$8.599314 \cdot 10^{-18}$	$2.00829 \cdot 10^{-21}$
a_7	$-1.18355 \cdot 10^{-18}$	$1.23682 \cdot 10^{-18}$	$-1.895706 \cdot 10^{-20}$	$-7.79232 \cdot 10^{-25}$
a_8	$8.12689 \cdot 10^{-22}$	$-1.22014 \cdot 10^{-21}$	$2.123069 \cdot 10^{-23}$	$-1.80029 \cdot 10^{-27}$
a_9	$-2.39907 \cdot 10^{-25}$	$4.88208 \cdot 10^{-25}$	$-9.523259 \cdot 10^{-27}$	$1.62365 \cdot 10^{-30}$

9.9.3 Longitudinal Heat Conduction in a Plane Partition Wall

The simplified heat balance for the partition element of a heat exchanger leads with the local temperatures of the fluids t and t' as well as a mean wall temperature Θ with neglect of the thermal resistance in the partition wall to the equation

$$(t - \Theta)\alpha = (\Theta - t')\alpha' = (t - t')k . \qquad (9.212)$$

In the case of counterflow, this results with the fluid temperatures according to Eqs. (9.177) and (9.178) in the following equation for the wall temperature

$$\Theta = t_1 - \frac{t_1 - t_1'}{1 - \frac{b}{a}e^{b-a}} \left\{ 1 - \left[1 - \left(1 - \frac{b}{a}\right)\frac{k}{\alpha}\right] e^{(b-a)x} \right\} . \qquad (9.213)$$

Fig. 9.27 shows for a counterflow heat exchanger the temperature curves t and t' of the fluids and Θ in the partition wall. It can be seen that neglecting the longitudinal heat conduction in the partition wall in the x direction is problematic, because wherever there is a temperature gradient, thermal conduction also occurs [109].

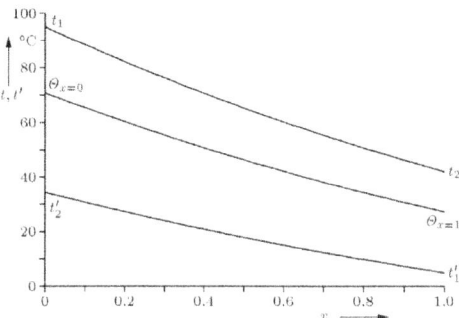

Figure 9.27 Temperature curve in a counterflow heat exchanger without taking into account the longitudinal heat conduction in the partition wall. The strong change of the mean partition wall temperature Θ in x-direction is visible. This is particularly pronounced in a counterflow heat exchanger. In an unidirectional flow heat exchanger, on the other hand, this fact only plays a subordinate role.

In a heat exchanger, it is assumed that only heat from the heat-emitting fluid passes through the partition wall to the heat-receiving fluid; there should be no heat transfer between the partition wall and the environment. This is contradicted by the temperature gradient in the partition wall at the left and right edges of the heat exchanger in Fig. 9.27. If the longitudinal heat conduction in the partition wall is taken into account, the temperature gradient here should have a horizontal tangent.

In [109], the mathematical treatment of the problem of longitudinal heat conduction in the partition wall is discussed in detail. In the following, the most important results are summarised. The following quantities are used: x_0 as the length of the plane partition wall in x-direction, the heat transfer coefficients α and α' and the over-all heat transfer coefficient k. The surface temperatures ϑ_0 und ϑ'_0 are expressed with the help of the mean wall temperature

$$\Theta = \frac{\vartheta_0 + \vartheta'_0}{2} \; ; \tag{9.214 a}$$

thus, the new auxiliary quantities ω und ω' are introduced as

$$t - \vartheta_0 = (t - \Theta)\omega \quad (9.214\,\text{b}) \qquad \vartheta'_0 - t' = (\Theta - t')\omega' \; . \quad (9.214\,\text{c})$$

These are given by

$$\omega = \frac{2\dfrac{k}{\alpha}}{1 - \left(\dfrac{k}{\alpha'} - \dfrac{k}{\alpha}\right)} \quad (9.215\,\text{a}) \qquad \omega' = \frac{2\dfrac{k}{\alpha'}}{1 + \left(\dfrac{k}{\alpha'} - \dfrac{k}{\alpha}\right)} \quad (9.215\,\text{b})$$

After several transformaions, the new auxiliary quantity z is used:

$$\frac{k}{\alpha\,\omega} = z \quad (9.216\,\text{a}) \qquad \frac{k}{\alpha'\,\omega'} = 1 - z \quad (9.216\,\text{b})$$

As an additional dimensionless ratio, in [109]

$$u = k\frac{\delta}{\lambda}\left(\frac{x_0}{\delta}\right)^2 \tag{9.217}$$

is introduced, where the thermal conductivity coefficient λ and the wall thickness δ appear.

The results are shown graphically in Fig. 9.28. For $u > 100$, one can disregard the heat conduction in the partition wall parallel to the partition surface. For $u < 100$, there is a remarkable influence on the operating characteristic which depends on z according to Eq. (9.216). The longitudinal heat conduction in the plane partition wall in the plane of the fluid flows reduces the heat transfer performance of a counterflow heat exchanger, which is described by the ratio Ψ:

$$\Psi = \frac{\text{operating characteristic with consideration of longidudinal heat conduction}}{\text{operating characteristic without consideration of longidudinal heat conduction}}$$

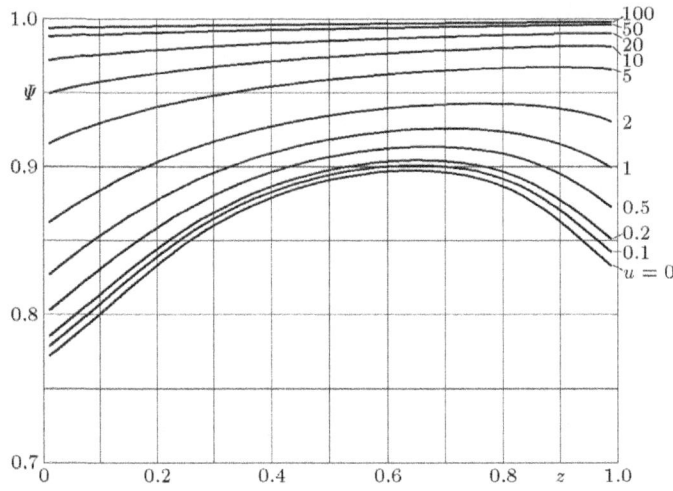

Figure 9.28 Ratio of operating characteristic with consideration of the heat conduction in longitudinal direction (= perpendicular to the heat transfer direction) in a counterflow heat exchanger to the operating characteristic without consideration of the heat conduction in longitudinal direction

Taking into account the longitudinal heat conduction in the partition wall, the temperature curve in a counterflow heat exchanger is qualitatively as shown in Fig. 9.29.

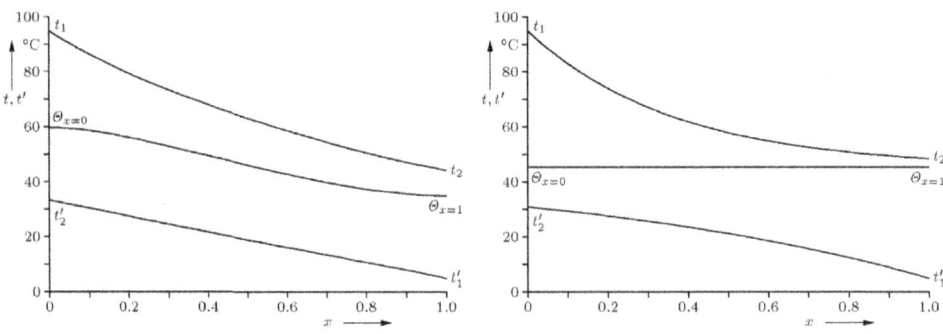

Figure 9.29 Temperature profile in a counterflow heat exchanger considering the longitudinal heat conduction in the partition wall

Fig. 9.30 Temperature profile in a counterflow heat exchanger with perfect temperature equalisation in the partition wall

The often practised neglect of the heat conduction resistance through the wall means $\lambda \to \infty$ in the direction of heat flow and $\lambda = 0$ in the perpendicular direction. In

the case where $\lambda \to \infty$ is in all directions of the partition wall, this leads to perfect temperature equilibrium in the partition wall, which leads to the borderline case $u = 0$ (Fig. 9.30). The following equations apply to the temperature curves of the warm and of the cold fluid:

$$t = t_1 - (t_1 - \Theta)\left(1 - e^{-\frac{\alpha A}{W}x}\right) \qquad (9.218\,\text{a})$$

$$t' = t'_1 + (\Theta - t'_1)\left(1 - e^{-\frac{\alpha' A}{W'}(1-x)}\right) \qquad (9.218\,\text{b})$$

From the equality of the heat flows on both sides of the partition wall

$$(t_1 - \Theta)W\left(1 - e^{-\frac{\alpha A}{W}}\right) = (\Theta - t'_1)W'\left(1 - e^{-\frac{\alpha' A}{W'}}\right) \qquad (9.218\,\text{c})$$

follows for the partition wall temperature and for the operating characteristic:

$$\Theta = \frac{t_1 W\left(1 - e^{-\frac{\alpha A}{W}}\right) + t'_1 W'\left(1 - e^{-\frac{\alpha' A}{W'}}\right)}{W\left(1 - e^{-\frac{\alpha A}{W}}\right) + W'\left(1 - e^{-\frac{\alpha' A}{W'}}\right)} \qquad (9.219)$$

$$\Phi \frac{W_1}{W} = \frac{1}{\dfrac{1}{1 - e^{-\frac{\alpha A}{W}}} + \dfrac{\dfrac{W}{W'}}{1 - e^{-\frac{\alpha' A}{W'}}}} \qquad (9.220)$$

Since heat conduction perpendicular to the heat transfer direction reduces the operating characteristic of a counterflow heat exchanger, this is the heat exchanger with the smallest operating characteristic. The equations for the fluid temperatures and the operating characteristic apply to all partition wall heat exchangers, because with an uniform partition wall temperature, the differences caused by the flow paths are eliminated.

9.9.4 Design Diagram

In the following, a crossflow heat exchanger is considered in which water flows through within smooth pipes and is cooled by air flowing over the finned outsides of the pipes - such as a motor vehicle heat exchanger. First, one needs empirical data on the heat transfer and pressure drop of the two fluids from measurements on similar heat exchangers. Data on heat transfer and pressure drop can be found in [66], [67], [68], [69], [77] and [92], but all conditions are not always published there.

Measured data are the inlet and outlet temperatures of the substance flows. In addition, measured data are the mass flows, from which the heat capacity flows of the mass flows result. All measurement data are averaged over the flow cross-sections. The calculation of the heat flows is a control option, since the heat fow released by the warmer fluid $\dot{Q}$ should be equal to the heat flow $\dot{Q}'$ absorbed by the colder fluid:

$$\dot{Q} = \dot{Q}' = (t_1 - t_2)W = (t'_2 - t'_1)W' \qquad (9.221)$$

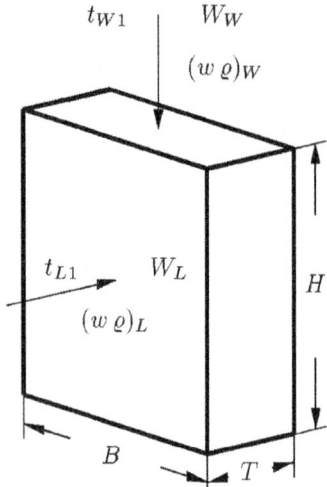

Figure 9.31 Block dimensions of a crossflow heat exchanger

Figure 9.32 Specific heat transfer data of a cross flow heat transfer system

Using the operating characteristic Φ or the mean temperature difference $(t-t')_m$, kA is obtained. This value depends on the size of the measured heat exchanger. The desire to separate the over-all heat transfer coefficient k and the area A leads to a problem: The area A is not physically specifiable. Any separation of k and A made for different surfaces on the air and water sides is arbitrary and should therefore be avoided. However, the elimination of the size and the formation of a specific heat transfer value remains desirable. Therefore, one should relate the quantity kA to the block volume $V = BHT$ according to Fig. 9.31.

As an intermediate step in the evaluation of measurement results, a diagram is obtained with the application of kA/V over the mass flow density (= the product of velocity and density) of the air flow $(w\varrho)_L$, with the mass flow density of the liquid water flow $(w\varrho)_W$ as a parameter (Fig. 9.32). It is advantageous to relate the velocities in the expressions $(w\varrho)_L$ and $(w\varrho)_W$ to the corresponding frontal areas BH and BT [109].

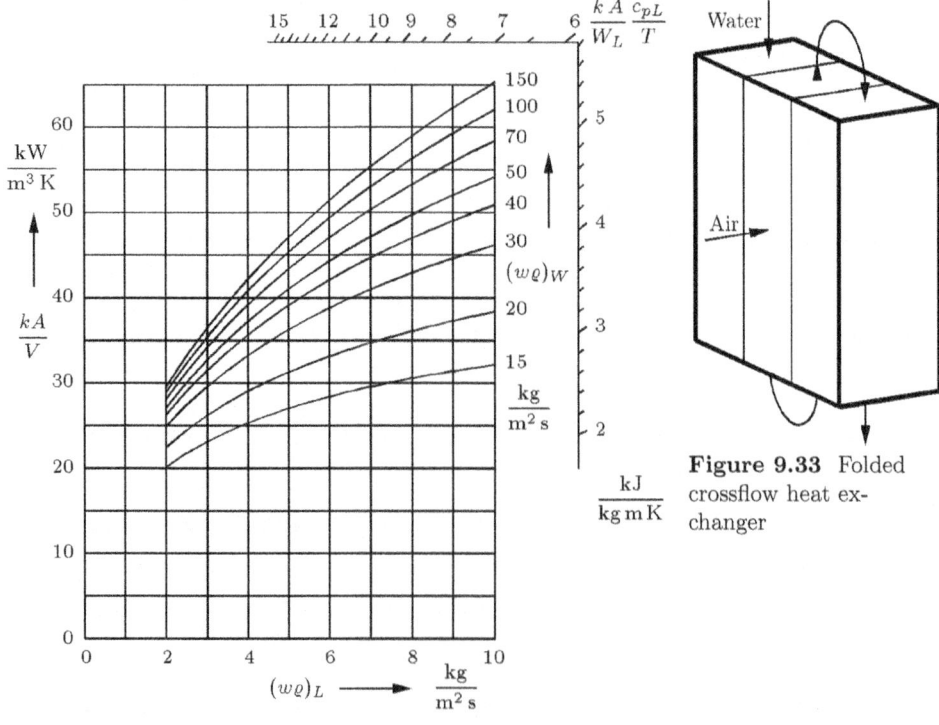

Figure 9.33 Folded crossflow heat exchanger

In a x, y coordinate system with x as abscissa and y as ordinate

$$\frac{y}{x} = \text{const}$$

is a straight line of origin. In a $\frac{kA}{V}, (w\varrho)_L$ coordinate system, to a straight line of origin with $V = BHT$, $(w\varrho)_L = \frac{\dot{m}_L}{BH}$ and $W_L = \dot{m}_L c_{pL}$, a value K belongs and is respresented by the following expression: $K = \dfrac{\frac{kA}{V}}{(w\varrho)_L}$

$$K = \frac{\frac{kA}{V}}{(w\varrho)_L} = \frac{kA}{BHT} \frac{BH}{\dot{m}_L} \frac{c_{pL}}{c_{pL}} = \frac{kA}{W_L} \frac{c_{pL}}{T} . \quad (9.222)$$

This expression indicates a design task.

To determine the dimensions of a crossflow heat exchanger, information about the desired heat flow $\dot{Q}$, the two mass flows $\dot{m}_L$ and $\dot{m}_W$ as well as their inlet temperatures t_{L1} and t_{W1} are required. Eq. (9.159) gives the operating characteristic Φ with $\dot{Q}$, $t_1 = t_{W1}$, $t'_1 = t_{L1}$ and W_1. The operating characteristic and the ratio of the heat capacity flows provides graphically or mathematically $\dfrac{kA}{W_L}$ with W_L as the heat capacity flow of the airflow. With the specific heat capacity c_{pL} and the block depth T one can calculate the value of the quantity K according to Eq. (9.222).

Each point on the straight line of origin in the $\frac{kA}{V}, (w\varrho)_L$ diagram provides the assignment of a value $(w\varrho)_L$ to a value of $(w\varrho)_W$ and leads via the equations of the heat capacity flows of the water flow and the air flow

$$W_W = BT(w\varrho)_W c_{pW} \qquad\qquad W_L = BH(w\varrho)_L c_{pL} \qquad (9.223)$$

to the equations of determination for the width B and the height H:

$$B = \frac{W_W}{T(w\varrho)_W c_{pW}} \qquad\qquad H = \frac{W_L}{B(w\varrho)_L c_{pL}} \qquad (9.224)$$

If the result of the calculation is not satisfactory, choosing a different point on the straight line of origin in Fig. 9.32 provides a new solution. In the case of unfavourably narrow and high block dimensions, a fold of the heat exchanger can possibly remedy the situation (Fig. 9.33).

Further solution variants result from the choice of a new number of tube or pipe rows and thus a different block depth T. The overview of the various possible solutions for a design task should be improved by using margin scales on the side of the diagram for the individual block depths (Fig. 9.34). The easy access to a multitude of solutions makes it possible to understand the design of a heat exchanger as an optimisation task.

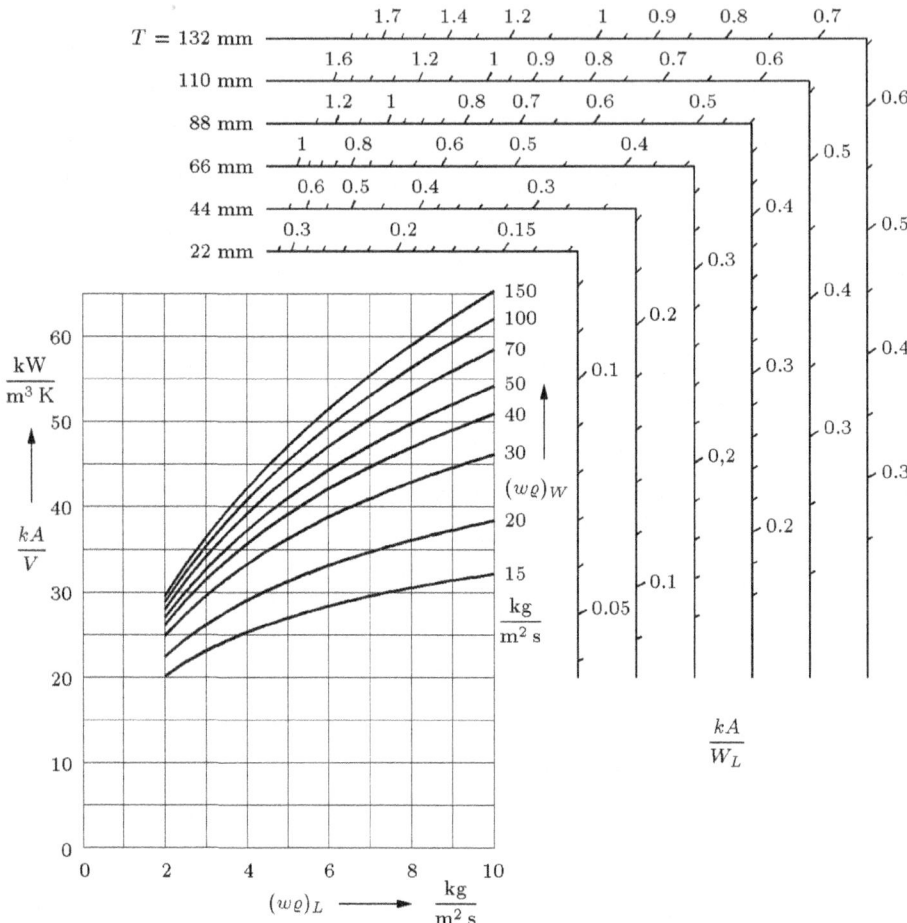

Figure 9.34 Design diagram for crossflow heat exchangers

Tasks for Section 9

1. A 200 mm thick plane wall of unknown thermal conductivity separates two rooms with temperatures 24.0 °C and 6.0 °C. The surface temperatures of the wall are 17.7 °C and 7.5 °C.

On one side of the wall an insulating layer with a thickness of 10 mm and a thermal conductivity of 0.0407 W/(m K) is applied. Afterwards, with unchanged room temperatures and heat transfer coefficients, the wall surface temperatures are 19.8 °C and 7.0 °C.

a) How does the heat flow after the implementation of the insulating layer relate to the heat flow through the uninsulated wall?

b) What is the thermal conductivity of the wall without an insulating layer?

c) What are the heat transfer coefficients at the two wall surfaces?

2. At an outside temperature of −15 °C, there should be a temperature of 20 °C in a room. The wall consists of an outer concrete layer with a thickness of 200 mm and a thermal conductivity

of 1.51 W/(m K) as well as an inner insulating layer of foamed material with a thermal conductivity of 0.041 W/(m K). The heat transfer coefficients outside and inside are 23 W/(m² K) and 5 W/(m² K). For comfort reasons, the surface temperature on the inside of the wall should not be below 16 °C.

a) What is the minimum thickness of the insulating layer?

b) What is the heat loss per square metre of wall area?

c) What would be the heat loss per square metre of wall area and the inside wall temperature without an insulating layer?

3. In a counterflow heat exchanger, a gas flow is to be cooled from 700 °C to 200 °C. Water with 15 °C is available as a coolant. The heat transfer coefficient on the water side is twenty times as large as the heat transfer coefficient on the gas side. The thermal resistance in the partition wall can be neglected. The water flow rate is chosen so that the highest partition wall temperature is 80 °C. What kA/W_1 value is required?

4. In a company brochure from the nineties, the over-all heat transfer coefficient $k = 1.6$ W/(m² K) is given for a new thermal insulating glass. As a particular advantage, it is mentioned that the room-side surface of the heat-insulating pane remains warmer in winter than with insulating glass of older designs, which eliminates the unhealthy veil of cold near the window. With an outside temperature of -10 °C and a room temperature of 21 °C, the temperature of the room-side pane surface is 15 °C. With an insulating glass of older design, this temperature is only 9.5 °C with the same heat transfer coefficients. The thermal insulation effect of the new thermal insulating glass corresponds to that of a 37 cm thick solid brick wall with a thermal conductivity $\lambda = 0.79$ W/(m K).

a) What heat flow passes through 1 m² of the thermal insulating glass at the temperatures given?

b) What is the heat transfer coefficient on the room side?

c) What heat flow passes through 1 m² of the older type of insulating glass at the temperatures stated?

d) What is the value of the heat transfer coefficient on the outside?

5. If one looks at the temperature curve for a heat exchanger with solely crossflow, one notices that the flow filament of the warm fluid, which is at the inlet cross-section of the cold fluid behaves as if the cold fluid would vaporise. The flow filament of the cold fluid at the inlet cross-section of the warm fluid has a temperature profile as if the warm fluid would condense. From this it is very easy to determine kA/W_1 and kA/W_2. What are these values if the inlet temperatures of both fluids are 0 °C and 100 °C, the highest temperature of the outlet profile of the cold fluid is 86.5 °C and the lowest temperature of the outlet profile of the warm fluid is 36.8 °C?

6. A copper pipe with an outer diameter of 50 mm, a wall thickness of 2 mm, a length of 3.4 m and a thermal conductivity of 370 W/(m K) is flowed through by a liquid with a heat transfer coefficient of 2000 W/(m² K). The heat transfer coefficient on the outside is 10 W/(m² K). Enveloping the pipe with a 50 mm thick thermal insulating layer, on the one hand causes an increase in the thermal resistance (insulating effect), on the other hand causes an increase in the surface area (increase in heat transfer with unchanged heat transfer coefficient).

a) What is the product of the over-all heat transfer coefficient k and the surface area A for the copper pipe without insulating layer?

b) What is the value of k without insulating layer if it is related in a first consideration to the outer surface and then related in a second consideration to the inner surface?

c) What thermal conductivity must the insulating layer have if the two effects mentioned cancel each other out and kA remains unchanged?

d) What value for kA is obtained if the thermal conductivity of the insulating layer is only 10 % of the value in c)?

e) What value for kA results if the thermal conductivity of the thermal insulating layer is as in sub-task d), but the thermal insulating layer is only 25 mm thick?

10 Humid Air

10.1 State Variables of Humid Air

Humid air (also named moist air) is a homogeneous mixture of dry air and water. Dry air is also called clean air. The water in the mixture may be vapor, liquid or solid. When calculating changes of state, the dry air is taken as the reference value. The phase of the dry air remains unchanged, while the water can undergo phase changes. The phase changes are called evaporation during the transition from the liquid to the gaseous aggregate state and as thawing during the phase transition in the opposite direction; other possible phase transitions are described in section 5.1.1.

As long as the water is only contained in the air as a vapor, the behaviour of ideal gases can be assumed for both mixing partners. For humid air then the law of *Dalton* applies. According to the thermal equation of state of ideal gases (4.25), to dry air and water vapor applies

$$p_L V = m_L R_L T \quad \text{with} \quad R_L = 287.06 \text{ J/(kg K)} \quad \text{(dry air)} \tag{10.1}$$

$$p_D V = m_D R_D T \quad \text{with} \quad R_D = 461.52 \text{ J/(kg K)} \quad \text{(water vapor)}. \tag{10.2}$$

For the humid air pressure (= total pressure) as sum of the partial pressures one gets

$$p = p_L + p_D . \tag{10.3}$$

Compared to other mixtures of ideal gases, humid air has the peculiarity that water vapor cannot be mixed in arbitrary quantities with dry air. The air can only contain such a quantity of water vapor until the partial pressure of the water vapor p_D reaches the saturation pressure p_{DS}.

The saturation pressure p_{DS} is a function of temperature. The saturation pressure of water vapor p_{DS} is independent of the atmospheric pressure. The relationship between saturation pressure and saturation temperature is represented by the empirical vapor pressure curve (cf. e.g. example 5.1 and Appendix: Tables A.8, A.9a and A.9b).

10.1.1 Relative Humidity

If the partial pressure of water vapor p_D is less than the saturation pressure p_{DS}, then the humid air is unsaturated. To describe the state of the unsaturated air one uses the relative humidity φ:

$$\varphi = \frac{p_D}{p_{DS}} = \frac{\varrho_D}{\varrho_{DS}} \tag{10.4}$$

ϱ_D and ϱ_{DS} are the densities of water vapor in unsaturated and saturated air, respectively.

10.1.2 Humidity Ratio and Saturation

In an air treatment process, the mass of clean air m_L is a constant quantity. The aim of an air treatment process may be to change the mass of water vapor m_D by humidification or dehumidification. To characterise the mixture composition, one uses the humidity ratio x, which is also called the moisture content or the water content:

$$x = \frac{m_D}{m_L} \qquad (10.5)$$

The humidity ratio x indicates the amount of water contained in 1 kg of dry air or in $(1+x)$ kg of humid air. The total mass of humid air is $m_{fL} = m_L(1+x)$.

Completely dry air has a humidity ratio of $x = 0$, for pure water x becomes $x = \infty$. In practice, however, only low x values up to about 0.04 kg/kg dry air are of interest. The humidity ratio x has the unit „kg/kg" and thus seems to be dimensionless. However, since the numerator contains "kg of water" and the denominator "kg of dry air", "water" cannot be shortened to "dry air", this should be taken into account when calculating the humidity ratio and carrying out a dimensional test of such a calculation.

Using from Eqs. (10.1) and (10.2) the masses of clean air and water vapor and using in addition Eq. (10.3,), one obtains for the humidity ratio

$$x = \frac{m_D}{m_L} = \frac{R_L}{R_D}\frac{p_D}{p_L} = 0.622\frac{p_D}{p - p_D} \ . \qquad (10.6)$$

If Eq. (10.6) is transformed to p_D, one obtains

$$p_D = \frac{p\,x}{0.622 + x} \ . \qquad (10.7)$$

According to Eq. (10.6), the maximum amount of moisture that can be contained in air in the form of vapor is

$$x_S = 0.622\frac{p_{DS}}{p - p_{DS}} \ . \qquad (10.8)$$

x_S depends on temperature and total pressure. Air containing x_S kg of water per kg of dry air, is called saturated humid air. At $x < x_S$ the air is unsaturated, at $x > x_S$ the air is supersaturated, which leads to fog formation or ice fog formation. The water vapor in the unsaturated humid air is superheated vapor. The water vapor in the saturated humid air is saturated vapor. In the supersaturated humid air there is a mixture of saturated vapor and liquid water (liquid-vapor mixture, wet vapor, also named fog) resp. below the triple temperature of water (0.01 °C) a mixture of saturated vapor and ice (ice fog). If one sets the degree of humidity x of the air in relation to the humidity ratio x_S of the saturated air, then one obtains the degree of saturation ψ:

$$\psi = \frac{x}{x_S} \qquad (10.9)$$

The relationship between the degree of saturation ψ and the relative humidity φ is obtained by inserting the Eqs. (10.6) and (10.8) into Eq. (10.9):

$$\psi = \frac{p_D}{p_{DS}}\frac{p - p_{DS}}{p - p_D} = \varphi\frac{p - p_{DS}}{p - p_D} \qquad (10.10)$$

At low temperatures, φ and ψ differ only little from each other. In the saturation state $\varphi = \psi = 1$ is given. For dry air $\varphi = \psi = 0$ is valid.

From the state of saturation, which is referred to in relative humidity, humidity ratio and degree of saturation, the dew point is clearly distinguishable: A change of the state of the unsaturated humid air towards the saturation state occurs at constant temperature by increasing the humidity ratio x. A change of the state of the unsaturated humid air in the direction of the dew point takes place with the humidity ratio x by lowering the temperature.

10.1.3 Specific Enthalpy

Air treatment processes in air conditioning involve isobaric changes of state. Here, the heat transferred in such processes is equal to the change in specific enthalpy. To get the specific enthalpy of both unsaturated and saturated humid air h, the specific enthalpy of dry air h_L and the product of the humidity ratio and the specific enthalpy of water vapor $x\,h_D$ are to be added: $h = h_L + x\,h_D$.

As a reference point for $h = 0$ kJ/kg the temperature $t_0 = 0\,°\text{C}$ is chosen; for practical calculations the difference to the triple point as reference point with $t_{Tr} = 0.01\,°\text{C}$ is meaningless for practical calculations (cf. sections 4.1.3 and 5.1.5). The specific enthalpy of dry air is

$$h_L = c_{pL}\,t \qquad \text{with} \qquad c_{pL} = 1.004 \text{ kJ/(kg K)}. \tag{10.11}$$

The specific enthalpy of water vapor is (cf. [128])

$$h_D = c_{pD}\,t + r_0 \quad \text{with} \quad c_{pD} = 1.905 \text{ kJ/(kg K)} \quad \text{and} \quad r_0 = 2500.9 \text{ kJ/kg}. \tag{10.12}$$

The specific enthalpy of liquid water is (cf. [128])

$$h_F = c_F\,t \qquad \text{with} \qquad c_F = 4.19 \text{ kJ/(kg K)}. \tag{10.13}$$

The specific enthalpy of ice is

$$h_E = c_E\,t - r_S \quad \text{with} \quad c_E = 2.05 \text{ kJ/(kg K)} \quad \text{and} \quad r_S = 333.0 \text{ kJ/kg}. \tag{10.14}$$

r_0 is the specific enthalpy of vaporisation of water and r_S is the specific enthalpy of ice formation (in each case at $t_0 = 0\,°\text{C}$).

The specific enthalpy h of humid air refers to 1 kg of dry air or, which is the same thing, to $(1+x)$ kg of humid air. To get the enthalpy H, one has to multiply the specific enthalpy h with the mass of dry air m_L: $H = m_L\,h$.

$(1+x)$ kg of unsaturated humid air consists of 1 kg of dry air with the specific enthalpy h_L and x kg of water vapor with the specific enthalpy $x\,h_D$. For unsaturated humid air, therfore

$$h = h_L + x\,h_D \tag{10.15}$$

results. For saturated humid air, in Eq. (10.15) x is replaced by x_S.

$(1+x)$ kg of supersaturated humid air consists of 1 kg of dry air with the specific enthalpy h_L, x_S kg of water vapor with the specific enthalpy $x_S\,h_D$ and $(x - x_S)$ kg water as fog or ice fog with the specific enthalpy $(x - x_S)\,h_F$ or $(x - x_S)\,h_E$.

The following applies to humid air in the fog region:

$$h = h_L + x_S\,h_D + (x - x_S)\,h_F \tag{10.16}$$

For humid air in the ice fog region the following applies:

$$h = h_L + x_S\,h_D + (x - x_S)\,h_E \tag{10.17}$$

10.1.4 Specific Volume and Density

The specific volume related to the volume of humid air is

$$v = \frac{V}{m_L + m_D} \,. \tag{10.18}$$

The density of the humid air is

$$\varrho = \frac{m_L + m_D}{V} \,. \tag{10.19}$$

Eqs. (10.1) and (10.2) are added, and using Eq. (10.3) one obtains

$$p_L V + p_D V = p V = (m_L R_L + m_D R_D) T = m_L R_D \left(\frac{R_L}{R_D} + \frac{m_D}{m_L} \right) T$$

$$V = m_L \left(0.622 + x \right) \frac{R_D T}{p} \,. \tag{10.20}$$

Eq. (10.20) is inserted into Eqs. (10.18) and (10.19). For unsaturated humid air this leads to

$$v = \frac{0.622 + x}{1 + x} \frac{R_D T}{p} \tag{10.21}$$

$$\varrho = \frac{1 + x}{0.622 + x} \frac{p}{R_D T} \,. \tag{10.22}$$

In the case of saturated humid air, in Eqs. (10.21) and (10.22) x is to be replaced by x_S. When humid air is supersaturated, in Eq. (10.20) one can neglect the volume of liquid or frozen water with respect to the volume of vapor. In this case, the value x_S valid for saturated vapor is to be used in Eq. (10.20). In the case of the mass of supersaturated humid air, one must also take into account the liquid or icy mass of the water in addition to the vapor mass, and calculate with the actual degree of humidity x. Therefore, in the following equations for the specific volume v and the density ϱ of supersaturated humid air both x_S and x are found:

$$v = \frac{0.622 + x_S}{1 + x} \frac{R_D T}{p} \qquad \varrho = \frac{1 + x}{0.622 + x_S} \frac{p}{R_D T} \tag{10.23}$$

Humid air has a smaller density than dry air at the same temperature and pressure.

10.2 Changes of State of Humid Air

Changes of state of humid air occur during heating or cooling, during humidification or dehumidification, and when mixing quantities of humid air, each with different state properties.

10.2.1 Temperature Change

The heat input or the heat output is considered as an isobaric change of state. Then the supply or release of heat is equal to the change in enthalpy. When humid air is heated, the humidity ratio x remains unchanged. When humid air is cooled to a temperature below the dew point, some of the moisture falls out; therefore, an amount of precipitation has to be noted. There are two possibilities:

a) The humidity precipitation can remain part of the humid air in the form of small droplets or ice crystals. The humid air is supersaturated, the humidity level x has not changed. An example is cloud formation.

b) The humidity precipitation, depending on the special condition, can separate as liquid or ice. The humidity ratio x becomes smaller, the humid air has the saturation state. An example is the cooling of humid air in a heat exchanger.

If the initial temperature t_1 and the specific enthalpy h_1 are known and the specific enthalpy h_2 is given by the transferred heat or by humidification or dehumidification, it can be difficult to determine the final temperature t_2. If the endpoint of the change of state is in the unsaturated region, it follows immediately

$$t_2 = \frac{h_2 - r_0\, x_2}{c_{pL} + c_{pD}\, x_2} \ . \tag{10.24}$$

If the endpoint of the change of state is in the fog region, then

$$t_2 = \frac{h_2 - r_0\, x_{S2}}{c_{pL} - (c_F - c_{pD})\, x_{S2} + c_F\, x_2} \tag{10.25}$$

follows. Since x_{S2} is a function of t_2, Eq. (10.25) can only be solved iteratively. If the endpoint of the change of state is in the ice fog region, then

$$t_2 = \frac{h_2 - (r_0 + r_S)\, x_{S2} + r_S\, x_2}{c_{pL} - (c_E - c_{pD})\, x_{S2} + c_E\, x_2} \tag{10.26}$$

follows. Eq. (10.26) can only be solved iteratively.

10.2.2 Humidification and Dehumidification

The humidity ratio (moisture content) x of the air can be increased by adding liquid water or water vapor:

$$x_2 = x_1 + \Delta x \tag{10.27}$$

The amount of clean air m_L remains constant. When adding liquid water with the mass Δm_F,

$$\Delta x = \frac{\Delta m_F}{m_L} \tag{10.28}$$

results. Correspondingly, when water vapor of the mass Δm_D is added, it follows

$$\Delta x = \frac{\Delta m_D}{m_L} \ . \tag{10.29}$$

The specific enthalpy of humid air changes during humidification:

$$h_2 = h_1 + \Delta h \tag{10.30}$$

When liquid water of mass Δm_F and specific enthalpy $h_F = c_F t$ is added,

$$\Delta h = \Delta x \, h_F = \frac{\Delta m_F}{m_L} c_F t \tag{10.31}$$

holds. Accordingly, when water vapor of mass Δm_D and specific enthalpy $h_D = c_{pD} t + r_0$ is added, and assuming water vapor to be an ideal gas,

$$\Delta h = \Delta x \, h_D = \frac{\Delta m_D}{m_L} (c_{pD} t + r_0) \tag{10.32}$$

follows. At higher temperatures and higher pressures, the specific enthalpy must be taken from the water vapor table. After the specific enthalpy h_2 has been determined via Eq. (10.30), with Eqs. (10.24) to (10.26) the temperature t_2 can be determined. The mass and enthalpy balances given here do not answer the question of the final temperatures. For their calculation, Eqs. (10.27) to (10.32) must be consulted. The relation between Δh and Δx is determined by the „drawing point'"(cf. section 10.6). Eqs. (10.27) to (10.32) apply mutatis mutandis to the dehumidification of humid air when the humidity separates according to section 10.2.1 b).

10.2.3 Mixing of Two Humid Air Quantities

When two quantities of humid air are mixed, a state is established which differs from the two initial states. The laws of conservation of mass and conservation of energy apply to the determination of the mixing state. From the law of conservation of mass it follows for the humidity ratio of the mixture x_M with $a_{12} = m_{L1}/m_{L2}$:

$$m_{L1} x_1 + m_{L2} x_2 = (m_{L1} + m_{L2}) x_M \quad x_M = \frac{m_{L1} x_1 + m_{L2} x_2}{m_{L1} + m_{L2}} = \frac{a_{12} x_1 + x_2}{a_{12} + 1} \tag{10.33}$$

From the law of conservation of energy it follows for the specific enthalpy of the mixture, if adiabatic mixing is assumed:

$$m_{L1} h_1 + m_{L2} h_2 = (m_{L1} + m_{L2}) h_M \quad h_M = \frac{m_{L1} h_1 + m_{L2} h_2}{m_{L1} + m_{L2}} = \frac{a_{12} h_1 + h_2}{a_{12} + 1} \tag{10.34}$$

Eqs. (10.33) and (10.34) give

$$a_{12} = \frac{x_2 - x_M}{x_M - x_1} = \frac{h_2 - h_M}{h_M - h_1}. \tag{10.35}$$

Eq. (10.35) is transformed to h_M:

$$h_M = \frac{h_2 - h_1}{x_2 - x_1} x_M + \frac{h_1 x_2 - h_2 x_1}{x_2 - x_1} \tag{10.36}$$

The temperature t_M of the mixed state (h_M, x_M) is calculated using Eqs. (10.24) to (10.26), where the subscript „2" is to be replaced by the subscript „M". If the state of the mixture is given and one component is known, the other component can be calculated:

$$x_2 = \frac{m_{LM} x_M - m_{L1} x_1}{m_{LM} - m_{L1}} \qquad h_2 = \frac{m_{LM} h_M - m_{L1} h_1}{m_{LM} - m_{L1}} \tag{10.37}$$

10.3 The h,x Diagram of *Mollier*

The computational treatment of the changes of state of humid air is, without special computational aids, associated with great effort because many problems can only be solved iteratively.

Very advantageous is the h, x diagram of *Mollier*, which allows graphical solutions in a very clear way. The specific enthalpy h is plotted as the ordinate and the humidity ratio x as the abscissa.

The two most important areas in the h, x diagram are the area of unsaturated humid air and the area of supersaturated humid air. Between the two areas lies the boundary case of saturated humid air. It is represented by the saturation curve. At $0\,°C$ (strictly speaking at the triple temperature $0.01\,°C$) the saturation curve has a slight kink. For the saturation curve, the equation is

$$h_S = c_{pL}\, t + (c_{pD}\, t + r_0)\, x_S \ . \tag{10.38}$$

h_S is plotted as a function of x_S. According to Eq. (10.8), x_S is a function of the partial pressure of saturation p_{DS} of the water vapor and the total pressure p. The partial pressure of saturation p_{DS} is a function of the temperature t. Thus, the saturation curve according to Eq. 10.38 at constant total pressure p is a function of temperature t. The limiting case of dry air is represented by the ordinate.

The set of curves of the isotherms in the unsaturated region is represented by

$$h = c_{pL}\, t + (c_{pD}\, t + r_0)\, x \ . \tag{10.39}$$

The isotherms given by Eq. (10.39) are straight lines with the ascent

$$\left(\frac{\partial h}{\partial x}\right)_t = c_{pD}\, t + r_0 \ . \tag{10.40}$$

The straight lines according to Eq. (10.39) are all very steep. At higher temperature the ascent increases. The set of curves of the isotherms in the fog region is represented by the following equation:

$$h = c_{pL}\, t + (c_{pD}\, t + r_0)\, x_S + (x - x_S)\, c_F\, t \tag{10.41}$$

The isotherms given by Eq. (10.41) are straight lines with the ascent

$$\left(\frac{\partial h}{\partial x}\right)_t = c_F\, t \ . \tag{10.42}$$

The increase of the fog isotherms is small compared to the increase of the isotherms in the unsaturated region. At $0\,°C$ the fog isotherm is horizontal in a rectangular h, x diagram. At higher temperatures the increase gets greater.

The set of curves of the isotherms in the ice fog region is

$$h = c_{pL}\, t + (c_{pD}\, t + r_0)\, x_S + (x - x_S)\, (c_E\, t - r_S) \ . \tag{10.43}$$

The isotherms according to Eq. (10.43) are straight lines with the ascent

$$\left(\frac{\partial h}{\partial x}\right)_t = c_E\, t - r_S \ . \tag{10.44}$$

The ascent of the ice fog isotherms is negative. At $0\,°C$ the slope is

$$\left(\frac{\partial h}{\partial x}\right)_{t=0\,°C} = -r_S \ . \tag{10.45}$$

At lower temperatures the isotherms drop more steeply.

In the region of supersaturated humid air there is not only one curve for all states with the temperature of 0 °C (more precisely: the triple temperature $t_{Tr} = 0.01\,°\mathrm{C}$): Between the fog isotherm 0 °C and the ice fog isotherm 0 °C there is a 0 °C region, where fog and ice fog occur in any mixing ratio.

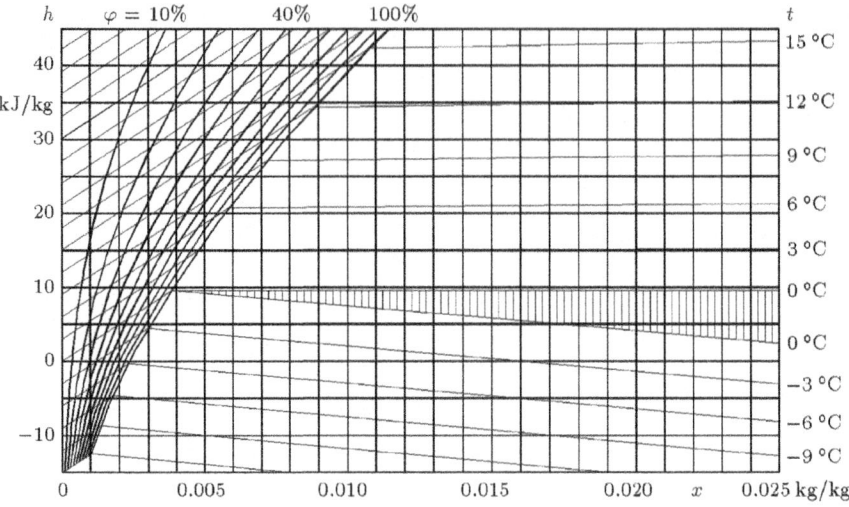

Figure 10.1 Rectangular h, x diagram (thin lines are isotherms)

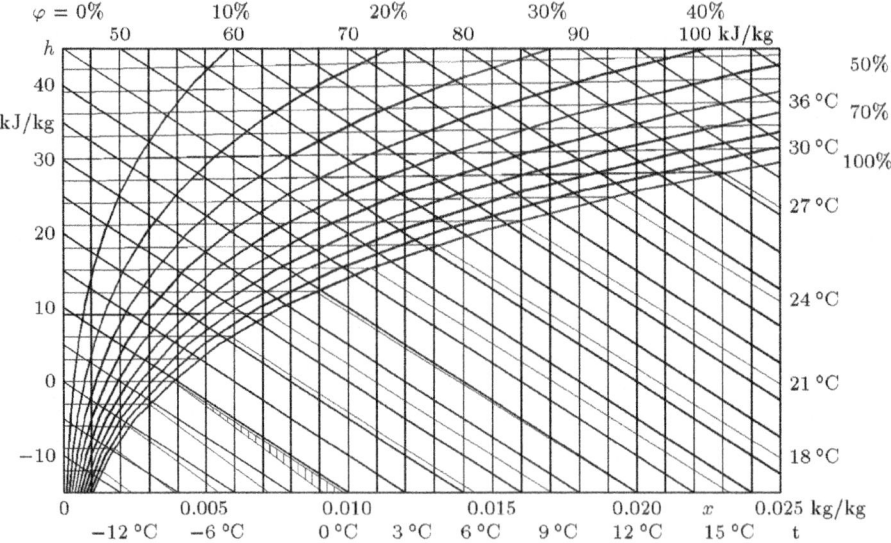

Figure 10.2 Oblique-angled h, x diagram (thin lines are isotherms, thick inclined lines are isenthalps)

The rectangular h, x diagram described so far (Fig. 10.1) has the disadvantage that the region of unsaturated humid air is very small. The largest part of the diagram is occupied by the area of supersaturated humid air. This does not correspond to the respective importance of the two areas. Therefore, in 1923 *Mollier* introduced an

oblique-angled h, x diagram in which the x axis (humidity ratio axis) has been rotated to the right so that the 0 °C isotherm in the unsaturated region is horizontal (Fig. 10.2).

The lines of constant humidity ratio x continue to run vertically. The lines of constant specific enthalpy h are straight lines with a negative slope, all parallel to the 0 °C fog isotherm (cf. [112], [17], [39]).

To indicate the state of the air in the unsaturated region, the curves of constant relative humidity φ are plotted. For $\varphi = 1$ the saturation curve is obtained.

If one wants to calculate a point of a curve $\varphi = $ const, one first chooses a temperature t and then finds the corresponding saturation pressure p_{DS}. Then one goes with a value $p_D = \varphi p_{DS}$ into Eq. (10.6) and calculates x. Eq. (10.15) yields h and thus a point on the curve $\varphi = $ const.

A curve of constant density $\varrho = $ const in the region of unsaturated vapor is obtained by solving Eq. (10.22) to t:

$$t = \frac{1+x}{0.622+x} \frac{p}{R_D \varrho} - T_0 \qquad (10.46)$$

In this equation, the temperature t is calculated as a function of ϱ and x. Eq. (10.15) gives the corresponding specific enthalpy.

To make it easier to follow the humidification and dehumidification processes in the oblique-angled h, x diagram according to Fig. 10.2, a boundary scale dh/dx is helpful. It indicates the direction of the change of state in relation to the zero point of the coordinate system. The zero point of the coordinate system ($h = 0$, $x = 0$) corresponds to dry air with 0 °C. This boundary scale is added to the oblique-angled h, x diagram given on page 464.

Example 10.1 Into the oblique-angled h, x diagram, some states are to be drawn: For the given temperature $t = 20$ °C, dry air state 1 ($\varphi = 0$; $x = 0$; $h = 20.080$ kJ/kg), humid air state 2 with relative humidity $\varphi = 0.4$ ($x = 0.0058750$ kg/kg; $h = 34.991$ kJ/kg), the corresponding saturation state 3 ($\varphi = 1.0$; $x_S = 0.014898$ kg/kg; $h_S = 57.894$ kJ/kg) and the dew point 4 ($t_{Tp} = 6.0051$ °C; $\varphi = 1.0$; $x_{Tp} = 0.0058750$ kg/kg; $h_{Tp} = 20.787$ kJ/kg) are shown. Point 5 is the zero point of the h, x diagram (Fig. 10.3).

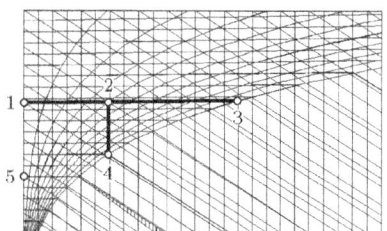

Figure 10.3 To example 10.1

10.3.1 Temperature Change

For a temperature change, all points of state above the dew point lie on a perpendicular $x = $ const. Below the dew point this is only true if the condensed liquid remains part of the humid air, that is, if the end point is in the fog region or in the ice-fog region. If one considers the end point of the temperature change leading into the supersaturated region as an intermediate state, whereupon this is followed by a deposition of the condensed liquid, then this deposition takes place at unchanged temperature on a fog isotherm. When the saturation curve is reached, the separation process is finished. This also applies analogously to the formation of frost (ice fog isotherm).

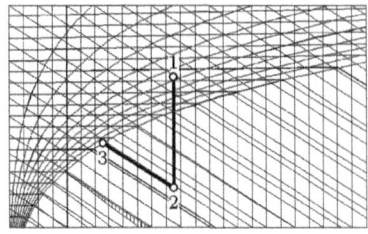

Figure 10.4 To example 10.2

Example 10.2 Humid air with the temperature $25\,°C$ and the humidity ratio $x = 0.0115$ kg/kg ($\varphi = 0.5727$; $h = 54.395$ kJ/kg; point 1) is cooled below the dew point temperature from $15.970\,°C$ to $7.647\,°C$ and enters the fog region ($h = 24.395$ kJ/kg; point 2). If the state is not stable and the liquid separates, the final state moves on a fog isotherm until the saturation state ($x = 0.0065842$ kg/kg; $\varphi = 1.0$; $h = 24.237$ kJ/kg; point 3) is reached (Fig. 10.4).

10.3.2 Humidification and Dehumidification

According to Eeqs. (10.31) and (10.32) the direction of the change of state $\Delta h/\Delta x$ is given by the specific enthalpy of the added or removed humidity h_F or h_D. The direction of the change of state in relation to the zero point of the coordinate system can be taken from the boundary scale. By parallel shifting one obtains the direction of the change of state in relation to the respective initial point.

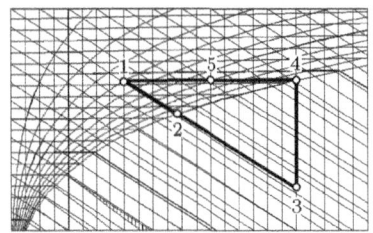

Figure 10.5 To example 10.3

Example 10.3 Humid air with $25\,°C$ and relative humidity 0.4 (state 1) is humidified with liquid water of $20\,°C$ first up to saturation (state 2) and then up to the humidity ratio 0.02 (state 3). Subsequently, heating takes place until the saturation state is reached (state 4). Finally, the air of state 4 is mixed with humid air in state 1 to form state 5, whereby the dry air masses of the mixing partners are equal (Fig. 10.5)

10.3.3 Mixing of Two Humid Air Quantities

From Eq. (10.36) it can be seen that all the mixing states in the h, x diagram lie on the straight line connecting the states of the two mixing partners. If, following Eqs. (10.27) and (10.30),

$$x_M = x_1 + \Delta x \qquad \text{and} \qquad h_M = h_1 + \Delta h \qquad (10.47)$$

holds, then one gets

$$\Delta h = \frac{h_M - h_1}{x_M - x_1} \Delta x \ . \qquad (10.48)$$

The dimensions of h_F and h_D on the one hand and $(h_M - h_1)/(x_M - x_1)$ on the other hand are equal.

10.4 Evaporation Model

10.4.1 Evaporation Coefficient

In an adiabatic evaporation channel, an air flow $\dot{m}_L$ with the substance values h_1, x_1, t_1 flows over an open water surface with the surface temperature ϑ_0 (Fig. 10.6). In the deeper layers of the water, where the processes at the surface do not have an immediate effect, the temperature is ϑ. In the air directly above the surface of the water, a boundary layer is formed, which adopts the surface temperature ϑ_0 of the water and which is saturated with humidity. The humidity ratio of the boundary

layer shall be denoted by x_S. Between the saturated boundary layer and the air flow, a diffusion process occurs in which the differential vapor flow $d\dot{m}_D$ diffuses into the air flow with the humidity ratio x. The drive for the diffusion flow is caused by the difference of the humidity ratios $x_S - x$. The differential vapor flow $d\dot{m}_D$ is proportional to the differential evaporation area dA. The evaporation coefficient σ is introduced as a proportionality factor:

$$d\dot{m}_D = \sigma (x_S - x)\, dA \tag{10.49}$$

The evaporation coefficient has the dimension of a clean air flow related to the area unit $(\text{kg dry air})/(\text{m}^2\,\text{h})$. The evaporation coefficient σ represents the hourly quantity of clean air of the humid air flow that would have to be saturated with water vapor on 1 m² evaporation surface in order to be able to remove the evaporated water quantity.

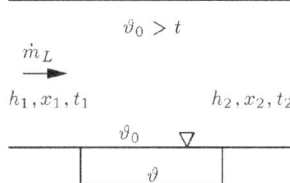

Figure 10.6 Evaporation channel

Due to the evaporation above the water surface dA, the substance values of the air change are:

$$h_2 = h_1 + dh \qquad x_2 = x_1 + dx \qquad t_2 = t_1 + dt \tag{10.50}$$

From Eqs. (10.5) and (10.49) follows

$$d\dot{m}_D = \dot{m}_L\, dx = \sigma (x_S - x)\, dA \qquad dx = \frac{\sigma}{\dot{m}_L}(x_S - x)\, dA\,. \tag{10.51}$$

The specific enthalpy change dh results from two energy balances:

10.4.2 Energy Balances

A system boundary is introduced, which is located in the area of the surface between air and water either slightly above the water surface or slightly below the water surface (Fig. 10.7).

The system boundary is crossed by substance flows and energy flows. To convert the differential water flow $d\dot{m}_F$ into the differential vapor flow $d\dot{m}_D$, the differential heat flow $d\dot{Q}_V$ must be supplied to the water surface for evaporation:

$$d\dot{Q}_V = d\dot{m}_D\, r \tag{10.52}$$

Here, $r = h_D - h_F$ is the specific heat of evaporation of the water. If the water surface has a higher temperature than the air flow ($\vartheta_0 > t$), then the differential heat flow $d\dot{Q}_L$ is transferred from the water surface to the air:

$$d\dot{Q}_L = \alpha_L (\vartheta_0 - t)\, dA \tag{10.53}$$

If one imagines the water surface as a boundary layer like a solid thin surface, then in the case of $\vartheta_0 < \vartheta$ the heat transfer from the deeper water layers with the temperature ϑ to the surface with the temperature ϑ_0 can be considered as convective heat transfer:

$$d\dot{Q}_W = \alpha_W (\vartheta - \vartheta_0)\, dA \tag{10.54}$$

The following energy balances are obtained:

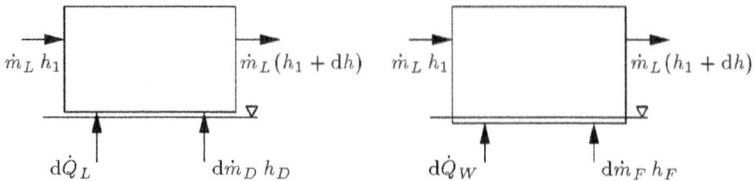

Figure 10.7 Energy balances in evaporation

For h_D and h_F, Eqs. (10.12) and (10.13) with temperature ϑ_0 hold:
$$h_D = c_{pD}\,\vartheta_0 + r_0 \qquad\qquad h_F = c_F\,\vartheta_0 \qquad (10.55)$$

The first balance gives
$$\dot{m}_L\,h_1 + \mathrm{d}\dot{Q}_L + \mathrm{d}\dot{m}_D\,h_D = \dot{m}_L\,(h_1 + \mathrm{d}h) \qquad \mathrm{d}\dot{Q}_L + \mathrm{d}\dot{m}_D\,h_D = \dot{m}_L\,\mathrm{d}h \,. \qquad (10.56)$$

The second balance gives
$$\dot{m}_L\,h_1 + \mathrm{d}\dot{Q}_W + \mathrm{d}\dot{m}_F\,h_F = \dot{m}_L\,(h_1 + \mathrm{d}h) \qquad \mathrm{d}\dot{Q}_W + \mathrm{d}\dot{m}_F\,h_F = \dot{m}_L\,\mathrm{d}h \,. \qquad (10.57)$$

From Eqs. (10.56) and (10.57), $\dot{m}_L\,\mathrm{d}h$ is eliminated. Using $r = h_D - h_F$ and Eq. (10.52), one obtains
$$\mathrm{d}\dot{Q}_L + \mathrm{d}\dot{Q}_V = \mathrm{d}\dot{Q}_W \,. \qquad (10.58)$$

The specific enthalpy of unsaturated humid air according to Eq. (10.15) with Eqs. (10.11) and (10.12) is a function of t and x:
$$h = c_{pL}\,t + c_{pD}\,t\,x + r_0\,x \qquad (10.59)$$

h is a specific state variable with the total differential $\mathrm{d}h$:
$$\mathrm{d}h = \left(\frac{\partial h}{\partial t}\right)_x \mathrm{d}t + \left(\frac{\partial h}{\partial x}\right)_t \mathrm{d}x = (c_{pL} + x\,c_{pD})\,\mathrm{d}t + (c_{pD}\,t + r_0)\,\mathrm{d}x = c_p\,\mathrm{d}t + h_D\,\mathrm{d}x \qquad (10.60)$$

$c_p = c_{pL} + x\,c_{pD}$ is the specific heat capacity of humid air related to 1 kg dry air or related to $(1+x)$ kg humid air.

10.4.3 *Lewis* Relationship

According to *Lewis*[1], evaporation can be described by the following model: An air flow $\dot{m}_L$ with the initial temperature t_1 is to be considered. This air flow $\dot{m}_L$, entering the evaporation surface A, is split into a fraction $z\,\dot{m}_L$ and a fraction $(1-z)\,\dot{m}_L$. The partial flow $z\,\dot{m}_L$ is heated from its initial temperature t_1 to the water surface temperature ϑ_0 (nearly wet bulb temperature) and reaches the saturation state of the boundary layer. The heat supplied to the partial flow z is (cf. [75])
$$\dot{Q}_L = z\,\dot{m}_L\,c_p\,(\vartheta_0 - t_1)\,, \qquad (10.61)$$

the humidity supplied to the partial flow z is
$$\dot{m}_D = z\,\dot{m}_L\,(x_S - x_1)\,. \qquad (10.62)$$

---
[1] *Gilbert Newton Lewis*, North American chemist, 1875 to 1946, worked in various fields of chemistry, physics and thermodynamics.

The saturated partial flow $z\dot{m}_L$ with the state h_S, x_S, ϑ_0 mixes during further flow with the unchanged partial flow $(1-z)\dot{m}_L$. This results in the final state of the mixed air h_2, x_2, t_2 leaving the evaporation surface A. The enthalpy balance of the mixture results in

$$(1-z)\dot{m}_L h_L + z\dot{m}_L h_S = \dot{m}_L h_2 \qquad h_2 = (1-z)h_1 + z h_S \qquad z = \frac{h_2 - h_1}{h_S - h_1}. \qquad (10.63)$$

The humidity balance of the mixture gives

$$(1-z)\dot{m}_L x_1 + z\dot{m}_L x_S = \dot{m}_L x_2 \qquad x_2 = (1-z)x_1 + z x_S \qquad z = \frac{x_2 - x_1}{x_S - x_1}. \qquad (10.64)$$

Assuming that during heating ϑ_0 and t_1 remain unchanged, the heat flow $\dot{Q}_L$ according to Eq. (10.53) can also be expressed as follows:

$$\dot{Q}_L = \alpha_L A (\vartheta_0 - t_1) \qquad (10.65)$$

From Eqs. (10.61) and (10.65) follows

$$z\dot{m}_L c_p = \alpha_L A \qquad z = \frac{\alpha_L A}{\dot{m}_L c_p}. \qquad (10.66)$$

Assuming that during heating x_S and x_1 remain unchanged, the humidity flow $\dot{m}_D$ taken up by the partial flow $z\dot{m}_L$ can also be expressed as follows:

$$\dot{m}_D = \sigma A (x_S - x_1) \qquad (10.67)$$

From Eqs. (10.62) and (10.67) follows

$$z\dot{m}_L = \sigma A \qquad z = \frac{\sigma A}{\dot{m}_L}. \qquad (10.68)$$

From Eqs. (10.66) and (10.68) one gets the *Lewis* relation:

$$z = \frac{\sigma A}{\dot{m}_L} = \frac{\alpha_L A}{\dot{m}_L c_p} \qquad \sigma = \frac{\alpha_L}{c_p} \qquad \frac{\sigma c_p}{\alpha_L} = 1 \qquad (10.69)$$

Eq. (10.69) is the expression for the fact that the same partial flow $z\dot{m}_L$ transfers the heat flow $\dot{Q}_L$ and the mass flow $\dot{m}_D$.

The *Lewis* relation is not a strictly valid law and does not always hold. However, it can be used, calculating air-conditioning problems.

10.5 Cooling Limit

When considering the state of the air $L(t,x)$ above a water surface in relation to the state of the evaporating water, seven cases can be distinguished with respect to the water surface states A to G (Fig. 10.8):

If one assumes that the state of the air above the water surface is practically unchanging, while the water surface can assume different temperatures, then the stationary state is state D. This state is called the adiabatic steady state. The water has the same temperature in all layers. The temperature of the water in state D is the lowest temperature that can be reached when the water is cooled by a flow of air from state L. This is the cooling limit. At the cooling limit $d\dot{Q}_W = 0$ is given.

According to Eq. (10.58), $d\dot Q_V = -d\dot Q_L$ holds, and with Eqs. (10.52) and (10.53) one obtains $d\dot m_D\, r = \alpha_L(t-\vartheta_0)\,dA$ as well as with Eq. (10.67) and the *Lewis* relation (cf. [112], [114])

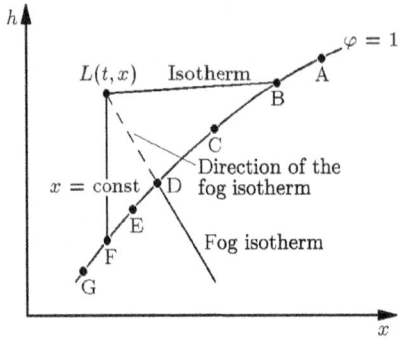

Figure 10.8 Water surface states during evaporation

$$\sigma(x_S - x)\,dA\,r = \sigma c_p(t-\vartheta_0)\,dA$$

$$\frac{c_p(t-\vartheta_0)}{x_S - x} = r . \tag{10.70}$$

Eq. (10.70) can be transformed into the equation

$$\frac{h_S - h}{x_S - x} = c_F\,\vartheta_0 \tag{10.71}$$

as shown below. With

$$r = h_D - h_F = c_{pD}\,\vartheta_0 + r_0 - c_F\,\vartheta_0$$

follows from Eq. (10.70):

$$c_p t - (c_{pL} + x\,c_{pD})\,\vartheta_0 = (c_{pD}\,\vartheta_0 + r_0)(x_S - x) - c_F\,\vartheta_0\,(x_S - x)$$
$$c_p t - c_{pL}\,\vartheta_0 - c_{pD}\,\vartheta_0\,x = c_{pD}\,\vartheta_0\,x_S + r_0\,x_S - c_{pD}\,\vartheta_0\,x - r_0\,x - c_F\,\vartheta_0\,(x_S - x)$$
$$c_p t + r_0\,x - (c_{pL}\,\vartheta_0 + c_{pD}\,\vartheta_0\,x_S + r_0\,x_S) = -c_F\,\vartheta_0\,(x_S - x)$$
$$h - h_S = -c_F\,\vartheta_0\,(x_S - x) \qquad \frac{h_S - h}{x_S - x} = c_F\,\vartheta_0$$

This is the slope of a fog isotherm. h_S and x_S refer to the cooling limit, h and x refer to the air state.

$(h_S - h)/(x_S - x)$ is the direction from the air state to the cooling limit. This direction is equal to the direction of the fog isotherm (Fig. 10.9).

Table 10.1 Heat and mass flows at the water surface

Case			$\dot m_D$	$\dot Q_V$	$\dot Q_L$	$\dot Q_W$	Release / Supply	
A	$t < \vartheta_0 < \vartheta$	$x < x_S$	↑	↑	↑	↑	$\dot Q_W = \dot Q_V + \dot Q_L$	
B	$t = \vartheta_0 < \vartheta$	$x < x_S$	↑	↑	0	↑	$\dot Q_W = \dot Q_V$	Unchanged temperature
C	$t > \vartheta_0 < \vartheta$	$x < x_S$	↑	↑	↓	↑	$\dot Q_L + \dot Q_W = \dot Q_V$	
D	$t > \vartheta_0 = \vartheta$	$x < x_S$	↑	↑	↓	0	$\dot Q_L = \dot Q_V$	Cooling limit
E	$t > \vartheta_0 > \vartheta$	$x < x_S$	↑	↑	↓	↓	$\dot Q_L = \dot Q_V + \dot Q_W$	
F	$t > \vartheta_0 > \vartheta$	$x = x_S$	0	0	↓	↓	$\dot Q_L = \dot Q_W$	No evaporation
G	$t > \vartheta_0 > \vartheta$	$x > x_S$	↓	↓	↓	↓	$\dot Q_V + \dot Q_L = \dot Q_W$	Condensation

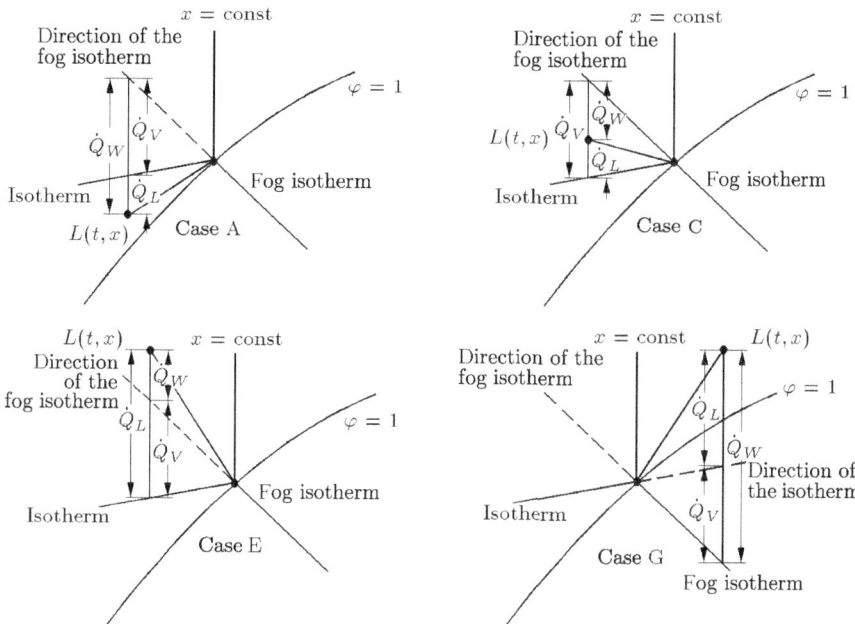

Figure 10.9 Relative quantities of the heat flows $\dot{Q}_L$, $\dot{Q}_W$ and $\dot{Q}_V$ at a water surface as a function of the air state $L(t,x)$

10.6 Evaporation and Dew Precipitation

The interaction between air and water surface has not only a great influence on processes in nature, but also in technical systems (e.g. in scrubbers in air-conditioning systems and in cooling towers with open cooling circuits). This interaction depends on the relative position of the air state to the state at the water surface in the h, x diagram.

If the air state is e.g. close to the saturation curve, then fog forms. If one assumes that the state of the water surface is practically unchanging, while the air under the influence of the water surface undergoes a change of state, then the question arises in which direction in the h, x diagram the change of state of the air takes place (Fig. 10.10).

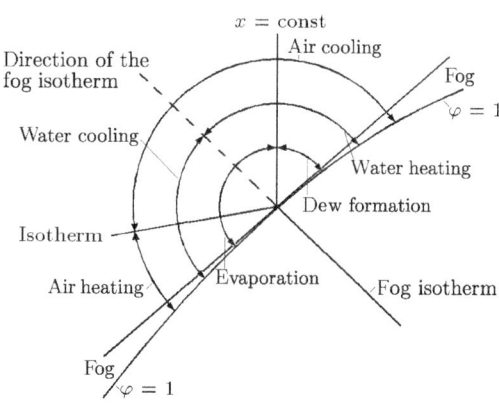

Figure 10.10 Effect of the water surface on the air state and the heat flows at the water surface

According to the first balance, from Eq. (10.56),

$$dh = h_D \frac{d\dot{m}_D}{\dot{m}_L} + \frac{d\dot{Q}_L}{\dot{m}_L} \ . \tag{10.72}$$

holds. From Eq. (10.51) follows

$$\frac{\dot{m}_D}{\dot{m}_L} = dx \ . \tag{10.73}$$

The transformed Eq. (10.51) is inserted into Eq. (10.53):

$$\frac{dA}{\dot{m}_L} = \frac{dx}{\sigma(x_S - x)} \tag{10.74}$$

Using the *Lewis* relation according to Eq. (10.69), one obtains

$$\frac{d\dot{Q}_L}{\dot{m}_L} = \frac{\alpha_L}{\sigma} \frac{\vartheta_0 - t}{x_S - x} dx = \frac{c_p(\vartheta_0 - t)}{x_S - x} dx \ . \tag{10.75}$$

The Eqs. (10.12), (10.73) and (10.75) are inserted into Eq. (10.72):

$$\frac{dh}{dx} = c_{pD} \vartheta_0 + r_0 + \frac{c_p(\vartheta_0 - t)}{x_S - x} \tag{10.76}$$

Eq. (10.76) can be transformed to

$$\frac{dh}{dx} = \frac{h_S - h}{x_S - x} \tag{10.77}$$

as shown below. Eq. (10.77) states that the air state moves in a straight line towards the boundary layer state. With $c_p = c_{pL} + x\, c_{pD}$ follows

$$c_{pD} \vartheta_0 + r_0 + \frac{c_p(\vartheta_0 - t)}{x_S - x} =$$

$$= \frac{1}{x_S - x}(c_{pD} \vartheta_0 x_S - c_{pD} \vartheta_0 x + r_0 x_S - r_0 x + c_{pL} \vartheta_0 + c_{pD} \vartheta_0 x - c_p t) =$$

$$= \frac{1}{x_S - x}(c_{pL} \vartheta_0 + c_{pD} \vartheta_0 x_S + r_0 x_S - c_p t - r_0 x) = \frac{h_S - h}{x_S - x} \ . \tag{10.78}$$

The point in the h, x diagram that marks the boundary layer state (h_S, x_S, ϑ_0) is called the "pulling point". Towards this point the state of the air moves.
The position of the "pulling point" on the saturation line depends on the condition of the surroundings for the heat transfer.

10.7 Water Vapor Diffusion Through Walls

Water vapor diffusion through a layer of air is described by *Fick*'s law (cf. [112]):

$$\dot{m}_{DD} = -D A \frac{dC}{dx} \tag{10.79}$$

D = Diffusion coefficient
A = Cross-sectional area
C = Water vapor concentration $C = \varrho_D = \frac{p_D}{R_D T}$

Thus, *Fick*'s law is given in the form

$$\dot{m}_{DD} = -\frac{D A}{R_D T} \frac{dp_D}{dx} \ . \tag{10.80}$$

For the vapor to diffuse in the x direction, $\dfrac{\mathrm{d}p_D}{\mathrm{d}x}$ must be negative. Integrating Eq. (10.80) gives

$$\dot{m}_{DD} = \frac{D\,A}{R_D\,T}\,\frac{p_{D0} - p'_{D0}}{\delta} \tag{10.81}$$

with the layer thickness δ. The expression $\dfrac{D}{R_D\,T}$ is temperature dependent, whereby also the diffusion coefficient D depends on the temperature: D increases with increasing temperature. However, the temperature dependence of $\dfrac{D}{R_D\,T}$ is not very strong, so that the expression can be regarded as approximately constant. If one summarises $\dfrac{D\,A}{R_D\,T}$ to a constant K

$$K = \frac{D\,A}{R_D\,T}, \tag{10.82}$$

thus the diffusion flow is

$$\dot{m}_{DD} = \frac{K}{\delta}(p_{D0} - p'_{D0}). \tag{10.83}$$

Diffusion through a wall is less than through a layer of air. The greater resistance of the wall to diffusion is expressed by the diffusion resistance factor μ. The diffusion flow through the wall is

$$\dot{m}_{DD} = \frac{K}{\mu\,\delta}(p_{D0} - p'_{D0}). \tag{10.84}$$

Analogous to the thermal resistance R_L according to Eq. (9.46), the vapor conduction resistance S can be introduced:

$$S = \frac{\mu\,\delta}{K} \tag{10.85}$$

In the case of a multi-layer wall, from Eq. (10.84) follows

$$\dot{m}_{DD} = \frac{K}{\sum \mu_i\,\delta_i}(p_{D0} - p'_{D0}). \tag{10.86}$$

In the stationary case is

$$\dot{m}_{DD} = \frac{K}{\sum \mu\,\delta}(p_{D0} - p'_{D0}) =$$
$$= \frac{K}{\mu_1\,\delta_1}(p_{D1} - p'_{D1}) = \frac{K}{\mu_2\,\delta_2}(p_{D2} - p'_{D2}) = \frac{K}{\mu_3\,\delta_3}(p_{D3} - p'_{D3}) = \ldots \tag{10.87}$$

At the contact surfaces of the layers are

$$p'_{D1} = p_{D2} \qquad p'_{D2} = p_{D3} \qquad p'_{D3} = p_{D4} \ldots \tag{10.88}$$

The vapor transfer resistance at the contact surfaces of the air spaces is negligible. Thus are

$$p_{D0} = p_{D1} \qquad p'_{Dn} = p'_{D0}. \tag{10.89}$$

The vapor pressures at the contacting surfaces of the layers are given by Eq. (10.87)

$$p'_{D1} = p_{D1} - \frac{\mu_1\,\delta_1}{\sum \mu_i\,\delta_i}(p_{D0} - p'_{D0}) \tag{10.90}$$

$$p'_{D2} = p_{D2} - \frac{\mu_2 \delta_2}{\Sigma \mu_i \delta_i}(p_{D0} - p'_{D0}) \qquad (10.91)$$

$$p'_{D3} = p_{D3} - \frac{\mu_3 \delta_3}{\Sigma \mu_i \delta_i}(p_{D0} - p'_{D0}) \ . \qquad (10.92)$$

The partial pressure curve calculated in this way is only valid for the case where no condensation occurs in the wall. To investigate whether condensation occurs, the course of the saturation partial pressure p_{DS} is determined. For this purpose, the temperature curve in the wall is required:

$$\vartheta_0 = t - \frac{\dot{Q}}{\alpha A} = \vartheta_1 \qquad (10.93)$$

$$\vartheta'_1 = \vartheta_1 - \frac{\delta_1 \dot{Q}}{\lambda_1 A} = \vartheta_2 \qquad (10.94)$$

$$\vartheta'_2 = \vartheta_2 - \frac{\delta_2 \dot{Q}}{\lambda_2 A} = \vartheta_3 \qquad (10.95)$$

The partial pressure of the water vapor cannot be higher than the saturation pressure. From Eq. (10.87) follows:

$$\frac{\dot{m}_{DD}}{K} = \frac{p_{D0} - p'_{D0}}{\Sigma \mu_i \delta_i} = \frac{p_{D1} - p'_{D1}}{\mu_1 \delta_1} = \frac{p_{D2} - p'_{D2}}{\mu_2 \delta_2} = \frac{p_{D3} - p'_{D3}}{\mu_3 \delta_3} = \ldots \qquad (10.96)$$

According to Eq. (10.96), a straight line is obtained if the partial pressure of water vapor is plotted over the products $\mu_i \delta_i$ and if no condensation occurs in the wall.

To avoid condensation, one uses barrier layers. A barrier layer is usually a very thin layer that has a very high diffusion resistance factor μ. The thermal resistance is insignificant.

Rules to prevent condensation:

The thermal resistance R_L should be large on the outside and small on the inside. The vapor conduction resistance S should be large on the inside and small on the outside. (Inside means warm, outside means cold: $\rightarrow$ buildings).

The product $\mu \lambda$ shall decrease from the warm to the cold side.

A barrier layer shall be placed on the side of higher water vapor partial pressure.

Example 10.4 A house wall consists of 5 layers with the following data:

No.	1	2	3	4	5
Material	Interior plaster	Foil	Cork	Masonry	Exterior plaster
Heat conductivity λ in W/(m K)	0.814	0.174	0.047	0.79	0.814
Diffusion resistance factor μ	10	6000	7	10	10
Layer thickness δ in m	0.02	0.001	0.04	0.24	0.02

On the left is a room with temperature 22 °C and relative humidity 80 %, on the right is outside air with temperature 0 °C and relative humidity 90 %. The heat transfer coefficients on the inside and outside are 8 W/(m² K) and 23 W/(m² K). It is to be investigated whether the water vapor diffusion flow through the wall condenses in the wall.

First calculation:

Column 1: number of the interface
Column 2: x-coordinate in m
Column 3: temperature in °C
Column 4: vapor pressure p_D in mbar
Column 5: saturation pressure p_{DS} in mbar

Second calculation:

Column 1: number of the interface
Column 2: x-coordinate in m
Column 3: temperature in °C
Column 4: vapor pressure p_D in mbar
Column 5: saturation pressure p_{DS} in mbar

1	2	3	4	5		1	2	3	4	5
		22.000	21.162	26.452				22.000	21.162	26.452
0	0.000	20.005	21.162	23.399		0	0.000	20.013	21.162	23.411
1	0.020	19.612	20.817	22.837		1	0.001	19.922	14.930	23.279
2	0.021	19.521	10.468	22.707		2	0.021	19.531	14.723	22.721
3	0.061	5.936	9.985	9.312 !		3	0.022	19.440	8.491	22.592
4	0.301	1.086	5.845	6.612		4	0.062	5.911	8.201	9.296
5	0.321	0.694	5.500	6.427		5	0.302	1.082	5.708	6.610
		0.000	5.500	6.112		6	0.322	0.691	5.500	6.426
								0.000	5.500	6.112

The first calculation shows a higher water vapor partial pressure p_D than the saturation pressure p_{DS} at $x = 0.061$ m (marked by an exclamation mark !) (Fig. 10.11). At this point condensation of the water vapor occurs. To prevent this, an additional foil of the same type as foil No. 2 is applied as a barrier layer on the left side — on the room side — of the wall. The second calculation shows that this measure is successful (Fig. 11.12).

Figure 10.11 Water vapor diffusion through a five-layer wall

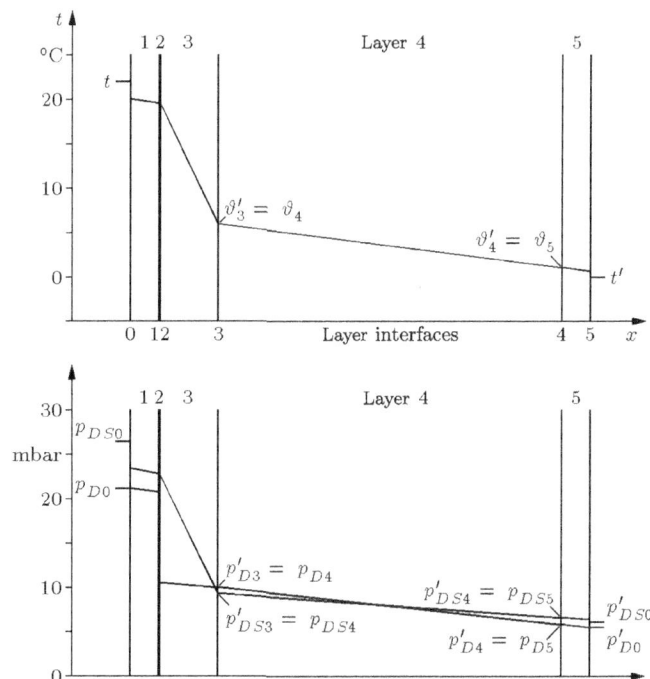

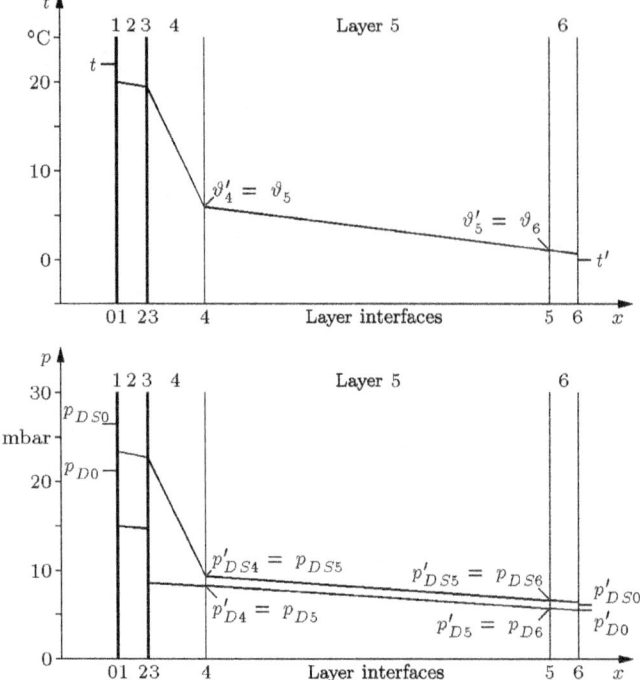

Figure 10.12 Water vapor diffusion through a six-layer wall

Fig. 10.11 and Fig. 10.12 each contain two diagrams. The upper diagram shows the temperature curve in a multi-layer wall (interior plaster, foil, cork, masonry, exterior plaster). The lower diagram shows the saturation pressure of the water vapor, which is dependent on the temperature, and the curve of the pressure of the water vapor, which depends on the vapor pressures of the air in the room to the left of the wall and in the outside air to the right of the wall.

Fig. 10.11 shows a condensation zone at the boundary between layers 3 and 4. This is remedied by a barrier layer on the inside of the wall, which turns the five-layer wall into a six-layer wall (Fig. 10.12). Everywhere in the wall, the pressure of the saturated vapor is now higher than the partial pressure of the water vapor.

Tasks for Section 10

1. Saturated humid air at 27 °C is mixed with dry air. At a barometric pressure of 1020 mbar, 3200 m³ of saturated humid air with 12 °C is produced.

a) What is the mass of the mixture and the mass of the dry air of the mixture?

b) What is the mass of the dry air and the temperature of the dry air?

2. At a barometer pressure of 925 mbar, 2200 m³ of humid air with a temperature of 28 °C and a relative humidity of 20 %, after an air treatment process shall have a temperature of 32 °C and a relative humidity of 33 %. The air treatment process is carried out in two stages. First, the air is humidified with liquid water of 18 °C, then it is heated.

a) What is the dry air mass of the humid air?

b) What is the total amount of water needed for humidification?

c) What is the temperature and the relative humidity of the air after humidification?

d) What specific enthalpy would water vapor have to have in order to reach the desired end state directly when humidifying not with liquid water but with water vapor?

3. On the Jungfraujoch at an altitude of 3454 m, at an air pressure of 653 mbar a temperature of 6 °C is measured. The dew point is -1 °C. What are the humidity ratio and the relative humidity of the air?

4. At a pressure of 1 bar and a temperature of 20 °C, humid air containing 6 g of water per kg of dry air, is isothermally compressed.

a) At what ratio of the final volume to the initial volume is saturation reached?

b) How much water per kg of dry air condenses during isothermal compression to 5 bar?

c) What temperature would humid air have, to reach at a pressure of 5 bar a humidity ratio of 6 g/kg and a relative humidity of 85 %?

5. At a barometric pressure of 800 mbar, unsaturated humid air with a dew point temperature of 5 °C flows over a water surface and gradually cools it to a cooling limit temperature of 14 °C.

a) What is the temperature of the unsaturated humid air?

b) What temperature would dry air have if the barometer pressure and the cooling limit temperature would remain unchanged?

c) What are the results of questions a) and b) at a barometer pressure of 1013 mbar?

6. Foggy air is converted into saturated humid air by the addition of heat. Further addition of the same amount of heat results in unsaturated humid air with 28.6 °C and a relative humidity of 34 %. All changes of state are isobaric at a total pressure of 945 mbar.

a) What is the humidity ratio of the air?

b) What is the temperature of the air in the intermediate state of saturation?

c) What is the temperature of the foggy air?

d) What is the total heat to be added if the humid air has a volume of 16.2 m^3 in the final state?

7. At an outside temperature of -5 °C and 100 % relative humidity, the air in a windowless hall should have a temperature of 27 °C and 60 % relative humidity. The two-layer wall of the hall consists of an outer concrete layer (thickness 20 cm; thermal conductivity 1.5 W/(m K); diffusion resistance factor 34) and an inner foam layer (thickness 2.5 cm; thermal conductivity 0.041 W/(m K); diffusion resistance factor 40). The heat transfer coefficients are 8 W/(m^2 K) inside and 23 W/(m^2 K) outside.

a) What are the wall temperatures at the surfaces of the layers?

b) How large must the product of diffusion resistance factor and thickness of the barrier layer be at least so that condensation in the wall is avoided?

c) How do all the results change if the outdoor air condition is assumed to be -15 °C with a relative humidity of 100 % ?

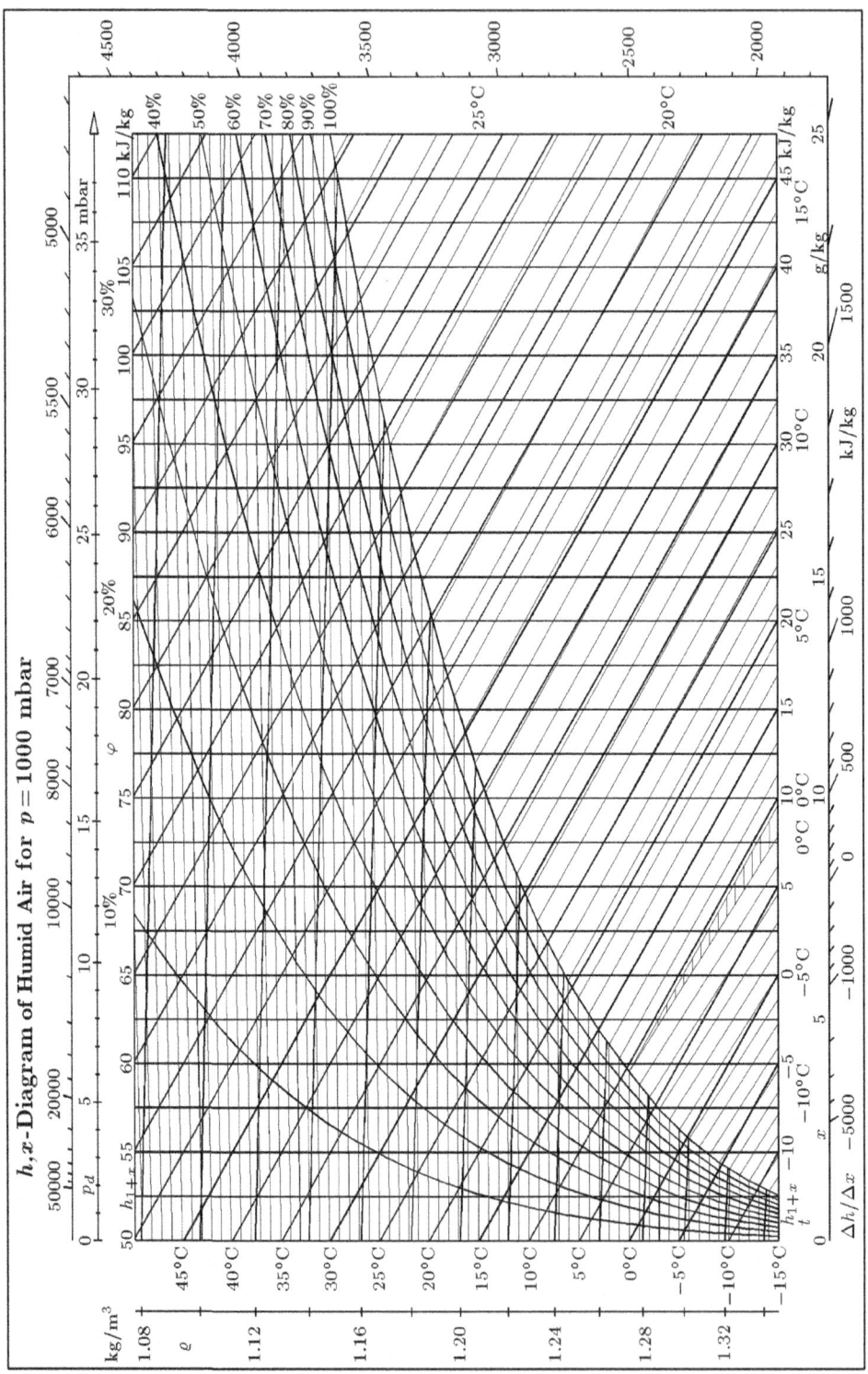

11 Combustion

Heat is predominantly produced by burning fossil fuels such as coal, lignite, petroleum products and natural gas; in addition, nuclear fission, liquids and gases generated from renewable energies, and the use of biomass and solar energy are also important for heating processes. This section focuses on the calculation of combustion processes of fossil fuels, of liquids and gases generated from renewable energies, and of biomass.

Fuels may be composed of substances with different constituents and with different chemical compounds. Often they consist of carbon C, hydrogen H and oxygen O.

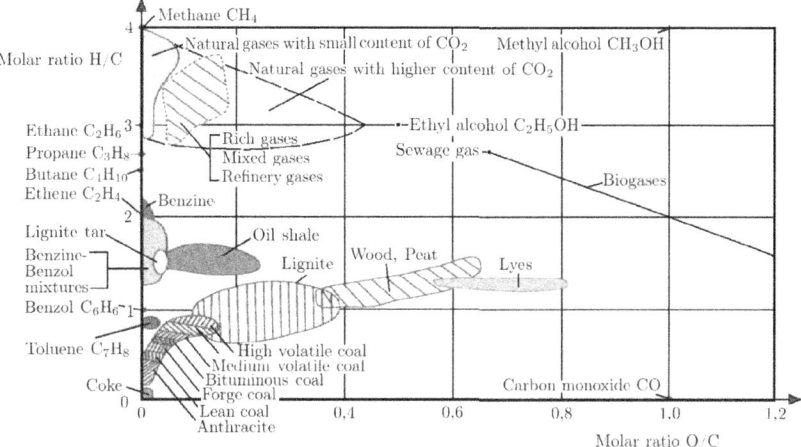

Figure 11.1 Carbon-based atomic hydrogen and oxygen contents of various fuels (molar ratio H/C, molar ratio O/C)

In Fig. 11.1 (e.g. [72]), several dry fuels are listed according to their molar ratio H/C (ordinate) and molar ratio O/C (abscissa):

• Gaseous fuels: methane CH_4, carbon monoxide CO, natural gases with small content of CO_2, natural gases with higher content of CO_2, rich gases, mixed gases, refinery gases, ethane C_2H_6, ethene C_2H_4, propane C_3H_8, butane C_4H_{10}. Hydrogen H_2 is also of technical interest as a gaseous fuel. It cannot be shown in Fig. 11.1.

• Liquid fuels: benzine. benzine-benzol mixtures, benzol C_6H_6, toluene C_7H_8, methyl alcohole CH_3OH, ethyl alcohole C_2H_5OH, lyes

• Solid fuels: oil shale, lignite tar, lignite, wood, peat, coke, hard coal (anthracite, lean coal, forge coal, bituminous coal, medium volatile coal, high volatile coal)

Fuels are burnt in a reaction chamber (e.g. the combustion chamber of a boiler, the cylinder of an internal combustion engine, etc.) (cf. Fig. 11.2). The chemical reactions that take place represent oxidations of the fuel or fuel mixture, whereby the required oxygen is usually added to the fuel. Oxygen is usually taken from the supplied air. Depending on the fuel and the amount of air supplied, combustion products of different compositions are formed; in addition, unburnt or only partially burnt substances may occur. These can be described as combustion gas (exhaust gas, flue gas), soot and ash or slag. To reproduce the combustion process mathematically, a quantity balance and an energy balance are required.

Figure 11.2 Combustion process

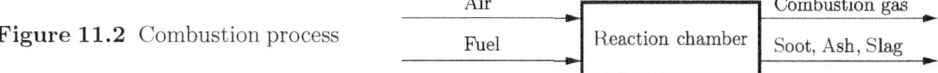

11.1 Fuels

In heat and energy technology, a distinction is made between gaseous, liquid and solid fuels as indicated above. For the description of combustion processes it is useful to relate the respective state variables to the quantities of the fuels used. For solid and liquid fuels, the mass m in kg is customary used as a quantity parameter. In the case of gaseous fuels, the volume at physical norm state V_0 (norm volume V_0 (section 4.1.3)) has proven itself as a further quantity. To distinguish from volumes at arbitrary states (indicated in m^3), in the following norm volumes V_0 are noted in the — in practice often used — unit Nm^3. Furthermore, e.g. for the recording of combustion reactions, the molar quantity (substance quantity) n in kmol is useful (section 4.1.6, section 12).

11.1.1 Gaseous Fuels

Gaseous fuels (fuel gases) may be technically pure substances or — more often — mixtures of combustible, carbonaceous and hydrogenous components and non-combustible (inert) components. Combustible components are e.g. hydrogen (H_2), carbon monoxide (CO), as the simplest hydrocarbon methane (CH_4) as well as higher hydrocarbons ($C_n H_m$) in different compositions (indicated as $\Sigma\, C_n H_m$), non-combustible components such as e.g. carbon dioxide (CO_2) and nitrogen (N_2, usually counted together with incombustible noble gases such as e.g. argon) as well as oxygen (O_2) (cf. e.g. [9], [74], [91]). As examples of the composition in volume fractions and of substance values, special natural gases are given in Table 11.1.

Table 11.1 Composition and substance values of natural gas L and natural gas H [107]

	r_{CH_4}	$r_{C_2H_6}$	$r_{C_3H_8}$	$r_{C_4H_{10}}$	r_{N_2}	r_{CO_2}	ϱ_0	M	R	c_p	$H_{s,0}$	$H_{i,0}$	c_{Zu}	c_{Zo}
	Vol.-%	Vol.-%	Vol.-%	Vol.-%	Vol.-%	Vol.-%	kg/Nm^3	kg/kmol	kJ/(kg K)	kJ/(kg K)	MJ/Nm^3	MJ/Nm^3	Vol.-%	Vol.-%
Nat. gas L[1]	83.8	3.6	0.8	0.4	9.8	1.6	0.83	18.6	0.448	1.91	37.26	33.66	4.7	16.3
Nat. gas H[2]	87.6	7.2	1.3	0.6	2.3	1.0	0.82	18.2	0.456	2.02	41.98	37.94	4.2	16.2
Nat. gas H[3]	98.2	0.6	0.2	0.1	0.8	0.1	0.73	16.4	0.509	2.15	39.85	35.93	4.4	16.5

[1] Netherlands   [2] Mixed gas H   [3] Siberian gas

Oxygen is required for the combustion of the combustible components. The combustion equations for complete combustion of these components are:

$$H_2 + 0.5\, O_2 = H_2O \tag{11.1}$$

$$CO + 0.5\, O_2 = CO_2 \tag{11.2}$$

$$CH_4 + 2\, O_2 = CO_2 + 2\, H_2O \tag{11.3}$$

$$C_n H_m + \left(n + \frac{m}{4}\right) O_2 = n\, CO_2 + \frac{m}{2} H_2O \tag{11.4}$$

The combustion equations give the ratio of the molecules reacting with each other during the chemical reaction and the ratio of the molar quantities as well as — assuming the presence of ideal gases — the ratio of the physical norm volumes (see section 4.1.6). Since the volume ratios in the volume fractions of the individual gases in a mixture of ideal gases are at the same time also the ratios of the physical norm volumes (section 4.5.3), the minimum oxygen demand per Nm^3 fuel gas (physical norm volume-related minimum oxygen demand) can be determined. If the fuel gas in volume fractions $r_i = r_i^B$ has the composition

$$r_{H_2} + r_{CO} + r_{CH_4} + \Sigma\, r_{C_n H_m} + r_{O_2} + r_{CO_2} + r_{N_2} = 1 \,, \tag{11.5}$$

then the minimum oxygen demand (more precisely: the norm volume-related minimum oxygen demand) o_{min}, measured in Nm^3 oxygen per Nm^3 fuel gas, is

$$o_{min} = \frac{V_{0\,Omin}}{V_{0\,B}} = 0.5\, r_{H_2} + 0.5\, r_{CO} + 2\, r_{CH_4} + \Sigma \left(n + \frac{m}{4} \right) r_{C_n H_m} - r_{O_2} \,. \tag{11.6}$$

The volume fraction of oxygen in the air is approximately $r_{O_2} = 0.21$, that of nitrogen simplified as approximately $r_{N_2} = 0.79$. The minimum (stoichiometric) air requirement (norm volume-related minimum air requirement) l_{min} in Nm^3 air per Nm^3 fuel gas is therefore

$$l_{min} = \frac{V_{0\,Lmin}}{V_{0\,B}} = \frac{1}{0.21}\, o_{min} = 4.76\, o_{min} \tag{11.7}$$

and, in conjunction with Eq. (11.6),

$$l_{min} = 4.76\, [0.5\, (r_{H_2} + r_{CO}) + 2\, r_{CH_4} + \Sigma \left(n + \frac{m}{4} \right) r_{C_n H_m} - r_{O_2}] \,. \tag{11.8}$$

Combustion never leads to a complete transformation of the reacting substances (educts). As chemical thermodynamics shows, an equilibrium is reached between the educts and the products. In order to shift this equilibrium to the side of the products, and in order to achieve more than the minimum amount of air required for the mixing of fuel gas and air, the oxygen content is increased above the minimum requirement.

The ratio of the amount of air l supplied per Nm^3 of fuel gas to the norm volume-related minimum air requirement l_{min} is called the air/fuel equivalent ratio (air ratio) λ:

$$\lambda = \frac{l}{l_{min}} \tag{11.9}$$

Combustion is described as complete when the proportion of combustible substances after combustion is negligibly small. The resulting combustion gases (exhaust gases, flue gases) contain only the components water (H_2O), carbon dioxide (CO_2), oxygen (O_2) and nitrogen (N_2). The following gas quantities in Nm^3 gas per Nm^3 fuel gas must be taken into account, if dry fuel gas and dry air are assumed:

$$v_{0\,H_2O} = \frac{V_{0\,H_2O}}{V_{0\,B}} = r_{H_2} + 2\, r_{CH_4} + \Sigma \frac{m}{2}\, r_{C_n H_m} \tag{11.10}$$

$$v_{0\,CO_2} = \frac{V_{0\,CO_2}}{V_{0\,B}} = r_{CO} + r_{CH_4} + \Sigma\, n\, r_{C_n H_m} + r_{CO_2} \tag{11.11}$$

$$\begin{aligned}
v_{0\,O_2} = \frac{V_{0\,O_2}}{V_{0\,B}} &= r_{O_2} + 0.21\, l - 0.5\, (r_{H_2} + r_{CO}) - 2\, r_{CH_4} - \Sigma \left(n + \frac{m}{4} \right) r_{C_n H_m} \\
&= 0.21\, (\lambda - 1)\, l_{min}
\end{aligned} \tag{11.12}$$

$$v_{0\,N_2} = \frac{V_{0\,N_2}}{V_{0\,B}} = r_{N_2} + 0.79\, \lambda\, l_{min} \tag{11.13}$$

The following norm volume-related moist combustion gas volume (moist exhaust gas volume) in Nm^3 combustion gas per Nm^3 fuel gas thus results in the case of complete combustion:

$$v_{0\,Af} = \left(\frac{V_{0\,A}}{V_{0\,B}}\right)_f = \frac{V_{0\,H_2O}}{V_{0\,B}} + \frac{V_{0\,CO_2}}{V_{0\,B}} + \frac{V_{0\,O_2}}{V_{0\,B}} + \frac{V_{0\,N_2}}{V_{0\,B}} = v_{0\,H_2O} + v_{0\,CO_2} + v_{0\,O_2} + v_{0\,N_2}$$
(11.14)

In the experimental determination of the composition of the combustion gas, the combustion gas is usually cooled down to the temperature of the surroundings. As a result, the water vapor contained in it is almost completely condensed; the dry combustion gas is therefore examined.

In this context, knowledge of the norm volume-related dry combustion gas volume (exhaust gas volume) $v_{0\,A\,t}$ in Nm^3 combustion gas per Nm^3 fuel gas is of interest. For this purpose, complete combustion results in

$$\begin{aligned} v_{0\,At} &= \left(\frac{V_{0\,A}}{V_{0\,B}}\right)_t = \left(\frac{V_{0\,A}}{V_{0\,B}}\right)_f - \frac{V_{0\,H_2O}}{V_{0\,B}} = \frac{V_{0\,CO_2}}{V_{0\,B}} + \frac{V_{0\,O_2}}{V_{0\,B}} + \frac{V_{0\,N_2}}{V_{0\,B}} \\ &= v_{0\,Af} - v_{0\,H_2O} = v_{0\,CO_2} + v_{0\,O_2} + v_{0\,N_2} \end{aligned}$$
(11.15)

If in Eqs. (11.12) and (11.13) $\lambda = 1$ is assumed, i.e. the minimum air requirement $l = l_{min}$ is used, then one obtains with Eqs. (11.14) and (11.15) the norm volume-related moist minimum combustion gas volume (moist minimum exhaust gas volume) $v_{0\,Amin\,f}$ or the norm volume-related dry minimum combustion gas volume (dry minimum exhaust gas volume) $v_{0\,Amin\,t}$.

In the case of gaseous fuels, Fig. 11.3 can be used to approximate the norm volume-related minimum air requirement l_{min}, the norm volume-related minimum moist combustion gas volume $v_{0\,Amin\,f}$ and the norm volume-related moist combustion gas volume $v_{0\,A\,f}$ as a function of the air/fuel equivalent ratio λ; on the abscissa, the norm volume-related lower heating value (also called norm volume-related lower calorific value) $H_{i,0}$ in MJ/Nm^3 and in kWh/Nm^3 is plotted (cf. section 11.3).

A volume change can occur during the combustion of gaseous fuels. The volume change can be read from the respective combustion equation. For example, if the combustion of H_2 is considered according to Eq. (11.1), it is obtained from

$$H_2 + 0.5\,O_2 = H_2O$$

$$1\,Nm^3 + 0.5\,Nm^3 = 1\,Nm^3$$

a volume reduction from 1.5 to 1 norm volume unit. According to Eqs. (11.1) to (11.4), a volume change occurs during the combustion of H_2, CO and in general also of $\Sigma\,C_nH_m$. If the total volume V of air and fuel gas is added to the reaction chamber at state (p, t)

$$V = V_B\,(1 + l)$$
(11.16)

where V_B is the fuel gas volume at state (p, t), then the moist combustion gas volume $V_{A\,f}$ (same state (p, t) hypothetically assumed) is

$$V_{Af} = V_B \left[1 + l - 0.5\,(r_{CO} + r_{H_2}) + \Sigma \left(\frac{m}{4} - 1\right) r_{C_nH_m}\right].$$
(11.17)

Related to the norm volume, the following results accordingly

$$v_{0\,Af} = \left(\frac{V_{0\,A}}{V_{0\,B}}\right)_f = 1 + l - 0.5\,(r_{CO} + r_{H_2}) + \Sigma\left(\frac{m}{4} - 1\right) r_{C_n H_m} \qquad (11.18)$$

(norm volume related norm volume of the moist combustion gas, complete combustion).

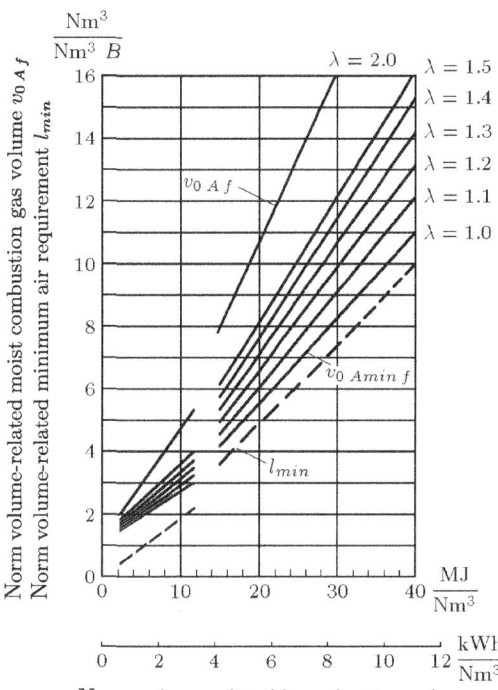

Figure 11.3 Gaseous fuels: norm volume-related minimum air requirement l_{min} and norm volume-related moist combustion gas volume $v_{0\,Af}$

Example 11.1 In a burner, ethyne (acetylene) as a fuel gas with air at an air/fuel equivalent ratio of $\lambda = 1.02$ is completely burnt. Ethyne and air are to be treated as ideal gases. With a gas meter, $\dot{V}_B = 5.3\ \mathrm{m^3/h}$ of ethyne is measured at the temperature $t_B = 15\,°C$ and the gage pressure $p_e = 11.8$ mbar, measured at the atmospheric pressure $p_{amb} = 0.963$ bar. How much air with the temperature $t_L = 20\,°C$ and the pressure $p_L = 0.963$ bar is required for combustion? How much combustion gas with the temperature $t_A = 360\,°C$ and the pressure $p_A = 0.963$ bar is produced, and what is its composition?

For the complete combustion of ethyne, the following holds:

$$C_2H_2 + 2.5\,O_2 = 2\,CO_2 + H_2O$$

$$1\,\mathrm{Nm^3}\ C_2H_2 + 2.5\,\mathrm{Nm^3}\ O_2 = 2\,\mathrm{Nm^3}\ CO_2 + 1\,\mathrm{Nm^3}\ H_2O$$

The norm volume-related minimum oxygen requirement is according to Eq. (11.6) with $r_{C_n H_m} = r_{C_2 H_2} = 1$

$$o_{min} = \left(n + \frac{m}{4}\right) r_{C_n H_m} = 2.5\,\frac{\mathrm{Nm^3\ O_2}}{\mathrm{Nm^3\ C_2H_2}}$$

and the norm volume-related minimum air requirement according to Eq. (11.7) is

$$l_{min} = \frac{2.5}{0.21}\,\frac{\mathrm{Nm^3\ }air}{\mathrm{Nm^3\ C_2H_2}} = 11.90\,\frac{\mathrm{Nm^3\ }air}{\mathrm{Nm^3\ C_2H_2}}\ .$$

At an air/fuel equivalent ratio $\lambda = 1.02$, according to Eq. (11.9)

$$l = \lambda l_{min} = \frac{1.02 \cdot 2.5}{0.21} \frac{\text{Nm}^3 \, air}{\text{Nm}^3 \, \text{C}_2\text{H}_2} = 12.14 \frac{\text{Nm}^3 \, air}{\text{Nm}^3 \, \text{C}_2\text{H}_2}$$

is added. During combustion, the following results according to Eqs. (11.10) to (11.13):

$$v_{0\,\text{H}_2\text{O}} = \frac{V_{0\,\text{H}_2\text{O}}}{V_{0\,B}} = \frac{m}{2} r_{\text{C}_n\text{H}_m} = 1 \frac{\text{Nm}^3 \, \text{H}_2\text{O}}{\text{Nm}^3 \, \text{C}_2\text{H}_2}$$

$$v_{0\,\text{CO}_2} = \frac{V_{0\,\text{CO}_2}}{V_{0\,B}} = n \, r_{\text{C}_n\text{H}_m} = 2 \frac{\text{Nm}^3 \, \text{CO}_2}{\text{Nm}^3 \, \text{C}_2\text{H}_2}$$

$$v_{0\,\text{O}_2} = \frac{V_{0\,\text{O}_2}}{V_{0\,B}} = 0.21 \, (\lambda - 1) \, l_{min} = 0.05 \frac{\text{Nm}^3 \, \text{O}_2}{\text{Nm}^3 \, \text{C}_2\text{H}_2}$$

$$v_{0\,\text{N}_2} = \frac{V_{0\,\text{N}_2}}{V_{0\,B}} = 0.79 \, \lambda \, l_{min} = 9.59 \frac{\text{Nm}^3 \, \text{N}_2}{\text{Nm}^3 \, \text{C}_2\text{H}_2}$$

Thus, according to Eq. (11.14), 12.64 Nm³ of moist combustion gas are produced per Nm³ C$_2$H$_2$. The norm volumes behave like the volume fractions:

$$r_{\text{H}_2\text{O}} : r_{\text{CO}_2} : r_{\text{O}_2} : r_{\text{N}_2} = 0.079 : 0.158 : 0.004 : 0.759$$

The combustion gas thus contains

$$7.9\,\% \, \text{H}_2\text{O}, \, 15.8\,\% \, \text{CO}_2, \, 0.4\,\% \, \text{O}_2, \, 75.9\,\% \, \text{N}_2 \, .$$

The volume of ethyne added per hour, expressed as a norm volume flow, is as follows

$$\dot{V}_{0\,B} = \frac{\dot{V}_B \, p \, T_0}{p_0 \, T} = \frac{5.3 \, \text{m}^3/\text{h} \cdot 0.975 \, \text{bar} \cdot 273.15 \, \text{K}}{1.01325 \, \text{bar} \cdot 288.15 \, \text{K}} = 4.834 \, \frac{\text{Nm}^3}{\text{h}} \, .$$

The volume flow of air to be supplied, given as norm volume per hour as well as actual volume per hour, is

$$\dot{V}_{0\,L} = 12.14 \, \dot{V}_{0\,B} = 58.69 \, \frac{\text{Nm}^3}{\text{h}} \qquad \dot{V}_L = \frac{\dot{V}_{0\,L} \, p_0 \, T_L}{p_L \, T_0} = 66.27 \, \frac{\text{m}^3}{\text{h}} \, .$$

The combustion gas volume flow, stated as norm volume per hour as well as actual volume per hour, is

$$\dot{V}_{0\,A} = 12.64 \, \dot{V}_{0\,B} = 61.10 \, \frac{\text{Nm}^3}{\text{h}} \qquad \dot{V}_A = \frac{\dot{V}_{0\,A} \, p_0 \, T_A}{p_A \, T_0} = 149.02 \, \frac{\text{m}^3}{\text{h}} \, .$$

11.1.2 Solid and Liquid Fuels

The composition of solid and liquid fuels shall be expressed in mass fractions. It is customary to denote these mass fractions, determined by elemental analysis, as follows:

carbon c; hydrogen h; nitrogen n; water w; oxygen o; sulphur s; ash a

It follows that

$$c + h + n + w + o + s + a = 1 \, . \tag{11.19}$$

The mass fractions c, h and s of carbon, hydrogen and sulphur are to be considered with regard to the combustion. For complete combustion the following equations apply:

$$C + O_2 = CO_2 \tag{11.20}$$

$$H_2 + 0.5\,O_2 = H_2O$$

$$S + O_2 = SO_2 \tag{11.21}$$

This gives the following mass balances with the respective molar masses M_i:

$$12.01 \text{ kg C} + 32.00 \text{ kg } O_2 = 44.01 \text{ kg } CO_2$$

$$2.02 \text{ kg } H_2 + 16.00 \text{ kg } O_2 = 18.02 \text{ kg } H_2O$$

$$32.06 \text{ kg S} + 32.00 \text{ kg } O_2 = 64.06 \text{ kg } SO_2$$

$$c \text{ kg C} + 2.66\,c \text{ kg } O_2 = 3.66\,c \text{ kg } CO_2 \tag{11.22}$$

$$h \text{ kg } H_2 + 7.94\,h \text{ kg } O_2 = 8.94\,h \text{ kg } H_2O \tag{11.23}$$

$$s \text{ kg S} + 1.00\,s \text{ kg } O_2 = 2.00\,s \text{ kg } SO_2 \tag{11.24}$$

For the minimum oxygen requirement (more precisely: for the specific minimum oxygen requirement) $\bar{o}_{min}$ in kg oxygen per kg fuel, one gets

$$\bar{o}_{min} = 2.66\,c + 7.94\,h + 1.00\,s - o \;. \tag{11.25}$$

The oxygen requirement characteristic σ is introduced as a key figure of a fuel [17]:

$$\bar{o}_{min} = 2.66\,c\,[1 + 2.98\,h/c + 0.375\,(s-o)/c] = 2.66\,c\,\sigma \tag{11.26}$$

$$\sigma = \frac{\bar{o}_{min}}{2.66\,c} = 1 + 2.98\,h/c + 0.375\,(s-o)/c \tag{11.27}$$

σ gives the ratio of the specific minimum oxygen requirement in kmol O_2 per kg fuel to the carbon content of the fuel in kmol C per kg of fuel. For pure carbon as fuel, $\sigma = 1$ kmol O_2/kmol C.

For practical purposes, instead of using the specific minimum oxygen requirement $\bar{o}_{min}$ in kg oxygen per kg fuel, the specific minimum norm volume oxygen requirement o^*_{min} in Nm^3 oxygen per kg fuel is used, for which with $\varrho_{0\,O_2} = 1.43$ kg O_2/Nm^3 O_2 holds:

$$o^*_{min} = \frac{\bar{o}_{min}}{\varrho_{0\,O_2}} = 1.86\,c\,\sigma \tag{11.28}$$

From this, the specific minimum norm volume air requirement is calculated in Nm^3 of air per kg of fuel:

$$l^*_{min} = \frac{o^*_{min}}{0.21} = 8.89\,c\,\sigma \tag{11.29}$$

If the specific norm volume of air l^* given in Nm^3 of air per kg of fuel is supplied to the combustion chamber, then the air/fuel equivalent ratio is

$$\lambda = \frac{l^*}{l^*_{min}} \;. \tag{11.30}$$

The nitrogen characteristic ν [17] is used to record the nitrogen content of a fuel. It indicates the ratio of the nitrogen content in kmol N_2 per kg of fuel to the carbon content in kmol C per kg of fuel:

$$\nu = \frac{n}{c} \frac{M_C}{M_{N_2}} = \frac{n}{c} \frac{12.01}{28.02} = 0.429 \frac{n}{c} \tag{11.31}$$

Both ratios σ and ν are also used to characterise fuel gases. The ratio ν is generally negligible for solid and liquid fuels.

In the complete combustion of a solid or liquid fuel, the combustion gases produced contain CO_2, H_2O, SO_2, O_2 and N_2.

Thus, for each kg of fuel, and with $\varrho_{0\,CO_2} = 1.964\,(kg\,CO_2)/(Nm^3\,CO_2)$, $\varrho_{0\,H_2O} = 0.804\,(kg\,H_2O)/(Nm^3\,H_2O)$, $\varrho_{0\,SO_2} = 2.858\,(kg\,SO_2)/(Nm^3\,SO_2)$ as well as $\varrho_{0\,N_2} = 1.250\,(kg\,N_2)/(Nm^3\,N_2)$, according to Eqs. (11.22) to (11.24) the following quantities of Nm^3 gas per kg of fuel are produced when the water (H_2O) is gaseous:

$$v^*_{0\,CO_2} = \frac{V_{0\,CO_2}}{m_B} = 3.66\,c/\varrho_{0\,CO_2} = 1.86\,c \tag{11.32}$$

$$v^*_{0\,H_2O} = \frac{V_{0\,H_2O}}{m_B} = (8.94\,h + w)/\varrho_{0\,H_2O} = 11.11\,h + 1.24\,w \tag{11.33}$$

$$v^*_{0\,SO_2} = \frac{V_{0\,SO_2}}{m_B} = 2.00\,s/\varrho_{0\,SO_2} = 0.68\,s \tag{11.34}$$

$$v^*_{0\,O_2} = \frac{V_{0\,O_2}}{m_B} = 0.21\,l^* - o^*_{min} = 0.21\,(\lambda - 1)\,l^*_{min} \tag{11.35}$$

$$v^*_{0\,N_2} = \frac{V_{0\,N_2}}{m_B} = 0.79\,l^* + n/\varrho_{0\,N_2} = 0.79\,l^* + 0.8\,n \tag{11.36}$$

With respect to complete combustion, the specific norm volume of the moist combustion gas in Nm^3 of combustion gas per kg of fuel is thus given as

$$\begin{aligned} v^*_{0\,Af} &= \left(\frac{V_{0\,A}}{m_B}\right)_f = \frac{V_{0\,H_2O}}{m_B} + \frac{V_{0\,CO_2}}{m_B} + \frac{V_{0\,SO_2}}{m_B} + \frac{V_{0\,O_2}}{m_B} + \frac{V_{0\,N_2}}{m_B} \\ &= v^*_{0\,H_2O} + v^*_{0\,CO_2} + v^*_{0\,SO_2} + v^*_{0\,O_2} + v^*_{0\,N_2}\,. \end{aligned} \tag{11.37}$$

With respect to complete combustion, the specific norm volume of the dry combustion gas $v^*_{0\,At}$ in Nm^3 combustion gas per kg fuel is thus given as

$$\begin{aligned} v^*_{0\,At} &= \left(\frac{V_{0\,A}}{m_B}\right)_t = \left(\frac{V_{0\,A}}{m_B}\right)_f - \frac{V_{0\,H_2O}}{m_B} - \frac{V_{0\,SO_2}}{m_B} = \frac{V_{0\,CO_2}}{m_B} + \frac{V_{0\,O_2}}{m_B} + \frac{V_{0\,N_2}}{m_B} \\ &= v^*_{0\,Af} - v^*_{0\,H_2O} - v^*_{0\,SO_2} = v^*_{0\,CO_2} + v^*_{0\,O_2} + v^*_{0\,N_2}\,. \end{aligned} \tag{11.38}$$

If in Eqs. (11.35) and (11.36) $\lambda = 1$ is assumed, i.e. with the specific minimum air requirement $l^* = l^*_{min}$, one obtains with Eqs. (11.37) and (11.38) the specific minimum norm volume of the moist combustion gas $v^*_{0\,Amin\,f}$ and the specific minimum norm volume of the dry combustion gas $v^*_{0\,Amin\,t}$, respectively.

The increase of the norm gas volume in combustion can be thought of as follows: With the combustion of carbon and sulphur, according to Eqs. (11.20) and (11.21) the amount of air supplied to the combustion chamber is just the same as the amount of gas flowing out of the combustion chamber. A change in the norm volume does not occur through the combustion of these components.

The amount of water, oxygen and nitrogen in the fuel is converted into the gaseous state by combustion. Thus, the norm volume per kg of fuel (in the case of water vapor, the hypothetic norm volume) for these components increases as follows:

$$\Delta v^*_{0\,H_2O} + \Delta v^*_{0\,O_2} + \Delta v^*_{0\,N_2} = \Delta \frac{V_{0\,H_2O}}{m_B} + \Delta \frac{V_{0\,O_2}}{m_B} + \Delta \frac{V_{0\,N_2}}{m_B} = \frac{w}{\varrho_{0\,H_2O}} + \frac{o}{\varrho_{0\,O_2}} + \frac{n}{\varrho_{0\,N_2}}$$
$$= 1.24\,w + 0.70\,o + 0.80\,n \qquad (11.39)$$

During the combustion of the hydrogen content which is not contained in gaseous form in the fuel, according to Eq. (11.1) and taking into account Eq. (11.33), an additional increase in norm volume per kg of fuel occurs as a result of the amount of oxygen required for this purpose:

$$\underline{\Delta} v^*_{0\,H_2O} = \underline{\Delta} \frac{V_{0\,H_2O}}{m_B} = 0.5 \cdot 11.11\,h = 5.56\,h \qquad (11.40)$$

Thus, from the fuel and the supplied specific air volume in the norm state l^*, the specific norm volume of the moist combustion gas in Nm^3 per kg fuel is

$$v^*_{0\,Af} = \left(\frac{V_{0\,A}}{m_B}\right)_f = l^* + 1.24\,w + 0.70\,o + 0.80\,n + 5.56\,h \;. \qquad (11.41)$$

11.1.3 Composition of the Combustion Gas, Combustion Triangles, Combustion Control

The importance of the norm volume-related volume of the dry combustion gas (dry exhaust gas) $v_{0\,At}$ in Nm^3 combustion gas per Nm^3 fuel gas and the the specific norm volume of the dry combustion gas (dry exhaust gas) $v^*_{0\,At}$ in Nm^3 combustion gas per kg of fuel has been already pointed out [23].

With them it is possible — when using Eqs. (11.10) to (11.13) for fuel gases and using Eqs. (11.32) to (11.36) for solid or liquid fuels — to calculate the composition of the combustion gas. If, for example, the volume fraction of carbon dioxide (carbon dioxide content) in the dry combustion gas $r^{at}_{CO_2}$ is needed, the following applies in the combustion

- of gaseous fuels
$$r^{at}_{CO_2} = \frac{v_{0\,CO_2}}{v_{0\,At}}, \qquad (11.42)$$

- of solid or liquid fuels
$$r^{at}_{CO_2} = \frac{v^*_{0\,CO_2}}{v^*_{0\,At}}. \qquad (11.43)$$

The superscript at indicates that a volume fraction in the dry combustion gas (exhaust gas, flue gas) is concerned. The volume fractions $r^{at}_{SO_2}$, $r^{at}_{O_2}$ and $r^{at}_{N_2}$ can be determined analogously.

This gives the sum of the volume fractions of the dry combustion gas during the complete combustion of gaseous fuels or of solid fuels or of liquid fuels:

$$r^{at}_{CO_2} + r^{at}_{SO_2} + r^{at}_{O_2} + r^{at}_{N_2} = 1 \;. \qquad (11.44)$$

In the case of incomplete combustion of fuel gases or of solid or liquid fuels, to CO_2, SO_2, O_2 and N_2 in addition CO and, in the case of very unfavourable combustion processes, e.g. also CH_4 and H_2 occur. Thus, the sum of the volume fractions of the dry combustion gas is generally given as

$$r^{at}_{CO_2} + r^{at}_{SO_2} + r^{at}_{O_2} + r^{at}_{N_2} + r^{at}_{CO} + r^{at}_{CH_4} + r^{at}_{H_2} = 1 \,. \tag{11.45}$$

If the combustion gas can be treated as an ideal gas, the volume fractions are equal to the mole fractions.

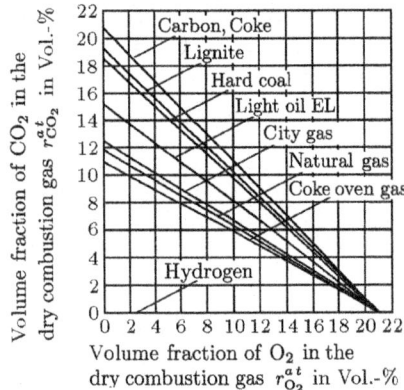

Volume fraction of O_2 in the dry combustion gas $r^{at}_{O_2}$ in Vol.-%

Figure 11.4 *Bunte* triangle for various solid, liquid and gaseous fuels [23]

A combustion triangle can be used to represent the relationship between the volume fractions of CO_2, O_2 and CO in in a dry combustion gas. A distinction is made between a *Bunte*[1] triangle and a *Ostwald*[2] triangle:

In the case of the *Bunte* triangle [24], complete combustion is assumed; CO and other unburnt components are not contained in the dry combustion gas. The volume fraction of O_2 in the dry combustion gas $r^{at}_{O_2}$ is plotted on the abscissa and the volume fraction of CO_2 in the dry combustion gas $r^{at}_{CO_2}$ is plotted on the ordinate.

In the case of complete combustion with $\lambda = 1$, the highest possible CO_2 content $r^{at}_{CO_2\,max}$ is found in the combustion gas; at the same time holds $r^{at}_{O_2} = 0$ (top left point). If $\lambda = \infty$ is assumed, then the combustion gas consists of pure air with $r^{at}_{CO_2} = 0$ in the combustion gas; at the same time holds $r^{at}_{O_2} = 0.21$ (point on the lower right). Both points are connected by a straight line. For different fuels — depending on their carbon content — different values result for $r^{at}_{CO_2\,max}$.

In Fig. 11.4, the *Bunte* triangles for technically important fuels are combined into one diagram.

In the *Ostwald* triangle [24] – in the same way as in the *Bunte* triangle – the volume fraction of O_2 is plotted on the abscissa and the volume fraction of CO_2 in the dry combustion gas on the ordinate. The information of the *Bunte* triangle is included as a straight line bounding to the upper right; for this line $r^{at}_{CO} = 0$ holds. Parallel to this

---
[1] *Hans Hugo Christian Bunte*, 1848 to 1928, was a German chemist and university lecturer.
[2] *Friedrich Wilhelm Ostwald*, 1853 to 1932, was a German chemist, philosopher and historian. One of the founders of physical chemistry, he was awarded to the Nobel Prize in 1909 for his work on catalysis, equilibrium conditions and reaction rates.

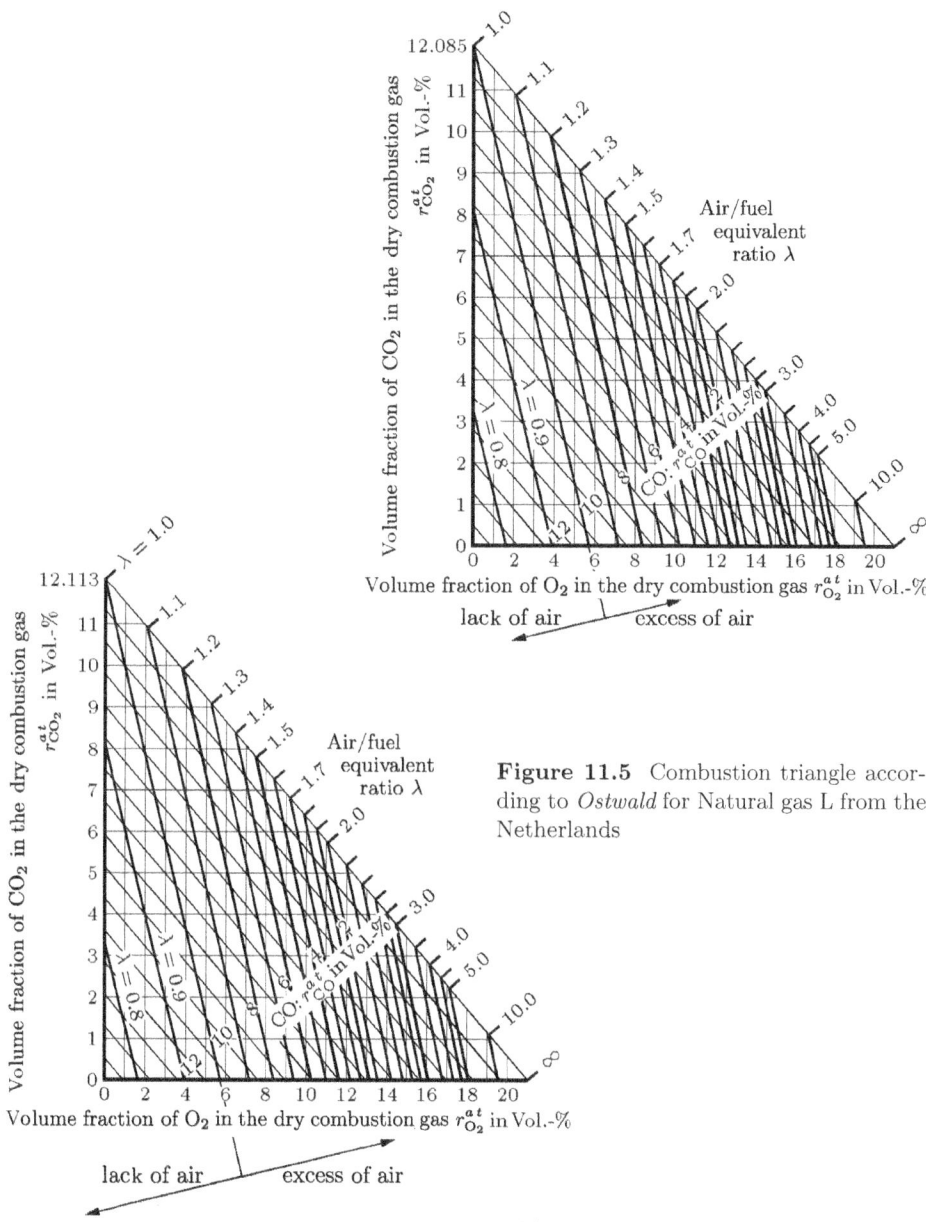

Figure 11.5 Combustion triangle according to *Ostwald* for Natural gas L from the Netherlands

Figure 11.6 Combustion triangle according to *Ostwald* for natural gas H (mixed gas)

other straight lines are shown, with which the conditions for incomplete combustion are recorded, where each line represents a constant CO-content. It is assumed here that the circumstance of imperfect combustion is only reflected in the formation of CO, i.e. that, for example, neither CH_4 or H_2 is present in the combustion gas nor solid carbon in the ash or slag, nor soot has formed. Furthermore, in the case of complete combus-

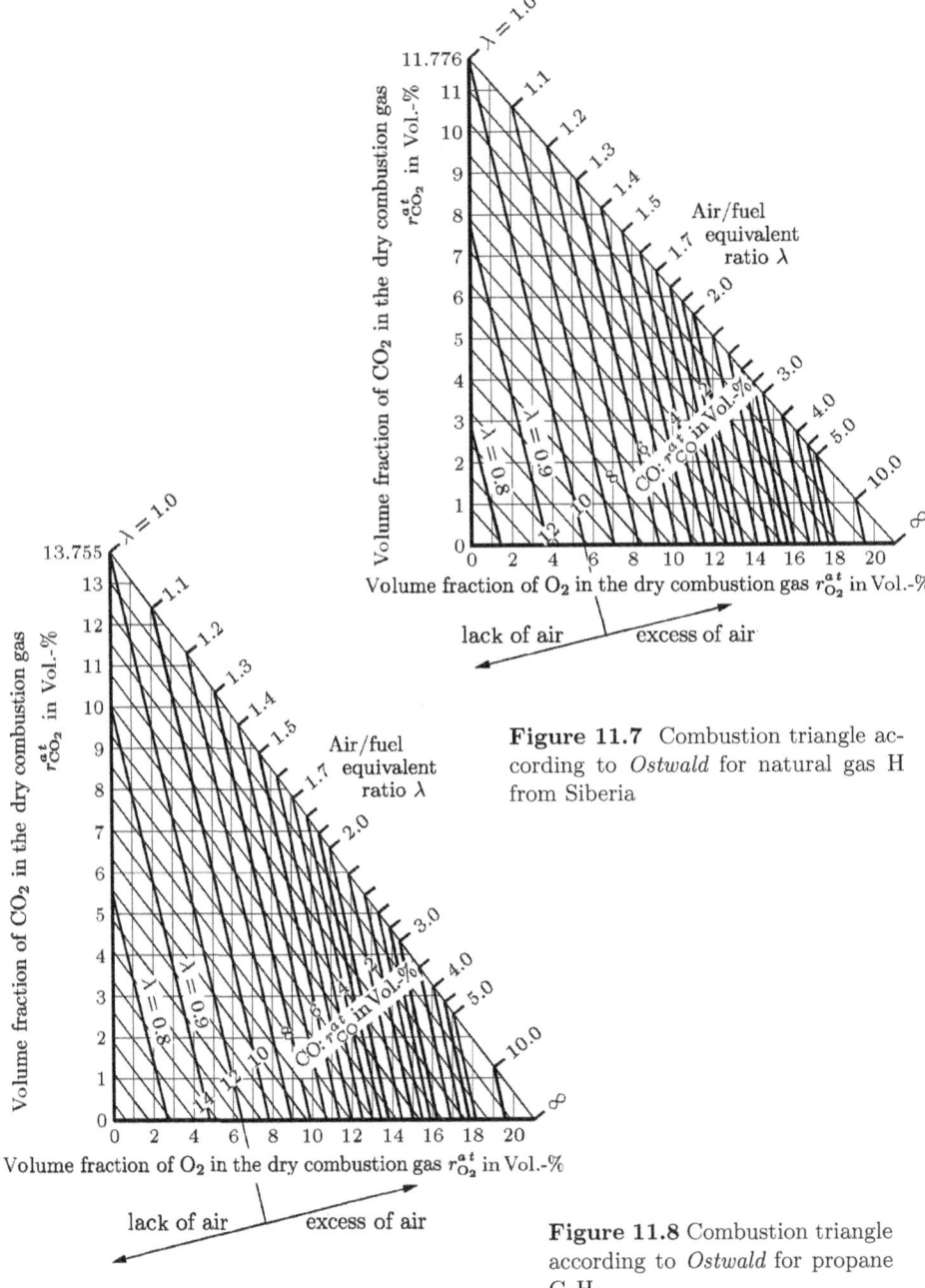

Figure 11.7 Combustion triangle according to *Ostwald* for natural gas H from Siberia

Figure 11.8 Combustion triangle according to *Ostwald* for propane C_3H_8

tion of natural gas L or natural gas H, it is assumed that the CO_2 contained in the fuel gas at high temperatures to a small extent decomposes according to $CO_2 = CO + 0.5\,O_2$ (Eq. (11.2)), resulting in the same proportions of $r^{at}_{CO_2}$, r^{at}_{CO} and $r^{at}_{O_2}$ as in incomplete combustion. Further, straight lines are drawn for constant air/fuel equivalent ratios in each case $\lambda = \text{const}$; the line for $\lambda = 1$ is particularly highlighted.

11.1 Fuels   477

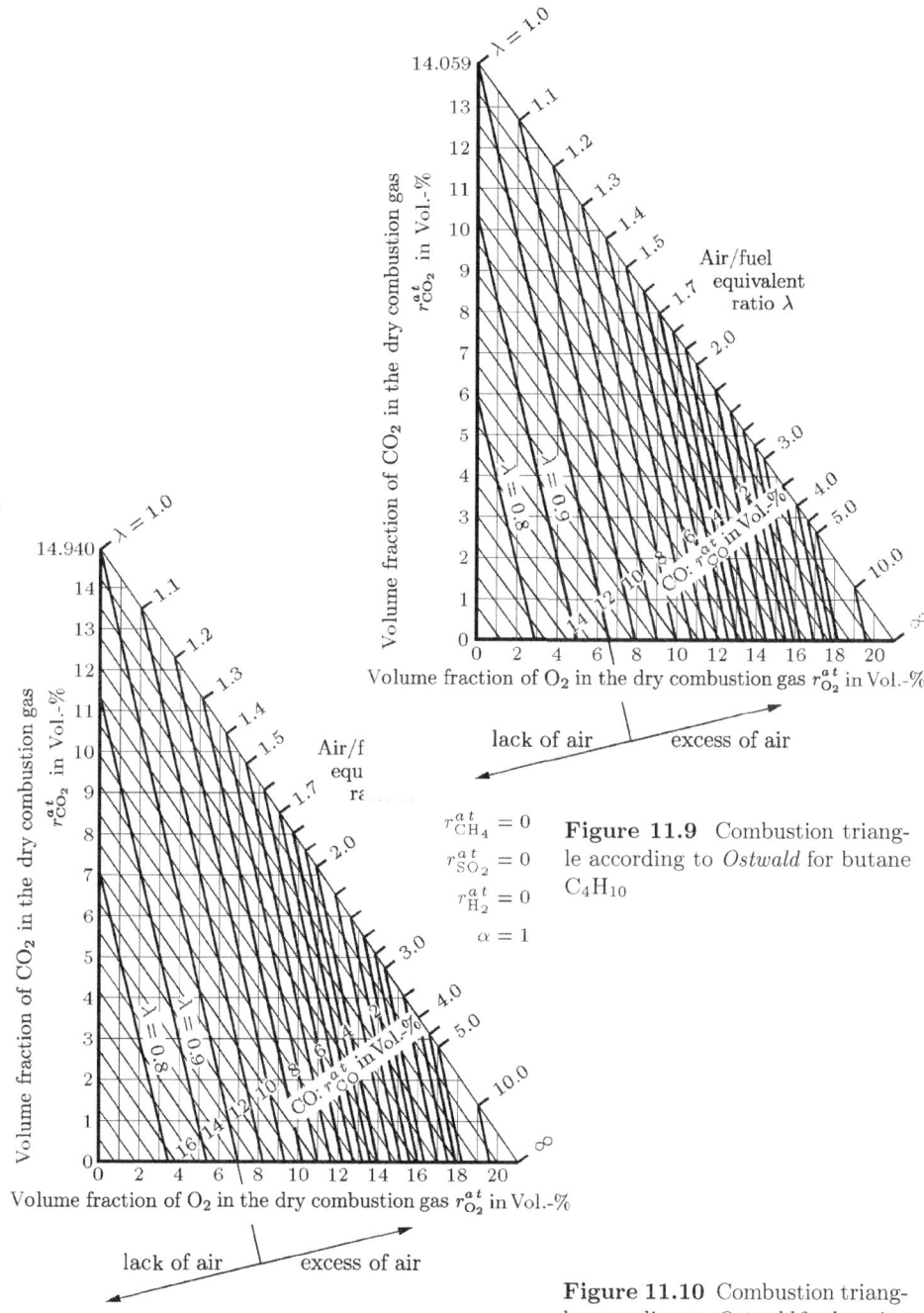

$r_{CH_4}^{a\,t} = 0$
$r_{SO_2}^{a\,t} = 0$
$r_{H_2}^{a\,t} = 0$
$\alpha = 1$

Figure 11.9 Combustion triangle according to *Ostwald* for butane C_4H_{10}

Figure 11.10 Combustion triangle according to *Ostwald* for benzine (petrol)

478   11 Combustion

In Figs. 11.5 to 11.12 *Ostwald* triangles are shown for natural gas L and natural gas H, propane C_3H_8 and butane C_4H_{10}, benzine (petrol) and *Diesel* fuel, light oil EL and heavy oil S.

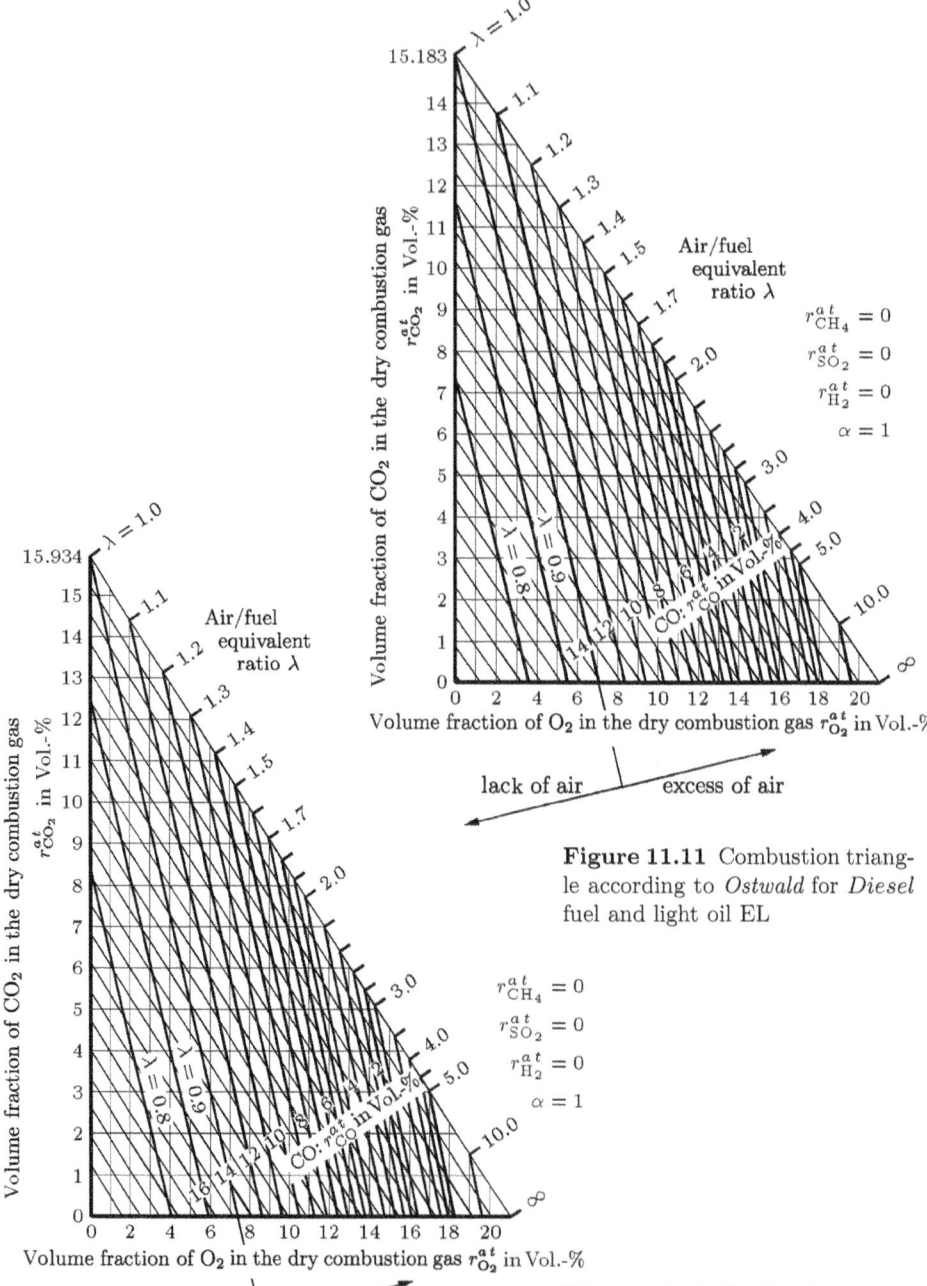

Figure 11.11 Combustion triangle according to *Ostwald* for *Diesel* fuel and light oil EL

Figure 11.12 Combustion triangle according to *Ostwald* for heavy oil S

For a check of the combustion quality (combustion control [23], [40], [133]) it is often necessary to know the information of the air/fuel equivalent ratio λ; for this purpose a measurement of the values of CO_2 or O_2 (in case of incomplete combustion also CO) in the combustion gas is helpful.

For a more precise determination, knowledge of the values of l_{min} or l^*_{min}, $v_{0\,Amin\,t}$ or $v^*_{0\,Amin\,t}$ and $r^{a\,t}_{CO_2\,max}$ is additionally required, which have the character of state variables for the respective fuels used.

The following equations are in the case of complete combustion exact, in the case of incomplete combustion also exact, insofar as incomplete combustion only results in the formation of CO (leading to $r^{a\,t}_{CO}$) and not in the formation of CH_4, H_2 and solid carbon C contained e.g. in ash, slag or soot:

- for gaseous fuels

$$\lambda = 1 + \frac{v_{0\,Amin\,t}}{l_{min}} \left[\frac{r^{a\,t}_{CO_2\,max}}{r^{a\,t}_{CO_2} + r^{a\,t}_{CO}} \left(1 - \frac{r^{a\,t}_{CO}}{2}\right) - 1 \right] \tag{11.46}$$

$$\lambda = 1 + \frac{v_{0\,Amin\,t}}{l_{min}} \frac{r^{a\,t}_{O_2} - \frac{r^{a\,t}_{CO}}{2}}{0.21 - r^{a\,t}_{O_2} + 0.79\frac{r^{a\,t}_{CO}}{2}} \tag{11.47}$$

- for solid and liquid fuels

$$\lambda = 1 + \frac{v^*_{0\,Amin\,t}}{l^*_{min}} \left[\frac{r^{a\,t}_{CO_2\,max}}{r^{a\,t}_{CO_2} + r^{a\,t}_{CO}} \left(1 - \frac{r^{a\,t}_{CO}}{2}\right) - 1 \right] \tag{11.48}$$

$$\lambda = 1 + \frac{v^*_{0\,Amin\,t}}{l^*_{min}} \frac{r^{a\,t}_{O_2} - \frac{r^{a\,t}_{CO}}{2}}{0.21 - r^{a\,t}_{O_2} + 0.79\frac{r^{a\,t}_{CO}}{2}} \tag{11.49}$$

If it can be assumed that $r^{a\,t}_{CO}/2$ is very small, then for complete combustion it holds exactly, for incomplete combustion to a good approximation:

- for gaseous fuels

$$\lambda = 1 + \frac{v_{0\,Amin\,t}}{l_{min}} \left(\frac{r^{a\,t}_{CO_2\,max}}{r^{a\,t}_{CO_2} + r^{a\,t}_{CO}} - 1 \right) \tag{11.50}$$

$$\lambda = 1 + \frac{v_{0\,Amin\,t}}{l_{min}} \frac{r^{a\,t}_{O_2}}{0.21 - r^{a\,t}_{O_2}} \tag{11.51}$$

- for solid and liquid fuels

$$\lambda = 1 + \frac{v^*_{0\,Amin\,t}}{l^*_{min}} \left(\frac{r^{a\,t}_{CO_2\,max}}{r^{a\,t}_{CO_2} + r^{a\,t}_{CO}} - 1 \right) \tag{11.52}$$

$$\lambda = 1 + \frac{v^*_{0\,Amin\,t}}{l^*_{min}} \frac{r^{a\,t}_{O_2}}{0.21 - r^{a\,t}_{O_2}} \tag{11.53}$$

For gaseous, solid and liquid fuels with a high upper calorific value (upper heating value) (e.g. for natural gas H and natural gas L) the following equations can be used for rough calculations:

$$\lambda \approx \frac{r^{a\,t}_{CO_2\,max}}{r^{a\,t}_{CO_2} + r^{a\,t}_{CO}} \tag{11.54}$$

$$\lambda \approx \frac{0{,}21}{0.21 - r_{O_2}^{at}} \tag{11.55}$$

Example 11.2 Natural gas H (mixed gas H according to Table 11.1) is completely combusted with the air/fuel equivalent ratio $\lambda = 1.3$. The following shall be calculated: o_{min}, l_{min}, l, $v_{0\,Af}$, $v_{0\,A\,min\,f}$, $v_{0\,At}$, $v_{0\,A\,min\,t}$, $r_{CO_2\,max}^{at}$ as well as the composition of the dry and moist combustion gas. The results for the proportions of CO_2 and O_2 in the dry combustion gas shall be verified with the help of Fig. 11.6.

For the complete combustion of natural gas H, the result is:

$$o_{min} = 2\,r_{CH_4} + 3.5\,r_{C_2H_6} + 5\,r_{C_3H_8} + 6.5\,r_{C_4H_{10}} = 2.108\,\frac{\text{Nm}^3\,O_2}{\text{Nm}^3\,B}$$

$$l_{min} = \frac{o_{min}}{0.21} = 10.04\,\frac{\text{Nm}^3\,air}{\text{Nm}^3\,B}$$

$$l = \lambda\,l_{min} = 1.3 \cdot 10.04\,\frac{\text{Nm}^3\,air}{\text{Nm}^3\,B} = 13.05\,\frac{\text{Nm}^3\,air}{\text{Nm}^3\,B}$$

$$v_{0\,H_2O} = 2\,r_{CH_4} + 3\,r_{C_2H_6} + 4\,r_{C_3H_8} + 5\,r_{C_4H_{10}} = 2.05\,\frac{\text{Nm}^3\,H_2O}{\text{Nm}^3\,B}$$

$$v_{0\,CO_2} = r_{CH_4} + 2\,r_{C_2H_6} + 3\,r_{C_3H_8} + 4\,r_{C_4H_{10}} + r_{CO_2} = 1.09\,\frac{\text{Nm}^3\,CO_2}{\text{Nm}^3\,B}$$

$$v_{0\,O_2} = 0.21\,l - 2\,r_{CH_4} - 3.5\,r_{C_2H_6} - 5\,r_{C_3H_8} - 6.5\,r_{C_4H_{10}} = 0.63\,\frac{\text{Nm}^3\,O_2}{\text{Nm}^3\,B}$$

$$v_{0\,N_2} = r_{N_2} + 0.79\,l = 10.32\,\frac{\text{Nm}^3\,N_2}{\text{Nm}^3\,B}$$

$$v_{0\,Af} = v_{0\,H_2O} + v_{0\,CO_2} + v_{0\,O_2} + v_{0\,N_2} = 14.09\,\frac{\text{Nm}^3\,Af}{\text{Nm}^3\,B}$$

$$v_{0\,A\,min\,f} = v_{0\,H_2O} + v_{0\,CO_2} + v_{0\,min\,N_2} = 11.08\,\frac{\text{Nm}^3\,Af}{\text{Nm}^3\,B}$$

$$v_{0\,At} = v_{0\,Af} - v_{0\,H_2O} = 12.04\,\frac{\text{Nm}^3\,At}{\text{Nm}^3\,B}$$

$$v_{0\,A\,min\,t} = v_{0\,A\,min\,f} - v_{0\,H_2O} = 9.03\,\frac{\text{Nm}^3\,At}{\text{Nm}^3\,B}$$

$$r_{CO_2\,max}^{at} = \frac{v_{0\,CO_2}}{v_{0\,A\,min\,t}} = 0.121\,\frac{\text{Nm}^3\,CO_2}{\text{Nm}^3\,At}$$

The following applies to the composition of the dry combustion gas:

$$r_{CO_2}^{at} = \frac{v_{0\,CO_2}}{v_{0\,At}} = 0.091\,\frac{\text{Nm}^3\,CO_2}{\text{Nm}^3\,At} \quad r_{O_2}^{at} = \frac{v_{0\,O_2}}{v_{0\,At}} = 0.052\,\frac{\text{Nm}^3\,O_2}{\text{Nm}^3\,At} \quad r_{N_2}^{at} = \frac{v_{0\,N_2}}{v_{0\,At}} = 0.857\,\frac{\text{Nm}^3\,N_2}{\text{Nm}^3\,At}$$

The following applies to the composition of the moist combustion gas:

$$r_{H_2O}^{af} = \frac{v_{0\,H_2O}}{v_{0\,Af}} = 0.145\,\frac{\text{Nm}^3\,H_2O}{\text{Nm}^3\,Af} \quad r_{CO_2}^{af} = \frac{v_{0\,CO_2}}{v_{0\,Af}} = 0.078\,\frac{\text{Nm}^3\,CO_2}{\text{Nm}^3\,Af}$$

$$r_{O_2}^{af} = \frac{v_{0\,O_2}}{v_{0\,Af}} = 0.045\,\frac{\text{Nm}^3\,O_2}{\text{Nm}^3\,Af} \quad r_{N_2}^{af} = \frac{v_{0\,N_2}}{v_{0\,Af}} = 0.732\,\frac{\text{Nm}^3\,N_2}{\text{Nm}^3\,Af}$$

In Fig. 11.6, the values of CO_2 and O_2 in the dry combustion gas can be found as $r_{CO_2}^{at} = 0.091$ and $r_{O_2}^{at} = 0.052$. This matches very satisfactory with the calculated results.

11.2 Technical Aspects of Combustion

11.2.1 Initiation and Progression of Combustion

In order to initiate combustion, it is not sufficient to bring the reaction substances fuel (i.e. solid, liquid or gaseous fuel) and atmospheric oxygen into contact with one another, because the velocity of the oxidation reactions of fuels is generally very small at ambient temperature. Rather, the fuel must be brought to a higher temperature — the ignition temperature — at least at one point in order to achieve faster oxidation. The minimum ignition energy required for this is only a few mJ for fuel gases in the case of spark ignition, but is considerably higher in the case of ignition with glow wire or auxiliary flame.

The ignition of a fuel gas-air mixture or a fuel gas-oxygen mixture at the respective ignition point is only successful within certain limits of mixing ratios. The highest possible still ignitable volume fraction (volume concentration) of a fuel gas in a fuel gas-air or fuel gas-oxygen mixture is called the upper ignition limit c_{Zo}, the lowest possible still ignitable volume fraction is called the lower ignition limit c_{Zu}. The ignition limits depend on pressure and temperature. At physical norm pressure p_0 and at temperature $t = 20\,°C$, these ignition limits for natural gas-air mixtures are approx. 4.2 % and 16.5 % (Table 11.1), for hydrogen-air mixtures about 4.0 % and 77.0 %.

Tp propagate the initiated combustion of the mixture of fuel and air, the heat generation must be at least as great as the heat dissipation by heat conduction, convective heat transfer or heat radiation. If the heat dissipation is greater, combustion ceases; if the heat generated is greater, the temperature of the burning material increases. At ambient temperature, spontaneous ignition of an oxidising substance can occur if the heat released during oxidation is prevented from flowing away: The substance heats up more and more until the ignition temperature is reached at one point. In this way, coal piles or even haystacks can self-ignite at ambient temperature.

The velocity at which combustion spreads is called the ignition speed (also flame speed). It can be increased by mixing the fuel well with the air supplied for combustion. For solid and liquid fuels, the increase in surface area (e.g. dust finely ground coal, atomisation or vaporisation of liquid fuels), for fuel gases the turbulent movement of the gas flows increases the ignition velocity. In the case of solid or liquid fuels, the transfer of combustible components into the gaseous phase by heating can promote the velocity of combustion.

11.2.2 Complete and Incomplete Combustion

In complete combustion, the flows leaving the reaction chamber contain no unburnt or only partially burnt substances. A combustion is called incomplete if carbon is still present in the ash or slag or if carbon is still present as soot, or if combustible components (CO, H_2, CH_4, C_nH_m, etc.) are still contained in the combustion gas (exhaust gas). However, a hydrogen-containing combustion gas is only observed in case of poor combustion control.

One therefore describes the completeness of the combustion of a solid, liquid or gaseous fuel by the quotient α [18]. α is that fraction of the carbon present in the fuel that is bound to CO_2, CO etc. in the combustion gas after the combustion process:

$$\alpha = \frac{\text{to } CO_2, CO \text{ etc. bound amount of carbon in the combustion gas}}{\text{amount of carbon in the fuel}} \tag{11.56}$$

The fraction $(1 - \alpha)$ of the carbon of the fuel thus remains unburnt and is discharged from the reaction chamber or the furnace as soot or discharged with ash or slag.

The main causes of incomplete combustion are lack of air ($\lambda < 1$) or also unfavourable air supply to parts of the fuel, excessive cooling of the fuel or of parts of the combustion gases during the combustion process below the ignition temperature as a result of heat release.

Example 11.3 A light fuel oil EL with mass fractions 86.7 % c, 13.2 % h and 0.1 % s burns at a fuel/air equivalent ratio $\lambda = 1.2$. There are 10.6 % CO_2 in the moist combustion gas. The combustion is incomplete in spite of excess air because the combustion gases cool down considerably in the combustion chamber before they could completely oxidise. The specific norm volume of the moist combustion gas $v^*_{0\,A\,unv\,f}$ is to be determined. What is the value of α?

First, the oxygen requirement characteristic σ of the fuel is to be calculated according to Eq. (11.27):

$$\sigma = 1 + 2.98 \frac{h}{c} + 0.375 \frac{s}{c} = 1 + 2.98 \cdot \frac{0.132}{0.867} + 0.375 \cdot \frac{0.001}{0.867} = 1 + 0.454 = 1.454$$

The specific norm volume of air l^* to be supplied to the combustion chamber is given by Eqs. (11.29) and (11.30)

$$l^* = \lambda\, l^*_{min} = \lambda\, 8.89\, c\, \sigma = 1.2 \cdot 8.89 \cdot 0.867 \cdot 1.454 \; \frac{Nm^3\, air}{kg\, B} = 13.45 \; \frac{Nm^3\, air}{kg\, B}.$$

For complete combustion, according to Eq. (11.32), the specific volume of CO_2 in the norm state would be

$$v^*_{0\,CO_2} = \frac{V_{0\,CO_2}}{m_B} = 1.86\, c = 1.86 \cdot 0.867 \; \frac{Nm^3\, CO_2}{kg\, B} = 1.61 \; \frac{Nm^3\, CO_2}{kg\, B}$$

and the specific norm volume of the moist combustion gas according to Eq. (11.41)

$$v^*_{0\,A\,f} = \left(\frac{V_{0\,A}}{m_B}\right)_f = l^* + 5.56\, h = (13.45 + 5.56 \cdot 0.132) \; \frac{Nm^3\, A\, f}{kg\, B} = 14.18 \; \frac{Nm^3\, A\, f}{kg\, B}.$$

The volume fraction of CO_2 in the moist combustion gas would thus be

$$r^{a\,f}_{CO_2} = \frac{1.61}{14.18} \frac{Nm^3\, CO_2}{Nm^3\, A\, f} = 0.1135 \; \frac{Nm^3\, CO_2}{Nm^3\, A\, f} \mathrel{\hat=} 11.35\,\%.$$

However, due to the incomplete combustion, the volume fraction of CO_2 is only $r^{a\,f}_{CO_2} = 0.106$ = 10.6 %. For the further calculation process, the knowledge of $v^*_{0\,A\,unv\,f} = (V_{0\,A}/m_B)_{unv\,f}$ is required for the real incomplete combustion.

If one assumes that the part of the volume fraction of CO_2 missing compared to complete combustion is present as the volume fraction of CO, the CO volume fraction does not lead to an increase in volume of the combustion gas, but the unreacted volume fraction of oxygen, which, according to the equation $CO + 0.5\, O_2 = CO_2$, must be half as large as the missing part of the volume fraction of CO_2. (Incomplete combustion ist denoted with the index "unv" because the German word for that is "unvollständig".) This results in an increase in volume of

$$\Delta v^*_{0\,A\,unv\,f} = \Delta \left(\frac{V_{0\,A}}{m_B}\right)_{unv\,f} = 0.5 \cdot (0.1135 - 0.106) \cdot 14.18 \; \frac{Nm^3\, A\, unv\, f}{kg\, B} = 0.05 \; \frac{Nm^3\, A\, unv\, f}{kg\, B},$$

and thus the specific norm volume of the combustion gas in the case of incomplete combustion becomes

$$v^*_{0\,A\,unv\,f} = \left(\frac{V_{0\,A}}{m_B}\right)_{unv\,f} = \left(\frac{V_{0\,A}}{m_B}\right)_f + \Delta\left(\frac{V_{0\,A}}{m_B}\right)_{unv\,f} = (14.18 + 0.05)\,\frac{\mathrm{Nm}^3\,A\,unv\,f}{\mathrm{kg}\,B}$$

$$= 14.23\,\frac{\mathrm{Nm}^3\,A\,unv\,f}{\mathrm{kg}\,B}.$$

According to Eq. (11.32) and analogous to Eq. (11.43), the carbon content m_c in CO_2 and CO of the combustion gas must still be determined. The specific norm volume of CO_2 as well as CO in the combustion gas are:

$$v^*_{0\,CO_2} = \frac{V_{0\,CO_2}}{m_B} = r^{a\,f}_{CO_2}\left(\frac{V_{0\,A}}{m_B}\right)_{unv\,f} = 0.106 \cdot 14.23\,\frac{\mathrm{Nm}^3\,CO_2}{\mathrm{kg}\,B} = 1.508\,\frac{\mathrm{Nm}^3\,CO_2}{\mathrm{kg}\,B}.$$

$$v^*_{0\,CO} = \frac{V_{0\,CO}}{m_B} = r^{a\,f}_{CO}\left(\frac{V_{0\,A}}{m_B}\right)_{unv\,f} = 0.0075 \cdot 14.23\,\frac{\mathrm{Nm}^3\,CO}{\mathrm{kg}\,B} = 0.107\,\frac{\mathrm{Nm}^3\,CO}{\mathrm{kg}\,B}.$$

The mass fractions of CO_2 as well as CO in kg CO_2 per kg of fuel as wells as in kg CO per kg of fuel are:

$$\frac{m_{CO_2}}{m_B} = \frac{V_{0\,CO_2}}{m_B}\varrho_{0\,CO_2} = 1.508\,\frac{\mathrm{Nm}^3\,CO_2}{\mathrm{kg}\,B} \cdot 1.9767\,\frac{\mathrm{kg}\,CO_2}{\mathrm{Nm}^3\,CO_2} = 2.981\,\frac{\mathrm{kg}\,CO_2}{\mathrm{kg}\,B}$$

$$\frac{m_{CO}}{m_B} = \frac{V_{0\,CO}}{m_B}\varrho_{0\,CO} = 0.107\,\frac{\mathrm{Nm}^3\,CO}{\mathrm{kg}\,B} \cdot 1.2505\,\frac{\mathrm{kg}\,CO}{\mathrm{Nm}^3\,CO} = 0.134\,\frac{\mathrm{kg}\,CO}{\mathrm{kg}\,B}$$

The mass fractions of carbon in CO_2 and CO can be calculated from the ratio of the molar masses of C and CO_2 as well as of C and CO and are as follows

$$\frac{M_C}{M_{CO_2}} = \frac{12.01\,\mathrm{kg\,C/kmol}}{44.01\,\mathrm{kg\,CO_2/kmol}} = 0.2729\,\frac{\mathrm{kg\,C}}{\mathrm{kg\,CO_2}} \qquad \frac{M_C}{M_{CO}} = \frac{12.01\,\mathrm{kg\,C/kmol}}{28.01\,\mathrm{kg\,CO/kmol}} = 0.429\,\frac{\mathrm{kg\,C}}{\mathrm{kg\,CO}}.$$

The carbon content in the CO_2 as well as in the CO of the combustion gas (expressed as mass of carbon in CO_2 and in CO, respectively, related to the mass of the fuel) is

$$\frac{m_C}{m_B} = \frac{M_C}{M_{CO_2}}\frac{m_{CO_2}}{m_B} + \frac{M_C}{M_{CO}}\frac{m_{CO}}{m_B}$$

$$= 0.2729\,\frac{\mathrm{kg\,C}}{\mathrm{kg\,CO_2}} \cdot 2.981\,\frac{\mathrm{kg\,CO_2}}{\mathrm{kg}\,B} + 0.429\,\frac{\mathrm{kg\,C}}{\mathrm{kg\,CO}} \cdot 0.134\,\frac{\mathrm{kg\,CO}}{\mathrm{kg}\,B}$$

$$= (0.813 + 0.057)\,\frac{\mathrm{kg\,C}}{\mathrm{kg}\,B}.$$

Thus, within the scope of the calculation accuracy,

$$\alpha = \frac{m_C/m_B}{c} = \frac{0.813 + 0.057}{0.867} = 1$$

results. So 100 % of the carbon in the fuel has been oxidised to CO_2 and CO. This also results without calculation from the definition of α and the assumptions made in the task.

11.2.3 Dew Point of Combustion Gases

The combustion gases formed during combustion generally contain water in the state of vapor and thus constitute a mixture of gas and water vapor (see section 10). The water vapor begins to condense when the temperature falls below the dew point. If the combustion gases contain sulphur dioxide (SO_2), this can dissolve in the liquefied water to form sulphurous acid, which can lead to the corrosion of a number of metallic materials. Nitrogen oxides (NO_x) in the combustion gas can lead to the formation of corrosive nitrous acid, among other reaction products. In conventional combustion systems (e.g. special boilers), constructive measures are therefore made to move this dew point into the free atmosphere.

On the other hand, for the provision of heat at low temperature levels (e. g. for room heating and for drinking water heating), especially when burning low-sulphur energy sources such as natural gas H, natural gas L, coke oven gas, propane, butane or low-sulphur light oil EL and when using corrosion-resistant materials, to cool down the combustion gases below the dew point may be desirable. This allows the enthalpy of condensation to be used energetically during the partial condensation of water vapor (condensing technology; condensing boilers).

To calculate the dew point, the partial pressure of the water vapor $p_{p\,H_2O}$ must be determined. If the volume fraction of the water vapor in the moist combustion gas is denoted by $r^{a\,f}_{H_2O}$ and the total pressure of the combustion gas is denoted by p, then Eq. (4.203) applies accordingly:

$$p_{p\,H_2O} = r^{a\,f}_{H_2O}\, p \tag{11.57}$$

According to Eq. (11.10), Eq. (11.14) and Eq. (11.18) respectively, for complete combustion of dry fuel gases with dry air follows

$$r^{a\,f}_{H_2O} = \frac{v_{0\,H_2O}}{v_{0\,A\,f}} = \frac{r_{H_2} + 2\,r_{CH_4} + \Sigma \dfrac{m}{2} r_{C_nH_m}}{1 + l - 0.5\,(r_{CO} + r_{H_2}) + \Sigma \left(\dfrac{m}{4} - 1\right) r_{C_nH_m}}. \tag{11.58}$$

According to Eqs. (11.33), (11.37) and (11.41) respectively, for complete combustion of solid as well as liquid fuels follows

$$r^{a\,f}_{H_2O} = \frac{v^*_{0\,H_2O}}{v^*_{0\,A\,f}} = \frac{11.11\,h + 1.24\,w}{l^* + 1.24\,w + 0.70\,o + 0.80\,n + 5.56\,h}. \tag{11.59}$$

11.2.4 Chimney Draught

Compared to the cold outside air, a buoyancy of the hot combustion gases in the chimney or in the stack is caused equal to the weight of the displaced air volume G_L, resulting from the higher temperature of the combustion gases. The difference between the buoyancy G_L and the weight of the combustion gas G_A is the driving force for the flow of the combustion gas through the chimney or the stack, respectively. If the cross-section of the chimney is A and its height is z, then analogous to Eq. (2.4), the — positively counted — upward force F is

$$F = G_L - G_A = A\,z\,(\gamma_L - \gamma_A) = A\,z\,g\,(\varrho_L - \varrho_A)\,. \tag{11.60}$$

The quotient of the force F and the cross-section A is called chimney draught Δp:

$$\Delta p = \frac{F}{A} = z\, g\, (\varrho_L - \varrho_A) \approx z\, g\, p \left(\frac{1}{R_L\, T_L} - \frac{1}{R_A\, T_A} \right) \tag{11.61}$$

It is larger the higher the stack and the higher the combustion gas temperature T_A are. The chimney draught compensates the friction losses of the combustion gas flow. In condensing boilers, as a consequence of the energetically advantageous low combustion gas temperature T_A, the chimney draught is very small, so that the combustion gases must be removed from the boiler with the aid of an additional fan.

Example 11.4 What is the dew point temperature for the combustion gas calculated in example 11.3 at a total pressure of $p = 0.981$ bar?

In the case of complete combustion, Eqs. (11.59) or (11.33) and (11.37) would be applicable. In the case of incomplete combustion, the results of Example 11.3 can be used:

$$r^{a\,f}_{H_2O} = \frac{v^*_{0\,H_2O}}{v^*_{0\,A\,unv\,f}} = \frac{11.11 \cdot 0.132}{14.23} = 0.103 = 10.3\,\%$$

According to Eq. (11.57), the partial pressure of water vapor is

$$p_{p\,H_2O} = r^{a\,f}_{H_2O}\, p = 0.103 \cdot 0.981 \text{ bar} = 0.101 \text{ bar}\,.$$

The condensation temperature is therefore according to Table A.9b in the appendix $t_S = 46\,°C$.

11.3 Upper Calorific Value and Lower Calorific Value

During combustion in an open system, heat is released from the enthalpies of the reactants. Depending on the substances involved, a single chemical reaction or several chemical reactions occur. A chemical reaction that takes place results in a difference in the chemical energies bound in the molecular structure, which are part of the enthalpies of the substances involved. This difference in enthalpies is released as heat in irreversible combustion processes — e.g. in furnaces.

In the following, it is assumed that the fuel with the mass m_B and the enthalpy H_B as well as the enthalpy H_L flow steadily into the combustion chamber at a pressure p_1 and a temperature t_1 (Fig. 11.13). In addition to the temperature-dependent enthalpy, the enthalpy H_B of the fuel also includes the chemically bound energy of the fuel.

After complete combustion, the products of combustion at the same pressure p_1 have a much higher temperature t_2. In the following, one has to think of the combustion products as being cooled back to the initial temperature of the substances at the beginning of the process t_1. It is to be assumed that the water vapor formed as a combustion product is practically completely liquefied by the recooling and that enthalpy of condensation (heat of condensation) is also released in the process. Thus, the combustion gas has the enthalpy H_A and the ash or slag has the enthalpy H_S.

Eq. (2.67) is now applied to combustion with recooling, neglecting the differences in potential and kinetic energies; moreover, W_e and W_{RA} can be set to zero:

$$Q_{11'} = (H_A + H_S)_{1'} - (H_B + H_L)_1 = H_{1'} - H_1 \tag{11.62}$$

The heat $Q_{11'}$ is negative because it is removed from the combustion chamber. The amount of specific heat $q_{11'}$ (related to the mass of 1 kg of fuel) is called the upper calorific value (also known as upper heating value UHV) H_s; the term specific upper calorific value is more accurate.

$$H_s = \frac{|Q_{11'}|}{m_B} = \frac{H_1 - H_{1'}}{m_B} \qquad (11.63)$$

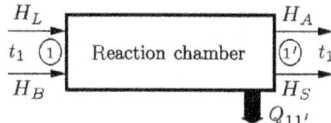

Figure 11.13 To explain the terms "upper calorific value" and "lower calorific value"

For table specifications, the physical norm state ($p_1 = p_0 = 1.01325$ bar; $t_1 = t_0 = 0\,°C$) is chosen as the reference state. In some cases H_s is also specified in such a way that the temperature of the starting substances (educts) as well as the recooled products is $t_1 = 15\,°C$ or $t_1 = 25\,°C$.

For fuel gases, the upper calorific value is related to the physical norm volume

$$H_{s,0} = \frac{|Q_{11'}|}{V_{0B}} = \frac{H_1 - H_{1'}}{V_{0B}} \; ; \qquad (11.64)$$

the term norm volume-related upper calorific value is more accurate (cf. Table 11.1).

In contrast to the specific upper calorific value H_s or the norm volume-related upper calorific value $H_{s,0}$, the specific lower calorific value H_i (also known as specific lower heating value LHV) or the norm volume-related lower calorific value $H_{i,0}$ (cf. Table 11.1) is characterized as follows: The combustion products are cooled down to the temperature of the starting substances (educts) t_1 but the entire amount of water produced is hypothetically thought of as water vapor.

Between the specific upper calorific value H_s and the specific lower calorific value H_i there is the relation

$$H_s - H_i = \frac{m_{H_2O}}{m_B} r \; , \qquad (11.65)$$

if m_{H_2O} is the mass of water vapor contained in the combustion gases, m_B is the mass of burnt fuel and r is the specific enthalpy of condensation (specific heat of condensation) of H_2O at the temperature t_1. For example, if a solid or liquid fuel with hydrogen content h and water content w is considered, then according to Eqs. (11.19) and (11.23), for each kg of fuel the following holds:

$$\frac{m_{H_2O}}{m_B} = 8.94\,h + w \qquad (11.66)$$

The difference between the specific upper calorific value H_s and the specific lower calorific value H_i is obtained (at the standard temperature of $t_1 = 25\,°C$ for combustion)

$$H_s - H_i = \frac{m_{H_2O}}{m_B} r = (8.94\,h + w) \cdot 2442.5 \text{ kJ/kg} \; . \qquad (11.67)$$

The difference between the norm volume-related upper calorific value $H_{s,0}$ and the norm volume-related lower calorific value $H_{i,0}$ is correspondingly

$$H_{s,0} - H_{i,0} = \frac{V_{0\,H_2O}}{V_{0B}} r_0 \qquad (11.68)$$

with $v_{0\,H_2O} = V_{0\,H_2O}/V_{0\,B}$ as the amount of water vapor produced by the combustion reactions in Nm^3 H_2O per Nm^3 fuel gas and r_0 as the norm volume-related enthalpy of condensation of H_2O. With Eq. (11.10), at the standard temperature $t_1 = 25\,°C$, for combustion the following holds:

$$H_{s,0} - H_{i,0} = \left(r_{H_2} + 2\,r_{CH_4} + \Sigma\,\frac{m}{2}\,r_{C_nH_m}\right) \cdot 1963.5\,\frac{kJ}{Nm^3} \tag{11.69}$$

The specific upper calorific value or specific lower calorific value of a mixture of solid or of liquid fuels can be calculated with the help of the mass fractions μ_i of the components. The norm volume-related upper calorific value or norm volume-related lower calorific value of a fuel gas mixture of ideal gases can be calculated with the aid of the volume fractions r_i of the components (cf. section 4.5.3):

$$H_s = \sum_{i=1}^{n} \mu_i\,H_{s\,i} \tag{11.70}$$

$$H_i = \sum_{i=1}^{n} \mu_i\,H_{i\,i} \tag{11.71}$$

$$H_{s,0} = \sum_{i=1}^{n} r_i\,H_{s,0\,i} \tag{11.72}$$

$$H_{i,0} = \sum_{i=1}^{n} r_i\,H_{i,0\,i} \tag{11.73}$$

The specific lower calorific values H_i of different liquid and solid fuels — given in MJ/kg — are contained in the Appendix of this book where the results of task 10 of section 11 are listed.

11.4 Theoretical Combustion Temperature

During combustion, the heat released increases the temperature of the combustion gases. If the process is adiabatic, i.e. no heat is released to a working fluid or to the environment, the combustion gases will reach the theoretical combustion temperature during combustion (adiabatic combustion temperature) t_{Amax}.

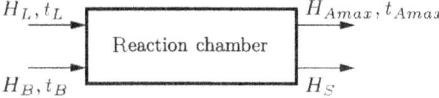

Figure 11.14 Adiabatic combustion

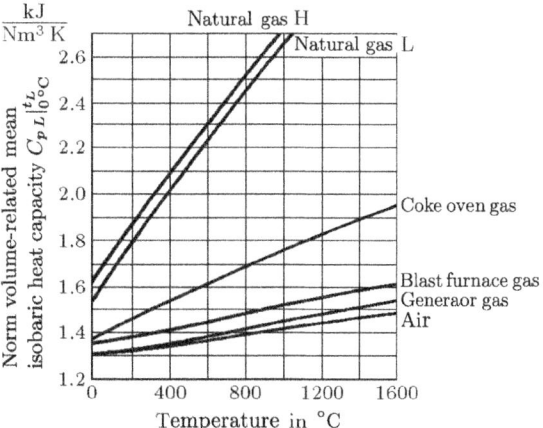

Figure 11.15 Norm volume related mean isobaric heat capacity of fuel gases (not dissociated) and of air

In addition to the temperature-dependent enthalpy, the enthalpy H_B of the fuel also includes the chemical energy bound in the fuel. Analogous to Eq. (11.62), the following applies according to Fig. 11.14

$$H_B + H_L = H_{Amax} + H_S \ . \tag{11.74}$$

If the complete combustion of a fuel gas is assumed, for the calculation of the theoretical combustion temperature t_{Amax} the following state variables are required:

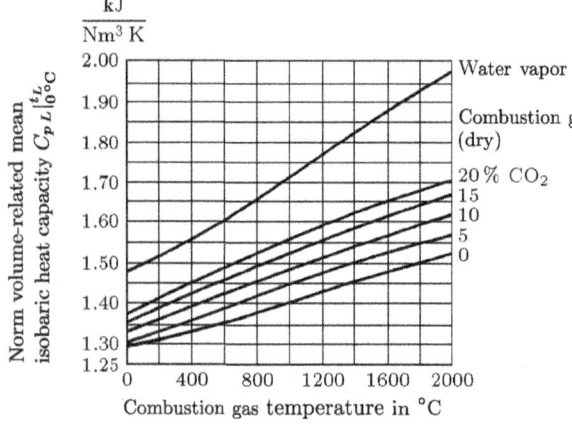

Figure 11.16 Norm volume-related mean isobaric heat capacity of water vapor and of dry combustion gas (not dissociated) as a function of the CO_2 content

- the norm volume-related mean isobaric heat capacity of the respective combustion gas $C_{pB}|_{0°C}^{t_B}$ in kJ/(Nm³ K) (cf. Fig. 11.15)
- the norm volume-related mean isobaric heat capacities of air $C_{pL}|_{0°C}^{t_L}$, of CO_2 $C_{pCO_2}|_{0°C}^{t_{Amax}}$, of H_2O (steam) $C_{pH_2O}|_{0°C}^{t_{Amax}}$, of O_2 $C_{pO_2}|_{0°C}^{t_{Amax}}$ and of N_2 $C_{pN_2}|_{0°C}^{t_{Amax}}$ in kJ/(Nm³ K) (cf. Fig. 11.15 for $C_{pL}|_{0°C}^{t_L}$ and Fig. 11.16, in which not only $C_{pH_2O}|_{0°C}^{t_L}$, but in the sense of a summary also $C_{pAt}|_{0°C}^{t_A}$ in kJ/(Nm³ K) of the dry combustion gas is shown, taking into account the respective value of $r_{CO_2}^{a\,t}$ in the dry combustion gas)

Because of the lower limit $t_{0'} = 0°C$ for the norm volume-related mean isobaric heat capacities $C_{pi}|_{0°C}^{t_{Amax}}$, $H_{0'B}$, $H_{0'L}$ and $H_{0'A}$ result as constant reference values for the enthalpies at $t_{0'} = 0°C$. Thus the enthalpy of the incoming fuel gas-air mixture is

$$H_B + H_L = V_{0B} C_{pB}|_{0°C}^{t_B} t_B + V_{0B} h_{0'B} + V_{0B} \lambda l_{min} (C_{pL}|_{0°C}^{t_L} t_L + h_{0'L}) \ . \tag{11.75}$$

The ash or soot content in the combustion gas can be neglected ($H_S = 0$); thus one obtains for the enthalpy of the combustion products at the same temperature limits as given above:

$$H_{Amax} = V_{0H_2O} C_{pH_2O}|_{0°C}^{t_{Amax}} t_{Amax} + V_{0CO_2} C_{pCO_2}|_{0°C}^{t_{Amax}} t_{Amax} +$$
$$V_{0O_2} C_{pO_2}|_{0°C}^{t_{Amax}} t_{Amax} + V_{0N_2} C_{pN_2}|_{0°C}^{t_{Amax}} t_{Amax} + H_{0'A} \tag{11.76}$$

If the norm temperature $t_{0'} = 0°C$ is again chosen as the reference temperature, the enthalpies H_B, H_L and H_{Amax} in Fig. 11.14 become $H_{0'B}$, $H_{0'L}$ and $H_{0'A}$. Thus, applying Eq. (11.64) mutatis mutandis and with a balance according to Fig. 11.13, the norm volume-related lower calorific value $H_{i,0}$ results from the following relation:

$$V_{0B} H_{i,0} = (H_{0'B} + H_{0'L}) - H_{0'A} = V_{0B} h_{0'B} + V_{0B} \lambda l_{min} h_{0'L} - H_{0'A} \tag{11.77}$$

If Eqs. (11.75) to (11.77) are inserted into Eq. (11.74) and if this is divided by $V_{0\,B}$, one gets for the theoretical combustion temperature t_{Amax}:

$$t_{Amax} = \frac{H_{i,0} + \lambda l_{min} C_{p\,L}|_{0\,°C}^{t_L} t_L + C_{p\,B}|_{0\,°C}^{t_B} t_B}{\sum \frac{V_{0\,i}}{V_{0\,B}} C_{p\,i}|_{0\,°C}^{t_{Amax}}}$$

$$= \frac{H_{i,0} + \lambda l_{min} C_{p\,L}|_{0\,°C}^{t_L} t_L + C_{p\,B}|_{0\,°C}^{t_B} t_B}{\frac{V_{0\,H_2O}}{V_{0\,B}} C_{p\,H_2O}|_{0\,°C}^{t_{Amax}} + \left(\frac{V_{0\,A}}{V_{0\,B}}\right)_t C_{p\,A\,t}|_{0\,°C}^{t_{Amax}}} \quad (11.78)$$

Into the sum in the denominator the expressions $V_{0\,i}/V_{0\,B} = v_{0\,i}$ for the respective components of the combustion gas according to Eqs. (11.10) to (11.13) or Eqs. (11.10) and (11.15) respectively, are to be inserted.

If a solid or a liquid fuel is used as fuel, for the calculation of the theoretical combustion temperature t_{Amax} the following state variables are required:

- the specific mean isobaric heat capacity of the respective fuel $c_{p\,B}|_{0\,°C}^{t_B}$ in kJ/(kg K)

- the norm volume-related mean isobaric heat capacities of air $C_{p\,L}|_{0\,°C}^{t_L}$, of CO_2 $C_{p\,CO_2}|_{0\,°C}^{t_{Amax}}$, of H_2O (vapor) $C_{p\,H_2O}|_{0\,°C}^{t_{Amax}}$, of SO_2 $C_{p\,SO_2}|_{0\,°C}^{t_{Amax}}$, of O_2 $C_{p\,O_2}|_{0\,°C}^{t_{Amax}}$ and of N_2 $C_{p\,N_2}|_{0\,°C}^{t_{Amax}}$ in kJ/(Nm^3 K) (cf. Fig. 11.15 for $C_{p\,L}|_{0\,°C}^{t_L}$ and Fig. 11.16, in which not only $C_{p\,H_2O}|_{0\,°C}^{t_A}$, but in the sense of a summary also $C_{p\,A\,t}|_{0\,°C}^{t_A}$ in kJ/(Nm^3 K) of the dry combustion gas is shown, taking into account the respective value of $r_{CO_2}^{a\,t}$ in the dry combustion gas)

By an analogous derivation, one obtains for the theoretical combustion temperature t_{Amax}:

$$t_{Amax} = \frac{H_i + \lambda l_{min}^* C_{p\,L}|_{0\,°C}^{t_L} t_L + c_{p\,B}|_{0\,°C}^{t_B} t_B}{\sum \frac{V_{0\,i}}{m_B} C_{p\,i}|_{0\,°C}^{t_{Amax}}}$$

$$= \frac{H_i + \lambda l_{min}^* C_{p\,L}|_{0\,°C}^{t_L} t_L + c_{p\,B}|_{0\,°C}^{t_B}}{\frac{V_{0\,H_2O}}{m_B} C_{p\,H_2O}|_{0\,°C}^{t_{Amax}} + \left(\frac{V_{0\,A}}{m_B}\right)_t C_{p\,A\,t}|_{0\,°C}^{t_{Amax}}} \quad (11.79)$$

Into the sum in the denominator the expressions $V_{0\,i}/m_B = v_{0\,i}^*$ for the respective components of the combustion gas according to Eqs. (11.32) to (11.36), or Eqs. (11.33) and (11.38) respectively, are to be inserted. The ash or soot content in the combustion gas can be neglected ($H_S = 0$).

The Eqs. (11.78) and (11.79) do not take into account the dissociation of the components H_2O, CO_2 and SO_2 in the combustion gas that becomes noticeable at temperatures above about 1500 °C. This refers to the shift of the reaction equilibrium in the combustion equations (11.1), (11.2) and (11.21) to the left; at even higher temperatures, in addition, H_2 and O_2 can partially decompose into their atoms. These processes require energy; this leads to a reduction of the theoretical combustion temperature t_{Amax} [9], [43].

The theoretical combustion temperature t_{Amax} becomes smaller the larger the air/fuel equivalent ratio λ is. If the highest possible combustion temperatures are to be achieved, efforts are made to keep the air/fuel equivalent ratio as small as possible: so small that complete combustion is just achieved. As a guideline for λ, the following can generally be considered: manually operated firing systems for solid fuels: $\lambda = 1.4 \ldots 2.5$; automatic grate firing systems

for solid fuels: $\lambda = 1.25 \ldots 1.4$; pulverised coal firing systems: $\lambda = 1.05 \ldots 1.2$; oil and gas firing systems: $\lambda = 1.02 \ldots 1.35$; *Otto* engines without stratified combustion resp. without exhaust gas recirculation with three-way catalytic converter: $\lambda = 0.99 \ldots 1.05$; *Diesel* engines: $\lambda = 1.0 \ldots 1.5$.

The theoretical combustion temperature can be increased by preheating the fuel and/or the combustion air.

In furnaces, the combustion temperature (e.g. $t \leq 1200\,°C$) is lower than the theoretical combustion temperature as a result of the transfer of usable heat to a working fluid (e.g. in steam generators) or to other substances to be heated (e.g. in furnaces to heat and/or melt metals).

If in Eq. (11.76) the temperature t_{Amax} is replaced by the temperature t_{St} at which the combustion gases enter the stack, H_{Amax} becomes H_A, and the difference $H_A - H_{0'A}$ represents the heat that could still be recovered if the combustion gases were cooled to $0\,°C$. This enthalpy difference is called combustion gas loss or stack loss:

$$H_A - H_{0'A} = t_{St} \Sigma V_{0\,i} C_{p\,i}|_{0\,°C}^{t_{St}} = t_{St}(V_{0\,H_2O} C_{p\,H_2O}|_{0\,°C}^{t_{St}} + V_{0\,At} C_{p\,At}|_{0\,°C}^{t_{St}}) \qquad (11.80)$$

To evaluate the rather inconveniently handled Eqs. (11.74) to (11.78) and (11.79) respectively, one makes use of the h_a, t_A diagram of *Rosin* and *Fehling* [24], which can be used with an accuracy of about $\pm 1.5\,\%$ for the combustion gases of practically all solid, liquid and gaseous fuels.

In terms of a good usable approximation, the h_a, t_A diagram of *Rosin* and *Fehling* (Fig. 11.17) is of good practical use. In the case of gaseous fuels, the enthalpy of the combustion gas related to the norm volume of the gaseous fuel h_A at any temperature t_A is divided by $v_{0\,Af}$, which is the volume of the moist combustion gas related to the norm volume of the gaseous fuel:

$$h_a = \frac{h_A}{v_{0\,Af}} \quad \text{in MJ/Nm}^3 \text{ of moist combustion gas} \qquad (11.81\,a)$$

The subscript 'a' is chosen to clarify that h_a refers to the norm volume of the moist combustion gas (exhaust gas) and not to the norm volume of the fuel gas.

In the same way, h_a results when considering the combustion gases of solid and liquid fuels. Here, the enthalpy of the combustion gas related to the mass of the fuel h_A^* is divided by the moist combustion gas related to the mass of the fuel $v_{0\,Af}^*$:

$$h_a = \frac{h_A^*}{v_{0\,Af}^*} \quad \text{in MJ/Nm}^3 \text{ of moist combustion gas} \qquad (11.81\,b)$$

The following applies to determine the theoretical combustion temperature:

$$h_{amax} = \frac{h_{Amax}}{v_{0\,Af}} \quad \text{resp.} \quad h_{amax} = \frac{h_{Amax}^*}{v_{0\,Af}^*} \quad \text{in MJ/Nm}^3 \text{ of moist combustion gas} \qquad (11.82)$$

It can thus be determined from the h_a, t_A diagram of *Rosin* and *Fehling*. In this diagram, the influence of dissociation is already taken into account. In Fig. 11.17 the air content

$$l_a = \frac{(\lambda - 1)\,l_{min}}{v_{0\,Af}} \quad \text{resp.} \quad l_a = \frac{(\lambda - 1)\,l_{min}^*}{v_{0\,Af}^*} \qquad (11.83)$$

is chosen as the parameter; the air content is the volume fraction of excess supplied air in the combustion gas. l_a can be taken from the upper diagrams of Fig. 11.17 depending on the specific lower calorific value H_i or depending on the norm volume-related lower calorific value $H_{i,0}$ and the air/fuel equivalent ratio λ.

11.4 Theoretical Combustion Temperature

The theoretical combustion temperatures for solid, liquid and gaseous fuels — depending on the air/fuel equivalent ratio λ — can also be taken from Figure 11.18 [21].

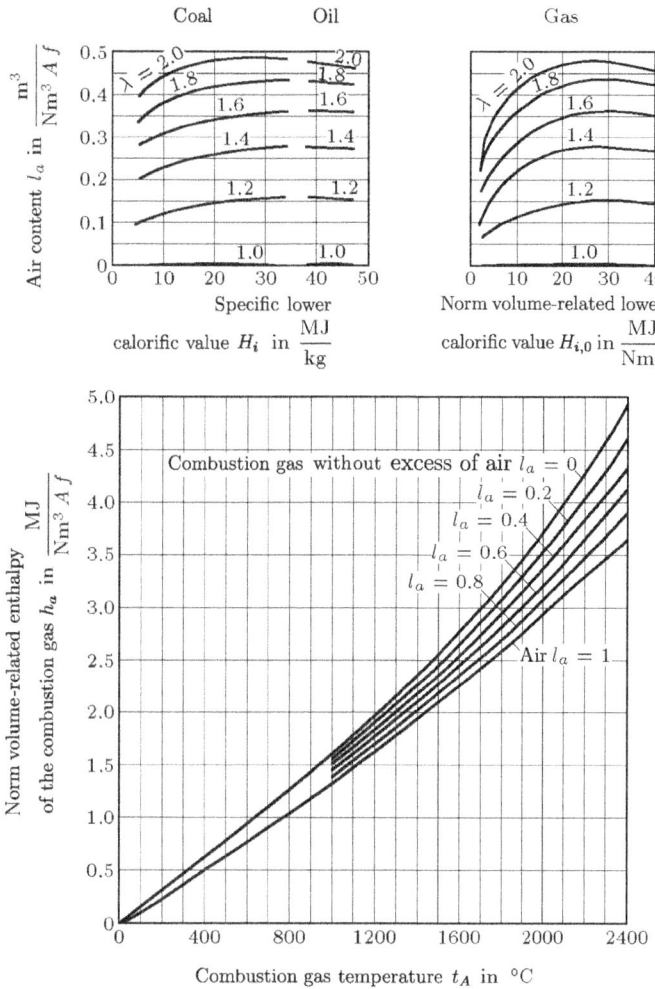

Figure 11.17
h_a, t_A diagram after
Rosin and *Fehling* [24]

Further diagrams (cf. e.g. [78]), in particular h, t_A diagrams, are used for fuels, combustion air and combustion gases, depending on the respective solid, liquid and gaseous fuels. In the following, only combustion gases are considered. For the norm volume-related enthalpies of the starting substances (reactants), if $t_B = t_L$ is set as a simplification, one obtains

$$h_B + h_L = \frac{H_B + H_L}{V_{0B}} = (C_{pB}|_{0°C}^{t_B} + \lambda l_{min} C_{pL}|_{0°C}^{t_L}) t_L + h_{0'B} + \lambda l_{min} h_{0'L} \quad (11.84)$$

and for the combustion products when the function $h_A = H_A/V_{0B} = f(t)$ is formed,

$$h_A = \frac{H_A}{V_{0B}} = t_A \Sigma \frac{V_{0i}}{V_{0B}} C_{pi}|_{0°C}^{t_A} + \frac{H_{0'A}}{V_{0B}} \ . \quad (11.85)$$

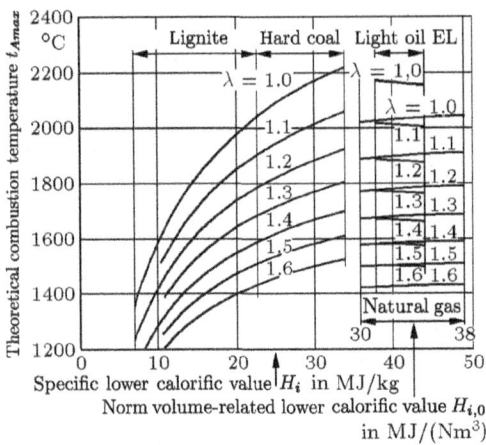

Figure 11.18 Theoretical combustion temperature for hard coal, lignite, fuel oil and natural gas as a function of lower calorific value H_i or $H_{i,0}$ respectively, and air/fuel equivalent ratio λ, for $t_B = t_L = 15\,°C$ [21]

Analogous to Eq. (11.74), under the condition $h_S = 0$ holds

$$h_B + h_L = h_{Amax} \,. \tag{11.86}$$

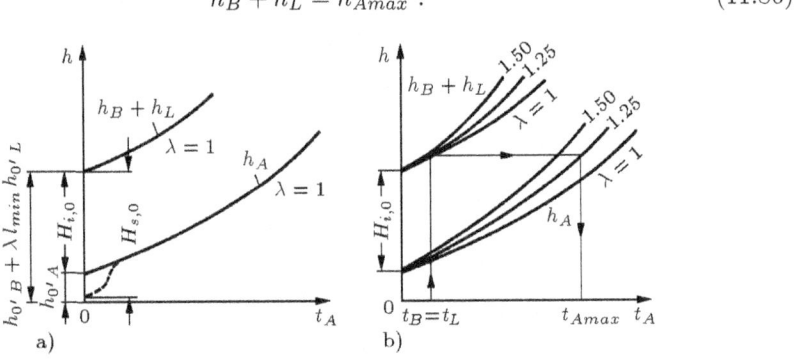

Figure 11.19 h, t_A diagram:
a) Basic set-up   b) Determination of the theoretical combustion temperature t_{Amax}

If the enthalpies per norm cubic meter of fuel gas are calculated according to Eqs. (11.84) and (11.85) in a common h, t_A diagram, the theoretical combustion temperature can be calculated with the condition $h_A = h_{Amax}$ from this diagram: This is found according to Fig. 11.19 b by calculating $h_B + h_L$ at the temperature of the incoming substances ($t_B = t_L$) and, according to Eq. (11.86), for the same norm volume-related enthalpy value on the curve for $h_A = h_{Amax}$ the temperature is read off. This temperature is then the theoretical combustion temperature according to Eq. (11.78) or Eq. (11.79).

The norm volume-related lower calorific value $H_{i,0}$ appears in the h, t_A diagram (Fig. 11.19 a) according to Eq. (11.77) divided by $V_{0\,B}$ on the ordinate axis for $t_{0'} = 0\,°C$. If the condensation of the water vapor at low temperatures is taken into account, the curve of the combustion products follows the dashed line, and at $t_{0'} = 0\,°C$ one obtains the norm volume-related upper calorific value $H_{s,0}$.

Example 11.5 What is the theoretical combustion temperature t_{Amax} at complete combustion of the light fuel oil EL in example 11.3 with the air ratio $\lambda = 1.2$, if its specific calorific value can be assumed as the lower specific calorific value $H_i = 42\,700$ kJ/kg? The light fuel oil EL and the combustion air enter the burner with the temperature $t = 20\,°C$. The specific isobaric heat capacity of the light fuel oil EL is $c_p = 2.01$ kJ/(kg K).

Using Eqs. (11.32) to (11.36), one obtains:

$$v^*_{0\,CO_2} = \frac{V_{0\,CO_2}}{m_B} = 1.86\ c = 1.61\ \frac{\text{Nm}^3\ CO_2}{\text{Nm}^3\ B}$$

$$v^*_{0\,H_2O} = \frac{V_{0\,H_2O}}{m_B} = 11.11\ h = 1.47\ \frac{\text{Nm}^3\ H_2O}{\text{Nm}^3\ B}$$

$$v^*_{0\,SO_2} = \frac{V_{0\,SO_2}}{m_B} = 0.68\ s \approx 0\ \frac{\text{Nm}^3\ SO_2}{\text{Nm}^3\ B}$$

$$v^*_{0\,O_2} = \frac{V_{0\,O_2}}{m_B} = 0.21\ (\lambda - 1)\ l^*_{min} = 0.47\ \frac{\text{Nm}^3\ O_2}{\text{Nm}^3\ B}$$

$$v^*_{0\,N_2} = \frac{V_{0\,N_2}}{m_B} = 0.79\,\lambda\,l^*_{min} = 10.63\ \frac{\text{Nm}^3\ N_2}{\text{Nm}^3\ B}$$

This gives the specific norm volume of the dry combustion gas:

$$v^*_{0\,A\,t} = \left(\frac{V_{0\,A}}{m_B}\right)_t = v^*_{0\,CO_2} + v^*_{0\,O_2} + v^*_{0\,N_2} = 12.71\ \frac{\text{Nm}^3\ A\ t}{\text{Nm}^3\ B}$$

For the determination of the mean norm volume-related isobaric heat capacities of air, of water vapor and of the norm volume of the dry combustion gas, Fig. 11.15 and Fig. 11.16 are used. For the two last-mentioned quantities, an estimate must first be made for the upper temperature limit t_{Amax}. It is estimated at $t_{Amax} = 1900\,°C$. Thus result

$$C_{p\,L}|^{t_L}_{0\,°C} = 1.31\ \frac{\text{kJ}}{\text{Nm}^3\ air\ K},\ C_{p\,H_2O}|^{t_{Amax}}_{0\,°C} = 1.95\ \frac{\text{kJ}}{\text{Nm}^3\ H_2O\ K}\ \text{and}\ C_{p\,A\,t} = 1.60\ \frac{\text{kJ}}{\text{Nm}^3\ A\ t\ K}.$$

With Eq. (11.79) one gets

$$t_{Amax} = \frac{H_i + \lambda\,l^*_{min}\,C_{p\,L}|^{t_L}_{0\,°C}\,t_l + c_{p\,B}\,t_B}{\Sigma\,\frac{V_{0\,i}}{m_B}\,C_{p\,i}|^{t_{Amax}}_{0\,°C}} = \frac{H_i + \lambda\,l^*_{min}\,C_{p\,L}|^{t_L}_{0\,°C}\,t_l + c_{p\,B}\,t_B}{\frac{V_{0\,H_2O}}{m_B}\,C_{p\,H_2O}|^{t_{Amax}}_{0\,°C} + \left(\frac{V_{0\,A}}{m_B}\right)_t C_{p\,A\,t}|^{t_{Amax}}_{0\,°C}}$$

$$= \frac{42700\,\frac{\text{kJ}}{\text{kg}\ B} + 13.45\,\frac{\text{Nm}^3\ air}{\text{kg}\ B} \cdot 1.31\,\frac{\text{kJ}}{\text{Nm}^3\ air\ K} \cdot 20\ K + 2.01\,\frac{\text{kJ}}{\text{kg}\ B\ K} \cdot 20\ K}{1.47\,\frac{\text{Nm}^3\ H_2O}{\text{kg}\ B} \cdot 1.95\,\frac{\text{kJ}}{\text{Nm}^3\ H_2O\ K} + 12.71\,\frac{\text{Nm}^3\ A\ t}{\text{kg}\ B} \cdot 1.60\,\frac{\text{kJ}}{\text{Nm}^3\ A\ t\ K}}$$

$$= 1857\,°C.$$

Because the result agrees well with the assumption made for t_{Amax}, a second iteration step is not necessary. For a graphical determination, Fig. 11.18 can be used; from this one obtains $t_{Amax} = 1870\,°C$.

Tasks for Section 11

1. In an information leaflet of a company producing protective gases it is stated that a protective gas for furnaces consisting of

$$r_{H_2} = 0.14; \quad r_{CO} = 0.09; \quad r_{CO_2} = 0.05; \quad r_{N_2} = 0.72$$

reacts completely when mixed with air in the following volume ratio: 64.6 % protective gas to 35.4 % air. Is this information correct?

2. Propane C_3H_8 is completely burnt with dry air at an air/fuel equivalent ratio of $\lambda = 1.2$. The gas can be regarded as an ideal gas. o_{min}, l_{min}, l, $v_{0\,A\,min\,f}$, $v_{0\,A\,f}$, $v_{0\,A\,min\,t}$, $v_{0\,A\,t}$, $r_{CO_2\,max}^{a\,t}$, the composition of the dry combustion gas and the composition of the moist combustion gas are to be calculated. The results for the volume fractions of CO_2 and O_2 in the dry combustion gas shall be checked with the aid of Fig. 11.8.

3. In a sewage sludge treatment plant, sewage sludge is inertised. In the gasifier, as a by-product a combustible lean gas with the following composition in volume fractions r_i is produced:

$$r_{CH_4} = 0.04; \quad r_{H_2} = 0.15; \quad r_{CO} = 0.17; \quad r_{CO_2} = 0.15; \quad r_{N_2} = 0.49$$

The lean gas is an ideal gas and is to be approximated as dry. It is used in a combustion engine for the combined production of electricity and heat. The lean gas is combusted with dry air; the air/fuel equivalent ratio is $\lambda = 1.5$ (lean-burn operation). o_{min}, l_{min}, l, $v_{0\,A\,min\,f}$, $v_{0\,A\,f}$, $v_{0\,A\,min\,t}$, $v_{0\,A\,t}$ and $r_{CO_2\,max}^{a\,t}$ are to be calculated.
The norm volume-related lower calorific value is $H_{i,0} = 5.2$ MJ/Nm^3. The norm volume-related minimum air requirement l_{min} and the norm volume-related norm volume of the moist exhaust gas $v_{0\,A\,f}$ is to be determined graphically using Fig. 11.3.

4. The lean gas according to task 3 is now combusted with a reduced air/fuel equivalent ratio λ. Within the scope of a combustion control, in the dry combustion gas $r_{CO_2}^{a\,t} = 0.181$ Nm^3 CO_2/Nm^3 $A\,t$ and $r_{CO}^{a\,t} = 0.008$ Nm^3 CO/Nm^3 $A\,t$ are measured; other unburnt or only partially burnt components do not occur in the combustion gas. λ shall be determined exactly according to Eq. (11.46) and approximately according to Eq. (11.50) and Eq. (11.54).

5. 1 kg rapeseed oil methyl ester (RME), in the combustion engine of a combined heat and power unit is completely burnt with the air/fuel equivalent ratio $\lambda = 1.2$. The combustion air is dry. Rapeseed oil methyl ester has the following composition in mass fractions:

$$c = 0.7760; \quad h = 0.1360; \quad s = 0.00007; \quad o = 0.0879; \quad n = 0.00003$$

The oxygen requirement characteristic σ, the nitrogen characteristic ν, the specific minimum oxygen requirement o_{min}^*, the specific minimum air requirement l_{min}^*, the specific air requirement l^*, the specific minimum norm volume of the moist combustion gas $v_{0\,A\,min\,f}^*$, the specific norm volume of the moist combustion gas $v_{0\,A\,f}^*$, the specific minimum norm volume of the dry combustion gas $v_{0\,A\,min\,t}^*$ and the specific norm volume of the dry combustion gas $v_{0\,A\,t}^*$ are to be determined.
What is the composition of the moist combustion gas $v_{0\,A\,f}^*$ as well as the dry combustion gas $v_{0\,A\,t}^*$, respectively? The calculation of $v_{0\,A\,f}^*$ using Eq. 11.41 is to be checked. What is the actual moist combustion gas quantity $v_{A\,f}^*$ if the combustion gas leaves the combustion engine with the state $p = 1.0$ bar and $t = 320\,°C$? The molar mass M and the specific gas constant R of the combustion gas is to be determined.
The exhaust pipe of the combined heat and power unit ends at a height of 10 m above the combustion engine as a chimney (stack) into the open air. What is the pressure difference Δp — i.e. the chimney draught — of the combustion gas flow between the combustion engine and the top of the exhaust pipe?

For the calculation of chemical reactions, the fuel wood (waf) is given with the fictitious molecule $C_1H_{1.44}O_{0.66}$. Analogously, the fictitious molecule $C_1H_pO_qS_rN_s$ of rapeseed oil methyl ester (RME) is to be calculated from the composition in mass fractions. The corresponding atomic masses in Table A.7 of the appendix are to be used.

6. Lignite briquettes with the composition in mass fractions

$$c = 0.517; \quad h = 0.040; \quad s = 0.006; \quad o = 0.207; \quad n = 0.010; \quad w = 0.160; \quad a = 0.060$$

are completely burnt with the air/fuel equivalent ratio $\lambda = 1.5$. The combustion air has the condition $p = 1.0$ bar and $t = 20\,°C$. The oxygen requirement characteristic σ, the nitrogen characteristic ν, the specific norm volume air requirement l^*, the specific norm volume of the moist combustion gas $v^*_{0\,A\,f}$ and the specific norm volume of the dry combustion gas $v^*_{0\,A\,t}$ are to be determined.
What is the composition of the specific norm volume of the moist combustion gas $v^*_{0\,A\,f}$ and the specific norm volume of the dry combustion gas $v^*_{0\,A\,t}$, respectively? The maximum volume fraction CO_2 in the specific norm volume of the minimum dry combustion gas $r^{a\,t}_{CO_2\,max}$ shall be investigated. What is the actual moist combustion gas quantity $v^*_{A\,f}$, if the combustion gas leaves the combustion boiler with the condition $p = 0.98$ bar and $t = 240\,°C$?

7. 1 kg of hard coal (forge coal) with the composition in mass fractions

$$c = 0.807; \quad h = 0.040; \quad s = 0.009; \quad o = 0.026; \quad n = 0.013; \quad w = 0.062; \quad a = 0.043$$

is completely burnt with the air/fuel equivalent ratio $\lambda = 1.3$. The combustion air has the condition $p = 0.98$ bar and $t = 5\,°C$. The oxygen requirement characteristic σ, the nitrogen characteristic ν, the specific norm volume air requirement l^*, the specific norm volume of the moist combustion gas $v^*_{0\,A\,f}$ and the specific norm volume of the dry combustion gas $v^*_{0\,A\,t}$ are to be determined. At the air/fuel equivalent ratio $\lambda = 1$, what are the specific minimum norm volume air requirement $l*_{min}$, the specific minimum norm volume of the moist combustion gas $v^*_{0\,Amin\,f}$ and the specific minimum norm volume of the dry combustion gas $v^*_{0\,Amin\,t}$? What is the composition of the moist and the dry minimum norm volume of the combustion gas $v^*_{0\,Amin\,f}$ and $v^*_{0\,Amin\,t}$, respectively? The maximum volume fraction $r^{a\,t}_{CO_2\,max}$ in the specific norm volume of the minimum dry combustion gas is to be investigated.

8. Wood pellets have the composition in mass fractions

$$c = 0.4450; \quad h = 0.0534; \quad s = 0.0030; \quad o = 0.3916; \quad n = 0.0220; \quad w = 0.0800; \quad a = 0.0050$$

They are completely burnt with the air/fuel equivalent ratio $\lambda = 1.0$ and, additionally, with the air/fuel equivalent ratio $\lambda = 1.5$. The oxygen requirement characteristic σ, the nitrogen characteristic ν, the specific minimum norm volume air requirement l^*_{min}, the specific norm volume air requirement l^*, the specific minimum norm volume of the moist combustion gas $v^*_{0\,Amin\,f}$, the specific norm volume of the moist combustion gas $v^*_{0\,A\,f}$, the specific minimum norm volume of the dry combustion gas $v^*_{0\,Amin\,t}$ and the specific norm volume of the dry combustion gas $v^*_{0\,A\,t}$ are to be determined. What is the composition of the specific minimum norm volume of the moist combustion gas $v^*_{0\,Amin\,f}$, of the specific minimum norm volume of the dry combustion gas $v^*_{0\,Amin\,t}$, of the specific norm volume of the moist combustion gas $v^*_{0\,A\,f}$ and of the specific norm volume of the dry combustion gas $v^*_{0\,A\,t}$, respectively?

9. It shall be investigated which thermodynamic effects, according to the assessment parameters in accordance with section 7.5, the operation of a further developed *Otto* engine and a further developed *Diesel* engine with an increased oxygen content in the combustion air will have. Compared to normal air (79 % N_2, 21 % O_2), the air enriched with 50 % more O_2 has the composition 68.5 % N_2, 31.5 % O_2.
The *Otto* process and the *Diesel* process are to be taken as the respective reversible comparison process, whereby the air can be simplified in each case as an ideal gas with $R = 0{,}2872$ kJ/(kg K); $c_p = 1{,}005$ kJ/(kg K); $c_v = 0{,}718$ kJ/(kg K); $\kappa = 1{,}4$ can be assumed.

a) The maximum combustion temperature for complete combustion with usual air at the air/fuel equivalent ratio $\lambda = 1.0$ is given with petrol and *Diesel* fuel at about $t_3 = 2100\,°C$ (cf. Fig. 11.18). What maximum combustion temperature can hypothetically be achieved in a first approximation if complete combustion with air — enriched to 31.5 % O_2 — at an air/fuel equivalent ratio $\lambda = 1.0$ is assumed and the influence of dissociation is neglected? The compositions of petrol with $C_{16}H_{34}$ and of *Diesel* fuel with $C_{21}H_{40}$ are to be fictitiously summarised. The resulting specific norm volumes of the combustion gases of both processes with usual air as well as with enriched air $v_{0\,Af}^*$ are to be calculated, and are to be compared with each other to conclude on the maximum combustion gas temperatures.

b) The initial state 1 for the processes to be considered is $p_1 = 1.0$ bar, $t_1 = 25\,°C$ and $s_1 = 6.867$ kJ/kg K. The following numerical values are to be calculated: the thermal and the exergetic efficiency η_{th} and ζ, the mechanical effort ratios a_{wp} and a_{wv}, the thermal effort ratios a_{qT} and a_{qs} as well as the total effort ratio a_g. The following processes shall be calculated:
— an *Otto* process with the compression ratio $\varepsilon = 13$ and the highest process state $p_3 = 70.0$ bar and $T_3 = 3273.15$ K
— a *Diesel* process with the highest process state of $p_3 = 300.0$ bar and $T_3 = 3273.15$ K
— a *Diesel* process with the highest process state of $p_3 = 300.0$ bar and $T_3 = 2423.15$ K

10. To determine the specific lower calorific value H_i of different fuels — given in MJ/kg — the usefulness of the so called Association Formula (Formula of the German Hard Coal Association, especially for solid fuels such as hard coal and lignite) shall be checked. In addition, *Boie*'s Formula shall be used.

$$H_i = 33.9\,c + 121\,h + 10.5\,s - 15.2\,o - 2.5\,w \qquad (Association\ Formula)$$
$$H_i = 34.8\,c + 93.9\,h + 10.5\,s + 6.3\,n - 10.8\,o - 2.5\,w \qquad (Boie's Formula)$$

Fuels: Xylene C_8H_{10}; Methane CH_4; Ethyl alcohol C_2H_5OH; fuels according to the following table:

	c	h	s	o	n	w	a
Wood (water- and ash-free; waf))	0.500	0.060	0	0.440	0	0	0
Wood pellets	0.4450	0.0534	0.0030	0.3916	0.0220	0.0800	0.0050
Landscape maint. wood (green waste)	0.2965	0.0356	0.0042	0.2609	0.0150	0.3578	0.0300
Digestate (from biogas production)	0.3280	0.0280	0.0040	0.2000	0.0160	0.3770	0.0470
Waste paper sludge	0.1201	0.0140	0.0004	0.0850	0.0010	0.5500	0.2295
Old bread (dry)	0.4010	0.0580	0.0020	0.3860	0.0220	0.1010	0.030
Hard coal (Forge coal)	0.8070	0.0400	0.0090	0.0260	0.0130	0.0622	0.0430
Raw lignite (moist)	0.2620	0.0200	0.0030	0.1050	0.0050	0.5750	0.0300
Lignite (dry)	0.5550	0.0420	0.0060	0.2220	0.0110	0.1000	0.0640
Rapeseed oil methyl ester (RME)	0.7760	0.1360	0.00007	0.0879	0.00003	0	0
Fuel Oil (extra light EL)	0.8600	0.1100	0.0100	0.02000	0	0	0

The calculated values are to be compared with data from lower calorific value tables for the fuels.

11. 1 kg of a mixture of g = 90 % ethyl alcohol (C_2H_5OH) and (1 - g = 10 %) water (H_2O) is completely burnt with the air/fuel equivalent ratio $\lambda = 1.0$. For variable g, the reaction equation is to be given as a mass balance and a norm volume balance and the lower calorific value H_i. Furthermore, for g = 0.9, the reaction equation is to be given as a mass balance and as a norm volume balance, and in addition the lower calorific value H_i is to be calculated.

12 Chemical Thermodynamics

12.1 Systems Involving Chemical Reactions

To describe systems in which chemical reactions take place, it makes sense to use the corresponding molar quantities n_i (given e.g. in mol or kmol) rather than the masses m_i (given e.g. in g or kg) for the substances i involved. If, for example, the molar quantity n_i of the i-th substance increases as a result of a chemical reaction, the molar quantities of other substances decrease at the same time. Examples of corresponding chemical reactions are

$$CH_4 + H_2O \rightleftharpoons CO + 3\,H_2 \tag{12.1}$$

$$CO + H_2O \rightleftharpoons CO_2 + H_2 \tag{12.2}$$

$$N_2 + 3\,H_2 \rightleftharpoons 2\,NH_3 \tag{12.3}$$

$$2\,H_2 + O_2 \rightleftharpoons 2\,H_2O \tag{12.4}$$

$$2\,CO + O_2 \rightleftharpoons 2\,CO_2 \tag{12.5}$$

These relations describe - read from left to right - the formation of carbon monoxide (CO) and hydrogen (H_2) from methane (CH_4) and water vapor (H_2O), the formation of carbon dioxide (CO_2) and hydrogen (H_2) from carbon monoxide (CO) and water vapor (H_2O), the formation of ammonia (NH_3) from nitrogen (N_2) and hydrogen (H_2), the formation of water vapor (H_2O) from hydrogen (H_2) and oxygen (O_2), and the formation of carbon dioxide (CO_2) from carbon monoxide (CO) and oxygen (O_2). The arrows in both directions indicate that the reactions can basically proceed in both directions; depending on e.g. pressure and temperature, a state of equilibrium can be reached in which all substances involved are present together.

Relationships 12.1 and 12.2 are important e.g. for hydrogen supply for fuel cells, relationship 12.3 for ammonia synthesis as a preliminary stage for the production of nitrogenous fertilisers, and relationships 12.4 and 12.5 e.g. as combustion reactions to provide heat. Relationship 12.4 is also a main reaction in most fuel cell designs.

It will first be assumed that all substances exist as gases, e.g. H_2O as water vapor and not as a liquid. The relations (12.1) to (12.5) not only clarify the qualitative statement that certain starting substances (e.g. according to Eq. 12.1 CH_4 and H_2O) are transformed into new substances (here CO and H_2) as a result of a chemical reaction, but also a quantitative statement about the quantities of decreasing and increasing substances - for example, that from 1 kilomole (1 kmol) methane (CH_4) and 1 kilomole (1 kmol) water vapor (H_2O) 1 kilomole (1 kmol) carbon monoxide (CO) and 3 kilomoles (3 kmol) hydrogen (H_2) are formed. With the help of the molar masses M_i, this quantitative fact can also be represented by the masses m_i of the reaction partners according to section 11.1: 16 kg methane (CH_4) and 18 kg water vapor (H_2O) produce 28 kg carbon monoxide (CO) and 6 kg hydrogen (H_2).

In this context, it is of interest how systems with chemically reacting substances can be thermodynamically described.

This is to be shown again with the help of Eq. (12.1). For a complete description, it is important which quantities of methane (CH_4), water vapor (H_2O), carbon monoxide

(CO) and hydrogen (H_2) are contained in the system under consideration. It is important that not only those substances are recorded that are present in pure form (here hydrogen H in the form of H_2), but also those substances that are bound to other substances (e.g. hydrogen H in CH_4 and in H_2O).

The thermodynamic state of a chemical system can be described in such a way that two of the three thermal state variables pressure p, volume V and temperature T are known. It proves to be useful that, for example, in a closed system with fixed walls, the state variables V and T are kept constant, whereas, for example, in a closed system with a moving wall and in an open system, the state variables p and T are kept constant, because in the simplest case chemical reactions take place either at constant volume (Fig. 12.1) or at constant pressure (Fig. 12.2) [9], [42].

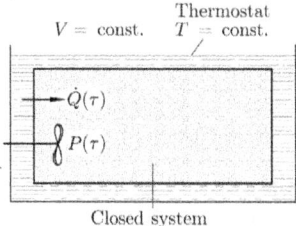

Figure 12.1 Chemical reaction in a closed system with solid walls

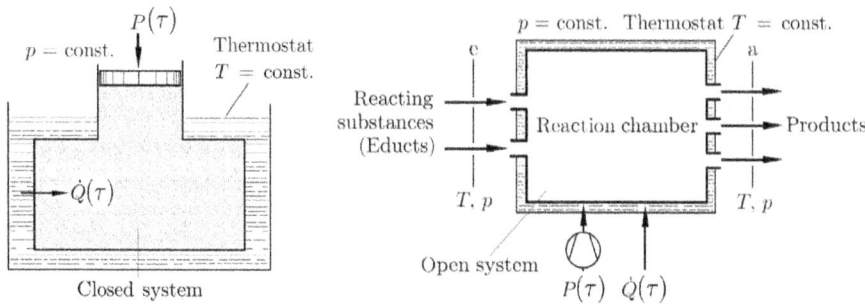

Figure 12.2 Chemical reaction in a closed system with a moving wall (left) and in an open system (right)

In the case of a considerable increase in temperature due to the addition of heat from outside, part of the methane (CH_4) present decomposes (dissociates) into the components carbon (C) and hydrogen (H_2) or part of the water vapor (H_2O) into the components hydrogen (H_2) and oxygen (O_2). Conversely, a decrease in temperature leads to a partial recombination of the dissociated substances into methane (CH_4) and water vapor (H_2O). In a closed system with fixed walls, the total volume V remains constant, but there is a change in pressure. In a closed system with a moving wall and in an open system, the imposed pressure p often remains constant, while the volume V changes.

It is assumed in the following that the state variables V and T or p and T have fixed values. Thus, the equilibrium composition of the system - i.e. the amounts of CH_4, H_2O, CO and H_2 as well as, if existing, the dissociated components - is also unambiguously

fixed. This state of equilibrium is called chemical equilibrium; this represents a special case of thermodynamic equilibrium. In the following, chemical equilibrium states are considered, among others. In a general formulation of the equilibrium conditions, it is obvious to represent all partners of a chemical reaction on one side of the equation; thus the stoichiometric relations of equations (12.1) to (12.5) become

$$CO + 3\,H_2 - CH_4 - H_2O = 0 \tag{12.6}$$

$$CO_2 + H_2 - CO - H_2O = 0 \tag{12.7}$$

$$2\,NH_3 - N_2 - 3\,H_2 = 0 \tag{12.8}$$

$$2\,H_2O - 2\,H_2 - O_2 = 0 \tag{12.9}$$

$$2\,CO_2 - 2\,CO - O_2 = 0 \;. \tag{12.10}$$

In relations (12.6) to (12.10), no arrows appear, but equals signs. Accordingly, any chemical reaction with N participants can be described by the relation

$$\nu_1\,A_1 + \nu_2\,A_2 + \nu_3\,A_3 + ... \nu_N\,A_N = 0$$

respectively

$$\sum_{i=1}^{N} \nu_i\,A_i = 0 \;, \tag{12.11}$$

where A_i is the symbol of the i-th substance and ν_i is a positive or negative number. The quantities ν_i are called stoichiometric factors. If one now assumes a small change in the composition and a differential change in the molar quantity n_i (amount of molar substance n_i) of the i-th substance by dn_i in the system under consideration, then, in accordance with the above considerations, a further differential change dn_j also occurs for another substance j or further differential changes dn_j occur for other substances j.

For example, according to the reaction corresponding to Eq. (12.6) it can be said: Whenever 1 kmol of CH_4 has reacted, 1 kmol of H_2O has also reacted, and 1 kmol of CO and 3 kmol of H_2 are produced. This relationship can be represented in the following way:

$$\frac{dn_1}{\nu_1} = \frac{dn_2}{\nu_2} = \frac{dn_3}{\nu_3} = ... = \frac{dn_i}{\nu_i} = ... = \frac{dn_N}{\nu_N} \tag{12.12}$$

12.2 Reaction Turnover and Reaction Rate

In Eq. (12.12), for all substances, the quotient dn_i/ν_i is constant; this is set to dz and represents the differential of a quantity z called the reaction turnover or the reaction run number:

$$dz = \frac{dn_i}{\nu_i} \tag{12.13}$$

The reaction turnover z has the dimension of a molar quantity (amount of molar substance) and thus the unit mol or kmol.

If z is related on a time interval, its time change $\dot{z} = dz/d\tau$ is the reaction rate in mol per second or in kmol per second (also known as turnover rate):

$$\dot{z} = \frac{dz}{d\tau} = \frac{dn_i}{\nu_i \, d\tau} \qquad (12.13\,\text{a})$$

The integration of Eq. (12.13) gives for the component i

$$n_i = n_i^0 + \nu_i \, z \quad . \qquad (12.14)$$

This also includes inert substances for which no reaction turnover occurs and $\nu_i = 0$ holds. The molar quantities n_i^0 represent the amounts of molar substances of the respective components at a certain state 0, in which the reacting mixture of substances is currently. According to Eq. (12.14), all molar quantities of the components depend on only one variable - the reaction turnover z. The molar quantity n of the mixture is obtained by adding up all n_i of the individual components i:

$$n = \sum_{i=1}^{N} n_i = \sum_{i=1}^{N} n_i^0 + \sum_{i=1}^{N} \nu_i \, z = n_0 + \sum_{i=1}^{N} \nu_i \, z \qquad (12.15)$$

n_0 here is the molar quantity in the state 0 for which $z = 0$ holds. The molar quantity n does not have to remain constant but can change; n is invariant only in equimolar reactions for which $\sum_{i=1}^{N} \nu_i = 0$ holds.

A reaction takes place only as long as all reactants are present. The reaction turnover

$$z = \frac{n_i - n_i^0}{\nu_i} \qquad (12.16)$$

reaches its maximum value z_{max} when one of the starting substances (reactants, also called educts) ($\nu_i < 0$) is completely converted. The minimum value z_{min} results when one of the products ($\nu_i > 0$) is no longer present. Let the starting substance that is completely consumed be denoted by $i = j$. Then holds:

$$z_{max} = -\frac{n_j^0}{\nu_j} = \frac{n_j^0}{|\nu_j|}$$

Let that product which is no longer present be denoted by $i = m$. Then holds:

$$z_{min} = -\frac{n_m^0}{\nu_m}$$

If, for this product, not $n_m^0 = 0$ holds, z_{min} becomes negative. With z_{max} and z_{min}, the degree of turnover (also called degree of reaction) ϵ is defined as a new quantity whose values lie between 0 and 1:

$$\epsilon = \frac{z - z_{min}}{z_{max} - z_{min}} \qquad 0 \leq \epsilon \leq 1 \qquad (12.17)$$

Hereby the molar quantities n_i as well as the molar fractions n_i/n can be given as functions of ϵ. The following applies to the extreme values of ϵ: If $\epsilon = 0$, the mixture is as close as possible to the starting substances (reactants, educts). If, on the other hand, $\epsilon = 1$, the mixture is as close as possible to the products. Whether these extreme values are actually achievable in the various reactions must, however, be clarified with the help of further considerations on the question of reaction equilibria; this is examined in section 12.9 (mass action law).

12.2 Reaction Turnover and Degree of Turnover

Example 12.1 For the synthesis of ammonia (NH_3) from nitrogen (N_2) and hydrogen (H_2), the following holds:
$$2\,NH_3 - N_2 - 3\,H_2 = 0 \tag{12.8}$$
From the analysis of a gas mixture sample taken during the course of this reaction, the following composition in mole fractions was obtained: $n^0_{NH_3}/n_0 = 0.25$; $n^0_{N_2}/n_0 = 0.17$; $n^0_{H_2}/n_0 = 0.58$. In these relations, n_0 is the unknown molar quantity of the mixture sample. The three mole fractions are to be plotted as a function of the degree of reaction ϵ. Further, the degree of reaction ϵ at which the mixture sample was taken shall be determined.

With Eq. (12.14) follow

$n_{NH_3} = 0.25\,n_0 + 2\,z$,

$n_{N_2} = 0.17\,n_0 - 1\,z$ and

$n_{H_2} = 0.58\,n_0 - 3\,z$.

Furthermore, for the molar quantity n of the reacting mixture, which depends on the reaction turnover z, Eq. (12.15) gives the relation

$n = n_0 - 2\,z$.

To calculate the degree of reaction ϵ, z_{min} and z_{max} have to be determined first. z_{min} results for the product ammonia (NH_3) with $n_{NH_3} = 0$ to $z_{min} = -0.125\,n_0$.
The substance of the reactants that runs out first is nitrogen (N_2), since $n^0_{N_2}/n_0 = 0.17 < (n^0_{H_2}/n_0)/3 = 0.58/3 = 0.1933$.
Thus, with $n_{N_2} = 0$ for z_{max}, the value $z_{max} = 0.17\,n_0$ is obtained.
From this follows with Eq. (12.17)

$$\epsilon = \frac{z - z_{min}}{z_{max} - z_{min}} = \frac{z + 0.125\,n_0}{(0.17 + 0.125)\,n_0} = \frac{z/n_0 + 0.125}{0.295}$$

and

$z/n_0 = 0.295\,\epsilon - 0.125$.

For the mole fractions n_i/n, the relations

$$\frac{n_{NH_3}}{n} = \frac{0.25\,n_0 + 2\,z}{n_0 - 2\,z} = \frac{0.25 + 2\,\dfrac{z}{n_0}}{1 - 2\,\dfrac{z}{n_0}} = \frac{0.25 + 2\,(0.295\,\epsilon - 0.125)}{1 - 2\,(0.295\,\epsilon - 0.125)} = \frac{0.59\,\epsilon}{1.25 - 0.59\,\epsilon}$$

$$\frac{n_{N_2}}{n} = \frac{0.17\,n_0 - 1\,z}{n_0 - 2\,z} = \frac{0.17 - 1\,\dfrac{z}{n_0}}{1 - 2\,\dfrac{z}{n_0}} = \frac{0.17 - 1\,(0.295\,\epsilon - 0.125)}{1 - 2\,(0.295\,\epsilon - 0.125)} = \frac{0.295 - 0.295\,\epsilon}{1.25 - 0.59\,\epsilon}$$

$$\frac{n_{H_2}}{n} = \frac{0.58\,n_0 - 3\,z}{n_0 - 2\,z} = \frac{0.58 - 3\,\dfrac{z}{n_0}}{1 - 2\,\dfrac{z}{n_0}} = \frac{0.58 - 3\,(0.295\,\epsilon - 0.125)}{1 - 2\,(0.295\,\epsilon - 0.125)} = \frac{0.955 - 0.885\,\epsilon}{1.25 - 0.59\,\epsilon}$$

follow. Fig. 12.3 shows the course of the mole fractions for N_2, H_2 and NH_3 in the mixture as a function of the degree of reaction ϵ. The mole fractions of N_2 and H_2 do not completely show the required stoichiometric ratio 1 : 3.

Since in the ammonia synthesis the reaction equilibrium is by no means entirely on the side of the product NH_3, the course shown in the right area of Fig. 12.3 is partly hypothetical.

The degree of reaction ϵ that was present in the mixture sample drawn can be calculated from the condition $z = 0$. From the relation $z/n_0 = 0.295\,\epsilon - 0.125$ follows $\epsilon = 0.125/0.295 = 0.4237$.

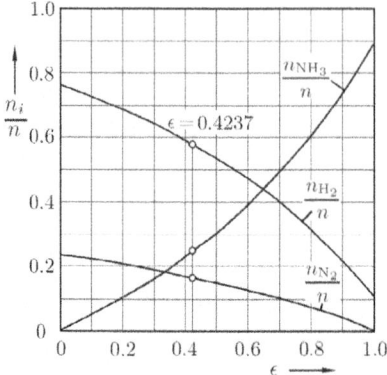

Figure 12.3 Course of the mole fractions for nitrogen (N_2), hydrogen (H_2) and ammonia (NH_3) in the mixture as a function of the degree of reaction ϵ

12.3 Molar Enthalpies of Reaction and Standard Molar Enthalpies of Formation; Theorem of *Hess*

12.3.1 Molar Enthalpies of Reaction

When calculating changes of state according to the first law as it is applied to sections 2 to 11, differences of the internal energy or of the enthalpy are formed for one and the same substance or for one and the same mixture of substances; in this case, the indeterminate constant of the internal energy U_{0i} or of the enthalpy H_{0i} is omitted, so that the knowledge of such a constant is not necessary.

In contrast, the knowledge of these constants is necessary when applying the first law to chemical reactions, since differences of the internal energies or of the enthalpies of different substances are to be calculated here and the undetermined constants cannot be shortened mathematically. This problem is discussed in more detail below for enthalpies. These are to be matched in such a way that the energy and power balances lead to results that agree with results from experiments.

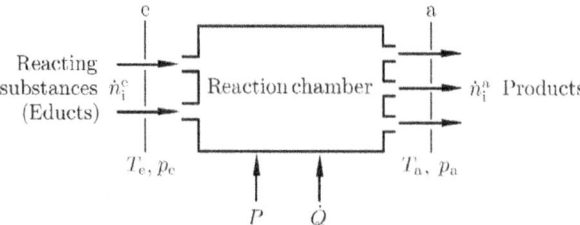

Figure 12.4 Chemical reaction in an open system: balancing according to the first law

Fig. 12.4 shows a stationary flow process (steady-state flow process) in which a chemical reaction takes place; individual reactive educts or a reactive mixture of educts (index

12.3 Molar Enthalpies of Reaction and Standard Molar Enthalpies of Formation

'e') with pressure p_e and temperature T_e enter across the system boundary, and a reaction according to

$$\sum_{i=1}^{N} \nu_i A_i = 0 \tag{12.11}$$

takes place. The reaction products (index 'a') exit with pressure p_a and temperature T_a across the system boundary. The entering and exiting subtance flows are either each separately occurring pure substance flows or ideal mixture flows, so that no mixture enthalpies have to be considered; this is expressed by the index '0'. The first law, neglecting changes in the time-related potential and kinetic energies as well as any external friction power that may occur, is as follows according to Eq. (2.77):

$$\dot{Q} + P = \sum_{i=1}^{N} \dot{n}_i^a H_{m,0i}(T_a, p_a) - \sum_{i=1}^{N} \dot{n}_i^e H_{m,0i}(T_e, p_e) \tag{12.18}$$

Here, $\dot{n}_i$ represents the molar flow rate and $H_{m,0i}$ the molar enthalpy of the respective component i.

According to Eq. (12.14), the following applies to the respective flows of the molar quantities of the components:

$$\dot{n}_i = \dot{n}_i^0 + \nu_i \dot{z}$$

In this, $\dot{z}$ represents as a time-related variable the reaction rate: The entry state e is chosen as state 0 before the start of the chemical reaction ($\dot{z} = 0$). Thus, with $\dot{n}_i^0 = \dot{n}_i^e$ one gets

$$\dot{n}_i^a = \dot{n}_i^e + \nu_i \dot{z}_a \quad , \tag{12.19}$$

where hereby all flows of the molar quantities $\dot{n}_i$ are recorded and their entry and exit states are mathematically connected with each other via the reaction rate $\dot{z}_a$.

$\dot{z}_a$ then becomes $\dot{z}_{max}$ when one of the entering mass flows i is completely consumed in the reaction chamber; then $\dot{n}_i^a = 0$ holds. Mostly, however, a state of equilibrium results at the end of the chemical reaction at a reaction rate $\dot{z}_a = \dot{z}_{equilib}$ (cf. section 12.9). Eq. (12.19) also includes those flows of the molar quantities which do not react, because for them $\nu_i = 0$ holds, so that here is $\dot{n}_i^a = \dot{n}_i^e$.

The right-hand side of Eq. (12.18) can be modified by introducing a reference state (e.g. a defined norm, standard or reference state) with the pressure p_0 and the temperature T_0 in such a way that enthalpy differences result which can be calculated by caloric equations of state or by tables of the molar enthalpies of the respective substances [9]:

$$\begin{aligned} \dot{Q} + P &= \sum_{i=1}^{N} \dot{n}_i^a \left[H_{m,0i}(T_a, p_a) - H_{m,0i}(T_0, p_0) \right] \\ &- \sum_{i=1}^{N} \dot{n}_i^e \left[H_{m,0i}(T_e, p_e) - H_{m,0i}(T_0, p_0) \right] \\ &+ \sum_{i=1}^{N} (\dot{n}_i^a - \dot{n}_i^e) H_{m,0i}(T_0, p_0) \end{aligned} \tag{12.20}$$

The third summand in Eq. (12.20) follows from the difference in the first two summands.

This can be given by the transformed Eq. (12.19) as follows:

$$\sum_{i=1}^{N} (\dot{n}_i^a - \dot{n}_i^e) H_{m,0i}(T_0, p_0) = \dot{z}_a \sum_{i=1}^{N} \nu_i H_{m,0i}(T_0, p_0) = \dot{z}_a \Delta^R H_m(T_0, p_0) \quad (12.21)$$

In this equation

$$\Delta^R H_m(T_0, p_0) = \sum_{i=1}^{N} \nu_i H_{m,0i}(T_0, p_0) \quad (12.22)$$

is called the molar enthalpy of reaction at pressure p_0 and temperature T_0. This represents the enthalpy change of the chemical reaction $\sum_{i=1}^{N} \nu_i A_i = 0$ taking place at pressure p_0 and temperature T_0 isothermally and isobarically, assuming complete conversion of the starting substances (educts) into the products.

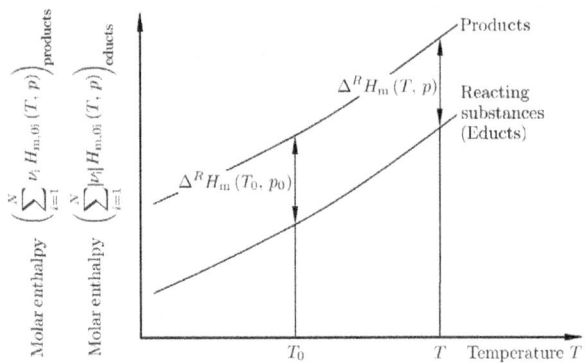

Figure 12.5 Dependence of molar enthalpy of reaction on temperature

The molar enthalpy of reaction $\Delta^R H_m(T_0, p_0)$ can be determined by calorimetric measurement of the chemical reaction under investigation. The molar enthalpy of reaction occurring at arbitrary temperatures and pressures

$$\Delta^R H_m(T, p) = \sum_{i=1}^{N} \nu_i H_{m,0i}(T, p) \quad (12.22\,a)$$

is hardly dependent on pressure and temperature (cf. Fig. 12.5).

If the isothermal and isobaric reaction is at a chosen reference state $T = T_a = T_e$ and $p = p_a = p_e$, the first two sum expressions in Eq. (12.20) each become zero, and the following is valid:

$$\dot{Q} + P = \sum_{i=1}^{N} (\dot{n}_i^a - \dot{n}_i^e) H_{m,0i}(T, p) = \dot{z}_a \Delta^R H_m(T, p) \quad (12.20\,a)$$

For a reference state $T_0 = T_a = T_e$ and $p_0 = p_a = p_e$, the following applies accordingly:

$$\dot{Q} + P = \sum_{i=1}^{N} (\dot{n}_i^a - \dot{n}_i^e) H_{m,0i}(T_0, p_0) = \dot{z}_a \Delta^R H_m(T_0, p_0) \quad (12.20\,b)$$

Often, within the framework of a simplified reaction control — e.g. for an oxidation reaction in a boiler, for example — the time-related work output P is omitted and only the time-related heat output $\dot{Q}$ is of interest, which is proportional to the molar reaction enthalpy $\Delta^R H_m (T_0, p_0)$ at a selected reference state $T_0 = T_a = T_e$ and $p_0 = p_a = p_e$ according to Eq. (12.20 b).

Reactions with $\Delta^R H_m > 0$ are called endothermic reactions. Because the sum of the enthalpy flows of the products then exceeds the sum of the enthalpy flows of the educts in Eq. (12.20 a), the sum of the two process variables $\dot{Q} + P$ — mathematically speaking — must be supplied.

Reactions with $\Delta^R H_m < 0$ are called exothermic reactions. Because the sum of the enthalpy flows of the educts then outweighs the sum of the enthalpy flows of the products in Eq. (12.20 a), the sum of the two process variables $\dot{Q} + P$ — mathematically speaking — is to be released. Since the time-related work P is exergetically more valuable than the time-related heat $\dot{Q}$, it is of interest — especially in oxidation reactions, e.g. in combustion engines or in fuel cells — to increase the amount of the time-related work output $|P|$ at the expense of the time-related heat output $|\dot{Q}|$ by a suitable process control and to approach a reversible reaction.

In the molar enthalpy of reaction $\Delta^R H_{\mathrm{m}}(T_0, p_0)$ the connection of the molar enthalpies of the reactants at the reference state (T_0, p_0) becomes visible. However, the numerical values of these molar enthalpies are still characterised by an arbitrarily chosen constant initial value (integration constant). With the large number of possible chemical reactions, it is important to avoid these arbitrary constants and to replace them with constants for the molar enthalpies of the individual reactants in such a way that they are transformed into a binding relationship.

If this goal is achieved, the number of molar reaction enthalpies to be measured $\Delta^R H_{\mathrm{m}}(T_0, p_0)$ can remain limited, since each substance in a fixed reference state has only one, unique substance-specific value of the molar enthalpy $H_{\mathrm{m}}(T_0, p_0)$, which is the same for all reactions of this substance.

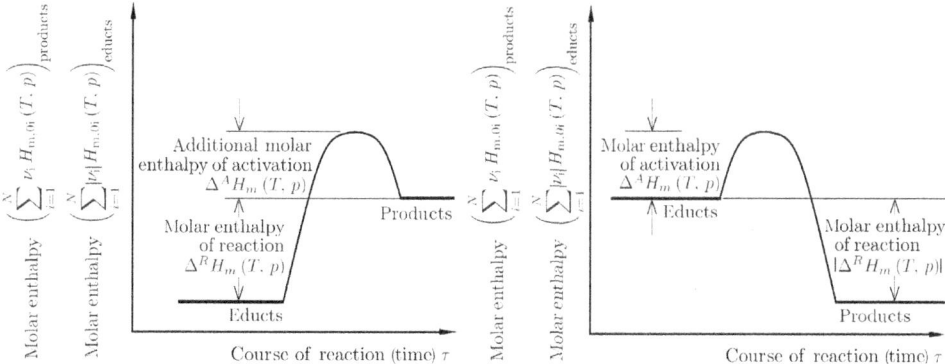

Figure 12.6 Molar activation enthalpy and molar enthalpy of reaction of an endothermic chemical reaction (left) and of an exothermic chemical reaction (right)

In order for isothermal-isobaric chemical reactions to proceed sufficiently fast, it is often necessary to carry them out not at a low temperature T, but at a higher tem-

perature T_R. Then, in endothermic reactions, in addition to the molar enthalpy of reaction $\Delta^R H_m$, an additional molar enthalpy of activation $\Delta^A H_m$ must be supplied, which is released again after the reaction when the system cools down to the original temperature T (Fig. 12.6 left). This applies equally to the molar activation enthalpy $\Delta^A H_m$ in exothermic reactions (Fig. 12.6 right). The molar activation enthalpy $\Delta^A H_m$ can be reduced in certain chemical reactions with the aid of a catalyst, whereby a rapidly proceeding reaction at a lower temperature T_K is possible; the catalyst itself remains unchanged. This is illustrated schematically in Fig. 12.7 using the example of an exothermic reaction.

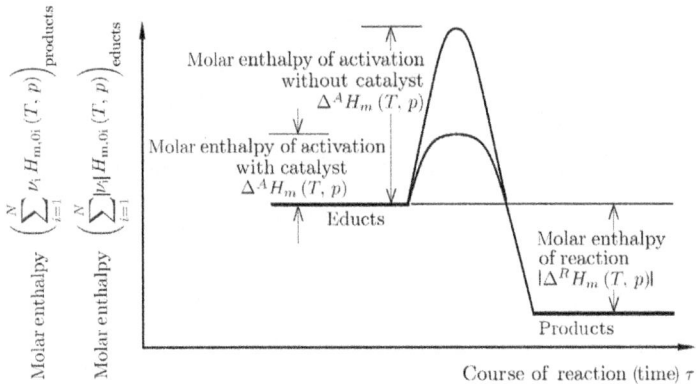

Fig. 12.7 Molar enthalpy of activation with and without catalyst, and molar enthalpy of reaction in an exothermic chemical reaction

12.3.2 Standard Molar Enthalpies of Formation; Theorem of *Hess*

Because the individual chemical elements from which a substance present as a chemical compound is formed remain unchanged in all reactions, it is suggested that the molar enthalpy of this substance be linked to the molar enthalpies of the respective elements from which the substance is formed. For this purpose, the isothermal and isobaric chemical reaction is used by which the chemical compound A_i is formed from a number of M elements E_k. For this reaction, called the formation reaction, the general equation is:

$$A_i = \sum_{k=1}^{M} |\nu_{ki}| E_k = \sum_{k=1}^{M} a_{ki} E_k \qquad (12.23)$$

Their molar enthalpy of reaction is called molar enthalpy of formation $H_{m,i}^f(T, p)$:

$$H_{m,i}^f(T, p) = H_{m,0i}(T, p) - \sum_{k=1}^{M} |\nu_{ki}| H_{m,0k}(T, p) \qquad (12.24)$$

Using the transformed Eq. (12.24), it follows according to Eq. (12.22 a) for the molar enthalpy of reaction $\Delta^R H_m(T, p)$ of any isothermal and isobaric chemical reaction involving a number of N educts and products taken together, each composed of a number M of elements:

$$\Delta^R H_m(T, p) = \sum_{i=1}^{N} \nu_i H_{m,0i}(T, p) = \sum_{i=1}^{N} \nu_i H_{m,i}^f(T, p) + \sum_{i=1}^{N} \nu_i \sum_{k=1}^{M} |\nu_{ki}| H_{m,0k}(T, p) \quad (12.25)$$

The order of the summations in the last term of Eq. (12.25) can be reversed. Looking at the sum in the brackets in Eq. (12.26), it is clear that this becomes zero, since the quantity of the respective element E_k — whether in the original state or within the substances formed with it — remains unchanged [9]:

$$\sum_{k=1}^{M} \left(\sum_{i=1}^{N} \nu_i |\nu_{ki}| \right) H_{m,0k}(T, p) = \sum_{k=1}^{M} \left(\sum_{i=1}^{N} \nu_i a_{ki} \right) H_{m,0k}(T, p) = 0 \quad (12.26)$$

It follows from Eq. (12.25):

$$\Delta^R H_m(T, p) = \sum_{i=1}^{N} \nu_i H_{m,i}^f(T, p) \quad (12.27)$$

Thus, any molar enthalpy of reaction $\Delta^R H_m(T, p)$ can be determined independently using the values of the molar enthalpies $H_{m,0k}(T, p)$ of the elements. Rather, it can be calculated solely from the molar enthalpies of formation of the compounds involved $H_{m,i}^f(T, p)$, which can be determined experimentally, whereby the respective stoichiometric coefficients ν_i must also be taken into account according to Eq. (12.27). Measurements of further molar enthalpies of reaction are therefore not necessary [9].

These results apply not only to general values for temperature T and pressure p, but also to a fixed reference state. The standard thermochemical state ($T = T_0 = 298.15$ K $= 25\,°C$, $p = p_0 = 1.0$ bar $= 100$ kPa) is chosen as the reference state: All substances present in the gaseous state are taken to be ideal gases.

Furthermore, those gaseous elements that are not noble gases are not treated as monatomic, but are treated as molecules in their appearance. For all other substances - liquid and solid elements and compounds - the stable phase is taken as the basis. Table A.7 in the appendix contains the following solid, liquid and gaseous elements in the forms

O_2, H_2, He, Ne, Ar, Kr, Xe, F_2, Cl_2, S (rhombic), C (graphite), C (diamond), N_2 .

Since in the calculation of the molar enthalpies of reaction according to Eq. (12.27) the molar enthalpies of the elements can remain irrelevant, these are set to zero in the standard state: $H_{m,0k}(T_0, p_0) = 0$. The molar enthalpy of formation of a compound in the standard state is called the standard molar enthalpy of formation and is given the name $H_i^{f\square} = H_{m,i}^{f\square}(T_0, p_0)$, often omitting the epithet 'molar'. The standard molar enthalpy of formation $H_i^{f\square}$ can be measured directly or calculated from other standard molar enthalpies of reaction.

From Eq. (12.27) follows:

$$\Delta^R H_m(T_0, p_0) = \sum_{i=1}^{N} \nu_i H_{m,i}^{f\square}(T_0, p_0) \quad (12.28)$$

The molar enthalpy of a pure substance is thus given by

$$H^*_{m,0i}(T, p) = H_i^{f\square} + [H_{m,0i}(T, p) - H_{m,0i}(T_0, p_0)] \ . \qquad (12.29)$$

The expression in the square bracket is the difference of the molar enthalpies of the substance i between the arbitrary state (T, p) and the standard state (T_0, p_0). It can be calculated with a caloric equation of state or taken from tables. The molar enthalpies matched using the standard molar enthalpies of reaction are marked with an asterisk and called molar conventional enthalpies $H^*_{m,0i}(T, p)$ [9]. They make it possible to easily formulate equations according to the first law also for chemical reactions without having to make further adjustments. The determination of molar enthalpies of reaction is no longer necessary. For equation (12.20), it can thus be written [9]:

$$\dot{Q} + P = \sum_{i=1}^{N} \dot{n}_i^a H^*_{m,0i}(T_a, p_a) - \sum_{i=1}^{N} \dot{n}_i^e H^*_{m,0i}(T_e, p_e) \qquad (12.30)$$

If one does not make use of adding or removing a time-related molar work P_m or a molar work W_m in chemical reactions, the reaction is irreversible to a great extent.

If the reaction then proceeds isothermally-isochorically in the standard chemical state ($dT = 0$, $dV = 0$), the molar heat of reaction $Q_m = Q_{mV}$ is equal to the change in molar internal energy ΔU_m according to the first law: $Q_{mV} = \Delta U_m$ If a reaction proceeds isothermally-isobarically ($dT = 0$, $dp = 0$) in the standard chemical state, the molar heat of reaction $Q_m = Q_{mp}$ is equal to the change in molar enthalpy ΔH_m according to the first law: $Q_{mp} = \Delta H_m = \Delta^R H_m(T_0, p_0)$

Since the molar enthalpy and the molar internal energy are state variables, the molar heats of reaction Q_{mV} and Q_{mp} depend, under the above conditions, only on the initial and final states of the system; they are thus independent of any intermediate steps.

Example 12.2 The sum of the two reactions $C + O_2 \rightleftharpoons CO_2$ (a) and $CO_2 \rightleftharpoons CO + 0.5\ O_2$ (b) is to be considered, which can be combined to give $C + O_2 \rightleftharpoons CO + 0.5\ O_2$ and hence the reaction $C + 0.5\ O_2 \rightleftharpoons CO$ (c). Using Eq. (12.28) and the values of Table A.7 in the appendix, computationally is to be shown that the sum of the standard molar enthalpies of reaction (a) and (b) is equal to the standard molar enthalpy of reaction (c).

Reaction (a): $CO_2 - C - O_2 = 0$ with $\Delta_a^R H_m(T_0, p_0) = (-393.509 - 0 - 0)$ MJ/kmol $= -393.509$ MJ/kmol

Reaction (b): $CO + 0.5\ O_2 - CO_2 = 0$ with $\Delta_b^R H_m(T_0, p_0) = (-110.525 + 0.5 \cdot 0 - (-393.509))$ MJ/kmol $= +282.984$ MJ/kmol

Reaction (c): $CO - C - 0.5\ O_2 = 0$ with $\Delta_c^R H_m(T_0, p_0) = (-110.525 - 0 - 0.5 \cdot 0)$ MJ/kmol $= -110.525$ MJ/kmol

For reaction (c) also holds: $\Delta_c^R H_m(T_0, p_0) = \Delta_a^R H_m(T_0, p_0) + \Delta_b^R H_m(T_0, p_0) = (-393.509 + 282.984)$ MJ/kmol $= -110.525$ MJ/kmol

In example 12.2, the molar enthalpies of reaction can therefore be added in the same way as the two stoichiometric equations (a) and (b). This law, which reflects the first law for chemical reactions, was developed by *Hess*[1] and was discovered in 1840 - i.e. before the formulation of the first law by *Robert Mayer*, *James Prescott Joule* and *Hermann von Helmholtz* (cf. section 2.1); it is called the law of constant heat sums.

An advantage of this law is that unknown molar enthalpies of reaction can be traced back to molar enthalpies of reaction of known reactions. For example, the molar enthalpy of reaction of incomplete combustion according to $C + 0.5\,O_2 \rightleftharpoons CO$ (c) can hardly be determined by an experiment because CO_2 is formed in addition to CO. In contrast, measurements of the respective molar reaction enthalpies are possible for both reactions (a) and (b).

From example 12.2 it is visible that according to Eq. (12. 28) the standard molar enthalpies of formation $H_i^{f\square} = H_{m,i}^{f\square}(T_0, p_0)$, multiplied by the respective stoichiometric coefficients ν_i according to the underlying reaction equation, can be added to compute the corresponding standard molar enthalpy of reaction $\Delta^R H_{m,0i}(T, p)$.

Example 12.3

a) A fuel cell is operated with Siberian natural gas H (cf. Table 11.1), which can be simplified as consisting entirely of methane (CH_4). Hydrogen (H_2) is produced from this with the help of the endothermic steam reforming reaction

$CH_4 + H_2O \rightleftharpoons CO + 3\,H_2$ (a)

and the exothermic carbon monoxide conversion reaction

$CO + H_2O \rightleftharpoons CO_2 + H_2$ (b) .

The sum of both reactions (a) and (b) is described by the reaction equation

$CH_4 + 2\,H_2O \rightleftharpoons CO_2 + 4\,H_2$ (c) .

Hydrogen H_2 is then used in the exothermic fuel cell reaction

$4\,H_2 + 2\,O_2 \rightleftharpoons 4\,H_2O$ (d)

to generate electricity.

If reactions (c) and (d) are combined, the result is

$CH_4 + 2\,O_2 + 2\,H_2O \rightleftharpoons CO_2 + 4\,H_2O$ (e)

or simplified

$CH_4 + 2\,O_2 \rightleftharpoons CO_2 + 2\,H_2O$ (f) .

For these reactions, using the values of the standard molar enthalpies of formation $H_i^{f\square}$ from Table A.7 in the appendix, the standard molar enthalpies of reaction $\Delta^R H_m(T_0, p_0)$ are to be calculated.

b) With the Power-to-Gas concept (Fig. 12.8), electrical energy from wind turbines is to be transformed into hydrogen (H_2) energy using water electrolysis and, if necessary, in a second step in methane (CH_4). Hydrogen or methane should then be fed into the natural gas grid. The two reactions required are the reverse reaction of (d)

$4\,H_2O \rightleftharpoons 4\,H_2 + 2\,O_2$ (g)

---

[1] *Hermann Heinrich Hess (Germain Henri Hess)*, 1802 to 1850, Swiss natural scientist, concerned himself among other fields with thermochemical investigations.

and the reverse reaction of (c)

4 H$_2$ + CO$_2$ ⇌ CH$_4$ + 2 H$_2$O . (h)

These reactions can be calculated to the total reaction

4 H$_2$O + CO$_2$ ⇌ 2 O$_2$ + CH$_4$ + 2 H$_2$O (i)

which is the reverse reaction of (e), which in simplified form is

2 H$_2$O + CO$_2$ ⇌ 2 O$_2$ + CH$_4$ (k)

and represents the reverse reaction of (f).

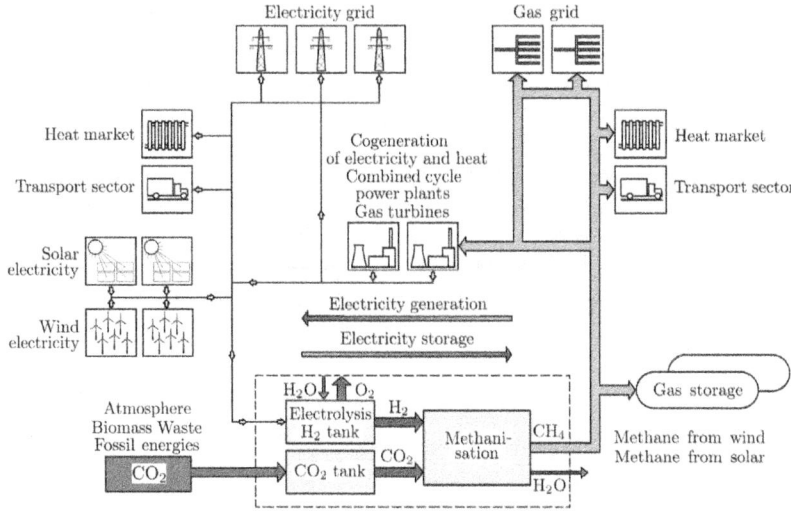

Figure 12.8 Power-to-Gas concept: generation, storage and use of hydrogen (H$_2$) and possibly methane (CH$_4$) using electrical energy from wind and solar power plants. The lower part of the diagram shows the water electrolysis for hydrogen production (H$_2$) and, in a second step, a possible conversion of hydrogen into methane.

For reactions (g), (h), (i) and (k), using the standard molar enthalpies of formation $H_i^{f\square}$ from Table A.7 in the appendix, the standard molar enthalpies of reaction $\Delta^R H_m(T_0, p_0)$ are to be calculated.

a) According to Eq. (12.28), the following holds for the respective chemical reactions:

CO + 3 H$_2$ - CH$_4$ - H$_2$O = 0 (a*)

$(-110.53 + 3 \cdot 0 + 74.87 + 241.83)$ MJ/kmol $= +206.17$ MJ/kmol

$= +206.17$ MJ/(kmol CH$_4$)

CO$_2$ + H$_2$ - CO - H$_2$O = 0 (b*)

$(-393.51 + 0 + 110.53 + 241.83)$ MJ/kmol $= -41.15$ MJ/kmol $= -41.15$ MJ/(kmol CO)

CO$_2$ + 4 H$_2$ - CH$_4$ - 2 H$_2$O = 0 (c*)

$(-393.51 + 4 \cdot 0 + 74.87 + 2 \cdot 241.83)$ MJ/kmol $= +165.02$ MJ/kmol

$= +165.02$ MJ/(kmol CH$_4$)

4 H$_2$O - 4 H$_2$ - 2 O$_2$ = 0 (d*)

$(-4 \cdot 241.83 - 4 \cdot 0 - 2 \cdot 0)$ MJ/(4 kmol H_2) $= -967.32$ MJ/(4 kmol H_2)

(related to 4 kmol H_2 formed with the help of 1 kmol CH_4).

Considering the equation $H_2O - H_2 - 0.5\, O_2 = 0$ (d**), one gets:

$(-241.83 - 0 - 0.5 \cdot 0)$ MJ/(kmol H_2) $= -241.83$ MJ/(kmol H_2)

$CO_2 + 4\, H_2O - CH_4 - 2\, O_2 - 2\, H_2O = 0$ (e*)

$(-393.51 - 4 \cdot 241.83 + 74.87 - 2 \cdot 0 + 2 \cdot 241.83)$ MJ/kmol $= -802.30$ MJ/kmol
$= -802.30$ MJ/(kmol CH_4)

$CO_2 + 2\, H_2O - CH_4 - 2\, O_2 = 0$ (f*)

$(-393.51 + 0 - 2 \cdot 241.83 + 74.87 - 2 \cdot 0)$ MJ/kmol $= -802.30$ MJ/kmol
$= -802.30$ MJ/(kmol $(-393.51 + 0 - 2 \cdot 241.83 + 74.87 - 2 \cdot 0)$ MJ/kmol $= -802.30$ MJ/kmol
$= -802.30$ MJ/(kmol CH_4)

Water is always understood as water vapor. If water is accounted for as a liquid, this results, for example, according to equations (d*) and (d**):

$4\, H_2O - 4\, H_2 - 2\, O_2 = 0$ (d*)

$(-4 \cdot 285.84 - 4 \cdot 0 - 2 \cdot 0)$ MJ/(4 kmol H_2) $= -1143.36$ MJ/(4 kmol H_2)

(related to 4 kmol H_2 formed with the help of 1 kmol CH_4), resp.

$H_2O - H_2 - 0.5\, O_2 = 0$ (d**)

$(-285.84 - 0 - 0.5 \cdot 0)$ MJ/(kmol H_2) $= -285.84$ MJ/(kmol H_2)

b) For reactions (g), (h), (i) and (k), one gets according to Eq. (12.28):

$4\, H_2 + 2\, O_2 - 4\, H_2O = 0$ (g*)

$(4 \cdot 0 + 2 \cdot 0 + 4 \cdot 241.83)$ MJ/(4 kmol H_2) $= +967.32$ MJ/(4 kmol H_2)

$CH_4 + 2\, H_2O - CO_2 - 4\, H_2 = 0$ (h*)

$(-74.87 - 2 \cdot 241.83 + 393.51 - 4 \cdot 0)$ MJ/kmol $= -165.02$ MJ/kmol
$= -165.02$ MJ/(kmol CH_4)

For the total reaction (i) summarised by calculation, one finds:

$CH_4 + 2\, O_2 + 2\, H_2O - 4\, H_2O - CO_2 = 0$ (i*)

$(-74.87 + 2 \cdot 0 - 2 \cdot 241.83 + 4 \cdot 241.83 + 393.51)$ MJ/kmol $= +802.30$ MJ/kmol $= +802.30$ MJ/(kmol CH_4)

$CH_4 + 2\, O_2 - 2\, H_2O - CO_2 = 0$ (k*)

$(-74.87 + 2 \cdot 0 + 2 \cdot 241.83 + 393.51)$ MJ/kmol $= +802.30$ MJ/kmol
$= +802.30$ MJ/(kmol CH_4)

In each case, water is understood as water vapor. If, for example, water is treated as a liquid in the water electrolysis according to (g*), it follows:

$4\, H_2 + 2\, O_2 - 4\, H_2O = 0$ (g*)

$(4 \cdot 0 + 2 \cdot 0 + 4 \cdot 285.84)$ MJ/(4 kmol H_2) $= +1143.36$ MJ/(4 kmol H_2)

If the equation $H_2 + 0.5\, O_2 - H_2O = 0$ (g**) is considered, one gets:

$(0 + 0.5 \cdot 0 + 285.84)$ MJ/kmol $= +285.84$ MJ/(kmol H_2)

12.4 Absolute Molar Entropies; Third Law of Thermodynamics

The real course of a chemical reaction — like the course of a real solely physical change of state — is irreversible. When quantifying irreversibilities, the corresponding reversible substitute process can serve as a standard of comparison. For this, an entropy balance is required according to the second law, whereby the sum of all entropy changes of the reaction participants as well as the entropy change of the environment must be considered.

Similarly to section 12.3, which discusses the preparation of molar enthalpy balances according to the first law, the preparation of molar entropy balances of the individual pure substances also requires the adjustment of constants. This is an essential difference compared to the calculation of entropy differences for solely physical changes of state of pure substances without chemical reactions, where the respective constant S_{0i} can be shortened mathematically. For an isothermal-isobaric chemical reaction, the molar reaction entropy is defined as follows:

$$\Delta^R S_m(T, p) = \sum_{i=1}^{N} \nu_i \, S_{m,0i}(T, p) \tag{12.31}$$

Here, the individual constants do not disappear; rather, the molar entropies of the individual substances have to be harmonised to each other. This is also essential when defining the molar reaction *Gibbs*[2] function, discussed in further sections:

$$\Delta^R G_m(T, p) = \sum_{i=1}^{N} \nu_i \, G_{m,0i}(T, p) = \Delta^R H_m(T, p) - T \, \Delta^R S_m(T, p) \tag{12.32}$$

The adjustment of the molar entropy constants of the substances participating in a reaction to each other cannot — in contrast to molar enthalpies — be done by calorimetric measurements.

In 1906, *Nernst*[3] solved this problem [9]: In his postulate, known as the 'New Heat Theorem', he assumed that for all reactions between solid substances, as absolute zero is approached, the molar reaction *Gibbs* function $\Delta^R G_m(T, p)$ does not change with temperature, i.e.

$$\lim_{T \to 0} \frac{d \Delta^R G_m}{dT} = 0 \tag{12.33}$$

should hold. This means that at zero thermodynamic temperature, the molar reaction entropies of all reactions between solid substances become zero:

$$\lim_{T \to 0} \Delta^R S_m(T, p) = 0 \tag{12.34}$$

This is the prerequisite for calculating molar reaction entropies and molar reaction *Gibbs* functions at other temperatures if the temperature-dependent molar isobaric heat capacities $C_{m,p,0i}$ of the substances involved in the reaction are known [9].

---
[2] *Josiah Willard Gibbs*, 1839 to 1903, North American natural scientist, made important contributions to the thermodynamics of equilibria.
[3] *Walther Nernst*, 1864 to 1941, German physicist and chemist. He was awarded the Nobel Prize in Chemistry in 1920 for his work in thermochemistry.

The *Nernst* heat theorem is considered a universal valid law about the behaviour of entropy at the zero point of thermodynamic temperature; it is also called the third law of thermodynamics. It can be expressed — as with the second law — in various ways. The version formulated by *Max Planck* in terms of harmonising the entropies of various pure substances reads:

'The entropy of any pure condensed substance which is in internal equilibrium has its smallest value at $T = 0$, which is independent of the other intensive state variables and can be set equal to zero.'

This formulation satisfies Eq.(12.34) as well as *Nernst*'s postulate. No indeterminate constant occurs in the determination of entropy, and the molar entropy at constant pressure p for a substance i can be calculated by integrating the entropy differential, starting at $T = 0$, following Eq. (5.154) as follows [9]:

$$S_{m,0i}(T,p) = \int_0^1 \frac{C_{mp,0i}(T,p)}{T} \, \mathrm{d}T \; ; \; p = \text{const} \qquad (12.35)$$

The molar entropies determined with Eq. (12.35) with the starting point at $T = 0$ are called conventional molar entropies; the designation absolute molar entropies is also common. For finite entropy values to result, the molar isobaric heat capacity $C_{m,p,0i}$ must also be zero at $T = 0$: Thus holds, as a consequence of *Nernst*'s postulate,

$$\lim_{T \to 0} C_{mp,0i}(T,p) = 0 \; . \qquad (12.36)$$

The left-hand part of Fig. 12.9 shows the temperature dependence of the expression $C_{mp,0i}/T$, while the right-hand part of Figure 12.9 shows the plot of $S_{m,0i}$ versus temperature at $p = \text{const}$, obtained by integrating $C_{m,p}/T$ according to Eq. (12.35).

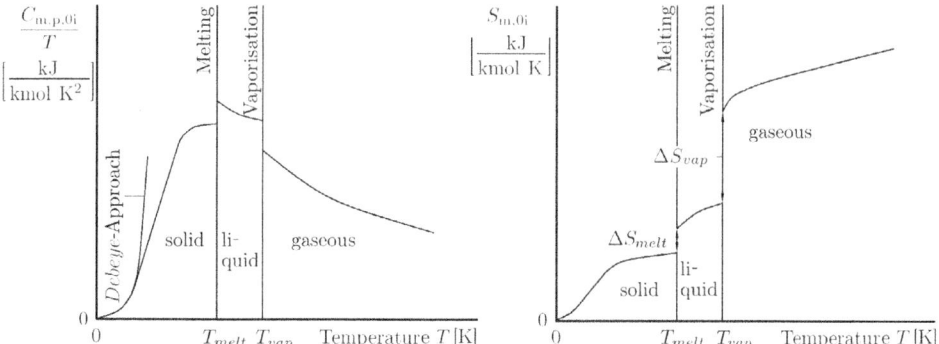

Figure 12.9 Determining the absolute molar entropy of a substance using the molar isobaric heat capacity

The *Nernst* postulate is confirmed by quantum theory and statistical thermodynamics. In particular, *P. Debeye* [9] found that the molar heat capacity and molar entropy of solid substances approach zero in a way that they are proportional to T^3 near absolute zero.

The entropy of ideal gases can be calculated from natural constants and spectroscopic measurements with the instruments of quantum theory and statistical thermodynamics, and the results thus obtained confirm the calculations according to the integration given in Eq.(12.35).

The results of these entropy calculations include standard molar entropies, which are given as $S_{m,0i}^{\square} = S_{m,0i}^{\square}(T_0, p_0) = S_{m,0i}(T_0, p_0)$. For them, $T_0 = 298.15$ K (25 °C) and $p_0 = 1$ bar $= 100$ kPa holds. For important pure substances, standard molar entropies $S_{m,0i}^{\square}$ are given in Table A.7 of the appendix.

For the calculation of the standard molar entropy of reaction, Eq. (12.31) states:

$$\Delta^R S_m(T_0, p_0) = \sum_{i=1}^{N} \nu_i S_{m,0i}(T_0, p_0) \tag{12.37}$$

With molar entropies in the standard state, entropies at states deviating from it can be calculated by determining entropy differences with respect to the standard state using the equations available for this purpose — e.g. for ideal gases according to section 4 or for real gases according to section 5.

Example 12.4

a) For reactions (a) to (f) given in example 12.3 a), the values of standard molar entropies $S_{m,0i}^{\square}$ from Table A.7 in the appendix are to be used to calculate the standard molar entropies $\Delta^R S_m(T_0, p_0)$.

b) For reactions (g) to (k) given in Example 12.3(b), the values of the standard molar entropies $S_{m,0i}^{\square}$ from Table A.7 in the appendix are to be used to calculate the standard molar reaction entropies $\Delta^R S_m(T_0, p_0)$.

a) According to Eq. (12.37), the following holds for the respective chemical reactions:

$CO + 3 H_2 - CH_4 - H_2O = 0$ (a*)

$(197.660 + 3 \cdot 130.680 - 186.250 - 188.835)$ kJ/(kmol K) $= +214.615$ kJ/(kmol K) $=$
$+214.615$ kJ/(kmol CH_4 K)

$CO_2 + H_2 - CO - H_2O = 0$ (b*)

$(213.785 + 130.680 - 197.660 - 188.835)$ kJ/(kmol K) $= -42.030$ kJ/(kmol K)
$= -42.030$ kJ/(kmol CO K)

$CO_2 + 4 H_2 - CH_4 - 2 H_2O = 0$ (c*)

$(213.785 + 4 \cdot 130.680 - 186.250 - 2 \cdot 188.835)$ kJ/(kmol K) $= +172.585$ kJ/(kmol K)
$= +172.585$ kJ/(kmol CH_4 K)

$4 H_2O - 4 H_2 - 2 O_2 = 0$ (d*)

$(4 \cdot 188.835 - 4 \cdot 130.680 - 2 \cdot 205.152)$ kJ/(4 kmol H_2 K) $= -177.684$ MJ/(4 kmol H_2 K)

(related to 4 kmol H_2 formed with the help of 1 kmol CH_4).

If the equation $H_2O - H_2 - 0.5 O_2 = 0$ (d**) is considered, one gets:

$(188.835 - 130.680 - 0.5 \cdot 205.152)$ kJ/(kmol H_2 K) $= -44.421$ kJ/(kmol H_2 K)

$CO_2 + 4 H_2O - CH_4 - 2 O_2 - 2 H_2O = 0$ (e*)

$(213.785 + 4 \cdot 188.835 - 186.25 - 2 \cdot 205.152 - 2 \cdot 188.835)$ kJ/(kmol K) $= -5.099$ kJ/(kmol K)
$= -5.099$ kJ/(kmol CH_4 K)

$CO_2 + 2\,H_2O - CH_4 - 2\,O_2 = 0$ (f*)

$(213.785 + 2 \cdot 188.835 - 186.25 - 2 \cdot 205.152)\,kJ/(kmol\,K)$
$= -5.099\,kJ/(kmol\,K) = -5.099\,kJ/(kmol\,CH_4\,K)$

Water is always understood as water vapor. If water is accounted for as a liquid, this results, for example, according to equations (d*) and (d**):

$4\,H_2O - 4\,H_2 - 2\,O_2 = 0$ (d*)

$(4 \cdot 69.93 - 4 \cdot 130.680 - 2 \cdot 205.152)\,kJ/(4\,kmol\,H_2\,K) = -653.304\,kJ\,(4\,kmol\,H_2\,K)$

(related to 4 kmol H_2 formed with the help of 1 kmol CH_4), resp.

$H_2O - H_2 - 0.5\,O_2 = 0$ (d**)

$(69.93 - 130.680 - 0.5 \cdot 205.152)\,kJ/(kmol\,H_2\,K) = -163.326\,kJ/(kmol\,H_2\,K)$

b) For reactions (g), (h), (i) and (k) follows, according to Eq. (12.37):

$4\,H_2 + 2\,O_2 - 4\,H_2O = 0$ (g*)

$(4 \cdot 130.680 + 2 \cdot 205.152 - 4 \cdot 188.835)\,kJ/(4\,kmol\,H_2\,K) = +177.684\,kJ/(4\,kmol\,H_2\,K)$

For reaction (h), the following applies accordingly

$CH_4 + 2\,H_2O - CO_2 - 4\,H_2 = 0$ (h*)

$(186.25 + 2 \cdot 188.835 - 213.785 - 4 \cdot 130.680)\,kJ/(kmol\,K) = -172.585\,kJ/(kmol\,K)$
$= -172.585\,kJ/(kmol\,CH_4\,K)$

$CH_4 + 2\,O_2 + 2\,H_2O - 4\,H_2O - CO_2 = 0$ (i*)

$(186.25 + 2 \cdot 205.152 + 2 \cdot 188.835 - 4 \cdot 188.835 - 213.785)\,kJ/(kmol\,K) = +5.099\,kJ/(kmol\,K)$
$= +5.099\,kJ/(kmol\,CH_4\,K)$

$CH_4 + 2\,O_2 - 2\,H_2O - CO_2 = 0$ (k*)

$(186.25 + 2 \cdot 205.152 - 2 \cdot 188.835 - 213.785)\,kJ/(kmol\,K) = +5.099\,kJ/(kmol\,K)$
$= +5.099\,kJ/(kmol\,CH_4\,K)$

In each case, water is understood as water vapor. If, for example, water is treated as a liquid in the water electrolysis according to (g*), it follows:

$4\,H_2 + 2\,O_2 - 4\,H_2O = 0$ (g*)

$(4 \cdot 130.680 + 2 \cdot 205.152 - 4 \cdot 69.93)\,kJ/(4\,kmol\,H_2\,K) = +653.304\,kJ/(4\,kmol\,H_2\,K)$

If the equation $H_2 + 0.5\,O_2 - H_2O = 0$ (g**) is considered, one gets:

$(130.680 + 0.5 \cdot 205.152 - 69.93)\,kJ/(kmol\,H_2\,K) = +163.326\,kJ/(kmol\,H_2\,K)$

Via the standard molar entropy $S_{m,0i}^{\square}$ and the standard molar enthalpy of formation $H_{m,0i}^{f\square}$ of a substance i, its molar *Gibbs*-function for pure substances is also

$$G_{m,0i}^{\square} = H_{m,0i}^{f\square} - T\,S_{m,0i}^{\square} \qquad (12.38)$$

fixed in the standard state (cf. Table A.7 in the appendix), so no further adjustment is required in the formation of the molar reaction *Gibbs* function in the standard state according to Eq. (12.32):

$$\Delta^R G_m(T_0, p_0) = \sum_{i=1}^{N} \nu_i G_{m,0i}(T_0, p_0) = \Delta^R H_m(T_0, p_0) - T_0 \Delta^R S_m(T_0, p_0) \quad (12.39)$$

The postulate of *Nernst* is also the basis for the theorem of the unreachability of the (absolute) zero of the thermodynamic temperature:

'It is impossible to reach the zero point of thermodynamic temperature ($T = 0$) in a finite number of process steps.'

In the meantime, the results of numerous attempts to reach very low temperatures are available. Thermodynamic temperatures down to about 0.0005 K could be achieved; the temperature $T = 0$ K was not reached [9].

12.5 The Importance of the Second Law for Chemical Reactions

The values of standard molar entropy provide absolute molar entropies that are useful for applying the second law to chemically reacting systems. In the following, a stationary flow process (steady-state flow process) is considered in which an isothermal-isobaric reaction takes place inside the reaction chamber according to Fig. 12.10, for which the general reaction equation

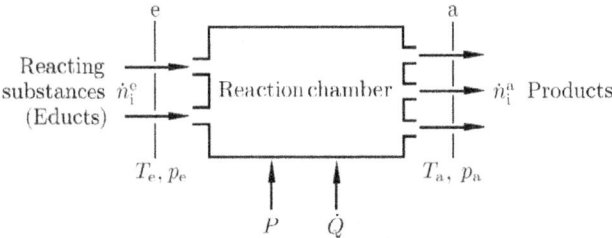

Figure 12.10 Stationary flow process of an isothermal-isobaric chemical reaction with educts separately entering and products separately leaving the reaction chamber

$$\sum_{i=1}^{N} \nu_i A_i = 0 \quad (12.11)$$

holds. In order not to have to take mixing effects into account, the reactants should flow into the reaction chamber separately from each other with the same temperature $T = T_e = T_a$ as well as the same pressure $p = p_e = p_a$ and the products should leave the reaction chamber separately with the same temperature $T = T_e = T_a$ and the same pressure $p = p_e = p_a$. For the respective molar mass flows $\dot{n}_i^e$ on entry and $\dot{n}_i^a$ on exit, the following applies for a substance i according to sections 12.2 and 12.3

$$\dot{n}_i^a = \dot{n}_i^e + \nu_i \dot{z}_a \; ; \quad (12.19)$$

where $\dot{z}_a$ is the reaction rate.

Using the first law, neglecting changes in the time-related kinetic and potential energies

according to Eq. (12.20 a) yields the power balance equation

$$\dot{Q} + P = \sum_{i=1}^{N} (\dot{n}_i^a - \dot{n}_i^e) H_{m,0i}(T, p) = \dot{z}_a \sum_{i=1}^{N} \nu_i H_{m,0i}(T, p) = \dot{z}_a \Delta^R H_m(T, p) \ . \tag{12.40}$$

In determining the time-related entropy change $\Delta \dot{S}(T, p)$, on the one hand, the entropy flow change resulting from the reaction $\sum_{i=1}^{N} (\dot{n}_i^a - \dot{n}_i^e) S_{m,0i}(T, p)$ and, on the other hand, the increase in the time-related entropy $\dot{S}_{irr}$ caused by the irreversibilities must be accounted for. When calculating the associated heat flow $\dot{Q} = T \Delta \dot{S}(T, p)$, it is important to note that, on the one hand, the heat flow $T \sum_{i=1}^{N} (\dot{n}_i^a - \dot{n}_i^e) S_{m,0i}(T, p)$ occurring in connection with the reaction, and on the other hand the heat flow $T\dot{S}_{irr}$ caused by the irreversibilities have to be taken into account, which are, due to the isothermal reaction, in connection with the environment.

With $\Delta^R S_m(T, p)$ one gets:

$$\dot{Q} = \dot{z}_a T \sum_{i=1}^{N} \nu_i S_{m,0i}(T, p) - T \dot{S}_{irr} = \dot{z}_a T \Delta^R S_m(T, p) - T \dot{S}_{irr} \tag{12.42}$$

From Eq. (12.40) follow

$$P = \dot{z}_a \Delta^R H_m(T, p) - \dot{Q} = \dot{z}_a \Delta^R H_m(T, p) - \dot{z}_a T \Delta^R S_m(T, p) + T \dot{S}_{irr} \tag{12.43}$$

and

$$P = \dot{z}_a \Delta^R G_m(T, p) + T \dot{S}_{irr} \ . \tag{12.44}$$

In this equation, $\Delta^R G_m(T, p)$ is the molar reaction *Gibbs* function of the isothermal-isobaric reaction according to Eq.(12.32). If a negative value is obtained for $\Delta^R G_m(T, p)$, not only heat but also work can be obtained from the reaction. If $\Delta^R G_m(T, p)$ is positive, work must be supplied to carry out the reaction.

If the boundary case of the reversible chemical reaction is considered, the molar reversible reaction work is obtained from Eq.(12.44) because $\dot{S}_{irr} = 0$ is given:

$$(W_m)_{rev} = \frac{P_{rev}}{\dot{z}_a} = \Delta^R G_m(T, p) \tag{12.45}$$

The molar reaction *Gibbs* function $\Delta^R G_m(T, p)$ is thus characteristic of a reversible isothermal-isobaric reaction.

(In parallel, in Section 5, to Eq. (5.214) and and to Eq. (5.215) is referred, which together reveal the characteristic of the *Gibbs* function $G_2 - G_1 = W_{p12}$ for capturing the pressure change work in solely physical reversible isothermal changes of state).

$\Delta^R G_m(T, p)$ gives the molar work that must at least be supplied or at the most can be gained in a formula reaction. The molar reaction *Gibbs* function is used in section 12.6 to determine the chemical exergy.

Fig. 12.11 shows two cases of a reversible isothermal-isobaric exothermic reaction ($\Delta^R H_m < 0$):

If the molar reaction entropy decreases ($\Delta^R S_m < 0$; Fig. 12.11 left), the molar reversible heat $(Q_m)_{rev} = T \Delta^R S_m < 0$ is simultaneously released from the system to the environment, whereby the molar reversible work, given as the absolute value of the molar reversible work, is less than the absolute value of the molar reaction enthalpy: $|(W_m)_{rev}| = |\Delta^R G_m| < |\Delta^R H_m|$.

If, on the other hand, the molar reaction entropy increases ($\Delta^R S_m > 0$; Fig. 12.11 right), the molar reversible heat $(Q_m)_{rev} = T \Delta^R S_m > 0$ is simultaneously supplied to the system from the environment, where the molar reversible work, given as the absolute value of the molar reversible work, is greater than the absolute value of the molar reaction enthalpy: $|(W_m)_{rev}| = |\Delta^R G_m| > |\Delta^R H_m|$.

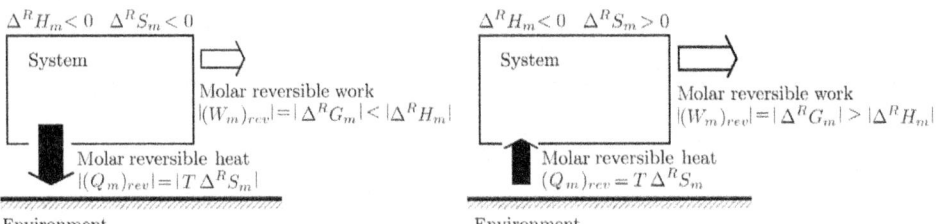

Figure 12.11 Molar reaction *Gibbs* function: work gain in a reversible isothermal-isobaric exothermic reaction

Often, in a chemical reaction it is refrained from supplying or extracting mechanical or electrical power. With the condition $P = 0$, one then obtains from Eq.(12.44) the time-related entropy increase due to irreversibilities:

$$\dot{S}_{irr} = -\frac{\dot{z}_a}{T} \Delta^R G_m (T, p) > 0. \qquad (12.46)$$

The molar reaction *Gibbs* function can then also be used to make statements about the direction of the reaction process: If

$\Delta^R G_m (T, p) < 0$ holds, a positive reaction rate results: $\dot{z}_a > 0$

Thus, the isothermal-isobaric reaction proceeds in the direction assumed when the chemical reaction equation was written down. If, on the other hand,

$\Delta^R G_m (T, p) > 0$ holds, the reaction rate is negative: $\dot{z}_a < 0$.

The reaction takes the opposite course: An isothermal-isobaric reaction with a positive molar reaction *Gibbs* function cannot proceed in the direction assumed when writing down the chemical reaction equation if $P = 0$ holds.

Eq. (12.44), however, opens up the possibility of forcing the desired reaction direction against the positive molar reaction *Gibbs* function $\Delta^R G_m (T, p) > 0$ and the negative reaction rate $\dot{z}_a < 0$ with $P > 0$. This can be managed by e.g. supplying mechanical or electrical power or an – exergy flow containing – high temperature heat flow (i.e. a heat flow with the ability to deliver useful work e.g. by a cyclic process (see sections 5 and 8)).

Here it must be ensured that the supplied power P is not simply used in an electrothermal process in an Ohm^4 resistance and, being irreversibly converted into an increase of an enthalpy flow, merely has the effect of increasing the temperature of the chemical reaction partners.

Instead, an electrochemical reaction is required in which the chemical reaction is carried out with the help of electrically charged electrons or ions. A reversible supply of mechanical power is theoretically also possible in the *van 't Hoff*[5] reaction model (cf. section 12.11).

Example 12.5

a) For reactions (a) to (f) given in Example 12.3 a), using the values of the molar standard *Gibbs* functions $G_{m,0i}^{\square}$ from Table A.7 in the appendix, the molar standard reaction *Gibbs* functions $\Delta^R G_m(T_0, p_0)$ are to be calculated.

b) For reactions (g) to (k) given in Example 12.3 b), using the values of the molar standard *Gibbs* functions $G_{m,0i}^{\square}$ from Table A.7 in the appendix, the molar standard reaction *Gibbs* molar functions $\Delta^R G_m(T_0, p_0)$ are to be calculated.

a) It is valid for the respective chemical reactions:

$CO + 3\,H_2 - CH_4 - H_2O = 0$ (a*)

$(-169.46 - 3 \cdot 38.962 + 130.40 + 298.13)\,MJ/kmol = +142.184\,MJ/kmol$
$= +142.184\,MJ/(kmol\,CH_4)$

$CO_2 + H_2 - CO - H_2O = 0$ (b*)

$(-457.25 - 38.962 + 169.46 + 298.13)\,MJ/kmol = -28.622\,MJ/kmol$
$= -28.622\,MJ/(kmol\,CO)$

$CO_2 + 4\,H_2 - CH_4 - 2\,H_2O = 0$ (c*)

$(-457.25 - 4 \cdot 38.962 + 130.40 + 2 \cdot 298.13)\,MJ/kmol = +113.562\,MJ/(kmol)$
$= +113.562\,MJ/(kmol\,CH_4)$

$4\,H_2O - 4\,H_2 - 2\,O_2 = 0$ (d*)

$(-4 \cdot 298.13 + 4 \cdot 38.962 + 2 \cdot 61.166)\,MJ/(4\,kmol\,H_2) = -914.34\,MJ/(4\,kmol\,H_2)$

If the equation $H_2O - H_2 - 0.5\,O_2 = 0$ (d**) is considered, one gets:

$(-298.13 + 38.962 + 0.5 \cdot 61.166\,MJ/(kmol\,H_2) = -228.585\,MJ/(kmol\,H_2)$

$CO_2 + 4\,H_2O - CH_4 - 2\,O_2 - 2\,H_2O = 0$ (e*)

$(-457.25 - 4 \cdot 298.13 + 130.40 + 2 \cdot 61.166 + 2 \cdot 298.13)\,MJ/kmol$
$= -800.778\,MJ/kmol = -800.778\,MJ/(kmol\,CH_4)$

$CO_2 + 2\,H_2O - CH_4 - 2\,O_2 = 0$ (f*)

$(-457.25 - 2 \cdot 298.13 + 130.40 + 2 \cdot 61.166)\,MJ/kmol = -800.778\,MJ/kmol$
$= -800.778\,MJ/(kmol\,CH_4)$

In each case, water is considered as water vapor. If water is accounted for as a liquid, this results as follows, for example, according to equations (d*) and (d**):

---

[4] *Georg Simon Ohm*, 1789 to 1854, German physicist. His main interest was in electricity, which was still little researched at the time.
[5] *Jacobus Henricus van 't Hoff*, 1852 to 1911, Dutch chemist

4 H$_2$O - 4 H$_2$ - 2 O$_2$ = 0 (d*)

($-4 \cdot 306.69 + 4 \cdot 38.962 + 2 \cdot 61.166$) MJ/(4 kmol H$_2$) = -948.58 MJ/(4 kmol H$_2$) respectively.

H$_2$O - H$_2$ - 0.5 O$_2$ = 0 (d**)

($-306.69 + 38.962 + 0.5 \cdot 61.166$) MJ/(kmol H$_2$) = -237.145 MJ/(kmol H$_2$)

b) For reactions (g), (h), (i) and (k), the following applies:

4 H$_2$ + 2 O$_2$ - 4 H$_2$O = 0 (g*)

($-4 \cdot 38.962 - 2 \cdot 61.166 + 4 \cdot 298.13$) MJ/(4 kmol H$_2$) = $+914.34$ MJ/(4 kmol H$_2$)

CH$_4$ + 2 H$_2$O - CO$_2$ - 4 H$_2$ = 0 (h*)

($-130.40 - 2 \cdot 298.13 + 457.25 + 4 \cdot 38.962$) MJ/kmol = -113.562 MJ/kmol
= -113.562 MJ/(kmol CH$_4$)

CH$_4$ + 2 O$_2$ + 2 H$_2$O - 4 H$_2$O - CO$_2$ = 0 (i*)

($-130.40 - 2 \cdot 61.166 - 2 \cdot 298.13 + 4 \cdot 298.13 + 457.25$) MJ/kmol = $+800.778$ MJ/kmol
= $+800.778$ MJ/(kmol CH$_4$)

CH$_4$ + 2 O$_2$ - 2 H$_2$O - CO$_2$ = 0 (k*)

($-130.40 - 2 \cdot 61.166 + 2 \cdot 298.13 + 457.25$) MJ/kmol = $+800.778$ MJ/kmol
= $+800.778$ MJ/(kmol CH$_4$)

In each case, water is understood as water vapor. If, for example, water is treated as a liquid in the water electrolysis according to (g*), it follows:

4 H$_2$ + 2 O$_2$ - 4 H$_2$O = 0 (g*)

($-4 \cdot 38.962 - 2 \cdot 61.166 + 4 \cdot 306.69$) MJ/(4 kmol H$_2$) = $+948.58$ MJ/(4 kmol H$_2$)

If the equation H$_2$ + 0.5 O$_2$ - H$_2$O = 0 (g**) is considered, one gets:

($-38.962 - 0.5 \cdot 61.166 + 306.69$) MJ/(kmol H$_2$) = $+237.145$ MJ/(kmol H$_2$)

These values can also be obtained using Eq. (12.39) from the results of examples 12.3 and 12.4.

Example 12.6

A utility company operates a demonstration plant to use surplus electrical energy from wind turbines to produce hydrogen. In each of six containers, four alkaline water electrolysis units are installed, operating at ambient pressure $p = 1.0$ bar and temperature $t = 80\,^\circ$C. Each unit has a hydrogen output of 15 Nm^3/h H$_2$ at full load operation; thus, the total output is 360 Nm^3/h H$_2$. The hydrogen produced is fed into a long-distance natural gas pipeline passing nearby; for this purpose it is compressed to a pressure of 55 bar.

Fig. 12.12 shows the schematic of an electrolysis cell: Two electrodes are placed in an electrolyte bath with an aqueous potassium hydroxide (KOH) solution with a 30 % KOH concentration. A semi-permeable membrane (diaphragm) is placed between the electrodes to prevent the gases H$_2$ and O$_2$ formed at the electrodes from mixing; on the other hand, the membrane is permeable to the aqueous solution and the OH$^-$ ions formed. By supplying electrical work, the following reactions take place:

Cathode: 2 H$_2$O + 2 e$^-$ $\rightleftharpoons$ H$_2$ + 2 OH$^-$,

Anode: 2 OH$^-$ $\rightleftharpoons$ H$_2$O + 0.5 O$_2$ + 2 e$^-$.

Summing up, one gets the reaction equation

H$_2$ + 0.5 O$_2$ - H$_2$O = 0

with $\Delta^R G_m(T_0, p_0) = +237.145$ MJ/kmol $= +237.145$ MJ/(kmol H_2O) $= +237.145$ MJ/(kmol H_2) under standard conditions.

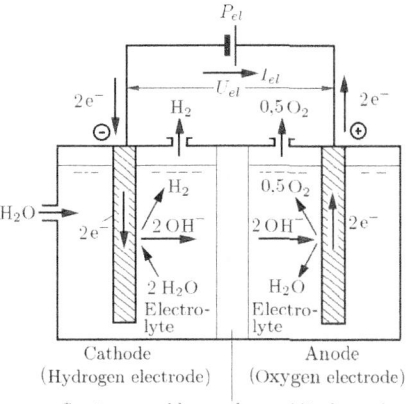

Figure 12.12 Schematic representation of a cell for water electrolysis

For an electrolysis cell operated with the cell voltage $U_{el} = 2.02$ V, the quantities $W_m = P/\dot{n}_{H_2}$ and $Q_m = \dot{Q}/\dot{n}_{H_2}$ related to the molar quantity flow rate $\dot{n}_{H_2}$ of the generated H_2 are to be calculated. For simplicity, it shall be assumed that the process takes place at the standard electrochemical state (T_0, p_0).

Thus the values for the reaction (g**) in examples 12.3, 12.4 and 12.5 can be used. These are, considering the liquid state of water, $\Delta^R H_m(T_0, p_0) = +285.84$ MJ/(kmol H_2), $\Delta^R S_m(T_0, p_0) = +163.326$ kJ/(kmol H_2 K) and $\Delta^R G_m(T_0, p_0) = +237.145$ MJ/(kmol H_2). Because only water is fed into the electrolysis cell, $\dot{n}_{H_2}^e = 0$ holds, and the equation $\dot{n}_{H_2}^a - \dot{n}_{H_2}^e = \dot{z}_a$ becomes $\dot{z}_a = \dot{n}_{H_2}^a = \dot{n}_{H_2}$: The reaction rate $\dot{z}_a$ of the water electrolysis is identical to the molar quantity flow rate $\dot{n}_{H_2}$ of the hydrogen produced. The electric power supplied to the electrolysis cell electrical power is

$$P = U_{el} I_{el} = U_{el} \dot{n}_{el} F .$$

Herein, $F = e N_A = 9.64853 \cdot 10^7$ As/kmol is the *Faraday* constant; it represents the product of the elementary electric charge e and the *Avogadro* constant N_A. According to the reaction equation, the molar quantity flow rate $\dot{n}_{el}$ of the electrons at the cathode has twice the magnitude of the molar quantity flow rate $\dot{n}_{H_2}$ of the hydrogen obtained, because two electrons are taken up into the ions 2 OH^- during the formation of each H_2 molecule: $\dot{n}_{el} = 2\dot{n}_{H_2}$.

This leads to

$$W_m = P/\dot{n}_{H_2} = 2 F U_{el} = 2 \cdot 9.64853 \cdot 10^7 \frac{\text{A s}}{\text{kmol}} \cdot 2.02 \text{ V} \cdot \frac{\text{MJ}}{10^6 \text{ V A s}} = 389.801 \text{ MJ/kmol} .$$

In the following, the electric power is related to the norm volume flow $\dot{V}_n$ of the hydrogen obtained. With the molar norm volume $(V_m)_0 = 22.414$ Nm³/kmol of all ideal gases (cf. section 4.1.6), the reaction work related to the norm volume in Nm³ is

$$W_n = \frac{P}{\dot{V}_n} = \frac{P}{\dot{n}_{H_2}} \frac{\dot{n}_{H_2}}{\dot{V}_n} = \frac{W_m}{(V_m)_0} = \frac{389.801 \text{ MJ kmol}}{\text{kmol } 22.414 \text{ Nm}^3} \cdot \frac{1 \text{ kWh}}{3.6 \text{ MJ}} = 4.831 \frac{\text{kWh}}{\text{Nm}^3} .$$

Due to the irreversible processes in the cell, the molar reaction work to be supplied is higher than the minimum necessary reversible molar reaction work $(W_m)_{rev} = \Delta^R G_m(T_0, p_0) = +237.145$ MJ/kmol.

According to Eqs.(12.44) and (12.45), the difference $W_m - (W_m)_{rev}$ corresponds to the dissipated molar electrical energy:

$$T_0 \frac{\dot{S}_{irr}}{\dot{n}_{H_2}} = W_m - (W_m)_{rev} = W_m - \Delta^R G_m(T_0, p_0)$$
$$= (389.801 - 237.145) \text{ MJ/kmol} = 152.656 \text{ MJ/kmol}.$$

Thus, for the increase in molar entropy one gets
$$S_{m\,irr} = \frac{\dot{S}_{irr}}{\dot{n}_{H_2}} = \frac{152.656 \text{ MJ}}{\text{kmol } 298.15 \text{ K}} = 0.5120 \frac{\text{MJ}}{\text{kmol K}}$$

According to Eq.(11.42) applied to the standard state, the dissipated molar energy
$$T_0 \frac{\dot{S}_{irr}}{\dot{n}_{H_2}} = T_0 \Delta^R S_m(T_0, p_0) - Q_m$$

partly consists of the molar energy for the chemical reaction and partly of the molar waste heat dissipated to the environment, since $\dot{S}_{irr}/\dot{n}_{H_2} > \Delta^R S_m(T_0, p_0)$ holds. From this, the molar waste heat can be determined:

$$Q_m = T_0 \Delta^R S_m(T_0, p_0) - T_0 \frac{\dot{S}_{irr}}{\dot{n}_{H_2}}$$
$$= 298.15 \text{ K} \cdot 0.163326 \frac{\text{MJ}}{\text{kmol K}} - 298.15 \text{ K} \cdot 0.5120 \frac{\text{MJ}}{\text{kmol K}} = -103.957 \frac{\text{MJ}}{\text{kmol}}.$$

In contrast, it should be noted that the reversible cell takes up useful heat from the environment:

$$Q_m = (Q_m)_{rev} = T_0 \Delta^R S_m(T_0, p_0) = 298.15 \text{ K} \cdot 0.163326 \frac{\text{MJ}}{\text{kmol K}} = 48.696 \frac{\text{MJ}}{\text{kmol}},$$

with $(W_m)_{rev}$, considerably less molar electrical energy is required. Its voltage $(U_{el})_{rev}$, called reversible cell voltage or equilibrium voltage, is smaller than U_{el}; it is only

$$(U_{el})_{rev} = \frac{\Delta^R G_m(T_0, p_0)}{2F} = \frac{237.145 \text{ MJ kmol}}{\text{kmol } 2 \cdot 9.64853 \cdot 10^7 \text{A s}} \cdot \frac{10^6 \text{ V A s}}{\text{MJ}} = 1.2289 \text{ V}.$$

The following relationship is thermodynamically useful for expressing the efficiency of the electrolytic cell:

$$\eta_{ELZ} = \frac{P_{rev}}{P} = \frac{\dot{n}_{H_2} \Delta^R G_m(T, p)}{P} = \frac{(U_{el})_{rev}}{U_{el}} = 1 - \frac{T \dot{S}_{irr}}{P}$$

With $T = T_0$ one gets:
$$\eta_{ELZ} = \frac{\dot{n}_{H_2} \Delta^R G_m(T_0, p_0)}{P} = \frac{237.145 \text{ MJ kmol}}{\text{kmol } 389.801 \text{ MJ}} = 0.6084$$

Furthermore, it is not uncommon to use the following efficiency definitions, where the norm volume-related upper calorific value $H_{s,0\,H_2}$ or the norm-volume-related lower calorific value $H_{i,0\,H_2}$ is divided by the norm volume-related electrical work:

$$\eta^*_{ELZ} = \frac{H_{s,0\,H_2}}{W_n} = \frac{3.540 \text{ kWh Nm}^3}{\text{Nm}^3 \, 4.831 \text{ kWh}} = 0.7328$$

$$\eta^{**}_{ELZ} = \frac{H_{i,0\,H_2}}{W_n} = \frac{2.995 \text{ kWh Nm}^3}{\text{Nm}^3 \, 4.831 \text{ kWh}} = 0.6200$$

In addition, it is of interest to consider the total norm volume-related electrical work consumed, which in the case of the demonstration unit also includes the auxiliary units; this was determined to be $W_{n\,ges} = 5.2 \text{ kWh/Nm}^3$. From this follows:

$$\eta^*_{ges} = \frac{H_{s,0\,H_2}}{W_{n\,ges}} = \frac{3.540 \text{ kWh Nm}^3}{\text{Nm}^3 \, 5.2 \text{ kWh}} = 0.6808$$

$$\eta^{**}_{ges} = \frac{H_{i,0\,H_2}}{W_{n\,ges}} = \frac{2.995 \text{ kWh Nm}^3}{\text{Nm}^3 \, 5.2 \text{ kWh}} = 0.5760$$

12.6 Chemical Exergies

In order to be able to determine the exergy (i.e. the ability of useful work) of a substance or a substance flow, it must be reversibly transferred from a state of thermal or mechanical disequilibrium to its thermal or mechanical equilibrium with the environment (see Section 8). Strictly speaking, a substance or a substance flow has exergy if it ist not in thermal or mechanical equilibrium with the environment.

An additional condition — the chemical equilibrium — is not considered in section 8, because this is not relevant for many technical processes. The maximum work that can be gained during a transition (at $T = T_0$ and $p = p_0$) into equilibrium of a substance or chemical equilibrium is called chemical energy in section 2; it is given in section 11 (e.g. in Table 11.1) for combustion processes for some selected fuels in approximate form as e.g. norm volume-related upper calorific value $H_{s,0}$ or as norm volume-related lower calorific value $H_{i,0}$.

In the following, the chemical exergy of a substance flow is considered. For its calculation, it is necessary to determine the chemical composition of the environment. This composition must be based on the composition of the natural environment. The thermodynamic environment must be an equilibrium environment so that the exergy balances are congruent with the statements of the first and second law. This environment is characterised by the fact that no mixing or segregation processes and no chemical reactions take place between its components at $T = T_0$ and $p = p_0$.

However, the terrestrial environment is not completely in thermodynamic equilibrium due to kinetic obstacles. In this respect, a compromise has to be found when calculating exergies, in which an equilibrium environment is determined and there is a similarity to the earthly atmosphere, to sea water and to the earth's crust in its material composition.

J. Ahrendts and subsequently *Ch. Diederichsen* [9] have taken such an earth-similarity into account by calculating the mass of each chemical element occurring in the equilibrium environment from the geochemical data, making plausible determinations regarding, among other things, the thickness of the earth layer and the depth of the ocean layer and its material stocks. Here, the ambient temperature is set equal to the thermochemical standard temperature $T_0 = 298.15$ K.

By introducing an earth-similarity criterion, practical disadvantages of the thermodynamically correctly determined equilibrium environments were largely eliminated; accordingly, from the thermodynamically equal model environments, that equilibrium environment was selected for whose gas phase the equilibrium pressure p_0 is close to the terrestrial atmospheric pressure of about 1 bar = 100 kPa and whose oxygen and

nitrogen fractions are close to the terrestrial values with $n_{O_2}/n = 0.23$ and $n_{N_2}/n = 0.75$ [9].

According to [9], one of the equilibrium environments named by *Ch. Diederichsen* comprises the terrestrial substance stock of the earth's atmosphere, an earth layer with a thickness of 0.1 m and the substance stock of the oceans with a depth of 100 m, taking into account the 17 most common elements on earth, from which 971 chemical compounds are formed. The gas phase with 5.820 % of the total mass and the ambient pressure $p_0 = 0.91771$ bar includes the components N_2, O_2, H_2O, Ar, CO_2 and traces of Cl_2, HCl and HNO_3; the liquid phase with 94.176 % of the total mass has 24 substances dissolved in water, and four pure solid phases together contain 0.004 % of the total mass of the equilibrium environment.

To be able to give the chemical exergy, the molar exergies $Ex_{m,0k} = Ex_{m,0k}(T_0, p_0)$ of the 17 elements (k = 1, 2, ... 17) in the standard state are used, which are listed in Table 12.1; they can be used to calculate the standard molar chemical exergies of all substances formed from the 17 elements and whose molar *Gibbs* functions in the standard state $G_{m,0K}^{\square}$ are known.

This is based on the isothermal-isobaric formation reaction with which the compound A_i is formed from the elements E_k according to the reaction equation

$$A_i - \sum_{k=1}^{M} |\nu_{ki}| E_k = 0 \ . \tag{12.23}$$

Assuming a reversible isothermal-isobaric formation reaction, the exergy balance according to Fig. 12.13 is given by

$$Ex_{m,0i}(T, p) = \sum_{k=1}^{M} |\nu_{ki}| Ex_{m,0k}(T, p) + (W_m)_{rev} + Ex_{m,(Q_m)_{rev}} \ . \tag{12.47}$$

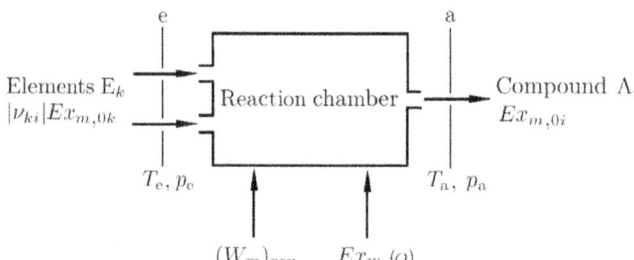

Figure 12.13 Reversible, isothermal-isobaric formation reaction of compound i using Eq. (12.23) to determine the exergy balance ($T_e = T_a = T$; $p_e = p_a = p$)

The molar reversible reaction work is equal to the molar reaction *Gibbs* function according to Eq.(12.45):

$$(W_m)_{rev} = \Delta^R G_m(T, p) = G_{m,0i}(T, p) - \sum_{k=1}^{M} |\nu_{ki}| G_{m,0k}(T, p) \ . \tag{12.48}$$

The molar exergy of the heat of reaction supplied or removed is calculated with the

help of the *Carnot* factor to be

$$Ex_{m,(Q_m)_{rev}} = \frac{T-T_0}{T}(Q_m)_{rev} = \frac{T-T_0}{T} T\Delta^R S_m(T,p) = (T-T_0)\Delta^R S_m(T,p) \ ; \quad (12.49)$$

for this, the molar reaction entropy of the formation reaction is expressed in the form

$$\Delta^R S_m(T,p) = S_{m,0i}(T,p) - \sum_{k=1}^{M} |\nu_{ki}| S_{m,0k}(T,p) \quad (12.50)$$

is used. Thus, the molar exergy of the compound A_i is given by

$$Ex_{m,0i}(T,p) = \sum_{k=1}^{M} |\nu_{ki}| Ex_{m,0k}(T,p) + G_{m,0i}(T,p)$$

$$- \sum_{k=1}^{M} |\nu_{ki}| G_{m,0k}(T,p) + (T-T_0)\Delta^R S_m(T,p) \ . \quad (12.51)$$

If the reaction is carried out at the standard temperature $T = T_0$ and at the standard pressure $p = p_0$, the standard molar exergy of the compound A_i is obtained from Eq. (12.51)

$$Ex_{m,0i}^{\square} = Ex_{m,0i}^{\square}(T_0, p_0) = G_{m,0i}^{\square} + \sum_{k=1}^{M} |\nu_{ki}| (Ex_{m,0k}^{\square} - G_{m,0k}^{\square}) \ . \quad (12.52)$$

Table 12.1 Standard molar exergies $Ex_{m,0k}^{\square}$ and standard molar *Gibbs* functions $-G_{m,0k}^{\square}$ and differences $Ex_{m,0k}^{\square} - G_{m,0k}^{\square}$ in MJ/kmol of the 17 most common elements on earth for the equilibrium environment calculated considering 971 compounds (cf. [9])

Element	-	$Ex_{m,0k}^{\square}$ $\frac{MJ}{kmol}$	$-G_{m,0k}^{\square}$ $\frac{MJ}{kmol}$	$Ex_{m,0k}^{\square} - G_{m,0k}^{\square}$ $\frac{MJ}{kmol}$	Element	-	$Ex_{m,0k}^{\square}$ $\frac{MJ}{kmol}$	$-G_{m,0k}^{\square}$ $\frac{MJ}{kmol}$	$Ex_{m,0k}^{\square} - G_{m,0k}^{\square}$ $\frac{MJ}{kmol}$
O_2	g	4.967	61.166	66.133	Al	so	844.53	8.44	852.97
H_2	g	234.683	38.962	273.645	Fe	so	367.49	8.13	375.62
Ar	g	11.642	46.167	57.809	Mn	so	482.91	9.54	492.45
Cl_2	g	50.235	66.512	110.747	Ti	so	884.45	9.16	893.61
S	so	531.524	9.557	541.081	Mg	so	688.18	9.74	697.92
N_2	g	0.743	57.128	57.871	Ca	so	790.94	12.40	803.34
P	so	864.97	12.25	877.22	Na	so	376.47	15.29	391.76
C	so	405.552	1.711	407.263	K	so	407.24	19.28	426.52
Si	so	853.19	5.61	858.80					

(g: gaseous; so: solid) The uncertainty of the tabulated values is several units of the last digit given.

Table 12.1 gives the standard molar exergies $Ex_{m,0K}^{\square}$ and standard molar *Gibbs* functions $G_{m,0k}^{\square}$ of the 17 elements required to evaluate this equation. Table A.7 in the appendix contains molar *Gibbs* functions $G_{m,0i}^{\square}$ in the standard state.

Example 12.7 What is the standard molar exergy of methane (CH_4)?

For the calculation the formation reaction

$CH_4 - C - 2\,H_2 = 0$

is to be used. With $G_{m,0\,CH_4}^{\square} = -130.40$ MJ/kmol according to Table A.7 in the appendix and the differences to be taken from Table 12.1, one obtains according to Eq. (12.52):

$$Ex_{m,0\,CH_4}^{\square} = G_{m,0\,CH_4}^{\square} + (Ex_{m,0\,C}^{\square} - G_{m,0\,C}^{\square}) + 2\,(Ex_{m,0\,H_2}^{\square} - G_{m,0\,H_2}^{\square})\ .$$

$$Ex_{m,0\,CH_4}^{\square} = (-130.40 + 407.263 + 2 \cdot 273.645)\,\frac{\text{MJ}}{\text{kmol}} = 824.153\,\frac{\text{MJ}}{\text{kmol}}\ .$$

12.7 Fuel Exergies

In combustion plants, the chemical energy released during combustion is used as heat or as internal energy or enthalpy of the hot combustion gases. In engines with internal combustion - such as reciprocating engines or gas turbines - work can also be generated.

With the help of the first and second law, it can be shown in the following how much technical useful work can be obtained from the chemical energy in the best case and what extents irreversibilities reach in a combustion process. For open systems, the exergy of a fuel is to be calculated. In addition, the amounts of exergy converted by the combustion process into the exergy of other forms of energy — heat, work, enthalpy of the combustion gases — are to be determined. The additional exergy loss quantifies the irreversibility of the combustion process.

If a fuel is in thermal and mechanical equilibrium with the environment at ambient temperature $T = T_0$ and ambient pressure $p = p_0$, it has no physical exergy; on the other hand, it has a considerable chemical exergy. This can be released as technical useful work when the fuel is converted by one or more reversible chemical reactions into substances such as, in particular, carbon dioxide (CO_2), water (H_2O) and sulphur dioxide (SO_2), which are contained in the environment. These substances are brought there by reversible mixing processes into the exergy-free equilibrium state with the thermodynamic environment.

As stated in the previous section, the exergy of chemically uniquely defined fuels (e.g. the standard molar exergy of methane (CH_4) according to example 12.7) can be determined from the standard molar exergies of the elements according to Table 12.1, if $T_0 = 298.15$ K and $p_0 = 100$ kPa $= 1$ bar is assumed, i.e. the ambient state is equated with the standard state [9]. For the elementary available fuels carbon (C), hydrogen (H_2) and sulphur (S), their exergy can be read directly from Table 12.1:

$Ex_{m,0\,C}^{\square}(T_0, p_0) = 405.552$ MJ/kmol

$Ex_{m,0\,H_2}^{\square}(T_0, p_0) = 234.683$ MJ/kmol

$Ex_{m,0\,S}^{\square}(T_0, p_0) = 531.524$ MJ/kmol

In the following, $T_0 = 298.15$ K and $p_0 = 100$ kPa are always assumed.

Compared to chemically clearly defined fuels such as methane (CH_4), carbon (C), hydrogen (H_2) and sulphur (S), natural gases, coals and oils can have very different compositions. In this respect, Table 12.1 cannot simply be used to calculate the respective fuel exergy; rather, the reversible isothermal-isobaric oxidation of the respective fuel must be accounted for according to Fig. 12.14: The fuel and oxygen (O_2) are brought

into the reaction chamber at ambient condition ($T = T_0$ and $p = p_0$). The reaction products (combustion gases) leave the reaction chamber unmixed and each with ambient state ($T = T_0$ and $p = p_0$) [9]; it is further assumed that a heat transfer into or out of the environment takes place at the temperature T_0. For the reversible reaction, the molar exergy does not change; thus, the molar exergy balance for a stationary system (steady-state system) becomes

$$Ex_{m,B}(T_0, p_0) + O_{min} Ex_{m,O_2}(T_0, p_0) + (W_m)_{rev} = \sum_{i=1}^{N} \nu_i Ex_{m,i}(T_0, p_0) \ . \quad (12.53)$$

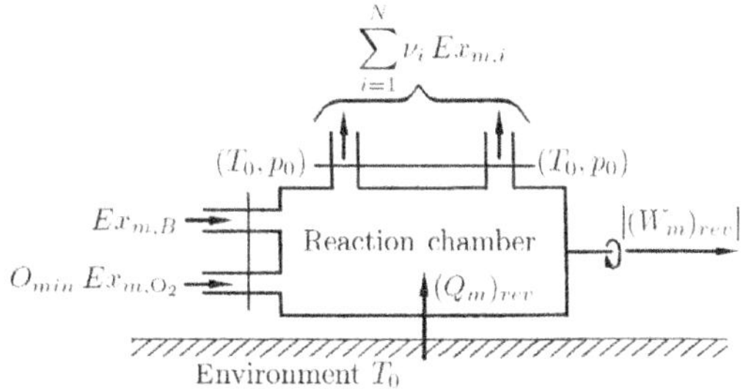

Figure 12.14 Exergy balance for determining the fuel exergy in a reversible chemical reaction in an open, steady-state flow system

The molar energy exchanged with the environment as molar heat $(Q_m)_{rev}$ consists only of exergetically worthless molar anergy; therefore, it does not need to be considered in the exergy balance.

In Eq. (12.53), O_{min} denotes the molar amount n_{O_2} of the minimum oxygen demand required for combustion, related to the molar quantity n_B of the fuel; $Ex_{m,B}$, Ex_{m,O_2} and $Ex_{m,i}$ are the molar exergies of the fuel, the oxygen and the reaction products, CO_2, H_2O and SO_2 respectively; with ν_i their stoichiometric factors are named in the reaction equation.

The molar work $(W_m)_{rev}$ related to the molar quantity of fuel is the molar reversible reaction work of the isothermal-isobaric oxidation reaction, which is given off in all oxidation reactions and is therefore negative; the molar upper calorific value of the fuel is denoted by $H_{sm}(T_0) = -\Delta^R H_m(T_0, p_0)$.

$$(W_m)_{rev} = \Delta^R G_m(T_0, p_0) = \Delta^R H_m(T_0, p_0) - T_0 \Delta^R S_m(T_0, p_0)$$

$$= -H_{sm}(T_0) - T_0 \Delta^R S_m(T_0, p_0) \quad (12.54)$$

This gives the following for the molar fuel exergy

$$Ex_{m,B}(T_0, p_0) = -(W_m)_{rev} + \Delta Ex_m(T_0, p_0)$$

$$= H_{sm}(T_0) + T_0 \Delta^R S_m(T_0, p_0) + \Delta Ex_m(T_0, p_0) \ , \quad (12.55)$$

in which the abbreviation is

$$\Delta Ex_m(T_0, p_0) = \sum_{i=1}^{N} \nu_i Ex_{m,i}(T_0, p_0) - O_{min} Ex_{m,O_2}(T_0, p_0) \ . \quad (12.56)$$

The term $\Delta Ex_m (T_0, p_0)$ represents a molar exergy and is non-zero, since the oxygen as well as the reaction products also have a chemical exergy at T_0 and p_0; according to [9], however, it comprises only a few percent of the molar fuel exergy.

The molar fuel exergy is predominantly characterised by the amount of reversible molar reaction work of its oxidation reaction, whereby this differs only slightly from the molar upper calorific value H_{sm}, since the term with the molar reaction entropy $\Delta^R S_m (T_0, p_0)$ is also comparatively small. This becomes visible e.g. for carbon (C); for hydrogen (H_2) this is only approximately true. According to Table 12.1, for carbon (C) with $H_{sm} = H_{im} = -\Delta^R H_m (T_0, p_0) = 393.51$ MJ/kmol during oxidation to carbon dioxide (CO_2) one obtains

$$-(W_m)_{rev} = H_{sm} + T_0 (S_{m\,CO_2}^{\square} - S_{m\,O_2}^{\square} - S_{m\,C}^{\square}) \tag{12.57}$$

$= 393.51$ MJ/kmol $+ 298.15$ K$\cdot(213.785 - 205.152 - 5.74)$ kJ/(kmol K) $= (393.51 + 0.863)$ MJ/kmol $= 394.37$ MJ/kmol $= 1.0022\, H_{sm} = 1.0022\, H_{im}$.

For hydrogen (H_2) one gets with $H_{sm} = -\Delta^R H_m (T_0, p_0) = 285.83$ MJ/kmol during oxidation to liquid water (H_2O)

$$-(W_m)_{rev} = H_{sm} + T_0 (S_{m\,H_2O}^{fl\,\square} - 0.5\, S_{m\,O_2}^{\square} - S_{m\,H_2}^{\square}) \tag{12.58}$$

$= 285.83$ MJ/kmol $+ 298.15$ K$\cdot(69.93 - 102.576 - 130.680)$ kJ/(kmol K) $= (285.83 - 48.696)$ MJ/kmol $= 237.134$ MJ/kmol $= 0.8296\, H_{sm} = 0.98072\, H_{im}$.

In the reversible oxidation of carbon, even a little more molar work could be gained than the upper molar calorific value indicates, because the molar reaction entropy results as a positive value. A little molar heat would be supplied from the environment, which makes a small contribution to the molar reversible reaction work.

In the reversible oxidation of hydrogen, on the other hand, the most significant difference of all fuels between the upper molar calorific value and the molar reversible reaction work can be stated, because two gases with comparatively large molar entropies are used as educts and liquid water is formed as a reaction product with lower molar entropy.

In the same way the molar reversible reaction works can be calculated which are characteristic for other oxidation reactions. In most cases, the differences in the values for $(W_m)_{rev}$ and H_{sm} are small; accordingly, the molar chemical energy reflected via the molar upper calorific value can be understood predominantly as molar exergy and thus as convertible molar energy.

The molar reversible reaction work $(W_m)_{rev} = \Delta^R G_m (T_0, p_0)$ is a property of the respective fuel; it represents the largest share of the molar exergy $Ex_{m,B} (T_0, p_0)$ and is independent of the underlying environment. The properties of the chosen thermodynamic environment only influence the comparatively small term $\Delta Ex_m (T_0, p_0)$ according to Eq. (12.56) [9]; this is based on the environment model described above. The molar exergies of the combustion products and oxygen contained in Eq. (12.56) take the following values for the standard state as ambient state: $Ex_{m,CO_2} = 16.15$ MJ/kmol; $Ex_{m,H_2O_{fl}} = 0.022$ MJ/kmol; $Ex_{m,SO_2} = 236.4$ MJ/kmol; $Ex_{m,O_2} = 4.967$ MJ/kmol [9].

The values of $\Delta Ex_m (T_0, p_0)$ determined hereby are less than $0.016\, H_{sm}$ for all hydrocarbons; carbon (C) has the value $0.0284\, H_{sm}$; hydrogen (H_2) is characterised by

a negative $\Delta Ex_m(T_0, p_0) = -0.00857\ H_{sm}$. The values for sulphur and the sulphur compounds are much higher.

The molar exergies of chemically uniform fuels thus calculated according to Eq. (12.55) are summarised in Table 12.2. The molar exergies of gaseous fuels come to about 95 % of the respective molar upper calorific value; an exception is hydrogen (H_2) with $Ex_{m,H_2} = 0.8211\ H_{sm}$. The molar exergies of the liquid fuels come to about 98 % of their respective molar upper calorific value. In contrast, the sulphur compounds have molar exergy values that far exceed the respective molar upper calorific value.

In Table 12.2, carbon (C) and sulphur (S) as technically important and chemically uniform solid fuel elements are listed. In addition, the fuel gases hydrogen (H_2), carbon monoxide (CO), methane (CH_4), propane (C_3H_8) and butane (C_4H_{10}) as well as for the liquid fuels hexane (C_6H_{14}), octane (C_8H_{18}), methyl alcohol (CH_3OH) and ethyl alcohol (C_2H_5OH) are listed. For them, the molar calorific value H_{im}, the molar calorific value H_{sm}, the negative value of the molar reversible reaction work $-(W_m)_{rev}$, the molar chemical exergy Ex_B and the ratios $Ex_{m,B}/H_{sm}$ and $Ex_{m,B}/(-(W_m)_{rev})$ at the standard chemical state are reported.

Table 12.2 Molar lower calorific value H_{im}, molar upper calorific value H_{sm}, negative molar reversible reaction work $-(W_m)_{rev}$ and molar chemical exergy $Ex_{m,B}$ of chemically uniform fuels at the standard chemical state. The molar chemical exergy $Ex_{m,B}$ is reported according to the ambient model given in section 12.6 [9].

Brennstoff	H_{im}	H_{sm}	$-(W_m)_{rev}$	$Ex_{m,B}$	$Ex_{m,B}/H_{sm}$	$Ex_{m,B}/(-(W_m)_{rev})$
	$\frac{MJ}{kmol}$	$\frac{MJ}{kmol}$	$\frac{MJ}{kmol}$	$\frac{MJ}{kmol}$	-	-
C	393.51	393.51	394.37	405.55	1.0306	1.0283
S	296.8	296.8	300.1	531.5	1.791	1.772
H_2	241.81	285.83	237.15	234.68	0.8211	0.9896
CO	282.98	282.98	257.21	270.88	0.9572	1.0535
CH_4	802.30	890.32	817.90	824.16	0.9257	1.0077
C_3H_8	2044.0	2220.0	2108.3	2132.0	0.9604	1.0112
CH_4H_{10}	2658.5	2878.5	2747.7	2780.1	0.9658	1.0118
C_6H_{14}	3855	4163	4023	4073	0.9784	1.0124
C_8H_{18}	5075	5471	5296	5363	0.9803	1.0127
CH_3OH	637.7	725.7	701.7	710.4	0.9789	1.0124
C_2H_5OH	1235.5	1367.6	1326.6	1343.6	0.9825	1.0132

With the information contained therein, the exergies of natural gases can also be determined if their composition is known in mole fractions n_i/n. In the following, it is assumed that natural gases can be regarded as ideal gases. Their molar exergy $Ex_m(T, p)$ is composed of the sum of the molar exergies of the respective components $\sum_{i=1}^{N} \frac{n_i}{n} Ex_{m,0i}(T, p)$ as well as of the negatively calculated molar work $-(W_m)_{rev}(T, p)$ and the molar heat $(Q_m)_{rev}(T, p)$, which occur during the reversible mixing process at constant temperature and constant pressure. If the molar exergy is considered at the standard chemical state and its state is agreed as the ambient state ($T = T_0$, $p = p_0$), the molar energy exchanged with the environment as molar heat $(Q_m)_{rev}(T_0, p_0)$ consists only of exergetically worthless molar anergy; it is therefore not to be considered

in the molar exergy balance. Thus it becomes

$$Ex_{m,B}(T_0, p_0) = \sum_{i=1}^{N} \frac{n_i}{n} Ex_{m,0i}(T_0, p_0) - (W_m)_{rev}(T_0, p_0) . \qquad (12.59)$$

Since the molar reversible work of the mixing process (index 'M') occurs at constant temperature, it results - analogous to Eqs. (5.213), (5.214), (5.215) and (12.48) - as a change in the molar free enthalpy to $(W_m)_{rev}(T_0, p_0) = -\Delta^M G_m(T_0, p_0) = \Delta^M H_m(T_0, p_0) - T_0 \Delta^M S_m(T_0, p_0)$. Since the molar enthalpy of an ideal gas does not change during an isothermal change of state, $\Delta^M H_m(T_0, p_0) = 0$ follows.

The specific entropy of mixing of ideal gases can be calculated according to Eq. (4.192 b). For $T = T_i = T_0$ its first summand becomes zero. For the molar entropy of mixing, taking into account $r_i = n_i/n$ according to Eq. (4.203) one can write:

$$\Delta^M S_m(T_0, p_0) = -R_m \sum_{i=1}^{N} \frac{n_i}{n} \ln r_i = -R_m \sum_{i=1}^{N} \frac{n_i}{n} \ln \frac{n_i}{n}. \quad \text{This gives}$$

$$Ex_{m,B}(T_0, p_0) = \sum_{i=1}^{N} \frac{n_i}{n} Ex_{m,0i}(T_0, p_0) + R_m T_0 \sum_{i=1}^{N} \frac{n_i}{n} \ln \frac{n_i}{n}$$

$$= \sum_{i=1}^{N} \frac{n_i}{n} [Ex_{m,0i}(T_0, p_0) + R_m T_0 \ln \frac{n_i}{n}] . \qquad (12.60)$$

In [9], it is found that, to a very good approximation, there is a linear relationship between the molar exergy of natural gases $Ex_{m,B}$ and their molar lower calorific value H_{im} or their molar upper calorific value H_{sm}, where the following relationships hold:

$Ex_{m,B}/H_{im} = 1.0313$ - $(3.3968 \text{ MJ/kmol})/H_{im}$
(at 600 MJ/kmol $< H_{im} <$ 875 MJ/kmol)

$Ex_{m,B}/H_{sm} = 0.9389$ - $(10.4465 \text{ MJ/kmol})/H_{sm}$
(at 660 MJ/kmol $< H_{sm} <$ 970 MJ/kmol)

On average, to a good approximation: $Ex_{m,B}/H_{im} = 1.027$; $Ex_{m,B}/H_{sm} = 0.934$.

The molar exergy of chemically undefined fuels - especially those of coal and fuel oil - cannot be readily calculated using Eq. (12.55) because the conventional entropy (absolute entropy) of the fuel required to determine the molar entropy of reaction $\Delta^R S_m$ is not known.

In [9], reference is made to a procedure to avoid this problem, with which linear relationships between the specific exergy ex_B and the specific lower calorific value H_i or between the specific exergy ex_B and the specific upper calorific value H_s can be specified for coal and fuel oil. The following equations are given for the ratios ex_B/H_i and ex_B/H_s [9]:

Coal:

$ex_B/H_i = 0.967 + (2.389 \text{ MJ/kg})/H_i$ (at $H_i <$ 33 MJ/kg)

$ex_B/H_s = 1.007 + (0.155 \text{ MJ/kg})/H_s$ (at $H_s <$ 34 MJ/kg)

Fuel oil :

$ex_B/H_i = 1.075 - (1.150 \text{ MJ/kg})/H_i$ (at 38 MJ/kg $< H_i <$ 44 MJ/kg)

$ex_B/H_s = 0.911 + (3.307 \text{ MJ/kg})/H_s$ (at 40 MJ/kg $< H_s <$ 47 MJ/kg)

The ratio ex_B/H_s is hardly different from the value 1.0; thus, in a first approximation, the specific exergy of chemically undefined fuels can be set equal to the upper specific calorific value.

In contrast to the reversible reactions assumed here, all real technical combustion processes in which the chemical energy is only converted into heat or internal energy or enthalpy - and not also partly into reaction work - are strongly irreversible. According to the statement of the second law, a considerable exergy loss occurs during combustion.

The exergy loss that occurs during the combustion of energy in a combustion chamber can be assigned to two different exergy losses: the exergy loss of the chemical reaction of combustion assumed to be adiabatic and the exergy loss during the cooling of the combustion gas and the associated heat transfer to the useful fluid (e.g. to process steam or heating water).

12.8 Chemical Potentials

The total entropy S of a system, which is an extensive quantity, can be given by summing up all products of the respective molar quantity n_i and the respective molar entropy $S_{m,i}$ of the individual constituents, as far as the individual constituents are taken to be unmixed with the other constituents:

$$S = \sum_{i=1}^{N} n_i S_{m,i} \qquad (12.61)$$

A differential change of the total entropy dS results generally according to the rule for the differentiation of the product $n_i S_{m,i}$ as follows:

$$dS = \sum_{i=1}^{N} n_i \, dS_{m,i} + \sum_{i=1}^{N} S_{m,i} \, dn_i \qquad (12.62)$$

Here $dS_{m,i}$ can be represented as $dS_{m,i} = (dU_{m,i} + p \, dV_{m,i})/T$ according to the combination of the first and second law:

$$dS = \sum_{i=1}^{N} n_i \left[(dU_{m,i} + p \, dV_{m,i})/T \right] + \sum_{i=1}^{N} S_{m,i} \, dn_i$$

$$dS = \sum_{i=1}^{N} \left[(n_i \, dU_{m,i} + p \, n_i \, dV_{m,i})/T \right] + \sum_{i=1}^{N} S_{m,i} \, dn_i$$

$$dS = \left(\sum_{i=1}^{N} n_i \, dU_{m,i} + \sum_{i=1}^{N} p \, n_i \, dV_{m,i} \right)/T + \sum_{i=1}^{N} S_{m,i} \, dn_i$$

$$T \, dS = \sum_{i=1}^{N} n_i \, dU_{m,i} + \sum_{i=1}^{N} p \, n_i \, dV_{m,i} + T \sum_{i=1}^{N} S_{m,i} \, dn_i$$

$n_i\, U_{m,i} = U_i$ as well as $n_i\, V_{m,i} = V_i$ and — in the case of a gas mixture produced before the reaction in a closed system without temperature and pressure change with unchanged total volume — according to section 4.5.3 also $\sum_{i=1}^{N} n_i\, V_{m,i} = \sum_{i=1}^{N} V_i = V$. Further $\Sigma U_i = U$ holds.

According to the rule for the differentiation of the products $U_i = n_i\, U_{m,i}$ and $V_i = n_i\, V_{m,i}$, $dU_i = n_i\, dU_{m,i} + U_{m,i}\, dn_i$ and $dV_i = n_i\, dV_{m,i} + V_{m,i}\, dn_i$ hold. From this follow $n_i\, dU_{m,i} = -U_{m,i}\, dn_i + dU_i$ and $n_i\, dV_{m,i} = -V_{m,i}\, dn_i + dV_i$. Thus

$$T\, dS = -\sum_{i=1}^{N} U_{m,i}\, dn_i + \sum_{i=1}^{N} dU_i - \sum_{i=1}^{N} p\, V_{m,i}\, dn_i + p\sum_{i=1}^{N} dV_i + \sum_{i=1}^{N} T\, S_{m,i}\, dn_i$$

becomes. With $\sum_{i=1}^{N} dU_i = dU$ and $p\sum_{i=1}^{N} dV_i = p\, dV$ follows:

$$T\, dS = -\sum_{i=1}^{N} U_{m,i}\, dn_i - \sum_{i=1}^{N} p\, V_{m,i}\, dn_i + \sum_{i=1}^{N} T\, S_{m,i}\, dn_i + dU + p\, dV$$

$$T\, dS = dU + p\, dV - \sum_{i=1}^{N} (U_{m,i} + p\, V_{m,i} - T\, S_{m,i})\, dn_i$$

The sum expression on the right-hand side of this equation represents for the individual components i the respective molar free enthalpy (i.e. the respective molar *Gibbs*-function) $G_{m,i} = U_{m,i} + p\, V_{m,i} - T\, S_{m,i} = H_{m,i} - T\, S_{m,i}$, multiplied by the differential change dn_i of the respective component i occurring during the chemical reaction (cf. Eq. (5.217)):

$$T\, dS = dU + p\, dV - \sum_{i=1}^{N} G_{m,i}\, dn_i$$

$$T\, dS - p\, dV + \sum_{i=1}^{N} G_{m,i}\, dn_i = dU$$

The molar free enthalpies $G_{m,i}$ of the individual constituents i are also called molar potentials $G_{m,i}$; because of their importance in the thermodynamics of chemical equilibrium, they are also called molar chemical potentials or only chemical potentials and named separately with $G_{m,i} = \mu_i$. Thus

$$T\, dS - p\, dV + \sum_{i=1}^{N} \mu_i\, dn_i = dU \qquad (12.63)$$

holds. dU is the - thus extended by the expression $\sum_{i=1}^{N} \mu_i\, dn_i$ - total differential of the thermodynamic potential for pure substances $U = U(S, V)$. Equations of the form $U = U(S, V)$ are called canonical equations of state or fundamental equations for pure substances; they contain all thermodynamic information of pure substances.

If one further considers

$$dU = dH - d(pV) = dH - p\,dV - V\,dp$$
$$dU = dF + d(TS) = dF + T\,dS + S\,dT$$
$$dU = dH - p\,dV - V\,dp = dG + d(TS) - p\,dV - V\,dp = dG + T\,dS + S\,dT - p\,dV - V\,dp$$

(cf. Eqs. (3.77), (5.220) and (5.221)), then inserting dU into Eq. (12.63) also yields the total differentials of the thermodynamic potentials for pure substances $H = H(S, p)$, $F = F(T, V)$ and $G = G(T, p)$, extended by the expression $\sum_{i=1}^{N} \mu_i\,dn_i$:

$$T\,dS + V\,dp + \sum_{i=1}^{N} \mu_i\,dn_i = dH \tag{12.64}$$

$$-S\,dT - p\,dV + \sum_{i=1}^{N} \mu_i\,dn_i = dF \tag{12.65}$$

$$-S\,dT + V\,dp + \sum_{i=1}^{N} \mu_i\,dn_i = dG \tag{12.66}$$

Compared to the fundamental relations for pure substances, Eqs. (12.63) to (12.66) thus have the additional term $\sum_{i=1}^{N} \mu_i\,dn_i$. From Eqs. (12.63) to (12.66) also follows if, apart from the molar quantity n_i of component i, the molar quantities n_j of all other components j are assumed to be constant as well as, for example, S and V, S and p, T and V and T and p respectively:

$$\mu_i = \left(\frac{\partial U}{\partial n_i}\right)_{S,V,n_j} = \left(\frac{\partial H}{\partial n_i}\right)_{S,p,n_j} = \left(\frac{\partial F}{\partial n_i}\right)_{T,V,n_j} = \left(\frac{\partial G}{\partial n_i}\right)_{T,p,n_j} \tag{12.67}$$

12.9 The Law of Mass Action

It is of particular interest in chemical reactions how a given system is composed in the state of chemical equilibrium. In the following, the state of equilibrium is to be defined by a constant temperature T and by a constant pressure p.

Under these conditions, the equilibrium state is most simply described by the free enthalpy G: Equation (12.66) $dG = -S\,dT + V\,dp + \sum_{i=1}^{N} \mu_i\,dn_i$ can be expressed by $dn_i = \nu_i\,dz$ in the form $dG = -S\,dT + V\,dp + \sum_{i=1}^{N} \mu_i\,\nu_i\,dz$ from which it follows for $dp = 0$ and $dT = 0$:

$$dG = \sum_{i=1}^{N} \mu_i\,\nu_i\,dz = \sum_{i=1}^{N} \mu_i\,dn_i \tag{12.68}$$

It follows:

$$\left(\frac{\partial G}{\partial z}\right)_{T,p} = \sum_{i=1}^{N} \nu_i \mu_i \qquad (12.69)$$

Eq. (12.69) represents the partial molar reaction *Gibbs* function.

It can be shown that for pure substances the thermodynamic potentials $S = S(U, V)$, $U = U(S, V)$, $H = H(S, p)$, $F = F(V, T)$ and $G = G(p, T)$ (cf. section 5.4.5.1) each assume an extreme value in the equilibrium state if the independent variables are chosen accordingly:

- The entropy S of a closed system assumes a maximum at constant internal energy ($dU = 0$) and constant volume ($dV = 0$), since according to Eq. (3.77): $T\,dS - p\,dV = dU$ holds. With $dU = 0$ and $dV = 0$, $dS = 0$ follows; i.e. the entropy S has reached its maximum value and cannot increase any further.

- The internal energy U assumes a minimum at constant volume ($dV = 0$) and at constant entropy ($dS = 0$).

- The enthalpy H assumes a minimum at constant entropy ($dS = 0$) and at constant pressure ($dp = 0$).

- The free energy F assumes a minimum at constant volume ($dV = 0$) and at constant temperature ($dT = 0$).

- The free enthalpy G assumes a minimum at constant pressure ($dp = 0$) and at constant temperature ($dT = 0$), since according to Eq. (3.77) holds: $T\,dS + V\,dp = dH = dG + d(TS) = dG + T\,dS + S\,dT$ and thus $V\,dp = dG + S\,dT$. With $dp = 0$ and $dT = 0$, also $dG = 0$ holds; i.e. the free enthalpy G has reached its lowest value and cannot decrease any further.

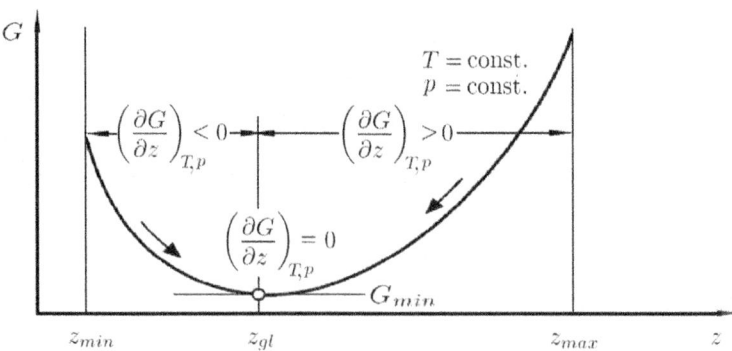

Figure 12.15 Partial molar reaction *Gibbs* function in relation to reaction turnover: possible directions of a chemical reaction

This shows how the equilibrium state of thermodynamic systems can be calculated under the above externally imposed conditions.

For the following considerations, the last statement is of interest: In chemical reactions, G according to Eq. (12.66) depends not only on T and p, but also on n_i as variables.

If, for example, the temperature T and the pressure p are kept constant in a closed system (cf. Figure 12.2 on the left), G tends towards a minimum:

$$dG = \sum_{i=1}^{N} \mu_i \, dn_i = \sum_{i=1}^{N} \mu_i \nu_i \, dz \leq 0 \qquad (12.70)$$

If the reaction is to proceed in the direction written in the reaction equation, $(\partial G/\partial z)_{T,p} < 0$ must hold. On the other hand, if $(\partial G/\partial z)_{T,p} > 0$ is valid, the reaction proceeds in the opposite direction. This is shown in Fig. 12.15 with respect to the reaction turnover z. If $(\partial G/\partial z)_{T,p} = 0$ is valid, the reaction has reached its equilibrium state and the reaction in the closed system has stopped.

The equation

$$\left(\frac{\partial G}{\partial z}\right)_{T,p} = \sum_{i=1}^{N} \mu_i \nu_i = 0 \qquad (12.71)$$

can be used for further statements if the respective chemical potential μ_i is known as a function of pressure and temperature. For this purpose, an ideal gas is assumed in the sense of a simplification and the equation for G is the basis: $G = H - TS$.

In this, the entropy S is to be determined in more detail. Since a mixture of ideal gases is to be thought of as being generated by a mixing process, the resulting mixing entropy must also be taken into account. According to Eq.(4.192 a), the following applies to a component i of the mixture:

$$\Delta S_i = (m_i \, c_{vi} \ln \frac{T}{T_i} + m_i \, R_i \ln \frac{V}{V_i}) \; .$$

In this, because $T = $ const, the first term becomes zero; further, because $T = $ const and $p = $ const, it follows $V/V_i = n/n_i$; also, $m_i R_i = n_i R_m$ is considered:

$$\Delta S_i = n_i \, R_m \ln \frac{n}{n_i} \; .$$

This gives for a mixture of ideal gases with the individual components i:

$$S = \sum_{i=1}^{N} n_i \, S_{m,i} + R_m \sum_{i=1}^{N} n_i \ln \frac{n}{n_i}$$

For the free enthalpy it follows:

$$G = \sum_{i=1}^{N} n_i \, [H_{m,i} - T \, S_{m,i} - R_m \, T \ln \frac{n}{n_i}] \qquad (12.72)$$

By $G_{m,i}^0 = H_{m,i}^0 - T \, S_{m,i}^0$ the molar free enthalpy of the pure component i in the state of the ideal gas is denoted. Thus Eq. (12.72) becomes

$$G = \sum_{i=1}^{N} n_i \, [G_{m,i}^0 - R_m \, T \ln \frac{n}{n_i}] \; . \qquad (12.73)$$

Now Eq. (12.68) is considered: $dG = \sum_{i=1}^{N} \mu_i \, dn_i$

Differentiating G according to n_i, using Eqs. (12.68) and (12.73), one gets

$$\mu_i = \left(\frac{\partial G}{\partial n_i}\right)_{p,T,n_j} = G^0_{m,i} - R_m T \ln \frac{n}{n_i} \quad \text{with} \quad j \neq i \ .$$

This differentiation takes into account that due to $n = n_1 + n_2 + n_3 + \ldots + n_N$ the differential quotient of $\sum_{i=1}^{N} n_i \ln \frac{n}{n_i}$ according to n_i becomes $\ln \frac{n}{n_i}$. From this follows for the equilibrium condition according to Eq. (12.71) $\sum_{i=1}^{N} \mu_i \nu_i = 0$:

$$\sum_{i=1}^{N} \mu_i \nu_i = \sum_{i=1}^{N} \nu_i \left(G^0_{m,i} - R_m T \ln \frac{n}{n_i}\right) = 0 \qquad (()12.74)$$

With $\ln \frac{n}{n_i} = -\ln \frac{n_i}{n}$ follows

$$\sum_{i=1}^{N} \nu_i G^0_{m,i} = \sum_{i=1}^{N} \nu_i R_m T \ln \frac{n}{n_i} = -\sum_{i=1}^{N} \nu_i R_m T \ln \frac{n_i}{n} \qquad (12.75)$$

$$-\frac{1}{R_m T} \sum_{i=1}^{N} \nu_i G^0_{m,i} = \sum_{i=1}^{N} \nu_i \ln \frac{n_i}{n} = \sum_{i=1}^{N} \ln \left(\frac{n_i}{n}\right)^{\nu_i} = \ln \left[\Pi \left(\frac{n_i}{n}\right)^{\nu_i}\right] \qquad (12.76)$$

or with the abbreviation $\sum_{i=1}^{N} = \Sigma$

$$\Pi \left(\frac{n_i}{n}\right)^{\nu_i} = e^{-\frac{\Sigma \nu_i G^0_{m,i}}{R_m T}} = e^{-\frac{\Delta^R G^0_m(T)}{R_m T}} \qquad ((12.77)$$

The right-hand side of this equation depends on p and T, since according to Eqs. (12.72) and (12.73) $G^0_{m,i}$ is a function of T and p and T appears in the denominator. $\Delta^R G^0_m(T)$ is the molar reaction *Gibbs* function of the ideal gas at standard pressure p_0.

If one introduces for the right side the abbreviation $K(T, p)$

$$K(T, p) = e^{\frac{-\Sigma \nu_i G^0_{m,i}}{R_m T}}, \qquad ((12.78)$$

then can be written

$$\Pi \left(\frac{n_i}{n}\right)^{\nu_i} = K(T, p) \ . \qquad (12.79)$$

n_i/n represents the ratio of the molar quantity n_i of component i to the total molar quantity n of the mixture of ideal gases, which according to Eq. (4.197) can also be expressed as the ratio of the partial pressure p_{pi} of component i to the total pressure

p of the mixture of ideal gases $n_i/n = p_{pi}/p$, since all components occupy the same volume and have the same temperature:

$$\Pi \left(\frac{p_{pi}}{p}\right)^{\nu_i} = \Pi \left(\frac{p_{pi}^{\nu_i}}{p^{\nu_i}}\right) = K(T, p) \tag{12.80}$$

With the new abbreviation

$$K'(T) = K(T, p) \Pi p^{\nu_i} \tag{12.81}$$

becomes

$$\Pi (p_{pi}^{\nu_i}) = K'(T) , \tag{12.82}$$

i.e.

$$p_{p1}^{\nu_1} p_{p2}^{\nu_2} p_{p3}^{\nu_3} \cdots p_{pN}^{\nu_N} = K'(T) . \tag{12.83}$$

Eqs. (12.77) to (12.83) are known as the law of mass action that was found by *Guldberg* and *Waage*[6]. In it, the stoichiometric coefficients ν_i of the educts are negative and those of the products are positive.

If the absolute values of the stoichiometric coefficients of the total b educts denoted by $\nu_1^e, \nu_2^e, \ldots \nu_b^e$ and those of the total $N - b$ products characterising the reaction with $\nu_{b+1}^a, \nu_{b+2}^a, \ldots \nu_N^a$, then the law of mass action can be written as follows according to Eq. (12.83):

$$\frac{p_{p(b+1)}^{\nu_{b+1}^a} p_{p(b+2)}^{\nu_{b+2}^a} \cdots p_{pN}^{\nu_N^a}}{p_{p1}^{\nu_1^e} p_{p2}^{\nu_2^e} \cdots p_{pb}^{\nu_b^e}} = K'(T) \tag{12.84}$$

Example 12.6 The constant K' is to be used in general for the reactions

$$CH_4 + H_2O \rightleftharpoons CO + 3H_2 \tag{12.1}$$

$$2H_2 + O_2 \rightleftharpoons 2H_2O \tag{12.4}$$

$$H_2 + 0.5 O_2 \rightleftharpoons H_2O \tag{12.4a}$$

and to be specified.

$$K'_{12.1} = \frac{p_{pCO} \, p_{pH_2}^3}{p_{pCH_4} \, p_{pH_2O}}$$

$$K'_{12.4} = \frac{p_{pH_2O}^2}{p_{pH_2}^2 \, p_{pO_2}}$$

$$K'_{12.4a} = \frac{p_{pH_2O}}{p_{pH_2} \, p_{pO_2}^{0,5}}$$

Depending on the formulation of the reaction of hydrogen and oxygen to water vapor according to Eqs. (12.4) and (12.4 a), different equilibrium constants $K'_{12.4}$ and $K'_{12.4a}$ result, between which the relation $K'_{12.4} = (K'_{12.4a})^2$ is valid.

---

[6] *Cato Maximilian Guldberg*, 1836 to 1902, and *Peter Waage*, 1833 to 1900, two Norwegian chemists, jointly formulated the law of mass action.

12.10 Pressure and Temperature Dependence of the Constants of the Law of Mass Action; Law of *Le Chatelier* and *Braun*

The law of mass action according to Eqs. (12.77) to (12.80) indicates under which pressure and temperature conditions the equilibrium of a chemical reaction occurs. For ideal gases the corresponding quantities are marked with a superscript '^{0}'. On one side of Eq. (12.79) and Eq. (12.80), respectively, are the expressions $\Pi\left(\frac{n_i}{n}\right)^{\nu_i}$ and $\Pi\left(\frac{p_{pi}}{p}\right)^{\nu_i}$, respectively. while on the other side there is the temperature- and pressure-dependent quantity K, which can be understood as a constant dependent on T and p:

$$K(p, T) = e^{-\frac{\Sigma \nu_i G^0_{m,i}}{R_m T}} \qquad (12.78)$$

It is of interest how the dependence of the constant K on T and p can be represented. In order to show the dependence on the pressure, Eq. (12.78) is first logarithmised and then differentiated according to p at constant temperature T:

$$\left(\frac{\partial \ln K}{\partial p}\right)_T = -\frac{1}{R_m T} \sum_{i=1}^{N} \nu_i \left(\frac{\partial G^0_{m,i}}{\partial p}\right)_T,$$

where is

$$\left(\frac{\partial G^0_{m,i}}{\partial p}\right)_T = \left(\frac{V^0_{m,i}\,\partial p - S^0_{m,i}\,\partial T}{\partial p}\right)_T = V^0_{m,i}.$$

Thus one gets

$$\left(\frac{\partial \ln K}{\partial p}\right)_T = -\frac{1}{R_m T} \sum_{i=1}^{N} \nu_i V^0_{m,i} = -\frac{\Delta^R V^0_m}{R_m T}. \qquad (12.85)$$

The trem $\sum_{i=1}^{N} \nu_i V^0_{m,i} = \Delta^R V^0_m$ is called the molar reaction volume of the ideal gas.

If the molar volume of the reaction products is greater than the molar volume of the educts, then because of the negative sign $(\partial \ln K/\partial p)_T < 0$ is given. This means that the molar quantity of the educts increases with an increase in pressure.

On the other hand, if the molar quantity of the reaction products is less than the molar quantity of the educts, then $(\partial \ln K/\partial p)_T > 0$ is given. This expresses that the molar quantity of the reaction products increases when the pressure is increased.

This shows that the system is evasive to the 'external constraint' created by an increase in pressure by decreasing the total molar quantity and thus the total volume V^0_m; this is also called the 'principle of least constraint'.

To investigate the dependence on temperature, Eq. (12.78) is first logarithmised and then differentiated according to T at constant pressure p:

$$\left(\frac{\partial \ln K}{\partial T}\right)_p = -\left(\frac{\partial \Sigma \nu_i\, (G^0_{m,i}/R_m T)}{\partial T}\right)_p \qquad (12.86)$$

Inserting the equation for the molar free enthalpies of the components i $G_{m,i}^0 = H_{m,i}^0 - T S_{m,i}^0$ into Eq. (12.86) and using the simplifying assumption that the molar enthalpies of the components $H_{m,i}^0$ do not depend on temperature, it follows by differentiating according to the quotient rule:

$$\left(\frac{\partial \ln K}{\partial T}\right)_p = \frac{1}{R_m T^2} \sum_{i=1}^{N} \nu_i G_{m,i}^0 + \frac{1}{R_m T} \sum_{i=1}^{N} \nu_i S_{m,i}^0 = \frac{1}{R_m T^2} \sum_{i=1}^{N} \nu_i (G_{m,i}^0 + T S_{m,i}^0)$$
(12.87)

Thus, using the equation $H_{m,i}^0 = G_{m,i}^0 + T S_{m,i}^0$, finally one gets

$$\left(\frac{\partial \ln K}{\partial T}\right)_p = \frac{1}{R_m T^2} \sum_{i=1}^{N} \nu_i H_{m,i}^0 .$$
(12.88)

In this equation, the expression $\sum_{i=1}^{N} \nu_i H_{m,i}^0 = \Delta^R H_m^0$ as the molar enthalpy of reaction represents the change in the molar enthalpies of all the ideal gases involved:

$$\left(\frac{\partial \ln K}{\partial T}\right)_p = \frac{\Delta^R H_m^0}{R_m T^2}$$
(12.89)

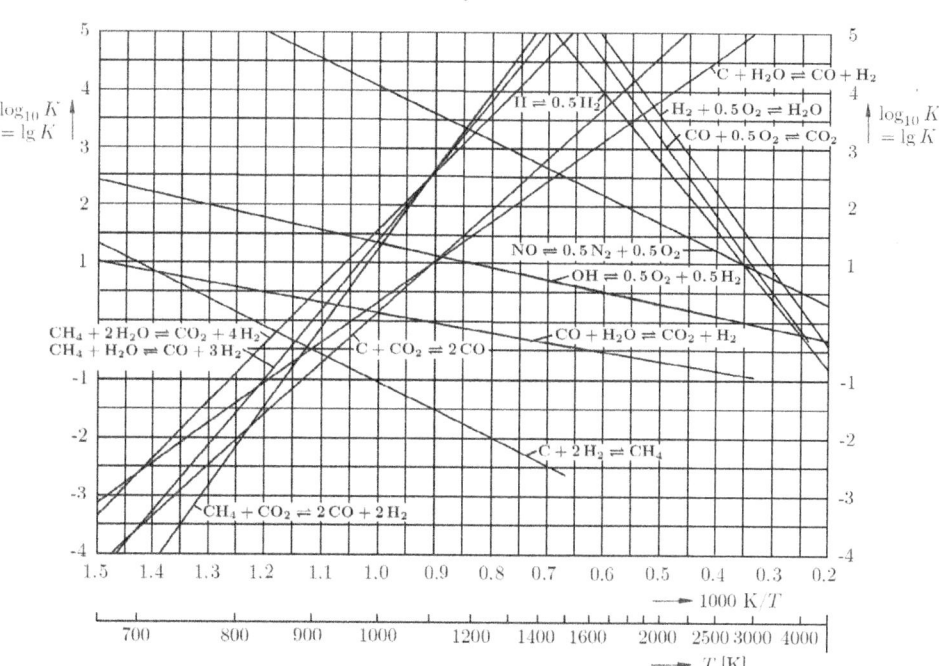

Figure 12.16 Temperature dependence of the equilibrium constant K for various chemical reactions at pressure $p_0 = 1.01325$ bar (cf. inter alia [112])

As far as there are no devices for the conversion of molar mechanical or molar electrical work in the technical arrangement for carrying out the chemical reaction, according to the first law $(Q_m)_{rev} + W_m = \Delta^R H_m^0$ the molar work becomes $W_m = 0$ and thus

one gets $(Q_m)_{rev} = \Delta^R H_m^0$. The occurring molar heat, which is then equal to the molar enthalpy of reaction, is called molar heat of reaction or sometimes molar heat tint, too.

Fig. 12.16 shows the temperature dependence of the equilibrium constant K for various chemical reactions at pressure $p_0 = 1.01325$ bar. Instead of the natural logarithm, the logarithm to the base 10 is plotted, where $\log_{10} K = \lg K = 0.434295 \ln K$ holds.

A reaction in which the molar enthalpy of reaction $\Delta^R H_m^0$ is negative and molar heat or, if necessary, molar work is released is called an exothermic reaction; according to Eq. (12.89), $(\partial \ln K / \partial T)_p$ is negative. This means that an increase in temperature shifts the chemical equilibrium towards the initial substances (educts).

A reaction in which the molar enthalpy of reaction $\Delta^R H_m^0$ is positive and molar heat or, if necessary, molar work must be supplied is called an endothermic reaction; according to Eq. (12.89), $(\partial \ln K / \partial T)_p$ is positive; this means that a temperature increase shifts the chemical equilibrium towards the products.

Fig. 12.16 shows the temperature dependence of the exothermic reactions

$H_2 + 0.5\,O_2 \rightleftharpoons H_2O$, $\quad CO + H_2O \rightleftharpoons CO_2 + H_2$, $\quad H \rightleftharpoons 0.5\,H_2$, $\quad CO + 0.5\,O_2 \rightleftharpoons CO_2$, $NO \rightleftharpoons 0.5\,N_2 + 0.5\,O_2$, $\quad OH \rightleftharpoons 0.5\,O_2 + 0.5\,H_2$ $\quad$ and $\quad C + 2\,H_2 \rightleftharpoons CH_4$

in the form of — with increasing temperature — decreasing straight lines.

In contrast, the endothermic reactions

$CH_4 + H_2O \rightleftharpoons CO + 3\,H_2$, $\quad CH_4 + 2\,H_2O \rightleftharpoons CO_2 + 4\,H_2$, $\quad C + CO_2 \rightleftharpoons 2\,CO$, $CH_4 + CO_2 \rightleftharpoons 2\,CO + 2\,H_2$ $\quad$ and $\quad C + H_2O \rightleftharpoons CO + H_2$

are represented by straight lines which increase with increasing temperatures.

Eq. (12.89) can be used to obtain an approximate formula showing the temperature dependence of the equilibrium constant K. If the molar enthalpy of reaction $\Delta^R H_m^0$ is assumed to be constant, it follows by integration over the temperature T

$$\ln K = -\frac{\Delta^R H_m^0}{R_m T} + \text{const} \;. \tag{12.90}$$

With Eq. (12.90), the temperature dependence of the molar enthalpy of reaction $\Delta^R H_m$ can be determined graphically by plotting measured values for $\ln K$ over $1/(R_m T)$; a straight line is obtained whose slope has the value - $\Delta^R H_m$. The assumption used here that the molar enthalpy of reaction $\Delta^R H_m$ is constant is generally only valid for smaller temperature intervals $T - T_0$; for larger temperature ranges, the temperature dependence of $\Delta^R H_m$ must be taken into account.

If an integration constant K_0 at temperature T_0 is known, the following applies to it:

$$\ln K_0 = -\frac{\Delta^R H_m^0}{R_m T_0} + \text{const.} \tag{12.91}$$

Subtracting this equation from Eq. (12.90) leads to

$$\ln\left(\frac{K}{K_0}\right) = \frac{\Delta^R H_m^0}{R_m} \frac{T - T_0}{T T_0} \;. \tag{12.92}$$

With the described procedure the course of the straight lines in Fig. 12.16 was determined.

From this it can be seen, for example, that with increasing temperature T according to the reaction equation $H_2 + 0.5\ O_2 \rightleftharpoons H_2O$, the exothermic conversion of hydrogen (H_2) and oxygen (0.5 O_2) to water vapor (H_2O) decreases, i.e. water vapor is decomposed more and more at very high, further increasing temperatures (dissociation, thermolysis). If heat is added to water vapor, the reaction moves in the direction of the 'heat-consuming' dissociation and by this reduces the temperature increase.

For example, for the endothermic reaction of methane (CH_4) with water vapor (H_2O) to carbon monoxide (CO) and hydrogen (3 H_2) the following can be seen: With increasing temperature, increasing amounts of the reaction products (CO) and (3 H_2) are produced; thus, the temperature increase is also reduced here.

With regard to the direction of the shift of an equilibrium due to the change of an influencing variable, the following general law was formulated by *Le Chatelier* and *Braun*[7]:

'If one of the quantities influencing the equilibrium is changed, the equilibrium shifts in such a way that the effect of the change is thereby reduced.'

If, according to Eq. (12.89), the temperature of an equilibrium mixture is increased by the addition of heat, the equilibrium shifts to that side which is associated with a positive heat demand, i.e. in the sense of a reduction of the temperature increase.

The law of *Le Chatelier* and *Braun* is also shown in Eq. (12.85): If the pressure is increased on the equilibrium mixture of a reaction which proceeds with a decrease in the molar quantity ($\nu_a < \nu_e$), then the equilibrium shifts to the right side of the reaction equation, whereby the molar quantity becomes smaller; in this case, the pressure increases less strongly due to the reduction in volume than would be the case without the shift in equilibrium. Such an equilibrium mixture can therefore be compressed with less work than an ideal gas that does not react chemically.

Furthermore, if the molar quantity of any component is increased, the equilibrium shifts in the sense of a consumption of this reactant, i.e. to the side of the reaction equation where this component is not present.

Example 12.8 For the production of hydrogen, carbon monoxide and water vapor is used. What is the temperature range in which a sufficient conversion will take place?

If, in the homogeneous water gas reaction, at a sufficient rate of reaction the educts water vapor (H_2O) and carbon monoxide (CO) are combined in equal molar quantities according to Eq. (12.2), the following equation holds:

$CO + H_2O \rightleftharpoons CO_2 + H_2$ resp. $CO_2 + H_2 - CO - H_2O = 0$

Here, $\sum_{i=1}^{N} \nu_i = 0$ is given. At the beginning of the reaction, here is assumed that there are not yet any reaction products in the reaction chamber. The equilibrium constant results with

---

[7] *Henri Louis Le Chatelier*, 1850 to 1936, was a French chemist, physicist and metallurgist. *Karl Ferdinand Braun*, 1850 to 1918, German physicist and electrical engineer, received the Nobel Prize in Physics in 1909.

the same initial or final pressures of the reactants, e.g. with $p_{CO} = p_{H_2O} = p_{CO_2} = p_{H_2} = 1.01325$ bar as follows:

$$K' = K = \frac{p_{pCO_2} p_{pH_2}}{p_{pCO} p_{pH_2O}}$$

This is an exothermic reaction with
$\Delta^R H_m^0 = -41.586$ MJ/kmol .

The equilibrium constant of this equilibrium is given in Fig. 12.15 and in Table 12.3 as a function of temperature. Eq. (12.14) applies accordingly to the balance of the molar quantities of the reaction:

$$n_{CO} = n_{CO}^0 - z \; ; \qquad n_{H_2O} = n_{H_2O}^0 - z \; ; \qquad n_{H_2} = z \; ; \qquad n_{CO_2} = z \; ;$$

The total molar quantity n does not change before, during and after the reaction, because with decreasing molar amounts of (CO + H_2O) the molar amounts of (CO_2 + H_2) arise to the same extent: $n = n_{CO}^0 + n_{H_2O}^0 = n_{CO} + n_{H_2O} + n_{CO_2} + n_{H_2}$

Thus, using Eq. (4.190) $p_{pi}/p = n_i/n$, the partial pressures can be given as follows:

$$\frac{p_{pCO_2}}{p} = \frac{z}{n_{CO}^0 + n_{H_2O}^0} \qquad p_{pCO_2} = p\frac{z}{n_{CO}^0 + n_{H_2O}^0}$$

$$\frac{p_{pH_2}}{p} = \frac{z}{n_{CO}^0 + n_{H_2O}^0} \qquad p_{pH_2} = p\frac{z}{n_{CO}^0 + n_{H_2O}^0}$$

$$\frac{p_{pCO}}{p} = \frac{n_{CO}^0 - z}{n_{CO}^0 + n_{H_2O}^0} \qquad p_{pH_2} = p\frac{n_{CO}^0 - z}{n_{CO}^0 + n_{H_2O}^0}$$

$$\frac{p_{pH_2O}}{p} = \frac{n_{H_2O}^0 - z}{n_{CO}^0 + n_{H_2O}^0} \qquad p_{pH_2O} = p\frac{n_{H_2O}^0 - z}{n_{CO}^0 + n_{H_2O}^0}$$

For the equilibrium constant $K' = K$ one obtains, according to Eq. (12.84), the relation

$$K = \frac{z^2}{(n_{CO}^0 - z)(n_{H_2O}^0 - z)} \; .$$

The value of the reaction turnover z is obtained by solving this equation for z if K is known. If the reaction is understood as a process for the production of hydrogen, then the quantity α

$$\alpha = \frac{p_{pH_2} + p_{pCO_2}}{p_{pH_2} + p_{pCO_2} + p_{pCO} + p_{pH_2O}}$$

represents the yield of the chemical reaction. α is determined for equal molar quantities of the two reactants $n_{CO}^0 = n_{H_2O}^0$ using the values of Table 12.3 and is shown as a function of temperature T in Fig. 12.17.

At a temperature of approximately 1100 K, the gas mixture in reaction equilibrium consists equally of the reactants $n_{CO} + n_{H_2O}$ and the products $n_{H_2} + n_{CO_2}$. At 300 K, the products almost completely predominate; on the other hand, at a temperature of 2000 K, the fraction of products $n_{H_2} + n_{CO_2}$ is less than one third.

Since hydrogen (H_2) and not carbon monoxide (CO) is often of interest for energy technology processes, the reaction should take place at temperatures that are below about 500 K. Regardless of the temperature, the proportion of combustible gas, i.e. the sum of the CO and H_2 content, is always 50 %, because for every CO molecule decreasing on the left side of the reaction equation, an H_2 molecule is formed on the right side. The gas mixture is heated — if hydrogen formation is the goal — as result of the chemical reaction because the molar enthalpy of reaction has a negative sign.

Table 12.3 Equilibrium constant $K = K'$ of the reaction $CO + H_2O = CO_2 + H_2$ as a function of temperature T [112]

Temp. T in K	298.15	300	400	500	600	700	800	900	1000
$\log_{10} K$	+4.9968	+4.9530	+3.1700	+2.1004	+1.4326	+0.9551	+0.6062	+0.3432	+0.1380
K	99260	89750	1479	126.0	27.08	9.017	4.038	2.204	1.374

Temp. T in K	1100	1200	1300	1400	1500	1750	2000	2500	3000
$\log_{10} K$	- 0.0248	- 0.1570	- 0.2648	- 0.3560	- 0.4313	- 0.5778	- 0.6790	- 0.8083	- 0.8422
K	0.9444	0.6966	0.5435	0.4406	0.3704	0.2644	0.2094	0.1555	0.1438

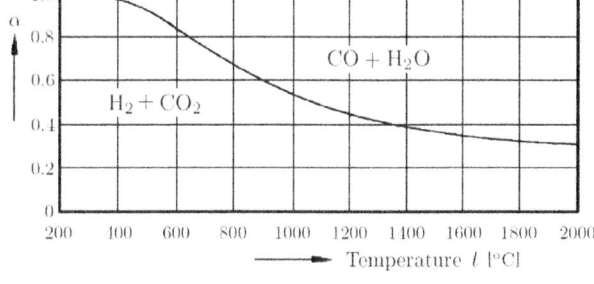

Bild 12.17 Water gas equilibrium as a function of temperature

12.11 Model of Isothermal-Isobaric Reversible Chemical Reactions

12.11.1 Model of Reversible Oxidation of Hydrogen

A reversible chemical reaction represents the ideal and thermodynamically best case of a chemical reaction. It is assumed that the reaction can be reversed without permanent changes (irreversibilities) occurring under inclusion of the environment. If the reaction is to be reversible, there must be no overall devaluation of energy: For example, there must be no additional molar work to be added compared to the reversible molar reaction work, which after the reaction becomes e.g. an exergetically less valuable additional heat compared to the reversible reaction.

The following model of thought by *Jacobus Henricus van't Hoff* makes use of a closed overall system (cf. Figure 12.2 left), whereby, among other things, semi-permeable walls are assumed (cf. e.g. [112]).

It is assumed that all substances involved are ideal gases and that, for example, the reaction $H_2 + 0.5\, O_2 \rightleftharpoons H_2O$ is carried out (Fig. 12.18). If the relation $H_2 + 0.5\, O_2 \rightleftharpoons H_2O$ is used, the thermodynamic quantities required for the description can be given in terms of 1 kmol H_2 and thus in terms of molar quantities. The educts H_2 and O_2 are to be present in pure form in the separately arranged containers a_1 and a_2 at pressures remaining unchanged in each case according to Fig. 12.18; the gaseous reaction product H_2O is likewise to be introduced in pure form into the container a_3 with constant pressure. The pressures p_{H_2}, p_{O_2} and p_{H_2O} indicated by the pistons can, for example, each be of equal magnitude and, for example, have the value $p_1 = p_2 = p_3 = 1.0$ bar - i.e. the pressure in the standard state - and also be equal to the ambient pressure p_L of the outside air. (If the temperature $T_1 = T_2 = T_3$ would be

the temperature of the standard state, the pressure of H_2O as an ideal gas $p_{H_2O} = p_3$ would be fictitious). Three machines to perform work (e.g. piston or fluid machines) are designated with b_1, b_2 and b_3, which are to be used optionally both as compressors and as expansion machines. In order to be able to balance the work occurring during the compression or the expansion of the ideal gases against each other, the machines are connected to each other via a common shaft with which the total molar reversible work $(W_{m,ges})_{rev}$ can be supplied or released.

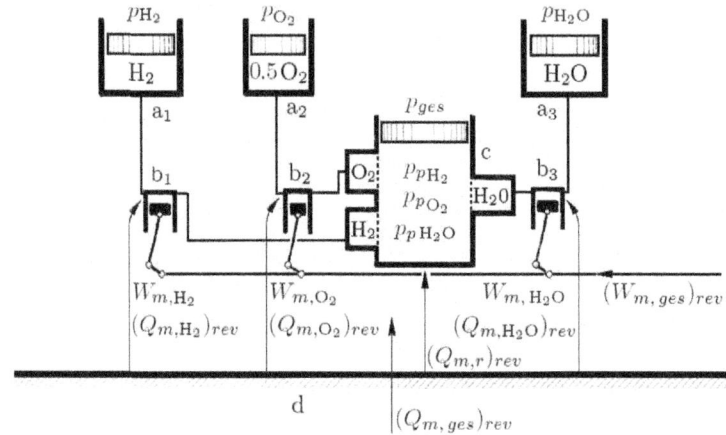

Figure 12.18 System for reversibly carrying out the reaction $H_2 + 0.5\,O_2 \rightleftharpoons H_2O$

The chamber for the chemical reaction c is to have, for example, with the aid of a piston, the always constant pressure p_{ges}, which results as the sum of the respective partial pressures $p_{ges} = p_{p1} + p_{p2} + p_{p3}$ of the reactants contained in the reaction chamber c; the reversible and isothermal reaction is thus also an isobaric reaction.

These pressures p_{ges}, p_{p1}, p_{p2} and p_{p3} are supposed to correspond to the chemical equilibrium between the reactants. The reaction chambeer is connected via three semi-permeable membranes to the three containers a_1, a_2 and a_3 containing the pure gases; the membranes are indicated by dashed lines. The pipelines are shown with solid lines.

The three machines and the reaction chamber c are each connected to the same infinitely large heat reservoir d in ideal heat conduction, which has the constant temperature T; this is to ensure that the processes both in the chemical reaction and in the three machines take place at the same constant temperature T isothermally [112].

For the reversible execution of the reaction, three individual steps are to be considered:
1. 1 kmol H_2 at the pressure p_{H_2} and 0.5 kmol O_2 at the pressure p_{O_2} are taken from the containers a_1 and a_2 by means of the machines b_1 and b_2 and are isothermally brought to the equilibrium pressure $p_{p\,H_2}$ and to the equilibrium pressure $p_{p\,O_2}$, then brought into the two storage chambers upstream of the reaction chamber and then brought into the reaction chamber c through the semi-permeable membranes. For this purpose, the following reversible-isothermal molar works related to 1 kmol H_2 are to be taken into account according to section 4.3.3:

$$W_{m,H_2} + W_{m,O_2} = R_m T \left(1 \ln \frac{p_{p\,H_2}}{p_{H_2}} + 0{,}5 \ln \frac{p_{p\,O_2}}{p_{O_2}}\right) \qquad (12.93)$$

In Eq. (12.93), $m_{H_2} R_{H_2} = n_{H_2} R_m$, $m_{O_2} R_{O_2} = n_{O_2} R_m$, $n_{H_2} = 1$ kmol and $n_{O_2} = 0.5$ kmol are considered.

The oppositely equal molar heat

$$(Q_{m,H_2})_{rev} + (Q_{m,O_2})_{rev} = -R_m T \left(1 \ln \frac{p_{p\,H_2}}{p_{H_2}} + 0, 5 \ln \frac{p_{p\,O_2}}{p_{H_2}} \right) \qquad (12.94)$$

has to be taken into account.

2. In the reaction chamber, the chemical reaction takes place at constant pressure p_{ges}, whereby the molar heat of reaction $(Q_{m,r})_{rev}$ has to be taken into account. Here, no molar work occurs in the reaction chamber itself, because the molar works take place in the upstream machines b_1 and b_2 and, at the same time, the resulting water vapor is removed with the molar work of the downstream machine b_3, as described in the following:

3. 1 kmol of the reaction product H_2O at the partial pressure $p_{p\,H_2O}$ is removed from the reaction chamber via the semi-permeable membrane into the storage chamber and then brought to the pressure p_{H_2O} by the machine b_3; the reaction product is then pushed into the container a_3. The molar work is

$$W_{m,\,H_2O} = R_m T \left(1 \ln \frac{p_{H_2O}}{p_{p\,H_2O}} \right). \qquad (12.95)$$

Here, $m_{H_2O} R_{H_2O} = n_{H_2O} R_m$ and $n_{H_2O} = 1$ kmol are considered.

The oppositely equal molar heat is thereby

$$(Q_{m,H_2O})_{rev} = -R_m T \left(1 \ln \frac{p_{H_2O}}{p_{p\,H_2O}} \right). \qquad (12.96)$$

This leads to the following total reversible molar work and total reversible molar heat:

$$(W_{m,\,ges})_{rev} = R_m T \left(1 \ln \frac{p_{p\,H_2}}{p_{H_2}} + 0, 5 \ln \frac{p_{p\,O_2}}{p_{O_2}} + 1 \ln \frac{p_{H_2O}}{p_{p\,H_2O}} \right) \qquad (12.97)$$

$$(Q_{m,\,ges})_{rev} = -R_m T \left(1 \ln \frac{p_{p\,H_2}}{p_{H_2}} + 0, 5 \ln \frac{p_{p\,O_2}}{p_{O_2}} + 1 \ln \frac{p_{H_2O}}{p_{p\,H_2O}} \right) - (Q_{m,r})_{rev} \qquad (12.98)$$

By transforming one obtains

$$(W_{m,\,ges})_{rev} = R_m T \left(\ln \frac{p_{H_2O}}{p_{H_2} p_{O_2}^{0,5}} - \ln \frac{p_{p\,H_2O}}{p_{p\,H_2} p_{p\,O_2}^{0,5}} \right) \qquad (12.99)$$

$$(Q_{m,\,ges})_{rev} = -R_m T \left(\ln \frac{p_{H_2O}}{p_{H_2} p_{O_2}^{0,5}} - \ln \frac{p_{p\,H_2O}}{p_{p\,H_2} p_{p\,O_2}^{0,5}} \right) - (Q_{m,r})_{rev} \qquad (12.100)$$

In this, the reaction constant K' according to Eq. (12.82) is

$$K' = \ln \frac{p_{p\,H_2O}}{p_{p\,H_2} p_{p\,O_2}^{0,5}}, \qquad (12.101)$$

so that also results

$$(W_{m,ges})_{rev} = R_m T \left(\ln \frac{p_{H_2O}}{p_{H_2}\, p_{O_2}^{0,5}} - \ln K' \right) \qquad (12.102)$$

$$(Q_{m,ges})_{rev} = - R_m T \left(\ln \frac{p_{H_2O}}{p_{H_2}\, p_{O_2}^{0,5}} - \ln K' \right) - (Q_{m,r})_{rev}. \qquad (12.103)$$

If the shift works in correspondence with the ambient air are also considered, in the molar work balance related to 1 kmol H_2 or to 1 kmol H_2O, one has in addition to take into account the — mostly quantitatively minor — expression $R_m T (1 + 0.5 - 1) = 0.5\, R_m T$, which is generally neglected [112].

The total reversible work becomes zero if holds:

$$\ln \frac{p_{H_2O}}{p_{H_2}\, p_{O_2}^{0,5}} = \ln \frac{p_{p\,H_2O}}{p_{p\,H_2}\, p_{p\,O_2}^{0,5}} \qquad (12.104)$$

This equation describes, among other things, the special case where the respective pressures of the individual gases before the reaction are equal to the respective partial pressures of the chemical equilibrium. In this case, no reaction takes place, as its result is already present. The reversible mixing of the individual gases is also not accompanied by a molar work — released or supplied — because the pressures in the storage containers are equal to the partial pressures of the respective gases in the reaction chamber.

In the case of

$$\ln \frac{p_{H_2O}}{p_{H_2}\, p_{O_2}^{0,5}} < \ln \frac{p_{p\,H_2O}}{p_{p\,H_2}\, p_{p\,O_2}^{0,5}} \qquad (12.105)$$

the molar work of the reversible reaction according to Eq. (12.102) is negative, i.e. is given off to the outside; this corresponds to the oxidation of hydrogen to water vapor. If the reaction proceeds irreversibly, it will appear partly or entirely as molar heat, and the reaction is exothermic. At

$$\ln \frac{p_{H_2O}}{p_{H_2}\, p_{O_2}^{0,5}} > \ln \frac{p_{p\,H_2O}}{p_{p\,H_2}\, p_{p\,O_2}^{0,5}} \qquad (12.106)$$

the reaction cannot proceed by itself, as molar work must be supplied. It is an endothermic reaction, i.e. the splitting of water vapor into oxygen and hydrogen.

The equilibrium constant K' has the very high value $K' = 1.114 \cdot 10^{40}$ at 25 °C. This matches in the case of $p_{ges} \approx p_{H_2O} \approx p_{p\,H_2O} \approx 1.0$ bar with the partial pressures of $p_{p\,H_2} = 2.50 \cdot 10^{-27}$ bar and $p_{p\,O_2} = 0.5\, p_{p\,H_2} = 1.25 \cdot 10^{-27}$ bar. Accordingly, of the approximately $6 \cdot 10^{26}$ molecules of a kilomole of water vapor, only about one molecule is dissociated. At 1300 K, K' reaches the value $K' = 1.158 \cdot 10^7$ and the partial pressures the values $p_{p\,H_2} = 2.43 \cdot 10^{-5}$ bar as well as $p_{p\,O_2} = 1.215 \cdot 10^{-5}$ bar.

The molar enthalpy of reaction at 25 °C is $\Delta^R H_{m,0} = -\,241.818$ MJ/kmol. The reversible molar work, according to Eq. (12.99) in the case of equal pressures of the educts and the final product $p_{H_2} = p_{O_2} = p_{H_2O} = 1.0$ bar and thus $\ln \left(\frac{p_{H_2O}}{p_{H_2} \cdot p_{O_2}^{0.5}} \right) = 0$, results in

$$(W_{m,\,ges})_{rev} = -R_m T \ln \left(\frac{p_{p\mathrm{H_2O}}}{p_{p\mathrm{H_2}} \, p_{p\mathrm{O_2}}^{0.5}} \right) = -R_m T \ln K'$$

$$= -8.31451 \,\mathrm{kJ/(kmol\,K)} \cdot 298.15\,\mathrm{K} \cdot \ln(1.11 \cdot 10^{40}) = -228.58\,\mathrm{MJ/kmol} \,. \quad (12.107)$$

This value is also obtained with Eqs. (12.32) and (12.45) and the numerical values of Table A.7 in the appendix:

$$(W_m)_{rev} = \Delta^R G_m^0 (T_0,\,p_0) = \Delta^R H_m^0 (T_0,\,p_0) - T_0 \Delta^R S_m^0 (T_0,\,p_0)$$

$$= (-241.83 - 1 \cdot 0 - 0.5 \cdot 0)\,\mathrm{MJ/kmol}$$

$$- 298.15\,K \cdot (188.835 - 1 \cdot 130.680 - 0.5 \cdot 205.152)\,\mathrm{kJ/(kmol\,K)}$$

$$= -228.58\,\mathrm{MJ/kmol} \,. \quad (12.107\,\mathrm{a})$$

Only -13.24 MJ/kmol are to be released as heat to the environment. This can be explained by the fact that the educts H_2 and O_2, according to the *van 't Hoff* model, are released from their storage tanks from the pressure of 1.0 bar to the very small pressures of $p_{p\,H_2} = 2.50 \cdot 10^{-27}$ bar and $p_{p\,O_2} = 1.25 \cdot 10^{-27}$ bar isothermally and hereupon enter the reaction chamber through the semi-permeable membranes without any work (cf. [112]).

12.11.2 Model of Arbitrary Homogeneous Reversible Chemical Reactions of Ideal Gases

For a general reversible chemical reaction according to Eq. (12.11), which has, for example, two educts 1 and 2 and two reaction products 3 and 4, holds:

$$\nu_3 A_3 + \nu_4 A_4 - |\nu_1| A_1 - |\nu_2| A_2 = 0 \quad (12.11)$$

The following are the general equations [112]

$$K' = \ln \frac{p_{p3}^{\nu_3} \, p_{p4}^{\nu_4}}{p_{p1}^{|\nu_1|} \, p_{p2}^{|\nu_2|}} \,, \quad (12,108)$$

$$(W_{m,\,ges})_{rev} = R_m T \left(\ln \frac{p_3^{\nu_3} \, p_4^{\nu_4}}{p_1^{|\nu_1|} \, p_2^{|\nu_2|}} - \ln K' \right) , \quad (12.109)$$

$$(Q_{m,\,ges})_{rev} = -R_m T \left(\ln \frac{p_3^{\nu_3} \, p_4^{\nu_4}}{p_1^{|\nu_1|} \, p_2^{|\nu_2|}} - \ln K' \right) - (Q_{m,\,r})_{rev} \,. \quad (12.110)$$

In the case of

$$\ln \frac{p_3^{\nu_3} \, p_4^{\nu_4}}{p_1^{|\nu_1|} \, p_2^{|\nu_2|}} < \ln \frac{p_{p3}^{\nu_3} \, p_{p4}^{\nu_4}}{p_{p1}^{|\nu_1|} \, p_{p2}^{|\nu_2|}} \quad (12.111)$$

the molar work requirement is negative; the reaction can therefore release molar work to the outside. To the extent that all or part of this useful molar work is converted into molar heat, the reaction is irreversible. If

$$\ln \frac{p_3^{\nu_3} \, p_4^{\nu_4}}{p_1^{|\nu_1|} \, p_2^{|\nu_2|}} > \ln \frac{p_{p3}^{\nu_3} \, p_{p4}^{\nu_4}}{p_{p1}^{|\nu_1|} \, p_{p2}^{|\nu_2|}} \quad (12.112)$$

holds, the molar work requirement is positive; molar work must be supplied to the reaction from outside for it to proceed.

By

$$K(T, p) = \Pi \left(\frac{p_{pi}}{p}\right)^{\nu_i} \qquad (12.113)$$

becomes from Eq. (12.109)

$$(W_{m,ges})_{rev} = -R_m T \ln \left[\Pi \left(\frac{p_{pi}}{p}\right)^{\nu_i}\right]. \qquad (12.114)$$

Eq. (12.77) can also be used with Eq. (4.190) and with $\sum_{i=1}^{N} = \Sigma$ can be written in the form

$$\ln \left[\Pi \left(\frac{n_i}{n}\right)^{\nu_i}\right] = \ln \left[\Pi \left(\frac{p_{pi}}{p}\right)^{\nu_i}\right] = -\frac{\Sigma \nu_i G^0_{m,i}}{R_m T} = -\frac{\Delta^R G^0_m(T)}{R_m T} \qquad (12.115)$$

A further transformation gives

$$\Delta^R G^0_m(T) = -R_m T \ln \left[\Pi \left(\frac{p_{pi}}{p}\right)^{\nu_i}\right]. \qquad (12.116)$$

The right-hand sides of Eqs. (12.114) and (12.116) coincide. Thus

$$(W_{m,ges})_{rev} = \Delta^R G^0_m(T) \qquad (12.117)$$

holds. Eq. (12.117) applies like Eq. (12.114) as well as Eq. (12.116) for a reversible isothermal-isobaric chemical reaction of ideal gases. Eq. (12.117) has been derived here for the special case of ideal gases and agrees with Eq. (12.45), which is generally valid for a reversible isothermal-isobaric chemical reaction. Thus, the *van 't Hoff* reaction model for reversible isothermal-isobaric chemical reactions of ideal gases is confirmed.

12.11.3 Reversible Storage of Heat and Work in the Form of Chemical Energy

Fig. 12.19 is intended to illustrate that a chemical reaction carried out isothermally, isobarically and reversibly can be used to carry out the storage of heat and work without losses in the latent form of chemically bound energy. An example of this is the decomposition of the molar quantity of 1 kmol water vapor (H_2O) into 1 kmol hydrogen (H_2) and 0.5 kmol oxygen (O_2).

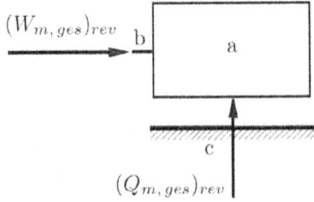

Bild 12.19 Arrangement for the reversible execution of arbitrary chemical reactions

12.11 Model of Isothermal-Isobaric Reversible Chemical Reactions

For this purpose, the system consisting of containers, reaction chamber and machines as shown in Fig. 12.18 shall be installed in the arrangement a of Fig. 12.19. With the shaft b a molar work $(W_{m,ges})_{rev}$ can be released or added to the arrangement. Further, the arrangement a is in perfect thermal connection with the large heat reservoir c of constant temperature T, from which the arrangement a can reversibly extract or to which the arrangement a can reversibly supply the molar heat $(Q_{m,ges})_{rev}$.

Insofar as the chemical reaction is characterised by a positive demand for molar work and molar heat, these process variables are stored as chemically bound molar energy, where the sum of the molar enthalpies of the reaction products is greater than the sum of the molar enthalpies of the educts by the molar enthalpy of reaction $\Delta_R H$:

$$(W_{m,ges})_{rev} + (Q_{m,ges})_{rev} = \Delta_R H \tag{12.118}$$

The molar heat input leads at the same time to the molar entropy increase $\Delta_R S = (Q_{m,ges})_{rev}/T$, which is associated with the molar entropy decrease of the same amount $\Delta S_{m,Beh} = -(Q_{m,ges})_{rev}/T$ of the heat reservoir c, because in the reversible process the sum of the molar entropies remains unchanged.

A chemical reaction can also proceed in such a way that a molar work input is associated with a molar heat removal or a molar work removal is associated with a molar heat input.

The same lossless storage of molar heat and molar work can also be realised without a chemical reaction if, for example, the molar quantity 1 kmol of gas is removed from a container a_1 with the constant pressure p_1, reversibly isothermally compressed to the pressure p_2 and stored in a second container a_2. Here, analogous to section 4.3.3, the reversible molar work

$$(W_{m,ges})_{rev} = (W_{m12})_{rev} = R_m T (\ln p_2 - \ln p_1) = R_m T \ln \frac{p_2}{p_1} \tag{12.119}$$

will be supplied and the molar work

$$(Q_{m,ges})_{rev} = (Q_{m,12})_{rev} = -R_m T (\ln p_2 - \ln p_1) = -R_m T \ln \frac{p_2}{p_1} \tag{12.120}$$

will be released. Equations (12.119) and (12.120) have the same structure as Eqs. (12.109) and (12.110). Here, ordinary pressures have replaced the power number products of the type of equilibrium constants K'; further, the molar heat of reaction $(Q_{m,r})_{rev}$ is absent, so that the molar work is equal to the absolute value of the molar heat which occurs [112].

However, the arrangement to store heat and work without loss according to Fig. 12.19 can also be interpreted that the molar heat of reaction $(Q_{m,r})_{rev}$ is considered. This can be done by a process of isothermal vaporisation of 1 kmol of a pure substance in a closed system according to Fig. 12.2 a) which is also an isobaric vaporisation. Thereby, the molar heat $(Q_{m,r})_{rev}$ is supplied without a molar pressure change work; the energy of the molar heat is stored in the vapor [112].

By linking isothermal vaporisation to isothermal compression, a complete thermal-mechanical analogue can be formed for the isothermal-isobaric chemical reaction, as Fig. 12.20 shows: Here a_1 and a_2 are two gas containers whose gases are under pressures

p_1 and p_2, respectively. b is an isothermally operating compressor, from which the molar heat of compression $(Q_{m,12})_{rev}$ is released at the same time when the molar compressor work $(W_{m,ges})_{rev} = (W_{m,12})_{rev}$ is supplied isothermally.

The molar heat of compression $(Q_{m,12})_{rev}$ is released isothermally in whole or in part to the vaporisation container c, which is reversibly connected to it in a heat-conducting manner. The container contains a suitable liquid which is in thermodynamic equilibrium with its vapor at the temperature T (wet vapor). By raising the piston in the vaporisation container more or less strongly, the molar heat from the compressor cylinder b $(Q_{m,12})_{rev}$ can be completely or partially bound. In addition, in this way the amount of molar heat $(Q_{m,ges})_{rev}$ that is to be supplied or to be absorbed by the heat reservoir d can be changed.

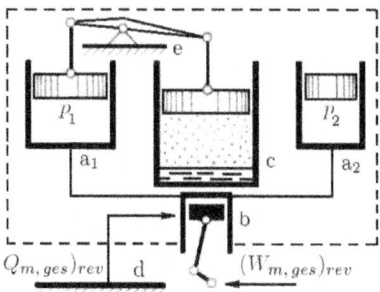

Figure 12.20 Thermal-mechanical model of a reversible chemical reaction: compression or expansion of a pure substance in the cylinder b and vaporisation or condensation of another pure substance in the container c as a form of a simplified reversible chemical reaction

A desired ratio between $(Q_{m,ges})_{rev}$ and $(W_{m,ges})_{rev}$ can be achieved by bringing the displacement of the piston in the vaporisator container c and the displacement of one of the two pistons in the gas container a_1 or a_2 into a certain ratio; this is indicated in Fig. 12.20 by the lever mechanism.

If the arrangement is surrounded by a system boundary, it represents a closed system which is connected to the environment only by the reversible molar heat $(Q_{m,ges})_{rev}$ and by the reversible molar work $(W_{m,ges})_{rev}$. Thus, the arrangement according to Fig. 12.20 corresponds to the arrangement for a chemical reaction according to Fig. 12.19 (cf. [112]).

12.12 Fuel Cells

In fuel cells, the chemical energy of a fuel is converted directly into electrical energy by means of an electrochemical process. This differs fundamentally from a machine which converts a part of the thermal energy — generated by combustion at higher temperatures — into mechanical energy.

In the fuel cells available today, the chemical energy of hydrogen (H_2) is mostly used. The summary electrochemical reaction takes place at the electrodes of the fuel cell; in the process, primarily the hydrogen component (H_2) of the fuel and the oxygen component (O_2) of the air are converted according to the reaction equation

$$H_2 + 0.5\,O_2 \rightleftharpoons H_2O \tag{12.4}$$

(cf. Fig. 12.21). In general, hydrogen (H$_2$) must first be obtained for this in an upstream process — the reforming process. The product of the reaction of hydrogen with oxygen is water vapor (H$_2$O); an electrolyte enables the conduction of ions, while electrons e$^-$ (electricity) are not conducted in it and are thus forced to take a different path through a consumer of electricity. The process can be seen as the reverse electrolysis of water.

The conversion of the fuel and the extraction of electrical work is possible with high efficiency because it is not limited by the ideal thermal efficiency of the Carnot process (cf. section 7.3.3) which indicates the upper limit of the efficiency of reversibly delivering mechanical work (and subsequently electrical work) via a combustion process with heat release at higher temperatures (cf. Fig. 12.22).

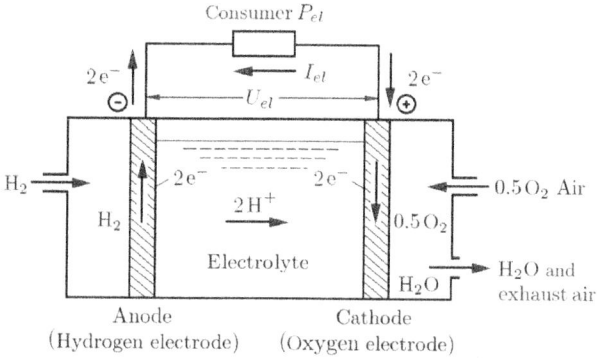

Figure 12.21 Fuel cell with polymer membrane (PEMFC)

Fuel cells have two electrodes: the anode, where the fuel is supplied, and the cathode, which is used to take up the oxygen contained in the air. The electrolyte — which conducts ions — is located between these electrodes. Different electrolytes are characteristic of the fuel cell types that are currently used and could be of importance in the future:

- Alkaline liquid electrolytes (fuel cell type: Alkaline Fuel Cell AFC)

- Acidic liquid electrolytes (common fuel cell type: Phosphoric Acid Fuel Cell (PAFC); further type: Direct Methanol Fuel Cell (DMFC))

- Molten carbonate electrolytes (fuel cell type: Molten Carbonate Fuel Cell (MCFC))

- Solid oxide electrolytes such as zirconium (fuel cell type: Solid Oxide Fuel Cell (SOFC))

- Plastic polymer membranes (fuel cell type: Proton Exchange Membrane Fuel Cell PEMFC) that can conduct protons

The gaseous hydrogen (H$_2$) is split into protons (H$^+$) and electrons (e$^-$) by the action of a catalyst:

$$H_2 \rightleftharpoons 2\,H^+ + 2\,e^- \quad (12.121)$$

For example, in the PEM fuel cell (Fig. 12.21), the protons (H$^+$) go through the electrolyte to the cathode, where they react — again with the assistance of a catalyst

— with the oxygen (O_2) supplied and the electrons (e^-) coming from the external circuit in the following manner:

$$0.5\,O_2 + 2\,H^+ + 2\,e^- \rightleftharpoons H_2O \tag{12.122}$$

The electrochemical process takes place at the level of individual cells, which, however, have only a low electrical cell voltage U_{el} as well as a low current I_{el} and therefore only a limited electrical power $P_{el} = -U_{el}\,I_{el}$; they are therefore combined into a stack to produce the desired voltage and power. The direct current obtained is converted into alternating current before it can be used directly or fed into the public grid. The cell voltage U_{el} decreases with increasing current I_{el}.

The power balance according to the first law gives according to Eq. (12.18)

$$\dot{Q} + P = \sum_{i=1}^{N} \dot{n}_i^a\, H_{m,0i}(T_a, p_a) - \sum_{i=1}^{N} \dot{n}_i^e\, H_{m,0i}(T_e, p_e) \;. \tag{12.18}$$

In the case of $T_a = T_e$ and $p_a = p_e$, using the reaction enthalpy $\Delta^R H_m(T)$ as the negative molar lower calorific value of hydrogen $H_{i\,m\,H_2}(T)$ (Table 12.2), considering the reaction rate $\dot{z} = \dot{n}_{H_2}$ and Eq. (12.20 a), one gets

$$\dot{Q} + P = \dot{n}_{H_2}\, \Delta^R H_m(T) = -\dot{n}_{H_2}\, H_{i\,m\,H_2}(T) \;. \tag{12.123}$$

In the reversible isothermal-isobaric reaction, the fuel cell simultaneously emits the reversible electrical power P_{rev} and the thermal power $\dot{Q}$; it is thus a cogeneration plant. The reversible electrical power is

$$P_{rev} = \dot{n}_{H_2}\, \Delta^R G_m(T, p) = -U_{rev}\, I \tag{12.124}$$

with the molar reaction *Gibbs* function $\Delta^R G_m(T, p)$ according to Eq. (12.45).

The ideal efficiency η_{id} of a fuel cell assuming a reversible function can be defined as the ratio of the molar reaction *Gibbs* function $\Delta^R G_m(T, p)$ to the negative molar lower calorific value $\Delta^R H_m(T) = -H_{i\,m\,H_2}(T)$:

$$\eta_{id} = \frac{\Delta^R G_m(T, p)}{\Delta^R H_m(T)} = -\frac{\Delta^R G_m(T, p)}{H_{i\,m\,H_2}(T)} \tag{12.125}$$

For an ideally operating fuel cell with hydrogen (H_2) as the fuel, the ideal efficiency η_{id} with the values of Table 12.4 as a function of Celsius temperature t is shown in Fig. 12.22.

The real efficiency η_{BZ} of the fuel cell, in which the influence of irreversible processes is also taken into account, is defined by the ratio of the actually recoverable molar reaction work $W_{BZm}(T)$ to the reversible molar reaction work $(W_{BZm})_{rev}(T) = \Delta^R G_m(T, p)$:

$$\eta_{BZ} = \frac{W_{BZm}(T)}{\Delta^R G_m(T, p)} \tag{12.126}$$

In addition, a definition is also possible via the ratio of the actual power P_{BZ} to the reversible power $(P_{BZ})_{rev}$ as well as via the ratio of the actually generated electric voltage U_{el} to the reversible electric voltage $(U_{el})_{rev}$ (so-called voltage efficiency)

$$\eta_{BZ} = \frac{W_{BZm}(T)}{\Delta^R G_m(T, p)} = \frac{P_{BZ}}{(P_{BZ})_{rev}} = \frac{U_{el}}{(U_{el})_{rev}} \tag{12.127}$$

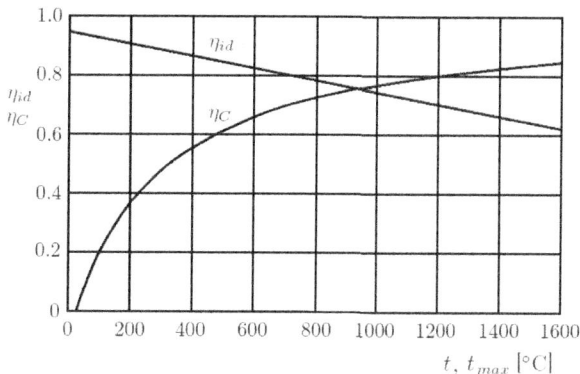

Figure 12.22 Temperature dependence of the thermal efficiency of the reversible *Carnot* process η_C; temperature dependence of the ideal fuel cell efficiency η_{id} based on the reaction equation $H_2 + 0.5\, O_2 \rightleftharpoons H_2O$

Thus, the overall efficiency of the fuel cell is:

$$\eta_{BZ\,ges} = \eta_{id}\,\eta_{BZ} = \frac{W_{BZm}(T)}{\Delta^R H_m(T)} = -\frac{W_{BZm}(T)}{H_{i\,m\,H_2}(T)} \qquad (12.128)$$

This does not yet take into account the energy demand of the additional techniques to serve the fuel cell.

Table 12.4 Molar enthalpy of reaction $\Delta^R H_m$, molar free enthalpy of reaction (molar reaction *Gibbs* function) $\Delta^R G_m$, standard molar entropy S_m at $p = 1.0$ bar [24] and logarithm to base 10 of the equilibrium constant K' of the reaction $H_2 + 0.5\, O_2 \rightleftharpoons H_2O$ [112]

Temp. T in K	298,15	300	400	500	600	700	800	900	1000	1100	1200
$\Delta^R H_m$	-241.81	-241.84	-242.85	-243.83	-244.76	-245.63	-246.44	-247.19	-247.86	-248.46	-249.00
$\Delta^R G_m$	-228.58	-228.50	-223.90	-219.05	-214.01	-208.81	-203.50	-198.08	-192.59	-187.03	-181.43
S_m	188.83	189.04	198.79	206.53	213.05	218.74	223.83	228.46	232.74	236.73	240.49
$\log_{10} K'$	40.047	39.787	29.240	22.886	18.632	15.583	13.289	11.498	10.061	8.883	7.899

The upstream reforming process (cf. Fig. 12.23) is easiest to carry out for natural gas as a fuel; it is more complex for oil and especially for coal.

Taking methane (CH_4) - the main component of natural gas - as an example, the addition of water vapor (H_2O) and heat results in the endothermic reaction

$$CH_4 + H_2O \rightleftharpoons CO + 3\,H_2 \; . \qquad (12.1)$$

The required heat is provided by the combustion of part of the CH_4 to water vapor (H_2O) and carbon dioxide (CO_2) in the case of an external reforming or fed from the heat of reaction released during the downstream electrochemical oxidation of the hydrogen in the case of an internal reforming. In a second, exothermic step — the carbon monoxide conversion reaction (also called homogeneous water gas reaction; cf. example 12.8) — the resulting carbon monoxide (CO) is converted into hydrogen (H_2) with water vapor (H_2O):

$$CO + H_2O \rightleftharpoons CO_2 + H_2 \qquad (12.2)$$

Thus, the sum of both reaction equations yields only hydrogen (H_2) as the fuel:

$$CH_4 + 2\,H_2O \rightleftharpoons 4\,H_2 + CO_2 \tag{12.129}$$

These processes are generally carried out at higher temperatures. In some fuel cell types, they are integrated into the fuel cell system to simplify the process: An 'internal reforming' refers to those cases where the two reactions of Eqs. (12.1) and (12.2) at the fuel cell electrodes almost simultaneously with the electrochemical reactions of Eqs. (12.121) and (12.122) take place. In addition, there are processes with 'indirect internal reforming', 'partial internal reforming' and 'external reforming' (Fig. 12.23).

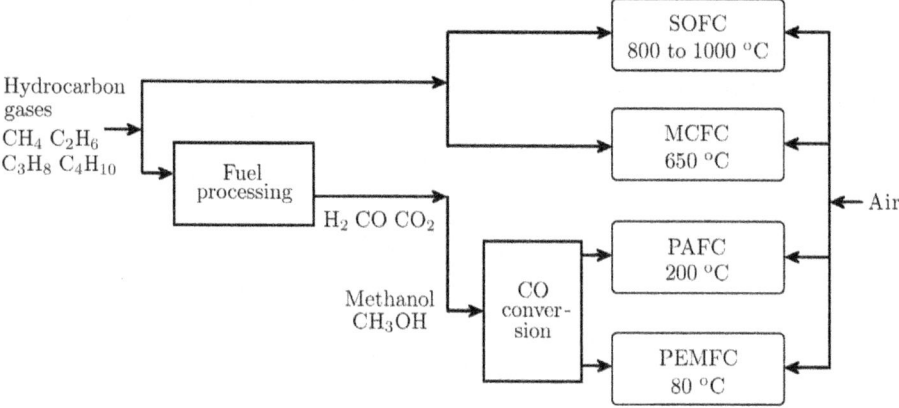

Figure 12.23 Fuel cell designs with external and internal fuel processing (reforming)

Pressure and temperature are significant for the velocity and the extent of electrochemical reactions. Higher temperatures reduce the efficiency but have a positive effect on the velocity of the reactions and thus on fuel cell performance. Higher pressures also have a positive influence on fuel cell performance.

Fuel cell systems usually consist of the sub-sections fuel treatment (reforming and conversion), power generation in the fuel cell stacks, electricity conversion, and heat recovery and/or additional power generation in an integrated gas and/or steam turbine process.

Appendix

Results of the Tasks

Section 1 (page 8)
1. ethine: $n = 0.13126$ kmol   oxygen: $n = 0.32813$ kmol
2. $\dot{m} = 1.843$ kg/min $\tau = 2$ h
3. $T = 111.65$ K

Section 2 (pages 41 to 42)
1. Eqs. (2.96) and (2.98), Table 2.6: $c = 4.1823$ kJ/(kg K)

Eq. (2.99): $\dot{W}_e = |\dot{W}_{RI}| = 1102.45$ kW

2. Eqs. (2.96) and (2.98), Table 2.6: $c = 4.1815$ kJ/(kg K)

$\tau = \dfrac{m\,c\,(t_2 - t_1)}{\dot{Q}} = 1.3855$ h

3. water: $\Delta T = 0.468$ K   mercury: $\Delta T = 14.084$ K
4. $\dot{V} = 56.549$ m^3/h $\dot{m} = 56\,549$ kg/h $P_e = 33.8910$ kW $(P_e)_{id} = 28.5537$ kW
5. $c_{pm}|_{50\,°C}^{600\,°C} = 0.9992$ kJ/(kg K) $\dot{Q}_{12} = 5495.6$ kW

$c_{pm}|_{50\,°C}^{700\,°C} = 1.0113$ kJ/(kg K) $\dot{Q}_{12} = 6573.8$ kW

6. $\dot{Q}_{12} = 16.598$ kW   2 heat exchangers

Section 3 (page 68)
1. $m\,g\,\Delta z = m_{Hg}\,c\,\Delta t$ $\Delta t = 1.06$ K

Eqs. (3.28) and (3.81): $S_2 - S_1 = m_{Hg}\,c\,\ln\dfrac{T_2}{T_1} = 0.256$ J/K

2. heat given off by the heat source 19 440 kJ

entropy change of the heat source -41.09 kJ/K

entropy change of the water $m\,c\,\ln\dfrac{T_2}{T_1} = 48.72$ kJ/K

heat absorbed by the surroundings 4477 kJ

entropy change of the surroundings 15.38 kJ/K

total entropy change 23.01 kJ/K

3. Eq. (3.77): $dS = \dfrac{dH - V\,dp}{T}$ $dH = 0$, $V = $ const : $dS = -\dfrac{V}{T}dp$

$\dot{S}_2 - \dot{S}_1 = \dfrac{\dot{V}}{T}(p_1 - p_2) = 4.419$ kJ/(h K)

4. $\left(\dfrac{\partial^2 z}{\partial x\,\partial y}\right) = \left(\dfrac{\partial^2 z}{\partial y\,\partial x}\right) = 30\,x\,y^4$ dz : total differential

$\left(\dfrac{\partial^2 z}{\partial x\,\partial y}\right) = \left(\dfrac{\partial^2 z}{\partial y\,\partial x}\right) = 24\,x^3\,y^5$ dz : total differential

5. $\Delta\dot{S} = 1.142$ W/K

Section 4 (pages 147 to 150)
1. Eq. (4.25): $p = 47.61$ bar

2. Eq. (4.36): $R = 518.68$ J/(kg K)   Eq. (4.25): $m = 31.91$ kg   price 14.28 Euro
a) price per kilogram 0.447 Euro/kg   b) $m = 29.19$ kg
price per kilogram 0.489 Euro/kg   price difference 0.042 Euro/kg

3. Gl. (2.4): $F = G - A = 0$ (hover state)   Gl. (2.1): $G = (V \varrho_{He} + m) g$
$m = 1350$ kg   Eq. (2.3): $A = V \varrho_L g$ with Eqs. (4.27) and (4.36): $V = \dfrac{m T R_m}{p (M_L - M_{He})}$
$V = 2456.57$ m^3 a) $d = 16.74$ m   b) Eq.(4.25): $m_{He} = 216.5$ kg
c) Eq. (4.96): $V_{St} = 1345.69$ m^3 d) $A = \dfrac{m M_L g}{M_L - M_{He}} = 15\,362$ N

In the equation for the static buoyancy A in the hover state, the properties p and T dependent from the height, are not found. Therefore, in the hover state the static buoyancy is not dependent from the height and is constant.

4. Eq. (2.96): $c_{pm}|_{20\,°C}^{150\,°C} = 0.9005$ kJ/(kg K)   Eq. (4.25): $\dot{m} = 1.4039$ kg/s
a) Eq. (3.82): $(\dot{Q}_{12})_{rev} = 164.35$ kW $= 0.92\,P_{el}$ $P_{el} = 178.6$ kW   b) $t_2 = 161.3\,°C$

5. a) Eq. (4.144): $n = 1.2351$ b) volume of the air content 21.991 cm^3
Eq. (4.116): air volume in the height 29.628 cm^3
changed height in one dose 6.7364 mm   adjustment path 24.31 mm
c) Eq. (4.25): air mass in the carburettor can 0.02693 g
Eqs. (2.96), (4.44) and (4.137): $c_p = 1.0034$ kJ/(kg K) $\kappa = 1.4007$
$c_n = -0.5048$ kJ/(kg K)   Eq. (2.99): heat 0.2651 J

6. Eq. (4.25): $m_1 = 5.4462$ kg   a) Eq. (4.66): $t_2 = 486.52\,°C$
Eqs. (2.96) and (2.98): $c_v = 0.8303$ kJ/(kg K)   Eq. (3.83): $(Q_{12})_{rev} = 2109.6$ kJ
b) Eq. (4.25): $m_3 = 4.7384$ kg at $600\,°C$ $\Delta m = 0.7078$ kg   c) Eqs. (2.96) and (2.98):
$c_p = 1.1745$ kJ/(kg K)   example 4.13: $(Q_{23})_{rev} = \dfrac{p_2 V}{R} c_p \ln \dfrac{T_3}{T_2} = 676.5$ kJ
d) Eq. (4.66): gage pressure 3.347 bar   e) Eqs. (2.96) and (2.98): $c_v = 0.8607$ kJ/(kg K)
Eq. (3.83): $(Q_{34})_{rev} = 2365.3$ kJ

7. a) isentrope $(F + p_1 A) x = \dfrac{p_1 V_1}{\kappa - 1} \left[\left(\dfrac{V_1}{V_2} \right)^{\kappa-1} - 1 \right]$ $\dfrac{V_1}{V_2} = \varepsilon$ $V_2 = V_1 - A x$
$F = \Delta p\,A$ $A = 0.008847$ m^2 $\Delta p = 15.298$ bar $p_1 + \Delta p = p_3$ $\kappa = 1.4$
$\dfrac{p_3}{p_1} = \dfrac{\varepsilon^\kappa - \varepsilon}{\varepsilon - 1}$ $p_1 = 6.45$ bar $p_3 = 21.748$ bar   iteration: $\varepsilon = 6.7751$
b) Eq. (4.94): $p_2 = p_1\,\varepsilon^\kappa = 93.937$ bar   c) $V_2 = \dfrac{V_1}{\kappa} = 0.32472$ ℓ
$x = \dfrac{V_1 - V_2}{A} = 212.5$ mm   d) Eq. (4.98): $t_2 = 357.02\,°C$
e) Eq. (4.94): $V_3 = 0.92338$ ℓ $\bar{x} = \dfrac{V_1 - V_3}{A} = 144.7$ mm

8. a) Eq. (4.27) and (4.182): $\varrho_0 = 1.2922$ kg/m^3 Eqs. (4.213) and (4.24): $R = 287.07$ J/(kg K)
b) Eqs. (4.27) and (4.182): $\varrho_{N_2} = 1.2562$ kg/m^3 (correct value 1.2498 kg/m^3)

9. Eqs. (4.27), (4.36) and (4.182): exhaust gas densitiy in the physical norm state
$\varrho_0 = 1.2150$ kg/m^3
Eq. (4.22): exhaust gas volume flow in the physical norm state $\dot{V}_0 = 0.0016004$ m^3/s
Eq. (4.207): CO volume flow in the physical norm state $\dot{V}_{0\,CO} = 0.000080018$ m^3/s
a) absorbed CO volume flow $\dot{V}_{0\,CO_{abs}}$

$$\frac{1.5}{100} = \frac{\dot{V}_{0\,CO} - \dot{V}_{0\,CO\,abs}}{\dot{V}_0 - \dot{V}_{0\,CO\,abs}} \qquad \dot{V}_{0\,CO\,abs} = \frac{\dot{V}_{0\,CO} - 0.015\,\dot{V}_0}{1 - 0.015} = 0.000056866 \text{ m}^3/\text{s}$$

Eq. (4.25): absorbed CO mass flow 0.071064 g/s

b) 76.726 % N_2 9.332 % H_2O 7.258 % CO_2 1.500 % CO   4.147 % H_2 1.037 % O_2

10. volume flow of the moist exhaust gas 0.33674 m³/s

Eq. (4.96): volume flow of the moist exhaust gas in the physical norm state
($T_0 = 273.15$ K, $p_0 = 1.01325$ bar) 0.17671 m³/s   water vapor flow 9.3658 g/s

Eq. (4.27): density of the water vapor in the physical norm state 0.80375 kg/m³

Eq. (4.22): volume flow of the water vapor in the physical norm state 0.011653 m³/s

Eq. (4.20): volume fraction of the water vapor 6.594 %

Eq. (4.201): volume fraction of the dry exhaust gas 93.406 %

Eqs. (4.27) and (4.182): density of the dry exhaust gas in the physical norm state 1.3408 kg/m³

Eq. (4.182): density of the moist exhaust gas in the physical norm state 1.3054 kg/m³

Eq. (4.22): mass flow of the moist exhaust gas 0.23068 kg/s

11. components of the coke gas in the norm state   38.25 m³/h CO_2 130.5 m³/h CO
381 m³/h H_2 129 m³/h CH_4 71.25 m³/h N_2

result according to the first equation of reaction 76.5 m³/h CO   76.5 m³/h H_2

result according to the second equation of reaction 90.75 m³/h CO   181.5 m³/h H_2

a) 45.375 m³/h O_2 + 170.696 m³/h N_2 = 216.071 m³/h air

b) 297.75 m³/h CO + 639 m³/h H_2 + 241.95 m³/h N_2 = 1178.7 m³/h endo gas

c) 25.26 % CO   54.21 % H_2 20.53 % N_2

12. reduced pressure in the vessel 0.27 bar   Eq. (4.25): air $m_L = 0.32085$ kg

Eqs. (4.204), (4.36) and (4.25): water gas $M_W = 16.073$ kg/kmol $R_W = 517.29$ J/(kg K)
$m_W = 0.52755$ kg

a) total mass $m = 0.84840$ kg

b) Eq. (4.203): 36.635 % H_2 32.150 % CO   3.738 % CO_2 22.178 % N_2 5.299 % O_2

volume fraction of air 25.234%   volume fraction of water gas 74.766%

13. heat release of the exhaust gas $(Q_{12})_{rev} = -1.2375$ kW

heat supplied to the false air $(\dot{Q}_{32})_{rev} = 0.3797$ kW

heat supplied to the combustion air $(\dot{Q}_{34})_{rev} = 0.8578$ kW

final temperature of the preheated air 267.39 °C

14. $m = 0.0694$ kg $n = 2.396$ mol $p_2 = 2.932$ bar

15. $p_2 = 1.583$ bar $G = 497.16$ N   balance of forces: $p_3 A + G = p_1 A$

$p_3 = 2\,p_1 - p_2 = 0.317$ bar $h_3 = 750$ mm $N = 4.609 \cdot 10^{22}$ molecules

16. $A_1 = 2897.1$ kN $A_2 = 2768.7$ kN $\Delta A = -128.4$ kN

17. a) $\Delta V = -490.87$ m³ $W_{VL} = 49087$ kJ

b) $G_D = 787\,800$ N $p_D = 0.01599$ bar $p_{ges} = 1.01599$ bar

c) $R = 0.51826$ kJ/(kg K) $\kappa = 1.31624$

d) $m_1 = 23375$ kg $T_1 = 284.034$ K

e) $U_2 - U_1 = -157625$ kJ $H_2 - H_1 = -207473$ kJ $Q_{12} = -207473$ kJ

18. a) $R = 0.29345$ kJ/(kg K) $\varrho_0 = 1.26408$ kg/m³

$\mu_{N_2} = 0.7119$ $\mu_{H_2O} = 0.0890$ $\mu_{CO_2} = 0.1553$ $\mu_{CO} = 0.0099$ $\mu_{O_2} = 0.0339$
b) CO_2: $\dot{m} = 1.2424$ kg/h H_2O: $\dot{m} = 0.7120$ kg/h
$\bar{r}_{N_2} = 0.9474$ $\bar{r}_{CO} = 0.0132$ $\bar{r}_{O_2} = 0.0394$

Section 5 (pages 210 to 212)

1. $p_1 = 2.19$ bar $t_1 = 123.11\,°C$ $x_1 = 0$ $v_1 = 0.0010632$ m^3/kg
$h_1 = h_2 = 516.99$ kJ/kg
$s_1 = 1.5612$ kJ/(kg K) $p_2 = 0.05$ bar $t_2 = 32.88\,°C$ $x_2 = 0.1565$ kg/kg
$v_2 = 4.4123$ m^3/kg $s_2 = 1.7155$ kJ/(kg K)

2. $\dfrac{\dot{m}(h_2 - h_1)}{H_i\,\eta} = \dfrac{2300 \text{ kg/h} \cdot (2702.24 \text{ kJ/kg} - 807.57 \text{ kJ/kg})}{28\,500 \text{ kJ/kg} \cdot 0.76} = 201.19$ kg/h

3. specific enthalpy in the final state $h_2 = h' + (h_1 - h')\dfrac{m_1}{m_2}$

$h_2 = 781.20$ kJ/kg $+ (2870.69$ kJ/kg $- 781.20$ kJ/kg$) \cdot \dfrac{1800 \text{ kg/h}}{1600 \text{ kg/h}} = 3131.88$ kJ/kg

$t_2 = 338.63\,°C$

4. state 1: $p_1 = 70$ bar $t_1 = 300\,°C$ $v_1 = 0.029494$ m^3/kg
$h_1 = 2839.83$ kJ/kg $s_1 = 5.9335$ kJ/(kg K) $m_1 = 0.9493$ kg
state 2: $p_2 = 120$ bar $t_2 = 553.95\,°C$ $v_2 = v_1$ $h_2 = 3491.87$ kJ/kg
$s_2 = 6.6677$ kJ/(kg K) $m_2 = m_1$
a) $(Q_{12})_{rev} = m_1(u_2 - u_1) = m_1[h_2 - h_1 - v_1(p_2 - p_1)] = 478.65$ kJ
b) $u_3 = u_2 = 3137.95$ kJ/kg   state 3: $p_3 = 103.31$ bar $t_3 = 547.48\,°C$
$v_3 = 0.03430$ m^3/kg $h_3 = 3492.27$ kJ/kg $s_3 = 6.7325$ kJ/(kg K) $m_3 = 0.8163$ kg

5. $h_1 = 3145.97$ kJ/kg $s_1 = 7.0713$ kJ/(kg K) $h_W = 167.54$ kJ/kg
$s_W = 0.5724$ kJ/(kg K) $h_2 = 2990.21$ kJ/kg $s_2 = 6.8060$ kJ/(kg K)
$\dfrac{\dot{m}_W}{\dot{m}_D} = \dfrac{h_1 - h_2}{h_2 - h_W} = 0.05518$ (kg water)/(kg superheated steam) $\Delta s = 0{,}07906$ kJ/(kg K)

6. $v = \dfrac{RT}{2p} + \sqrt{\left(\dfrac{RT}{2p}\right)^2 + \dfrac{1}{p}\left(B_0 + \dfrac{B_1}{T}\right)}$ $c_v^0 = R\left[c_1 + c_2\dfrac{T}{T_1} + c_3\left(\dfrac{T}{T_1}\right)^2\right]$ $T_1 = 1$ K

Eqs. (5.159) and (2.41): $h = RT\left[1 + c_1 + \dfrac{c_2}{2}\dfrac{T}{T_1} + \dfrac{c_3}{3}\left(\dfrac{T}{T_1}\right)^2\right] + \dfrac{1}{v}\left(2B_0 + 3\dfrac{B_1}{T}\right) + h^*$

Gl. (5.161): $s = R\left[c_1 \ln\dfrac{T}{T_1} + c_2\dfrac{T}{T_1} + \dfrac{c_3}{2}\left(\dfrac{T}{T_1}\right)^2\right] + R\ln\dfrac{v}{v_1} + \dfrac{B_1}{vT^2} + s^*$ $v_1 = 1$ dm^3/kg

7. a) $m_1 = 5.0$ kg $p_1 = 0.15$ bar $h_1 = 2600$ kJ/kg $s_1 = 8.01$ kJ/(kg K)
b) $p_2 = 0.247$ bar $h_2 = 3000$ kJ/kg $s_2 = 8.73$ kJ/(kg K) $Q_{12} = 1515$ kJ
ideal gas: permissible assumption, since hereby $p_2 = 0.247$ bar results
c) $x_3 = 0.69$ $p_3 = 0.1$ bar $h_3 = 1840$ kJ/kg $s_3 = 5.81$ kJ/(kg K) $Q_{13} = -\,3550$ kJ
d) $S_2 - S_1 = 3.61$ kJ/K $S_3 - S_1 = -11.0$ kJ/K

8. a) $x_1 = 0.20$ $m_{N1} = 200$ kg $m_{F1} = 160$ kg $m_{D1} = 40$ kg
$p_1 = 44.7$ bar $h_1 = 1460$ kJ/kg $s_1 = 3.50$ kJ/(kg K)
b) $t_2 = 544\,°C$ $h_2 = 3257$ kJ/kg $s_2 = 6.01$ kJ/(kg K)
$S_2 - S_1 = 503$ kJ/K $H_2 - H_1 = 360$ MJ $U_2 - U_1 = 309$ MJ
path A: $Q_{12} = 309$ MJ   path B: $Q_{12} = 348$ MJ

9. $p = p_k e^E \qquad E = \dfrac{T_k}{T}(b_1 \Theta + b_2 \Theta^2 + b_3 \Theta^3 + b_4 \Theta^4 + b_5 \Theta^5)$

Eq. (5.197): $r = -(v'' - v')p(E + b_1 + 2b_2\Theta + 3b_3\Theta^2 + 4b_4\Theta^3 + 5b_5\Theta^4)$
$h^* = 189.1935$ kJ/kg $\qquad s^* = 1.1658$ kJ/(kg K)

Section 6 (pages 242 to 244)

1. $\kappa = 1.2717 \qquad \bar\kappa = 1.2724$

Eq. (4.100): $T_2 = T_1 \left(\dfrac{p_2}{p_1}\right)^{\frac{\bar\kappa - 1}{\bar\kappa}} \qquad t_2 = 108.22\,°C$

a) Eqs. (6.20), (4.111) and (4.105):

$(P_e)_{id} = \dfrac{\kappa}{\kappa - 1}\dot m R T_1 \left[\left(\dfrac{p_2}{p_1}\right)^{\frac{\bar\kappa-1}{\bar\kappa}} - 1\right] = 12.262$ kW

$(\dot Q_{23})_{rev} = \dot m c_p (t_3 - t_2) = -12.262$ kW $\qquad c_p = \dfrac{\kappa}{\kappa-1} R$

$c_p = 0.8841$ kJ/(kg K) $\qquad c = 4.1886$ kJ/(kg K)

water flow 619.93 kg/h

b) Eqs. (6.20), (4.86) and (4.83): $(P_e)_{id} = 10.795$ kW

water outlet temperature $22.97\,°C$ c) Fig. A.1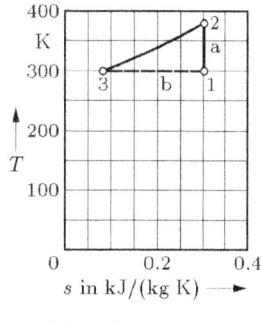

Fig. A.1

d) Eqs. (4.79) and (4.89):

gas $\Delta s_G = \bar c_p \ln \dfrac{T_3}{T_2} = R \ln \dfrac{p_1}{p_2} = -0.2172$ kJ/(kg K)

$\Delta \dot S_G = -0.03621$ kW/K $\qquad \bar c_p = \dfrac{\bar\kappa}{\bar\kappa-1}R = 0.8825$ kJ/(kg K)

Eq. (5.4): water $\qquad$ a) $\Delta \dot S_W = 0.04234$ kW/K $\qquad$ b) $\Delta \dot S_W = 0.03740$ kW/K

total change of entropy flow $\qquad$ a) $\Delta \dot S = 0.00613$ kW/K $\qquad$ b) $\Delta \dot S = 0.00119$ kW/K

2. $V_0 = 2.6923$ m^3 $\qquad$ a) Eq. (4.81): $V_a = 0.2991$ m^3 $\qquad V_e = 0.3846$ m^3

$V_N = 0.0855$ m^3 $\qquad$ b) Eq. (4.94): $V_a = 0.5604$ m^3 $\qquad V_e = 0.6706$ m^3

$V_N = 0.1102$ m^3 $\qquad$ c) case a): 17.09 s $\qquad$ case b): 22.04 s

d) Eq. (4.81): $V_a = 2.0940$ m^3 $\qquad V_e = V_0 \qquad V_N = 0.5983$ m^3

3. stroke volume $0.3927\,\ell$ $\qquad$ a) Eq. (4.25): $\Delta m = 0.4476$ g

b) mass in the air vessel at ambient pressure 5.129 kg

mass in the air vessel at final pressure 43.612 kg

air mass to be compressed 38.483 kg $\qquad$ c) 143.3 min

4. a) $(D^2 - d^2)\dfrac{\pi}{4}s = \dfrac{1}{4}D^2\dfrac{\pi}{4}s \qquad d = 69.3$ mm

b) intake volume flow 0.007288 m^3/s

Eq. (6.20), Table 4.2 Eq. (2): power of one stage
1.075 kW $\qquad$ total power 2.150 kW

c) Eqs. (6.20), (4.86) and (4.88): power of one stage 0.986 kW $\qquad$ total power 1.972 kW

d) Eq. (6.20), Table 4.2 Eq. (2): total power without intercooling 2.350 kW

saved power $(2.350 - 2.150)$ kW $= 0.200$ kW

e) Figure A.2

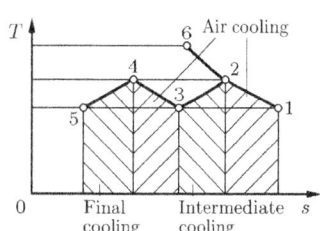

Figure A.2

5. a) Eq. (4.27): 618.8 mbar gage pressure   b) Gl. (4.100): 55.33 °C
c) Eq. (6.20), Table 4.2 Eq. (2): compressor power (= − turbine power) 697.7 kW
d) Table 6.1 Eq. (19), Eqs. (4.43), (4.49), (4.21) and (4.24): 6.766 kg/s
6. a) Table 6.2 Eq. (2), Table 6.5 Eq. (2): 61.41 °C
b) Eq. (4.144): $n = 1.4906$ c) Table 6.1 Eq. (22): internal power 811.3 kW
Table 6.3 Eq. (1): coupling power 836.4 kW
d) Table 6.2 Eq. (4), Table 6.5 Eq. (4): final temperature in the turbine 375.02 °C
coupling power − 836.4 kW   Table 6.3 Eq. (2): internal power − 862.2 kW
Table 6.1 Eq. (22), Eqs. (4.43), (4.49), (4.21) and (4.24): 9.723 kg/s
7. $h_1 = 3406.52$ kJ/kg $s_1 = 6.7694$ kJ/(kg K) $h_2 = 2829.26$ kJ/kg $h_3 = 2581.53$ kJ/kg
a) Table 6.2 Eq. (3),   Eq. (6.47),   Table 6.5 Eq. (3):   0.6997
b) Table 6.1 Eq. (22),   Table 6.3 Eq. (2): $W_e = -923.6$ kW
c) Table 6.1 Eq. (19),   Eq. (6.41): $(W_e)_{id} = -1375.0$ kW

Section 7 (pages 329 to 332)
1. a) Eqs. (4.94) and (7.55): $p_{2O} = 24.11$ bar $p_{2D} = p_3 = 72.72$ bar
$p_4 = 2.895$ bar   Eq. (4.66): $t_{2O} = 423.25$ °C   Eq. (4.100): $t_1 = 4.09$ °C
$t_{2D} = 681.46$ °C $t_4 = 562.93$ °C   b) Eq. (7.69): Otto cycle 0.6019
Eq. (7.70) or (7.59), (7.60), (7.36): Diesel cycle 0.6515
c) Eq. (4.25): mass of a cylinder filling 0.0005771 kg
Eq. (7.61): Otto cycle $(P_e)_{id} = -12.19$ kW   Diesel cycle $(P_e)_{id} = -15.08$ kW
power gain 2.89 kW
2. Eq. (4.100): $t_2 = 207.97$ °C $t_3 = t_5 = 291.25$ °C
a) Eq. (4.77): 76.22 kW
b) Eq. (7.90): $(P_e)_{id} = -157.2$ kW
c) Eq. (7.93): 0.4788   d) Eq. (7.87): 0.3886
e) Eq. (7.104): 0.6814   e) Fig. A.3

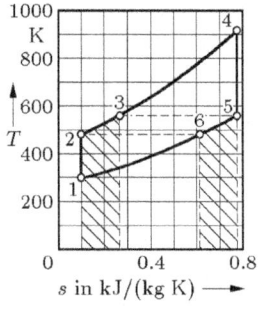

3. Eq. (4.81): $p_2 = 9.1$ bar
Eq. (4.66): $p_3 = 21$ bar   Eq. (4.81): $p_4 = 42$ bar
a) Eq. (4.25): $m = 0.0005092$ kg
Eq. (4.87): heat flow supply (= cooling capacity   4.5415 kW

Figure A.3

b) Eqs. (7.151) and (7.155): 0.7647   Eq. (7.151): 5.9389 kW
4. a) Fig. A.4 $h_1 = 3336.13$ kJ/kg $s_1 = 6.2651$ kJ/(kg K)
$h_2 = 2631.39$ kJ/kg $s_2 = 6.6395$ kJ/(kg K)
$h_3 = h_4 = 2225.08$ kJ/kg
$x_4 = 0.7595$ $h_5 = 2475.36$ kJ/kg $x_5 = 0.8769$
$h_6 = 2102.46$ kJ/kg $x_6 = 0.7987$
b) $(P_e)_{id\,HT} = -56\,379$ kW $(P_e)_{id\,NT} = -28\,171$ kW
c) 4515 kW   d) high pressure turbine 0.8187
low pressure turbine 0.7682

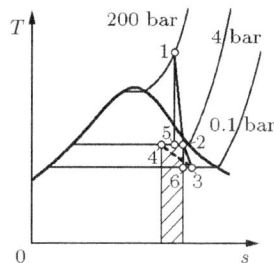

Figure A.4

Figure A.5

5. a) nominations according to Fig. A.5

		p	t	x	h
		bar	°C		kJ/kg
State	1	0.9312	−35	1	1715.0
	2	5	78.09		1947.9
	3	5	4.14	0	518.93
	4	0.9312	−35	0.1286	518.93
	5	4	−1.88	1	1758.7
	6	18	108.27		1979.7
	7	18	45.38	0	712.35
	8	4	−1.88	0.1743	712.35

b) $\dot{m}_\mathrm{I} = 0.06688$ kg/s $\dot{m}_\mathrm{II} = 0.09134$ kg/s

c) 95.575 kW d) 115.76 kW e) $(P_e)_{id\,\mathrm{I}} = 15.575$ kW $(P_e)_{id\,\mathrm{II}} = 20.189$ kW

$(P_e)_{id\,(\mathrm{I+II})} = 35.764$ kW f) 2.2369 g) 4.1358

6. $V_1 = 0.00171$ m³ $V_2 = 0.0001427$ m³ $T_2 = 805.58$ K $p_2 = 32.423$ bar

$T_3 = 2013.94$ K $p_3 = 81.058$ bar $V_3 = 0.0001427$ m³ $V_5 = 0.00171$ m³

$p_5 = 2.500$ bar $T_5 = 745.37$ K $Q_{23} = 1.735$ kJ $Q_{51} = -0.642$ kJ

$(W_e)_{id} = -1.093$ kJ $\eta_{th} = 0.6299$ $(P_e)_{id} = -54.65$ kW

7. $V_1 = 0.008563$ m³ $V_2 = 0.0004281$ m³ $T_2 = 988.2$ K $p_2 = 66.289$ bar

$T_4 = 2174.04$ K $V_4 = 0.0009418$ m³ $p_4 = 66.289$ bar $V_5 = 0.008563$ m³

$T_5 = 899.08$ K $p_5 = 3.015$ bar $Q_{34} = 11.918$ kJ $Q_{51} = -4.315$ kJ

$(W_e)_{id} = -7.603$ kJ $\eta_{th} = 0.6380$ $(P_e)_{id} = -190$ kW

8. Stirling cycle according to Fig. 7.23:

2 isothermal changes of state 12 and 34: $T_2 = T_1$ $T_4 = T_3$

$w_{p12} = w_{V12} = q_{s12} = RT_1 \ln(p_2/p_1)$

$w_{p34} = w_{V34} = -q_{s34} = RT_3 \ln(p_4/p_3) = RT_3 \ln(p_2/p_1)$

$p_4/p_3 = p_2/p_1$ follows from the isochoric changes of state 23 and 41, since $p_3 = p_2 T_3/T_1$ and $p_4 = p_1 T_3/T_1$ are valid.

2 isochoric changes of state: $v_3 = v_2$ $v_1 = v_4$

$w_{p23} = v_2(p_3 - p_2) = R(T_3 - T_1)$ $w_{V23} = 0$

$q_{s23} = c_v(T_3 - T_2) = c_v(T_3 - T_1) = R(T_3 - T_1)/(\kappa - 1)$

$w_{p41} = v_1(p_1 - p_4) = R(T_1 - T_3) = -R(T_3 - T_1)$ $w_{V41} = 0$

$q_{s41} = c_v(T_1 - T_4) = -c_v(T_3 - T_1) = -R(T_3 - T_1)/(\kappa - 1)$

562 Appendix

$$a_{wp} = \frac{\Sigma|w_{pij}|}{|\Sigma w_{pij}|} = \frac{RT_1 \ln(p_2/p_1) + RT_3 \ln(p_2/p_1) + 2R(T_3 - T_1)}{R(T_3 - T_1)\ln(p_2/p_1) + R(T_3 - T_1) - R(T_3 - T_1)}$$

$$a_{wp} = \frac{R(T_3 + T_1)\ln(p_2/p_1))}{R(T_3 - T_1)\ln(p_2/p_1)} + \frac{2R(T_3 - T_1)}{R(T_3 - T_1)\ln p_2/p_1}$$

$$a_{wp} = \frac{T_3 + T_1}{T_3 - T_1} + \frac{2R(T_3 - T_1)}{R(T_3 - T_1)\ln(p_2/p_1)}$$

$$a_{wv} = \frac{\Sigma|w_{Vij}|}{|w_{Vij}|} = \frac{RT_1 \ln(p_2/p_1) + RT_3 \ln(p_2/p_1)}{R(T_3 - T_1)\ln(p_2/p_1)} = \frac{T_3 + T_1}{T_3 - T_1}$$

$$a_{qs} = \frac{\Sigma|q_{sij}|}{|q_{sij}|} = \frac{RT_1 \ln(p_2/p_1) + RT_3 \ln(p_2/p_1) + 2R(T_3 - T_1)/(\kappa - 1)}{R(T_3 - T_1)\ln(p_2/p_1) + R(T_3 - T_1)/(\kappa - 1) - R(T_3 - T_1)/(\kappa - 1)}$$

$$a_{qs} = \frac{R(T_3 + T_1)\ln(p_2/p_1)}{R(T_3 - T_1)\ln(p_2/p_1)} + \frac{2R(T_3 - T_1)/(\kappa - 1)}{R(T_3 - T_1)\ln(p_2/p_1)}$$

$$a_{qs} = \frac{T_3 + T_1}{T_3 - T_1} + \frac{2}{(\kappa - 1)\ln(p_2/p_1)}$$

9. *Otto* cycle: $p_2 = 27.172$ bar $\quad p_3 = 70.0$ bar $\quad T_3 = 1973.15$ K $\quad \eta_{th} = 0.6107$
$\zeta = 0.7970 \quad a_{wp} = 4.0864 \quad a_{wv} = 2.2689 \quad a_{qT} = 46.2461 \quad a_{qs} = 2.2748 \quad a_g = 54.8762$

10. *Diesel* cycle: $p_2 = p_3 = 70.0$ bar $\quad T_3 = 1973.15$ K $\quad \eta_{th} = 0.6538 \quad \zeta = 0.8253$
$a_{wp} = 3.2265 \quad a_{wv} = 2.5904 \quad a_{qT} = 38.2585 \quad a_{qs} = 2.0593 \quad a_g = 46.1347$

11. *Joule* cycle: $p_2 = p_3 = 20.0$ bar $\quad T_3 = 1523.15$ K $\quad \eta_{th} = 0.5751 \quad \zeta = 0.8002$
$a_{wp} = 2.7084 \quad a_{wv} = 2.6425 \quad a_{qT} = 37.9486 \quad a_{qs} = 2.4776 \quad a_g = 45.7771$

12. a) *Stirling* cycle (air): $p_2 = 6.1275$ bar $p_3 = 20.0$ bar $T_3 = 973.15$ K $\eta_{th} = 0.6936$
$\zeta = 1 \quad a_{wp} = 2.9867 \quad a_{wv} = 1.8834 \quad a_{qT} = 27.3250 \quad a_{qs} = 4.6416 \quad a_g = 36.8367$

b) *Stirling* cycle (Helium): $p_2 = 6.1275$ bar $\quad p_3 = 20.0$ bar $\quad T_3 = 973.15$ K
$\eta_{th} = 0.6936 \quad \zeta = 1 \quad a_{wp} = 2.9867 \quad a_{wv} = 1.8834 \quad a_{qT} = 16.9061 \quad a_{qs} = 3.5382$
$a_g = 25.3144$

c) *Stirling* cycle (Helium): $p_2 = 45.9564$ bar $\quad p_3 = 150.0$ bar $\quad T_3 = 973.15$ K
$\eta_{th} = 0.6936 \quad \zeta = 1 \quad a_{wp} = 2.9867 \quad a_{wv} = 1.8834 \quad a_{qT} = 14.6831 \quad a_{qs} = 3.5382$
$a_g = 23.0914$

13. *Clausius Rankine* cycle: $p_2 = p_3 = 350.0$ bar $\quad T_3 = 873.15$ K $\quad \eta_{th} = 0.4732 \quad \zeta = 1$
$a_{wp} = 1.0451 \quad a_{wv} = 1.1254 \quad a_{qT} = 6.1900 \quad a_{qs} = 3.2264 \quad a_g = 11.5869$

Section 8 (page 374)

1. $EU = 146.78$ kJ $\qquad EH = 414.79$ kJ

2. $m\,c_v(t_2 - t_0) - T_0\,m\bigl(c_p \ln\dfrac{T_2}{T_0} - R \ln\dfrac{p}{p_0}\bigr) + p_0\,m\,R\bigl(\dfrac{T_2}{p} - \dfrac{T_0}{p_0}\bigr) = 2 \cdot 146.78$ kJ

a) iteration: $t_2 = 306.37\,°C$ $\qquad$ b) 757.11 kJ $\qquad$ c) 610.33 kJ $\qquad$ d) 1060.0 kJ
e) $(EQ_{12})_{rev} = 355.0$ kJ $\qquad (AQ_{12})_{rev} = 705.0$ kJ

3. polytropic exponent 2.4150 $\qquad E_V = A_G = 428.23$ kJ
reversible substitution process:

$\qquad\qquad\qquad (Q_{12})_{rev} + W_{p12} \quad = H_2 - H_1$
energy: $\qquad\quad$ 613.82 kJ + 584.19 kJ $= 1198.01$ kJ
exergy: $\qquad\quad$ 185.59 kJ + 584.19 kJ $= 769.78$ kJ
anergy: $\qquad\quad$ 428.23 kJ + 0 kJ $\quad = 428.23$ kJ

real process:
$$W_{p12} + |W_{RI}| = H_2 - H_1$$
energy: $584.19 \text{ kJ} + 613.82 \text{ kJ} = 1198.01 \text{ kJ}$
exergy: $584.19 \text{ kJ} + 613.82 \text{ kJ} = 769.78 \text{ kJ} + 428.23 \text{ kJ}$
anergy: $0 \text{ kJ} + 0 \text{ kJ} = 428.23 \text{ kJ} - 428.23 \text{ kJ}$

4. $\zeta = \dfrac{T_3 - T_1}{T_3 - T_0}$ $\zeta_1 = 0.9854$ $\zeta_2 = 0.9908$

Section 9 (pages 440 to 442)

1. a) $\dot{Q}$ = heat flow without insulation $\dot{Q}^*$ = heat flow with insulation
$\dot{Q}^*/\dot{Q} = 2/3$
b) $R_{L\,is}/R_L = 0.88235$ $\lambda = 0.7182 \text{ W/(m K)}$
c) $\alpha = 5.814 \text{ W/(m}^2 \text{ K)}$ $\alpha' = 24.42 \text{ W/(m}^2 \text{ K)}$
2. a) wall thickness = 56.3 mm b) with insulation: $\dot{q} = 20 \text{ W/m}^2$
c) without insulation: $\dot{q} = 93.1 \text{ W/m}^2$ inner wall temperature $1.38\,°C$
3. water outlet temperature $49\,°C$ $W_1/W_2 = 0.068$ $\Phi = 0.730$
$(t - t')_m = 370 \text{ K}$ $k\,A/W_1 = 1.351$
4. a) $\dot{q} = 49.6 \text{ W/m}^2$ b) $\alpha_i = 8.2667 \text{ W/(m}^2 \text{ K)}$ c) $\dot{q} = 95.067 \text{ W/m}^2$
d) window outside temperature $= -8.2304\,°C$ $\alpha_a = 28.029 \text{ W/(m}^2 \text{ K)}$
5. $k\,A/W_L = 2.002$ $k\,A/W_W = 1.000$
6. a) $k\,A = 5.3115 \text{ W/K}$ b) $A_i = 0.4913 \text{ m}^2$ $A_a = 0.5341 \text{ m}^2$
$k_{Aa} = 9.945 \text{ W/(m}^2 \text{ K)}$ $k_{Ai} = 10.811 \text{ W/(m}^2 \text{ K)}$ c) $\lambda_u = 0.41198 \text{ W/(m K)}$
d) $k\,A = 0.76236 \text{ W/K}$ e) $k\,A = 0.97007 \text{ W/K}$

Section 10 (pages 462 to 463)

1. a) $27\,°C$: $p_{DS} = 35.68 \text{ mbar}$ $x_1 = 0.02255 \text{ kg/kg}$ $h_1 = 84.66 \text{ kJ/kg}$
$12\,°C$: $p_{DS} = 14.028 \text{ mbar}$ $x_M = 0.00867 \text{ kg/kg}$ $h_M = 33.93 \text{ kJ/kg}$
$v_M = 0.8067 \text{ m}^3/\text{kg}$
$m = 3966.8 \text{ kg}$ $m_{LM} = 3932.7 \text{ kg}$ b) $m_{L2} = 2420.7 \text{ kg}$
Gl.(10.35): $h_2 = h_M - (x_M/(x_1 - x_M))(h_1 - h_M) = 2.242 \text{ kJ/kg}$
$t_2 = 2.242\,°C/1.004 = 2.23\,°C$
2. a) $28\,°C$: $p_{DS} = 37.83 \text{ mbar}$ $p_D = 7.566 \text{ mbar}$ $x = 0.0051296 \text{ kg/kg}$
$h = 41.214 \text{ kJ/kg}$ $v = 0.93747 \text{ m}^3/\text{kg}$ $m_L = (V/v)/(1+x) = 2334.8 \text{ kg}$
b) $m_D = x\,m_L = 11.976 \text{ kg}$ $32\,°C$: $p_{DS} = 47.59 \text{ mbar}$ $p_D = 15.705 \text{ mbar}$
$x = 0.010743 \text{ kg/kg}$ $h = 59.650 \text{ kJ/kg}$ $m_D = 25.083 \text{ kg}$
$m_W = (25.083 - 11.976) \text{ kg} = 13.107 \text{ kg}$
c) $\Delta x = 0.0056134 \text{ kg/kg}$ $\Delta h = 0.0056134 \cdot 4.19 \cdot 18 \text{ kJ/kg} = 0.423 \text{ kJ/kg}$
$h_2 = h_1 + \Delta h = 41.637 \text{ kJ/kg}$ $t_2 = 14.43\,°C$ $p_{DS} = 16.44 \text{ mbar}$
$\varphi = 15.705/16.44 = 0.9553$
d) $Q = (59.650 - 41.637) \cdot 2334.8 \text{ kJ} = 42056 \text{ kJ/kg}$
e) $h_D = (59.650 - 41.214)/0.0056134 \text{ kJ/kg} = 3284.3 \text{ kJ/kg}$

3. air pressure: 653 mbar   6 °C: $p_{DS} = 9.3535$ mbar
dew point -1 °C: $p_D = 5.6267$ mbar $x = 0.0054062$ kg/kg $\varphi = 0.60156$
4. a) $t = 20$ °C: $p_{DS} = 23.37$ mbar $x = 0.006$ kg/kg $p_D = 9.554$ mbar
b) $V_2/V_1 = 9.554/23.37 = 0.4085$ 5 bar: $x_S = 2.92$ g/kg
$\Delta x = 6$ g/kg $- 2.92$ g/kg $= 3.08$ g/kg
c) 5 bar: $p_D = 5 \cdot 9.554$ mbar $= 47.77$ mbar $\varphi = 0.85$ $p_S = p_D/\varphi = 56.2$ mbar
$t = 34.97$ °C
5. a) 5 °C: $p_{DS} = 8.726$ mbar $x_S = 0.0068593$ kg/kg
14 °C: $p_{DS} = 15.989$ mbar $x_S = 0.012685$ kg/kg
$h = 45.776$ kJ/kg $t = 28.14$ °C   b) $h = 45.374$ kJ/kg $t = 45.19$ °C
c) 5 °C: $x_S = 0.0054045$ kg/kg   14 °C: $x_S = 0.0099750$ kg/kg
$h = 39.000$ kJ/kg $t = 25.12$ °C $h = 38.683$ kJ/kg $t = 38.53$ °C
6. 28.6 °C: $p_{DS} = 39.171$ mbar $p_D = 13.318$ mbar   a) $x = 0.0088912$ kg/kg
$v = 0.92153$ m^3/kg $m = 17.580$ kg $m_L = 17.425$ kg $h_3 = 51.435$ kJ/kg
b) dew point: 11.22 °C $h_2 = 33.696$ kJ/kg $h_1 = 2h_2 - h_3 = 15.957$ kJ/kg
c) $t_1 = 3.13$ °C
d) $Q = (51.435 - 15.957) \cdot 17.425$ kJ $= 618.2$ kJ
7. a) $k = 1.09701$ W/(m^2 K) $\dot{q} = 35.104$ W/m^2 $\vartheta_0 = 22.612$ °C
$\vartheta_Z = 1.207$ °C
$\vartheta'_0 = -3.474$ °C   b) $t = 27$ °C: $p_{DS} = 35.679$ mbar
$p_D = \varphi \cdot p_{DS} = p_{D0} = p_{D1} = 21.407$ mbar $\vartheta_0 = 22.612$ °C $p_{DS0} = 27.456$ mbar
$\vartheta_Z = 1.207$ °C $p_{DSZ} = 6.670$ mbar $p_{DZ} = 19.177$ mbar $p_{DSZ} < p_{DZ}$ condensation !
$\vartheta'_0 = -3.474$ °C $p'_{DS0} = 4.574$ mbar $t' = -5$ °C $p'_{DS} = 4.018$ mbar
b) A barrier layer on the inside with $\mu \cdot \delta > 37$ m prevents condensation.
c) A barrier layer on the inside with $\mu \cdot \delta > 69$ m prevents condensation.

Section 11 (pages 494 bis 496)

1. $l_{min} = 0.5 \, (r_{CO} + r_{H_2})/0.21 = 0.547$ Nm^3 air/Nm^3 B
$V_{0B} : V_{0\,Lmin} = 1.000 : 0.547 = 64.6 : 35.4$
2. $o_{min} = 5.000$ Nm^3 O$_2$/Nm^3 B $l_{min} = 23.810$ Nm^3 air/Nm^3 B $l = 28.751$ Nm^3 air/Nm^3 B
$v_{0\,Amin\,f} = 25.810$ Nm^3 Af/Nm^3 B $v_{0\,Af} = 30.572$ Nm^3 Af/Nm^3 B
$v_{0\,Amin\,t} = 21.810$ Nm^3 At/Nm^3 B $v_{0\,At} = 26.572$ Nm^3 At/Nm^3 B
$r_{CO_2\,max}^{at} = 0.138$ Nm^3 CO$_2$/Nm^3 At $r_{CO_2}^{at} = 0.1129$ Nm^3 CO$_2$/Nm^3 At
$r_{O_2}^{at} = 0.0376$ Nm^3 O$_2$/Nm^3 At $r_{N_2}^{at} = 0.8495$ Nm^3 N$_2$/Nm^3 At
$r_{H_2O}^{af} = 0.1309$ Nm^3 H$_2$O/Nm^3 Af $r_{CO_2}^{af} = 0.0981$ Nm^3 CO$_2$/Nm^3 Af
$r_{O_2}^{af} = 0.0327$ Nm^3 O$_2$/Nm^3 Af $r_{N_2}^{af} = 0.7383$ Nm^3 N$_2$/Nm^3 Af
Fig. 11.6: $r_{CO_2}^{at} = 0.113$ Nm^3 CO$_2$/Nm^3 At $r_{O_2}^{at} = 0.037$ Nm^3 O$_2$/Nm^3 At
3. $o_{min} = 0.240$ Nm^3 O$_2$/Nm^3 B $l_{min} = 1.143$ Nm^3 air/Nm^3 B $l = 1.715$ Nm^3 air/Nm^3 B
$v_{0\,Amin\,f} = 1.983$ Nm^3 Af/Nm^3 B $v_{0\,Af} = 2.488$ Nm^3 Af/Nm^3 B
$v_{0\,Amin\,t} = 1.753$ Nm^3 At/Nm^3 B $v_{0\,At} = 2.325$ Nm^3 At/Nm^3 B
$r_{CO_2\,max}^{at} = 0.205$ Nm^3 CO$_2$/Nm^3 At

Fig. 11.3: $l_{min} = 0.98 \, \text{Nm}^3 \, air/\text{Nm}^3 \, B$ $v_{0\,Af} = 2.3 \, \text{Nm}^3 \, Af/\text{Nm}^3 \, B$

4. Eq. (11.46): $\lambda = 1.123$ Eq. (11.50): $\lambda = 1.130$ Eq. (11.54): $\lambda = 1.085$

5. $\sigma = 1.480 \, \text{kmol} \, O_2/\text{kmol} \, C$ $\nu = 0.000017 \, \text{kmol} \, N_2/\text{kmol} \, C$
$o^*_{min} = 2.1387 \, \text{Nm}^3 O_2/\text{kg} \, B$ $l^*_{min} = 10.1840 \, \text{Nm}^3 \, air/\text{kg} \, B$ $l^* = 12.2210 \, \text{Nm}^3 \, air/\text{kg} \, B$
$v^*_{0\,Aminf} = 11.0273 \, \text{Nm}^3 fA/\text{kg} \, B$ $v^*_{0\,Af} = 13.0741 \, \text{Nm}^3 fA/\text{kg} \, B$
$v^*_{0\,Amint} = 9.4888 \, \text{Nm}^3 tA/\text{kg} \, B$ $v^*_{0\,At} = 11.5256 \, \text{Nm}^3 tA/\text{kg} \, B$ $r^{af}_{CO_2} = 0.1104$
$r^{af}_{H_2O} = 0.1182$ $r^{af}_{SO_2} = 0.000004$ $r^{af}_{O_2} = 0.0327$ $r^{af}_{N_2} = 0.7386$
$r^{at}_{CO_2} = 0.1252$ $r^{at}_{O_2} = 0.0371$ $r^{at}_{N_2} = 0.8377$ $v^*_{Af} = 28.7647 \, \text{m}^3 fA/\text{kg} \, B$
$M_A = 28.7241 \, \text{kg}/\text{kmol}$ $R_A = 0.28946 \, \text{kg}/\text{kmol}$ $\Delta p = 0.6791 \, \text{mbar}$
$CH_{2.08399} O_{0.08503} S_{0.00004} N_{0.00002} \approx CH_{2.084} O_{0.085}$

6. $\sigma = 1.085 \, \text{kmol} \, O_2/\text{kmol} \, C$ $\nu = 0.0083 \, \text{kmol} \, N_2/\text{kmol} \, C$ $l^* = 7.458 \, \text{Nm}^3 \, air/\text{kg} \, B$
$v^*_{0\,Af} = 8.154 \, \text{Nm}^3 fA/\text{kg} \, B$ $v^*_{0\,At} = 7.384 \, \text{Nm}^3 tA/\text{kg} \, B$ $r^{af}_{CO_2} = 0.1180$
$r^{af}_{H_2O} = 0.0939$ $r^{af}_{SO_2} = 0.0005$ $r^{af}_{O_2} = 0.0640$ $r^{af}_{N_2} = 0.7236$
$r^{at}_{CO_2} = 0.1303$ $r^{at}_{O_2} = 0.0707$ $r^{at}_{N_2} = 0.7990$ $r^{at}_{CO_2\,max} = 0.1964$
$v^*_{Af} = 15.838 \, \text{m}^3 fA/\text{kg} \, B$

7. $\sigma = 1.140 \, \text{kmol} \, O_2/\text{kmol} \, C$ $\nu = 0.0069 \, \text{kmol} \, N_2/\text{kmol} \, C$ $l^* = 10.602 \, \text{Nm}^3 \, air/\text{kg} \, B$
$v^*_{0\,Af} = 10.9278 \, \text{Nm}^3 fA/\text{kg} \, B$ $v^*_{0\,At} = 10.4004 \, \text{Nm}^3 tA/\text{kg} \, B$ $l^*_{min} = 8.155 \, \text{Nm}^3 \, air/\text{kg} \, B$
$v^*_{0\,Amin\,f} = 8.4804 \, \text{Nm}^3 fA/\text{kg} \, B$ $v^*_{0\,Amin\,t} = 7.9530 \, \text{Nm}^3 tA/\text{kg} \, B$ $r^{af}_{CO_2} = 0.1770$
$r^{af}_{H_2O} = 0.0615$ $r^{af}_{SO_2} = 0.0007$ $r^{af}_{O_2} = 0$ $r^{af}_{N_2} = 0.7608$ $r^{at}_{CO_2} = 0.1887$
$r^{at}_{N_2} = 0.8113$ $r^{at}_{CO_2\,max} = 0.1887$

8. $\sigma = 1.0301 \, \text{kmol} \, O_2/\text{kmol} \, C$ $\nu = 0.0212 \, \text{kmol} \, N_2/\text{kmol} \, C$ $l^*_{min} = 4.0624 \, \text{Nm}^3 \, air/\text{kg} \, B$
$l^* = 6.0937 \, \text{kmol} \, L/\text{kmol} \, C$ $v^*_{0\,Amin\,f} = 4.7491 \, \text{Nm}^3 fA/\text{kg} \, B$
$v^*_{0\,Amin\,t} = 4.0546 \, \text{Nm}^3 tA/\text{kg} \, B$ $v^*_{0\,Af} = 6.7990 \, \text{Nm}^3 fA/\text{kg} \, B$
$v^*_{0\,At} = 6.0858 \, \text{Nm}^3 tA/\text{kg} \, B$

$\lambda = 1.0$:
$r^{af}_{CO_2} = 0.1743$ $r^{af}_{H_2O} = 0.1458$ $r^{af}_{SO_2} = 0.0004$ $r^{af}_{O_2} = 0$ $r^{af}_{N_2} = 0.6795$
$r^{at}_{CO_2} = r^{at}_{CO_2\,max} = 0.2041$ $r^{at}_{N_2} = 0.7959$

$\lambda = 1.5$:
$r^{af}_{CO_2} = 0.1221$ $r^{af}_{H_2O} = 0.1021$ $r^{af}_{SO_2} = 0.0003$ $r^{af}_{O_2} = 0.0629$ $r^{af}_{N_2} = 0.7126$
$r^{at}_{CO_2} = 0.1360$ $r^{at}_{O_2} = 0.0701$ $r^{at}_{N_2} = 0.7939$

9. a) petrol (*Otto* fuel): reduction of the amount of combustion gas to approx. $\frac{86}{125} \cdot v^*_{0\,Af}$; thereby increasing t_{max} from $2060\,°C$ to $\frac{125}{86} \cdot 2060\,°C \approx 3000\,°C$

gas oil (*Diesel* fuel): reduction of the amount of combustion gas to approx. $\frac{108}{157} \cdot v^*_{0\,Af}$; thereby increasing t_{max} from $2060\,°C$ to $\frac{157}{108} \cdot 2060\,°C \approx 3000\,°C$

b) *Otto* cycle: $p_2 = 36.268 \, \text{bar}$ $p_3 = 70.0 \, \text{bar}$ $T_3 = 3273.15 \, \text{K}$ $\eta_{th} = 0.6416$
$\zeta = 0.7705$ $a_{wp} = 3.2010$ $a_{wv} = 1.6814$ $a_{qT} = 39.9400$ $a_{qs} = 2.1174$ $a_g = 46.9399$
Diesel cycle 1: $p_2 = p_3 = 300.0 \, \text{bar}$ $T_3 = 3273.15 \, \text{K}$ $\eta_{th} = 0.7662$ $\zeta = 0.8796$
$a_{wp} = 2.8222$ $a_{wv} = 2.3015$ $a_{qT} = 32.4068$ $a_{qs} = 1.6103$ $a_g = 39.1404$
Diesel cycle 2: $p_2 = p_3 = 300.0 \, \text{bar}$ $T_3 = 2423.15 \, \text{K}$ $\eta_{th} = 0.7830$ $\zeta = 0.9254$
$a_{wp} = 4.4632$ $a_{wv} = 3.4737$ $a_{qT} = 42.7480$ $a_{qs} = 1.5542$ $a_g = 52.2391$

10.

	Table value MJ/kg B	Association formula MJ/kg B	Formula of Boie MJ/kg B	Deviation from table value	
				Association formula (%)	Formula of Boie (%)
Xylol C_8H_{10}	40.876	42.17	40.42	$+3.2$	-1.1
Methane CH_4	50.028	54.59	48.73	$+9.1$	-2.6
Ethyl alcohol C_2H_5OH	27.708	28.27	26.72	$+2.0$	-3.6
Wood (free of water and ash; waf)	18.3	17.52	18.28	-4.8	-0.6
Wood pellets	16.64	15.62	16.24	-6.1	-2.4
Landscape mainten. wood (green waste)	10.86	9.54	10.09	-12.2	-7.1
Digestate (from biogas production)	–	10.53	10.61	–	–
Waste paper sludge	3.10	3.10	3.21	± 0.0	$+3.5$
Old bread (dry)	16.20	14.51	15.14	-10.4	-6.5
Hard coal (forge coal)	31.23	31.74	31.58	$+1.6$	$+1.1$
Raw lignite (moist)	8.37	8.30	8.49	-0.9	$+1.3$
Lignite (dry)	20.80	20.34	20.74	$+2.2$	-0.3
Rapeseed oil methylester	37.20	41.43	38.83	$+11.1$	$+4.5$
Fuel oil (extra light EL)	42.60	45.35	42.55	$+6.4$	-0.1

11. $1\,\text{kg}\,B + g \cdot 2.082\,\text{kg}\,O_2 + g \cdot 6.866\,\text{kg}\,N_2 \to g \cdot 1.911\,\text{kg}\,CO_2$
$+ (1 + 0.173 \cdot g)\,\text{kg}\,H_2O + g \cdot 6.866\,\text{kg}\,N_2$
$1\,\text{kg}\,B + g \cdot 1.458\,\text{Nm}^3\,O_2 + g \cdot 5.491\,\text{Nm}^3\,N_2 \to g \cdot 0.967\,\text{Nm}^3\,CO_2$
$+ (1.244 + 0.215 \cdot g)\,\text{Nm}^3\,H_2O + g \cdot 5.491\,\text{Nm}^3\,N_2$
$g = 0.9$:
$1\,\text{kg}\,B + 1.8756\,\text{kg}\,O_2 + 6.1794\,\text{kg}\,N_2 \to 1.7199\,\text{kg}\,CO_2 + 1.1557\,\text{kg}\,H_2O$
$+ 6.1794\,\text{kg}\,N_2$
$1\,\text{kg}\,B + 1.3122\,\text{Nm}^3\,O_2 + 4.9419\,\text{Nm}^3\,N_2 \to 0.8703\,\text{Nm}^3\,CO_2 + 1.4375\,\text{Nm}^3\,H_2O$
$+ 4.9419\,\text{Nm}^3\,N_2$
$H_i = (g \cdot 27.708 - (1 - g) \cdot 2.501)\,\text{MJ/kg}\,B = 24.69\,\text{MJ/kg}\,B$

Properties of Substances

Universal Constants of Thermodynamics

$(V_m)_0$ = mol volume in norm state = 22.41410(19) m³/kmol
N_A = Avogadro number = 6.0221367(36)·10²⁶ molecules/kmol
R_m = universal gas constant = 8.314510(70) kJ/(kmol K)

The numbers in brackets describe the range of variation in units of the last digits and thus the uncertainty of the given values.

For example, 8.314510(70) means 8.314510 ± 0.000070

For example, 15.9994(3) means 15.9994 ± 0.0003.

Specific Heat Capacities and Isentropic Exponents of Several Ideal Gases

Table A.1 Coefficients of the balancing polynomias for c_p

	H₂	N₂	O₂	CO	SO₂
a_0	14.198835	1.03867	0.9148035	1.03977	0.608273
a_1	5.1219731 ·10⁻³	1.19697 ·10⁻⁵	1.1142613 ·10⁻⁴	2.05062 ·10⁻⁵	5.58701·10⁻⁴
a_2	− 3.9621966 ·10⁻⁵	− 3.08115 ·10⁻⁸	7.9873803 ·10⁻⁷	1.01911 ·10⁻⁷	2.65712 ·10⁻⁷
a_3	1.8541994 ·10⁻⁷	2.37993 ·10⁻⁹	7.2136812 ·10⁻¹⁰	2.57399 ·10⁻⁹	− 2.95578 ·10⁻⁹
a_4	− 5.6148295 ·10⁻¹⁰	− 6.36632 ·10⁻¹²	− 1.3372879 ·10⁻¹¹	− 8.38933 ·10⁻¹²	6.15134 ·10⁻¹²
a_5	1.1544510 ·10⁻¹²	7.92153 ·10⁻¹⁵	4.1958417 ·10⁻¹⁴	1.21949 ·10⁻¹⁴	− 6.85033 ·10⁻¹⁵
a_6	− 1.5977289 ·10⁻¹⁵	− 5.40751 ·10⁻¹⁸	− 7.1053002 ·10⁻¹⁷	− 9.66631 ·10⁻¹⁸	4.43085 ·10⁻¹⁸
a_7	1.4598684 ·10⁻¹⁸	1.96038 ·10⁻²¹	7.3611113 ·10⁻²⁰	4.06887 ·10⁻²¹	− 1.56474 ·10⁻²¹
a_8	− 8.4356364 ·10⁻²²	− 2.95746 ·10⁻²⁵	− 4.6619206 ·10⁻²³	− 7.13820 ·10⁻²⁵	2.33266 ·10⁻²⁵
a_9	2.7917387 ·10⁻²⁵		1.6600446 ·10⁻²⁶		
a_{10}	− 4.0306899 ·10⁻²⁹		− 2.5485300 ·10⁻³⁰		

	CO₂	H₂O	NO₂	N₂O	Luft
a_0	0.81650777	1.8583998	0.790444	0.850481	1.00326
a_1	1.0921781 ·10⁻³	1.8315303 ·10⁻⁴	6.46468 ·10⁻⁴	1.10819 ·10⁻³	3.31469 ·10⁻⁵
a_2	− 9.3659550 ·10⁻⁷	1.2200933 ·10⁻⁶	8.78716 ·10⁻⁷	− 1.59704 ·10⁻⁶	2.19458 ·10⁻⁷
a_3	− 1.9413723 ·10⁻⁹	2.0272951 ·10⁻⁹	− 4.26744 ·10⁻⁹	2.57993 ·10⁻⁹	1.37657 ·10⁻⁹
a_4	1.3865062 ·10⁻¹¹	− 2.4418519 ·10⁻¹¹	6.90645 ·10⁻¹²	− 3.65896 ·10⁻¹²	− 4.93196 ·10⁻¹²
a_5	− 3.9575880 ·10⁻¹⁴	8.0838394 ·10⁻¹⁴	− 6.03641 ·10⁻¹⁵	3.48688 ·10⁻¹⁵	7.11970 ·10⁻¹⁵
a_6	6.6933732 ·10⁻¹⁷	− 1.4590993 ·10⁻¹⁶	2.89216 ·10⁻¹⁸	− 1.97856 ·10⁻¹⁸	− 5.50786 ·10⁻¹⁸
a_7	− 7.0743313 ·10⁻²⁰	1.5854100 ·10⁻¹⁹	− 6.52474 ·10⁻²²	5.91706 ·10⁻²²	2.25552 ·10⁻²¹
a_8	4.5864721 ·10⁻²³	− 1.0374537 ·10⁻²²	3.77757 ·10⁻²⁶	− 6.91526 ·10⁻²⁶	− 3.85310 ·10⁻²⁵
a_9	− 1.6694210 ·10⁻²⁶	3.7772387 · 10⁻²⁶			
a_{10}	2.6128455 ·10⁻³⁰	− 5.8888202 ·10⁻³⁰			

The true specific heat capacity at constant pressure of air was determined for the composition 78.09 % N₂, 20.95 % O₂, 0.93 % Ar and 0.03 % CO₂ using Eqs. (4.171) and (4.206).

Table A.2 True values of c_p, κ, mean values of $c_{pm}|_{0°C}^t$, κ_m and $\overline{c}_{pm}|_{0°C}^t$, $\overline{\kappa}_m$ of H_2 and of N_2

t °C	c_p kJ/kgK	κ	c_{pm} kJ/kgK	κ_m	$\overline{c}_{pm}$ kJ/kgK	$\overline{\kappa}_m$	c_p kJ/kgK	κ	c_{pm} kJ/kgK	κ_m	$\overline{c}_{pm}$ kJ/kgK	$\overline{\kappa}_m$
0	14.199	1.4094	14.199	1.4094	14.199	1.4094	1.0387	1.4001	1.0387	1.4001	1.0387	1.4001
25	14.305	1.4051	14.255	1.4071	14.255	1.4072	1.0390	1.3999	1.0388	1.4000	1.0388	1.4000
50	14.376	1.4023	14.299	1.4054	14.297	1.4055	1.0395	1.3997	1.0390	1.3999	1.0390	1.3999
100	14.454	1.3993	14.360	1.4030	14.353	1.4032	1.0414	1.3986	1.0396	1.3996	1.0396	1.3996
150	14.489	1.3979	14.398	1.4015	14.388	1.4019	1.0451	1.3966	1.0408	1.3989	1.0406	1.3991
200	14.507	1.3972	14.423	1.4005	14.410	1.4010	1.0509	1.3936	1.0426	1.3980	1.0420	1.3983
250	14.521	1.3967	14.441	1.3998	14.426	1.4004	1.0586	1.3896	1.0450	1.3967	1.0440	1.3972
300	14.538	1.3961	14.456	1.3992	14.439	1.3999	1.0679	1.3849	1.0480	1.3951	1.0463	1.3960
350	14.559	1.3953	14.469	1.3987	14.450	1.3994	1.0784	1.3797	1.0516	1.3932	1.0490	1.3946
400	14.586	1.3942	14.482	1.3982	14.460	1.3990	1.0899	1.3743	1.0556	1.3911	1.0520	1.3930
450	14.622	1.3929	14.495	1.3977	14.471	1.3986	1.1017	1.3687	1.0601	1.3888	1.0552	1.3913
500	14.667	1.3912	14.510	1.3971	14.482	1.3982	1.1138	1.3633	1.0649	1.3864	1.0586	1.3896
550	14.722	1.3892	14.527	1.3965	14.494	1.3978	1.1257	1.3581	1.0699	1.3839	1.0621	1.3878
600	14.785	1.3869	14.546	1.3958	14.507	1.3973	1.1373	1.3531	1.0750	1.3814	1.0656	1.3861
650	14.857	1.3843	14.567	1.3950	14.521	1.3967	1.1486	1.3485	1.0802	1.3789	1.0691	1.3843
700	14.938	1.3814	14.590	1.3941	14.537	1.3961	1.1593	1.3441	1.0855	1.3763	1.0727	1.3826
750	15.025	1.3784	14.616	1.3931	14.554	1.3955	1.1695	1.3401	1.0908	1.3738	1.0761	1.3808
800	15.118	1.3752	14.645	1.3921	14.572	1.3948	1.1791	1.3364	1.0960	1.3714	1.0796	1.3792
900	15.318	1.3685	14.708	1.3897	14.611	1.3933	1.1967	1.3298	1.1062	1.3667	1.0862	1.3760
1000	15.530	1.3616	14.780	1.3871	14.654	1.3917	1.2121	1.3243	1.1160	1.3623	1.0925	1.3730
1100	15.746	1.3549	14.858	1.3843	14.700	1.3900	1.2256	1.3196	1.1254	1.3582	1.0984	1.3703
1200	15.964	1.3484	14.941	1.3813	14.748	1.3882	1.2373	1.3156	1.1342	1.3544	1.1039	1.3677

Table A.3 True values of c_p, κ, mean values of $c_{pm}|_{0°C}^t$, κ_m and $\overline{c}_{pm}|_{0°C}^t$, $\overline{\kappa}_m$ of O_2 and of CO

t °C	c_p kJ/kgK	κ	c_{pm} kJ/kgK	κ_m	$\overline{c}_{pm}$ kJ/kgK	$\overline{\kappa}_m$	c_p kJ/kgK	κ	c_{pm} kJ/kgK	κ_m	$\overline{c}_{pm}$ kJ/kgK	$\overline{\kappa}_m$
0	0.9148	1.3967	0.9148	1.3967	0.9148	1.3967	1.0398	1.3995	1.0398	1.3995	1.0398	1.3995
25	0.9181	1.3947	0.9164	1.3958	0.9163	1.3958	1.0404	1.3992	1.0401	1.3994	1.0401	1.3994
50	0.9224	1.3922	0.9183	1.3946	0.9182	1.3947	1.0413	1.3987	1.0404	1.3992	1.0404	1.3992
100	0.9337	1.3856	0.9230	1.3918	0.9225	1.3921	1.0447	1.3969	1.0416	1.3985	1.0415	1.3986
150	0.9476	1.3778	0.9288	1.3884	0.9276	1.3891	1.0504	1.3939	1.0435	1.3975	1.0432	1.3977
200	0.9631	1.3695	0.9355	1.3846	0.9332	1.3859	1.0585	1.3897	1.0462	1.3961	1.0454	1.3965
250	0.9791	1.3613	0.9426	1.3806	0.9391	1.3825	1.0685	1.3847	1.0496	1.3943	1.0482	1.3951
300	0.9948	1.3535	0.9500	1.3765	0.9450	1.3793	1.0801	1.3790	1.0537	1.3922	1.0514	1.3934
350	1.0098	1.3465	0.9575	1.3725	0.9508	1.3761	1.0926	1.3730	1.0584	1.3898	1.0549	1.3916
400	1.0238	1.3401	0.9649	1.3685	0.9564	1.3730	1.1057	1.3670	1.0635	1.3872	1.0587	1.3896
450	1.0367	1.3345	0.9722	1.3648	0.9618	1.3701	1.1189	1.3611	1.0689	1.3845	1.0626	1.3876
500	1.0485	1.3295	0.9792	1.3612	0.9670	1.3674	1.1320	1.3554	1.0746	1.3817	1.0666	1.3856
550	1.0592	1.3250	0.9860	1.3578	0.9720	1.3649	1.1446	1.3501	1.0804	1.3788	1.0707	1.3836
600	1.0690	1.3211	0.9925	1.3546	0.9766	1.3625	1.1567	1.3452	1.0862	1.3760	1.0748	1.3816
650	1.0779	1.3176	0.9988	1.3516	0.9811	1.3603	1.1682	1.3407	1.0921	1.3733	1.0788	1.3796
700	1.0860	1.3145	1.0047	1.3488	0.9853	1.3582	1,1790	1,3365	1.0979	1.3705	1.0827	1.3777
750	1.0934	1.3117	1.0104	1.3462	0.9892	1.3562	1.1891	1.3327	1.1037	1.3679	1.0866	1.3759
800	1.1002	1.3092	1.0158	1.3437	0.9930	1.3544	1.1986	1.3292	1.1093	1.3654	1.0903	1.3741
900	1.1122	1.3048	1.0258	1.3392	0.9999	1.3511	1.2157	1.3230	1.1202	1.3605	1.0975	1.3708
1000	1.1227	1.3011	1.0350	1.3352	1.0062	1.3482	1.2305	1.3179	1.1305	1.3561	1.1041	1.3677
1100	1.1321	1.2979	1.0434	1.3316	1.0118	1.3455	1.2433	1.3136	1.1402	1.3520	1.1104	1.3649
1200	1.1408	1.2950	1.0512	1.3284	1.0170	1.3432	1.2544	1.3100	1.1493	1.3482	1.1161	1.3623

Table A.4 True values of c_p, κ, mean values of $c_{pm}|_{0\,°C}^{t}$, κ_m and $\bar{c}_{pm}|_{0\,°C}^{t}$, $\bar{\kappa}_m$ of CO_2 and of H_2O

t	c_p	κ	c_{pm}	κ_m	$\bar{c}_{pm}$	$\bar{\kappa}_m$	c_p	κ	c_{pm}	κ_m	$\bar{c}_{pm}$	$\bar{\kappa}_m$
°C	$\frac{kJ}{kg\,K}$		$\frac{kJ}{kg\,K}$		$\frac{kJ}{kg\,K}$		$\frac{kJ}{kg\,K}$		$\frac{kJ}{kg\,K}$		$\frac{kJ}{kg\,K}$	
0	0.8165	1.3010	0.8165	1.3010	0.8165	1.3010	1.8584	1.3304	1.8584	1.3304	1.8584	1.3304
25	0.8432	1.2888	0.8300	1.2947	0.8298	1.2948	1.8638	1.3291	1.8609	1.3298	1.8609	1.3298
50	0.8686	1.2780	0.8430	1.2888	0.8423	1.2892	1.8707	1.3275	1.8640	1.3291	1.8639	1.3291
100	0.9155	1.2600	0.8677	1.2783	0.8652	1.2794	1.8892	1.3233	1.8718	1.3273	1.8710	1.3275
150	0.9574	1.2459	0.8908	1.2692	0.8856	1.2712	1.9125	1.3181	1.8813	1.3251	1.8794	1.3255
200	0.9950	1.2344	0.9122	1.2612	0.9040	1.2642	1.9393	1.3123	1.8924	1.3225	1.8887	1.3234
250	1.0289	1.2249	0.9322	1.2542	0.9207	1.2582	1.9683	1.3063	1.9047	1.3198	1.8987	1.3211
300	1.0596	1.2170	0.9509	1.2479	0.9359	1.2529	1.9987	1.3002	1.9178	1.3169	1.9091	1.3188
350	1.0876	1.2102	0.9685	1.2424	0.9499	1.2483	2.0303	1.2942	1.9316	1.3139	1.9198	1.3165
400	1.1132	1.2044	0.9850	1.2373	0.9628	1.2441	2.0627	1.2882	1.9460	1.3109	1.9306	1.3142
450	1.1365	1.1994	1.0005	1.2328	0.9747	1.2404	2.0960	1.2824	1.9608	1.3078	1.9415	1.3118
500	1.1579	1.1950	1.0152	1.2286	0.9858	1.2371	2.1300	1.2766	1.9760	1.3047	1.9525	1.3095
550	1.1774	1.1911	1.0291	1.2249	0.9961	1.2341	2.1647	1.2710	1.9915	1.3016	1.9636	1.3073
600	1.1952	1.1877	1.0422	1.2214	1.0058	1.2313	2.1999	1.2655	2.0074	1.2985	1.9747	1.3050
650	1.2115	1.1847	1.0546	1.2182	1.0148	1.2288	2.2354	1.2602	2.0236	1.2955	1.9858	1.3028
700	1.2265	1.1821	1.0664	1.2153	1.0233	1.2264	2.2710	1.2551	2.0400	1.2924	1.9969	1.3006
750	1.2402	1.1797	1.0775	1.2126	1.0313	1.2243	2.3066	1.2501	2.0566	1.2893	2.0079	1.2984
800	1.2528	1.1776	1.0881	1.2101	1.0388	1.2223	2.3419	1.2454	2.0733	1.2863	2.0190	1.2963
900	1.2749	1.1740	1.1076	1.2056	1.0525	1.2188	2.4113	1.2367	2.1070	1.2805	2.0408	1.2922
1000	1.2937	1.1710	1.1253	1.2018	1.0649	1.2157	2.4781	1.2289	2.1408	1.2748	2.0623	1.2883
1100	1.3097	1.1686	1.1414	1.1984	1.0759	1.2130	2.5414	1.2219	2.1744	1.2694	2.0832	1.2846
1200	1.3233	1.1665	1.1560	1.1954	1.0860	1.2106	2.6013	1.2157	2.2075	1.2643	2.1036	1.2811

Table A.5 True values of c_p, κ, mean values of $c_{pm}|_{0\,°C}^{t}$, κ_m and $\bar{c}_{pm}|_{0\,°C}^{t}$, $\bar{\kappa}_m$ of NO_2 and of N_2O

t	c_p	κ	c_{pm}	κ_m	$\bar{c}_{pm}$	$\bar{\kappa}_m$	c_p	κ	c_{pm}	κ_m	$\bar{c}_{pm}$	$\bar{\kappa}_m$
°C	$\frac{kJ}{kg\,K}$		$\frac{kJ}{kg\,K}$		$\frac{kJ}{kg\,K}$		$\frac{kJ}{kg\,K}$		$\frac{kJ}{kg\,K}$		$\frac{kJ}{kg\,K}$	
0	0.7904	1.2964	0.7904	1.2964	0.7904	1.2964	0.8505	1.2855	0.8505	1.2855	0.8505	1.2855
25	0.8071	1.2885	0.7987	1.2925	0.7986	1.2925	0.8772	1.2745	0.8640	1.2798	0.8638	1.2799
50	0.8245	1.2807	0.8072	1.2885	0.8067	1.2887	0.9022	1.2648	0.8769	1.2746	0.8762	1.2749
100	0.8602	1.2660	0.8248	1.2806	0.8229	1.2814	0.9476	1.2490	0.9011	1.2652	0.8986	1.2662
150	0.8959	1.2527	0.8425	1.2731	0.8387	1.2747	0.9879	1.2364	0.9235	1.2572	0.9185	1.2589
200	0.9300	1.2412	0.8602	1.2660	0.8537	1.2685	1.0240	1.2262	0.9442	1.2501	0.9362	1.2528
250	0.9621	1.2313	0.8774	1.2594	0.8680	1.2630	1.0567	1.2177	0.9635	1.2439	0.9523	1.2475
300	0.9915	1.2229	0.8940	1.2534	0.8814	1.2579	1.0864	1.2105	0.9815	1.2383	0.9670	1.2428
350	1.0182	1.2158	0.9099	1.2479	0.8939	1.2534	1.1135	1.2043	0.9985	1.2334	0.9805	1.2386
400	1.0423	1.2098	0.9249	1.2428	0.9056	1.2493	1.1382	1.1990	1.0144	1.2288	0.9929	1.2350
450	1.0639	1.2046	0.9392	1.2383	0.9165	1.2456	1.1608	1.1944	1.0294	1.2248	1.0045	1.2316
500	1.0832	1.2003	0.9527	1.2341	0.9266	1.2423	1.1815	1.1903	1.0436	1.2210	1.0152	1.2286
550	1.1004	1.1965	0.9653	1.2303	0.9359	1.2393	1.2004	1.1868	1.0570	1.2176	1.0252	1.2259
600	1.1159	1.1933	0.9772	1.2269	0.9447	1.2366	1.2178	1.1836	1.0697	1.2145	1.0345	1.2234
650	1.1297	1.1904	0.9884	1.2238	0.9528	1.2341	1.2336	1.1808	1.0817	1.2116	1.0432	1.2211
700	1.1422	1.1880	0.9990	1.2209	0.9604	1.2318	1.2481	1.1784	1.0931	1.2089	1.0514	1.2190
750	1.1535	1.1858	1.0089	1.2182	0.9676	1.2297	1.2614	1.1762	1.1039	1.2065	1.0592	1.2171
800	1.1636	1.1839	1.0183	1.2158	0.9742	1.2278	1.2735	1.1742	1.1141	1.2042	1.0664	1.2153
900	1.1811	1.1807	1.0354	1.2115	0.9863	1.2243	1.2951	1.1708	1.1330	1.2001	1.0798	1.2121
1000	1.1957	1.1781	1.0507	1.2077	0.9971	1.2214	1.3135	1.1680	1.1502	1.1965	1.0917	1.2093
1100	1.2080	1.1759	1.0645	1.2045	1.0067	1.2188	1.3292	1.1657	1.1658	1.1934	1.1025	1.2068
1200	1.2186	1.1741	1.0769	1.2017	1.0153	1.2166	1.3428	1.1637	1.1800	1.1906	1.1122	1.2046

Table A.6 True values of c_p, κ, mean values of $c_{pm}|_{0°C}^{t}$, κ_m and $\bar{c}_{pm}|_{0°C}^{t}$, $\bar{\kappa}_m$ of SO_2 and of air

t	c_p	κ	c_{pm}	κ_m	$\bar{c}_{pm}$	$\bar{\kappa}_m$	c_p	κ	c_{pm}	κ_m	$\bar{c}_{pm}$	$\bar{\kappa}_m$
°C	$\frac{kJ}{kg\,K}$		$\frac{kJ}{kg\,K}$		$\frac{kJ}{kg\,K}$		$\frac{kJ}{kg\,K}$		$\frac{kJ}{kg\,K}$		$\frac{kJ}{kg\,K}$	
0	0.6083	1.2712	0.6083	1.2712	0.6083	1.2712	1.0033	1.4008	1.0033	1.4008	1.0033	1.4008
25	0.6224	1.2635	0.6153	1.2673	0.6152	1.2674	1.0042	1.4003	1.0037	1.4006	1.0037	1.4006
50	0.6365	1.2561	0.6224	1.2635	0.6220	1.2637	1.0056	1.3995	1.0043	1.4002	1.0043	1.4002
100	0.6644	1.2428	0.6365	1.2561	0.6350	1.2569	1.0097	1.3972	1.0059	1.3993	1.0057	1.3994
150	0.6907	1.2314	0.6502	1.2494	0.6472	1.2508	1.0158	1.3939	1.0081	1.3981	1.0077	1.3983
200	0.7149	1.2218	0.6634	1.2432	0.6585	1.2455	1.0237	1.3897	1.0110	1.3965	1.0101	1.3970
250	0.7367	1.2138	0.6759	1.2376	0.6689	1.2407	1.0332	1.3847	1.0145	1.3946	1.0129	1.3955
300	0.7561	1.2072	0.6877	1.2326	0.6785	1.2365	1.0439	1.3793	1.0185	1.3925	1.0161	1.3938
350	0.7732	1.2017	0.6987	1.2281	0.6872	1.2328	1.0554	1.3736	1.0229	1.3901	1.0195	1.3920
400	0.7882	1.1971	0.7090	1.2241	0.6952	1.2295	1.0673	1.3679	1.0277	1.3876	1.0230	1.3900
450	0.8013	1.1933	0.7185	1.2204	0.7025	1.2266	1,0792	1.3624	1.0328	1.3849	1.0267	1.3881
500	0.8129	1.1900	0.7274	1.2172	0.7093	1.2240	1.0911	1.3570	1.0380	1.3823	1.0305	1.3861
550	0.8231	1.1872	0.7356	1.2142	0.7154	1.2216	1.1026	1.3520	1.0434	1.3796	1.0342	1.3842
600	0.8321	1.1848	0.7433	1.2115	0.7211	1.2195	1.1136	1.3473	1.0488	1.3769	1.0380	1.3823
650	0.8401	1.1827	0.7505	1.2091	0.7264	1.2175	1.1242	1.3429	1.0542	1.3742	1.0417	1.3804
700	0.8472	1.1809	0.7571	1.2069	0.7313	1.2158	1.1342	1.3389	1.0595	1.3716	1.0453	1.3786
750	0.8536	1.1793	0,7633	1.2048	0.7358	1.2142	1.1436	1.3352	1.0648	1.3691	1.0489	1.3768
800	0.8593	1.1779	0.7692	1.2030	0.7400	1.2127	1.1524	1.3317	1.0700	1.3666	1.0523	1.3751
900	0.8691	1.1755	0.7797	1.1997	0.7476	1.2101	1.1685	1.3257	1.0801	1.3620	1.0589	1.3719
1000	0.8773	1.1736	0.7891	1.1968	0.7543	1.2078	1.1825	1.3206	1.0896	1.3577	1.0651	1.3689
1100	0.8842	1.1720	0.7974	1.1944	0.7602	1,2059	1.1949	1.3162	1.0987	1.3537	1.0709	1.3662
1200	0.8901	1.1707	0.8049	1.1922	0.7655	1.2042	1.2057	1.3125	1.1071	1.3500	1.0763	1.3637

Additional Informations to Tables A.1 to A.6:

The dependence of the specific isobaric heat capacity c_p on the temperature is described for a series of ideal gases in the range from 10 K to 6000 K in the form of tables in [5]. For some gases, there is no temperature dependence. For example, $c_p/R = 2{,}5000 = $ const applies to the noble gas argon regardless of temperature.

For the temperature-dependent c_p values, the following approach is adopted:

$$c_p = \sum_{i=0}^{n} a_i\, t^i \tag{A.1}$$

Eq. (A.1) is a balancing polynomial of the nth degree. The coefficients a_i (Table A.1) are determined so that the sum of the squares of the errors becomes a minimum. The deviations are less than 0.01 % in the temperature range between 270 K and 1500 K. The function according to Eq. (A.1) gives the true values of the specific heat capacity at constant pressure. The mean specific heat capacity at constant pressure according to Eq. (2.96) is then

$$c_{pm}|_{0\,°C}^{t} = \sum_{i=0}^{n} \frac{a_i}{i+1} t^i\,. \tag{A.2}$$

Eqs. (4.48) to (4.50) are valid for both the true as well as the mean values of the specific isobaric heat capacity. This means that the isentropic exponent according to Eq. (4.44) is different for the true and for the mean specific heat capacity.

Using the true value for c_p according to Eq. (A.1) one obtains

$$c_v = c_p - R \tag{A.3}$$

$$\kappa = \frac{c_p}{c_p - R} \ . \tag{A.4}$$

With the mean value c_{pm} according to Eq. (A.2) one obtains

$$c_{vm} = c_{pm} - R \tag{A.5}$$

$$\kappa_m = \frac{c_{pm}}{c_{pm} - R} \ . \tag{A.6}$$

Also in the case of a polytropic change of state, a distinction is made between the true and the mean value of the specific heat capacity. According to Eq. (4.132)

$$c_n = c_p - \frac{n}{n-1} R \tag{A.7}$$

$$c_{nm} = c_{pm} - \frac{n}{n-1} R \ . \tag{A.8}$$

become. In example 5.2 it was shown that one has to distinguish between different averaging methods for the specific heat capacity. The mean values c_{pm}, c_{vm}, c_{nm} are needed for the calculation of the change of internal energy, the change of enthalpy and for the calculation of heat. The mean values $\bar{c}_{pm}$, $\bar{c}_{vm}$ $\bar{c}_{nm}$ are needed for the calculation of the change of entropy. The values given in Eqs. (2.90) and (2.96) are as follows

$$\bar{c}_{pm}|_{0\,°C}^{t} = \frac{1}{\ln \frac{T}{T_0}} \int_{T_0}^{T} c_p \frac{dT}{T} \tag{A.9}$$

$$\bar{c}_{pm}|_{t_1}^{t_2} = \frac{1}{\ln \frac{T_2}{T_1}} \left(\bar{c}_{pm}|_{0\,°C}^{t_2} \cdot \ln \frac{T_2}{T_0} - \bar{c}_{pm}|_{0\,°C}^{t_1} \cdot \ln \frac{T_1}{T_0} \right) \ . \tag{A.10}$$

With the approach according to Eq. (A.1) it follows from Eq. (A.9)

$$\bar{c}_{pm}|_{0\,°C}^{t} = \sum_{i=0}^{n} (-1)^i a_i T_0^i + \frac{1}{\ln \frac{T}{T_0}} \sum_{k=1}^{n} (T^k - T_0^k) \frac{1}{k} \sum_{i=k}^{n} (-1)^{i-k} \binom{i}{k} a_i T_0^{i-k}$$

$$= (a_0 - a_1 T_0 + a_2 T_0^2 - a_3 T_0^3 + a_4 T_0^4 - a_5 T_0^5 + a_6 T_0^6 - + \ldots) +$$

$$+ \frac{1}{\ln \frac{T}{T_0}} [(T - T_0)(a_1 - 2 a_2 T_0 + 3 a_3 T_0^2 - 4 a_4 T_0^3 + 5 a_5 T_0^4 - 6 a_6 T_0^5 + - \ldots) +$$

$$+ (T^2 - T_0^2) \frac{1}{2} (a_2 - 3 a_3 T_0 + 6 a_4 T_0^2 - 10 a_5 T_0^3 + 15 a_6 T_0^4 - + \ldots) +$$

$$+ (T^3 - T_0^3) \frac{1}{3} (a_3 - 4 a_4 T_0 + 10 a_5 T_0^2 - 20 a_6 T_0^3 + - \ldots) +$$

$$+ (T^4 - T_0^4) \frac{1}{4} (a_4 - 5 a_5 T_0 + 15 a_6 T_0^2 - + \ldots) +$$

$$+ (T^5 - T_0^5) \frac{1}{5} (a_5 - 6 a_6 T_0 + - \ldots) + \ldots] \ . \tag{A.11}$$

Analogous to Eqs. (A.5), (A.6) and (A.8) apply the equations

$$\bar{c}_{vm} = \bar{c}_{pm} - R \tag{A.12}$$

$$\overline{\kappa}_m = \frac{\overline{c}_{pm}}{\overline{c}_{pm} - R} \tag{A.13}$$

$$\overline{c}_{nm} = \overline{c}_{pm} - \frac{n}{n-1} R . \tag{A.14}$$

Into the equations of Tables 4.3 to 4.5 according to Eq. (A.6) κ_m is to be inserted.

According to section 4.3.5, the isentrope is a special case of the polytrope. The question arises which value for the isentropic exponent is to be used in Eq. (4.117). From Eqs. (4.129), (4.132) and (4.140)

$$\mathrm{d}Q_{rev} = m \left(c_p - \frac{n}{n-1} R \right) \mathrm{d}T \tag{A.15}$$

$$\mathrm{d}S = m \left(c_p - \frac{n}{n-1} R \right) \frac{\mathrm{d}T}{T} \tag{A.16}$$

one sees that at the transition to the isentrope, where Eqs. (3.87) and (3.88) are valid, n can no longer be constant at variable c_p. The polytropic exponent, which now becomes the isentropic exponent, must be variable and, when integrating Eqs. (A.15) and (A.16), has different values.

In the transition from Eq. (A.15) to Eq. (4.130), c_p becomes c_{pm} according to Eq. (2.96) and in the case of the isentropes n becomes κ_m according to Eq. (A.6). In the integration of Eq. (A.16) leading to Eq. (4.141), c_p passes into $\overline{c}_{pm}$ according to Eq. (A.10) and in the case of the isentrope n into $\overline{\kappa}_m$ according to Eq. (A.13).

In the case of an isentropic change of state, according to Eq. (4.58), if the index m is not used in the specific isobaric heat capacity and the isentropic exponent, follows

$$\overline{c}_{pm} \ln \frac{T_2}{T_1} = \frac{\overline{\kappa}_m}{\overline{\kappa}_m - 1} R \ln \frac{T_2}{T_1} = R \ln \left(\frac{T_2}{T_1} \right)^{\frac{\overline{\kappa}_m}{\overline{\kappa}_m - 1}} = R \ln \frac{p_2}{p_1} . \tag{A.17}$$

Accordingly, in Eqs. (4.94), (4.95) and (4.98) to (4.101) the isentropic exponent $\overline{\kappa}$ or $\overline{\kappa}_m$ is to be used ccording to Eq. (A.13).

If one considers a unique and constant polytropic exponent as a characteristic of a polytropic change of state, the isentrope can no longer be regarded as a special case of the polytrope if the specific isobaric heat capacity depends on the temperature.

Tables A.2 to A.6 contain the properties calculated according to Eqs. (A.1) to (A.6) and (A.11) to (A.13). If an isobar is plotted in a T, s diagram for an ideal gas with a temperature dependent specific isobaric heat capacity, one has to use Eq. (4.58). Here, for c_p the mean specific isobaric heat capacity $\overline{c}_{pm}$ is to be inserted according to Eq. (A.11):

$$s = s_1 + \overline{c}_{pm} \ln \frac{T}{T_0} - R \ln \frac{p}{p_0} \tag{A.18}$$

For an isobar according to Eq. (A.18), the entropy is only a function of the temperature. A check whether $\overline{c}_{pm}$ is to be used here can be done according to Eq. (5.50) via the relation

$$T \left(\frac{\partial s}{\partial T} \right)_p = c_p . \tag{A.19}$$

Here c_p is the true specific heat capacity at constant pressure at temperature T. One obtains from Eq. (A.18)

$$\left(\frac{\partial s}{\partial T} \right)_p = \frac{\mathrm{d}\overline{c}_{pm}}{\mathrm{d}T} \ln \frac{T}{T_0} + \frac{\overline{c}_{pm}}{T} . \tag{A.20}$$

Therefore, according to Eqs. (A.19) and (A.20) the following equation must hold:

$$T\frac{\mathrm{d}\bar{c}_{pm}}{\mathrm{d}T}\ln\frac{T}{T_0}+\bar{c}_{pm}=c_p \qquad (A.21)$$

The derivative of $\bar{c}_{pm}$ is

$$\frac{\mathrm{d}\bar{c}_{pm}}{\mathrm{d}T}=-\frac{1}{(\ln\frac{T}{T_0})^2}\frac{1}{T}\sum_{k=1}^{n}(T^k-T_0^k)\frac{1}{k}\sum_{i=k}^{n}(-1)^{i-k}\binom{i}{k}a_i T_0^{i-k}+$$

$$+\frac{1}{\ln\frac{T}{T_0}}\sum_{k=1}^{n}T^{k-1}\sum_{i=k}^{n}(-1)^{i-k}\binom{i}{k}a_i T_0^{i-k} \; . \qquad (A.22)$$

After multiplication with $T\ln(T/T_0)$ and addition of $\bar{c}_{pm}$, the first expression disappears:

$$T\left(\frac{\partial s}{\partial T}\right)_p=\sum_{i=0}^{n}(-1)^i a_s i\, T_0^i+\sum_{k=1}^{n}T^k\sum_{i=k}^{n}(-1)^{i-k}\binom{i}{k}a_i T_0^{i-k}$$

$$=a_0-a_1 T_0+a_2 T_0^2-a_3 T_0^3+a_4 T_0^4-+\cdots+T(a_1-2\,a_2\,T_0+3\,a_3\,T_0^2+4\,a_4\,T_0^3-+\ldots)+$$

$$+T^2(a_2-3\,a_3\,T_0+6\,a_4\,T_0^2-+\ldots)+T^3(a_3-4\,a_4\,T_0+10\,a_5\,T_0^2-+\ldots)+\ldots$$

$$T\left(\frac{\partial s}{\partial T}\right)_p=a_0+a_1(T-T_0)+a_2(T^2-2\,T\,T_0+T_0^2)+a_3(T^3-3\,T^2\,T_0+3\,T\,T_0^2-T_0^3)+$$

$$+a_4(T^4-4\,T^3\,T_0+6\,T^2\,T_0^2-4\,T\,T_0^3+T_0^4)+\ldots$$

$$=a_0+a_1(T-T_0)+a_2(T-T_0)^2+a_3(T-T_0)^3+a_4(T-T_0)^4+\ldots$$

$$=a_0+a_1\,t+a_2\,t^2+a_3\,t^3+a_4\,t^4+\ldots \qquad (A.23)$$

The right sides of Eqs. (A.1) and (A.23) agree. This proves the correctness of Eqs. (A.21) and (A.18).

The considerations described here have not always been consistently observed in the previous sections. Therefore the following examples are used to show the effects of the new approach:

Example 4.2 For the entropy change one needs $\bar{c}_{vm}|_{18\,°C}^{212\,°C}=0.7295$ kJ/(kg K). Thus one obtains in c) $S_{G2}-S_{G1}=m\,\bar{c}_{vm}\ln\frac{T_2}{T_1}=0{,}669$ kg $\cdot\, 0{,}7295$ kJ/(kg K) $\cdot\ln\frac{485\text{ K}}{291\text{ K}}=$
0,249 kJ/K . For the total entropy changes according to c) and d) one obtains 0.164 kJ/K and 0.080 kJ/K.

Example 4.3 The numerical values for c_{pm} and $\bar{c}_{pm}$ agree: There is no effect.

Example 4.8 $\bar{c}_{pm}|_{18\,°C}^{234\,°C}=1.0129$ kJ/(kg K) $\qquad \bar{\kappa}_m=\dfrac{1.0129\text{ kJ/(kg K)}}{(1.0129-0.2871)\text{ kJ/(kg K)}}=$
1.3956

$T_2=291\text{ K}\cdot 4^{0.3956}=504$ K $\qquad t_2=231\,°C \qquad p_2=0.9493$ bar $\cdot\, 4^{1.3956}=6.571$ bar

$W_{V12}=\dfrac{p_1 V_1}{\kappa_m-1}\left[\left(\dfrac{V_1}{V_2}\right)^{\bar{\kappa}_m-1}-1\right]=\dfrac{0.9493\text{ bar}\cdot 1.0053\cdot 10^{-3}\text{ m}^3}{1.3949-1}\left(4^{1.3956-1}-1\right)$

$=176.54$ J $\qquad$ The impact energy is $\quad 176.54$ J $-\,71.58$ J $=104.96$ J

Example 4.12 The numerical values for c_{pm} and $\bar{c}_{pm}$ match: There is no effect.

Table A.7 Molar mass M, specific gas constant R, specific isobaric heat capacity c_p^0 resp. c_p; molar enthalpy of formation $H_{m,0i}^{f\square}$, molar absolute entropy $S_{m,0i}^{\square}$, molar *Gibbs* function $G_{m,0i}^{\square}$ and phase Ph in the standard thermochemical state ($T_0 = 298.15$ K, $p_0 = 1$ bar $= 100$ kPa) [9].

Substance	M kg/kmol	R kJ/(kg K)	c_p^0 resp. c_p kJ/(kg K)	$H_{m,0i}^{f\square}$ MJ/kmol	$S_{m,0i}^{\square}$ kJ/(kmol K)	$G_{m,0i}^{\square}$ MJ/kmol	Ph
O	15.9994	0.51967	1.3696	249,18	161.059	201.16	g
O_2	31.9988	0.25984	0.9181	0	205.152	-61.166	g
H	1.00794	8.24897	20.622	217.908	114.717	183.795	g
H_2	2.01588	4.12449	14.304	0	130.680	-38.962	g
OH	17.0073	0.48888	1.7576	47.52	189.395	-8.95	g
H_2O	18.0153	0.46152	1.8646	-241.83	188.835	-298.13	g
H_2O	18.0153	0.46152	4.1819	-285.84	69.93	-306.69	fl
He	4.002602	2.07727	5.1932	0	126.153	-37.613	g
Ne	20.1797	0.41202	1.0300	0	146.328	-43.628	g
Ar	39.948	0.20813	0.5203	0	154.846	-46.167	g
Kr	83.80	0.09922	0.2480	0	164.085	-48.922	g
Xe	131.293	0.06333	0.1583	0	169.685	-50.592	g
F_2	37.99680	0.21882	0.8239	0	202.791	-60.462	g
HF	20.00634	0.41559	1.4564	-273.3	173.779	-325.1	g
Cl_2	70.906	0.11726	0.4788	0	223.081	-66.512	g
HCl	36.461	0.22804	0.7991	-92.31	186.902	-148.03	g
S	32.065	0.25930	0.7095	0	32.054	-9.557	rh
S	32.065	0.25930	0.7383	277.17	167.829	227.13	g
S_2	64.130	0.12965	0.5068	128.6	228.17	60.57	g
SO_2	64.064	0.12978	0.6219	-296.8	248.22	-370.8	g
H_2S	34.081	0.24396	1.0049	-20.6	205.81	-81.96	g
N	14.0067	0.59361	1.4840	472.7	153.301	427.0	g
N_2	28.0134	0.29681	1.0396	0	191.609	-57.128	g
NO	30.0061	0.27709	0.9965	90.25	210.76	27.41	g
NO_2	46.0055	0.18073	0.7938	33.10	240.04	-38.47	g
N_2O	44.0128	0.18891	0.8700	82.05	219.96	16.07	g
NH_3	17.0305	0.48821	2.0921	-45.94	192.77	-103.41	g
C	12.0107	0.69226	0.7091	0	5.74	-1.711	Gr
C	12.0107	0.69226	1.7350	716.7	158.10	669.5	g
CO	28.0101	0.29684	1.0404	-110.53	197.660	-169.46	g
CO_2	44.010	0.18892	0.8438	-393.51	213.785	-457.25	g
CH_4	16.042	0.51829	2.185	-74.87	186.25	-130.40	g
CH_3OH	32.042	0.25949	2.546	-239.45	126.61	-277.20	fl
CH_3OH	32.042	0.25949	1.370	-200.66	239.81	-272.16	g
COS	60.076	0.13840	0.672	-138.40	231.58	-207.52	g
HCN	27.025	0.30765	1.293	135.14	201.83	74.96	g
C_2H_2	26.037	0.31933	1.693	226.77	200.94	166.86	g
C_2H_4	28.053	0.29638	1.488	52.47	219.33	-12.92	g
C_2H_6	30.069	0.27651	1.730	-84.73	229.60	-153.18	g
C_2H_5OH	46.068	0.18048	2.434	-276.98	161.00	-324.98	fl
C_2H_5OH	46.068	0.18048	1.420	-235.10	282.7	-319.39	g
HCOOH	46.025	0.18065	2.154	-424.7	129.0	-463.2	fl
CH_2O	30.026	0.27691	1.167	-115.90	218.95	-181.18	g
C_3H_8	44.096	0.18855	1.667	-103.85	270.02	-184.36	g
n-C_4H_{10}	58.122	0.14305	1.690	-124.73	310.14	-217.20	g
n-C_5H_{12}	72.149	0.11524	2.297	-173.83	259.86	-251.31	fl
n-C_6H_{14}	86.175	0.09648	2.263	-198.8	292.5	-286.0	fl
C_6H_6	78.112	0.10644	1.7425	-49.04	171.54	-2.10	fl
n-C_7H_{16}	100.20	0.08298	2.242	-224.4	328.0	-322.2	fl
Air, dry	28.9654	0.28705	1.0047	-0.142	198.827	-59.42	g

Conversion of the molar absolute standard entropy $S_{m,0i}^{\square}$ into the specific absolute standard entropy $s_{0i}^{\square}$:
$s_{0i}^{\square} = S_{m,0i}^{\square}/M$

Table A.8 Steam table (water vapor table), liquid-vapor line and solid-vapor line [128], [129]

t °C	p_{DS} bar	t °C	p_{DS} bar	t °C	p_{DS} bar
−20.0	0.001033	25.0	0.031697	88.0	0.650174
−19.0	0.001136	26.0	0.033637	90.0	0.701824
−18.0	0.001249	27.0	0.035679	92.0	0.756849
−17.0	0.001372	28.0	0.037828	94.0	0.815420
−16.0	0.001507	29.0	0.040089	96.0	0.877711
−15.0	0.001653	30.0	0.042467	98.0	0.943902
−14.0	0.001812	31.0	0.044966	100.0	1.014180
−13.0	0.001985	32.0	0.047592	105.0	1.209021
−12.0	0.002173	33.0	0.050351	110.0	1.433760
−11.0	0.002377	34.0	0.053247	115.0	1.691770
−10.0	0.002599	35.0	0.056286	120.0	1.986654
−9.0	0.002839	36.0	0.059475	125.0	2.322242
−8.0	0.003100	37.0	0.062818	130.0	2.702596
−7.0	0.003382	38.0	0.066324	135.0	3.132010
−6.0	0.003687	39.0	0.069997	140.0	3.615010
−5.0	0.004018	40.0	0.073844	145.0	4.156349
−4.0	0.004375	41.0	0.077873	150.0	4.761014
−3.0	0.004761	42.0	0.082090	155.0	5.434216
−2.0	0.005177	43.0	0.086503	160.0	6.181392
−1.0	0.005627	44.0	0.091118	165.0	7.008204
0.0	0.006112	45.0	0.095944	170.0	7.920532
1.0	0.006571	46.0	0.100988	175.0	8.924475
2.0	0.007060	47.0	0.106259	180.0	10.026346
3.0	0.007581	48.0	0.111764	185.0	11.232669
4.0	0.008135	49.0	0.117512	190.0	12.550179
5.0	0.008726	50.0	0.123513	195.0	13.985815
6.0	0.009354	52.0	0.136305	200.0	15.546719
7.0	0.010021	54.0	0.150215	210.0	19.073907
8.0	0.010730	56.0	0.165322	220.0	23.192877
9.0	0.011483	58.0	0.181708	230.0	27.967925
10.0	0.012282	60.0	0.199458	240.0	33.466519
11.0	0.013129	62.0	0.218664	250.0	39.759391
12.0	0.014028	64.0	0.239421	260.0	46.920711
13.0	0.014981	66.0	0.261827	270.0	55.028395
14.0	0.015989	68.0	0.285986	280.0	64.164593
15.0	0.017057	70.0	0.312006	290.0	74.416425
16.0	0.018188	72.0	0.340001	300.0	85.877083
17.0	0.019383	74.0	0.370088	310.0	98.647456
18.0	0.020647	76.0	0.402389	320,0	112.838559
19.0	0.021982	78.0	0.437031	330.0	128.575219
20.0	0.023392	80.0	0.474147	340.0	146.001811
21.0	0.024881	82.0	0.513875	350.0	165.291643
22.0	0.026452	84.0	0.556355	360.0	186.664034
23.0	0.028109	86.0	0.601738	370.0	210.433673
24.0	0.029856				

Table A.9a Steam table (water vapor table), saturation state (temperature table) [128], [130]

t °C	p bar	v' $\frac{m^3}{kg}$	v'' $\frac{m^3}{kg}$	h' $\frac{kJ}{kg}$	h'' $\frac{kJ}{kg}$	r $\frac{kJ}{kg}$	s' $\frac{kJ}{kg\,K}$	s'' $\frac{kJ}{kg\,K}$
0.01	0.00611655	0.0010002	205.99122	0.00	2500.92	2500.92	0.0000	9.1555
1.00	0.00657086	0.0010001	192.43878	4.18	2502.73	2498.55	0.0153	9.1291
2.00	0.00705986	0.0010001	179.75781	8.39	2504.57	2496.18	0.0306	9.1027
3.00	0.00758081	0.0010001	168.00844	12.60	2506.40	2493.80	0.0459	9.0765
4.00	0.00813548	0.0010001	157.11576	16.81	2508.23	2491.42	0.0611	9.0505
5.00	0.00872575	0.0010001	147.01134	21.02	2510.06	2489.04	0.0763	9.0248
6.00	0.00935355	0.0010001	137.63268	25.22	2511.89	2486.67	0.0913	8.9993
7.00	0.01002091	0.0010001	128.92264	29.43	2513.72	2484.29	0.1064	8.9741
8.00	0.01072995	0.0010002	120.82895	33.63	2515.55	2481.92	0.1213	8.9491
9.00	0.01148288	0.0010003	113.30375	37.83	2517.38	2479.55	0.1362	8.9243
10.00	0.01228199	0.0010003	106.30323	42.02	2519.21	2477.19	0.1511	8.8998
11.00	0.01312969	0.0010004	99.787212	46.22	2521.04	2474.82	0.1659	8.8754
12.00	0.01402848	0.0010005	93.718867	50.41	2522.86	2472.45	0.1806	8.8513
13.00	0.01498096	0.0010007	88.064395	54.60	2524.69	2470.09	0.1953	8.8274
14.00	0.01598984	0.0010008	82.792764	58.79	2526.51	2467.72	0.2099	8.8037
15.00	0.01705793	0.0010009	77.875465	62.98	2528.33	2465.35	0.2245	8.7803
16.00	0.01818816	0.0010011	73.286294	67.17	2530.16	2462.99	0.2390	8.7570
17.00	0.01938358	0.0010013	69.001148	71.36	2531.98	2460.62	0.2534	8.7339
18.00	0.02064735	0.0010014	64.997848	75.54	2533.80	2458.26	0.2678	8.7111
1.009	0.02198275	0.0010016	61.255971	79.73	2535.62	2455.89	0.2822	8.6884
20.00	0.02339318	0.0010018	57.756701	83.91	2537.43	2453.52	0.2965	8.6660
21.00	0.02488219	0.0010021	54.482696	88.10	2539.25	2451.15	0.3107	8.6437
22.00	0.02645344	0.0010023	51.417958	92.28	2541.07	2448.79	0.3249	8.6217
23.00	0.02811072	0.0010025	48.547727	96.47	2542.88	2446.41	0.3391	8.5998
24.00	0.02985798	0.0010028	45.858375	100.65	2544.69	2444.04	0.3532	8.5781
25.00	0.03169929	0.0010030	43.337313	104.83	2546.51	2441.68	0.3672	8.5566
26.00	0.03363889	0.0010033	40.972903	109.01	2548.32	2439.31	0.3812	8.5353
27.00	0.03568112	0.0010035	38.754386	113.19	2550.13	2436.94	0.3952	8.5142
28.00	0.03783053	0.0010038	36.671808	117.37	2551.93	2434.56	0.4091	8.4933
29.00	0.04009178	0.0010041	34.715952	121.55	2553.74	2432.19	0.4229	8.4725
30.00	0.04246971	0.0010044	32.878285	125.73	2555.55	2429.82	0.4368	8.4520
31.00	0.04496931	0.0010047	31.150898	129.91	2557.35	2427.44	0.4505	8.4316
32.00	0.04765957	0.0010050	29.526461	134.09	2559.15	2425.06	0.4642	8.4113
33.00	0.05035433	0.0010054	27.998174	138.27	2560.95	2422.68	0.4779	8.3913
34.00	0.05325058	0.0010057	26.559728	142.45	2562.75	2420.30	0.4915	8.3714
35.00	0.05629016	0.0010060	25.205265	146.63	2564.55	2417.92	0.5051	8.3517
36.00	0.05947893	0.0010064	23.929344	150.81	2566.34	2415.53	0.5187	8.3321
37.00	0.06282292	0.0010068	22.726910	154.99	2568.14	2413.15	0.5322	8.3127
38.00	0.06632835	0.0010071	21.593263	159.17	2569.93	2410.76	0.5456	8.2935
39.00	0.07000164	0.0010075	20.524030	163.35	2571.72	2408.37	0.5590	8.2745
40.00	0.07384938	0.0010079	19.515144	167.53	2573.51	2405.98	0.5724	8.2555
41.00	0.07787838	0.0010083	18.562818	171.71	2575.30	2403.59	0.5857	8.2368
42.00	0.08209563	0.0010087	17.663524	175.89	2577.08	2401.19	0.5990	8.2182
43.00	0.08650835	0.0010091	16.813976	180.07	2578.87	2398.80	0.6123	8.1998
44.00	0.09112392	0.0010095	16.011111	184.25	2580.65	2396.40	0.6255	8.1815
45.00	0.09594999	0.0010099	15.252072	188.44	2582.43	2393.99	0.6386	8.1633
46.00	0.10099437	0.0010104	14.534192	192.62	2584.20	2391.58	0.6517	8.1453
47.00	0.10626513	0.0010108	13.854984	196.80	2585.98	2389.18	0.6648	8.1275
48.00	0.11177053	0.0010112	13.212125	200.98	2587.75	2386.77	0.6779	8.1098
49.00	0.11751906	0.0010117	12.603445	205.16	2589.52	2384.36	0.6908	8.0922

Table A.9a Steam table (water vapor table), saturation state (temperature table, continued) [128], [130]

t °C	p bar	v' $\frac{m^3}{kg}$	v'' $\frac{m^3}{kg}$	h' $\frac{kJ}{kg}$	h'' $\frac{kJ}{kg}$	r $\frac{kJ}{kg}$	s' $\frac{kJ}{kg\,K}$	s'' $\frac{kJ}{kg\,K}$
50.00	0.12351946	0.0010121	12.026915	209.34	2591.29	2381.95	0.7038	8.0748
51.00	0.12978067	0.0010126	11.480639	213.52	2593.06	2379.54	0.7167	8.0576
52.00	0.13631188	0.0010131	10.962845	217.71	2594.82	2377.11	0.7296	8.0404
53.00	0.14312253	0.0010136	10.471872	221.89	2596.58	2374.69	0.7425	8.0234
54.00	0.15022227	0.0010141	10.006565	226.07	2598.34	2372.27	0.7553	8.0066
55.00	0.15762102	0.0010146	9.5642779	230.26	2600.09	2369.83	0.7680	7.9898
56.00	0.16532893	0.0010151	9.1448392	234.44	2601.85	2367.41	0.7808	7.9732
57.00	0.17335643	0.0010156	8.7465752	238.63	2603.60	2364.97	0.7934	7.9568
58.00	0.18171417	0.0010161	8.3682890	242.81	2605.35	2362.54	0.8061	7.9404
59.00	0.19041308	0.0010166	8.0088579	247.00	2607.09	2360.09	0.8187	7.9242
60.00	0.19946434	0.0010171	7.6672286	251.18	2608.84	2357.66	0.8313	7.9081
61.00	0.20887940	0.0010177	7.3424128	255.37	2610.58	2355.21	0.8438	7.8922
62.00	0.21866997	0.0010182	7.0334822	259.55	2612.31	2352.76	0.8563	7.8764
63.00	0.22884804	0.0010188	6.7395651	263.74	2614.05	2350.31	0.8688	7.8607
64.00	0.23942587	0.0010193	6.4598427	267.93	2615.78	2347.85	0.8813	7.8451
65.00	0.25041598	0.0010199	6.1935449	272.12	2617.51	2345.39	0.8937	7.8296
66.00	0.26183120	0.0010204	5.9399481	276.31	2619.23	2342.92	0.9060	7.8142
67.00	0.27368461	0.0010210	5.6983713	280.49	2620.95	2340.46	0.9183	7.7990
68.00	0.28598961	0.0010216	5.4681741	284.68	2622.67	2337.99	0.9306	7.7839
69.00	0.29875985	0.0010222	5.2487535	288.87	2624.39	2335.52	0.9429	7.7689
70.00	0.31200930	0.0010228	5.0395418	293.07	2626.10	2333.03	0.9551	7.7540
71.00	0.32575221	0.0010234	4.8400043	297.26	2627.81	2330.55	0.9673	7.7392
72.00	0.34000313	0.0010240	4.6496373	301.45	2629.51	2328.06	0.9795	7.7246
73.00	0.35477691	0.0010246	4.4681549	305.64	2631.21	2325.57	0.9916	7.7100
74.00	0.37008870	0.0010252	4.2945425	309.84	2632.91	2323.07	1.0037	7.6955
75.00	0.38595396	0.0010258	4.1289448	314.03	2634.60	2320.57	1.0158	7.6812
76.00	0.40238844	0.0010265	3.9707746	318.23	2636.29	2318.06	1.0278	7.6670
77.00	0.41940822	0.0010271	3.8196559	322.42	2637.98	2315.56	1.0398	7.6528
78.00	0.43702968	0.0010277	3.6752339	326.62	2639.66	2313.04	1.0517	7.6388
79.00	0.45526951	0.0010284	3.5371734	330.81	2641.34	2310.53	1.0637	7.6249
80.00	0.47414474	0.0010291	3.4051579	335.01	2643.02	2308.01	1.0756	7.6111
81.00	0.49367269	0.0010297	3.2788883	339.21	2644.69	2305.48	1.0874	7.5973
82.00	0.51387103	0.0010304	3.1580817	343.41	2646.35	2302.94	1.0993	7.5837
83.00	0.53475772	0.0010311	3.0424709	347.61	2648.02	2300.41	1.1111	7.5702
84.00	0.55635107	0.0010317	2.9318031	351.81	2649.67	2297.86	1.1229	7.5567
85.00	0.57866972	0.0010324	2.8258389	356.02	2651.33	2295.31	1.1346	7.5434
86.00	0.60173262	0.0010331	2.7243520	360.22	2652.98	2292.76	1.1463	7.5302
87.00	0.62555907	0.0010338	2.6271282	364.42	2654.62	2290.20	1.1580	7.5170
88.00	0.65016869	0.0010345	2.5339644	368.63	2656.26	2287.63	1.1696	7.5040
90.00	0.70181766	0.0010360	2.3590584	377.04	2659.53	2282.49	1.1929	7.4781
92.00	0.75684329	0.0010374	2.1982139	385.46	2662.78	2277.32	1.2160	7.4526
94.00	0.81541481	0.0010389	2.0501548	393.88	2666.01	2272.13	1.2389	7.4275
96.00	0.87770695	0.0010404	1.9137330	402.30	2669.22	2266.92	1.2618	7.4027
98.00	0.94390004	0.0010419	1.7879143	410.73	2672.40	2261.67	1.2846	7.3783
100.00	1.0141800	0.0010435	1.6717661	419.17	2675.57	2256.40	1.3072	7.3541
105.00	1.2090309	0.0010474	1.4183787	440.27	2683.39	2243.12	1.3633	7.2952
110.00	1.4337871	0.0010516	1.2092928	461.42	2691.06	2229.64	1.4188	7.2381
115.00	1.6918238	0.0010559	1.0358406	482.59	2698.58	2215.99	1.4737	7.1828
120.00	1.9867442	0.0010603	0.8912122	503.81	2705.93	2202.12	1.5279	7.1291
125.00	2.3223815	0.0010649	0.7700260	525.07	2713.10	2188.03	1.5816	7.0770

Table A.9a Steam table (water vapor table), saturation state (temperature table, continued) [128], [130]

t °C	p bar	v' $\frac{m^3}{kg}$	v'' $\frac{m^3}{kg}$	h' $\frac{kJ}{kg}$	h'' $\frac{kJ}{kg}$	r $\frac{kJ}{kg}$	s' $\frac{kJ}{kg\,K}$	s'' $\frac{kJ}{kg\,K}$
130.00	2.7027998	0.0010697	0.6680045	546.38	2720.08	2173.70	1.6346	7.0264
135.00	3.1322942	0.0010746	0.5817293	567.75	2726.87	2159.12	1.6872	6.9772
140.00	3.6153910	0.0010798	0.5084543	589.16	2733.44	2144.28	1.7392	6.9293
145.00	4.1568464	0.0010850	0.4465962	610.64	2739.80	2129.16	1.7907	6.8826
150.00	4.7616454	0.0010905	0.3924528	632.18	2745.93	2113.75	1.8418	6.8371
155.00	5.4349998	0.0010962	0.3464599	653.79	2751.81	2098.02	1.8924	6.7926
160.00	6.1823462	0.0011020	0.3067820	675.47	2757.44	2081.97	1.9426	6.7491
165.00	7.0093435	0.0011080	0.2724306	697.24	2762.81	2065.57	1.9923	6.7066
170.00	7.9218701	0.0011143	0.2425894	719.08	2767.90	2048.82	2.0417	6.6650
175.00	8.9260210	0.0011207	0.2165812	741.02	2772.71	2031.69	2.0906	6.6241
180.00	10.028105	0.0011274	0.1938422	763.05	2777.22	2014.17	2.1392	6.5840
185.00	11.234643	0.0011343	0.1739009	785.19	2781.41	1996.22	2.1875	6.5447
190.00	12.552362	0.0011415	0.1563619	807.43	2785.28	1977.85	2.2355	6.5059
195.00	13.988195	0.0011489	0.1408922	829.79	2788.82	1959.03	2.2832	6.4678
200.00	15.549279	0.0011565	0.1272104	852.27	2792.01	1939.74	2.3305	6.4302
205.00	17.242952	0.0011645	0.1150779	874.88	2794.83	1919.95	2.3777	6.3930
210.00	19.076750	0.0011727	0.1042920	897.63	2797.27	1899.64	2.4245	6.3563
215.00	21.058409	0.0011813	0.0946795	920.53	2799.32	1878.79	2.4712	6.3200
220.00	23.195862	0.0011902	0.0860924	943.58	2800.95	1857.37	2.5177	6.2840
225.00	25.497240	0.0011994	0.0784035	966.80	2802.15	1835.35	2.5640	6.2483
230.00	27.970875	0.0012090	0.0715035	990.19	2802.90	1812.71	2.6101	6.2128
235.00	30.625299	0.0012190	0.0652979	1013.77	2803.17	1789.40	2.6561	6.1775
240.00	33.469251	0.0012295	0.0597050	1037.55	2802.96	1765.41	2.7020	6.1423
245.00	36.511680	0.0012403	0.0546539	1061.55	2802.22	1740.67	2.7478	6.1072
250.00	39.761749	0.0012517	0.0500830	1085.77	2800.93	1715.16	2.7935	6.0721
255.00	43.228851	0.0012636	0.0459385	1110.23	2799.07	1688.84	2.8392	6.0369
260.00	46.922610	0.0012761	0.0421733	1134.96	2796.60	1661.64	2.8849	6.0016
265.00	50.852902	0.0012892	0.0387462	1159.96	2793.49	1633.53	2.9307	5.9661
270.00	55.029868	0.0013030	0.0356210	1185.27	2789.69	1604.42	2.9765	5.9304
275.00	59.464393	0.0013175	0.0327658	1210.90	2785.17	1574.27	3.0224	5.8944
280.00	64.165829	0.0013328	0.0301526	1236.89	2779.87	1542.98	3.0685	5.8579
285.00	69.146631	0.0013491	0.2775627	1263.25	2773.73	1510.48	3.1147	5.8209
290.00	74.417783	0.0013663	0.0255549	1290.03	2766.70	1476.67	3.1612	5.7834
295.00	79.991147	0.0013846	0.0235286	1317.27	2758.70	1441.43	3.2080	5.7451
300.00	85.879049	0.0014042	0.0216601	1345.01	2749.64	1404.63	3.2552	5.7059
305.00	92.094349	0.0014252	0.0199335	1373.30	2739.43	1366.13	3.3028	5.6657
310.00	98.650512	0.0014479	0.0183347	1402.22	2727.95	1325.73	3.3510	5.6244
315.00	105.56171	0.0014724	0.0168510	1431.83	2715.06	1283.23	3.3998	5.5816
320.00	112.84293	0.0014990	0.0154707	1462.22	2700.59	1238.37	3.4494	5.5372
325.00	120.51002	0.0015283	0.0141832	1493.52	2684.33	1190.81	3.5000	5.4908
330.00	128.58052	0.0015606	0.0129785	1525.87	2666.03	1140.16	3.5518	5.4422
335.00	137.07261	0.0015967	0.0118474	1559.45	2645.35	1085.90	3.6050	5.3906
340.00	146.00677	0.0016376	0.0107807	1594.53	2621.85	1027.32	3.6601	5.3356
345.00	155.40554	0.0016846	0.0097690	1631.48	2594.90	963.42	3.7176	5.2762
350.00	165.29415	0.0017400	0.0088024	1670.89	2563.64	892.75	3.7784	5.2110
355.00	175.70123	0.0018079	0.0078684	1713.72	2526.65	812.93	3.8439	5.1380
360.00	186.66007	0.0018954	0.0069493	1761.67	2481.49	719.82	3.9167	5.0536
365.00	198.21364	0.0020172	0.0060115	1817.77	2422.95	605.18	4.0014	4.9497
370.00	210.43563	0.0022152	0.0049544	1890.69	2334.52	443.83	4.1112	4.8012
373.946	220.64000	0.0031056		2084.26		0.00	4.4070	

Table A.9b Steam table (water vapor table), saturation state (pressure table) [128], [130]

p	t	v'	v''	h'	h''	r	s'	s''
bar	°C	$\dfrac{\text{m}^3}{\text{kg}}$	$\dfrac{\text{m}^3}{\text{kg}}$	$\dfrac{\text{kJ}}{\text{kg}}$	$\dfrac{\text{kJ}}{\text{kg}}$	$\dfrac{\text{kJ}}{\text{kg}}$	$\dfrac{\text{kJ}}{\text{kg K}}$	$\dfrac{\text{kJ}}{\text{kg K}}$
0.00611655	0.010	0.0010002	205.99122	0.00	2500.92	2500.92	0.0000	9.1555
0.008	3.7614432	0.0010001	159.64022	15.81	2507.79	2491.98	0.0575	9.0567
0.010	6.9695702	0.0010001	129.17834	29.30	2513.67	2484.37	0.1059	8.9749
0.012	9.6538483	0.0010003	108.66968	40.57	2518.58	2478.01	0.1460	8.9082
0.014	11.969188	0.0010005	93.899499	50.28	2522.81	2472.53	0.1802	8.8521
0.016	14.009787	0.0010008	82.742965	58.83	2526.53	2467.70	0.2100	8.8035
0.018	15.837373	0.0010011	74.011302	66.49	2529.86	2463.37	0.2366	8.7608
0.020	17.494681	0.0010014	66.986876	73.43	2532.88	2459.45	0.2606	8.7226
0.025	21.076866	0.0010021	54.239891	88.42	2539.39	2450.97	0.3118	8.6420
0.030	24.079018	0.0010028	45.653200	100.98	2544.84	2443.86	0.3543	8.5764
0.035	26.672154	0.0010035	39.466269	111.82	2549.53	2437.71	0.3906	8.5211
0.040	28.960379	0.0010041	34.791142	121.39	2553.67	2432.28	0.4224	8.4734
0.045	31.011975	0.0010047	31.130850	129.96	2557.37	2427.41	0.4507	8.4313
0.050	32.874255	0.0010053	28.185285	137.75	2560.73	2422.98	0.4762	8.3939
0.055	34.581407	0.0010059	25.762372	144.88	2563.80	2418.92	0.4994	8.3599
0.060	36.158974	0.0010065	23.733402	151.48	2566.63	2415.15	0.5208	8.3290
0.065	37.626562	0.0010070	22.008847	157.61	2569.26	2411.65	0.5406	8.3007
0.070	38.999564	0.0010075	20.524482	163.35	2571.72	2408.37	0.5590	8.2745
0.075	40.290295	0.0010080	19.233023	168.75	2574.03	2405.28	0.5763	8.2501
0.080	41.508765	0.0010085	18.098873	173.84	2576.21	2402.37	0.5925	8.2273
0.085	42.663232	0.0010089	17.094716	178.67	2578.27	2399.60	0.6078	8.2060
0.090	43.760583	0.0010094	16.199237	183.25	2580.22	2396.97	0.6223	8.1858
0.095	44.806632	0.0010098	15.395556	187.63	2582.08	2394.45	0.6361	8.1668
0.100	45.806329	0.0010103	14.670130	191.81	2583.86	2392.05	0.6492	8.1488
0.110	47.683108	0.0010111	13.412043	199.65	2587.19	2387.54	0.6737	8.1154
0.120	49.418655	0.0010119	12.358277	206.91	2590.26	2383.35	0.6963	8.0849
0.130	51.034274	0.0010126	11.462431	213.67	2593.12	2379.45	0.7172	8.0570
0.140	52.546696	0.0010134	10.691210	219.99	2595.78	2375.79	0.7366	8.0311
0.150	53.969313	0.0010140	10.020098	225.94	2598.28	2372.34	0.7549	8.0071
0.160	55.313017	0.0010147	9.4306341	231.57	2600.64	2369.07	0.7720	7.9846
0.170	56.586794	0.0010154	8.9086517	236.90	2602.88	2365.98	0.7882	7.9636
0.180	57.798146	0.0010160	8.4430937	241.97	2604.99	2363.02	0.8035	7.9437
0.190	58.953403	0.0010166	8.0252030	246.80	2607.01	2360.21	0.8181	7.9254
0.200	60.057960	0.0010172	7.6479524	251.42	2608.94	2357.52	0.8320	7.9072
0.210	61.116453	0.0010177	7.3056367	255.85	2610.78	2354.93	0.8453	7.8903
0.220	62.132895	0.0010183	6.9935753	260.11	2612.54	2352.43	0.8580	7.8743
0.230	63.110786	0.0010188	6.7078911	264.20	2614.24	2350.04	0.8702	7.8589
0.240	64.053196	0.0010193	6.4453453	268.15	2615.87	2347.72	0.8822	7.8442
0.250	64.962834	0.0010198	6.2032105	271.96	2617.44	2345.48	0.8932	7.8302
0.260	65.842105	0.0010203	5.9791736	275.64	2618.96	2343.32	0.9041	7.8167
0.270	66.693151	0.0010208	5.7712593	279.21	2620.42	2341.21	0.9146	7.8037
0.280	67.517889	0.0010213	5.5777731	282.66	2621.84	2339.18	0.9247	7.7912
0.290	68.318045	0.0010218	5.3972476	286.02	2623.22	2337.20	0.9345	7.7791
0.300	69.095174	0.0010222	5.2284110	289.27	2624.55	2335.28	0.9441	7.7675
0.320	70.585865	0.0010231	4.9214992	295.52	2627.10	2331.58	0.9623	7.7453
0.340	71.999784	0.0010240	4.6496775	301.45	2629.51	2328.06	0.9795	7.7246
0.360	73.345137	0.0010248	4.4072009	307.09	2631.80	2324.71	0.9958	7.7050
0.380	74.628851	0.0010256	4.1895188	312.47	2633.97	2321.50	1.0113	7.6865
0.400	75.856830	0.0010264	3.9929787	317.62	2636.05	2318.43	1.0261	7.6690
0.450	78.714610	0.0010282	3.5759446	329.62	2640.86	2311.24	1.0603	7.6288

Table A.9b Steam table (water vapor table), saturation state (pressure table, continued) [128], [130]

p	t	v'	v''	h'	h''	r	s'	s''
bar	°C	$\dfrac{m^3}{kg}$	$\dfrac{m^3}{kg}$	$\dfrac{kJ}{kg}$	$\dfrac{kJ}{kg}$	$\dfrac{kJ}{kg}$	$\dfrac{kJ}{kg\,K}$	$\dfrac{kJ}{kg\,K}$
0.500	81.316893	0.0010299	3.2400272	340.54	2645.22	2304.68	1.0912	7.5930
0.550	83.709303	0.0010315	2.9634786	350.59	2649.19	2298.60	1.1194	7.5606
0.600	85.925998	0.0010331	2.7317136	359.91	2652.86	2292.95	1.1454	7.5311
0.650	87.993254	0.0010345	2.5345798	368.60	2656.25	2287.65	1.1696	7.5040
0.700	89.931734	0.0010359	2.3647890	376.75	2659.42	2282.67	1.1921	7.4790
0.750	91.757999	0.0010372	2.2169721	384.44	2662.39	2277.95	1.2132	7.4557
0.800	93.485536	0.0010385	2.0870847	391.71	2665.18	2273.47	1.2330	7.4339
0.850	95.125483	0.0010397	1.9720216	398.62	2667.82	2269.20	1.2518	7.4135
0.900	96.687148	0.0010409	1.8693584	405.20	2670.31	2265.11	1.2696	7.3943
0.950	98.178394	0.0010420	1.7771744	411.48	2672.69	2261.21	1.2866	7.3761
1.000	99.605929	0.0010432	1.6939277	417.50	2674.95	2257.45	1.3028	7.3588
1.100	102.29217	0.0010453	1.5494619	428.84	2679.17	2250.33	1.3330	7.3269
1.200	104.78355	0.0010473	1.4283600	439.36	2683.05	2243.69	1.3609	7.2977
1.300	107.10908	0.0010492	1.3253299	449.19	2686.64	2237.45	1.3868	7.2709
1.400	109.29159	0.0010510	1.2365717	458.42	2689.98	2231.56	1.4110	7.2461
1.500	111.34938	0.0010527	1.1592851	467.13	2693.11	2225.98	1.4337	7.2230
1.600	113.29738	0.0010544	1.0913606	475.38	2696.04	2220.66	1.4551	7.2014
1.700	115.14790	0.0010560	1.0317830	483.22	2698.80	2215.58	1.4753	7.1812
1.800	116.91127	0.0010576	0.9774731	490.70	2701.41	2210.71	1.4945	7.1621
1.900	118.59619	0.0010591	0.9292424	497.85	2703.88	2206.03	1.5127	7.1440
2.000	120.21009	0.0010605	0.8856817	504.70	2706.23	2201.53	1.5302	7.1269
2.100	121.75938	0.0010619	0.8461371	511.29	2708.47	2197.18	1.5469	7.1106
2.200	123.24960	0.0010633	0.8100720	517.63	2710.61	2192.98	1.5628	7.0951
2.300	124.68559	0.0010646	0.7770419	523.74	2712.65	2188.91	1.5782	7.0803
2.400	126.07160	0.0010659	0.7466752	529.64	2714.61	2184.97	1.5930	7.0661
2.500	127.41141	0.0010672	0.7186589	535.35	2716.49	2181.14	1.6072	7.0524
2.600	128.70833	0.0010685	0.6927273	540.87	2718.30	2177.43	1.6210	7.0394
2.700	129.96535	0.0010697	0.6686537	546.24	2720.03	2173.79	1.6343	7.0268
2.800	131.18515	0.0010709	0.6462431	551.44	2721.71	2170.27	1.6471	7.0146
3.000	133.52242	0.0010732	0.6057596	561.43	2724.88	2163.45	1.6717	6.9916
3.500	138.85715	0.0010786	0.5241795	584.26	2731.96	2147.70	1.7274	6.9401
4.000	143.60836	0.0010836	0.4623829	604.66	2738.05	2133.39	1.7765	6.8955
4.500	147.90340	0.0010882	0.4138974	623.14	2743.39	2120.25	1.8205	6.8560
5.000	151.83108	0.0010925	0.3748054	640.09	2748.11	2108.02	1.8604	6.8207
5.500	155.45595	0.0010967	0.3425964	655.76	2752.33	2096.57	1.8970	6.7886
6.000	158.82648	0.0011006	0.3155827	670.38	2756.14	2085.76	1.9308	6.7592
7.000	164.94620	0.0011080	0.2727749	697.00	2762.75	2065.75	1.9918	6.7071
8.000	170.40649	0.0011148	0.2403401	720.86	2768.30	2047.44	2.0457	6.6616
9.000	175.35045	0.0011212	0.2148868	742.56	2773.03	2030.47	2.0940	6.6213
10.000	179.87801	0.0011272	0.1943619	762.52	2777.11	2014.59	2.1381	6.5850
11.000	184.06188	0.0011330	0.1774483	781.03	2780.65	1999.62	2.1785	6.5520
12.000	187.95674	0.0011385	0.1632616	798.33	2783.74	1985.41	2.2159	6.5217
13.000	191.60481	0.0011438	0.1511857	814.60	2786.46	1971.86	2.2508	6.4936
14.000	195.03941	0.0011489	0.1407777	829.97	2788.85	1958.88	2.2835	6.4675
15.000	198.28733	0.0011539	0.1317112	844.56	2790.96	1946.40	2.3143	6.4430
16.000	201.37047	0.0011587	0.1237400	858.46	2792.82	1934.36	2.3435	6.4199
17.000	204.30695	0.0011634	0.1166749	871.74	2794.46	1922.72	2.3711	6.3981
18.000	207.11197	0.0011679	0.1103684	884.47	2795.91	1911.44	2.3975	6.3775
19.000	209.79839	0.0011724	0.1047031	896.71	2797.18	1900.47	2.4227	6.3578
20.000	212.37723	0.0011767	0.0995851	908.50	2798.29	1889.79	2.4468	6.3390

Table A.9b Steam table (water vapor table), saturation state (pressure table, continued) [128], [130]

p bar	t °C	v' m³/kg	v'' m³/kg	h' kJ/kg	h'' kJ/kg	r kJ/kg	s' kJ/kg K	s'' kJ/kg K
22.000	217.24882	0.0011852	0.0906985	930.88	2800.10	1869.22	2.4921	6.3038
24.000	221.78893	0.0011934	0.0832442	951.87	2801.43	1849.56	2.5343	6.2712
26.000	226.04560	0.0012014	0.0768987	971.67	2802.34	1830.67	2.5736	6.2409
28.000	230.05680	0.0012091	0.0714292	990.46	2802.90	1812.44	2.6106	6.2124
30.000	233.85311	0.0012167	0.0666644	1008.35	2803.15	1794.80	2.6455	6.1856
32.000	237.45949	0.0012241	0.0624748	1025.44	2803.13	1777.69	2.6787	6.1602
34.000	240.89668	0.0012314	0.0587612	1041.84	2802.86	1761.02	2.7102	6.1360
36.000	244.18204	0.0012385	0.0554459	1057.61	2802.38	1744.77	2.7403	6.1129
38.000	247.33029	0.0012456	0.0524673	1072.81	2801.69	1728.88	2.7691	6.0908
40.000	250.35405	0.0012526	0.0497761	1087.49	2800.82	1713.33	2.7968	6.0696
42.000	253.26417	0.0012594	0.0473321	1101.71	2799.79	1698.08	2.8234	6.0491
44.000	256.07011	0.0012663	0.0451022	1115.50	2798.60	1683.10	2.8490	6.0293
46.000	258.78013	0.0012730	0.0430592	1128.90	2797.26	1668.36	2.8738	6.0102
48.000	261.40152	0.0012797	0.0411802	1141.94	2795.80	1653.86	2.8978	5.9917
50.000	263.94072	0.0012864	0.0394459	1154.64	2794.21	1639.57	2.9210	5.9737
55.000	269.96529	0.0013029	0.0356417	1185.09	2789.72	1604.63	2.9762	5.9307
60.000	275.58499	0.0013193	0.0324481	1213.92	2784.59	1570.67	3.0278	5.8901
65.000	280.85759	0.0013356	0.0297268	1241.38	2778.88	1537.50	3.0764	5.8516
70.000	285.82881	0.0013519	0.0273784	1267.66	2772.63	1504.97	3.1224	5.8148
75.000	290.53548	0.0013682	0.0253298	1292.93	2765.89	1472.96	3.1662	5.7793
80.000	295.00773	0.0013847	0.0235256	1317.31	2758.68	1441.37	3.2081	5.7450
85.000	299.27057	0.0014013	0.0219235	1340.93	2751.03	1410.10	3.2483	5.7117
90.000	303.34498	0.0014181	0.0204902	1363.87	2742.94	1379.07	3.2870	5.6791
95.000	307.24877	0.0014352	0.0191994	1386.23	2734.43	1348.20	3.3244	5.6473
100.000	310.99715	0.0014526	0.0180300	1408.06	2725.49	1317.43	3.3606	5.6160
105.000	314.60326	0.0014703	0.0169648	1429.45	2716.13	1286.68	3.3959	5.5851
110.000	318.07851	0.0014885	0.0159896	1450.44	2706.35	1255.91	3.4303	5.5545
115.000	321.43289	0.0015071	0.0150927	1471.10	2696.12	1225.02	3.4638	5.5241
120.000	324.67518	0.0015263	0.0142642	1491.46	2685.45	1193.99	3.4967	5.4939
125.000	327.81314	0.0015461	0.0134958	1511.58	2674.31	1162.73	3.5290	5.4638
130.000	330.85366	0.0015665	0.0127804	1531.51	2662.68	1131.17	3.5608	5.4336
135.000	333.80291	0.0015877	0.0121120	1551.29	2650.54	1099.25	3.5921	5.4032
140.000	336.66639	0.0016097	0.0114852	1570.96	2637.86	1066.90	3.6232	5.3727
145.000	339.44907	0.0016328	0.0108953	1590.58	2624.59	1034.01	3.6539	5.3418
150.000	342.15539	0.0016570	0.0103384	1610.20	2610.70	1000.50	3.6846	5.3106
155.000	344.78942	0.0016824	0.0098107	1629.88	2596.12	966.24	3.7151	5.2788
160.000	347.35480	0.0017094	0.0093088	1649.69	2580.79	931.10	3.7457	5.2463
165.000	349.85489	0.0017383	0.0088299	1669.70	2564.62	894.92	3.7765	5.2130
170.000	352.29271	0.0017693	0.0083709	1690.03	2547.50	857.47	3.8077	5.1787
175.000	354.67106	0.0018029	0.0079292	1710.77	2529.30	818.53	3.8394	5.1431
180.000	356.99245	0.0018398	0.0075017	1732.09	2509.83	777.74	3.8718	5.1061
185.000	359.25916	0.0018807	0.0070856	1754.14	2488.85	734.71	3.9053	5.0670
190.000	361.47319	0.0019268	0.0066773	1777.15	2466.01	688.86	3.9401	5.0256
195.000	363.63618	0.0019792	0.0062725	1801.39	2440.78	639.39	3.9767	4.9808
200.000	365.74926	0.0020400	0.0058652	1827.21	2412.35	585.14	4.0156	4.9314
205.000	367.81289	0.0021126	0.0054457	1855.34	2379.25	523.91	4.0579	4.8753
210.000	369.82689	0.0022055	0.0049960	1887.56	2338.59	451.03	4.1064	4.8079
215.000	371.79103	0.0023468	0.0044734	1929.53	2283.12	353.59	4.1698	4.7181
220.000	373.70540	0.0027044	0.0036475	2011.34	2173.09	161.75	4.2945	4.5446
220.640	373.94600	0.0031056		2084.26		0.00	4.4070	

Table A.10 Steam table (water vapor table), superheated steam [128]

p bar	t °C	v m³/kg	h kJ/kg	s kJ/kgK	t °C	v m³/kg	h kJ/kg	s kJ/kgK
0.2	100	8.58569	2686.19	8.1262	350	14.37498	3177.35	9.1311
0.2	150	9.74880	2782.32	8.3680	400	15.52977	3279.78	9.2892
0.2	200	10.90743	2879.14	8.5842	450	16.68434	3383.84	9.4383
0.2	250	12.06412	2977.12	8.7811	500	17.83874	3489.57	9.5797
0.2	300	13.21983	3076.49	8.9624	600	20.14721	3706.19	9.8431
0.4	100	4.27999	2683.68	7.8009	350	7.18498	3176.97	8.8108
0.4	150	4.86636	2780.91	8.0455	400	7.76287	3279.47	8.9690
0.4	200	5.44813	2878.23	8.2629	450	8.34052	3383.58	9.1182
0.4	250	6.02793	2976.48	8.4602	500	8.91802	3489.35	9.2596
0.4	300	6.60673	3076.00	8.6419	600	10.07267	3706.04	9.5231
0.6	100	2.84460	2681.10	7.6083	350	4.78831	3176.59	8.6232
0.6	150	3.23883	2779.49	7.8557	400	5.17390	3279.16	8.7815
0.6	200	3.62835	2877.32	8.0743	450	5.55925	3383.32	8.9308
0.6	250	4.01586	2975.83	8.2722	500	5.94444	3489.14	9.0722
0.6	300	4.40237	3075.52	8.4541	600	6.71449	3705.88	9.3358
1.0	100	1.69596	2675.77	7.3610	350	2.87097	3175.82	8.3865
1.0	150	1.93673	2776.59	7.6147	400	3.10272	3278.54	8.5451
1.0	200	2.17249	2875.48	7.8356	450	3.33424	3382.81	8.6945
1.0	250	2.40619	2974.54	8.0346	500	3.56558	3488.71	8.8361
1.0	300	2.63887	3074.54	8.2171	600	4.02795	3705.57	9.0998
1.2	150	1.61116	2775.12	7.5278	400	2.58493	3278.23	8.4606
1.2	200	1.80852	2874.55	7.7499	450	2.77798	3382.56	8.6101
1.2	250	2.00377	2973.89	7.9495	500	2.97087	3488.49	8.7517
1.2	300	2.19799	3074.05	8.1323	550	3.16364	3596.10	8.8866
1.2	350	2.39164	3175.43	8.3019	600	3.35631	3705.42	9.0155
1.5	150	1.28557	2772.89	7.4207	400	2.06713	3277.76	8.3571
1.5	200	1.44453	2873.14	7.6447	450	2.22173	3382.17	8.5067
1.5	250	1.60134	2972.90	7.8451	500	2.37615	3488.17	8.6484
1.5	300	1.75711	3073.31	8.0284	550	2.53046	3595.83	8.7833
1.5	350	1.91230	3174.86	8.1983	600	2.68468	3705.18	8.9123
2.0	150	0.95989	2769.09	7.2809	400	1.54934	3276.98	8.2235
2.0	200	1.08052	2870.78	7.5081	450	1.66547	3381.53	8.3733
2.0	250	1.19891	2971.26	7.7100	500	1.78144	3487.64	8.5151
2.0	300	1.31623	3072.08	7.8940	550	1.89728	3595.37	8.6501
2.0	350	1.43296	3173.89	8.0643	600	2.01304	3704.79	8.7792
4.0	150	0.47089	2752.78	6.9305	400	0.77264	3273.86	7.9001
4.0	200	0.53434	2860.99	7.1724	450	0.83108	3378.96	8.0507
4.0	250	0.59520	2964.56	7.3805	500	0.88936	3485.49	8.1931
4.0	300	0.65488	3067.11	7.5677	550	0.94752	3593.55	8.3286
4.0	350	0.71395	3170.01	7.7398	600	1.00559	3703.24	8.4579
6.0	200	0.35212	2850.66	6.9684	450	0.55295	3376.38	7.8609
6.0	250	0.39390	2957.65	7.1834	500	0.59200	3483.33	8.0039
6.0	300	0.43441	3062.06	7.3740	550	0.63093	3591.73	8.1398
6.0	350	0.47426	3166.10	7.5480	600	0.66977	3701.68	8.2694
6.0	400	0.51373	3270.72	7.7095	650	0.70854	3813.24	8.3937
8.0	200	0.26087	2839.77	6.8176	450	0.41388	3373.79	7.7255
8.0	250	0.29320	2950.54	7.0403	500	0.44332	3481.17	7.8690
8.0	300	0.32415	3056.92	7.2345	550	0.47263	3589.90	8.0053
8.0	350	0.35441	3162.15	7.4106	600	0.50186	3700.12	8.1353
8.0	400	0.38427	3267.56	7.5733	650	0.53101	3811.90	8.2598

Table A.10 Steam table (water vapor table), superheated steam (continued) [128]

p bar	t °C	v m³/kg	h kJ/kg	s kJ/kg K	t °C	v m³/kg	h kJ/kg	s kJ/kg K
10.0	200	0.20600	2828.27	6.6955	450	0.33044	3371.19	7.6198
10.0	250	0.23274	2943.22	6.9266	500	0.35411	3479.00	7.7640
10.0	300	0.25798	3051.70	7.1247	550	0.37766	3588.07	7.9007
10.0	350	0.28249	3158.16	7.3028	600	0.40111	3698.56	8.0309
10.0	400	0.30659	3264.39	7.4668	650	0.42450	3810.55	8.1557
15.0	200	0.13244	2796.02	6.4537	450	0.21918	3364.65	7.4259
15.0	250	0.15200	2923.96	6.7111	500	0.23516	3473.57	7.5716
15.0	300	0.16970	3038.27	6.9199	550	0.25102	3583.49	7.7093
15.0	350	0.18658	3148.03	7.1035	600	0.26678	3694.64	7.8404
15.0	400	0.20301	3256.37	7.2708	650	0.28248	3807.17	7.9657
20.0	250	0.11148	2903.23	6.5474	500	0.17568	3468.09	7.4335
20.0	300	0.12550	3024.25	6.7685	550	0.18769	3578.88	7.5723
20.0	350	0.13859	3137.64	6.9582	600	0.19961	3690.71	7.7042
20.0	400	0.15121	3248.23	7.1290	650	0.21146	3803.79	7.8301
20.0	450	0.16354	3358.05	7.2863	700	0.22326	3918.24	7.9509
30.0	250	0.07062	2856.55	6.2893	500	0.11619	3457.04	7.2356
30.0	300	0.08118	2994.35	6.5412	550	0.12437	3569.59	7.3767
30.0	350	0.09056	3116.06	6.7449	600	0.13244	3682,81	7.5102
30.0	400	0.09938	3231.57	6.9233	650	0.14045	3796.99	7.6373
30.0	450	0.10788	3344.66	7.0853	700	0.14840	3912.34	7.7590
40.0	300	0.05887	2961.65	6.3638	550	0.09270	3560.22	7.2353
40.0	350	0.06647	3093.32	6.5843	600	0.09886	3674.85	7.3704
40.0	400	0.07343	3214.37	6.7712	650	0.10494	3790.15	7.4989
40.0	450	0.08004	3330.99	6.9383	700	0.11097	3906.41	7.6215
40.0	500	0.08644	3445.84	7.0919	750	0.11696	4023.80	7.7391
60.0	300	0.03619	2885.49	6.0702	550	0.06102	3541.19	7.0306
60.0	350	0.04225	3043.86	6.3356	600	0.06526	3658.76	7.1692
60.0	400	0.04742	3178.18	6.5431	650	0.06943	3776.36	7.3002
60.0	450	0.05217	3302.76	6.7216	700	0.07354	3894.47	7.4248
60.0	500	0.05667	3422.95	6.8824	750	0.07761	4013.37	7.5439
80.0	300	0.02428	2786.38	5.7935	550	0.04517	3521.77	6.8798
80.0	350	0.02998	2988.06	6.1319	600	0.04846	3642.42	7.0221
80.0	400	0.03435	3139.31	6.3657	650	0.05167	3762.42	7.1557
80.0	450	0.03820	3273.23	6.5577	700	0.05483	3882.42	7.2823
80.0	500	0.04177	3399.37	6.7264	750	0.05793	4002.86	7.4030
100.0	350	0.02244	2923.96	5.9458	600	0.03838	3625.84	6.9045
100.0	400	0.02644	3097.38	6.2139	650	0.04102	3748.32	7.0409
100.0	450	0.02978	3242.28	6.4217	700	0.04359	3870.27	7.1696
100.0	500	0.03281	3375.06	6.5993	750	0.04613	3992.28	7.2918
100.0	550	0.03566	3501.94	6.7584	800	0.04862	4114.73	7.4087
150.0	350	0.01148	2693.00	5.4435	600	0.02492	3583.31	6.6797
150.0	400	0.01567	2975.55	5.8817	650	0.02680	3712.41	6.8235
150.0	450	0.01848	3157.84	6.1433	700	0.02862	3839.48	6.9576
150.0	500	0.02083	3310.79	6.3479	750	0.03039	3965.56	7.0839
150.0	550	0.02295	3450.47	6.5230	800	0.03212	4091.33	7.2039
200.0	400	0.00995	2816.84	5.5525	650	0.01969	3675.59	6.6596
200.0	450	0.01272	3061.53	5.9041	700	0.02113	3808.15	6.7994
200.0	500	0.01479	3241.19	6.1445	750	0.02252	3938.52	6.9301
200.0	550	0.01657	3396.24	6.3390	800	0.02387	4067.73	7.0534
200.0	600	0.01818	3539.23	6.5077				

Table A.11 Properties of dry air at p = 1 bar (p = 0,1 MPa)

t	ρ	c_p	$10^3 \beta$	$10^3 \lambda$	$10^6 \eta$	$10^6 \nu$	$10^6 a$	Pr
°C	$\dfrac{\text{kg}}{\text{m}^3}$	$\dfrac{\text{kJ}}{\text{kg K}}$	$\dfrac{1}{\text{K}}$	$\dfrac{\text{W}}{\text{m K}}$	$\dfrac{\text{kg}}{\text{m s}}$	$\dfrac{\text{m}^2}{\text{s}}$	$\dfrac{\text{m}^2}{\text{s}}$	-
-200	5.106	1.186	17.24	6.886	4.997	0.979	1.137	0.8606
-180	3.851	1.071	11.83	8.775	6.623	1.720	2.127	0.8086
-160	3.126	1.036	9.293	10.64	7.994	2.558	3.286	0.7784
-140	2.639	1.010	7.726	12.47	9.294	3.522	4.677	0.7530
-120	2.287	1.014	6.657	14.26	10.55	4.614	6.150	0.7502
-100	2.019	1.011	5.852	16.02	11.77	5.829	7.851	0.7423
-80	1.807	1.009	5.227	17.74	12.94	7.159	9.730	0.7357
-60	1.636	1.007	4.725	19.41	14.07	8.598	11.78	0.7301
-40	1.495	1.007	4.313	21.04	15.16	10.14	13.97	0.7258
-30	1.433	1.007	4.133	21.84	15.70	10.95	15.13	0.7236
-20	1.377	1.007	3.968	22.63	16.22	11.78	16.33	0.7215
-10	1.324	1.006	3.815	23.41	16.74	12.64	17.57	0.7196
0	1.275	1.006	3.674	24.18	17.24	13.52	18.83	0.7179
10	1.230	1.007	3.543	24.94	17.74	14.42	20.14	0.7163
20	1.188	1.007	3.421	25.69	18.24	15.35	21.47	0.7148
30	1.149	1.007	3.307	26.43	18.72	16.30	22.84	0.7134
40	1.112	1.007	3.200	27.16	19.20	17.26	24.24	0.7122
60	1.045	1.009	3.007	28.60	20.14	19.27	27.13	0.7100
80	0.9859	1.010	2.836	30.01	21.05	21.35	30.14	0.7083
100	0.9329	1.012	2.683	31.39	21.94	23.51	33.26	0.7070
120	0.8854	1.014	2.546	32.75	22.80	25.75	36.48	0.7060
140	0.8425	1.016	2.422	34.08	23.65	28.07	39.80	0.7054
160	0.8036	1.019	2.310	35.39	24.48	30.46	43.21	0.7050
180	0.7681	1.022	2.208	36.68	25.29	32.93	46.71	0.7049
200	0.7356	1.026	2.115	37.95	26.09	35.47	50.30	0.7051
250	0.6653	1.035	1.912	41.06	28.02	42.11	59.62	0.7063
300	0.6072	1.046	1.745	44.09	29.86	49.18	69.43	0.7083
350	0.5585	1.057	1.605	47.05	31.64	56.65	79.68	0.7109
400	0.5170	1.069	1.486	49.96	33.35	64.51	90.38	0.7137
450	0.4813	1.081	1.383	52.82	35.01	72.74	101.5	0.7166
500	0.4502	1.093	1.293	55.64	36.62	81.35	113.1	0.7194
550	0.4228	1.105	1.215	58.41	38.19	90.31	125.1	0.7221
600	0.3986	1.116	1.145	61.14	39.71	99.63	137.5	0.7247
650	0.3770	1.126	1.083	63.83	41.20	109.3	150.3	0.7271
700	0.3576	1.137	1.027	66.46	42.66	119.3	163.5	0.7295
750	0.3402	1.146	0.9772	69.03	44.08	129.6	177.1	0.7318
800	0.3243	1.155	0.9317	71.54	45.48	140.2	191.0	0.7342
850	0.3099	1.163	0.8902	73.98	46.85	151.2	205.2	0.7368
900	0.2967	1.171	0.8523	76.33	48.19	162.4	219.7	0.7395
1000	0.2734	1.185	0.7853	80.77	50.82	185.9	249.2	0.7458

Table A.12 Properties for boiling liquid water (The properties for the density ρ, the specific isobaric heat capacity c_p, the thermal expansion coefficient β, the thermal conductivity λ, the dynamic viscosity η, the thermal diffusivity a and the Prandtl number Pr can also be used in a first approximation for liquid water at the same temperature and at higher pressures.)

t	p	ρ	c_p	$10^3\,\beta$	λ	$10^3\,\eta$	$10^6\,a$	Pr
°C	bar	$\dfrac{\text{kg}}{\text{m}^3}$	$\dfrac{\text{kJ}}{\text{kg K}}$	$\dfrac{1}{\text{K}}$	$\dfrac{\text{W}}{\text{m K}}$	$\dfrac{\text{kg}}{\text{m s}}$	$\dfrac{\text{m}^2}{\text{s}}$	-
0.01	0.00611655	999.7925	4.21991	- 0.06797	0.56104	1.79116	132.978	13.4724
10	0.01228199	999.6546	4.19554	0.08769	0.58000	1.30598	138.289	9.44709
20	0.02339318	998.1618	4.18436	0.20666	0.59842	1.00165	143.277	7.00390
30	0.04246971	995.6062	4.18008	0.30330	0.61546	0.79736	147.886	5.41550
40	0.07384938	992.1751	4.17965	0.38545	0.63058	0.65297	152.059	4.32802
50	0.12351946	987.9962	4.18155	0.45779	0.64355	0.55683	155.772	3.55310
60	0.19946434	983.1602	4.18513	0.52329	0.65435	0.46638	159.029	2.98290
70	0.31200930	977.7337	4.19022	0.58401	0.66309	0.40387	161.851	2.55217
80	0.47414474	971.7662	4.19687	0.64143	0.66999	0.35433	164.279	2.21958
90	0.70181766	965.2953	4.20528	0.69666	0.67525	0.31440	166.345	1.95799
100	1.0141800	958.3491	4.21567	0.75062	0.67909	0.28174	168.088	1.74900
110	1.4337871	950.9480	4.22833	0.80409	0.68169	0.25470	169.536	1.57982
120	1.9867442	943.1066	4.24351	0.85777	0.68319	0.23205	170.709	1.44135
130	2.7027998	934.8340	4.26150	0.91230	0.68370	0.21290	171.620	1.32698
140	3.6153910	926.1344	4.28258	0.96836	0.68330	0.19654	172.279	1.23182
150	4.7616454	917.0077	4.30708	1.02660	0.68204	0.18246	172.685	1.15223
160	6.1823462	907.4495	4.33535	1.08773	0.67996	0.17024	172.837	1.08544
170	7.9218701	897.4510	4.36782	1.15256	0.67705	0.15955	172.721	1.02931
180	10.028105	886.9990	4.40497	1.22196	0.67332	0.15014	172.328	0.98221
190	12.552362	876.0757	4.44740	1.29698	0.66875	0.14178	171.639	0.94289
200	15.549279	864.6581	4.49584	1.37882	0.66331	0.13432	170.632	0.91039
210	19.076750	852.7176	4.55121	1.46899	0.65697	0.12760	169.283	0.88398
220	23.195862	840.2190	4.61463	1.56929	0.64965	0.12152	167.552	0.86316
230	27.970875	827.1192	4.68756	1.68207	0.64131	0.11596	165.407	0.84761
240	33.469251	813.3656	4.77190	1.81027	0.63185	0.11085	162.793	0.83718
250	39.761749	798.8942	4.87013	1.95783	0.62119	0.10611	159.659	0.83192
260	46.922610	783.6257	4.98562	2.13005	0.60924	0.10168	155.941	0.83209
270	55.029868	767.4612	5.12302	2.33429	0.59591	0.09750	151.565	0.83818
280	64.165829	750.2752	5.28893	2.58112	0.58115	0.09351	146.454	0.85098
290	74.417783	731.9052	5.49310	2.88622	0.56496	0.08966	140.522	0.87174
300	85.879049	712.1356	5.75040	3.27392	0.54743	0.08590	133.680	0.90229
310	98.650512	690.6716	6.08478	3.78393	0.52875	0.08217	125.815	0.94558
320	112.84293	667.0939	6.53734	4.48561	0.50920	0.07841	116.762	1.00664
330	128.58052	640.7732	7.18634	5.51342	0.48907	0.07454	106.208	1.09521
340	146.00677	610.6676	8.20797	7.17536	0.46851	0.07043	93.4713	1.23392
350	165.29415	574.7065	10.1160	10.39303	0.44737	0.06588	76.9506	1.48962
360	186.66007	527.5916	15.0044	19.12076	0.42572	0.06033	53.7784	2.12627
370	210.43563	451.4257	45.1552	76.38429	0.42504	0.05207	20.8514	5.53165
373.946	220.64000	321.1999	∞	∞	0.83000	0.03943	0	∞

State Diagrams

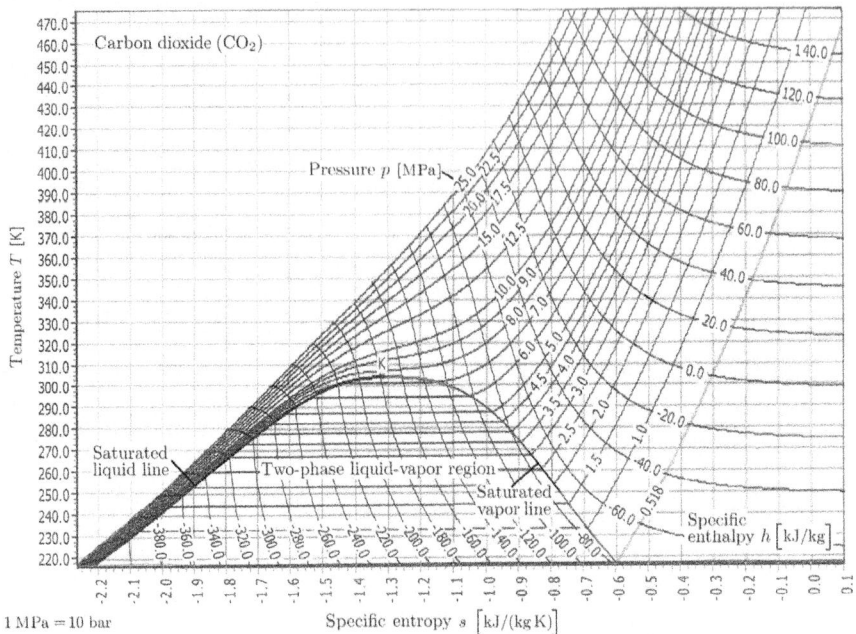

Diagram 1: T,s diagram of carbon dioxide (liquid and vapor) (CO_2)

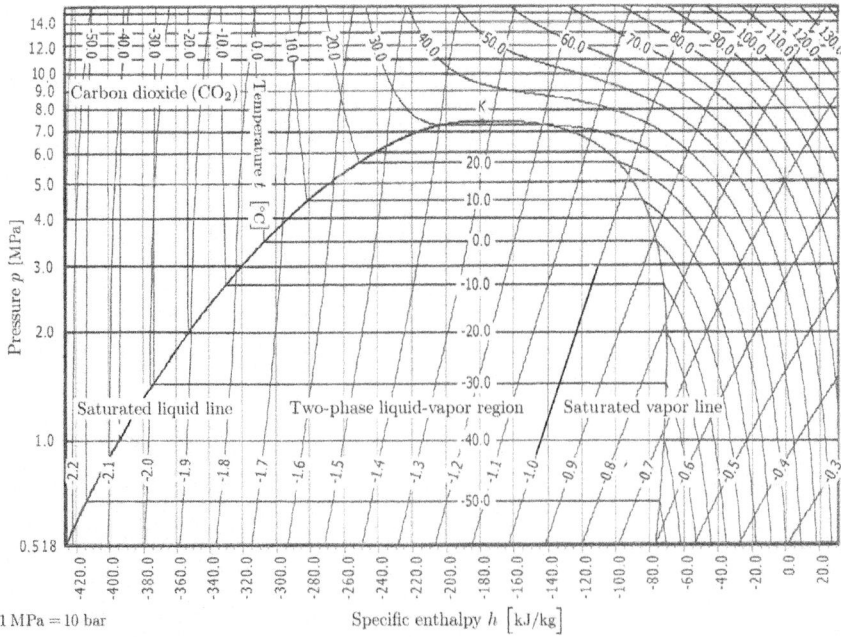

Diagram 2: lg p,h diagram of carbon dioxide (liquid and vapor) (CO_2)

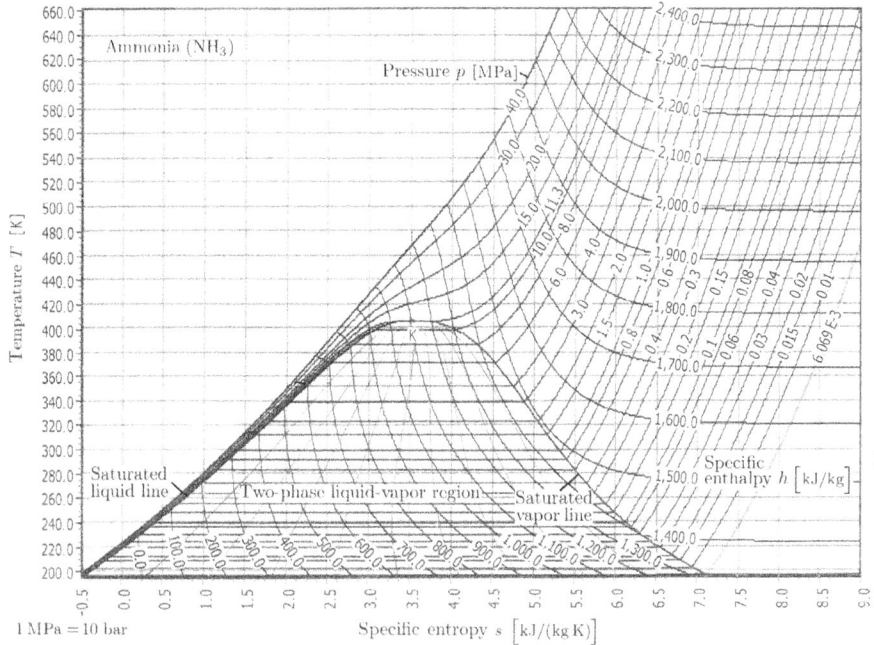

Diagram 3: T,s **diagram of ammonia (liquid and vapor) (NH$_3$)**

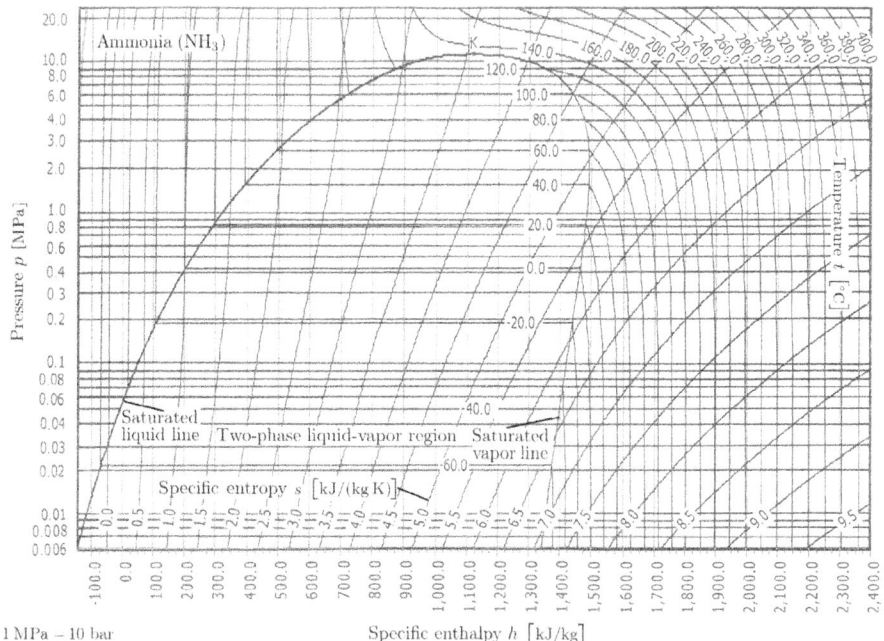

Diagram 4: lg p,h **diagram of ammonia (liquid and vapor) (NH$_3$)**

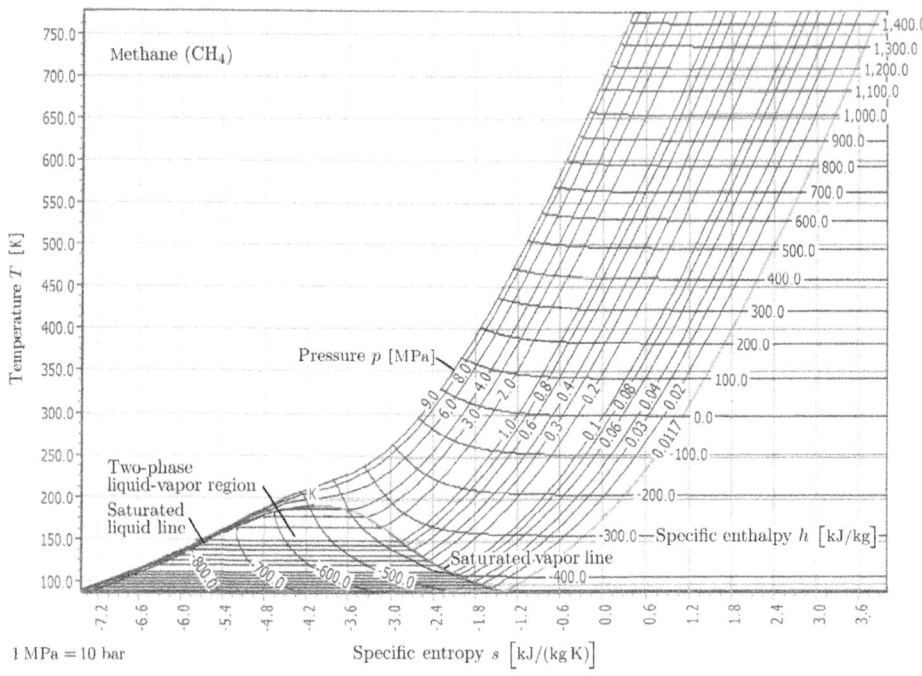

Diagram 5: T,s diagram of methane (liquid and vapor) (CH_4)

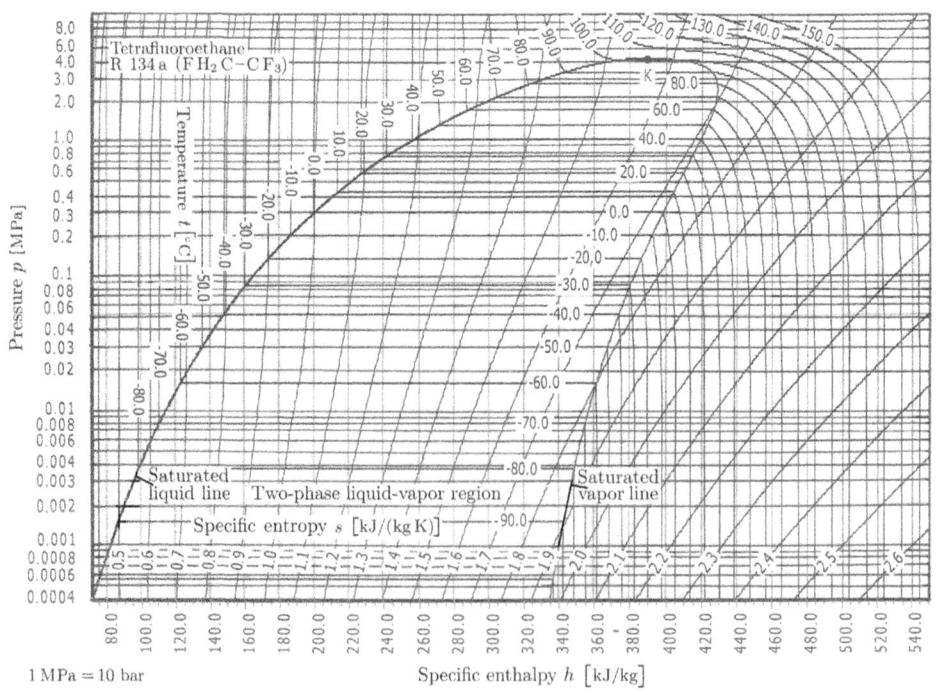

Diagram 6: lg p,h diagram of refrigerant tetrafluoroethane (liquid and vapor) (FH_2C-CF_3)

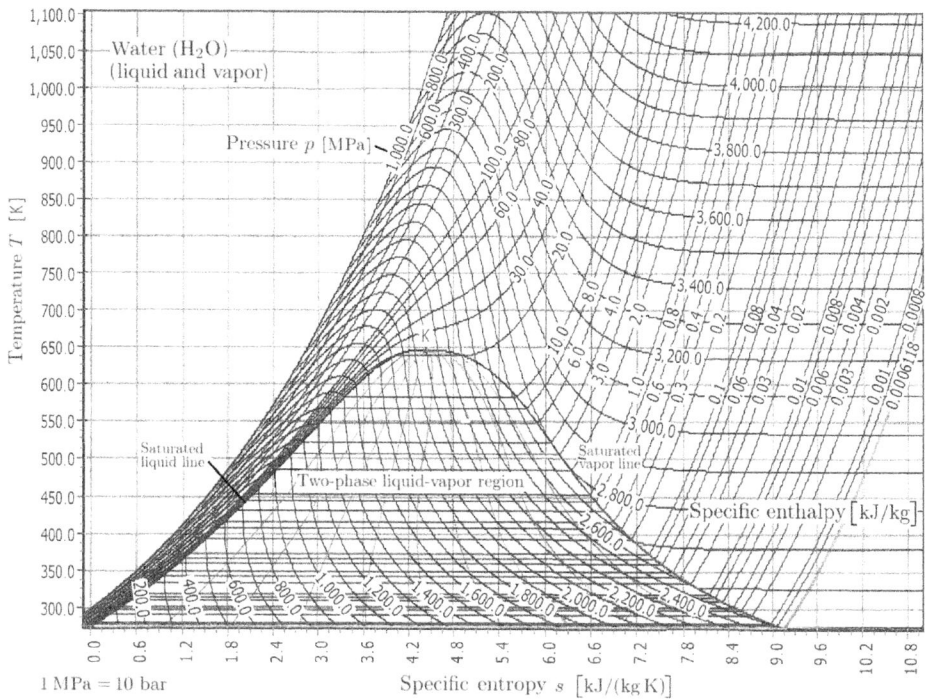

Diagram 7: T,s diagram of water (liquid and vapor) (H_2O)

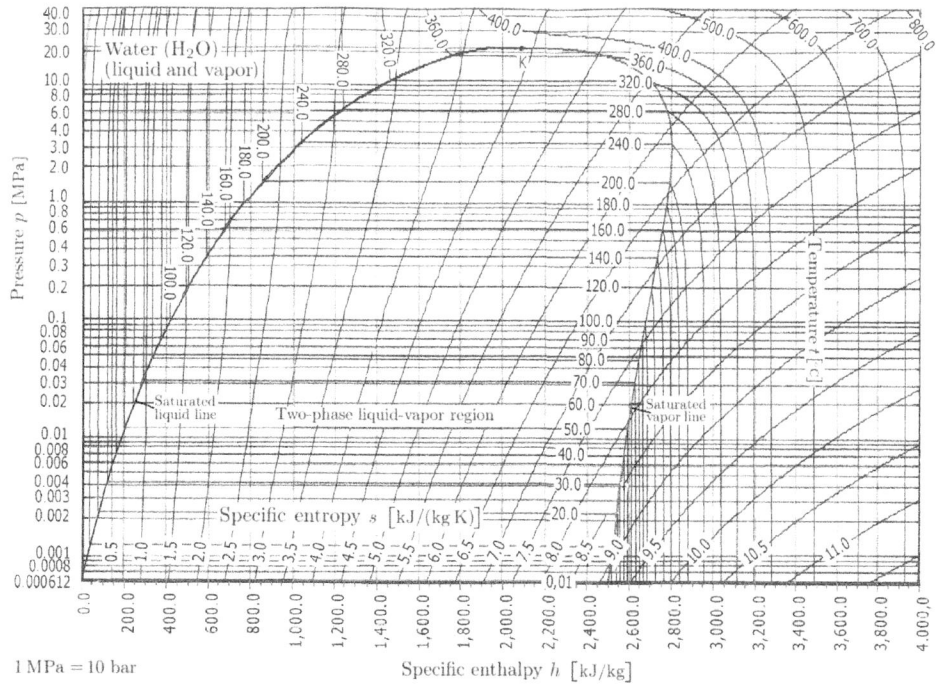

Diagram 8: lg p,h diagram of water (liquid and vapor) (H_2O)

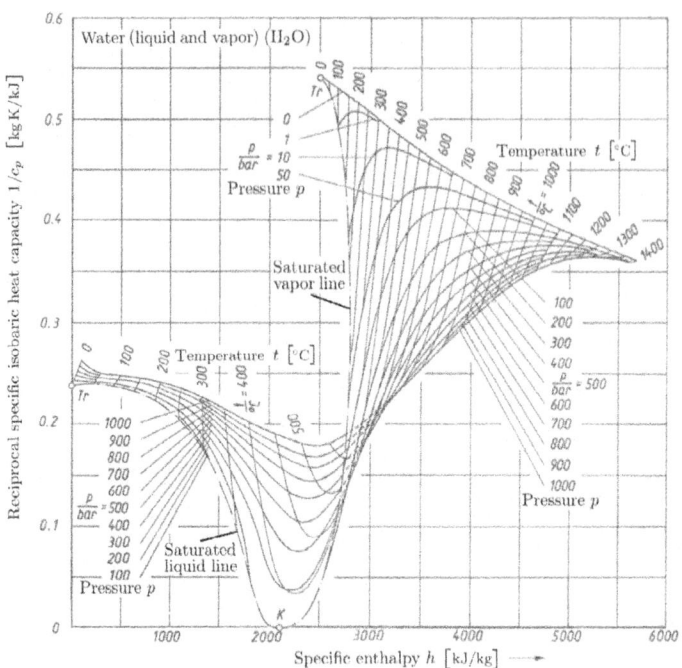

Diagram 9: $1/c_p, h$ diagram of water (liquid and vapor) (H_2O) [26]

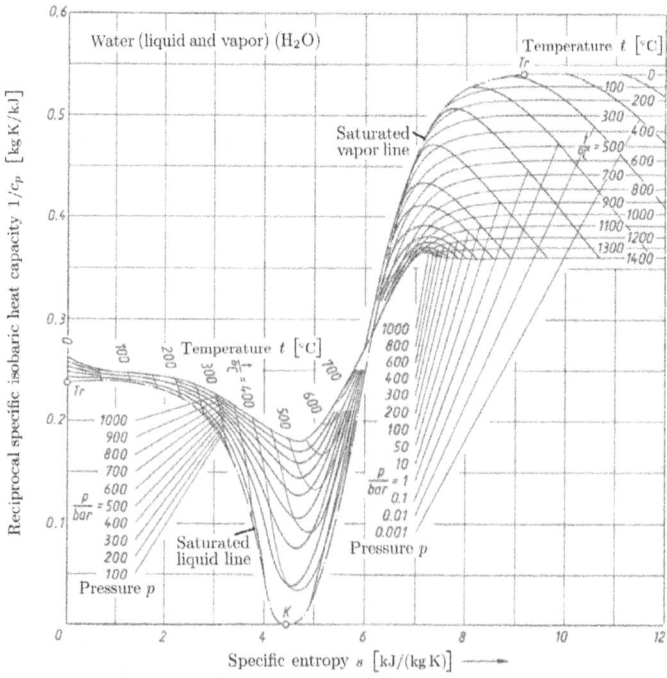

Diagram 10: $1/c_p, s$ diagram of water (liquid and vapor) (H_2O) [26]

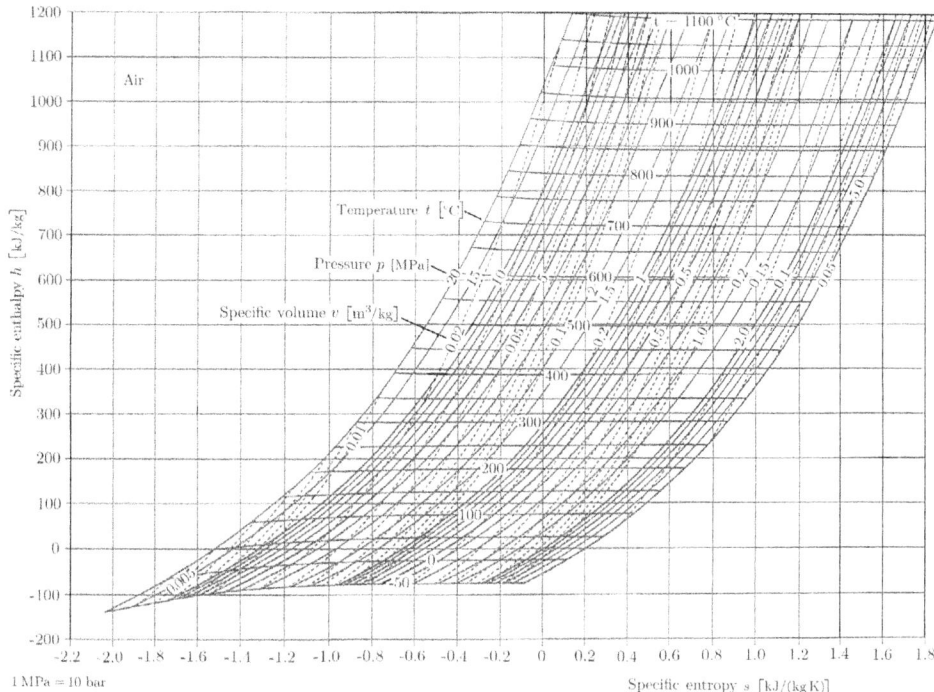

Diagram 11: h, s diagram of air

References

[1] *Asami, T.*: New method to determine the BWR coefficients in saturated regions. Cryogenics 28 (1988) Nr. 8 S. 521/526.

[2] *Baehr, H. D.* und *K. Schwier*: Die thermodynamischen Eigenschaften der Luft. Springer, Berlin 1961.

[3] *Baehr, H. D.* und *E. Hicken*: Die thermodynamischen Eigenschaften von CF_2Cl_2 (R 12) im kältetechnisch wichtigen Zustandsbereich. Kältetechnik 17 (1965) 143/150.

[4] *Baehr, H. D.*: Zur Definition exergetischer Wirkungsgrade. BWK 20 (1968) S. 197/200.

[5] *Baehr, H. D., H. Hartmann. H.-Chr. Pohl* und *H. Schomäcker*: Thermodynamische Funktionen idealer Gase für Temperaturen bis 6000 °K. Springer, Berlin 1968.

[6] *Baehr, H. D.*: Kanonische Zustandsgleichungen und ihre Bedeutung für die technischen Anwendungen der Thermodynamik. Kältetechnik 23 (1971) S. 78/81.

[7] *Baehr, H. D.*: Thermodynamik. 3. Aufl. Springer, Berlin 1973, S. 132.

[8] *Baehr, H. D.*: Thermodynamik. 7. Aufl. Springer, Berlin 1989, S. 162, 422.

[9] *Baehr, H. D.*: Thermodynamik. 16. Aufl. Springer, Berlin Heidelberg 2016.

[10] *Baehr, H. D.* und *S. Kabelac*: Vorläufige Zustandsgleichungen für das ozonunschädliche Kältemittel R 134 a. Ki Klima-Kälte-Heizung 2/1989 S. 69/71.

[11] *Bayer, C.* und *W. Koch-Emmery*: Das stationäre Betriebsverhalten von wasserbeheizten Lufterhitzern bei verschiedenen Luftzuständen. Gesundh.-Ing. 90 (1969) S. 87/93.

[12] *Beattie, J. A.* und *O. C. Bridgman*: A New Equation of State for Fluids.
I. Application to Gaseous Ethyl Ether and Carbon Dioxide.
J. Am. Chem. Soc. 49 (1927) S. 1665/1667.
II. Application to Helium, Neon, Argon, Hydrogen, Nitrogen, Oxygen, Air and Methane. J. Am. Chem. Soc. 50 (1928) S. 3133/3138.
III. The Normal Densities and the Compressibilities of several Gases at 0°.
J. Am. Chem. Soc. 50 (1928) S. 3151/3157.

[13] *Bender, E.*: Zur Aufstellung von Zustandsgleichungen, aus denen sich die Sättigungsgrößen exakt berechnen lassen — gezeigt am Beispiel des Methans. Kältetechnik 23 (1971) S. 258/264, Berichtigung S. 348.

[14] *Bender, E.*: Equations of State for Ethylene and Propylene. Cryogenics 15 (1975) S. 667/673.

[15] *Benedict, M., G. B. Webb* und *L. C. Rubin*: An Empirical Equation for Thermodynamic Properties of Light Hydrocarbons and Their Mixtures.
I. Methane, Ethane and n-Butane. J. Chem. Phys. 8 (1940) S. 334/345
II. Mixtures of Methane, Etane, Propane and n-Butane.
J. Chem. Phys. 10 (1942) S. 747/758.

[16] *Benedict, M., G. B. Webb* und *L. C. Rubin*: Constants for Twelve Hydrocarbons. Chem. Eng. Progr. 47 (1951) S. 419/422.

[17] Bošnjaković, F.: Technische Thermodynamik. Bd. 1 und 2. 4. Aufl. Steinkopff, Dresden 1965.

[18] Bošnjaković, F.: Technische Thermodynamik Teil 1. 8. korrigierte Aufl. Steinkopff, Darmstadt 1998.

[19] Bošnjaković, F., M. Viličić und B. Slipčević: Einheitliche Berechnung von Rekuperatoren. VDI-Forschungsheft 432 (1951) S. 5/26.

[20] Bowman, R. A., A. C. Mueller und W. M. Nagle: Mean temperature difference in design. Trans. ASME 62 (1940) S. 283/294.

[21] Brandt, F.: Brennstoffe und Verbrennungsrechnung. 2. Aufl. Vulkan-Verlag, Essen 1991.

[22] Carslaw, H. S. und J. C. Jaeger: Conduction of heat in solids. 2. Aufl. Clarendon Press, Oxford 1959.

[23] Cerbe, G., et al.: Grundlagen der Gastechnik. 7. Aufl. Carl Hanser, München Wien 2008.

[24] Cerbe, G. und H.-J. Hoffmann bzw. Cerbe, G. und G. Wilhelms: Einführung in die Thermodynamik. 12. bis 17. Aufl. Carl Hanser, München Wien 1999 bis 2013.

[25] CRC Handbook of Chemistry and Physics. Hrsg. D. R. Lide. 73. Aufl. Boca Raton (Florida) 1992–1993.

[26] Dehli, M.: Über kanonische Zustandsgleichungen und ihre Anwendungsmöglichkeiten in der Technik. VDI-Forschungsheft 570 (1975).

[27] Dehli, M.: Energieeinsparpotenziale bei der Drucklufterzeugung. Brennstoff-Wärme-Kraft BWK Bd. 54 (2002) Nr. 6 S. 57/62.

[28] Dehli, M.: Prozesse zur Verbesserung der Wirksamkeit von Gas-Expansionsanlagen unter thermodynamischen Gesichtspunkten. FHTE, Hochschule für Technik, Esslingen 1994.

[29] Dehli, M.: Energierückgewinnung mit Gas-Expansionsanlagen. Gas Erdgas gwf 137 (1996) Nr. 4 S. 196/206.

[30] Dehli, M.: Concepts for Gas Expansion at High Temperatures. Proceedings of the International Conference on Gas Expansion, Maastricht 1997.

[31] Dehli, M.: Energierückgewinnung in technischen Systemen.
Teil 1: Rückgewinnungskonzepte bei der Erdgasbereitstellung. Brennstoff-Wärme-Kraft BWK Bd. 52 (2000) Nr. 7/8 S. 44/51.
Teil 2: Rückgewinnungskonzept bei der Rauchgasreinigung von Steinkohlekraftwerken. Brennstoff-Wärme-Kraft BWK Bd. 52 (2000) Nr. 9 S. 49/51.

[32] Dehli, M.: g,s-Zustandsdiagramm für Wasser und Wasserdampf. Brennstoff-Wärme-Kraft BWK Bd. 57 (2005) Nr. 11 S. 63/67.

[33] Dehli, M.: Vergleichende Bewertung von thermodynamischen Kreisprozessen. Brennstoff-Wärme-Kraft BWK Bd. 51 (1999) Nr. 11/12 S. 48/58.

[34] Dehli, M.: Thermodynamische Kriterien für die Weiterentwicklung von Kraftmaschinen. Brennstoff-Wärme-Kraft BWK Bd. 57 (2005) Nr. 1/2 S. 61/66.

[35] Döring, R. und H. J. Löffler: Thermodynamische Eigenschaften von Trifluormethan (R 23). Kältetechnik 20 (1968) S. 342/348.

[36] Döring, R.: Thermodynamische Eigenschaften von Ammoniak (R 717). Klima- und Kälte-Ingenieur, Ki-extra Nr. 5. C. F. Müller, Karlsruhe 1978.

[37] Dvorák, Z. und J. Petrák: Beitrag zur Ermittlung von thermodynamischen Eigenschaften der Kältemittel R 22, R 505 und des Ammoniaks. Klima- und Kälte- Ingenieur 3 (1975) S. 319/324.

[38] Eckert, E.: Einführung in den Wärme- und Stoffaustausch. 2. Aufl. Springer, Berlin 1959.

[39] Elsner, N.: Grundlagen der Technischen Thermodynamik. Bertelsmann, Gütersloh 1973.

[40] Energietechnische Arbeitsmappe. VDI-Gesellschaft Energietechnik. 15., bearbeitete und erweiterte Aufl. Springer, Berlin Heidelberg 2000.

[41] Faltin, H.: Technische Wärmelehre. 5. Aufl. Akademie-Verlag, Berlin 1968.

[42] Glück, B.: Strahlungsheizung - Theorie und Praxis. Müller, Karlsruhe 1982.

[43] Gordon, S.: Thermodynamic and transport properties of hydrocarbon with air. Vol. 1. NASA Techn. Paper 1902, 1982.

[44] Gregorig, R.: Wärmeaustausch und Wärmeaustauscher. 2. Aufl. Sauerländer, Aarau 1973.

[45] Grigull, U. und H. Sandner: Wärmeleitung. Springer, Berlin 1979.

[46] Gröber, H., S. Erk und U. Grigull: Die Grundgesetze der Wärmeübertragung. 3. Aufl. Springer, Berlin 1957.

[47] Handbuch Maschinenbau. Grundlagen und Anwendungen der Maschinenbau-Technik. 24. Auflage. Part: Böge, G.; Böge, W.; Surek, D.: Abschnitt Hydrostatik, -dynamik, Gasdynamik S. 361/389. Springer Vieweg, Wiesbaden 2021.

[48] Hausen, H.: Wärmeübertragung durch Rippenrohre. Z. VDI-Beiheft Verfahrenstechnik (1940) Nr. 2 S. 55/57.

[49] Hausen, H.: Darstellung des Wärmeüberganges in Rohren durch verallgemeinerte Potenzbeziehungen, Z. VDI-Beiheft Verfahrenstechnik (1943) Nr. 4 S. 91/98.

[50] Hausen, H.: Wärmeübertragung im Gegenstrom, Gleichstrom und Kreuzstrom. Springer, Berlin 1950.

[51] Hausen, H.: Neue Gleichungen für die Wärmeübertragung bei freier und erzwungener Strömung. Allg. Wärmetechn. 9 (1959) S. 75/79.

[52] Hausen, H.: Erweiterte Gleichung für den Wärmeübergang bei turbulenter Strömung. Wärme- u. Stoffübertr. 7 (1974) S. 222/225.

[53] Hausen, H.: Wärmeübertragung im Gegenstrom, Gleichstrom und Kreuzstrom. 2. Aufl. Springer, Berlin 1976.

[54] Hausen, H. und H. Rögener: Warum stimmt die thermodynamische Temperaturskala mit den gasthermometrischen Messungen überein? PTB-Mitt. 87 (1977) Nr. 2 S. 97/102.

[55] Heckenberger, T. E. J.: Wärmeleitfähigkeit und Viskosität des neuen Kältemittels R 134a. Ki Klima-Kälte-Heizung 11/1990 S. 484/486.

[56] Hell, F.: Grundlagen der Wärmeübertragung. 3. Aufl. VDI-Verlag, Düsseldorf 1982.

[57] *Hofmann, E.*: Der Wärmeübergang bei der Strömung im Rohr. Z. ges. Kälte-Ind. Bd. 44 (1937) S. 99/107.

[58] *Holborn, L.* und *J. Otto*: Über die Isothermen einiger Gase zwischen $+400°$ und $-183°$. Z. Phys. Bd. 33 (1925) S. 1/11.

[59] *Hou, Y. C.* und *J. J. Martin*: Physical and Thermodynamic Properties of Trifluoromethane. AIChE Journal 5 (1959) S. 125/129.

[60] Internationale Praktische Temperaturskala von 1968. Verbesserte Ausg. von 1975. PTB-Mitteilungen 87 (1977) Nr. 6 S. 497/510. Internationale Praktische Temperaturskala von 1990 (ITS-90), Physikalisch-Technische Bundesanstalt (PTB), Braunschweig.

[61] *Jahnke, E., F. Emde* und *F. Lösch*: Tafeln höherer Funktionen. 6. Aufl. Teubner, Stuttgart 1960.

[62] *Jakob, M.*: Heat Transfer. Bd. 1 und 2. Wiley, New York 1959.

[63] *Jolls, K. R.* und *G. P. Willers*: Computer Generated Phase Diagram for Ethylene and Propylene. Cryogenics 18 (1978) S. 329/336.

[64] *Kamke, D.* und *K. Krämer*: Physikalische Grundlagen der Maßeinheiten. Teubner, Stuttgart 1977.

[65] *Kasparek, G.*: Der Energieaustausch durch Wärmestrahlung zwischen Feststoffoberflächen. BWK 24 (1972) S. 229/233.

[66] *Kays, W. M.*: Loss coefficients for abrupt changes in flow cross section with low Reynolds number flow in single and multi-tube systems.
Trans. ASME 72 (1950) S. 1067/1074.

[67] *Kays, W. M.* und *A. L. London*: Heat-transfer and flow-friction characteristics of some compact heat-exchanger surfaces.
Teil 1: Test system and procedure.
Teil 2: Design data for thirteen surfaces. Trans. ASME 72 (1950) S. 1075/1085 und 1087/1097.
Teil 3: Design data for five surfaces. Trans. ASME 74 (1952) S. 1167/1178.

[68] *Kays, W. M.* und *A. L. London*: Compact Heat Exchangers.
2. Aufl. McGraw-Hill, New York 1964.

[69] *Kays, W. M.* und *A. L. London*: Hochleistungswärmeübertrager.
Akademie-Verlag, Berlin 1973.

[70] *Küper, P.* und *H. J. Löffler*: Eine neue Dampftafel für das Kältemittel R 22. Kältetechnik 23 (1971) S. 47/51.

[71] *Küttner, K. H.*: Kolbenmaschinen. 5. Aufl. Teubner, Stuttgart 1984, S. 181.

[72] *Kugeler, K.* und *P.-W. Phlippen*: Energietechnik.
2. Aufl. Springer, Berlin Heidelberg 1993.

[73] *Landolt-Börnstein*: Zahlenwerte und Funktionen. Bd. IV/4a, Springer, Berlin 1967.

[74] *Leggewie, G.*: Flüssiggase. Carl Hanser, München Wien 1969.

[75] *Lewis, W. K.*: The evaporation of a liquid into a gas. Mech. Eng. 44 (1922) S. 445/446.

[76] *Löffler, H. J.* und *H. Hinrichsen*: „Azeotrope" Kältemittelgemische. Azeotrope Punkte, Kälteleistung, Kompressionsendtemperatur. Kältetechnik 21 (1969) S. 6/14.

[77] *London, A. L.* und *C. K. Ferguson*: Test results of high-performance heat-exchanger surface used in aircraft intercoolers and their significance for gas-turbine regenerator design. Trans. ASME 71 (1949) S. 17/26.

[78] *Lutz, O.* und *F. Wolf*: IS-Tafel für Luft und Verbrennungsgase. Springer, Berlin 1938.

[79] *Martin, J. J.* und *Y. C. Hou*: Development of an Equation of State for Gases. AIChE Journal 1 (1955) S. 142/151, Berichtigung S. 506.

[80] *Martin, J. J., R. M. Kapoor* und *N. de Nevers*: An Improved Equation of State for Gases. AIChE Journal 5 (1959) S. 159/160.

[81] *Martin, J. J.* und *R. C. Downing*: Thermodynamische Eigenschaften des Kältemittels R 502. Kältetechnik 23 (1971) S. 265/267.

[82] The Scientific Papers of *James Clerk Maxwell*. Hrsg. *W. D. Niven*. 2 Bände, Cambridge University Press 1890, Neudruck Dover Publications Inc., New York 1965, Bd. 2 S. 424.

[83] *McCarty, R. D.*: A Modified *Benedict-Webb-Rubin* Equation of State for Methane Using Recent Experimental Data. Cryogenics 14 (1974) S. 276/280.

[84] *Mears, W. H., J. V. Sinka, P. F. Malbrunot, P. A. Meunier, A. G. Dedit* und *G. M. Scatena*: Pressure-Volume-Temperature Behavior of a Mixture of Difluoromethane and Pentafluoromonochloroethane. J. Chem. Eng. Data 13 (1968) S. 344/347. Ref.: Kältetechnik 21 (1969) S. 48 und 81.

[85] *Morsy, T. E.*: Eine neue Dampftafel für Tetrafluordichloräthan (R 114). Kältetechnik 17 (1965) S. 86/89.

[86] *Morsy, T. E.*: Eine neue Zustandsgleichung für Trifluormethan (R 23). Kältetechnik 17 (1965) S. 272/275.

[87] *Morsy, T. E.*: Trifluormethan (R 23). Thermodynamische Eigenschaften und Dampftafel. Kältetechnik 18 (1966) S. 203/206.

[88] *Morsy, T. E.*: Ein *Mollier*-Diagramm für Trifluormethan (R 23). Kältetechnik 18 (1966) S. 347/349.

[89] *Morsy, T. E.*: Thermodynamische Eigenschaften des Kältemittels R 500. Kältetechnik 20 (1968) S. 94/101.

[90] *Morsy, T. E.* und *D. Straub*: Dampftafel und *Mollier*-Diagramm von Tetrafluormethan. Kältetechnik 20 (1968) S. 210/214.

[91] *Nunner, W.*: Wärmeübergang und Druckabfall in rauhen Rohren. VDI-Forschungsheft 455 (1956).

[92] *Nußelt, W.*: Eine neue Formel für den Wärmedurchgang im Kreuzstrom. Techn. Mech. Thermodyn. 1 (1930) S. 417/422.

[93] *Oswatitsch, K.*: Grundlagen der Gasdynamik. Springer, Wien 1976.

[94] *Planck, M.*: Theorie der Wärmestrahlung. 6. Aufl. Barth, Leipzig 1966.

[95] *Prandtl, L.*: Eine Beziehung zwischen Wärmeaustausch und Strömungswiderstand der Flüssigkeiten. Phys. Z. Bd. 11 (1910) S. 1072/1078.

[96] *Prandtl, L.*: Bemerkung über den Wärmeübergang im Rohr. Phys. Z. Bd. 29 (1928) S. 487/489.

[97] *Pitzer, K. S.*: The Volumetric and Thermodynamic Properties of Fluids.
I. Theoretical Basis and Virial Coefficients.
J. Am. Chem. Soc. 77 (1955) S. 3427/3433.
Pitzer, K. S., D. Z. Lippmann, R. F. Curl, C. M. Huggins und *D. E. Petersen*: The Volumetric and Thermodynamic Properties of Fluids.
II. Compressibility Factor, Vapor Pressure and Entropy of Vaporization.
J. Am. Chem. Soc. 77 (1955) S. 3433/3440.

[98] *Ražnjević, K.*: Thermodynamische Tabellen. VDI-Verlag, Düsseldorf 1977.

[99] *Redlich, O.* und *J. N. S. Kwong*: On the Thermodynamics of Solutions.
V. Chem. Rev. 44 (1949) S. 233/244.

[100] *Rehwald, W.*: Elementare Einführung in die Bessel-, Neumann- und Hankel-Funktionen. Hirzel, Stuttgart 1959.

[101] *Renz, U.*: Thermodynamische Eigenschaften des Eises. Kältetechnik 21 (1969) S. 266/269.

[102] *Riedel, L.*: Untersuchungen über eine Erweiterung des Theorems der übereinstimmenden Zustände.
Teil I Eine neue universelle Dampfdruckformel. Chemie-Ing.-Technik 26 (1954) S. 83/89.
Teil II Die Flüssigkeitsdichte im Sättigungszustand. Chemie-Ing.-Technik 26 (1954) S. 259/264.
Teil III Kritischer Koeffizient, Dichte des gesättigten Dampfes und Verdampfungswärme.
Chemie-Ing.-Technik 26 (1954) S. 679/683.

[103] *Rombusch, U. K.*: Ein erweitertes Korrespondenzprinzip zur Bestimmung von Zustandsgrößen.
Allg. Wärmetechnik 11 (1962) S. 41/50 und 133/145.

[104] *Rombusch, U. K.* und *H. Giesen*: Zur Berechnung von Dampftafeln.
Kältetechnik 16 (1964) S. 66/69.

[105] *Rombusch, U. K.*: Ein *Mollier-i*, log p-Diagramm für Trifluormonobrom-Methan (R 13 B 1).
Kältetechnik 16 (1964) S. 69/76.

[106] *Rombusch, U. K.* und *H. Giesen*: Neue *Mollier-i*, log p-Diagramme für die Kältemittel R 11, R 12, R 13 und R 21. Kältetechnik 18 (1966) S. 37/40.

[107] Angaben der Ruhrgas AG, Essen 2003.

[108] *Sawitzki, P.*: Zweidimensionale Temperaturfelder in geraden Rechteckrippen.
Wärme- und Stoffübertragung 5 (1972) S. 253/256.

[109] *Schedwill, H.*: Thermische Auslegung von Kreuzstromwärmeaustauschern.
Fortschr.-Ber. VDI-Z. Reihe 6 Nr. 19 (1968).

[110] *Schlünder, E. U.*: Einführung in die Wärmeübertragung.
3. Aufl. Vieweg, Braunschweig 1981.

[111] *Schmidt, E.*: Die Wärmeübertragung durch Rippen.
Z. VDI 70 (1926) S. 885/889 und 947/951.

[112] *Schmidt, E.*: Einführung in die Technische Thermodynamik. 7. Aufl. Springer, Berlin 1958.

[113] *Schmidt, E.* und *U. Grigull*: Properties of Water and Steam in SI-Units. 4. Aufl. Springer und Oldenbourg, Berlin und München 1989.

[114] *Schmidt, E.*: Verdunstung und Wärmeübergang. Gesundh.-Ing. 52 (1929) S. 525/529.

[115] *Schmidt, Th. E.*: Bestimmung der Wärmeübergangszahlen aus gemessenen Wärmedurchgangszahlen. Forsch. Ing.-Wes. 4 (1933) S. 183/186. Berichtigung S. 214.

[116] *Schmidt, Th. E.*: Die Wärmeleistung von berippten Oberflächen. Abh. des Deutschen Kältetechn. Vereins Nr. 4. Müller, Karlsruhe 1950.

[117] *Scholten, W.*: Vorläufige thermodynamische Eigenschaften von R 12 B 1 (Bromchlordifluormethan). Kältetechnik 24 (1972) S. 45/48.

[118] *Sievers, U.*: Die thermodynamischen Eigenschaften von Kohlendioxid. Fortschr.-Ber. VDI-Z. Reihe 6 Nr. 155 (1984).

[119] *Sinka, J. V.*: Physical Properties of a Refrigerant Mixture of Monofluoromonochloromethane and Tetrafluorodichloroethane. J. Chem. Eng. Data 15 (1970) S. 71/73.

[120] *Sinka, J. V., E. Rosenthal* und *R. P. Dixon*: Pressure-Volume-Temperature Relationship for a Mixture of Monochlorotrifluoromethane and Trifluoromethane. J. Chem. Eng. Data 15 (1970) S. 73/74. Ref.: Kältetechnik 22 (1970) S. 233.

[121] *Soave, G.*: Equilibrium constants from a modified *Redlich-Kwong* equation of state. Chem. Eng. Sci. 27 (1972) S. 1197/1203.

[122] *Stephan, P., Schaber, K., Stephan, K.* und *F. Mayinger*: Thermodynamik Grundlagen und technische Anwendungen – Band 2: Mehrstoffsysteme und chemische Reaktionen. 16. Aufl. Springer, Berlin Heidelberg 2017.

[123] *Surek, D.; Stempik, S.*: Dynamic Pressure Oscillations and Compression Shock of the Impeller in the Side Channel Compressor. Proceedings of the International Rotation Equipment Conference 2008, Düsseldorf, S. 101/110.

[124] *Teja, A. S.* und *A. Singh*: Equations of State for Ethane, Propane and n-Butane. Cryogenics 17 (1977) S. 591/596.

[125] *Tsonopoulos, C.* und *J. M. Prausnitz*: Equations of State. A Review for Engineering Applications. Cryogenics 9 (1969) S. 315/327.

[126] *Van Ness, H. C.*: Use of the *Redlich* and *Kwong* Equation of State in Calculating Thermodynamic Properties of Gases from Experimental Compressibility Data. AIChE Journal 1 (1955) S. 100/104.

[127] *Wagner, W., J. Ewers* und *R. Schmidt*: An equation of state for oxygen vapour — second and third virial coefficients. Cryogenics 24 (1984) Nr. 1 S. 37/43.

[128] *Wagner, W.* und *A. Kruse*: Zustandsgrößen von Wasser und Wasserdampf. Der Industrie-Standard IAPWS-IF97 für die thermodynamischen Zustandsgrößen und ergänzende Gleichungen für andere Eigenschaften. Springer, Berlin Heidelberg 1998.

[129] *Wagner, W., A. Saul* und *A. Pruß*: International Equations for the Pressure along the Melting and along the Sublimation Curve of Ordinary Water Substance. J. Phys. Chem. Ref. Data 23 (1994) Nr. 3 S. 515/527.

[130] *Wagner, W.* und *U. Overhoff*: ThermoFluids. Interaktive Software für die Berechnung thermodynamischer Eigenschaften für mehr als 60 Stoffe auf der Basis der Software Fluidcal. Lehrstuhl für Thermodynamik der Ruhr-Universität Bochum), Springer-Verlag, Heidelberg 2005.

[131] VDI-Wärmeatlas. VDI-Verlag, Düsseldorf 1954/1963.

[132] VDI-Wärmeatlas. 4. bis 10. Aufl. VDI-Verlag, Düsseldorf, bzw. Springer-Verlag, Berlin, 1984 bis 2006.

[133] VDI-Wärmeatlas. Berechnungsblätter für den Wärmeübergang. VDI-Gesellschaft Verfahrenstechnik und Chemieingenieurwesen (Hrsg.). 9. Aufl. Springer, Berlin Heidelberg 2002.

[134] *Wukalowitsch, M. P., W. N. Subarjev* und *P. G. Prusakov*: Zustandsgleichung für Wasserdampf bei 800 bis 1500 °C und 5 bis 1000 bar. Russische Originalarbeit: Teploenergetika 12 (1965) Nr. 9 S. 67/71. Englische Übersetzung: Thermal Engineering 12 (1965) S. 88/93. Deutsches Referat: BWK 18 (1966) S. 410/411.

[135] *Wukalowitsch, M. P., W. N. Subarjev, A. A. Alexandroff* und *P. G. Prusakov*: Eigenschaften von Wasserdampf bei 800 bis 1500 °C. Russische Originalarbeit: Teploenergetika 13 (1966) Nr. 3 S. 77/82. Englische Übersetzung: Thermal Engineering 13 (1966) S. 101/106. Deutsches Referat: BWK 19 (1967) S. 320.

Index

absolute zero temperature 6, 70, 513, 516
absorption ratio 375, 376
additional work 227, 228
adiabatic process 50
adiabatic system 5
air 95, 99, 128, 133, 467, 469, 570, 574, 584, 591
air content 490
air/fuel equivalent ratio 468, 472, 480, 481, 490, 491
air pressure 95
air requirement 467, 469, 471
air volume, specific 446
ambient state 333, 334
ammonia 446, 501, 502, 574, 587
anergy 358
anergy free energies 363
anergy gain 367
angular ratio 378
anthracite hard coal 465
area efficiency 416
argon 99, 467, 574
ash 465, 483, 482, 485
Avogadro law 74
Avogadro number 74

barrier layer 460
Bernoulli equation 35, 115-118
biomass 465, 566
black body 375, 376
black body radiation 376
boiling 153, 156, 412-414, 429
boiling line 152, 153
boiling temperature 154, 412-414
Boltzmann constant 377
boundary curve 174
boundary layer 396, 399, 401, 402, 452, 453
bound energy 246, 337
bound energy, exergy of 327
Boyle and *Mariotte* law 69
bubble vaporisation 157, 413, 414
Bunte triangle 474
buoyancy 11, 97, 484
butane 465, 477, 574

caloric equation of state 75
calorific value 466, 485-487, 527-531
carbon 465, 470, 481, 482, 574
carbon dioxide 99, 466, 473-478, 517, 522, 569, 574, 586
carbon monoxide 466, 468, 474, 479, 574
Carnot factor 258, 259
Carnot process 255-259, 284, 285, 298, 303, 314, 319, 319-320
Carnot process, single-polytropic 272, 285, 294, 298, 302
catalyst 506
cavity method 384, 385
change of state 5, 77
change of state of humid air 446
change of state, quasi-static 45
change of state with variable mass 96, 97
chimney 484
chimney draught 484, 485
Celsius scale 6, 66, 69-71
circular fin with rectangular cross-section 419, 420
circular integral 66, 249
Claude process 323
Clausius-Clapeyron equation 193, 194
Clausius-Rankine process 274-277, 293, 298, 319
clean air 443
closed system 4, 498
coefficient of thermal expansion 11, 70
coefficient of volume expansion 11, 70
coke oven gas 474, 484
cold vapor compression process 325-327
combustion chamber 465
combustion control 473, 479, 480
combustion engine 261
combustion equation 466, 468, 471
combustion gas 465, 473, 474, 481, 484
combustion gas quantity, dry 468, 472
combustion gas quantity (exhaust gas quantity) 467, 468
combustion gas quantity, moist 468, 472
combustion process 465
combustion product 465
combustion quality 479, 480
combustion rate 481
combustion temperature 487, 489, 492
combustion triangle 473-478
comparative evaluation of right-hand cyclic processes 277
comparison process 44, 236
comparison process, reversible 261
complete combustion 466-469, 472-473

compression 215-218
compression ratio 262
compression shock, normal 142-147
compression volume 219
compressor 95, 96, 215-218, 229-231
condensation 151-154, 411, 412, 416, 428, 429
condensing boiler 484
contact temperature 395
continuity equation 35, 111, 112
convection 396-398
convection, free 408-411
conversion of thermal energy into
  mechanical energy 253, 259
cooling limit 455-458
counterflow heat exchanger 422, 423
coupling work 16
critical point 155, 164-168, 170, 171, 204
critical state 118-120
crossflow correction factor 430-432
crossflow heat exchanger 424-428
cyclic process work 249

Dalton's law 99, 443
degree of humidity 443, 444
degree of saturation 444
dehumidification 447
dehumidification, h,x diagram 452
density 7, 72, 73, 100, 103, 111-113, 119,
  127, 129, 133, 137-139, 142-146, 179, 446
density of humid air 446
dependence of pressure and temperature
  of K 538, 539
design diagram, heat exchanger 437-440
desublimation line 153
desublimation 152
dew point 444, 484
dew point of combustion gases 484
dew point temperature 485
dew precipitation 457, 458
Diesel engine 260
Diesel fuel 478
Diesel process 260, 261, 289, 292, 296, 298,
  300, 301, 308, 316
Diesel process, generalised 261, 290, 291, 309
diffusion 453, 458-462
diffusion coefficient 458
displacement machine 214, 220-223, 231
dissipative energy 67
dissociation 487, 489, 539, 540, 546
dry air 443, 467
dynamics of ideal gases 111-147

educt 467, 486, 500
efficiency 235-242, 254-260, 284, 370-373
efficiency, exergetic 284, 370-373
efficiency, thermal 254-260
effort ratio, mechanical 279, 282, 298
effort ratio, thermal 279, 282, 298
electrolysis cell 520-523
elemental analysis 470
emission 375
emission to half-space 378
emission ratio 375
end product 467-467
energies without anergy 363
energy 9, 10, 43, 333-335
energy balance 24-30, 218, 220, 465, 503-509
energy balances of evaporation 453, 454
energy conversion 10, 43, 334-336
energy devaluation 43, 333-335
energy, dissipative 67
energy, internal 21, 22, 34, 97, 98, 151, 152,
  312
energy recovery 232, 233
energy, thermal 9, 20, 21, 97, 98
energy units 13
enthalpy 9, 23, 24, 27, 28, 35, 97, 98, 108,
  109, 312, 503-509
enthalpy, exergy of 347-350
enthalpy, molar 503-509
enthalpy of activation, molar 506
enthalpy of condensation, of vaporisation
  162, 411
enthalpy of formation, molar 506-509
enthalpy of reaction, molar 504
entropy 49-52, 55-67, 76, 77, 93
entropy, absolute molar 512-514
entropy change 52, 55-67
entropy change heat 206, 245
entropy change heat, specific 206, 278
entropy diagram 62, 63
entropy, molar 512-514
environment 4, 523, 524
environment, chemical 523, 524
equation of state, caloric 75
equation of state, thermal 6, 69, 73, 74,
  172-183
equilibrium, chemical 497, 533-543
equilibrium constant 536-543
equilibrium environment 523, 524
equilibrium, thermal 156
equilibrium, thermodynamic 156
equivalent diameter 401

equivalent pipe length 39
Ericsson process 268-270, 286, 294, 298, 299, 304, 314
ethyne 469, 470, 574
evaporation 155, 156, 443, 452-458
evaporation channel 452
evaporation coefficient 452, 453
evaporation model 452
excess air 475-478
exergy 333-373, 523-531
exergy efficiency 284, 370-373
exergy loss 363-369
exergy, molar 523-531
exergy of bound energy 337, 338
exergy of enthalpy 347-350
exergy of free energy 351
exergy of free enthalpy 351-353
exergy of heat 335, 336
exergy of internal energy 343-346
exergy of pressure change work 342, 343
exergy of shift work 341, 342
exergy of temperature change heat 338-340
exergy of volume change work 340, 341
exhaust gas 465

Fanno curve
Fick's law 458
film condensation 154, 413-414
film evaporation 157, 369-371
fin efficiency 416, 417
finned heat transfer surfaces 418-420
first constant of radiation 376
first law of thermodynamics 10
fix point, *Kelvin* scale 71
flame speed 481
flow, forced 406-408, 416
flow, free 408-411, 416
flow function 122-124
flow process, adiabatic 28
flow ratio 376
flow shapes 36
flow velocity, average 13
flue gas 465
fluid 5
fluid flow machine 214, 223
fluid mechanics 35-41
fog 444, 456, 457
fog isotherm 457, 458
free energy 197, 200-203, 351, 354-357
free energy as a thermodynamic potential 355-357

free energy, exergy of 351
free enthalpy 197, 200 201, 203-206, 351-353, 532-534
free enthalpy as a thermodynamic potential 355-357
free enthalpy, exergy of 351-353
free enthalpy, molar 512, 518, 532
frictional compressible pipe flow
 - adiabatic 131-135
 - isothermal 127-131
friction force 19
friction work 19, 20, 46, 48, 49, 55, 56, 58, 59, 223, 257, 258
friction work, external 20
friction work, internal 20
fuel 465, 466, 470, 479, 481, 526-531
fuel cell 258, 550-554
fuel exergy, molar 526-531
fuel gas 466, 481
fuel, gaseous 466-470, 479, 481, 486, 529-533
fuel mixture 465, 466

Galilei number 411
gas constant, molar 74
gas constant, specific 73
gas constant, universal 74
gas expansion process 273, 292, 297, 298, 311, 312, 316
gas expansion process as a refrigeration process 321-325
gas mixtures 99-110
gas phase 151, 152
gas thermometer 71, 72
gas turbine 231-235
gas turbine plant 231-235, 260, 261
Gaussian error function 394
Gay-Lussac's law 69
generalised *Diesel* process 261, 290, 291, 297, 298, 300, 309
generation of electrical energy 261
Grashof number 400
greenhouse effect 377, 378
grey body 376
g, s diagram 203-206

half-infinite body 394
hard coal 465
heat 9, 10, 22, 91, 245
heat capacity 30-35
heat capacity, mean specific 31-33, 60 567-573

Index 603

heat capacity, molar isobaric 513
heat capacity, specific 31-35, 76, 157-160,
    567-572, 574, 584, 585
heat coordinate 67
heat, exergy of 335, 336
heat flow 22, 23
heat flow equation 398, 429
heating loss 228, 230
heating of humid air 447
heating oil, heavy 478, 531
heating oil, light 478, 531
heat, molar reversible 518, 545-550
heat of vaporisation 162, 193, 411
heat of vaporisation, specific external 162
heat of vaporisation, specific internal 162
heat penetration coefficient 395, 396
heat pump 317-321, 325-328
heat radiation 375
heat sink 57, 252-256, 319
heat source 57, 252-256, 319
heat transfer 375
heat transfer coefficient 397-399, 416
heat transfer coefficient, over-all 414, 415
heat transfer, over-all 414
heat transfer resistance, over-all 415, 416
heat transfer with phase transition 428
heat transfer with temperature gradient
    47, 50, 56, 57, 255, 256, 319, 320
h, s diagram 62, 63, 170, 171
humid air 443
humid air, h, x-diagram 464
humidification 447, 448
humidity ratio 443, 444
h, x diagram 449-452, 464
h, x diagram of *Mollier* 449-452, 464
hydraulic diameter 401
hydrocarbon 466
hydrogen 100, 101, 105, 106, 465-474, 528,
    541-547, 550-554, 568, 574

ice fog 445
ideal gas 69-110, 111-147, 205, 443-464,
    535-550, 553
ignition limit 481
ignition speed 481
incomplete combustion 475, 481
indexed work 224
individual resistance coefficient 38
injection ratio 262
injection volume 262
instationary heat conduction 391-396

internal change of state 245
internal energy 9, 21, 22, 93, 102, 312
internal energy, exergy of 343-346
internal friction work 46, 223, 226-232
internal work 224, 226
intensity 376, 377
intensity distribution 376
inversion curve, differential 207, 210
inversion curve, integral 208, 210
irradiance number 378-382
irreversibility 46-52, 256-258, 363
irreversible process 46-49, 55-61, 256-258
isenthalpic throttling 207, 208
isentrope 84
isentropic change of state 61, 62, 84-88
isentropic exponent 75, 77, 85, 191-193
isobare 79
isobaric change of state 79-82
isochore 77
isochoric change of state 77-79
isolated system 5
isotherm 82
isothermal change of state 82-84
isothermal exponent 191-193

Joule process 265, 287, 295, 298, 300,
    306, 315
Joule-Thomson effect 207-210

Kelvin scale 6, 66, 70-72
kinetic energy 9, 13, 116, 225
Kirchhoff's law 375

lack of air 474-478
Lambert's cosine law 378
laminar flow
Laval nozzle 135-141
left-hand *Carnot* process 319, 320
left-hand *Clausius-Rankine* process 326
left-hand cyclic process 251, 252, 317-319
left-hand Gas expansion process 321-325
left-hand *Joule* process 320-321
Lewis relation 454, 455, 458
light heating oil EL 478, 482, 531
lignite 465
liquefaction 151-153, 156
liquefaction line 152
liquid 151-154, 157, 204
liquid, saturated 152, 154
longitudinal heat conduction 434-437
lower ignition limit 481

Mach number 114
mass 7
mass action law 533-537, 538-543
mass flow 7
mass fraction 100
Maxwell relationship 175, 176
mean temperature difference 23, 417, 418, 421-423, 428, 429
mechanical effort ratio 279, 282, 285-293, 298, 302-310
mechanical energy 9, 213, 225, 226, 253, 259
melting 151, 152
melting line 152, 153
melting pressure curve 152, 167, 194
methane 466, 574, 588
minimum air requirement 467-469, 471
minimum moist combustion gas volume 468
minimum oxygen requirement 467, 471
mixing of two humid air quantities 448
mixing temperature 102, 103, 109, 110
moist combustion gas volume 468, 469
molar flow rate 7
molar mass 7, 74, 100, 104, 106, 466, 574
molar mass of a mixture 104, 106
molar quantity 7, 74, 497
molar volume 74
mole fraction 104-106, 536
Mollier h, x-diagram 464
multi-stage compression 216-218, 266-269
multi-stage expansion 218, 268, 269
multi-substance system 155, 156

natural gas 465, 466, 475, 476, 484, 530
natural gas H 466, 475, 476, 484
natural gas L 466, 475, 484
nitrogen 99, 466-468, 470, 472, 473, 568, 574
nitrogen characteristic 472
nitrogen oxides 484, 569, 574
norm pressure 71
norm state, physical 70, 71, 466, 486
norm volume 71, 466, 486
nozzle 117, 124-126
nuclear energy, nuclear fission 9, 465
Nußelt number 399

Ohm's law 389
oil shale 465
open system 4, 5
operating characteristic 417, 418

Ostwald triangle 473-478
Otto engine 260, 264
Otto process 260-264, 288, 296, 298, 301, 307, 315, 316
oversaturated humid air 444, 445
oxygen 99, 465, 466, 468, 469, 568, 574
oxygen requirement characteristic 471

partial pressure 99, 537, 542, 544-548
partial pressure fraction 99, 103-108, 537, 542, 544, 545, 548
partition wall heat exchanger 421-440
peat 465
Peclet number 400
perpetual mobile of the first kind 10
perpetual mobile of the second kind 46, 254
petrol 477
phase transformation number 411
phase transition 152, 156, 411-414, 416, 428, 429
physical norm state 70, 71, 466, 486
physical norm volume 71, 466, 486
piston compressor 231
Planck's constant 377
Planck's elementary quantum of action 377
polytrope 89
polytropic change of state 89-96
polytropic exponent 89-96
potential, chemical 531-533
potential energy 9, 10-12
potential, thermodynamic 200, 201, 534, 535
power engine 213
power engine process 252
power machine 213
power number 405, 406
power-to-gas concept 509-511
power units 13
Prandtl analogy 402-405
Prandtl number 399
pressure change work 17, 18, 90, 91, 245, 246
pressure change work, exergy of 342, 343
pressure change work, specific 278
pressure equalisation 47, 58
pressure increase ratio 262
pressure loss in pipes 36-41, 129-131, 134-135
pressure units 12
process 8
process, irreversible 44
process, non-adiabatic 59-61
process, reversible 44